✦ *Understanding Our Universe*

SECOND EDITION

Understanding Our Universe

SECOND EDITION

Stacy Palen
Weber State University

Laura Kay
Barnard College

Brad Smith
Santa Fe, New Mexico

George Blumenthal
University of California—Santa Cruz

W. W. Norton & Company, Inc.
New York • London

W. W. NORTON & COMPANY has been independent since its founding in 1923, when William Warder Norton and Mary D. Herter Norton first published lectures delivered at the People's Institute, the adult education division of New York City's Cooper Union. The firm soon expanded its program beyond the Institute, publishing books by celebrated academics from America and abroad. By mid-century, the two major pillars of Norton's publishing program—trade books and college texts—were firmly established. In the 1950s, the Norton family transferred control of the company to its employees, and today—with a staff of four hundred and a comparable number of trade, college, and professional titles published each year—W. W. Norton & Company stands as the largest and oldest publishing house owned wholly by its employees.

Copyright ©2015, 2012 by W. W. Norton & Company, Inc.
All rights reserved.
Printed in the United States of America.

Editor: Erik Fahlgren
Project Editor: Diane Cipollone
Editorial Assistant: Arielle Holstein
Copy Editor: Christopher Curioli
Managing Editor, College: Marian Johnson
Managing Editor, College Digital Media: Kim Yi
Director of Production, College: Jane Searle
Media Editor: Rob Bellinger
Associate Media Editor: Julia Sammaritano
Media Project Editor: Kristen Sheerin, Danielle Belfiore
Assistant Media Editor: Paula Iborra
Marketing Manager: Stacy Loyal
Design Director: Hope Miller Goodell
Photo Editor: Stephanie Romeo
Permissions Associate: Bethany Salminen
Composition: Carole Desnoes
Illustrations: Precision Graphics
Manufacturing: LSC Communications, Kendallville

Permission to use copyrighted material is included in the credits section of this book, which begins on page C-1.

Library of Congress Cataloging-in-Publication Data
Palen, Stacy, author.
  Understanding our universe second edition / Stacy Palen, Weber State University, Laura Kay, Barnard College, Brad Smith, Santa Fe, New Mexico, George Blumenthal, University of California?Santa Cruz.
      pages cm
   Includes index.
   ISBN 978-0-393-93631-5 (pbk.)
   1. Astronomy--Textbooks. I. Kay, Laura, author. II. Smith, Brad, 1931- author. III. Blumenthal, George (George Ray), author. IV. Title.
   QB43.3.P35 2015
   520--dc23
                    2014031344

W. W. Norton & Company, Inc., 500 Fifth Avenue, New York, NY 10110-0017
wwnorton.com
W. W. Norton & Company Ltd., 15 Carlisle Street, London W1D 3BS
  5 6 7 8 9 0

Stacy Palen dedicates this book to Helene Detwiler, Everett Boles, Keith Palen, Dutchie Armstrong and "Miss T."

Laura Kay thanks her partner, M.P.M.

Brad Smith dedicates this book to his patient and understanding wife, Diane McGregor.

George Blumenthal gratefully thanks his wife, Kelly Weisberg, and his children, Aaron and Sarah Blumenthal, for their support during this project. He also wants to thank Professor Robert Greenler for stimulating his interest in all things related to physics.

BRIEF TABLE OF CONTENTS

Part I Introduction to Astronomy

Chapter 1 **Thinking Like an Astronomer** 2

Chapter 2 **Patterns in the Sky—Motions of Earth** 20

Chapter 3 **Laws of Motion** 46

Chapter 4 **Light and Telescopes** 72

Part II The Solar System

Chapter 5 **The Formation of Stars and Planets** 96

Chapter 6 **Terrestrial Worlds in the Inner Solar System** 126

Chapter 7 **Atmospheres of Venus, Earth, and Mars** 156

Chapter 8 **The Giant Planets** 182

Chapter 9 **Small Bodies of the Solar System** 214

Part III Stars and Stellar Evolution

Chapter 10 **Measuring the Stars** 244

Chapter 11 **Our Star: The Sun** 274

Chapter 12 **Evolution of Low-Mass Stars** 302

Chapter 13 **Evolution of High-Mass Stars** 328

Part IV Galaxies, the Universe, and Cosmology

Chapter 14 **Measuring Galaxies** 358

Chapter 15 **Our Galaxy: The Milky Way** 388

Chapter 16 **The Evolution of the Universe** 410

Chapter 17 **Formation of Structure** 440

Chapter 18 **Life in the Universe** 460

✦ CONTENTS

Preface xv
AstroTours xxix
Nebraska Simulations xxix
Visual Analogies xxx
Astronomy in Action Videos xxx
About the Authors xxxi

Part I Introduction to Astronomy

CHAPTER 1 Thinking Like an Astronomer 2

1.1 Astronomy Gives Us a Universal Context 4
1.2 Science Is a Way of Viewing the World 7
1.3 Astronomers Use Mathematics to Find Patterns 10
　　Working It Out 1.1 Units and Scientific Notation 13
　　Reading Astronomy News
　　　"Pluto Is Demoted to 'Dwarf Planet'" 14
　　Summary 16
　　Questions and Problems 17
　　Exploration: Logical Fallacies 19

CHAPTER 2 Patterns in the Sky—Motions of Earth and the Moon 20

2.1 Earth Spins on Its Axis 22
2.2 Revolution Around the Sun Leads to Changes during the Year 28
　　Working It Out 2.1 Manipulating Equations 29
2.3 The Moon's Appearance Changes as It Orbits Earth 35
2.4 Shadows Cause Eclipses 38
　　Reading Astronomy News
　　　"Two Eclipses, Two Stories" 40
　　Summary 42
　　Questions and Problems 43
　　Exploration: Phases of the Moon 45

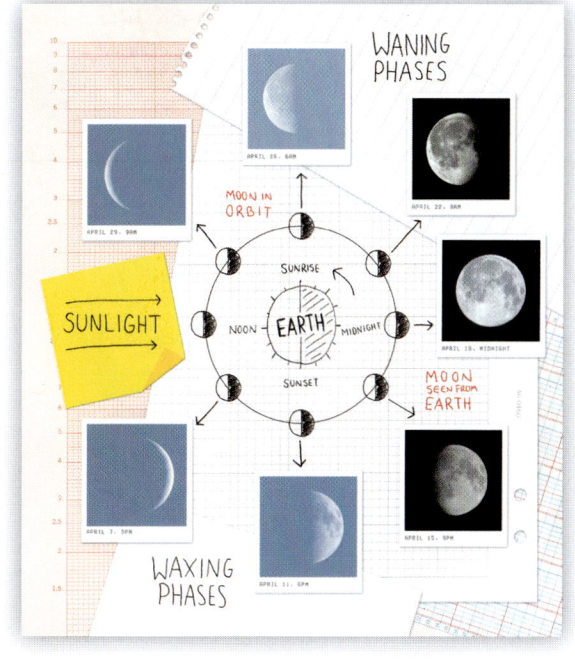

CHAPTER 3 Laws of Motion 46

3.1 Since Ancient Times Astronomers Have Studied the Motions of the Planets 48
3.2 Galileo Was the First Modern Scientist 51
　　Working It Out 3.1 Kepler's Law 52
3.3 Newton's Laws Govern Motion 53
　　Working It Out 3.2 Finding the Acceleration 56
3.4 Gravity Is a Force between Any Two Massive Objects 58

ix

x CONTENTS

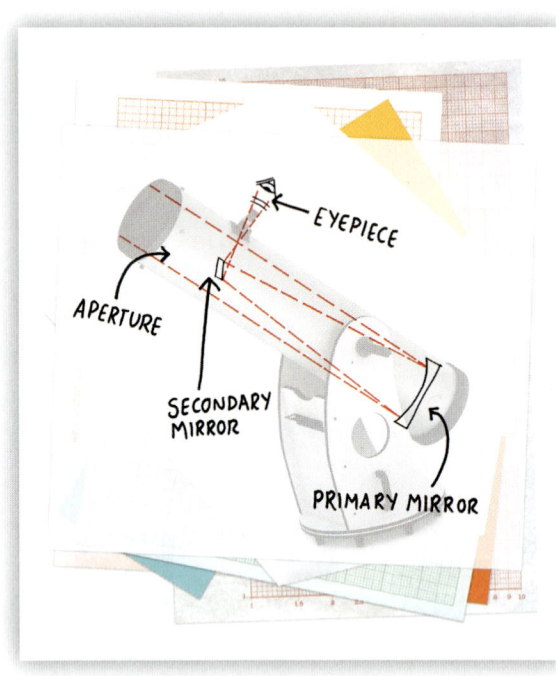

    Working It Out 3.3 Newton's Law of Gravity: Playing with Proportionality   60

3.5   Orbits Are One Body "Falling Around" Another   62
    Reading Astronomy News
       "Curiosity, Interrupted: Sun Makes Mars Go Dark"   66
    Summary   67
    Questions and Problems   68
    Exploration: Newtonian Features   71

CHAPTER 4   Light and Telescopes   72

4.1   What Is Light?   74
    Working It Out 4.1 Wavelength and Frequency   77
4.2   Cameras and Spectographs Record Astronomical Data   80
4.3   Telescopes Collect Light   84
    Working It Out 4.2 Diffraction Limit   89
    Reading Astronomy News
       "ATK Building World's Largest Space Telescope in Magna"   91
    Summary   92
    Questions and Problems   93
    Exploration: Light as a Wave   95

Part II The Solar System

CHAPTER 5   The Formation of Stars and Planets   96

5.1   Molecular Clouds Are the Cradles of Star Formation   98
5.2   The Protostar Becomes a Star   100
    Working It Out 5.1 The Stefan-Boltzmann Law and Wien's Law   101
5.3   Planets Form in a Disk around the Protostar   105
5.4   The Inner and Outer Disk Have Different Compositions   110
5.5   A Case Study: The Solar System   111
5.6   Planetary Systems Are Common   114
    Working It Out 5.2 Making Use of the Doppler Shift   116
    Reading Astronomy News
       "Kepler's Continuing Mission"   120
    Summary   121
    Questions and Problems   122
    Exploration: Exploring Extrasolar Planets   125

Chapter 6   Terrestrial Worlds in the Inner Solar System   126

6.1   Impacts Help Shape the Terrestrial Planets   128
    Working It Out 6.1 How to Read Cosmic Clocks   132
6.2   The Surfaces of Terrestrial Planets Are Affected by Processes in the Interior   133
    Working It Out 6.2 The Density of Earth   133
6.3   Planetary Surfaces Evolve through Tectonism   138
6.4   Volcanism Reveals a Geologically Active Planet   144
6.5   Wind and Water Modify Surfaces   147
    Reading Astronomy News
       "Moon Is Wetter, Chemically More Complex Than Thought, NASA Says"   151

Summary 152
Questions and Problems 153
Exploration: Earth's Tides 155

CHAPTER 7 Atmospheres of Venus, Earth, and Mars 156

7.1 Atmospheres Change over Time 158
7.2 Secondary Atmospheres Evolve 159
Working It Out 7.1 How Can We Find the Temperature of a Planet? 162
7.3 Earth's Atmosphere Has Detailed Structure 164
7.4 The Atmospheres of Venus and Mars Differ from Earth's 172
7.5 Greenhouse Gasses Affect Global Climates 174
Reading Astronomy News
"Curiosity Rover Sees Signs of Vanishing Martian Atmosphere" 177
Summary 178
Questions and Problems 179
Exploration: Climate Change 181

CHAPTER 8 The Giant Planets 182

8.1 Giant Planets Are Large, Cold, and Massive 184
Working It Out 8.1 Finding the Diameter of a Giant Planet 186
8.2 The Giant Planets Have Clouds and Weather 189
Working It Out 8.2 Measuring Wind Speeds on Distant Planets 195
8.3 The Interiors of the Giant Planets Are Hot and Dense 197
Working It Out 8.3 Internal Thermal Energy Heats the Giant Planets 198
8.4 The Giant Planets Are Magnetic Powerhouses 199
8.5 Rings Surround the Giant Planets 203
Reading Astronomy News
"Giant Propeller Structures Seen in Saturn's Rings" 208
Summary 210
Questions and Problems 211
Exploration: Estimating Rotation Periods of Giant Planets 213

CHAPTER 9 Small Bodies of the Solar System 214

9.1 Dwarf Planets May Outnumber Planets 216
9.2 Moons as Small Worlds 217
9.3 Asteroids Are Pieces of the Past 223
9.4 Comets Are Clumps of Ice 226
9.5 Comet Collisions Still Happen Today 232
Working It Out 9.1 Finding the Radius of a Meteoroid 234
9.6 Meteorites Are Remnants of the Early Solar System 234
Reading Astronomy News
"NASA's Asteroid Sample-Return Mission Moves into Development" 239
Summary 240
Questions and Problems 241
Exploration: Comparative Dwarf Planetology 243

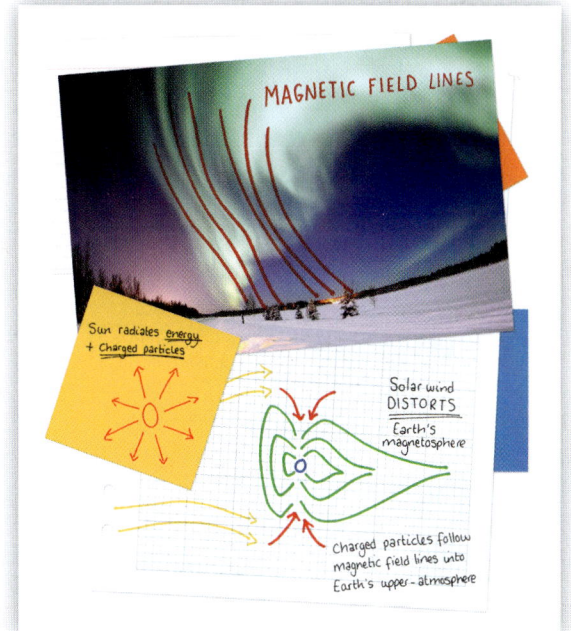

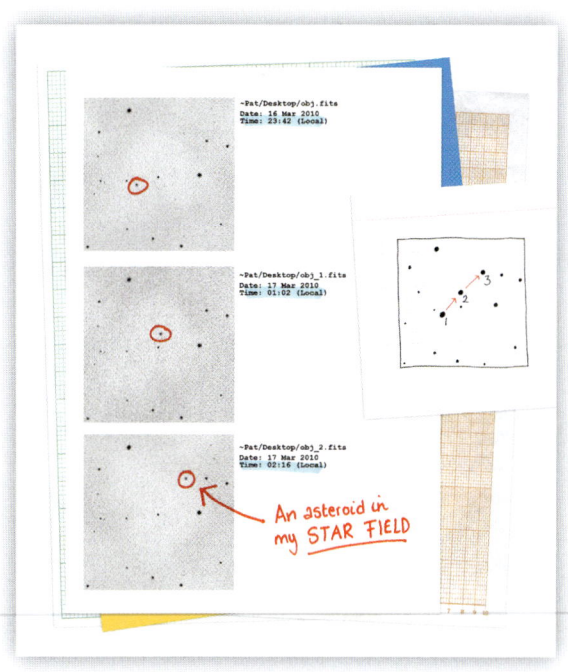

Part III Stars and Stellar Evolution

CHAPTER 10 Measuring the Stars 244

10.1 The Luminosity of a Star Can Be Found from the Brightness and Distance 246
  Working It Out 10.1 Parallax and Distance 249
10.2 Radiation Tells Us the Temperature, Size, and Composition of Stars 252
10.3 The Mass of a Star Can Be Determined in Some Binary Systems 262
10.4 The H-R Diagram Is the Key to Understanding Stars 265
  Reading Astronomy News
    "Mystery of Nearby SS Cygni Star System Finally Resolved" 269
  Summary 270
  Questions and Problems 271
  Exploration: The H-R Diagram 273

CHAPTER 11 Our Star: The Sun 274

11.1 The Structure of the Sun Is a Matter of Balance 276
  Working It Out 11.1 How Much Longer Will the Sun "Live"? 278
11.2 Energy in the Sun's Core Moves through Radiation and Convection 281
11.3 The Atmosphere of the Sun 285
11.4 The Atmosphere of the Sun Is Very Active 289
  Working It Out 11.2 Sunspots and Temperature 291
  Reading Astronomy News
    "Weather Forecast in Space: Not Sunny, with Solar Flares" 296
  Summary 297
  Questions and Problems 298
  Exploration: The Proton-Proton Chain 301

CHAPTER 12 Evolution of Low-Mass Stars 302

12.1 The Life and Times of a Main-Sequence Star Follow a Predictable Path 304
  Working It Out 12.1 Estimating Main-Sequence Lifetimes 306
12.2 A Star Runs Out of Hydrogen and Leaves the Main Sequence 306
12.3 Helium Begins to Burn in Degenerate Core 310
12.4 The Low-Mass Star Enters the Last Stages of Its Evolution 312
12.5 Star Clusters Are Snapshots of Stellar Evolution 317
12.6 Binary Stars Sometimes Share Mass, Resulting in Novae and Supernovae 319
  Reading Astronomy News
    "Scientists May Be Missing Many Star Explosions" 323
  Summary 324
  Questions and Problems 325
  Exploration: Evolution of Low-Mass Stars 327

CHAPTER 13 Evolution of High-Mass Stars 328

13.1 High-Mass Stars Follow Their Own Path 330
13.2 High-Mass Stars Go Out with a Bang 333
13.3 Supernovae Change the Galaxy 336

13.4 Einstein Moved Beyond Newtonian Physics 340
  Working It Out 13.1 The Boxcar Experiment 343
13.5 Gravity Is a Distortion of Spacetime 345
13.6 Black Holes Are a Natural Limit 350
  Working It Out 13.2 Finding the Schwarzschild Radius 350
  Reading Astronomy News
    "What a Scorcher—Hotter, Heavier, and Millions of Times Brighter Than the Sun" 353
  Summary 354
  Questions and Problems 355
  Exploration: The CNO Cycle 357

Part IV Galaxies, the Universe, and Cosmology

CHAPTER 14 Measuring Galaxies 358

14.1 Galaxies Come in Many Sizes and Shapes 360
14.2 Stars Form in the Spiral Arms of a Galaxy's Disk 365
14.3 Galaxies Are Mostly Dark Matter 367
14.4 A Supermassive Black Hole Exists at the Heart of Most Galaxies 371
14.5 We Live in an Expanding Universe 375
  Working It Out 14.1 Redshift: Calculating the Recession Velocity and Distance of Galaxies 377
  Reading Astronomy News
    "Colliding Galaxies Swirl in Dazzling New Photo" 382
  Summary 384
  Questions and Problems 385
  Exploration: Galaxy Classification 387

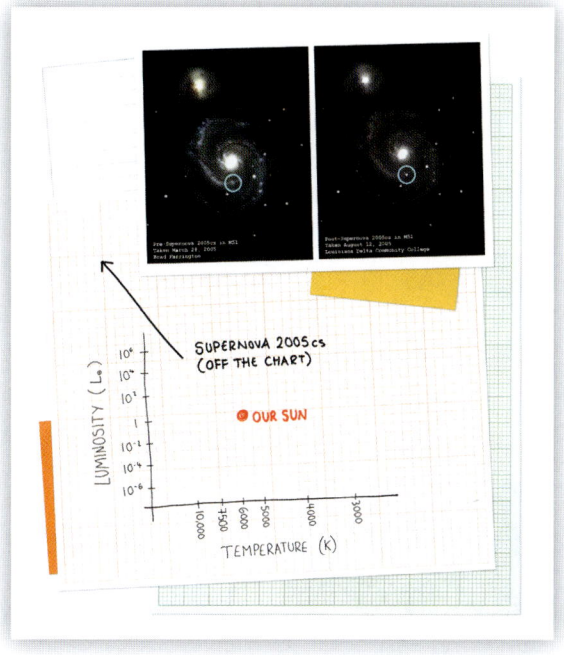

CHAPTER 15 Our Galaxy: The Milky Way 388

15.1 Measuring the Milky Way Is a Challenge 390
15.2 Components of the Milky Way Reveal Its Evolution 393
15.3 The Milky Way Is Mostly Dark Matter 399
  Working It Out 15.1 Finding the Mass of a Galaxy 400
15.4 The Milky Way Contains a Supermassive Black Hole 400
15.5 The Milky Way Offers Clues about How Galaxies Form 402
  Reading Astronomy News
    "Hyperfast Star Kicked Out of Milky Way" 404
  Summary 405
  Questions and Problems 406
  Exploration: The Center of the Milky Way 408

CHAPTER 16 The Evolution of the Universe 410

16.1 Hubble's Law Implies a Hot, Dense Beginning 412
  Working It Out 16.1 Expansion and the Age of the Universe 413
16.2 The Cosmic Microwave Background Confirms the Big Bang 416
16.3 The Expansion of the Universe Is Speeding Up 421
16.4 The Earliest Moments of the Universe Connect the Very Largest Size Scales to the Very Smallest 425

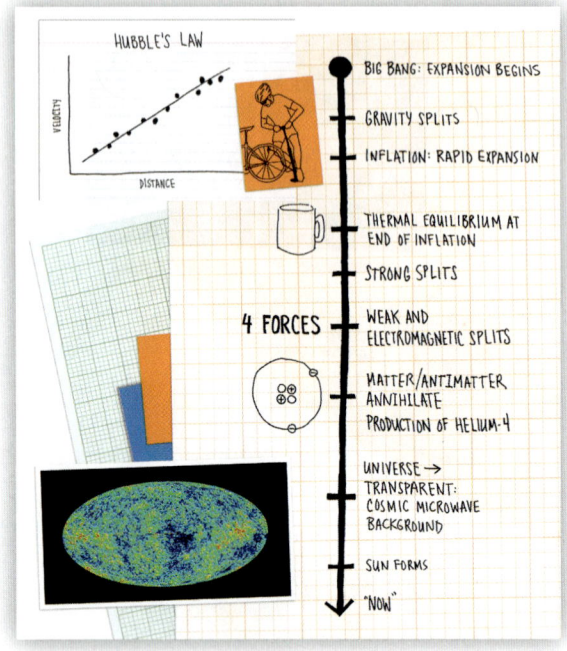

16.5 Inflation Solves Several Problems in Cosmology 430

16.6 Other Universes? 432

Reading Astronomy News
  "Planck: Big Bang's Afterglow Reveals Older Universe, More Matter" 434
Summary 435
Questions and Problems 436
Exploration: Hubble's Law for Balloons 438

CHAPTER 17 Formation of Structure 440

17.1 Galaxies Form Groups, Clusters, and Larger Structures 442

17.2 Gravity Forms Large-Scale Structure 444

17.3 The First Stars and Galaxies Form 449

17.4 Galaxies Evolve 451

17.5 Astronomers Think about the Deep Future 453

Reading Astronomy News
  "Massive Black Holes Sidle Up to Other Galaxies" 455
Summary 456
Questions and Problems 457
Exploration: The Story of a Proton 459

CHAPTER 18 Life in the Universe 460

18.1 Life on Earth Began Early and Evolved Over Time 462

18.2 Life beyond Earth Is Possible 466
  Working It Out 18.1 Exponential Growth 467

18.3 Scientists Search for Signs of Intelligent Life 473

18.4 The Fate of Life on Earth 477

Reading Astronomy News
  "Astronomer Uses Kepler Telescope's Data in Hunt for Spacecraft from Other Worlds" 480
Summary 482
Questions and Problems 483
Exploration: Fermi Problems and the Drake Equation 485

Appendix 1 Periodic Table of the Elements A-1
Appendix 2 Properties of Planets, Dwarf Planets, and Moons A-2
Appendix 3 Nearest and Brightest Stars A-5
Appendix 4 Star Maps A-8
Glossary G-1
Selected Answers SA-1
Credits C-1
Index I-1

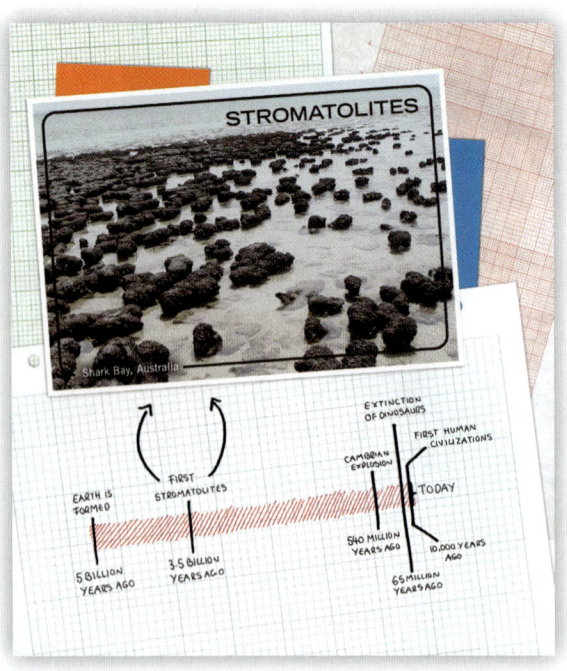

PREFACE

Dear Student,

You may wonder why it is a good idea to take a general-education science course. Throughout your education, you have been exposed to different ways of thinking—different approaches to solving problems, different definitions of *understanding*, and different meanings of the verb *to know*. Astronomy offers one example of the scientific viewpoint. Scientists, including astronomers, have a specific approach to problem solving (sometimes called the scientific method, although the common understanding of this term only skims the surface of the process). Astronomers "understand" when they can make correct predictions about what will happen next. Astronomers "know" when an idea has been tested dozens or even hundreds of times and that the idea has stood the test of time.

Your instructor likely has two basic goals in mind for you as you take this course. The first is to understand some basic physical concepts and be familiar with the night sky. The second is to think like a scientist and learn to use the scientific method not only to answer questions in this course but also to make decisions in your life. We have written the Second Edition of *Understanding Our Universe* with these two goals in mind.

Throughout this book, we emphasize not only the content of astronomy (the masses of the planets, the compositions of stellar atmospheres) but also *how* we know what we know. We believe the scientific method is a valuable tool that you can carry with you, and use, for the rest of your life.

Astronomy is one of the purest expressions of one of the more distinctive impulses of humanity—curiosity. Astronomy does not capture the public interest because it is profitable, will cure cancer, or build better bridges. People choose to learn about astronomy because they are curious about the universe.

The most effective way to learn something is to "do" it. Whether playing an instrument or a sport or becoming a good cook, reading "how" can only take you so far. The same is true of learning astronomy. This book helps you "do" as you learn. We start with the illustrations at the beginning of each chapter. These **chapter-opening figures** demonstrate different ways you might interact with the material. They are presented from the viewpoint of a student who is wondering about the universe, asking questions, and keeping a journal of experiments that investigate the answers. Your instructor may ask you to keep such a journal. Or you may choose to keep one on your own, as a useful way of investigating the world around you.

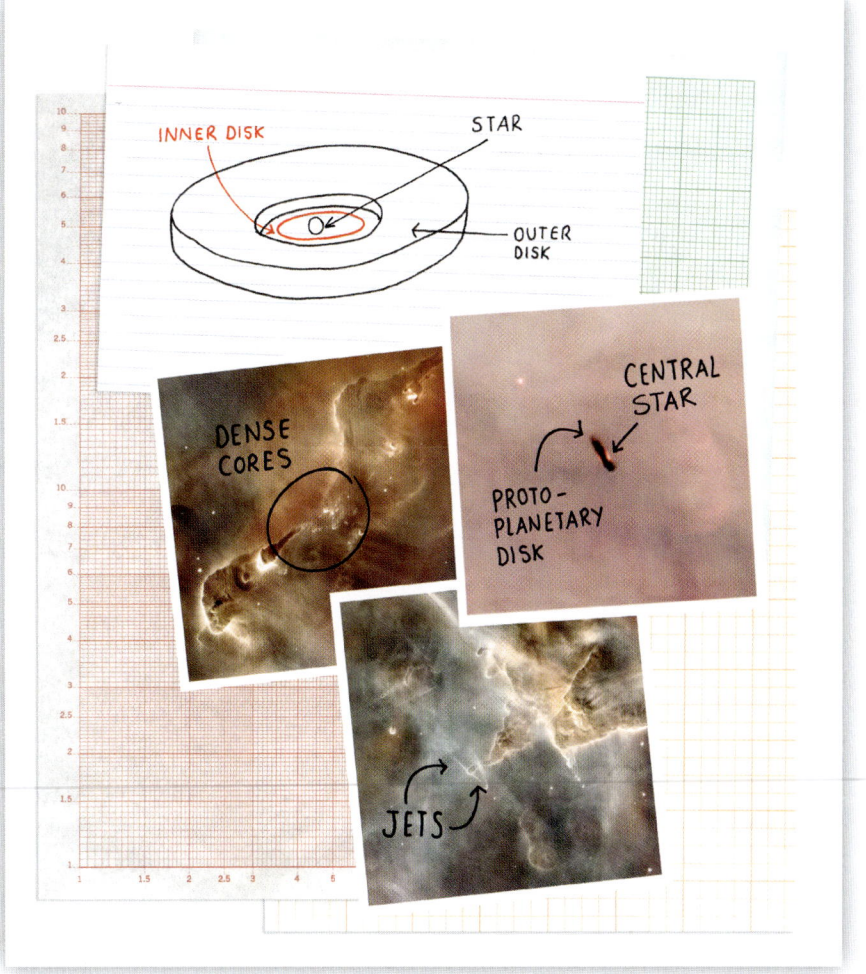

xv

Vocabulary Alert

pressure In common language, we often use *pressure* interchangeably with the word *force*. Astronomers specifically use pressure to mean the force per unit area that atoms or molecules exert as they speed around and collide with each other and their surroundings.

dense In common language, we use this word in many ways, some of which are metaphorical and unkind, as in "You can't understand this? You are so dense!" Astronomers specifically use *density* to mean "the amount of mass packed into a volume"; denser material contains more mass in the same amount of space. In practical terms, you are familiar with density by how heavy an object feels for its size: a pool ball and a tennis ball are roughly the same size, but the pool ball has greater mass and therefore feels heavier because it is denser.

As you learn any new subject, one of the stumbling blocks is often the language of the subject itself. This can be jargon—the specialized words unique to that subject—for example, *supernova* or *Cepheid variable*. But it can also be ordinary words that are used in a special way. As an example, the common word *inflation* usually applies to balloons or tires in everyday life, but economists use it very differently, and astronomers use it differently still. Throughout the book, we have included **Vocabulary Alerts** that point out the astronomical uses of common words to help you recognize how those terms are used by astronomers.

In learning science, there is another potential language issue. The language of science is mathematics, and it can be as challenging to learn as any other language. The choice to use mathematics as the language of science is not arbitrary; nature "speaks" math. To learn about nature, you will also need to speak its language. We don't want the language of math to obscure the concepts, so we have placed this book's mathematics in **Working It Out** boxes to make it clear when we are beginning and ending a mathematical argument, so that you can spend time with the concepts in the chapter text and then revisit the mathematics to study the formal language of the argument. Read through a Working It Out box once, then cover the worked example with a piece of paper, and work through the example until you can do it on your own. When you can do this, you will have learned a bit of the language of science. You will learn to work with data and identify when someone else's data isn't quite right. We want you to be comfortable reading, hearing, and speaking the language of science, and we will provide you with tools to make it easier.

Working It Out 5.2 | Making Use of the Doppler Shift

We noted in Section 4.2 that atoms and molecules emit and absorb light only at certain wavelengths. The spectrum of an atom or molecule has absorption or emission lines that look something like a bar code instead of a rainbow, and each type of atom or molecule has a unique set of lines. These lines are called *spectral lines*. A prominent spectral line of hydrogen atoms has a rest wavelength, λ_{rest}, of 656.3 nanometers (nm). Suppose that using a telescope, you measure the wavelength of this line in the spectrum of a distant object and find that instead of seeing the line at 656.3 nm, you see it at an observed wavelength, λ_{obs}, of 659.0 nm. The mathematical form of the Doppler effect shows that the object is moving at a radial velocity (v_r) of

$$v_r = \frac{\lambda_{obs} - \lambda_{rest}}{\lambda_{rest}} \times c$$

$$v_r = \frac{659.0\,\text{nm} - 656.3\,\text{nm}}{656.3\,\text{nm}} \times (3 \times 10^8\,\text{m/s})$$

$$v_r = 1.2 \times 10^6\,\text{m/s}$$

The object is moving away from you (because the wavelength became longer and redder) with a speed of 1.2×10^6 m/s, or 1,200 kilometers per second (km/s).

Now consider our stellar neighbor, Alpha Centauri, which is moving toward us with a radial velocity of -21.6 km/s (-2.16×10^4 m/s). (Negative velocity means the object is moving toward us.) What is the observed wavelength, λ_{obs}, of a magnesium line in Alpha Centauri's spectrum having a rest wavelength, λ_{rest}, of 517.27 nm? First, we need to manipulate the Doppler equation to get λ_{obs} all by itself. Then we can plug in all the numbers.

$$v_r = \frac{\lambda_{obs} - \lambda_{rest}}{\lambda_{rest}} \times c$$

Solve this equation for λ_{obs} to get

$$\lambda_{obs} = \lambda_{rest} + \frac{v_r}{c} \lambda_{rest}$$

Both terms on the right contain λ_{rest}. Factor it out to make the equation a little more convenient:

$$\lambda_{obs} = \left(1 + \frac{v_r}{c}\right) \lambda_{rest}$$

We are ready to plug in some numbers to solve for the observed wavelength:

$$\lambda_{obs} = \left(1 + \frac{-2.16 \times 10^4\,\text{m/s}}{3 \times 10^8\,\text{m/s}}\right) \times 517.27\,\text{nm}$$

$$\lambda_{obs} = 517.23\,\text{nm}$$

Although the observed Doppler blueshift (517.23 − 517.27) is only −0.04 nm, it is easily measurable with modern instrumentation.

Back in Section 7.2 . . .

. . . you learned that objects reach thermal equilibrium when they exchange energy and come to the same temperature.

There are a few physical concepts that are applicable in many astronomical situations. Rather than placing them all at the front of the text and asking you to remember them later, we have integrated them into the astronomy content, placing them where you first need them to understand the science. In later chapters, we have provided you with **Concept Connection** icons to remind you where you first saw the concept, so you can go back and review it as needed.

Many of these physical concepts, among others, are further explained in a series of short **Astronomy in Action** videos available on the student website. Those videos feature one of the authors (and several students) demonstrating physical concepts at work. Your instructor might assign these videos to you or you might choose to watch them on your own to create a better picture of each concept in your mind.

As a citizen of the world, you make judgments about science, distinguishing between good science and pseudoscience. You use these judgments to make decisions in the grocery store, pharmacy, car dealership, and voting booth. You base these decisions on the presentation of information you receive through the media, which is very different from the presentation in class. Recognizing what is credible and questioning what is not is an important skill. To help you hone this skill, we have provided **Reading Astronomy News** sections at the end of every chapter. These boxes include a news article with questions to help you make sense of how science is presented to you. It is important that you learn to be critical of the information you receive, and these boxes will help you do that.

At the end of each chapter, we have provided several types of questions, problems, and activities for you to practice your skills. The **Summary Self-Test** may be

READING ASTRONOMY News

NASA's Kepler mission may be disabled, but researchers say the best results are yet to come!

Kepler's Continuing Mission

By **RACHEL COURTLAND**, IEEE Spectrum

In early August, the moment that Bill Borucki had been dreading finally arrived. As the principal investigator of NASA's Kepler space telescope, Borucki had been working with his colleagues to restore the spacecraft's ability precisely to point itself. The planet-hunting telescope has four reaction wheels—essentially, electrically driven flywheels—and at least three must be functional to maintain positioning. But in the past few years, two of those wheels had been on the fritz. One went off line in July 2012 after showing elevated levels of friction, and a second followed suit in May 2013, effectively ending science operations. After a few months of recovery efforts, the telescope team was finally forced to call it quits, 6 months after the mission was originally scheduled to finish but years before they hoped it would.

The failures mark the end of an era for Kepler. With only two reaction wheels, the telescope can't steady itself well enough to ensure that light from each star hits the same fraction of a pixel on its charge-coupled devices for months on end without deviation. That's what Kepler needs in order to detect, with high precision, the transit of a planet: the slight dip in the brightness of a star that occurs when an orbiting planet crosses in front of it.

But the Kepler spacecraft might still have its uses, and the data it has already gathered almost certainly will. The telescope's managers are currently evaluating proposals for what might be done with a two-wheeled spacecraft. And the telescope's analysis team is gearing up for the rest of the science mission: a 2- to 3-year effort to crawl systematically through the 4 years of data that Kepler has collected since its launch in 2009.

That analysis effort, which will incorporate new machine-learning techniques and a bit of human experimentation, could yield a bounty of new potential planets on top of the 3,500 that Kepler has found so far. "We expect somewhere between several hundred more planets to maybe as many as a thousand," Borucki says. If all goes well, the revised hunt might even uncover the first handful of terrestrial twins—or at the very least, near cousins: roughly Earth-size planets on nearly yearlong orbits around Sun-like stars.

Uncovering those Earth analogues won't be easy. The orbits are slow and the planets themselves are small. "You're looking for a percent of a percent" dip in the brightness of a star, says Jon Jenkins, the telescope's analysis lead. "That's a very demanding and challenging measurement to make."

The task will be made even more difficult by an unexpected complication: Stars vary in brightness due to sunspots and flares, and Kepler's observations reveal that these variations are greater than scientists had previously estimated. Those fluctuations can hide the presence of a planet, reducing the telescope's sensitivity to terrestrial transits by 50 percent.

In April 2012, NASA granted Kepler a 4-year extension that would have compensated for the extra noise. But with the failure of the reaction wheels, Jenkins and his colleagues now must find a different way to uncover planetary signals.

Earlier this year, they moved the data processing from a set of computer clusters containing 700 microprocessors to the Pleiades supercomputer at the NASA Ames Research Center in Moffett Field, California, where they have the use of up to 15,000 of the machine's more than 160,000 cores. The team is also working on implementing a machine-learning process using an algorithm called the random forest, which will be trained with data already categorized by Kepler scientists. Once it's up and running, the software should be able to speedily differentiate false positives and data artifacts from promising candidates. Eventually, Jenkins says, the analysis team will insert fake data into the pipeline to test the performance of both the humans that ordinarily do the processing and the automated algorithms. "We need to know for every planet we detect how many we missed," Jenkins says.

No one can predict exactly how many planets Kepler will find. The telescope's main goal was to determine how common planets are in and around the habitable zones of stars—the areas around stars with the right temperature range for liquid water to be present. Such statistics could help astrophysicists decide how practical it would be to build a space telescope capable of directly detecting light from Earth-like planets, which is necessary to determine whether they have atmospheres that could support life.

For Earth-size planets in settings similar to our own, developing a good statistical estimate will be difficult. With small numbers, the uncertainty in the size of the overall population will be large. "The best-case scenario is that Kepler could still have, with very large error bars, a number for us at the end of the day," says Sara Seager, a professor of planetary

▶❙ **AstroTour:** Star Formation

▶▶ **Nebraska Simulation:** Exoplanet Radial Velocity Simulator

used to check your understanding. If you can answer these questions correctly, you have a basic grasp of the information in the chapter. Next, a separate group of true/false and multiple-choice questions focuses on more detailed facts and concepts from the chapter. Conceptual questions ask you to synthesize information and explain the "how" or "why" of a situation. Problems give you a chance to practice the quantitative skills you learned in the chapter and to work through a situation mathematically.

Each chapter has an **Exploration** activity that shows you how to use the concepts and skills you learned in an interactive way. About half of the book's Explorations ask you to use animations and simulations on the student website, while the others are hands-on, paper-and-pencil activities that use everyday objects such as ice cubes or balloons.

If you think of human knowledge as an island, each scientific experiment makes the island a little bigger by adding a pebble or a grain of sand to the shoreline. But each of those pebbles also increases our exposure to the ocean of the unknown: The bigger the island of knowledge, the longer the shoreline of ignorance. Throughout this book, we have tried to show clearly which pebbles

---

Exploration: Exploring Extrasolar Planets 125

Exploration | Exploring Extrasolar Planets

wwnpag.es/uou2

Visit the Student Site (wwnpag.es/uou2) and open the Exoplanet Radial Velocity Simulator Nebraska Simulation in Chapter 5. This applet has a number of different panels that allow you to experiment with the variables that are important for measurement of radial velocities. First, in the window labeled "Visualization Controls," check the box to show multiple views. Compare the views shown in panels 1–3 with the colored arrows in the last panel to see where an observer would stand to see the view shown. Start the animation (in the "Animation Controls" panel), and allow it to run while you watch the planet orbit its star from each of the views shown. Stop the animation, and in the "Presets" panel, select "Option A" and then click "set."

1. Is Earth's view of this system most nearly like the "side view" or most nearly like the "orbit view"?

2. Is the orbit of this planet circular or elongated?

3. Study the radial velocity graph in the upper right panel. The blue curve shows the radial velocity of the star over a full period. What is the maximum radial velocity of the star?

4. The horizontal axis of the graph shows the "phase," or fraction of the period. A phase of 0.5 is halfway through a period. The vertical red line indicates the phase shown in views in the upper left panel. Start the animation to see how the red line sweeps across the graph as the planet orbits the star. The period of this planet is 365 days. How many days pass between the minimum radial velocity and the maximum radial velocity?

5. When the planet moves away from Earth, the star moves toward Earth. The sign of the radial velocity tells the direction of the motion (toward or away). Is the radial velocity of the star positive or negative at this time in the orbit? If you could graph the radial velocity of the planet at this point in the orbit, would it be positive or negative?

In the "Presets" window, select "Option B" and then click "set."

6. What has changed about the orbit of the planet as shown in the views in the upper left panel?

7. When is the planet moving fastest: when it is close to the star or when it is far from the star?

8. When is the star moving fastest: when the planet is close to it or when it is far away?

9. Explain how an astronomer would determine, from a radial velocity graph of the star's motion, whether the orbit of the planet was in a circular or elongated orbit.

10. Study the Earth view panel at the top of the window. Would this planet be a good candidate for a transit observation? Why or why not?

In the "System Orientation" panel, change the inclination to 0.0.

11. Now is Earth's view of this system most nearly like the "side view" or most nearly like the "orbit view"?

12. How does the radial velocity of the star change as the planet orbits?

13. Click the box that says "show simulated measurements," and change the "noise" to 1.0 m/s. The gray dots are simulated data, and the blue line is the theoretical curve. Use the slider bar to change the inclination. What happens to the radial velocity as the inclination increases? (Hint: Pay attention to the vertical axis as you move the slider, not just the blue line.)

14. What is the smallest inclination for which you would find the data convincing? That is, what is the smallest inclination for which the theoretical curve is in good agreement with the data?

NORTON SMARTWORK • wwnorton.com/NSW

of knowledge are on the shore, which we are just catching a glimpse of under the water, and which are only thought to be there because of the way the water smoothes out as it passes over them. Sometimes the most speculative ideas are the most interesting because they show how astronomers approach unsolved problems and explore the unknown. As astronomers, we authors know that one of the greatest feelings in the world is to forge a pebble yourself and place it on the shoreline.

Astronomy gives you a sense of perspective that no other field of study offers. The universe is vast, fascinating, and beautiful, filled with a wealth of objects that, surprisingly, can be understood using only a handful of principles. By the end of this book, you will have gained a sense of your place in the universe—both how incredibly small and insignificant you are and how incredibly unique and important you are.

Dear Instructor,

We wrote this book with a few overarching goals: to inspire students, to make the material interactive, and to create a useful and flexible tool that can support multiple learning styles.

As scientists and as teachers, we are passionate about the work we do. We hope to share that passion with students and inspire them to engage in science on their own. As authors, one way we do this is through the "student notebook"–style sketches at the beginning of each chapter. These figures model student engagement, and the Learning Goals on the facing page challenge them to try something similar on their own. Elsewhere in a chapter, we remind students of this chapter-opening figure to encourage them to interact with the content and make it their own.

Through our own experience, familiarity with education research, and surveys of instructors, we have come to know a great deal about how students learn and what goals teachers have for their students. We have explicitly addressed many of these goals and learning styles in this book, sometimes in large, immediately visible ways such as the inclusion of feature boxes but also through less obvious efforts such as questions and problems that relate astronomical concepts to everyday situations or take fresh approaches to organizing material.

For example, many teachers state that they would like their students to become "educated scientific consumers" and "critical thinkers" or that their students should "be able to read a news story about science and understand its significance." We have specifically addressed these goals in our Reading Astronomy News feature, which presents a news article and a series of questions that guide a student's critical thinking about the article, the data presented, and the sources.

Many teachers want students to develop better spatial reasoning and visualization skills. We address this explicitly by teaching students to make and use spatial models. One example is in Chapter 2, where we ask students to use an orange and a lamp to understand the celestial sphere and the phases of the Moon. In nearly every chapter, we have Visual Analogy figures that compare astronomy concepts to everyday events or objects. Through these analogies, we strive to make the material more interesting, relevant, and memorable.

Education research shows that the most effective way to learn is by doing. Exploration activities at the end of each chapter are hands-on, asking students to take the concepts they've learned in the chapter and apply them as they interact with animations and simulations on the student website or work through pencil-and-paper activities. Many of these Explorations incorporate everyday objects and can be used either in your classroom or as activities at home.

To learn astronomy, students must also learn the language of science—not just the jargon, but the everyday words we scientists use in special ways. *Theory* is a famous example of a word that students think they understand, but their definition is very different from ours. The first time we use an ordinary word in a special way, a Vocabulary Alert in the margin calls attention to it, helping to reduce student confusion. This is in addition to the back-of-the-book Glossary, which includes all the text's boldface words in addition to other terms students may be unfamiliar with.

We also believe students should be fairly fluent in the more formal language of science—mathematics. We have placed the math in Working It Out boxes, so it does not interrupt the flow of the text or get in the way of students' understanding of conceptual material. But we've gone further by beginning with fundamental ideas in early math boxes and slowly building complexity in math boxes that appear later in the book. We've also worked to remove some of the stumbling blocks that crush student confidence by providing calculator hints, references to earlier boxes, and detailed, fully worked examples.

In our overall organization, we have made several efforts to encourage students to engage with the material and build confidence in their scientific skills as they proceed through the book. We organize the physical principles with a "just-in-time" approach; for example, we cover the Stefan-Boltzmann law in Chapter 6, when it is used for the first time in an astronomical context. For both stars and galaxies, we have organized the material to cover the general case first and then delve into more details with specific examples. Thus, you will find "stars" before the Sun, and "galaxies" before the Milky Way. This allows us to avoid frustrating students by making assumptions about what they know about stars or galaxies or forward-referencing to basic definitions and overarching concepts. This organization also implicitly helps students to understand their place in the universe: our galaxy and our star are each one of many. They are specific examples of a physical universe in which the same laws apply everywhere. Planets have been organized comparatively, to emphasize that science is a process of studying individual examples that lead to collective conclusions. All of these organizational choices were made with the student perspective in mind and a clear sense of the logical hierarchy of the material.

Even our layout has been designed to maximize student engagement—one wide text column is interrupted as seldom as possible.

SmartWork, an online tutorial and homework system, puts student assessment at your fingertips. SmartWork contains more than 1,300 questions and problems that are tied directly to this text, including the Summary Self-Test questions and versions of the Reading Astronomy News and Exploration questions. Any of these could be used as a reading quiz to be completed before class or as homework. Every question in SmartWork has hints and answer-specific feedback so that students are coached to work toward the correct answer. Instructors can easily modify any of the provided questions, answers, and feedback or can create their own questions.

We approached this text by asking: What do teachers want students to learn, and how can we best help students learn those things? Where possible, we consulted the education research to help guide us, and that guidance has led us down some previously unexplored paths. That research has continued to be useful in this second edition, but we have had another excellent resource to draw on.

In this edition, we have responded to commentary from you, our colleagues. You were concerned that students would not be able to find the "just-in-time" material, so we have added a Concept Connection icon in the margin, which points students back to the original explanation of these topics. They will be better able to find the material on Wien's law and the Stefan-Boltzmann law, which we moved to Chapter 5, where these concepts are first needed. There is no opportunity for students to forget the material before they actually need to use it in an astronomical context, and yet the Concept Connection icon allows them to find the material again and again. This icon reinforces the important fact that the universe is governed by a small number of physical concepts that appear again and again in very different contexts.

We revised each chapter, updating the science, to reflect the fast pace of astronomical research today. This is especially noticeable in the material on extrasolar planets and on the very latest results in cosmology; however, each chapter has been revised to reflect the progress in the field.

We better balanced the cognitive load between chapters, for example by moving material between Chapters 3 and 4 so that students are not grappling with both complex three-dimensional visualization skills and fundamental physics at the same time. We reorganized Chapter 6, moving the impact coverage forward for similar reasons. This also allowed us to streamline some of the text, reducing the need to remind students of earlier material.

We added further skill-building text with a new section on reading graphs in Chapter 1 and new Working It Out boxes that continue to build students' mathematical fluency throughout the text. We also developed new end-of-chapter problems that address student understanding at multiple skill levels. Even more skill-building content is available in the accompanying workbook, *Learning Astronomy by Doing Astronomy*.

We made descriptions of complex relationships even more accessible. For example, we have new visual analogies, such as the one of the solar wind shaping magnetospheres. We have revised figures to be more straightforward, such as the one showing the structure of the Sun. The Hertzsprung-Russell (H-R) diagram showing the evolution of low-mass stars has been split into multiple figures to match the narrative of the text better—each part of the evolution is shown separately, and then the entire sequence is shown in a culminating figure. This requires more space but is worth the effort, as it allows students to focus on each step individually and then put the whole picture together once each part is understood.

We reorganized Part IV: Galaxies, the Universe, and Cosmology to improve the logical flow of the cosmology concepts, balance cognitive load, and emphasize the process of science that has led to understanding that the universe began in the Big Bang. We begin in Chapter 14 by introducing galaxies as a whole and our measurements of them, including recession velocities. Then we address the Milky Way in Chapter 15—a specific example of a galaxy that we can discuss in detail. This follows the repeating motif of moving from the general to the specific that exists throughout the text and gives students a basic grounding in the concepts of spiral galaxies, supermassive black holes, and dark matter before they need to apply those concepts to the specific example of our own galaxy. In Chapter 16, we return to the implications of extragalactic recession velocities, showing how the Big Bang was derived from observational evidence, then was used to make predictions which have been later verified. In Chapter 17, we address issues of large-scale structure and the evolution of the universe over time.

Many professors find themselves under pressure from accrediting bodies or internal assessment offices to assess their courses in terms of learning goals and to update their teaching methods. To help you with this, we've revised each chapter's Learning Goals and organized the end-of-chapter Summary by Learning Goal. In SmartWork, questions and problems are tagged and can be sorted by Learning Goal.

We've also created a series of 23 videos explaining and demonstrating concepts from the text, accompanied by questions integrated into SmartWork. You might assign these videos prior to lecture—either as part of a flipped modality or as a "reading quiz." In either case, you can use SmartWork's diagnostic feedback from the questions to tailor your in-class discussions. Or you might show the videos in class to stimulate discussion. Or you might simply use them as a jumping-off point—to get ideas for activities to do with your own students.

We continue to look for better ways to engage students, so please let us know how these features work for your students.

Sincerely,
Stacy Palen
Laura Kay
Brad Smith
George Blumenthal

Ancillaries for Students

SmartWork

Steven Desch, Guilford Technical Community College; Violet Mager, Susquehanna University; Todd Young, Wayne State College

More than 1,700 questions support *Understanding Our Universe, Second Edition*—all with answer-specific feedback, hints, and ebook links. Questions include Summary Self-Tests and versions of the Explorations (based on AstroTours and the University of Nebraska simulations) and Reading Astronomy News questions. Image-labeling questions based on NASA images allow students to apply course knowledge to images that are not contained in the text. Astronomy in Action video questions focus on overcoming common misconceptions, while Process of Science questions take students through the steps of a discovery and ask them to participate in the decision-making process that leads to that discovery.

Student Website wwnpag.es/uou2

W. W. Norton's free and open student website features the following:

- Twenty-eight AstroTour animations. These animations, some of which are interactive, use art from the text to help students visualize important physical and astronomical concepts.
- Nebraska Simulations (sometimes called applets; or NAAPs, for Nebraska Astronomy Applet Program). These simulations allow students to manipulate variables and see how physical systems work.
- Astronomy in Action videos demonstrate the most important concepts in a visual, easy to understand, and memorable way.

Learning Astronomy by Doing Astronomy: Collaborative Lecture Activities

Stacy Palen, Weber State University, and Ana Larson, University of Washington

Many students learn best by doing. Devising, writing, testing, and revising suitable in-class activities that use real astronomical data, illuminate astronomical concepts, and ask probing questions requiring students to confront misconceptions can be challenging and time consuming. In this workbook, the authors draw on their experience teaching thousands of students in many different types of courses (large in-class, small in-class, hybrid, online, flipped, and so forth) to provide 30 field-tested activities that can be used in any classroom today. The activities have been designed to require no special software, materials, or equipment and to take no more than 50 minutes each to do.

Starry Night Planetarium Software (College Version 7) and Workbook

Steven Desch, Guilford Technical Community College

Starry Night is a realistic, user-friendly planetarium simulation program designed to allow students in urban areas to perform observational activities on a computer screen. Norton's unique accompanying workbook offers observation assignments that guide students' virtual explorations and help them apply what they've learned from the text reading assignments. The workbook is fully integrated with *Understanding Our Universe, Second Edition*.

For Instructors

Instructor's Manual

Ben Sugerman, Goucher College

This resource includes brief chapter overviews, suggested classroom discussions/activities, notes on the AstroTour animations and Nebraska Simulations contained on the Instructor's Resource Disk, teaching suggestions for how to use the Reading Astronomy News and the Exploration activity elements found in the textbook, and worked solutions to all end-of-chapter questions and problems.

Test Bank

Ray O'Neal, Florida A&M University; Todd Vaccaro, St. Cloud State University; Lisa M. Will, San Diego City College

The Test Bank has been revised using Bloom's Taxonomy and provides a quality bank of more than 900 items. Each chapter of the Test Bank consists of six question levels classified according to Bloom's Taxonomy:

Remembering
Understanding
Applying
Analyzing
Evaluating
Creating

Questions are further classified by section and difficulty, making it easy to construct tests and quizzes that are meaningful and diagnostic. The question types include short answer and multiple choice.

Norton Instructor's Resource Website
This Web resource contains the following resources to download:

- Test Bank, available in ExamView, Word RTF, and PDF formats.
- Instructor's Manual in PDF format.
- Lecture PowerPoint slides with lecture notes.
- All art and tables in JPEG and PPT formats.
- Starry Night College, W. W. Norton Edition, Instructor's Manual.
- AstroTour animations, some of which are interactive, use art from the text to help students visualize important physical and astronomical concepts.
- University of Nebraska simulations (sometimes called applets; or NAAPs, for Nebraska Astronomy Applet Program). Well known by introductory astronomy instructors, the simulations allow students to manipulate variables and see how physical systems work.
- Coursepacks, available in BlackBoard, Angel, Desire2Learn, and Moodle formats.

Coursepacks
Norton's Coursepacks, available for use in various Learning Management Systems (LMSs), feature all Quiz+ and Test Bank questions, links to the AstroTours and Applets, plus discussion questions from the Reading Astronomy News features, Astronomy in Action video quizzes, Explorations worksheets and pre- and post-tests from the *Learning Astronomy by Doing Astronomy* Workbook. Coursepacks are available in BlackBoard, Angel, Desire2Learn, and Moodle formats.

Instructor's Resource Folder
This two-disk set contains the Instructor's Resource DVD—which contains the same files as the Instructor's Resource Website—and the Test Bank on CD-ROM in ExamView format.

Acknowledgments

Projects of this magnitude always require a large number of participants, many of whom work hard in the background, keeping track of the thousands of small details that no one person could possibly remember. The authors would like to acknowledge the extraordinary efforts of the staff at W. W. Norton: Arielle Holstein, who kept the mail (and the schedule) flowing smoothly; Diane Cipollone, who shepherded the manuscript through the layout process and was very patient with all the late updates to the science material; the copy editor, Christopher Curioli, who made sure that all the grammar and punctuation survived the multiple rounds of the editing process. We would especially like to thank Becky Kohn, who brought years of expertise and a genuine appreciation of science to the developmental editing process; and Erik Fahlgren, who pushed us to consider new (sometimes uncomfortable) ideas and let go of outdated (often comfortable) ideas in our pursuit of the best book we could write.

We'd also like to thank Ron Proctor, of the Ott Planetarium, whose early efforts on the chapter openers were invaluable; and Kiss Me I'm Polish, who turned his sketches into a finished product. Jane Searle managed the production. Hope Miller Goodell was the design director. Rob Bellinger and Julia Sammaritano worked on the media and supplements, and Stacy Loyal will help get this book in the hands of people who can use it.

And we would like to thank the reviewers, whose input at every stage improved the final product:

Second Edition Reviewers
Loren Anderson, *West Virginia University*
Jonathan Barnes, *Salt Lake Community College*
Celso Batalha, *Evergreen Valley College*
Lloyd Black, *Rowan University*
Ann Bragg, *Marietta College*
Eric Bubar, *Marymount University*
Karen Castle, *Diablo Valley College–Pleasant Hill*
Kwang-Ping Cheng, *California State University–Fullerton*
James Cooney, *University of Central Florida*
Noella D'Cruz, *Joliet Junior College*
Declan De Paor, *Old Dominion University*
Ethan Dolle, *Northern Arizona University*
Jess Dowdy, *Abilene Christian University*
Hardin Dunham, *Angelo State University*
Michael Frey, *Cypress College*
Jeffrey Gillis-Davis, *University of Hawaii at Manoa*
Karl Haish, *Utah Valley University*
Anthony Heinzman, *Victor Valley College*
Olenka Hubickyj, *Foothill College*
Doug Ingram, *Texas Christian University*
Donald Isenhower, *Abilene Christian University*
Joe Jensen, *Utah Valley University*
Kishor Kapale, *Western Illinois University*
Viken Kiledjian, *East Los Angeles College*
Lauren Likkel, *University of Wisconsin–Eau Claire*
Paul Mason, *University of Texas–El Paso*
Kent Morrison, *The University of New Mexico*
David Nero, *University of Toledo*
Hon K. Ng, *Florida State University*
Ray O'Neal, *Florida A&M University*
Marina Papenkova, *East Los Angeles College*
Dan Robertson, *Monroe Community College–Rochester*
Dwight Russell, *Baylor University*
Ann Schmiedekamp, *Pennsylvania State University–Abington*
Haywood Smith, *University of Florida*
Mark Sonntag, *Angelo State University*
Ben Sugarman, *Goucher College*
Robert Sweetland, *Redlands Community College*
Don Terndrup, *The Ohio State University*
Tad Thurston, *Oklahoma City Community College*
Robert Tyson, *University of North Carolina–Charlotte*
Trina Van Ausdal, *Salt Lake Community College*
Scott Williams, *Angelo State University*
John Wilson, *Sam Houston State University*
Laura Woodney, *California State University–San Bernardino*
Amy White, *St. Charles Community College*
Kaisa Young, *Nicholls State University*

Previous Editions' Reviewers
James S. Brooks, *Florida State University*
Edward Brown, *Michigan State University*
James Cooney, *University of Central Florida*
Kelle Cruz, *Hunter College*
Robert Friedfeld, *Stephen F. Austin State University*
Steven A. Hawley, *University of Kansas*
Eric R. Hedin, *Ball State University*
Scott Hildreth, *Chabot College*
Emily S. Howard, *Broward College*
Dain Kavars, *Ball State University*
Kevin Krisciunas, *Texas A&M University*
Duncan Lorimer, *West Virginia University*
Jane H. MacGibbon, *University of North Florida*
James McAteer, *New Mexico State University*
Ian McLean, *University of California–Los Angeles*
Jo Ann Merrell, *Saddleback College*
Michele Montgomery, *University of Central Florida*
Christopher Palma, *Pennsylvania State University*
Nicolas A. Pereyra, *University of Texas–Pan American*
Vahe Peroomian, *University of California–Los Angeles*
Dwight Russell, *Baylor University*
Ulysses J. Sofia, *American University*
Michael Solontoi, *University of Washington*
Trina Van Ausdal, *Salt Lake Community College*
Nilakshi Veerabathina, *University of Texas–Arlington*

ASTROTOURS

- The Earth Spins and Revolves 22, 30
- The View from the Poles 26
- The Celestial Sphere and the Ecliptic 26, 30
- The Moon's Orbit: Eclipses and Phases 36
- Kepler's Laws 50
- Velocity, Acceleration, and Inertia 58
- Newton's Laws and Universal Gravitation 63
- Elliptical Orbits
- Light as a Wave, Light as a Photon 76, 95
- Geometric Optics and Lenses 86
- Star Formation 98
- Solar System Formation 105
- The Doppler Effect 115
- Processes That Shape the Planets 129
- Tides and the Moon 137
- Continental Drift 141
- Hot Spot Creating a Chain of Islands 144
- Atmospheres: Formation and Escape 159
- Greenhouse Effect 161
- Cometary Orbits 227
- Stellar Spectrum 252
- Atomic Energy Levels and the Bohr Model 254
- Atomic Energy Levels and Light Emission and Absorption 257
- H-R Diagram 266
- The Solar Core 279
- Dark Matter 369
- Active Galactic Nuclei 373
- Galaxy Interactions and Mergers 375
- Hubble's Law 377, 412
- Big Bang Nucleosynthesis 446

NEBRASKA SIMULATIONS

- Rotating Sky Explorer 22
- Motions of the Sun Simulator 22
- Seasons and Ecliptic Simulator 32
- Phases of the Moon Simulator 45
- Retrograde Motion Animation 48
- Planetary Orbit Simulator 49, 71
- Eccentricity Demonstrator 50
- EM Spectrum Module 79
- Three Views Spectrum Demonstrator 83
- Telescope Simulator 86
- Blackbody Curves and Filters Explorer 100, 252
- Exoplanet Radial Velocity Simulator 117, 125
- Exoplanet Transit Simulator 118, 472
- Solar System Properties Explorer 129
- Tidal Bulge Simulator 155
- Gas Retention Simulator 159
- Parallax Explorer 248
- Hydrogen Atom Simulation 255
- Eclipsing Binary Simulator 264
- Center of a Mass Simulator 264
- Hertzsprung-Russell Diagram Explorer 266, 273, 304, 327
- Spectroscopic Parallax Explorer 268
- Proton-Proton Animation 280, 301
- CNO Cycle 357
- Traffic Density Analogy 367
- Milky Way Rotational Velocity Explorer 368
- Galactic Redshift Simulator 377
- Circumstellar Habitable Zone Explorer 471
- Milky Way Habitability Explorer 472

✦ VISUAL ANALOGIES

Figure 1.2	Scale of the universe	5
Figure 2.3	Projecting the celestial sphere	23
Figure 2.12	Precession of the equinoxes	34
Figure 2.14	Phases of the Moon	36
Figure 3.15	Uniform circular motion	64
Figure 3.16	Gravity and orbits	65
Figure 4.2	Generation of electromagnetic waves	75
Figure 4.3	Propagation of electromagnetic waves	76
Figure 4.5	Quantization and intensity of light	77
Figure 5.6	Dynamic balance between pressure and gravity	103
Figure 5.11	Conservation of angular momentum	107
Figure 5.19	Doppler shift	115
Figure 6.10	Earth's magnetic field	137
Figure 6.13	Convection	140
Figure 7.2	Equilibrium temperature of a planet	160
Figure 8.5	Size of the Great Red Spot	189
Figure 8.15	Magnetic fields of the giant planets	200
Figure 8.16	Jupiter's magnetosphere	201
Figure 9.29	Radiant of a meteor shower	235
Figure 10.2	Stereoscopic vision and parallax	247
Figure 10.6	Energy states of an atom	253
Figure 10.7	Transition between energy states	254
Figure 11.7	Random deflection of photons within the Sun	283
Figure 12.5	Evolution of a red giant	309
Figure 13.6	Collapse of a massive star	334
Figure 13.13	Pulsars	337
Figure 14.1	Edge-on and face-on perspectives of galaxies	361
Figure 14.15	Inferring AGN size from changes in brightness	372
Figure 14.18	Hubble's law	379
Figure 16.4	Redshift due to expansion	416
Figure 16.7	Recombination	419
Figure 16.17	Inflation	431

✦ ASTRONOMY IN ACTION VIDEOS

Part I: Introduction to Astronomy
The Cause of Earth's Seasons
Phases of the Moon
The Celestial Sphere
The Earth-Moon-Sun System
Velocity, Force and Acceleration
Center of Mass

Part II: The Solar System
Changing Equilibrium
Doppler Shift
Tides
Wien's Law
Angular Momentum
Charged particles and Magnetic Forces

Part III: Stars and Stellar Evolution
Parallax
Inverse Square Law
Emission and Absorption
Random Walk
Type II Supernova
Pulsar rotation

Part IV: Galaxies, the Universe, and Cosmology
Expanding Balloon Universe
Size of Active Galactic Nuclei
Observable vs. Actual Universe
Infinity and the Number Line
Galaxy Shapes and Orientation

✦ ABOUT THE AUTHORS

Stacy Palen is an award-winning professor in the physics department and the director of the Ott Planetarium at Weber State University. She received her BS in physics from Rutgers University and her PhD in physics from the University of Iowa. As a lecturer and postdoc at the University of Washington, she taught Introductory Astronomy more than 20 times over 4 years. Since joining Weber State, she has been very active in science outreach activities ranging from star parties to running the state Science Olympiad. Stacy does research in formal and informal astronomy education and the death of Sun-like stars. She spends much of her time thinking, teaching, and writing about the applications of science in everyday life. She then puts that science to use on her small farm in Ogden, Utah.

Laura Kay is an Ann Whitney Olin professor in the Department of Physics and Astronomy at Barnard College, where she has taught since 1991. She received a BS degree in physics from Stanford University, and MS and PhD degrees in astronomy and astrophysics from the University of California–Santa Cruz. She studies active galactic nuclei, using ground-based and space telescopes. She teaches courses in astronomy, astrobiology, women and science, and polar exploration.

Brad Smith is a retired professor of planetary science. He has served as an associate professor of astronomy at New Mexico State University, a professor of planetary sciences and astronomy at the University of Arizona, and as a research astronomer at the University of Hawaii. Through his interest in Solar System astronomy, he has participated as a team member or imaging team leader on several U.S. and international space missions, including *Mars Mariners 6*, *7*, and *9*; *Viking*; *Voyagers 1* and *2*; and the Soviet *Vega* and *Phobos* missions. He later turned his interest to extrasolar planetary systems, investigating circumstellar debris disks as a member of the Hubble Space Telescope NICMOS experiment team. Brad has four times been awarded the NASA Medal for Exceptional Scientific Achievement. He is a member of the IAU Working Group for Planetary System Nomenclature and is Chair of the Task Group for Mars Nomenclature.

George Blumenthal is chancellor at the University of California–Santa Cruz, where he has been a professor of astronomy and astrophysics since 1972. He received his BS degree from the University of Wisconsin–Milwaukee and his PhD in physics from the University of California–San Diego. As a theoretical astrophysicist, George's research encompasses several broad areas, including the nature of the dark matter that constitutes most of the mass in the universe, the origin of galaxies and other large structures in the universe, the earliest moments in the universe, astrophysical radiation processes, and the structure of active galactic nuclei such as quasars. Besides teaching and conducting research, he has served as Chair of the UC–Santa Cruz Astronomy and Astrophysics Department, has chaired the Academic Senate for both the UC–Santa Cruz campus and the entire University of California system, and has served as the faculty representative to the UC Board of Regents.

1 Thinking Like an Astronomer

The illustration on the opposite page shows a pattern in the sky. Over the course of 6 months, a student in North America took photographs of the location where the Sun set along a ridgeline near his house and then compared them with a map of the area. He sees that the Sun gradually moves north as the year progresses from winter to summer. Understanding this pattern is part of what astronomy is all about. Loosely translated, the word **astronomy** means "finding patterns among the stars." However, modern astronomy is about far more than looking at the sky and cataloging the visible stars. What are the Sun and Moon made of? How far away are they? How do stars shine? How did the universe begin? How will it end? Astronomy is a living, dynamic science that seeks the answers to these and many other compelling questions. In this chapter, we will begin the study of astronomy by exploring our place in the universe and the methods of science.

◆ LEARNING GOALS

Scientists seek knowledge using a very specific set of processes, sometimes collectively called the scientific method. Part of this procedure stems from recognizing patterns in nature. Part of it stems from putting those patterns together to understand how they apply in different places and at different times. In the illustration on the opposite page, the student has noticed a pattern in his observations of the setting Sun. By the end of this chapter, you should understand how patterns of observations like these fit into the development of a scientific law and a scientific theory. You should also be able to:

LG 1 Relate our place in the universe to the rest of the universe.

LG 2 Explain how the patterns of our daily lives are connected to the larger universe.

LG 3 Describe our astronomical origins.

LG 4 Describe the scientific method.

LG 5 Extract meaning from a graph.

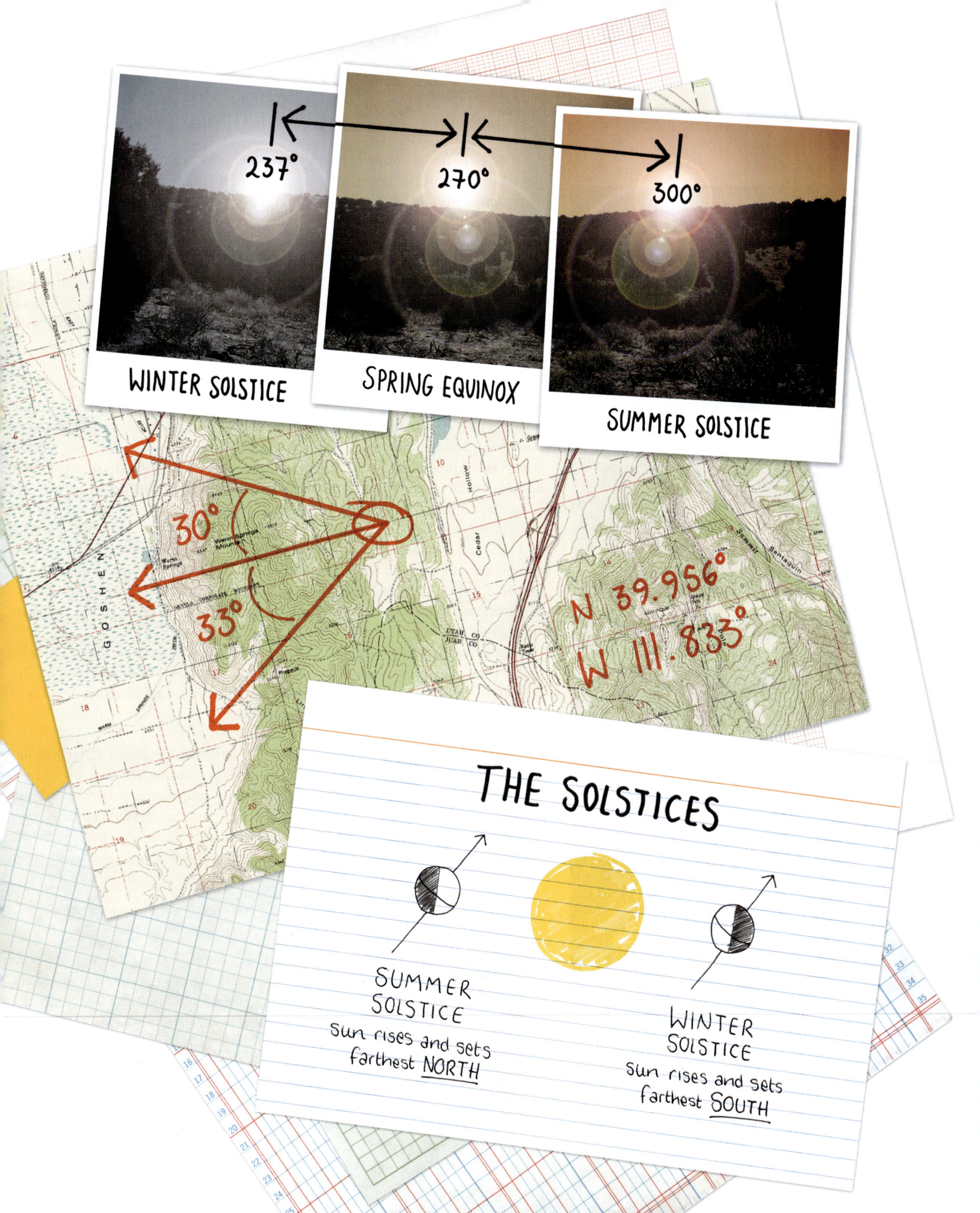

1.1 Astronomy Gives Us a Universal Context

Astronomers think of our place in the universe as both a location and a time. Locating Earth in the larger universe is the first step in learning the science of astronomy.

Our Place in the Universe

Most people have an address where they receive mail—street number, street, city, state, country. But we can expand our view to include the enormously vast universe we live in. What is our "cosmic address"? It might include: planet, star, galaxy, galaxy group, galaxy cluster.

We reside on a planet called Earth, which is orbiting under the influence of gravity around a star called the Sun. The Sun is an ordinary, middle-aged star, more massive and luminous than some stars but less massive and luminous than others. The Sun is extraordinary only because of its importance to us within our own Solar System. Our Solar System consists of eight planets—Mercury, Venus, Earth, Mars, Jupiter, Saturn, Uranus, and Neptune. It also contains many smaller bodies, which we will discuss in coming chapters, including *dwarf planets* (for example, Pluto, Ceres, or Eris), *asteroids* (for example, Ida or Eros), and *comets* (for example, Halley).

The Sun is located about halfway out from the center of the *Milky Way Galaxy*, a flattened collection of stars, gas, and dust. Our Sun is just one among several hundred billion stars scattered throughout our galaxy. Astronomers are discovering that many of these stars also have planets around them, which suggests that planetary systems are common.

The Milky Way, in turn, is part of a small collection of a few dozen galaxies called the Local Group. The Milky Way Galaxy and the Andromeda Galaxy are true giants within the Local Group. Most others are dwarf galaxies. The Local Group itself is part of a vastly larger collection of thousands of galaxies—a supercluster—called the Virgo Supercluster.

We can now define our cosmic address, illustrated in **Figure 1.1**: Earth, Solar System, Milky Way Galaxy, Local Group, Virgo Supercluster. Yet even this address is not complete because the vast structure we just described is only the local universe. The part of the universe that we can see extends much farther—a distance that light takes 13.8 billion years to travel. Within this volume, we estimate that there are *several hundred billion galaxies*—roughly as many galaxies as there are stars in the Milky Way.

The Scale of the Universe

One of the first challenges we face as we begin to think about the universe is its sheer size. A hill is big, and a mountain is really big. If a mountain is really big, then Earth is enormous. But where do we go from there? As the scale of the universe comes to dwarf our human experience, we run out of words. To develop a sense of scale, we can change from talking about distance to talking about time. **Figure 1.2** begins with Earth and progresses outward to the observable universe and illustrates that even relatively small distances in astronomy are so vast that they are measured in units of **light-years (ly)**: the distance light travels in 1 year.

To understand how astronomers use time as a measure of distance, think

Figure 1.1 Our place in the universe is given by our cosmic address: Earth, Solar System, Milky Way Galaxy, Local Group, Virgo Supercluster. We live on Earth, a planet orbiting the Sun in our Solar System, which is a star in the Milky Way Galaxy. The Milky Way is a large galaxy within the Local Group of galaxies, which in turn is located in the Virgo Supercluster.

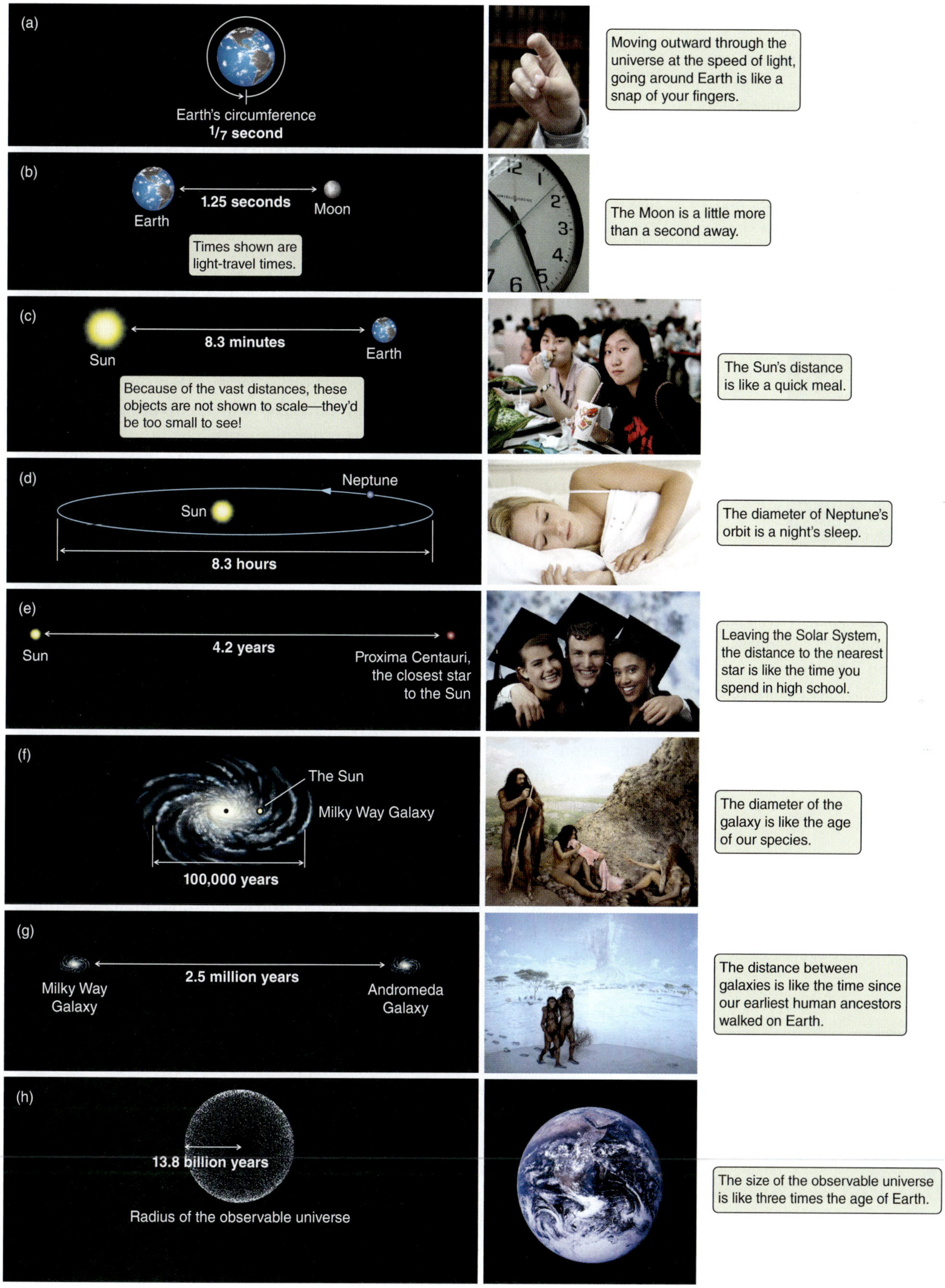

Figure 1.2 Thinking about the time it takes for light to travel between objects helps us to comprehend the vast distances in the observable universe.

(a) Earth's circumference — 1/7 second. Moving outward through the universe at the speed of light, going around Earth is like a snap of your fingers.

(b) Earth — 1.25 seconds — Moon. Times shown are light-travel times. The Moon is a little more than a second away.

(c) Sun — 8.3 minutes — Earth. Because of the vast distances, these objects are not shown to scale—they'd be too small to see! The Sun's distance is like a quick meal.

(d) Sun, Neptune — 8.3 hours. The diameter of Neptune's orbit is a night's sleep.

(e) Sun — 4.2 years — Proxima Centauri, the closest star to the Sun. Leaving the Solar System, the distance to the nearest star is like the time you spend in high school.

(f) The Sun, Milky Way Galaxy — 100,000 years. The diameter of the galaxy is like the age of our species.

(g) Milky Way Galaxy — 2.5 million years — Andromeda Galaxy. The distance between galaxies is like the time since our earliest human ancestors walked on Earth.

(h) 13.8 billion years — Radius of the observable universe. The size of the observable universe is like three times the age of Earth.

 **VISUAL ANALOGY**

Vocabulary Alert

massive In common language, *massive* can mean either "very large" or "very heavy." Astronomers specifically mean that more massive objects have more "stuff" in them.

satellite In common language, *satellite* typically refers to a human-made object. Astronomers use this word to describe any object, human-made or natural, that orbits another object.

about traveling in a car at 60 kilometers per hour (km/h). At 60 km/h, you travel 1 kilometer in 1 minute, or 60 kilometers in 1 hour. In 10 hours, you would travel 600 kilometers. To get a feel for the difference between 1 kilometer and 600 kilometers, you can think about the difference between 1 minute and 10 hours. In astronomy, the speed of a car on the highway is far too slow to be a useful measure of time. Instead, we use the fastest speed in the universe—the speed of light. Light travels at 300,000 kilometers per second (km/s). Light can circle Earth (a distance of 40,000 km) in just under one-seventh of a second—about the time it takes you to snap your fingers.

The Origin and Evolution of the Universe

As we will discuss in detail in Chapter 16, both theory and observation tell us that the universe began 13.8 billion years ago in an event known as the *Big Bang*. The only chemical elements in the early universe were hydrogen and helium, plus tiny amounts of lithium, beryllium, and boron. Yet we live on a planet with a core of iron and nickel, surrounded by an outer layer made up of rocks that contain large amounts of silicon and various other elements. The human body contains carbon, nitrogen, oxygen, sodium, phosphorus, and a host of other chemical elements. If these elements were not present in the early universe, where did they come from?

The answer to this question begins deep within stars. In the core of a star, less **massive** atoms, like hydrogen, combine to form more massive atoms, eventually leading to atoms such as carbon. (Terms in red signify a "Vocabulary Alert" in the margin of the text.) When a star nears the end of its life, it loses much of its material back into space—including some of these more massive atoms. This material combines with material lost from other stars, some of which produced even more massive atoms as they exploded, to form large clouds of dust and gas. Those clouds go on to make new stars and planets, like our Sun and Solar System. Prior "generations" of stars supplied the building blocks for the chemical processes, such as life, that go on around us (**Figure 1.3**). Look around you. Everything you see is made of atoms that were formed in stars long ago.

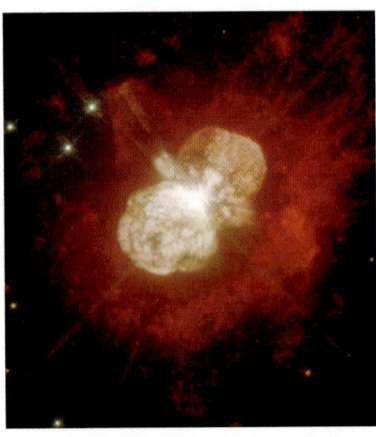

Figure 1.3 You and everything around you contain atoms that were forged in the interiors of stars that lived and died before the Sun and Earth were formed. The supermassive star Eta Carinae is currently ejecting a cloud of enriched material. This star is located about 7,500 light-years from Earth and emits 5 million times more light than the Sun.

An Astronomer's Toolkit

In 1957, the Soviet Union launched Sputnik, the first human-made **satellite.** Since that time, we have lived in an age of space exploration that has given us a new perspective on the universe. The atmosphere that shields us from harmful solar radiation also blocks much of the light that travels through space. Space astronomy shows views hidden from ground-based telescopes by our atmosphere. Satellite observatories have brought us discovery after surprising discovery. Each has forever altered our perception of the universe.

In addition to putting satellites into space around Earth, humans have walked on the Moon (**Figure 1.4**), and unmanned probes have visited all eight planets. Spacecraft have flown past asteroids, comets, and even the Sun. Spacecraft have also landed on Mars, Venus, Titan (Saturn's largest moon), and an asteroid and have plunged into both the atmosphere of Jupiter and the heart of a comet. Most of what we know of the Solar System has resulted from these past six decades of exploration since the space age began.

Astronomers collect information from many varieties of light, from highest-energy *gamma rays* (G) and *X-rays* (X), through *ultraviolet* (U), visible

Figure 1.4 *Apollo 15* (1971) was the fourth U.S. mission to land on the moon. Here astronaut James B. Irwin stands by the lunar rover during an excursion to explore and collect samples from the Moon.

(V), and *infrared* (I) radiation, down to the lowest-energy radio waves (R). **Figure 1.5** combines a visible-light image of the Parkes radio telescope and an image of the Milky Way in the radio part of the spectrum, illustrating the new perspectives we have gained from improved technology. The "R" beneath the photograph stands for radio waves (see the abbreviations defined earlier in this paragraph); in this text, the type of light used to obtain an image is indicated by the letter that is highlighted in the wave graphic appearing below the image.

Another tool of astronomy—telescopes—often comes to mind when we think of studying space. However, the 21st-century astronomer spends far more time staring at a computer screen than peering through the eyepiece of a telescope. Modern astronomers use computers to collect and analyze data from telescopes, calculate physical models of astronomical objects, and prepare reports on the results of their work. You may also be surprised to learn that much astronomy is now carried out in large physics facilities like the one shown in **Figure 1.6**. Astronomers work with scientists in related fields, such as physics, chemistry, geology, and planetary science, to develop a deeper understanding of physical laws and to make sense of their observations of the distant universe.

Figure 1.5 In the 20th century, advances in telescope technology opened new windows on the universe. This is the Milky Way as we would see it if our eyes were sensitive to radio waves, shown as a backdrop to the Parkes radio telescope in Australia. The bright blue object is an artist's impression of a fast radio burst.

1.2 Science Is a Way of Viewing the World

As we view the universe through the eyes of astronomers, we will also learn how science works. Science is a way of exploring the physical world through the scientific method.

The Scientific Method

The **scientific method** is a systematic way of testing new ideas or explanations. Often, the method begins with a fact—an observation or a measurement. For example, you might observe that the weather changes in a predictable way each year and wonder why that happens. You then create a **hypothesis**, a testable explanation of the observation: "I think that it is cold in the winter and warm in the summer because Earth is closer to the Sun in the summer." You and your colleagues come up with a test: if it is cold in the winter and warm in the summer because Earth is closer to the Sun in the summer, then it will be cold in the winter everywhere on the planet—Australia should have winter at the same time of year as the United States. This test can be used to falsify your hypothesis. You travel to the opposite hemisphere in the winter and find that it is summer there. Your hypothesis has just been **falsified**, which means that it has been proved incorrect. This is good! It means you know something you didn't know before. Now you must revise or completely change your hypothesis to be consistent with the new data.

Any idea that is not testable—that is not **falsifiable**—must be accepted or rejected based on intuition alone, so it is not scientific. A falsifiable hypothesis or idea does not have to be testable using current technology, but we must be able to imagine an experiment or observation that *could* prove the idea wrong if we could carry it out. As continuing tests support a hypothesis by failing to disprove it, scientists come to accept the hypothesis as a *theory*. A classic example is Einstein's theory of relativity, which has withstood more than a century of scientific efforts to disprove its predictions.

Figure 1.6 The high-energy particle collider at the European Organization for Nuclear Research (CERN), shown here as a circle drawn above the tunnels of the facility, has provided clues about the physical environment during the birth of the universe. Laboratory astrophysics, in which astronomers model important physical processes under controlled conditions, has become an important part of astronomy. The dashed line represents the boundary between France and Switzerland.

Vocabulary Alert

falsified/falsifiable: In common language, we are likely to think of "falsified" evidence as having been *manipulated* to misrepresent the truth. Astronomers (and scientists in general) use *falsifiable* in the sense of "being able to prove a hypothesis false," as we will throughout this book.

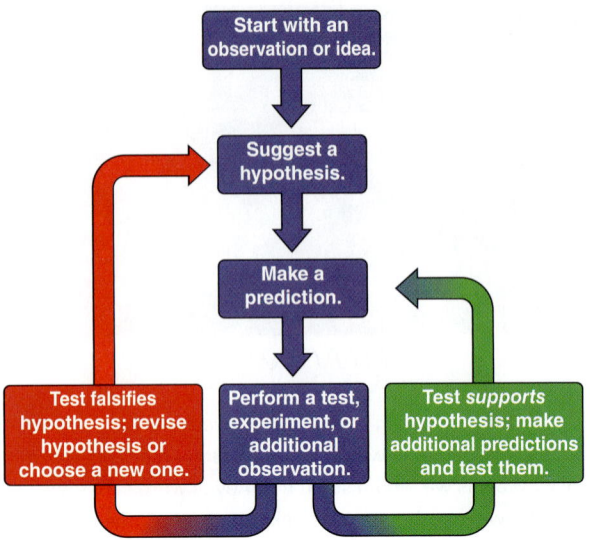

Figure 1.7 The scientific method is the path by which an idea or observation leads to a falsifiable hypothesis. The hypothesis is either accepted as a tested theory or rejected on the basis of observational or experimental tests of its predictions. The green loop goes on indefinitely as scientists continue to test the hypothesis.

Vocabulary Alert

theory In common language, a *theory* is weak—just an idea or a guess. Scientists use this word to label the most well-known, well-tested, and well-supported principles in science.

The path to scientific knowledge is solidly based on the scientific method. **Figure 1.7** illustrates the pathway of the scientific method. It begins with an observation or idea, followed by a hypothesis, a prediction, further observation or experiments to test the prediction, and perhaps ending as a tested theory. Look back at the chapter-opening illustration. Where does the activity represented in that figure fit onto this simplified flowchart?

Still, science can no more be said to be the scientific method than music can be said to be the rules for writing down a musical score. The scientific method provides the rules for testing whether an idea is false, but it offers no insight into where the idea came from in the first place or how an experiment was designed. Scientists discussing their work use words such as *insight*, *intuition*, and *creativity*. Scientists speak of a beautiful theory in the same way that an artist speaks of a beautiful painting or a musician speaks of a beautiful performance. Science has an aesthetic that is as human and as profound as any found in the arts.

The Language of Science

We have already seen that scientists often use everyday words in special ways. For example, in everyday language, *theory* may mean something that is little more than a guess: "Do you have a theory about who might have done it?" "My theory is that a third party could win the next election." In everyday language, a theory isn't something we take too seriously. "After all," we say, "it is only a theory."

In stark contrast, scientists use the word **theory** to mean a carefully constructed proposition that takes into account every piece of data as well as our entire understanding of how the world works. A theory has been used to make testable predictions, and all of those predictions have come true. Every attempt to prove it false has failed. A theory such as the theory of general relativity is not a mere speculation but is instead a crowning achievement of science. Even so, scientific theories are accepted only as long as their predictions are correct. A theory that fails only a single test is proved false. In this sense, all scientific knowledge is subject to challenge.

Theories are at the top of the loosely defined hierarchy of scientific knowledge. At the bottom is an *idea*—a notion about how something might be. Moving up the hierarchy we come to a *fact*, which is an observation or measurement. The radius of Earth is a fact, for example. A *hypothesis* is an idea that leads to testable predictions. A hypothesis may be the forerunner of a scientific theory, or it may be based on an existing theory, or both. At the top we reach a *theory*: an idea that has been examined carefully, is consistent with all existing theoretical and experimental knowledge, and makes testable predictions. Ultimately, the success of the predictions is the deciding factor between competing theories. A *law* is a series of observations that lead to an ability to make predictions but has no underlying explanation of why the phenomenon occurs. So we might have a "law of daytime" that says the Sun rises and sets once each day. And we could have a "theory of daytime" that says the Sun rises and sets once each day because Earth spins on its axis. Scientists themselves are sometimes sloppy about the way they use these words, and you will sometimes see them used differently than in these formal definitions.

Underlying this hierarchy of knowledge are scientific principles. A scientific principle is a general idea about how the universe is that guides our construction of new theories. For example, at the heart of modern astronomy is the cosmological principle. The **cosmological principle** is the testable assumption that the same physical laws that apply here and now also apply everywhere and at all times. This implies that there are no special locations or directions in the universe. The physical laws that act in laboratories also act in the centers of stars or in the hearts of distant galaxies. Each new theory that succeeds in explaining patterns and relationships among objects in the sky adds to our confidence in this cornerstone of our worldview.

This principle provides an example of Occam's razor, another guiding principle in science. **Occam's razor** states that when we are faced with two hypotheses that explain all the observations equally well, we should use the one that requires the fewest assumptions, until we have evidence to the contrary. For example, we might hypothesize that atoms are constructed differently in the Andromeda Galaxy than in the Milky Way Galaxy. This would be a violation of the cosmological principle. But that hypothesis would require a large number of assumptions about how the atoms are constructed and yet still appear to behave identically to atoms in the Milky Way. For example, we might assume that the center of the atom is negatively charged in Andromeda, opposite to the Milky Way, where the center of the atom is positively charged. Then we would need to make an assumption about where the boundary is between Andromeda-like matter and Milky Way–like matter. And then we would need to make an assumption about why atoms on the boundary between the two regions did not destroy each other. And we would need an assumption about *why* atoms in the two regions are constructed so differently; and so on. If reasonable experimental evidence is ever found that the cosmological principle is not true, scientists will construct a new description of the universe that takes the new data into account. Until then, it is the hypothesis that has the fewest assumptions, satisfying Occam's razor. To date, the cosmological principle has been repeatedly tested and remains unfalsified.

Scientific Revolutions

Limiting our definition of science to the existing theories fails to convey the dynamic nature of scientific inquiry. Scientists do not have all the answers and must constantly refine their ideas in response to new data and new insights. The vulnerability of knowledge may seem like a weakness. "Gee, you really don't know anything," the cynical person might say. But this vulnerability is actually science's great strength, because it means that science self-corrects. Wrong ideas are eventually overturned by new information. In science, even our most cherished ideas about the nature of the physical world remain fair game, subject to challenge by new evidence. Many of history's best scientists earned their status by falsifying a universally accepted idea. This is a powerful motivation for scientists to challenge old ideas constantly, inventing new explanations for our observations.

For example, the classical physics developed by Sir Isaac Newton in the 17th century withstood the scrutiny of scientists for more than 200 years. It seemed that little remained but cleanup work—filling in the details. Yet during the late 19th and early 20th centuries, a series of scientific revolutions completely changed our understanding of the nature of reality. Albert Einstein (**Figure 1.8**) is representative of these scientific revolutions. Einstein's special and general theories of relativity replaced Newton's mechanics. Einstein did not prove Newton wrong, but instead showed that Newton's theories were a special case of a far more general and powerful set of physical laws. Einstein's new ideas unified the concepts of mass and energy and destroyed the conventional notion of space and time as separate entities.

Figure 1.8 Albert Einstein is perhaps the most famous scientist of the 20th century. Einstein helped usher in two different scientific revolutions, one of which he was never able to accept.

Throughout this text, you will encounter many other discoveries and successful ideas that forced scientists to abandon accepted theories. Einstein himself never embraced the view of the world offered by *quantum mechanics*—a second revolution he helped start. Yet quantum mechanics, a statistical description of the behavior of particles smaller than atoms, has held up to challenges for more than 100 years. In science, all authorities are subject to challenge, even Einstein.

1.3 Astronomers Use Mathematics to Find Patterns

One of the primary advantages of scientific thinking is that it allows us to make predictions. Once a pattern, such as the rising and setting of the Sun, has been observed, scientists can predict what will happen next.

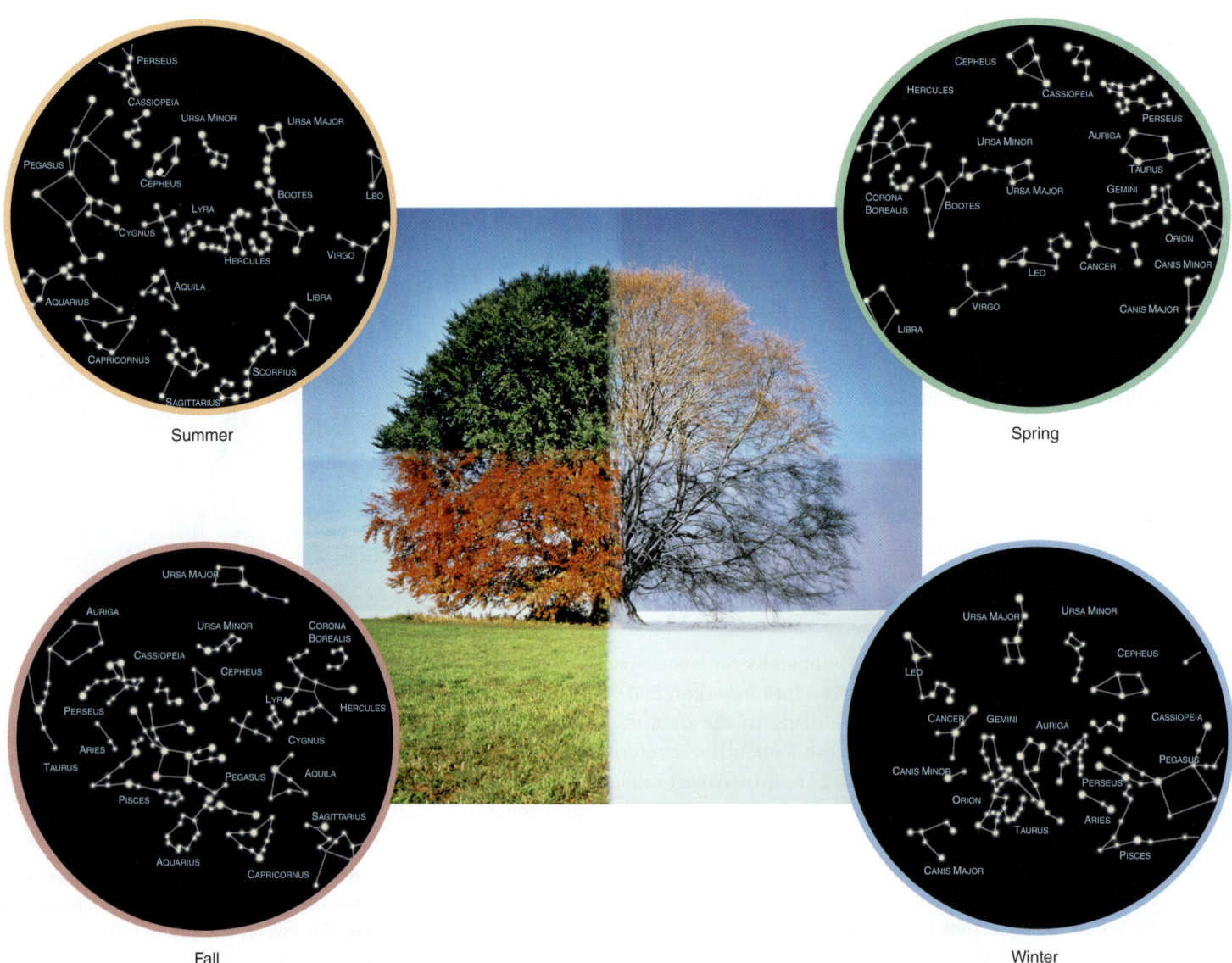

Figure 1.9 Since ancient times, our ancestors have recognized that patterns in the sky change with the seasons. These and other patterns shape our lives. These star maps show the sky in the Northern Hemisphere during each season.

Finding Patterns

Imagine that the patterns in your life became disrupted, so that things became entirely unpredictable. For example, what would life be like if sometimes when you let go of an object, it fell up instead of down? Or what if one day the Sun rose at noon and set at 1:00 P.M., the next day it rose at 6:00 A.M. and set at 10:00 P.M., and the day after that the Sun did not rise at all? In fact, objects do fall toward the ground. The Sun rises, sets, and then rises again at predictable times, and in predictable locations, as shown in the chapter-opening illustration. Spring turns into summer, summer turns into autumn, autumn turns into winter, and winter turns into spring. The rhythms of nature produce patterns in our lives, and these patterns give us clues about the nature of the physical world.

Astronomers identify and characterize these patterns and use them to understand the world around us. Some of the most easily identified patterns in nature are those we see in the sky. What in the sky will look different or the same a week from now? A month from now? A year from now? As you can see in **Figure 1.9**, patterns in the sky mark the changing of the seasons, which determines the planting and harvesting of crops. It is no surprise that astronomy, which studies these patterns that are so important to agriculture, is the oldest of all sciences.

Reading Graphs

Astronomers use mathematics to analyze patterns. There are many branches of mathematics, most of them dealing with more than just numbers. But all branches of mathematics share one thing: they deal with patterns.

Mathematical patterns are often shown in graphical form. Reading graphs is a skill that is important not only in astronomy but also in life. Economists, social and political scientists, mortgage brokers, financial analysts, retirement planners, doctors, and scientists all use graphs to evaluate and communicate important information. **Figure 1.10** shows the relationship between distance and time in a car trip. As with all graphs, there are three key pieces: the independent variable, the dependent variable, and the plotted data. The independent variable is typically the one the experimenter has control over. In this case, this is time: the experimenter chooses the times at which she will make a measurement. The independent variable is always shown on the horizontal (x) axis. The dependent variable *depends* on the independent variable. Here, the dependent variable is distance: the experimenter measures the distance at various times. The vertical (y) axis reflects the dependent variable.

When you see a graph, you should first look at the axes to find out what the variables are; that is, what information is plotted on the graph, and in what units. Here, we have distance in miles plotted against time in minutes.

The third key piece of information is the data plotted. Trends in these data show the behavior of the system. Study the graph. You can see that the first data point (a) is plotted at 10 minutes on the horizontal axis and 5 miles on the vertical axis. During those first 10 minutes, the distance becomes steadily larger, as the car travels 5 miles from its starting place. The upward rise means the distance is increasing with time.

The slope of the line gives the speed of the car and is calculated by dividing the "rise" (the change in the vertical axis) by the "run" (the change on the horizontal axis). To get a feel for what's happening, you need to understand the slope. A horizontal slope means that the distance is not changing: the car is parked.

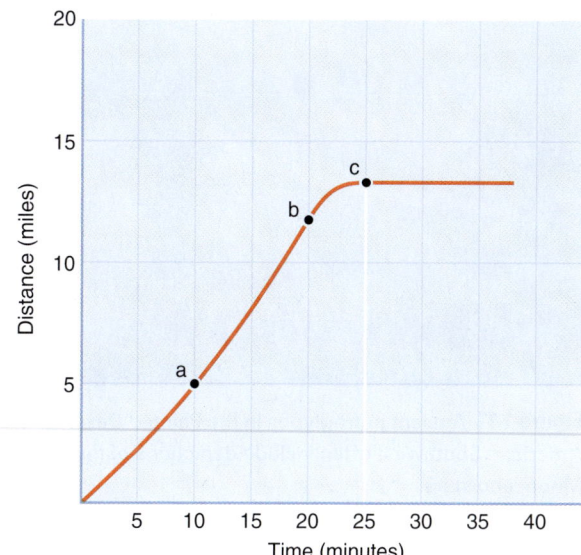

Figure 1.10 This graph of the relationship between distance and time in a car trip shows changes in speed.

A rising slope means the distance is increasing: the car is moving away. A falling slope means the distance is getting smaller: the car is approaching. A steeper slope means the car is moving faster.

During the second 10 minutes, between times (a) and (b), the car travels 7 miles. Notice that the line rises more sharply—the slope is steeper. The steeper slope tells you that the car is going faster, even if you don't calculate the speed exactly.

After 20 minutes, the car begins to slow down. The distance is still increasing, as the plotted data are still rising. But the slope becomes shallower, until the curve flattens out at 25 minutes (c), and the driver arrives at the destination and parks the car.

Many other patterns can be seen in graphs, and you will encounter graphs often in your life. Always remember to study four things when you see a new graph: What is plotted on the horizontal axis? What is plotted on the vertical axis? What is the trend of the data (up, down, or flat), and how fast is it changing?

If patterns are the heart of science, and mathematics is the language of patterns, it should come as no surprise that *mathematics is the language of science.* Trying to study science while avoiding mathematics is the practical equivalent of trying to study Shakespeare while avoiding the written or spoken word. It cannot be done meaningfully. Distaste for mathematics is one of the most common obstacles standing between a nonscientist and an appreciation of the world as seen through the eyes of a scientist.

Part of the responsibility for moving beyond this obstacle lies with us, the authors. It is our job to take on the role of translators, using words to express concepts as much as possible, even when these concepts are more concisely and accurately expressed mathematically. When we do use mathematics, we will explain in everyday language what the equations mean and show you how equations express concepts that you can connect to the world. We will also limit the mathematics to a few tools that we would like you to know. These mathematical tools, a few of which are described in **Working It Out 1.1**, enable scientists to convey complex information compactly and accurately.

Your responsibility is to accept the challenge and make an honest effort to think through the mathematical concepts that we use. The mathematics in this book are on a par with what it takes to balance a checkbook, build a bookshelf that stands up straight, check your gas mileage, estimate how long it will take to drive to another city, or buy enough food to feed an extra guest or two at dinner.

Almost everyone harbors a spark of interest in astronomy. Because you are reading this book, you probably share this spark as well. The prominence of the Sun, Moon, and stars in cave paintings and rock drawings, such as the one in **Figure 1.11**, which dates back a thousand years or more, tells us that these objects have long occupied the human imagination. Your initial interest may have grown over the years as you saw or read news reports about spectacular discoveries made in your lifetime—some so amazing that they seemed to blur the line between science fact and science fiction. If you are like many people, your understanding of astronomy may not yet have gone much beyond learning about the constellations and the names of the stars in them. We hope this book will take you to places you never imagined going and lead you to new insights and understandings.

Figure 1.11 Ancient petroglyphs in the Painted Desert of the American Southwest often include depictions of the Sun, Moon, and stars.

Working It Out 1.1 | Units and Scientific Notation

Mathematics is the language of patterns, and the universe "speaks" math. If you want to ask questions about the universe and understand the answers, you have to speak math too. Throughout the text, we present the math as needed, with worked examples to help you understand how to work it out on your own. We begin with two essential pieces: units, which relate numbers to the physical quantities they represent, and scientific notation, which makes it possible to express compactly the extremely large numbers that are common in astronomy.

Units

Scientists use the metric system of units because most units are related to each other by multiplying or dividing by 10 and so simply involve moving the decimal place to the right or left. Metric measurements also have prefixes that show the relationship of the units. There are 100 *centi*meters (cm) in a meter, 1,000 meters in a *kilo*meter (km), and 1,000 grams in a *kilo*gram (kg). To convert from meters to centimeters, multiply by 100 by moving the decimal place two spaces to the right. For example, 25 meters is $25 \times 100 = 2,500$ cm. But to convert from yards to inches, you must first multiply by 3 to convert from yards to feet and multiply that result by 12 to convert feet to inches.

Though many of the units we use will be unfamiliar, you can start to develop an intuition about them. For example:

- 1 meter is about 3 feet (ft).
- 1 cm is a bit less than half an inch.
- 1 km is approximately two-thirds of a mile.

- **Approximately how many feet are in 2.5 meters?** Because 1 meter is about 3 ft, 2.5 meters is about $3 \times 2.5 = 7.5$ ft. This is about the diameter of the Hubble Space Telescope's primary mirror.
- **Approximately how many centimeters are in a foot?** We can estimate this value in two ways. We know that 1 ft is 12 inches, and 1 inch is about 2.5 cm. Therefore, 1 ft = 12 inches × 2.5 cm / inch ≅ 30 cm. Here is another way to calculate an approximate conversion. Because 1 meter is approximately 3 ft, 1 ft is approximately ⅓ meter. We know that 1 meter is 100 cm, so there are roughly 33 cm in a foot. If we measure precisely, we find that 1 ft is 30.48 cm, which falls between our two estimates. The telescope used to discover Pluto was just over a foot in diameter.

- **How can you check your own work?** One way to check your own work is to think about whether the answer should be larger or smaller than the original number. The number of centimeters, for example, should always be larger than the number of meters, because there is more than 1 cm in a meter. Often, a quick estimate can also help. For example, 1 ft is about an "order of magnitude" (a power of 10) larger than a centimeter; in other words, there are a few tens—not hundreds—of centimeters in a foot. So if you have converted 1.2 ft to centimeters and calculated an answer of 3,200 cm, you should try again.

Scientific Notation

Scientists deal with numbers of vastly different sizes using scientific notation. Writing out 7,540,000,000,000,000,000,000 in standard notation is very inefficient. Scientific notation uses the first few digits (the "significant" ones) and counts the number of decimal places to create the condensed form 7.54×10^{21}. Similarly, rather than writing out 0.000000000005, we write 5×10^{-12}. The exponent on the 10 tells you where to move the decimal place. If it is positive, move the decimal place that many places to the right. If it's negative, move the decimal place that many places to the left. For example, the average distance to the Sun is 149,598,000 km, but astronomers usually express it as 1.49598×10^8 km.

- **Write the number 13.8×10^9 in standard notation.** Because the exponent of the 10 is 9, and it is positive, we must move the decimal point nine places to the right: 13,800,000,000. This number is the age of the universe in years. Usually, we would say "thirteen point eight billion years," where the word *billion* stands in for "times 10 to the 9."
- **Write the number 2×10^{-10} in standard notation.** The exponent of the 10 is −10. It is negative, so we must move the decimal point 10 places to the left: 0.0000000002. This is about the size of an atom in meters.
- **Write the number 50,000 in scientific notation.** To put this number in scientific notation, we must move the decimal place four places to the left: 5×10^4. This is about the radius of the Milky Way Galaxy in light-years (ly). To convert back to standard notation, we would move the decimal place to the right, so the exponent is positive.

Units and Scientific Notation CONTINUED

- **Write the number 0.000000570 in scientific notation.** To put this in scientific notation, we must move the decimal place seven places to the right: 5.70×10^{-7}. We could also write this as 570×10^{-9}. Why would we do that? Because we have special names for units every three decimal places. The special name for 10^{-9} meters is *nanometer* (nm). So 570×10^{-9} meters is 570 nm. This is the wavelength of yellow light. The *nano* part of this special name is called a prefix. Other useful prefixes can be found inside the front cover of this book.

Calculator hint: How do you put numbers in scientific notation into your calculator? Most scientific calculators have a button that says EXP or EE. These mean "times 10 to the." So for 4×10^{12}, you would type [4][EXP][1][2] or [4][EE][1][2] into your calculator. Usually, this number shows up in the window on your calculator either just as you see it written in this book or as a 4 with a smaller 12 all the way over on the right of the window.

READING ASTRONOMY News

This article from the New York Times describes the astronomical controversy that developed around the change of Pluto's classification from "planet" to "dwarf planet." Many people felt very strongly about this at the time.

Pluto Is Demoted to "Dwarf Planet"

By **DENNIS OVERBYE**, *New York Times*

Pluto got its walking papers today.

Throw away the placemats. Grab a magic marker for the classroom charts. Take a pair of scissors to the solar system mobile.

After years of wrangling and a week of bitter debate, astronomers voted on a sweeping reclassification of the solar system. In what many of them described as a triumph of science over sentiment, Pluto was demoted to the status of a "dwarf planet."

In the new solar system, there are eight planets, at least three dwarf planets, and tens of thousands of so-called "smaller solar system bodies," like comets and asteroids.

For now, the dwarf planets include, besides Pluto, Ceres, the largest asteroid, and an object known as UB 313, nicknamed "Xena," that is larger than Pluto and, like it, orbits out beyond Neptune in a zone of icy debris known as the Kuiper Belt. But there are dozens more potential dwarf planets known in that zone, planetary scientists say, and the number in that category could quickly swell.

In a nod to Pluto's fans, the astronomers declared Pluto to be the prototype for a new category of such "trans-Neptunian" objects but failed in a close vote to approve the name Plutonians for them.

"The new definition makes perfect sense in terms of the science we know," said Alan Boss, a planetary theorist at the Carnegie Institution of Washington, adding that it doesn't go too far in cultural terms. "We have a duty to satisfy the whole world."

The vote completed a stunning turnaround from only a week ago when the assembled astronomers had been presented with a proposal that would include 12 planets, including Pluto, Ceres, Xena, and even Pluto's moon Charon. Dr. Boss said today's decision spoke to the integrity of the planet-defining process. "The officers were willing to change their resolution and find something that would stand up under the highest scientific scrutiny and be approved," he said.

Jay Pasachoff, a Williams College astronomer who favored somehow keeping Pluto a planet, said, "The spirit of the meeting was of future discovery and activity in science rather than any respect for the past."

Mike Brown of the California Institute of Technology, who as the discoverer of Xena had the most to lose personally from Pluto's and Xena's downgrading, said he was relieved. "Through this whole crazy circus-like procedure, somehow the right answer was stumbled

SEE PLUTO

on," he said. "It's been a long time coming. Science is self-correcting eventually, even when strong emotions are involved."

It has long been clear that Pluto, discovered in 1930, stood apart from the previously discovered planets. Not only was it much smaller than them, only about 1,600 miles in diameter, smaller than the Moon, but its elongated orbit is tilted with respect to the other planets and it goes inside the orbit of Neptune part of its 248-year journey around the Sun.

Pluto makes a better match with the other ice balls that have since been discovered in the dark realms beyond Neptune, they have argued. In 2000, when the new Rose Center for Earth and Space opened at the American Museum of Natural History, Pluto was denoted in a display as a Kuiper Belt Object and not a planet.

Two years ago, the International Astronomical Union appointed a working group of astronomers to come up with a definition that would resolve this tension. The group, led by Iwan Williams of Queen Mary University in London, deadlocked. This year a new group with broader roots, led by Owen Gingerich of Harvard, took up the problem.

According to the new rules, a planet [must] meet three criteria: it must orbit the Sun, it must be big enough for gravity to squash it into a round ball, and it must have cleared other things out of the way in its orbital neighborhood. The latter measure knocks out Pluto and Xena, which orbit among the icy wrecks of the Kuiper Belt, and Ceres, which is in the asteroid belt.

Dwarf planets only have to be round.

"I think this is something we can all get used to as we find more Pluto-like objects in outer solar system," Dr. Pasachoff said.

The final voting came from about 400 to 500 of the 2,400 astronomers who were registered at the meeting of the International Astronomical [Union] in Prague. Many of the astronomers, Dr. Pasachoff explained, had already left, thinking there would be nothing but dry resolutions to decide in the union's final assembly.

It was hardly the first time that astronomers have rethought a planet. The asteroid Ceres was hailed as the eighth planet when it was first discovered in 1801 by Giovanni Piazzi floating in the space between Mars and Jupiter. It remained a "planet" for about half a century until the discovery of more and more things like it in the same part of space led astronomers to dub them asteroids.

In the aftermath, some astronomers pointed out that the new definition only applies to our own solar system and that there was so far no such thing as an extra-solar planet.

The decision was bound to have both a cultural and economic impact on the industry of astronomical artifacts and toys, publishing and education. The *World Book Encyclopedia*, for example, had been holding the presses for its new 2007 edition until Pluto's status could be clarified.

Neil deGrasse Tyson, director of the Hayden Planetarium in New York, said children are flexible when asked about the cultural impact of today's redefinition. He said that he had not bothered to watch the International Astronomical Union's vote in the Internet, as many astronomers did. "Counting planets is not an interesting exercise to me," he said. "I'm happy however they choose to define it. It doesn't really make any difference to me."

Dr. Tyson said the continuing preoccupation with what the public and schoolchildren would think about this was a concern and a troubling precedent. "I don't know any other science that says about its frontier, 'I wonder what the public thinks,'" he said. "The frontier should move in whatever way it needs to move."

Evaluating the News

1. Why was Pluto reclassified? Is Pluto any different now than it was before?
2. Mike Brown makes the statement in this article that "Science is self-correcting eventually, even when strong emotions are involved." What does he mean by this? How does it relate to what you learned about the nature of science in Section 1.2?
3. Is the reclassification of Pluto a unique event? Have similar "self-corrections" happened before? Are these self-corrections a weakness or a strength of science? (Consider Section 1.2.)
4. One way to evaluate scientific claims is to consider a what-if question. What if Pluto had not been reclassified? How would we classify Ceres, Charon, and the many other Pluto-like objects? How many planets would there be in the Solar System if we included all the ones that are Pluto-like?
5. How does this article fit into the scheme of the scientific method that you learned about in Section 1.2? Explain how the reclassification of Pluto is an example of the scientific method at work.

Dennis Ovesbye: "Vote Makes It Official: Pluto isn't What It Used to Be." From *The New York Times*, August 25, 2006. Copyright © 2006 The New York Times. All rights reserved. Used by permission and protected by the copyright laws of the United States. The printing. redistribution, or retransmission of this Content without express permission is prohibited.

SUMMARY

Astronomy is the study of patterns in the sky. This branch of science seeks answers to many compelling questions about the universe. It uses all available tools to follow the scientific method. Science is based on objective reality, physical evidence, and testable hypotheses. The great strength of science lies in its ability to improve continuously its description of objective reality, as new information becomes available.

LG 1 Earth is a small planet orbiting a more or less average star, halfway out in the disk of a spiral galaxy that is one among billions.

LG 2 The physical laws that apply to everyday life on Earth are the same physical laws that apply everywhere else in the universe.

LG 3 The atoms that make up Earth—and, indeed, you, yourself—come from particles that have been around since the Big Bang, processed through stars, and collected into the Sun and the Solar System.

LG 4 The scientific method is an approach to learning about the physical world. It includes observation, forming hypotheses, making predictions to enable the testing and refining of those hypotheses, and repeated testing of theories.

LG 5 Scientists often represent information graphically. Reading a graph means looking carefully at the axes and the trends in the data, which can be found from the slope.

SUMMARY SELF-TEST

1. Rank the following in order of increasing size: Sun, Virgo Supercluster, Earth, Solar System, Local Group, Milky Way Galaxy, universe.

2. If we compare our place in the universe with a very distant place,
   a. the laws of physics are different in each place.
   b. some laws of physics are different in each place.
   c. all of the laws of physics are the same in each place.
   d. some laws of physics are the same in each place, but we don't know about others.

3. The poetic statement "We are stardust" means that
   a. Earth exists because of the collision of two stars.
   b. the atoms in our bodies have passed through (and in many cases formed in) stars.
   c. Earth is primarily formed of material that used to be in the Sun.
   d. Earth and the other planets will eventually form a star.

4. The scientific method is a way of trying to _____ , not prove, ideas; thus all scientific knowledge is provisional.

5. In a graph of distance versus time (like the one in Figure 1.10), a line that went from upper left to lower right slope would mean that
   a. the car is approaching.
   b. the car is slowing down.
   c. the car is moving away but is behind you.
   d. the car is speeding up but is behind you.

6. Rank the following in order of size: a light-minute, a light-year, a light-hour, the radius of Earth, the distance from Earth to the Sun, the radius of the Solar System.

7. Our Solar System is to the universe as _____ is to three times the age of Earth.

8. The following astronomical events led to the formation of you. Place them in order of their occurrence over astronomical time.
   a. Stars die and distribute heavy elements into the space between the stars.
   b. Hydrogen and helium are made in the Big Bang.
   c. Enriched dust and gas gather into clouds in interstellar space.
   d. Stars are born and process light elements into heavier ones.
   e. The Sun and planets form from a cloud of interstellar dust and gas.

9. Write 1.60934×10^3 (the number of meters in a mile) and 9.154×10^{-3} (Earth's diameter compared to the Sun's) in standard notation.

10. Write 86,400 (the number of seconds in a day) and 0.0123 (the Moon's mass compared to Earth's) in scientific notation.

QUESTIONS AND PROBLEMS

Multiple Choice and True/False

11. **T/F:** A scientific theory can be tested by observations and proved to be true.

12. **T/F:** A pattern in nature can reveal an underlying physical law.

13. **T/F:** A theory, in science, is a guess about what might be true.

14. **T/F:** Once a theory is proved in science, scientists stop testing it.

15. **T/F:** Astronomers use instruments from many branches of science to investigate the universe.

16. The Solar System contains
    a. planets, dwarf planets, asteroids, and galaxies.
    b. planets, dwarf planets, comets, and billions of stars.
    c. planets, dwarf planets, asteroids, comets, and one star.
    d. planets, dwarf planets, one star, and many galaxies.

17. The Sun is part of
    a. the Solar System.
    b. the Milky Way Galaxy.
    c. the universe.
    d. all of the above

18. A light-year is a measure of
    a. distance.   c. speed.
    b. time.   d. mass.

19. Which of the following was *not* made in the Big Bang?
    a. hydrogen   c. beryllium
    b. lithium   d. carbon

20. The fact that scientific revolutions take place means that
    a. all the science we know is wrong.
    b. the science we know now is more correct than it was in the past.
    c. all knowledge about the universe is relative.
    d. the laws of physics that govern the universe keep changing.

21. Occam's razor states that
    a. the universe is expanding in all directions.
    b. the laws of nature are the same everywhere in the universe.
    c. if two hypotheses fit the facts equally well, choose the simpler one.
    d. patterns in nature are really manifestations of random occurrences.

22. The cosmological principle states
    a. on a large scale, the universe is the same everywhere at a given time.
    b. the universe is the same at all times.
    c. our location is special.
    d. all of the above

23. There are _____ nanometers (10^{-9}) in 1 micrometer (10^{-6}).
    a. 10   c. 1,000
    b. 100   d. 1,000,000

24. In Figure 1.10 during the time period from 20 to 25 minutes, the car
    a. is traveling away from the starting point and speeding up.
    b. is traveling away from the starting point and slowing down.
    c. is traveling toward the starting point and speeding up.
    d. is traveling toward the starting point and slowing down.

25. Suppose that you see a graph representing the relationship between the value of a stock you hold and the time you hold it. In this case, _____ would be plotted on the horizontal axis, and _____ would be plotted on the vertical axis.
    a. time; stock value
    b. stock value; time
    c. dollars; stock value
    d. time; dollars

Conceptual Questions

26. Suppose you lived on the planet named "Tau Ceti e" that orbits Tau Ceti, a nearby star in our galaxy. How would you write your cosmic address?

27. Draw a diagram representing your cosmic address. How can you best show the difference in size scales represented in the pieces of this address? How do these differences in scale compare to the differences in scale in your postal address?

28. Figure 1.2 gives examples of timescales that are analogous to the distance scales in the universe. Pick one of the distances in the figure, and think of a different analogy. That is, think of a different activity that takes the same amount of time.

29. Examine Figure 1.2. When an event occurs on the Sun, how long does it take us to know about it?

30. When a star explodes in the Andromeda Galaxy, how long does it take for us to see it on Earth? (Hint: Study Figure 1.2.)

31. Scientists say that we are "made of stardust." Explain what they mean.

32. What does the word *falsifiable* mean? Give an example of an idea that is not falsifiable. Give an example of an idea that is falsifiable.

33. Explain how the word *theory* is used differently by a scientist than in common everyday language.

34. What is the difference between *hypothesis* and *theory*?

35. Astronomers commonly work with scientists in other fields. Use what you have learned about the scientific method to predict what would happen if a discrepancy were found between fields.

36. Suppose the tabloid newspaper at your local supermarket claims that children born under a full Moon become better students than children born at other times.
    a. Is this theory falsifiable?
    b. If so, how could it be tested?

37. A textbook published in 1945 stated that it takes 800,000 years for light to reach us from the Andromeda Galaxy. In this book, we say that it takes 2,500,000 years. What does this tell you about a scientific fact and how our knowledge changes with time?

38. Astrology makes testable predictions. For example, it predicts that the horoscope for your star sign on any day should fit you better than horoscopes for other star signs. Without indicating which sign is which, read the daily horoscopes to a friend, and ask how many of them might describe your friend's experiences on the previous day. Repeat the experiment every day for a week and keep records. Did one particular horoscope sign consistently describe your friend's experiences?

39. Imagine yourself living on a planet orbiting a star in a very distant galaxy. What does the cosmological principle tell you about the way you would perceive the universe from this distant location?

40. Make a concept map (a diagram that shows all the links between different ideas) from the information presented in Section 1.2. Use this map to test your understanding of the scientific use of the word *theory*.

Problems

41. Estimate your height in meters.

42. Astronomers often express distances in terms of the amount of time it takes light to travel that distance. We call such distances light-minutes, light-hours, or light-years. Express the distance to a nearby town in terms of both the distance in miles and the time it takes to get there by car traveling at 60 miles per hour (car-hours). Express the distance to a nearby friend's house in terms of distance and in terms of time. Invent your own units for the travel-time measurement to your friend's house.

43. New York is 2,444 miles from Los Angeles. What is that distance in car-hours? In car-days? (Assume a reasonable number for the speed limit.) What is that distance in walking-hours? That is, how far is New York from Los Angeles in foot-days, foot-months, or foot-years?

44. (a) It takes about 8 minutes for light to travel from the Sun to Earth. Pluto is 40 times as far from us as the Sun when Pluto is closest to Earth. How long does it take light to reach Earth from Pluto? (b) Radio waves travel at the speed of light. What problems would you have if you tried to conduct a two-way conversation between Earth and a spacecraft orbiting Pluto?

45. Figure 1.2 gives the time it takes light to travel across the diameter of the Milky Way Galaxy, from the Milky Way Galaxy to the Andromeda Galaxy, and halfway across the universe. Express each of these travel times of light in standard notation and scientific notation.

46. Find the diameter of the Milky Way Galaxy in miles. You will need to reference Figure 1.2 to find the travel time of light across the galaxy. Do you have a "feel" for this number?

47. One way to get a feel for very large numbers is to "scale them" to everyday objects, much like making a map of a large area on a small piece of paper. Imagine the Sun is the size of a grain of sand and Earth a speck of dust 0.083 meters (83 millimeters) away. (Each light-minute of distance is represented by 10 millimeters.) On this scale, how many millimeters (mm) represent a light-second? How many represent a light-hour? On this scale, what is the distance from Earth to the Moon? From the Sun to Neptune?

48. The average distance from Earth to the Moon is 384,000 km. How many hours would it take, traveling at 800 kilometers per hour (km/h)—the typical speed of jet aircraft—to reach the Moon? How many days is this? How many months?

49. The average distance from Earth to the Moon is 384,000 km. In the late 1960s, astronauts reached the Moon in about 3 days. How fast (on average) must they have been traveling (in km/h) to cover this distance in this time? Compare this speed to the speed of a jet aircraft (800 km/h).

50. Estimate the number of stars in the Local Group. (Assume that the number of stars in all the dwarf galaxies in our Local Group is negligible compared to the number of stars in our Milky Way. Assume also that the Andromeda Galaxy and the Milky Way are similar.)

SMARTWORK

Norton's online homework system includes algorithmically generated versions of these questions, plus additional conceptual exercises. If your instructor assigns questions in SmartWork, log in at smartwork.wwnorton.com.

Exploration | Logical Fallacies

Logic is fundamental to the study of science and to scientific thinking. A logical fallacy is an error in reasoning, which good scientific thinking avoids. For example, "because Einstein said so" is not an adequate argument. No matter how famous the scientist is (even if he is Einstein), he must still supply a logical argument and evidence to support his claim. When someone claims that something must be true because Einstein said it, they have committed the logical fallacy known as an appeal to authority. There are many types of logical fallacies, but a few of them crop up often enough in discussions about science that you should be aware of them.

Ad hominem. In an *ad hominem* fallacy, you attack the person who is making the argument instead of the argument itself. Here is an extreme example of an ad hominem argument: "A famous politician says Earth is warming. But I think this politician is an idiot. So Earth can't be warming."

Appeal to belief. This fallacy has the general pattern "Most people believe X is true, therefore X is true." So, for example, "Most people believe Earth orbits the Sun. Therefore Earth orbits the Sun." Note that even if the conclusion is correct, you may still have committed a logical fallacy in your argument.

Begging the question. In this fallacy, also known as circular reasoning, you assume the claim is true and then use this assumption as your evidence to prove the claim is true. For example, "I am trustworthy, therefore I must be telling the truth." No real evidence is presented for the conclusion.

Biased sample. If a sample drawn from a larger pool has a bias, then conclusions about the sample cannot be applied to the larger pool. For example, imagine you poll students at your college and find that 30 percent of them visit the library one or more times per week. Then you conclude that 30 percent of Americans visit the library one or more times per week. You have committed the biased sample fallacy, because university students are not a representative sample of the American public.

Post hoc ergo propter hoc. *Post hoc ergo propter hoc* is Latin for "After this, therefore because of this." Just because one thing follows another doesn't mean that one caused the other. For example, "There was an eclipse and then the king died. Therefore, the eclipse killed the king." This fallacy is often connected to the inverse reasoning: "If we can prevent an eclipse, the king won't die."

Slippery slope. In this fallacy, you claim that a chain reaction of events will take place, inevitably leading to a conclusion that no one could want. For example, "If I fail astronomy, I will not ever be able to get a college degree, and then I won't ever be able to get a good job, and then I will be living in a van down by the river until I'm old and toothless and starve to death." None of these steps actually follows inevitably from the one before.

Following are some examples of logical fallacies. Identify the type of fallacy represented. Each of the fallacies is represented once.

1. You get a chain email threatening terrible consequences if you break the chain. You move it to your spam box. On the way home, you get in a car accident. The following morning, you retrieve the chain email and send it along.

2. If I get question number 1 on the assignment wrong, then I'll get question number 2 wrong as well, and before you know it, I will fail the assignment.

3. All my friends love the band Degenerate Electrons. Therefore, all people my age love this band.

4. Eighty percent of Americans believe in the Tooth Fairy. Therefore, the Tooth Fairy exists.

5. My professor says that the universe is expanding. But my professor is a nerd, and I don't like nerds. So the universe can't be expanding.

6. When applying for a job, you use a friend as a reference. Your prospective employer asks you how she can be sure your friend is trustworthy, and you say, "I can vouch for her."

SMARTWORK • smartwork.wwnorton.com

2 Patterns in the Sky—Motions of Earth and the Moon

Through careful observation, our distant ancestors learned that they could use patterns in the sky to predict the changing length of day, the change of seasons, and the rise and fall of tides. These patterns still draw our attention on dark, cloudless nights. But now we see these patterns with the perspective of centuries of modern science, and we can explain changes as a consequence of the motions of Earth and the Moon. Discovering the cause of these patterns has shown us the way outward into the universe. In the illustration on the opposite page, a student has matched photographs of the Moon to a sketch that relates the appearance of the Moon to the positions of Earth, the Moon, and the Sun. In the sketch, the viewpoint is from far above Earth's North Pole.

✦ LEARNING GOALS

In this chapter, we look at patterns in the sky and on Earth, and you learn about the underlying motions that cause these patterns. For example, in the illustration on the opposite page, a student is finding a pattern in the relationship between the appearance of the Moon and its location relative to Earth and the Sun. By the end of this chapter, you should be able to produce such an illustration or identify from an image of the Moon where the Moon must be relative to the Sun and Earth. You should also be able to:

LG 1 Understand how Earth's rotation and revolution affect our perception of celestial motions as seen from different places on Earth.

LG 2 Visualize how Earth's motion around the Sun and the tilt of Earth's axis relative to the plane of its orbit determine which stars are visible at night and which seasons are experienced in different locations through the year.

LG 3 Connect the motion of the Moon in its orbit around Earth to the phases we observe and to the spectacle of eclipses.

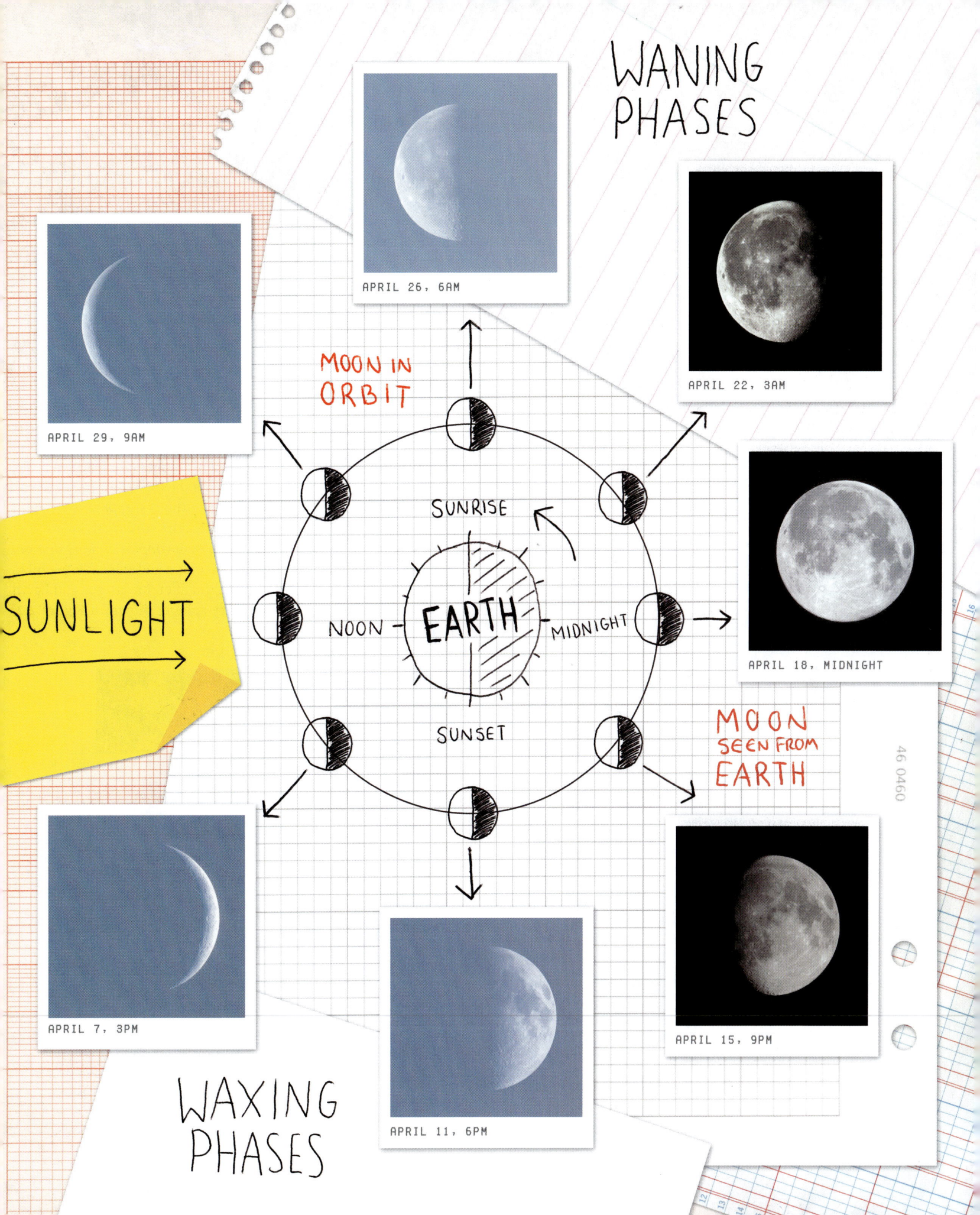

▶▶ **Nebraska Simulation:** Rotating Sky Explorer

▶‖ **AstroTour:** The Earth Spins and Revolves

▶▶ **Nebraska Simulation:** Sun Motions Demonstrator

2.1 Earth Spins on Its Axis

Long before Christopher Columbus journeyed to the New World, Aristotle and other Greek philosophers knew that Earth is a sphere. However, because Earth seems stationary, it was difficult for them to accept that the daily and annual patterns in the sky are caused by Earth's motions. As we will see in this section, Earth's rotation on its axis sets the very rhythm of life on Earth—the rising and setting of the Sun, Moon, and stars.

The Celestial Sphere

As Earth rotates, its surface is moving quite fast—about 1,670 kilometers per hour (km/h) at the equator. We do not "feel" that motion any more than we would feel the motion of a car with a perfectly smooth ride cruising down a straight highway. Nor do we feel the *direction* of Earth's spin, although the motion of the Sun, Moon, and stars across the sky reveals this direction. Earth's **North Pole** is at the north end of Earth's rotation axis. Imagine you are transported to a point in space far above the North Pole. From this vantage point, you would see Earth rotate counterclockwise, once each 24-hour period, as shown in **Figure 2.1**. As the rotating Earth carries us from west to east, objects in the sky *appear* to move in the other direction, from east to west. As seen from Earth's surface, the path each celestial body makes across the sky each day is called its *apparent daily motion*.

To help visualize the apparent daily motions of the Sun and stars, it is useful to think of the sky as a huge sphere with the stars painted on its surface and Earth at its center. Astronomers refer to this imaginary sphere, shown in **Figure 2.2**,

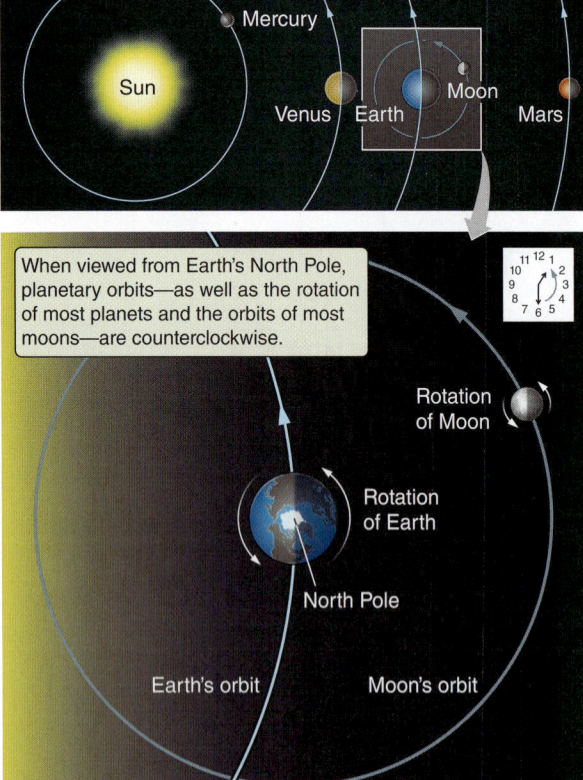

Figure 2.1 Nearly all of the large objects in the Solar System rotate and revolve in the same direction. (Not drawn to scale.)

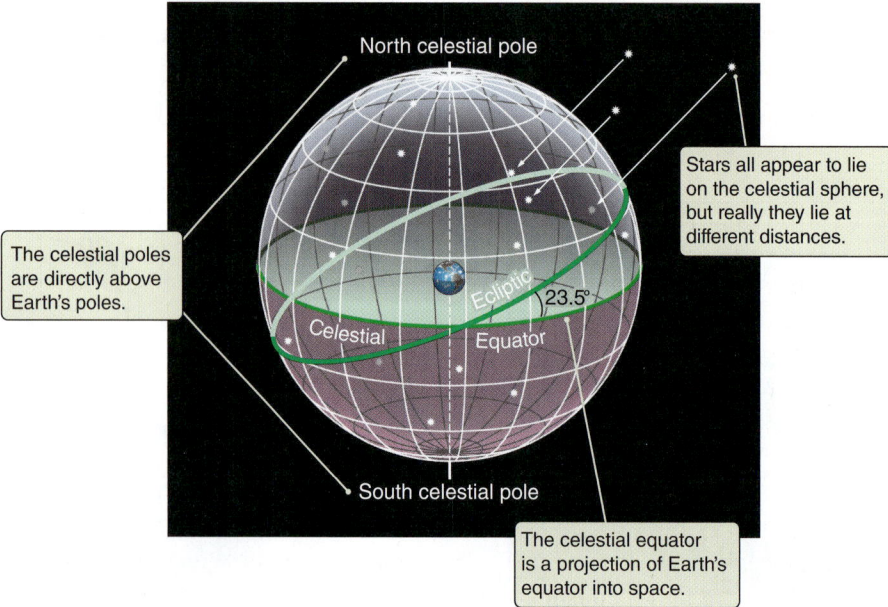

Figure 2.2 The celestial sphere is a useful fiction for thinking about the appearance and apparent motion of the stars in the sky. This imaginary sphere is centered on Earth and allows us to ignore, temporarily, the complication that stars are not all located at the same distance from Earth. The ecliptic will be discussed in Section 2.2.

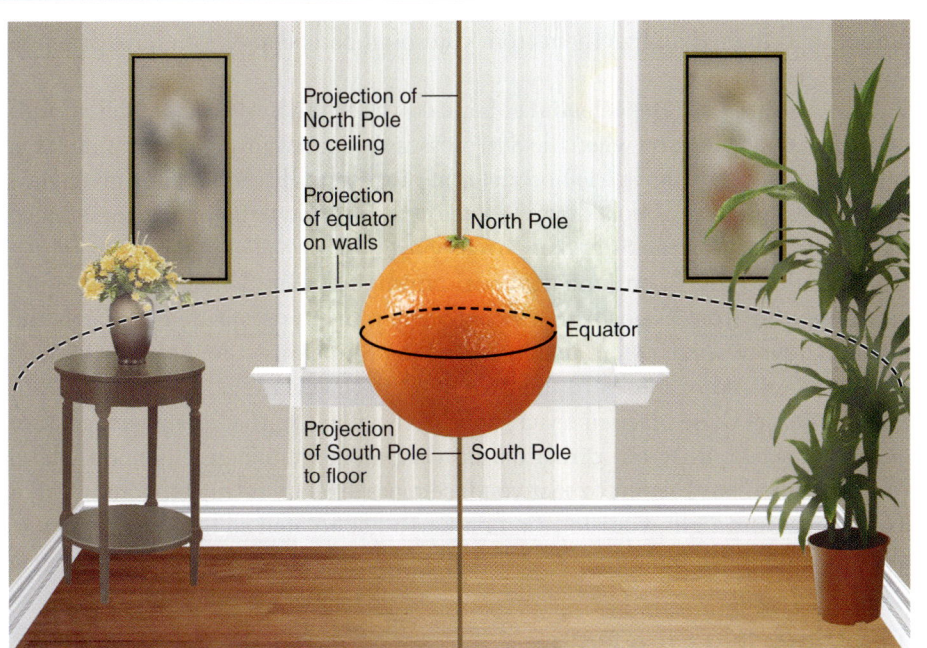

Figure 2.3 You can draw poles and an equator on an orange and imagine these points projected onto the walls of your room. Similarly, we imagine the poles and equator of Earth projected onto the celestial sphere. (Not drawn to scale.)

as the **celestial sphere**. The celestial sphere is useful because it is easy to draw and visualize, but don't forget that it is imaginary. Each point on the celestial sphere indicates a *direction* in space. Directly above Earth's North Pole is the **north celestial pole (NCP)**. Directly above Earth's **South Pole**, which is at the south end of Earth's rotation axis, is the **south celestial pole (SCP)**. The **ecliptic** is the path of the Sun in the sky throughout the year. (We'll return to the ecliptic in Section 2.2.) Directly above Earth's **equator** is the **celestial equator**. The celestial equator divides the sky into a northern half and a southern half. If you point one arm toward the celestial equator and one arm toward the north celestial pole, your arms will always form a right angle, so the north celestial pole is 90° away from the celestial equator. The angle between the celestial equator and the south celestial pole is also 90°. You might want to draw an equator and north and south poles on an orange and visualize those markings projected onto the walls of your room, as shown in **Figure 2.3**. (We will use this orange again throughout this chapter, so don't eat it!)

The **zenith** is the point in the sky directly above you wherever you are. You can find the **horizon** by standing up and pointing your right hand at the zenith and your left hand straight out from your side. Turn in a complete circle. Your left hand has traced out the entire horizon. You can divide the sky into an east half and a west half with a line that runs from the horizon at due north through the zenith to the horizon at due south. This imaginary north–south line is called the **meridian**, shown as a dashed line in **Figure 2.4**.

Take a moment to visualize all these locations in space. You may want to draw a little person on your orange and visualize that person's meridian and horizon on the walls of your room. Where is the zenith for that person? How is the meridian oriented relative to the celestial equator? How is the meridian oriented rela-

Vocabulary Alert

horizon In common language, the horizon is the place where the sky appears to meet the ground. For astronomers, however, it means the circle that is 90° from the zenith, ignoring obstructions. (This is the line you would see if you held your eyes perfectly level and turned all the way around.) A line to the horizon always makes a right angle with a line to the zenith.

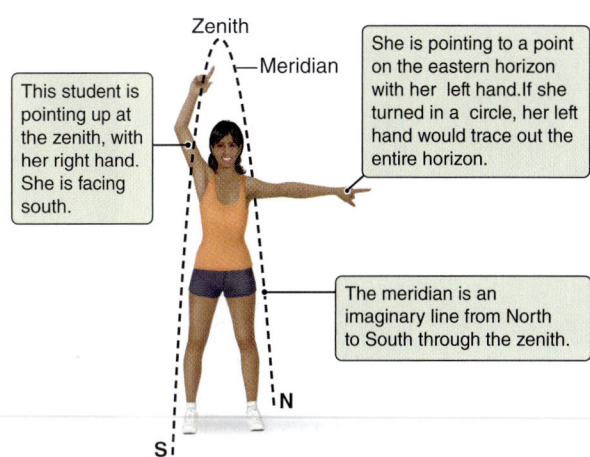

Figure 2.4 The meridian is a line on the celestial sphere that runs from north to south, dividing the sky into an east half and a west half.

Vocabulary Alert

altitude In common language, altitude is the height of an object, such as an airplane, above the ground. Astronomers use the word to refer to the angle formed between an imaginary line from an observer to an object and a second line from the observer to the point on the horizon directly below the object.

tive to the horizon? When you can answer these questions for the person on your orange, and then visualize them from your own perspective, you have oriented yourself to the sky.

To see how to use the celestial sphere, consider the Sun at noon and at midnight. Local noon occurs when the Sun crosses the meridian at your location. This is the highest point above the horizon that the Sun will reach on any given day. The highest point is almost never the zenith. You have to be in a specific place on a specific day for the Sun to be directly over your head at noon; for example, at latitude 23.5° north on June 21. Local midnight occurs when the Sun is precisely opposite from its position at local noon. From our perspective on Earth, the celestial sphere appears to rotate, carrying the Sun across the sky to its highest point at noon and over toward the west to set in the evening. In reality, the Sun remains in the same place in space, and Earth rotates so that any given location on Earth faces a different direction at every moment. When it is noon where you live, Earth has rotated so that you face most directly toward the Sun. Half a day later, at midnight, your location on Earth has rotated so that you face most directly away from the Sun.

The View from the Poles

The apparent daily motions of the stars and the Sun depend on where you live. The apparent daily motions of celestial objects in northern Canada, for example, are quite different from the apparent daily motions seen from an island in the Tropics. To understand why your location matters, let's examine the special case of the North Pole. This is known as a limiting case, meaning we've gone as far as we can go in some way. One way scientists test theories is by considering such limiting cases. In this example, we have gone as far north as possible. Here, the north celestial pole is directly overhead at the zenith.

Imagine that you are standing on the North Pole watching the sky, as in **Figure 2.5a**. (Ignore the Sun for the moment and pretend that you can always see stars in the sky.) You are standing where Earth's axis of rotation intersects its surface, which is like standing at the center of a rotating wheel. As Earth rotates, the spot directly above you remains fixed over your head while everything else in the sky appears to revolve in a counterclockwise direction around this spot. Figure 2.5b depicts this overhead view. If you have trouble visualizing this, spin your orange and imagine looking at the objects in your room from its "north pole."

Everywhere on Earth, all the time, half of the sky is "below the horizon," as shown in Figure 2.5c; the view is blocked by Earth itself. Everywhere except at the poles, this visible half of the sky changes as Earth rotates, because the zenith points at different locations in the sky as Earth carries you around. In contrast, if you are standing at the North Pole, the zenith is always in the same location in space, so the objects visible from the North Pole follow circular paths that always have the same **altitude**, or angle above the horizon. Objects close to the zenith appear to follow small circles, while objects near the horizon follow the largest circles. The

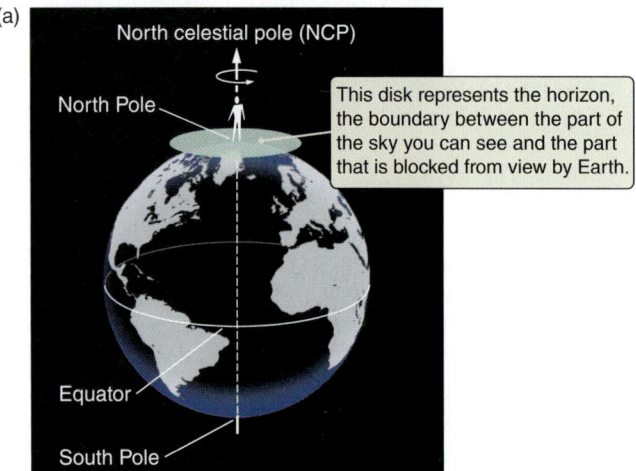
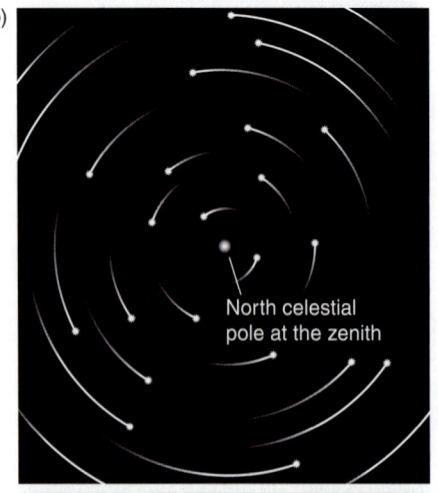
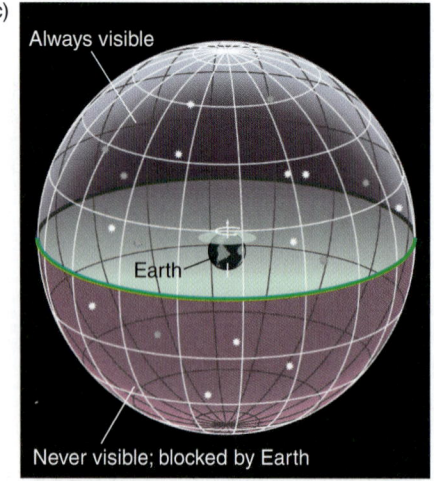

Figure 2.5 An observer (a) standing at the North Pole sees (b) stars moving throughout the night on counterclockwise, circular paths about the zenith. (c) The same half of the sky is always visible from the North Pole.

view from the North Pole is special because nothing rises or sets each day as Earth turns; from there we always see the *same* half of the celestial sphere.

The view from Earth's South Pole is much the same, but with two major differences. First, the South Pole is on the opposite side of Earth from the North Pole, so the visible half of the sky at the South Pole is precisely the half that is hidden from view at the North Pole. The second difference is that instead of appearing to move counterclockwise around the sky, stars appear to move clockwise around the south celestial pole. Try to visualize this movement. It might help to stand up and rotate around from right to left. As you look at the ceiling, things appear to move in a counterclockwise direction; but as you look at the floor, they appear to be moving clockwise.

The View Away from the Poles

Now imagine leaving the North Pole to travel south to lower latitudes. **Latitude** measures how far north or south of the equator you are on the surface of Earth. Imagine a line from the center of Earth to your location on the surface of the planet, as in **Figure 2.6**. Now imagine a second line from the center of Earth to the point on the equator closest to you. The angle between these two lines is your latitude. At the North Pole, these two imaginary lines form a 90° angle. At the equator, they form a 0° angle. So the latitude of the North Pole is 90° north, and the latitude of the equator is 0°. The South Pole is at latitude 90° south.

Your latitude determines the part of the sky that you can see throughout the year. As you move south, the zenith moves away from the north celestial pole, and so the horizon moves as well. At the North Pole, the horizon makes a 90° angle with the north celestial pole, which is at the zenith. At a latitude of 60° north, the horizon is tilted 60° from the north celestial pole. The angle between your horizon and the north celestial pole is equal to your latitude no matter where you are on Earth.

Probably the best way to solidify your understanding of the view of the sky at different latitudes is to draw pictures like the one in Figure 2.6. If you can draw a picture like this for any latitude—filling in the values for each angle in the drawing and imagining what the sky looks like from that location—then you will be well on your way to developing a working knowledge of the appearance of the sky. That knowledge will prove useful later when we discuss a variety of phenomena, such as the changing of the seasons.

The apparent motion of the stars about the celestial poles also differs from latitude to latitude. **Figure 2.7** shows two time-lapse views of the sky from different latitudes. The visible part of the sky constantly changes, as stars rise and set with Earth's rotation. From this perspective, the horizon appears fixed, and the stars appear to move. If we focus our attention on the north celestial pole, we see much the same thing we saw from Earth's North Pole. The north celestial pole remains fixed in the sky, and all of the stars appear to move throughout the night in counterclockwise, circular paths around that point.

From the vantage point of an observer in the Northern Hemisphere, stars located close enough to the north celestial pole

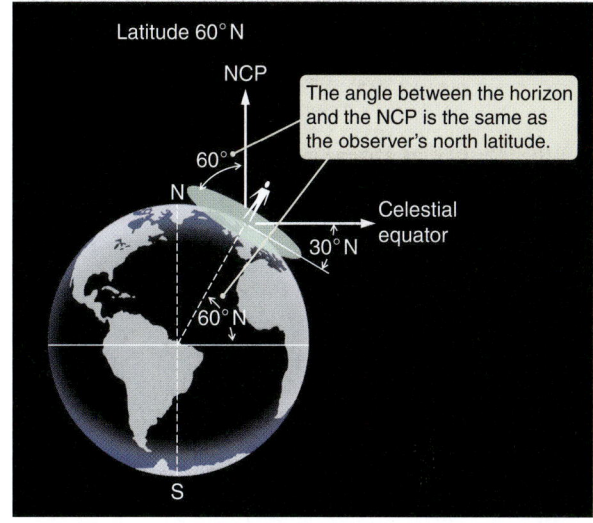

Figure 2.6 Our perspective on the sky depends on our location on Earth. The locations of the celestial poles and celestial equator in an observer's sky depend on the observer's latitude. In this case, an observer at latitude 60° north sees the north celestial pole at an altitude of 60° above the northern horizon and the celestial equator 30° above the southern horizon.

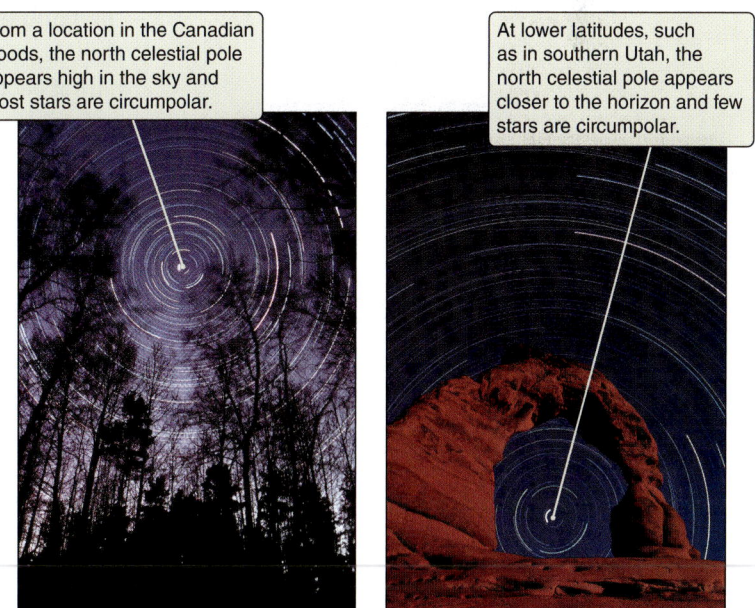

Figure 2.7 Time exposures of the sky show the apparent motions of stars through the night. Note the difference in the circumpolar portion of the sky as seen from the two different latitudes.

AstroTour: The View from the Poles

AstroTour: The Celestial Sphere and the Ecliptic

never dip below the horizon. How close is close enough? Recall from Figure 2.6 that the latitude is equal to the altitude of the north celestial pole. Stars closer to the north celestial pole than this angle never dip below the horizon as they complete their apparent paths around the pole. These stars are called **circumpolar** ("around the pole") stars, meaning that they can always be seen above the horizon by this observer. Another group of stars, close to the south celestial pole, never rise above the horizon and can *never* be seen by this observer. Stars between those that never rise and those in the circumpolar region can be seen for *only part of* each day. These stars appear to rise and set as Earth turns. The only place on Earth where you can see the entire sky over the course of 24 hours is the equator. From the equator, the north and south celestial poles sit on the northern and southern horizons, respectively, and the whole of the heavens passes through the sky each day. (Even though the Sun lights the sky for roughly half of this time, the stars are still there.)

Figure 2.8 shows the orientation of the sky as seen by observers at four different latitudes. For an observer at the North Pole (Figure 2.8a), the celestial equator lies exactly along the horizon. The north celestial pole is at the zenith, and the southern half of the sky is never visible. Stars neither rise nor set; their paths form circles parallel to the horizon.

At other latitudes, the celestial equator intersects the horizon due east and due west. Therefore, a star on the celestial equator rises due east and sets due west. Stars located north of the celestial equator rise north of east and set north of west. Stars located south of the celestial equator rise south of east and set south of west.

Regardless of where you are on Earth (with the exception of the poles), half of the celestial equator is always visible above the horizon. Therefore, any object located on the celestial equator is visible half of the time—above the horizon for 12 hours each day. Objects that are not on the celestial equator are above the horizon for differing amounts of time. Figure 2.8b and 2.8d show that stars in the observer's hemisphere are visible for more than half the day because more than half of each star's path in the sky is above the horizon. In contrast, stars in the opposite hemisphere are visible for less than half the day because less than half of each star's path in the sky is above the horizon.

For example, as seen from the Northern Hemisphere, stars north of the celestial equator remain above the horizon for more than 12 hours each day. The farther north the star is, the longer it stays up. Circumpolar stars are the extreme example of this phenomenon; they are always above the horizon. In contrast, stars south of the celestial equator are above the horizon for less than 12 hours a day. The farther south a star is, the less time it stays up. For an observer in the Northern Hemisphere, stars located close to the south celestial pole never rise above the horizon.

Since ancient times, travelers have used the stars for navigation. They would find the north or south celestial poles by recognizing the stars that surround them. In the Northern Hemisphere, a moderately bright star happens to be located close to the north celestial pole. This star is called Polaris, the "North Star." The altitude of Polaris is nearly equal to the latitude of the observer. If you are in Phoenix, Arizona, for example (latitude 33.5° north), the north celestial pole has an altitude of 33.5°. A navigator who has located the North Star can identify north and therefore also south, east, and west, as well as her latitude. This enables the navigator to determine which direction to travel. Figuring out your *longitude* (east–west location) is much more complicated because of Earth's rotation. Longitude cannot be determined from astronomical observation alone.

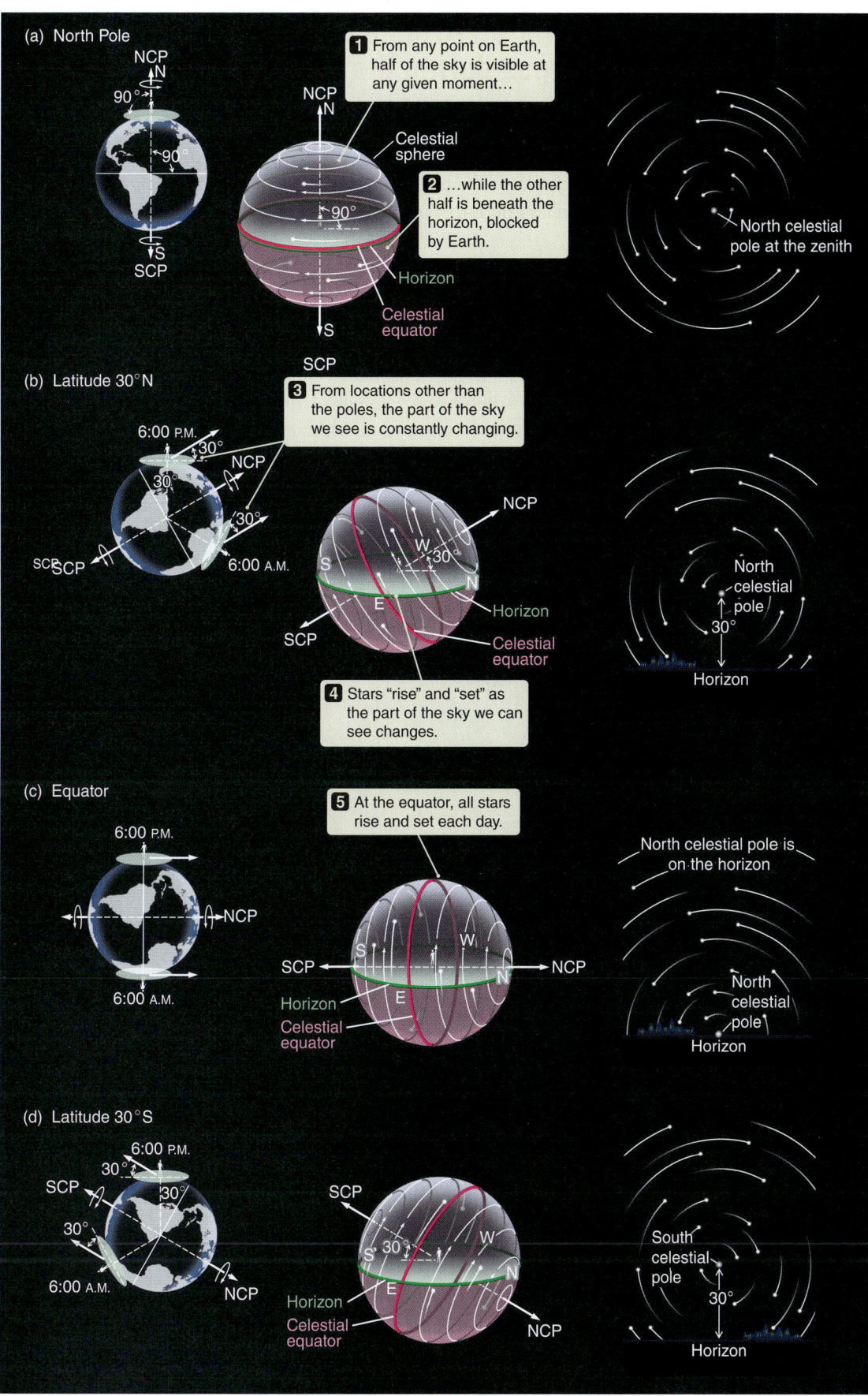

Figure 2.8 The celestial sphere is shown here as viewed by observers at four different latitudes. At all locations other than the poles, stars rise and set as the part of the celestial sphere that we see changes during the day.

2.2 Revolution Around the Sun Leads to Changes during the Year

Vocabulary Alert

revolve In common language, the words *revolve* and *rotate* are sometimes used synonymously to describe something that spins. Astronomers distinguish between the two terms, using *rotate* to mean that an object spins about an axis through its center and *revolve* to mean that one object orbits another. Earth rotates about its axis (causing our day) and revolves around the Sun (causing our year).

Earth **revolves** around (or orbits) the Sun in a nearly circular path in the same direction that Earth spins about its axis—counterclockwise as viewed from above Earth's North Pole. Because of this motion, the stars in the night sky change throughout the year, and Earth experiences seasons.

Earth Orbits the Sun

Earth's average distance from the Sun is 1.50×10^8 km—this distance is called an **astronomical unit (AU)**. The astronomical unit is handy for measuring distances in the Solar System. Earth orbits the Sun once in one **year**, by definition. Because Earth moves a distance over a period of time, it has a speed (**Working It Out 2.1**), which is necessary for motion. This motion is responsible for many of the patterns we see in the sky and on Earth, such as the night-to-night changes in the stars we see overhead. As shown in **Figure 2.9**, as Earth orbits the Sun, our view of the night sky changes. Six months from now, Earth will be on the other side of the Sun. The stars that are overhead at midnight 6 months from now are those that are overhead at noon today. Take a moment to visualize this motion. You can again use your orange as Earth and a table lamp with the shade

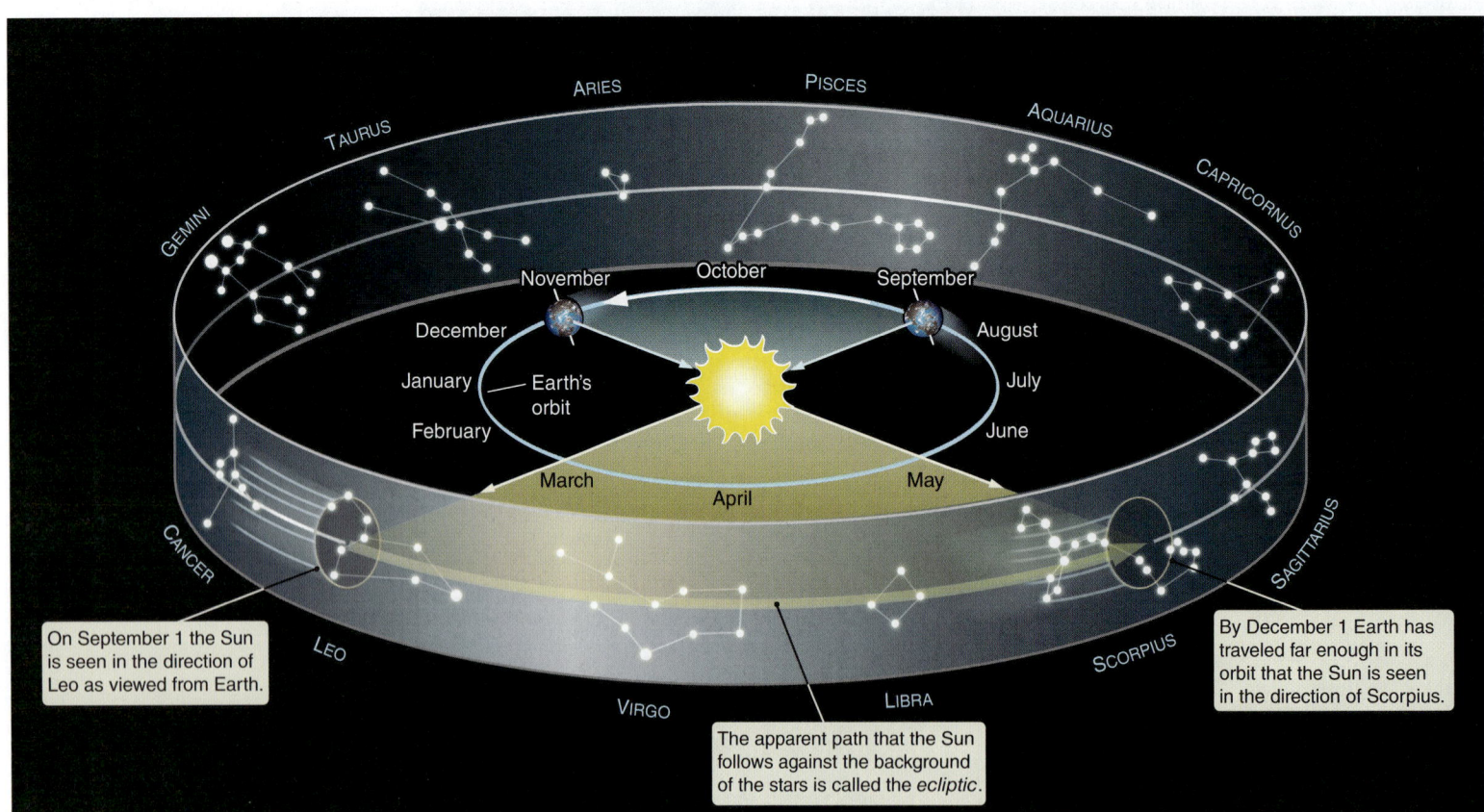

Figure 2.9 As Earth orbits the Sun, different stars appear in the night sky. Also, the Sun's apparent position against the background of stars changes. The imaginary circle traced by the annual path of the Sun is called the ecliptic. Constellations along the ecliptic form the zodiac.

Working It Out 2.1 | Manipulating Equations

So far, we have discussed scientific notation and units. Now we want to relate quantities to each other, such as distance and time. In Section 2.1, we mentioned the speed of the surface of Earth as it rotates. How do we know that speed?

How are distance, time, and speed related? If you travel a distance of 100 kilometers, and it takes you a time of 1 hour, you have traveled at a speed of 100 kilometers per hour. That's how we say it in English. How do we translate this sentence into math? We write the equation:

$$\text{Speed} = \frac{\text{Distance}}{\text{Time}}$$

But that takes up a lot of space, so we abbreviate it by using the letter s to represent speed, the letter d to represent distance, and the letter t to represent time:

$$s = \frac{d}{t}$$

To find the speed of the surface of Earth as it rotates, we imagine a spot on Earth's equator. The time it takes to go around once is 1 day, and the distance traveled is Earth's circumference. So the speed is

$$s = \frac{\text{Circumference}}{1 \text{ day}}$$

The circumference is given by $2 \times \pi \times r$, where r is the radius. Earth's radius is 6,378 kilometers (km). To find the circumference, multiply that by 2 and then by π to get 40,070 km. (Does it matter if you multiply first by π and then by 2? If you can't remember the rule, try it and see.)

Now we have the circumference, and we can find the speed:

$$s = \frac{\text{Circumference}}{1 \text{ day}} = \frac{40{,}074 \text{ km}}{1 \text{ day}} = 40{,}074 \text{ km/day}$$

But that's not the way we usually write a speed. We usually use km/h. One day = 24 hours, so (1 day) ÷ (24 hours) is equal to 1. We can always divide or multiply by 1, so let's multiply our speed by (1 day) ÷ (24 hours):

$$s = 40{,}074 \, \frac{\text{km}}{\cancel{\text{day}}} \times \frac{1 \, \cancel{\text{day}}}{24 \text{ hours}}$$

The unit of "day" divides (or cancels) out, and we have

$$s = \frac{40{,}074 \text{ km}}{24 \text{ hr}} = 1{,}670 \text{ km/hr}$$

This agrees with the value we learned earlier.

How can we manipulate this equation to solve for distance? We start with the original equation:

$$s = \frac{d}{t}$$

The d, which is what we are looking for, is "buried" on the right side of the equation. We want to get it all by itself on the left. That's what we mean when we say "solve for." The first thing to do is to flip the equation around:

$$\frac{d}{t} = s$$

We can do this because of the equals sign. It's like saying, "two quarters is equal to 50 cents" versus "50 cents is equal to two quarters." It has to be true on both sides. We still have not solved for d. To do this, we need to multiply by t. This cancels the t on the bottom of the left side of the equation. But if you do something only to one side of an equation, the two sides are not equal anymore. So whatever we do on the left, we also have to do on the right:

$$\frac{d}{\cancel{t}} \times \cancel{t} = s \times t$$

The t on the left cancels, to give us our final answer:

$$d = s \times t$$

We often write this as:

$$d = st$$

For simplicity, the multiplication symbol is left out of the equation. When two terms are written side by side, with no symbols between them, it means you should multiply them.

Use the information in Section 2.2 to find the circumference of Earth's orbit for yourself, assuming that the orbit is circular.

removed or a flashlight to represent the Sun. Move the orange around the "Sun" and notice which parts of your room walls are visible from the "nighttime side" of the orange at different points in the orbit.

Also take a moment to notice the location of your "Sun" relative to the walls of your room for the "observer" on your orange. If we correspondingly note the position of the Sun relative to the stars each day for a year, we find that it traces out a path against the background of the stars called the ecliptic (see Figure 2.9), and the plane of Earth's orbit around the Sun is called the **ecliptic plane**. On September 1, the Sun is in the direction of the constellation Leo. Six months later, on March 1, Earth is on the other side of the Sun, and the Sun is in the direction of the constellation Aquarius. The constellations that lie along the ecliptic are called the constellations of the **zodiac**. Ancient astrologers assigned special mystical significance to these stars because they lie along the path of the Sun. Actually, the constellations of the zodiac are nothing more than random patterns of distant stars that happen by chance to be located near the ecliptic.

▶❚ **AstroTour: The Earth Spins and Revolves**

Seasons and the Tilt of Earth's Axis

So far we have discussed the rotation of Earth on its axis and the revolution of Earth around the Sun. To understand why the seasons change, we need to consider the combined effects of these two motions. Many people believe that Earth is closer to the Sun in the summer and farther away in the winter and that this is the cause for the seasons. What if this idea were a hypothesis? Can it be falsified? Yes, we can make a prediction. If the distance from Earth to the Sun causes the seasons, all of Earth should experience summer at the same time of year. But the United States experiences summer in June, while Chile experiences summer in December. In modern times, we can directly measure the distance, and we find that Earth is actually closest to the Sun at the beginning of January. We have just falsified this hypothesis, and we need to go look for another one that explains *all* of the available facts. We will find an answer that fits all the facts by investigating Earth's axial tilt.

▶❚ **AstroTour: The Celestial Sphere and the Ecliptic**

Earth's axis of rotation is tilted 23.5° from the perpendicular to Earth's orbital plane, as shown in **Figure 2.10**. As Earth moves around the Sun, its axis always points toward Polaris, in the same direction in space. As Earth orbits the Sun, sometimes Earth is on one side of the Sun, and sometimes it is on the other side. Therefore, sometimes Earth's North Pole is tilted toward the Sun, and sometimes the South Pole is tilted toward the Sun. When Earth's North Pole is tilted toward the Sun, the Sun is *north* of the celestial equator; for observers in the Northern Hemisphere, the Sun is above the horizon more than 12 hours each **day**. Six months later, when Earth's North Pole is tilted away from the Sun, the Sun is *south* of the celestial equator; for observers in the Northern Hemisphere, the Sun is above the horizon less than 12 hours each day. If we look at the circle of the ecliptic in Figure 2.2, we see that it is tilted by 23.5° with respect to the celestial equator. Take a moment to visualize this with your orange and your table lamp. Tilt the orange so that its north pole no longer points at the ceiling but at some distant point in the sky through a wall or window. Keep the North Pole dot on your orange pointing in that direction, and "orbit" it around your "Sun." At one point, the tilt will be toward the "Sun." Halfway around the orbit, the tilt will be away from the "Sun."

Vocabulary Alert

day In common language, this word means both the time during which the Sun is up in the sky and the length of time it takes Earth to rotate once (from midnight to midnight). The context of the sentence tells us which meaning is intended. Unfortunately, astronomers use this word in both senses as well. You will have to consider the context of the sentence to know which meaning is being used in each instance.

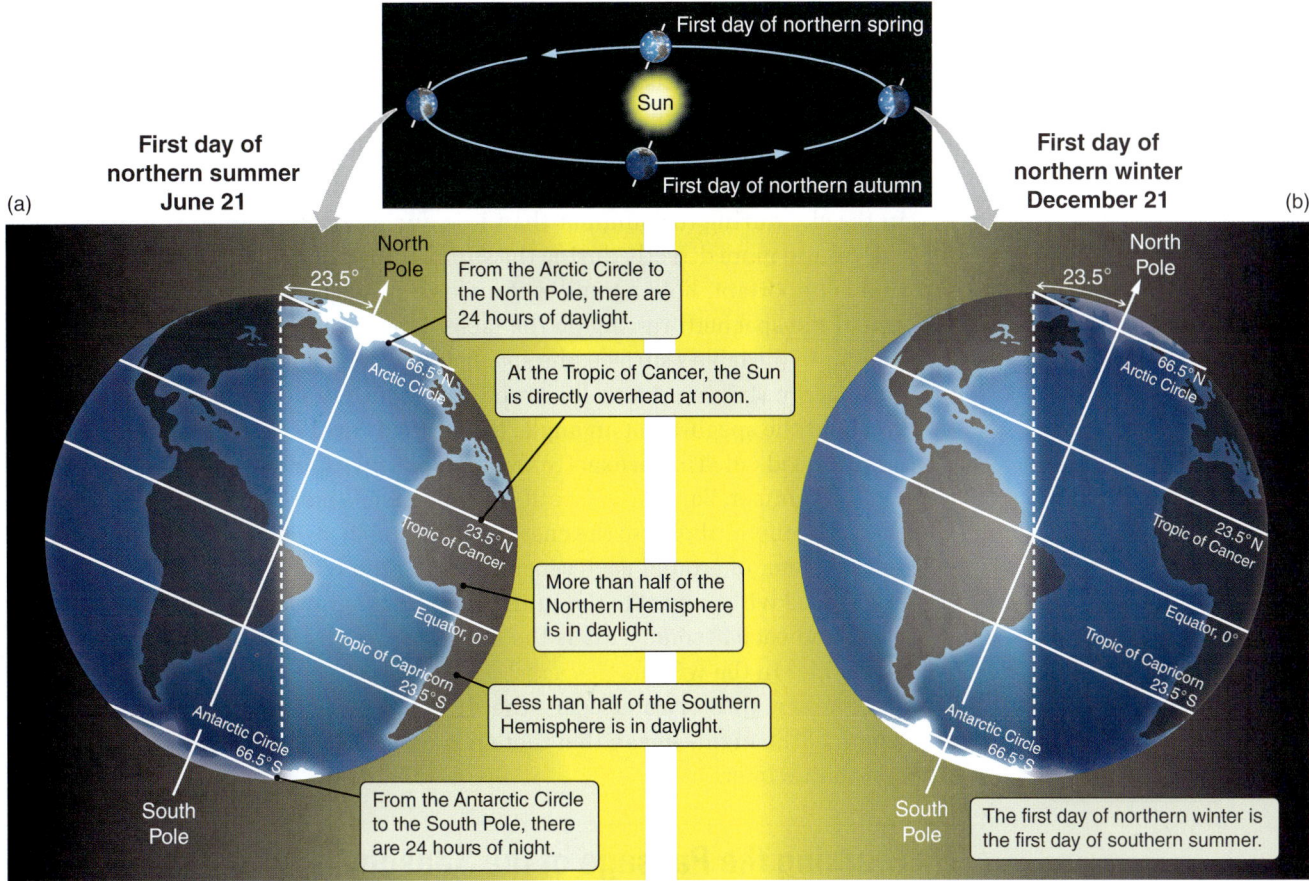

Figure 2.10 Earth's orbit around the Sun is nearly circular. Here it is shown "from the side," in perspective, and so it looks much more elliptical than it is. This perspective view is a common way to portray orbits of planets and commonly leads people to believe that the orbits are highly elongated. (a) On the first day of the northern summer (approximately June 21), the northern end of Earth's axis is tilted toward the Sun, while the southern end is tipped away. (b) Six months later, on the first day of the northern winter (approximately December 21), the situation is reversed. This explains why seasons are opposite in the Northern and Southern hemispheres.

In the preceding paragraph, we were careful to specify the *Northern* Hemisphere because seasons are opposite in the Southern Hemisphere. Look again at Figure 2.10. On June 21, while the Northern Hemisphere is enjoying the long days and short nights of summer, Earth's South Pole is tilted away from the Sun. It is winter in the Southern Hemisphere; the days are short and the nights are long there. But on December 21, Earth's South Pole is tilted toward the Sun. It is summer in the Southern Hemisphere; the days are long and the nights are short there.

To understand how the combination of Earth's axial tilt and its path around the Sun creates seasons, consider a limiting case. If Earth's spin axis were exactly perpendicular to the plane of its orbit, then the Sun would always be on the celestial equator. At every latitude, the Sun would follow the same path through the sky every day, rising due east each morning and setting due west each evening. The Sun would be above the horizon exactly half the time, and days and nights would always be exactly 12 hours long everywhere on Earth. In short, if Earth's axis were exactly perpendicular to the plane of Earth's orbit, there would be no seasons.

The differing length of the night through the year is part of the explanation for seasonal temperature changes, but it is not the whole story. Another important effect relates to the angle at which the Sun's rays strike Earth. The Sun is higher in the sky during the summer than it is during the winter, so sunlight strikes the ground more directly during the summer than during the winter. To see why this is important, hold a bundle of uncooked spaghetti in your hand, with one end of the spaghetti resting on the table. When you hold the spaghetti perpendicular to the table's surface, the spaghetti covers a smaller area. If the spaghetti was energy striking the table, it would be concentrated in a small area. But if you hold the spaghetti at an angle, the strands cover a larger area; the energy is more spread out. This is exactly what happens with the changing seasons. During the summer, Earth's surface at your location is more nearly perpendicular to the incoming sunlight, so the energy is more intense—more energy falls on each square meter of ground each second. During the winter, the surface of Earth is more tilted with respect to the sunlight, so less energy falls on each square meter of ground each second. This is the main reason why it is hotter in the summer and colder in the winter.

Together, these two effects—the directness of sunlight and the differing length of the night—mean that there is more heating from the Sun during summer than winter.

Marking the Passage of the Seasons

There are four special days during Earth's orbit that mark a unique moment in the year. As Earth orbits the Sun, the Sun moves along the ecliptic, which is tilted 23.5° with respect to the celestial equator. The day when the Sun is highest in the sky as it crosses the meridian (recall that the meridian is the line from due north to due south that passes overhead) is called the **summer solstice**. On this day, the Sun rises farthest north of east and sets farthest north of west. This occurs each year about June 21, the first day of summer in the Northern Hemisphere. This orientation of Earth and Sun is shown in Figure 2.10a.

Six months later, the North Pole is tilted away from the Sun. This day is the **winter solstice**, shown in Figure 2.10b. This occurs each year about December 21, the shortest day of the year and the first day of winter in the Northern Hemisphere. Almost all cultural traditions in the Northern Hemisphere include a major celebration of some sort in late December. These winter festivals celebrate the return of the source of Earth's light and warmth. The days have stopped growing shorter and are beginning to get longer. Spring will come again.

Between these two special days, there are days when the Sun lies directly above Earth's equator, so that the entire Earth experiences 12 hours of daylight and 12 hours of darkness. These are called the equinoxes (*equinox* means "equal night"). The **autumnal equinox** marks the beginning of fall, about September 22, which is halfway between summer solstice and winter solstice. The **vernal equinox** marks the beginning of spring, about March 20, which is halfway between winter solstice and summer solstice.

Figure 2.11 shows these four special dates from two perspectives. Figure 2.11a has a stationary Sun, which is what's actually happening, and Figure 2.11b

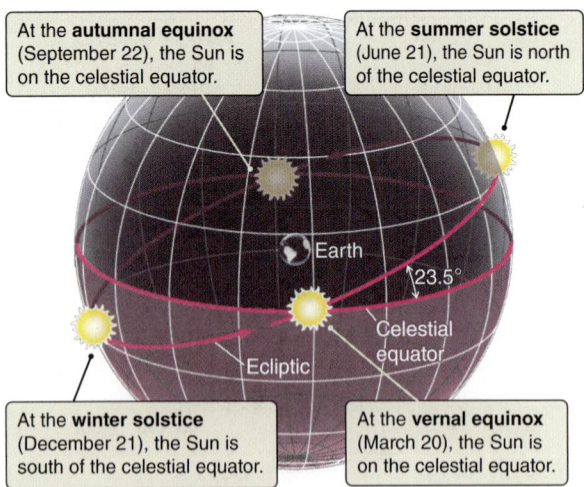

Figure 2.11 The motion of Earth around the Sun appears different when seen from the perspective where (a) the Sun or (b) Earth is stationary. Practice working back and forth between these points of view. Shifting back and forth between reference frames is a skill you will use often in your life. Note that part (a) is labeled with seasons in the Northern Hemisphere.

shows the Sun moving along the celestial sphere, which is how things appear to observers on Earth. In both cases, we are looking at the plane of Earth's orbit from "the side," so that it is shown in perspective and looks quite flattened. We have also tilted the picture so that the North Pole of Earth points straight up. Use your orange and your light source to reproduce these pictures and the motions implied by the arrows. Practice shifting between these two perspectives. Once you can look at one of the positions in Figure 2.11a and predict the corresponding positions of the Sun and Earth in Figure 2.11b (and vice versa), you will know that you really understand these differing perspectives.

Just as it takes time for a pot of water on a stove to heat up when the burner is turned up and time for the pot to cool off when the burner is turned down, it takes time for Earth to respond to changes in heating from the Sun. The hottest months of northern summer are usually July and August, which come *after* the summer solstice, when the days are growing shorter. Similarly, the coldest months of northern winter are usually January and February, which occur *after* the winter solstice, when the days are growing longer. Temperature changes on Earth follow the changes in the amount of heating we receive from the Sun.

In modern times, we use the calendar known as the **Gregorian calendar**, which is based on the *tropical year*. The **tropical year** measures the time from one vernal equinox to the next—from the start of Northern Hemisphere spring to the start of the next Northern Hemisphere spring—and is 365.242189 days long. Notice that the tropical year has approximately one-quarter "extra" day. To make up for that fraction of a day, nearly every fourth year is made a **leap year**, with an extra day in February. This prevents the seasons from becoming out of sync with the months. The Gregorian calendar also makes other adjustments on longer timescales.

▶▶ **Nebraska Simulation:** Seasons and Ecliptic Simulator

Seasons and Location

The variation in day length is extreme near Earth's poles. The **Arctic Circle** and the **Antarctic Circle** (see Figure 2.10) are regions close to the poles: farther than 66.5° from the equator. In these regions, the Sun is above the horizon 24 hours a day for part of the year, earning these polar regions the nickname "land of the midnight Sun." There is an equally long period during which the Sun never rises and the nights are 24 hours long. The Sun never rises very high in the sky at these latitudes, so the sunlight is never very direct. Even with the long days at the height of summer, the polar regions remain relatively cool.

In contrast, on the equator days and nights are 12 hours long throughout the year (see Figure 2.8c). The Sun passes directly overhead on the first day of spring and the first day of autumn because these are the days when the Sun is precisely on the celestial equator (see Figure 2.11b). Sunlight is perpendicular to the ground at the equator on these days. On the summer solstice, the Sun is at its northernmost point along the ecliptic. On this day, and on the winter solstice, the Sun is *farthest* from the zenith at noon, and therefore sunlight is *least* direct.

If you turn back to Figure 2.10, you can see a band between the latitudes of 23.5° south and 23.5° north—between Rio de Janeiro and Honolulu, for example. Here the Sun is directly overhead at noon twice during the year. The band between the latitudes of 23.5° south and 23.5° north is called the **Tropics**. The northern limit of this region is called the Tropic of Cancer; the southern limit is called the Tropic of Capricorn.

Figure 2.12 (a) Earth's axis of rotation changes orientation in the same way that the axis of a spinning top changes orientation. (b) This precession causes the projection of Earth's rotation axis to move in a circle with a radius of 23.5°, centered on the north ecliptic pole (orange cross), with a period of 26,000 years. The red cross shows the location of the north celestial pole in the early 21st century.

Earth's Axis Wobbles

When the ancient Egyptian astronomer Ptolemy (Claudius Ptolemaeus) and his associates were formalizing their knowledge of the positions and motions of objects in the sky 2,000 years ago, the Sun appeared in the constellation Cancer on the first day of northern summer and in the constellation Capricorn on the first day of northern winter. Today, the Sun is in Taurus on the first day of northern summer and in Sagittarius on the first day of northern winter. Why has this change occurred? There are *two* motions associated with Earth and its axis. Earth spins on its axis, but its axis also wobbles like the axis of a spinning top, shown in **Figure 2.12a**. The wobble is very slow; it takes about 26,000 years for the north celestial pole to make one trip around a large circle. In Section 2.1, we saw that the north celestial pole currently lies very near Polaris. However, if you could travel several thousand years into the past or future, you would find that the northern sky does not appear to rotate about a point near Polaris, but instead the stars rotate about another point on the path shown in Figure 2.12b. This figure shows the path of the north celestial pole through the sky during one *precession* cycle.

The celestial equator is perpendicular to Earth's axis. Therefore, as Earth's axis wobbles, the celestial equator must also wobble. As the celestial equator wobbles, the locations where it crosses the ecliptic—the equinoxes—change as

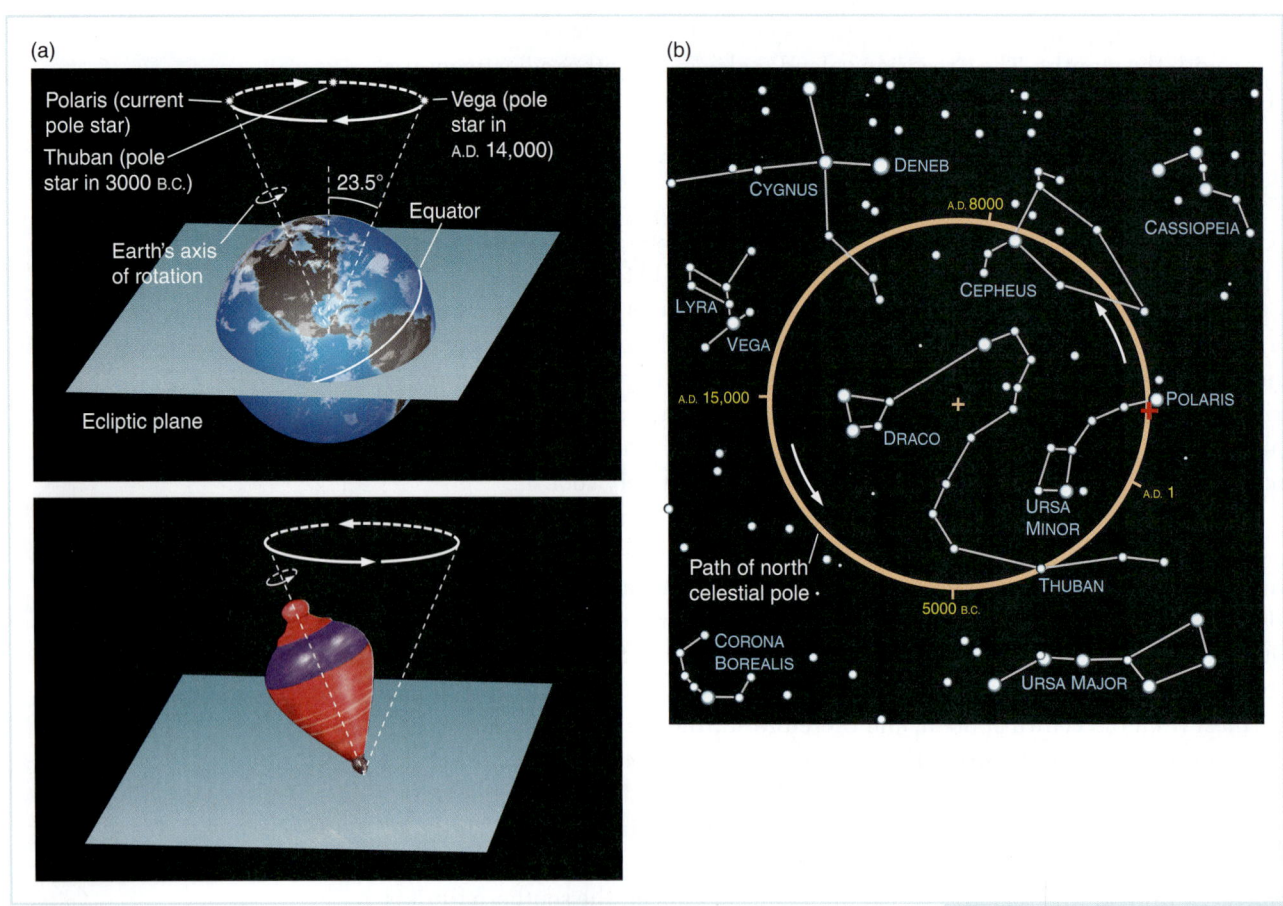

VISUAL ANALOGY

well. During each 26,000-year wobble of Earth's axis, the locations of the equinoxes make one complete circuit around the celestial equator. This change of the position of the equinox, due to the wobble of Earth's axis, is called the **precession of the equinoxes**.

2.3 The Moon's Appearance Changes as It Orbits Earth

After the Sun, the Moon is the most prominent object in the sky. Earth and the Moon orbit around each other, and together they orbit the Sun. The Moon takes just over 27 days to orbit Earth. One aspect of the Moon's appearance does not change: we always see the same side from Earth. However, as the Moon orbits, it is illuminated differently, and so its appearance changes constantly.

The Unchanging Face of the Moon

If you were to go outside next week or next month, or 20 years from now, or 20,000 *centuries* from now, you would still see the same patterns on the Moon. This observation is responsible for the common misconception that the Moon does not rotate. In fact, the Moon *does* rotate on its axis—exactly once for each revolution it makes about Earth.

Once again, use your orange to help you visualize this idea. This time, the orange represents the Moon. Use your chair or some other object to represent Earth. Face the person you drew on the orange toward the chair. First, make the orange "orbit" around the chair without rotating on its axis, keeping the person on the orange always facing the same wall of the room. When you do it this way, different sides of the orange face toward the chair at different points in the orbit. Now, make the orange orbit the chair, with the person on the orange always facing the chair. You will have to turn the orange, relative to the walls of the room, to make this happen, which means that the orange is rotating about its axis. By the time the orange completes one orbit, it will have rotated about its axis exactly once. The Moon does exactly the same thing, rotating on its axis once per revolution around Earth, always keeping the same face toward Earth, as shown in **Figure 2.13**. This phenomenon, where the same side of the Moon always faces toward Earth, is called **synchronous rotation** because the revolution and the rotation are synchronized (or in sync) with each other. The Moon's synchronous rotation occurs because it is elongated, which causes its *near side* always to fall toward Earth.

The Moon's *far side*, facing away from Earth, is often improperly called the dark side of the Moon. In fact, the far side spends just as much time in sunlight as the near side. The far side is not dark as in "unlit," but until the middle of the 20th century it was dark as in "unknown." Until spacecraft orbited the Moon, we had no knowledge of the far side.

The Changing Phases of the Moon

Unlike the Sun, the Moon has no light source of its own; it shines by reflected sunlight. Like Earth, half of the Moon is always in bright sunlight, and half is always in darkness. Our view of the illuminated portion of the Moon is constantly

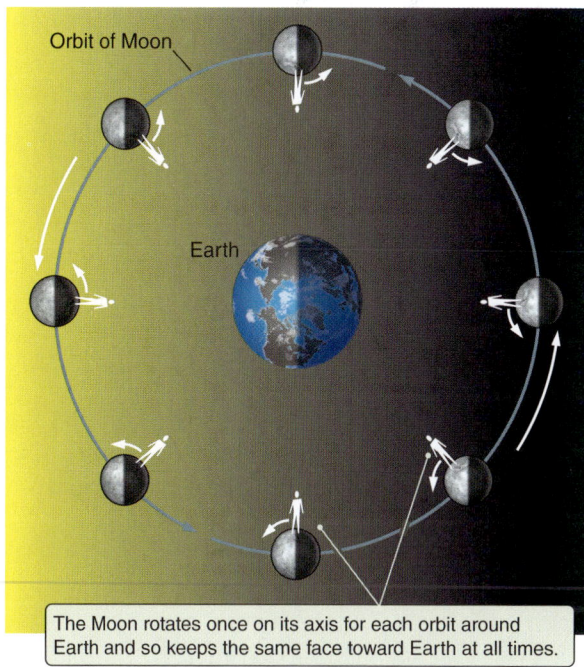

Figure 2.13 The Moon rotates once on its axis for each orbit around Earth—an effect called *synchronous rotation*. In this illustration, the Sun is far to the left of the Earth-Moon system.

Figure 2.14 You can experiment with illumination effects by using an orange as the Moon, a lamp with no shade as the Sun, and your own head as Earth. As you move the orange around your head, viewing it in different relative locations, you will see that the illuminated part of the orange mimics the phases of the Moon.

▶❙❙ AstroTour: The Moon's Orbit: Eclipses and Phases

changing, which explains the phases of the Moon. During a *new Moon*, when the Moon is between Earth and the Sun, the side facing away from us is illuminated, and during a *full Moon*, when Earth is between the Moon and the Sun, the side facing toward us is illuminated. The rest of the time, only a part of the illuminated portion can be seen from Earth. Sometimes the Moon appears as a circular disk in the sky. At other times, it is nothing more than a sliver.

To help you visualize the changing phases of the Moon, use your orange, your lamp, and your head, as shown in **Figure 2.14**. Your head is Earth, the orange is the Moon, and the lamp is the Sun. Turn off all the other lights in the room, and stand as far from the lamp as you can. Hold the orange slightly above your head so that it is illuminated from one side by the lamp. Move the orange clockwise and watch how the appearance of the orange changes. When you are between the orange and the lamp, the face of the orange that is toward you is fully illuminated. The orange appears to be a bright, circular disk. As the orange moves around its circle, you will see a progression of lighted shapes, depending on how much of the bright side and how much of the dark side of the orange you can see. This progression of shapes exactly mimics the changing phases of the Moon.

Figure 2.15 shows the changing phases of the Moon. The **new Moon** occurs when the Moon is between Earth and the Sun: the far side of the Moon is illuminated but the near side is in darkness and we cannot see it at night. At this phase, the Moon is up in the daytime, as it is in the direction of the Sun. A new Moon is never visible in the nighttime sky. It appears close to the Sun in the sky, so it rises in the east at sunrise, crosses the meridian near noon, and sets in the west near sunset.

A few days later, as the Moon orbits Earth, a sliver of its illuminated half, known as a **crescent**, becomes visible. Because the Moon appears to be "filling out" from night to night at this time, this phase of the Moon is called a **waxing** crescent Moon. (*Waxing* here means "growing in size and brilliance.") During the week that the Moon is in this phase, the Moon is visible east of the Sun. It is most noticeable just after sunset, near the western horizon.

As the Moon moves farther along in its orbit and the angle between the Sun and the Moon grows, more and more of the near side becomes illuminated, until half of the near side of the Moon is in brightness and half is in darkness—the phase called the **first quarter Moon** because the Moon is one-quarter of the way through its orbit. The first quarter Moon rises at noon, crosses the meridian at sunset, and sets at midnight.

As the Moon moves beyond first quarter, more than half of the near side is illuminated—the phase is called a waxing **gibbous** Moon. The gibbous Moon waxes until finally we see the entire near side of the Moon—a **full Moon**. The Sun and the Moon are now opposite each other in the sky. The full Moon rises as the Sun sets, crosses the meridian at midnight, and sets in the morning as the Sun rises.

The second half of the Moon's cycle of phases is the reverse of the first half. The Moon appears gibbous, but now the near side is becoming less illuminated—the phase called a **waning** gibbous Moon (*waning* means "becoming smaller"). When the Moon is waning, the left side—as viewed from the Northern Hemisphere—appears bright. A **third quarter Moon** occurs when half the near side is in sunlight and half is in darkness. A third quarter Moon rises at midnight, crosses the meridian near sunrise, and sets at noon. The cycle continues with a waning crescent Moon in the morning sky, west of the Sun, until the new Moon once more rises and sets with the Sun, and the cycle begins again. Notice that when

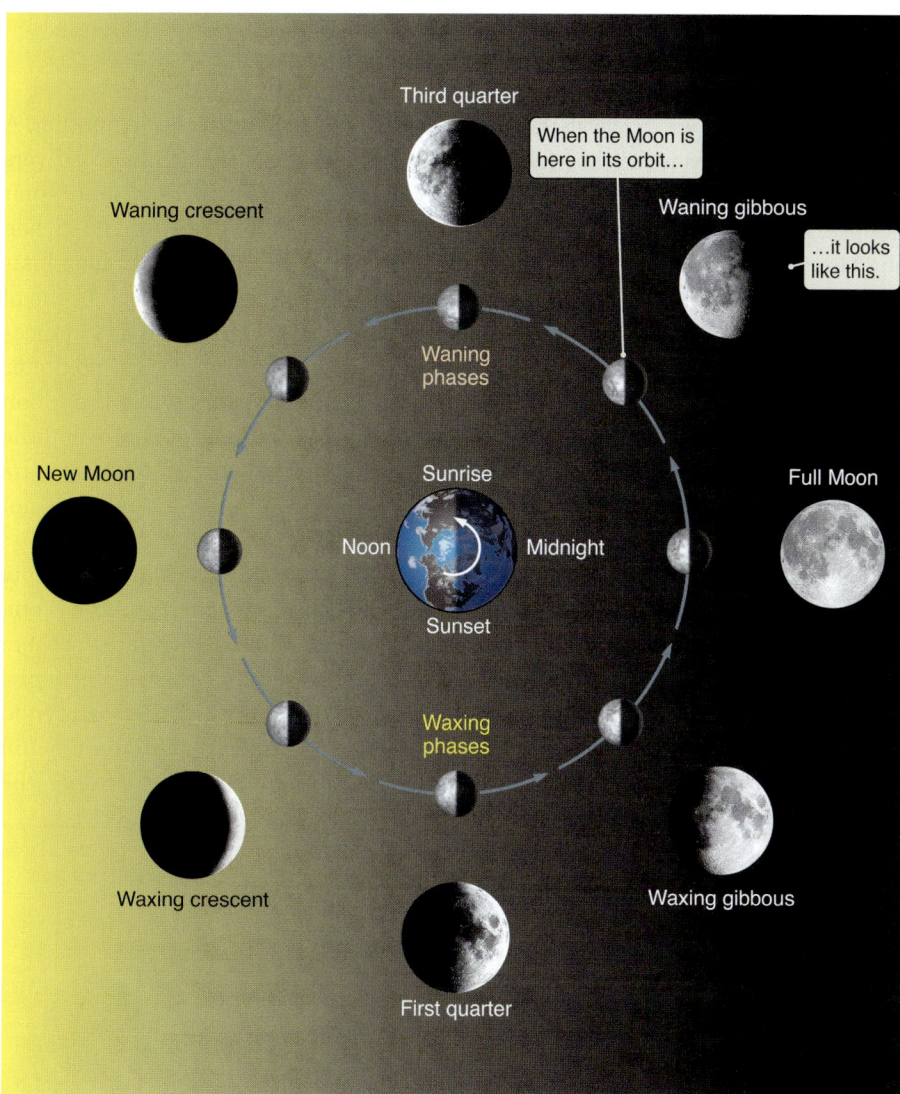

Figure 2.15 The inner circle of images (connected by blue arrows) shows the Moon as it orbits Earth as seen by an observer far above Earth's North Pole. The outer ring of images shows the corresponding phases of the Moon as seen from Earth.

the Moon is farther than Earth from the Sun, it is in gibbous phases. When it is closer than Earth to the Sun, it is in crescent phases.

Do not try to memorize all the possible combinations of where the Moon is in the sky at each phase and at every time of day. You do not have to. Instead, work on *understanding* the motion and phases of the Moon. Use your orange and your lamp or draw a picture like the chapter-opening illustration and follow the Moon around its orbit. From your drawing, figure out what phase you would see and where it would appear in the sky at a given time of day. Now return to this chapter's opening illustration and compare it to Figure 2.15, which is more compact but contains the same information. Does the earlier illustration make more sense to you now? As an extra test of your understanding, think about the phases of Earth an astronaut on the Moon would see when looking back at our planet.

▶▶ **Nebraska Simulation:** Lunar Phase Simulator

2.4 Shadows Cause Eclipses

An **eclipse** occurs when the shadow of one astronomical body falls on another. A **solar eclipse** occurs when the Moon passes between Earth and the Sun. Observers in the shadow of the Moon will see a solar eclipse. There are three types of solar eclipse: *total*, *partial*, and *annular*. A **total solar eclipse** (**Figure 2.16**) occurs when the Moon completely blocks the disk of the Sun. A total solar eclipse never lasts longer than 7½ minutes and is usually significantly shorter. Even so, it is one of the most amazing sights in nature. People all over the world travel great distances to see a total solar eclipse. The next one visible in the continental United States will occur on August 21, 2017. A **partial solar eclipse** occurs when the Moon partially covers the disk of the Sun. An **annular solar eclipse** occurs when the Moon is slightly farther away from Earth in its noncircular orbit, so it appears slightly smaller in the sky. It is centered over the disk of the Sun but does not block the entire disk. A ring is visible around the blocked portion.

Figure 2.17a shows the geometry of a solar eclipse, when the Moon's shadow falls on the surface of Earth. Figures like this usually show Earth and the Moon much closer together than they really are. The page is too small to draw them correctly and still see the critical details. Figure 2.17b shows the geometry of a solar eclipse with Earth, the Moon, and the separation between them drawn to the correct scale. Compare this drawing to Figure 2.17a, and you will understand why drawings of Earth and the Moon are rarely drawn to the correct scale. If the

Figure 2.16 During a total eclipse, the Sun produces a remarkable spectacle, as the disk is completely covered by the Moon.

(a) Solar eclipse geometry (not to scale)

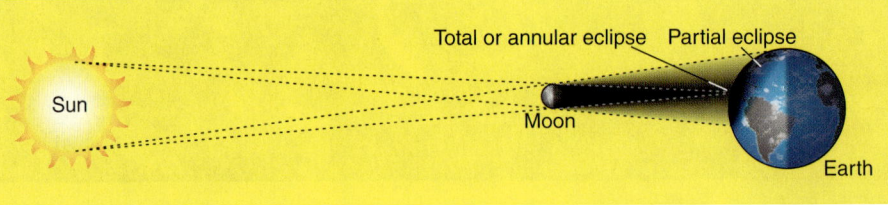

(b) Solar eclipse to scale

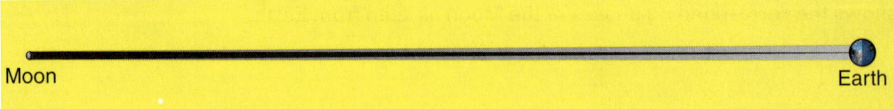

(c) Lunar eclipse geometry (not to scale)

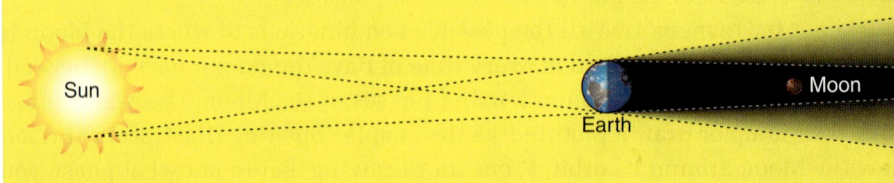

(d) Lunar eclipse to scale

Figure 2.17 (a, b) A solar eclipse occurs when the shadow of the Moon falls on the surface of Earth. (c, d) A lunar eclipse occurs when the Moon passes through Earth's shadow.

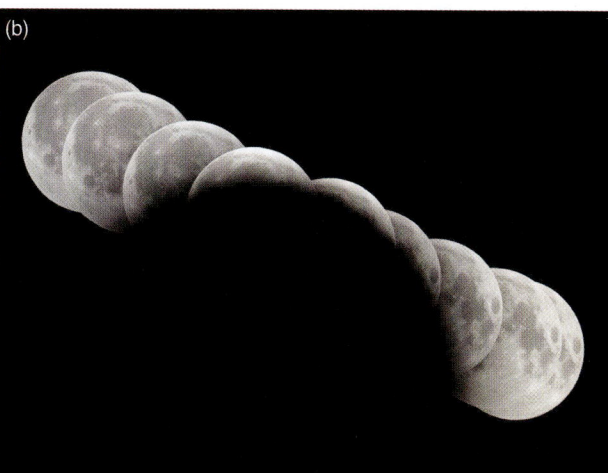

Figure 2.18 (a) During a total lunar eclipse, the Moon often appears blood red. (b) A time-lapse series of photographs of a partial lunar eclipse clearly shows Earth's shadow. Note the size of Earth's shadow compared to the size of the Moon.

Sun were drawn to scale in Figure 2.17b, it would be bigger than your head and located almost 64 meters off the left side of the page.

A **lunar eclipse**, which occurs when the Moon is partially or entirely in Earth's shadow, is very different in character from a solar eclipse. The geometry of a lunar eclipse is shown in Figure 2.17c and is drawn to scale in Figure 2.17d. Because Earth is much larger than the Moon, Earth's shadow at the distance of the Moon is more than twice the diameter of the Moon. A **total lunar eclipse**, when the Moon is entirely within Earth's shadow, lasts as long as 1 hour 40 minutes. A total lunar eclipse is often called a blood-red Moon in literature and poetry (**Figure 2.18a**): the Moon often appears red because it is being illuminated by red light from the Sun that is bent as it travels through Earth's atmosphere and hits the Moon (other colors of light are scattered away from the Moon and therefore don't illuminate it).

To see a total solar eclipse, you must be located within the very narrow band of the Moon's shadow as it moves across Earth's surface. In contrast, when the Moon is immersed in Earth's shadow, anyone located in the hemisphere of Earth that is facing the Moon can see it. Many more people have experienced a total lunar eclipse than have experienced a total solar eclipse.

If Earth's shadow incompletely covers the Moon, some of the disk of the Moon remains bright and some of it is in shadow. This is called a **partial lunar eclipse**. Figure 2.18b shows a composite of images taken at different times during a partial lunar eclipse. In the center image, the Moon is nearly completely eclipsed by Earth's shadow.

Imagine Earth, the Moon, and the Sun all sitting on the same flat tabletop. If the Moon's orbit were in exactly the same plane as the orbit of Earth, then the Moon would pass directly between Earth and the Sun at every new Moon. The Moon's shadow would pass across the face of Earth, and we would see a solar eclipse. Similarly, each full Moon would be marked by a lunar eclipse. But solar and lunar eclipses do not happen every month! This is because the Moon's orbit does not lie in exactly the same plane as the orbit of Earth. As you can see in **Figure 2.19**, the plane of the Moon's orbit around Earth is inclined by about 5° with respect to the plane of Earth's orbit around the Sun. Most of the time, the Moon is "above" or "below" the line between Earth and the Sun. About twice per year, the orbital planes line up at points called "nodes," and eclipses can occur.

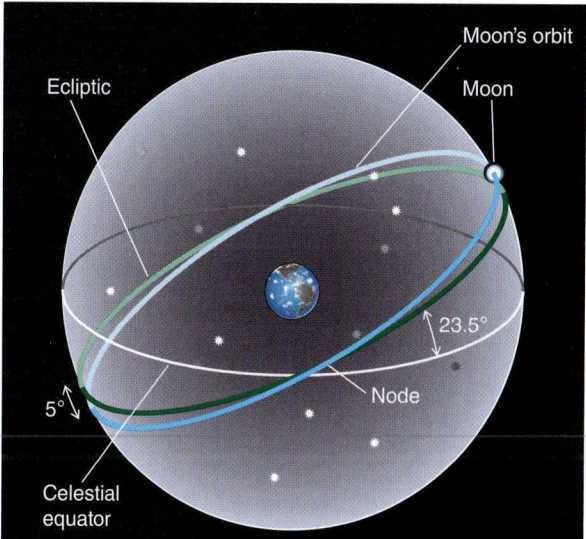

Figure 2.19 The orbit of the Moon is tilted with respect to the ecliptic, so we do not see eclipses every month.

READING ASTRONOMY News

Eclipses occur anytime the shadow of one astronomical object falls on another, even if one of those objects is man-made.

Two Eclipses, Two Stories

By **PHIL PLAIT,** *Slate.com*

For NASA's Solar Dynamics Observatory, it's eclipse season.

SDO circles the Earth in a geosynchronous orbit, meaning it makes one complete path around our planet every 24 hours. This is a special orbit, because it means to someone on Earth the satellite stays in one spot in the sky, making communication with it much easier.

But it means that twice a year the orbits of SDO and Earth line up, and the Earth irritatingly gets in the way of SDO's view of the Sun, partially blocking it. These times are called "eclipse seasons," and we're in the middle of one of them now. On March 2, the Earth got between SDO and the Sun . . . and not only that, a few hours later the Moon did as well! Here's the result [**Figure 2.20**].

I love the two different tales of these pictures. In the picture on the right the Moon's silhouette is sharp and distinct, but on the left the Earth's edge is fuzzy and distorted. The reason for that is pretty obvious: We have air! There is no sharp edge to our atmosphere, and so the amount of light it blocks from the Sun varies. Brighter stuff gets through, so you can see that bright, twisting filament in the Sun is more visible through Earth's air than the dimmer parts of the Sun's surface.

Note how the Earth's edge is nearly straight, too, while the Moon's is highly curved. The Earth is far bigger than the Moon (by a factor of 4 in diameter), so you'd expect that. But SDO is much closer to the Earth as well, so the curve of the Earth's edge isn't as obvious, looking more like a line.

One more thing: You might think SDO's

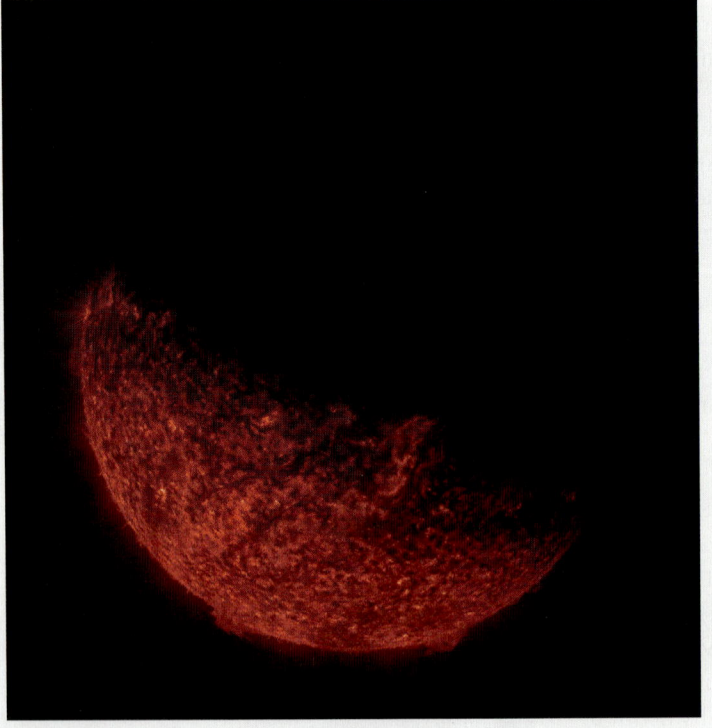 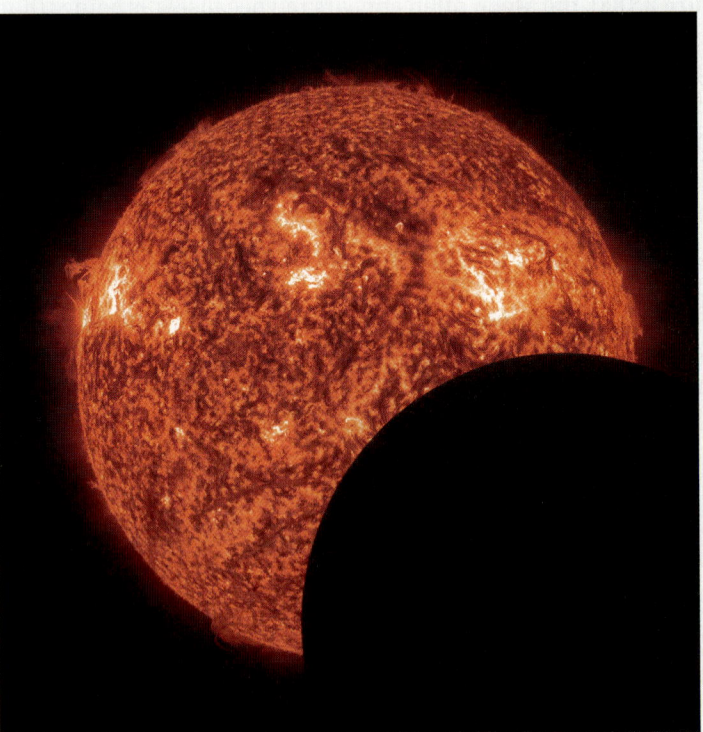

Figure 2.20 Earth (left) and the Moon (right) take turns taking a bite out of the Sun as seen by NASA's Solar Dynamics Observatory. (Image credit: NASA/SDO)

see TWO ECLIPSES

Two Eclipses (cont.)

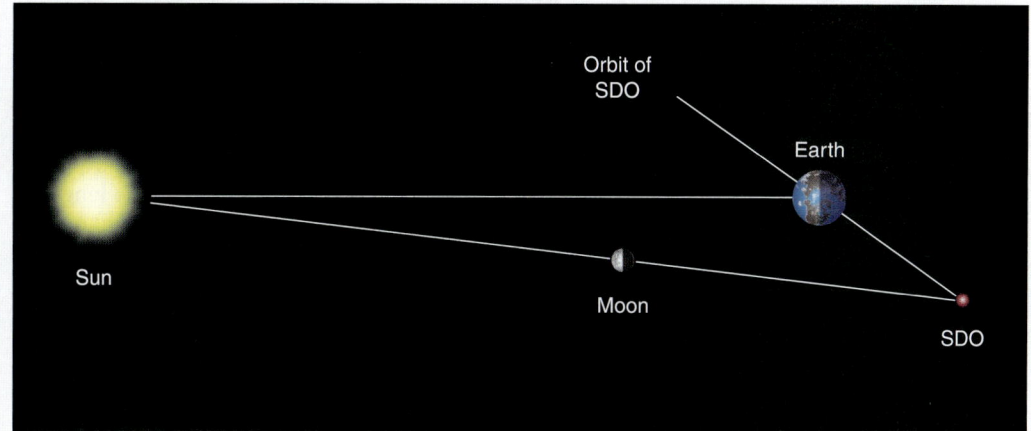

Figure 2.21 A diagram of the orbit of the Earth, Moon, and SDO (not to scale). (Image credit: Phil Plait)

view of the Sun would be blocked every orbit as the Earth got in the way, but the designers accounted for that by inclining the orbit by about 29° so the Earth usually stays out of the way [**Figure 2.21**].

This also has the advantage of keeping it close to Earth to accommodate its pretty bandwidth-intensive data stream back to Earth. Most of the time SDO has a clear view of the Sun, like the way I drew the diagram. You can see that the orbital plane of SDO intersects the orbit of Earth at two points, called the nodes. When SDO is at one of its nodes, twice per orbit, it is in the same plane as the Earth and Sun.

Normally that's no big deal. But as the Earth goes around the Sun, eventually those nodes line up with the line connecting the Earth and Sun. When SDO passes through its node on the outside part of its orbit, Earth gets in the way. That alignment can happen for about three weeks, twice per year, and that's when we have eclipse seasons.

The Moon's orbit is tilted with respect to the Earth's by about 5°, so it has two nodes as well. Over time the Moon will eventually be at a node when the node aligns with the Earth-Sun line, and we get a lunar or solar eclipse—that's why we don't get an eclipse every month! The Moon has to be at a node at the right time to get an eclipse.

So for SDO to see Earth eclipses the geometry has to be just right . . . and then to have the Moon get in the way of the Sun is even more rare, especially on the same day!

So this year, we hit the orbital alignment jackpot. I imagine the SDO scientists are lamenting so many big objects getting in their way of the Sun, but for the rest of us it makes a pretty picture, and a fun lesson in orbital mechanics.

Evaluating the News

1. In Figure 2.20, how would you describe these eclipses: total, partial, or annular? Solar or lunar?
2. Resketch Figure 2.21. Add in the shadows of Earth and the Moon, as are shown in Figure 2.17.
3. Figure 2.21 is a diagram of the system from "the side," in the plane of Earth's orbit. Redraw the figure as it would be seen from far above Earth's North Pole.
4. Figure 2.21 has been drawn during a different time of year, when SDO is not in eclipse season. Describe how this diagram would be different if it were drawn during eclipse season.
5. This special alignment of SDO lasts for "about three weeks." During this time, roughly how many eclipses of the Sun by Earth will SDO observe? Will there be as many, fewer, or more eclipses of the Sun by the Moon during this time period?

Phil Plait: "Two Eclipses, Two Stories." From Slate, March 27, 2013 © 2013 The Slate Group. All rights reserved. Used by permission and protected by the Copyright Laws of the United States. The printing, copying redistribution, or retransmission of this Content without express written permission is prohibited.

SUMMARY

The motions of Earth and the Moon are responsible for many of the repeating patterns that can be observed in the sky, and are directly connected to the calendar. Daily patterns of rising and setting are caused by Earth's rotation about its axis. Annual patterns of the stars in the sky and the passage of the seasons are caused by Earth's revolution around the Sun. The tilt of Earth on its axis changes both the length of daytime and the intensity of sunlight, causing the seasons. The pattern of the phases of the Moon lasts roughly a month, and is caused by the Moon's revolution about Earth. Occasionally, special alignments of Earth, the Moon and the Sun cause eclipses.

LG 1 As Earth rotates on its axis, stars follow circular arcs through the sky. At the equator, these are circular arcs from the eastern horizon to the western horizon. At higher latitudes, some stars move in complete circles around a celestial pole. At the North Pole or South Pole of Earth, all the stars move in complete circles in the sky.

LG 2 As Earth revolves around the Sun, the nighttime side of Earth faces different directions in space, and different stars are visible at night. Because Earth is tilted on its axis, the length of daytime and the intensity of sunlight change throughout the year, causing the seasons.

LG 3 The Moon's motion around Earth causes it to be illuminated differently at different times. When the Moon is farther than Earth from the Sun, it is in gibbous phases. When it is closer than Earth to the Sun, it is in crescent phases. Twice a year, at new or at full Moon, the Moon is exactly in line between Earth and the Sun. At these times, eclipses occur.

SUMMARY SELF-TEST

1. The Sun, Moon, and stars
   a. change their relative positions over time.
   b. appear to move each day because the celestial sphere rotates about Earth.
   c. rise north or south of west and set north or south of east, depending on their location on the celestial sphere.
   d. always remain in the same position relative to each other.

2. Which stars we see at night depend on (select all that apply)
   a. our location on Earth.
   b. Earth's location in its orbit.
   c. the time of the observation.
   d. the motion of stars relative to one another over the course of the year.

3. You see the Moon rising, just as the Sun is setting. What phase is it in?
   a. full
   b. new
   c. first quarter
   d. third quarter
   e. waning crescent

4. If you were standing at Earth's North Pole, where would you see the north celestial pole?
   a. at the zenith
   b. on the eastern horizon
   c. 23.5° south of the zenith
   d. none of the above; the north celestial pole can't be seen from there

5. Where on Earth can you stand and, over the course of a year, see the entire sky?
   a. only at the North Pole
   b. at either pole
   c. at the equator
   d. anywhere

6. The seasons are caused by _____.

7. The tilt of Jupiter's rotational axis with respect to its orbital plane is 3°. If Earth's axis had this tilt, then the seasons on Earth
   a. would be much more extreme.
   b. would be much less extreme.
   c. would be pretty much the same.
   d. would occur much differently every year, depending on the alignments.

8. You see the first quarter Moon on the meridian. Where is the Sun?
   a. on the western horizon
   b. on the eastern horizon
   c. below the horizon
   d. on the meridian

9. You do not see eclipses every month because
   a. all eclipses happen at night.
   b. the Sun, Earth, and the Moon line up only twice a year.
   c. the Sun, Earth, and the Moon line up only once a year.
   d. eclipses happen randomly and are unpredictable.

10. If the Moon were in the same orbital plane, but twice as far from Earth, which of the following would happen? (Choose all that apply.)
    a. The phases of the Moon would remain unchanged.
    b. Total eclipses of the Sun would not be possible.
    c. Total eclipses of the Moon would not be possible.
    d. The Moon's cycle would take longer.

QUESTIONS AND PROBLEMS

Multiple Choice and True/False

11. T/F: The celestial sphere is not an actual object in the sky.
12. T/F: Eclipses happen somewhere on Earth every month.
13. T/F: The phases of the Moon are caused by the relative position of Earth, the Moon, and the Sun.
14. T/F: If a star rises north of east, it will set south of west.
15. T/F: From the North Pole, all stars in the night sky are circumpolar stars.
16. The tilt of Earth's axis causes the seasons because
    a. one hemisphere of Earth is closer to the Sun in summer.
    b. the days are longer in summer.
    c. the rays of light strike the ground more directly in summer.
    d. both a and b
    e. both b and c
17. On the vernal and autumnal equinoxes,
    a. the entire Earth has 12 hours of daylight and 12 hours of darkness.
    b. the Sun rises due east and sets due west.
    c. the Sun is located on the celestial equator.
    d. all of the above
    e. none of the above
18. We always see the same side of the Moon because
    a. the Moon does not rotate on its axis.
    b. the Moon rotates once each revolution.
    c. when the other side of the Moon is facing toward us, it is unlit.
    d. when the other side of the Moon is facing Earth, it is on the opposite side of Earth.
    e. none of the above
19. You see the Moon on the meridian at sunrise. The phase of the Moon is
    a. waxing gibbous.   b. full.
    c. new.   d. first quarter.
    e. third quarter.
20. A lunar eclipse occurs when the _____ shadow falls on the _____.
    a. Earth's; Moon   b. Moon's; Earth
    c. Sun's; Moon   d. Sun's; Earth
21. You see the full Moon on the meridian. From this information, you can determine that the time where you are is
    a. noon.
    b. sunrise, about 6:00 A.M.
    c. sunset, about 6:00 P.M.
    d. midnight.
22. What do we call the group of constellations through which the Sun appears to move over the course of a year?
    a. the celestial equator
    b. the ecliptic
    c. the line of nodes
    d. the zodiac
23. If you were standing at Earth's South Pole, which stars would you see rising and setting?
    a. all of them
    b. all the stars north of the Arctic Circle
    c. all the stars south of the Arctic Circle
    d. none of them
24. On the summer solstice in the Northern Hemisphere, the Sun
    a. rises due east, passes through its highest point on the meridian, and sets due west.
    b. rises north of east, passes through its highest point on the meridian, and sets north of west.
    c. rises north of east, passes through its highest point on the meridian, and sets south of west.
    d. rises south of east, passes through its highest point on the meridian, and sets south of west.
    e. none of the above
25. In the Tropics,
    a. the Sun is directly overhead twice per year.
    b. the Sun's rays strike Earth exactly perpendicularly at some times of year.
    c. the seasons vary less than elsewhere on Earth.
    d. all of the above

Conceptual Questions

26. In your study group, two of your fellow students are discussing the phases of the Moon. One argues that the phases are caused by the shadow of Earth on the Moon. The other argues that the phases are caused by the orientation of Earth, the Moon, and the Sun. Explain how the photos in the chapter-opening illustration falsify one of these hypotheses.
27. Why is there no "east celestial pole" or "west celestial pole"?
28. Polaris was used for navigation by seafarers such as Columbus as they sailed from Europe to the New World. When Magellan sailed the South Seas, he could not use Polaris for navigation. Explain why.
29. We tend to associate certain constellations with certain times of year. For example, we see the zodiacal constellation Gemini in the Northern Hemisphere's winter (Southern Hemisphere's summer) and the zodiacal constellation Sagittarius in the Northern Hemisphere's summer. Why do we not see Sagittarius in the Northern Hemisphere's winter (Southern Hemisphere's summer) or Gemini in the Northern Hemisphere's summer?
30. Describe the Sun's apparent daily motion on the celestial sphere at the vernal equinox.
31. Why is winter solstice not the coldest time of year?
32. What is the approximate time of day when you see the full Moon near the meridian? At what time is the first quarter (waxing) Moon on the eastern horizon? Use a sketch to help explain your answers.

33. Assume that the Moon's orbit is circular. Suppose you are standing on the side of the Moon that faces Earth.
    a. How would Earth appear to move in the sky as the Moon made one revolution around Earth?
    b. How would the "phases of Earth" appear to you compared to the phases of the Moon as seen from Earth?

34. Astronomers are sometimes asked to serve as expert witnesses in court cases. Suppose you are called in as an expert witness, and the defendant states that he could not see the pedestrian because the full Moon was casting long shadows across the street at midnight. Is this claim credible? Why or why not?

35. From your own home, why are you more likely to witness a partial eclipse of the Sun rather than a total eclipse?

36. Why do we not see a lunar eclipse each time the Moon is full or witness a solar eclipse each time the Moon is new?

37. Why does the fully eclipsed Moon appear reddish?

38. In the Gregorian calendar, the length of a year is not 365 days, but actually about 365.24 days. How do we handle this extra quarter day to keep our calendars from getting out of sync?

39. Vampires are currently prevalent in popular fiction. These creatures have extreme responses to even a tiny amount of sunlight (the response depends on the author), but moonlight doesn't affect them at all. Is this logical? How is moonlight related to sunlight?

40. Suppose you are on a plane from Dallas, Texas, to Santiago, Chile. On the way there, you realize something amazing. You have just experienced the longest day of the year in the Northern Hemisphere and are about to experience the shortest day of the year in the Southern Hemisphere on the same day! On what day of the year are you flying? How do you explain this phenomenon to the person in the seat next to you?

Problems

41. Earth is spinning at 1,670 km/h at the equator. Use this number to find Earth's equatorial diameter.

42. The waxing crescent Moon appears to the east of the Sun and moves farther east each day. Does this mean it rises earlier each day or later? By how much?

43. Romance novelists sometimes say that as the hero rides off into the sunset, the full Moon is overhead. Is this correct? Why or why not? Draw a picture of the Sun, Moon, and Earth at full Moon phase to explain your answer.

44. Suppose you are on vacation in Australia, right on the Tropic of Capricorn. What is your latitude? What is the largest angle from the south celestial pole at which stars are circumpolar at your location?

45. The Moon's orbit is tilted by about 5° relative to Earth's orbit around the Sun. What is the highest altitude in the sky that the Moon can reach, as seen in Philadelphia (latitude 40° north)?

46. Imagine you are standing on the South Pole at the time of the southern summer solstice.
    a. How far above the horizon will the Sun be at noon?
    b. How far above (or below) the horizon will the Sun be at midnight?

47. Find out the latitude where you live. Draw and label a diagram showing that your latitude is the same as (a) the altitude of the north celestial pole and (b) the angle (along the meridian) between the celestial equator and your local zenith. What is the noontime altitude of the Sun as seen from your home at the times of winter solstice and summer solstice?

48. Let's say you use a protractor to estimate an angle of 40° between your zenith and Polaris. Are you in the continental United States or Canada?

49. Suppose the tilt of Earth's equator relative to its orbit were 10° instead of 23.5°. At what latitudes would the Arctic and Antarctic circles and the tropics of Cancer and Capricorn be located?

50. Carefully draw a diagram of the Moon and its shadow at both its current distance and twice as far from Earth. If the Moon were twice as far from Earth, which of the following would happen?
    a. Total eclipses of the Sun would not be possible.
    b. Total eclipses of the Moon would not be possible.

SMARTWORK

Norton's online homework system includes algorithmically generated versions of these questions, plus additional conceptual exercises. If your instructor assigns questions in SmartWork, log in at smartwork.wwnorton.com.

Exploration | Phases of the Moon

wwnpag.es/uou2

In this exploration, we will be examining the phases of the Moon. Visit the Student Site (wwnpag.es/uou2) and open the Phases of the Moon Nebraska Simulation in Chapter 2. This simulator animates the orbit of the Moon around Earth, allowing you to control the simulation speed and a number of other parameters.

Begin by starting the animation to explore how it works. Examine all three image frames. The large frame shows the Earth-Moon system, as looking down from far above Earth's North Pole. The upper right frame shows what the Moon looks like to the person on the ground. The lower right frame shows where the Moon appears in the person's sky. Stop the animation, and press "reset" in the upper menu bar.

1. What time of day is this for the person shown on Earth?

2. What phase is the Moon in?

3. Where is the Moon in this person's sky?

Run the animation until the Moon reaches waxing crescent phase.

4. As viewed from Earth, which side of the Moon is illuminated (the left or the right)?

5. The person shown on Earth will observe this waxing crescent Moon either after sunset or before sunrise. At which of these times can the person see the waxing crescent Moon?

Run the animation until the Moon reaches first quarter and the Sun is setting for the person on Earth. (Hint: You may want to slow the animation rate!)

6. How many full days have passed since new Moon?

7. At this instant, where is the first quarter Moon in the person's sky?

8. If an astronaut was standing on the near side of the Moon at this time, what phase of Earth would he see?

Three observations about the phases of the Moon are connected: the location of the Moon in the sky, the time for the observer, and the phase of the Moon. If you know two of these, you can figure out the third. Use the animation to fill in the missing pieces in the following situations:

9. An observer sees the Moon in _____ phase, overhead, at midnight.

10. An observer sees the Moon in third quarter phase, rising in the East, at _____.

11. An observer sees the Moon in full phase, _____, at 6 A.M.

SMARTWORK • smartwork.wwnorton.com

3 Laws of Motion

In the previous chapter, you learned that the planets, including Earth, orbit the Sun, but not the reason why. Gravity is the force that holds the planets in orbit. Because the Sun is far more massive than all the other parts of the Solar System combined, its gravity shapes the motions of every object in its vicinity, from the almost circular orbits of some planets to the extremely elongated orbits of comets.

As our understanding of the universe has expanded, we have come to realize that our Solar System is only one example of gravity at work. The illustration on the opposite page shows the launch of a space shuttle, in which the force of exhaust from the rocket pushes against, and overcomes, the force of Earth's gravity in order to boost the spacecraft into orbit. On the illustration, a student is working out the forces that are acting on the shuttle, in which direction they act, and how their strengths compare.

◆ LEARNING GOALS

By the end of this chapter, you should know how to think about and diagram forces such as gravity. You should be able to look at a photograph like the one on the postcard at right and see the fundamental laws of physics at work. You should understand how to draw action-reaction pairs of forces on objects such as Earth or a spacecraft. You should also be able to:

LG 1 Describe planetary orbits, and explain how astronomers came to know that these orbits are elliptical around the Sun.

LG 2 List the physical laws that govern the motion of all objects.

LG 3 Combine motion and gravitation to explain planetary orbits.

LG 4 Understand the concept of a frame of reference.

3.1 Since Ancient Times Astronomers Have Studied the Motions of the Planets

Astronomy challenges us to think in novel ways, to imagine ourselves in space, looking down at Earth, or even farther away, looking down at the Solar System. The struggle to understand our place in the universe begins with understanding the motions of Earth, the Sun, and the planets. The history of the progression of ideas—from Earth at the center of all things to Earth as a tiny, insignificant rock—is full of heroes and villains and is a wonderful example of the self-correcting nature of science.

Early Astronomy

In ancient times, astronomers and philosophers hypothesized that the Sun might be the center of the Solar System, but they did not have the tools to test the hypothesis or the mathematical insight to formulate a more complete and testable model. Because we can't feel Earth's motion through space, a **geocentric**—that is, Earth-centered—**model** of the Solar System prevailed. For nearly 1,500 years, most educated people believed that the Sun, the Moon, and the planets that are visible to the naked eye (Mercury, Venus, Mars, Jupiter, and Saturn) all moved in circles around a stationary Earth.

Ancient peoples were aware that planets move in a generally eastward direction among the "fixed stars." They also knew that these planets would occasionally exhibit **apparent retrograde motion**; that is, they would seem to turn around, move westward for a while, and then return to their normal eastward travel. **Figure 3.1** shows a time-lapse sequence of Mars going through its retrograde "loop." This odd behavior of the five known planets created a puzzling problem for the geocentric model; if the planets moved in circles around a stationary Earth, the only explanation for the retrograde motion would be that the planets

▶▶ **Nebraska Simulation:** Retrograde Motion

Vocabulary Alert

model In common language, a model is typically a scaled-down, three-dimensional version of a larger object. A model of a car, for example, is as close to the appearance of the real thing as possible, but typically is nonfunctional. In science, a model is a description of a system that accounts for its properties.

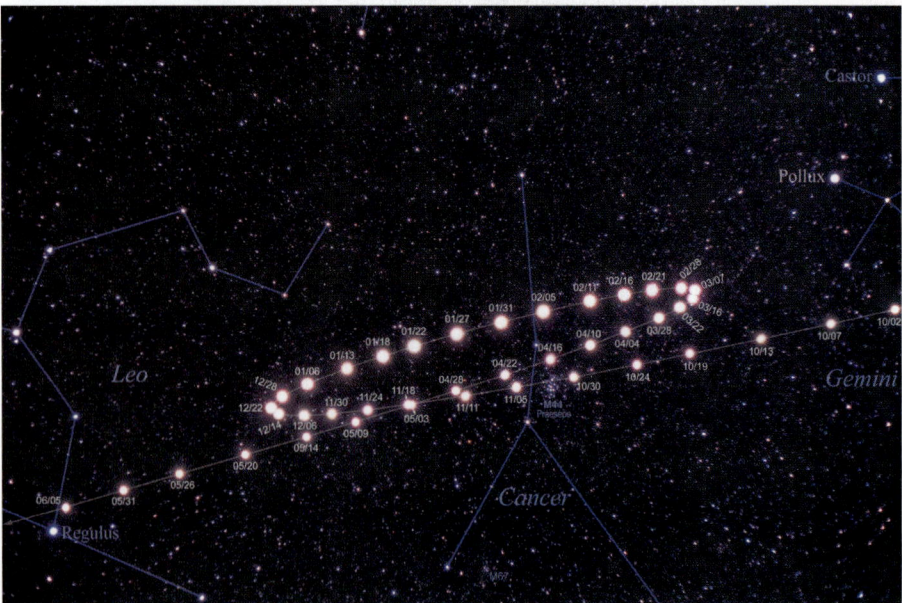

Figure 3.1 This time-lapse photographic series shows Mars as it moves in apparent retrograde motion.

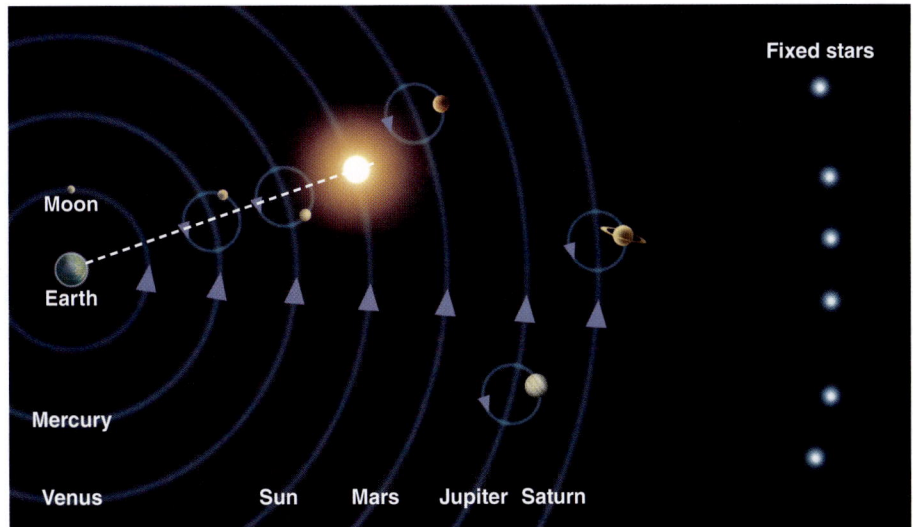

Figure 3.2 Ptolemy's model of the Solar System included a complex system of interconnected circles to explain retrograde motion. While traveling along its larger circle, a planet would at the same time be moving along its smaller circle. At times, the motions would be in opposite directions, creating the observed retrograde motion.

stop, turn around, and go the other way. The Greek astronomer Ptolemy (Claudius Ptolemaeus, 90–168 CE) modified the geocentric model with a complex system of interconnected circles to try to obtain more accurate results and to explain retrograde motion (**Figure 3.2**).

The Copernican Revolution

Nicolaus Copernicus (1473–1543) is famous for placing the Sun rather than Earth at the center of the Solar System. He was not the first person to consider the idea that Earth orbited the Sun, but he was the first to develop a mathematical model that made predictions that later astronomers would be able to test. This work was the beginning of what was later called the Copernican Revolution. Through the work of scientists such as Tycho Brahe (1546–1601), Galileo Galilei (1564–1642), Johannes Kepler (1571–1630), and Sir Isaac Newton (1642–1727), the **heliocentric**—Sun-centered—theory of the Solar System became one of the most well-corroborated theories in all of science.

In 1543, Copernicus published a heliocentric model that explained retrograde motion much more simply than the geocentric model. If we are in a car or train and we pass a slower-moving car or train, it can seem to us that the other vehicle is moving backward. It can be hard to tell which vehicle is moving and in which direction without an external frame of reference. A **frame of reference** is a system within which an observer measures positions and motions using coordinates such as distance and time. Copernicus provided this frame of reference for the Sun and its planets. In the Copernican model, the outer planets Mars, Jupiter, and Saturn undergo apparent retrograde motion when Earth overtakes them in their orbits. Likewise, the inner planets Mercury and Venus move in apparent retrograde motion when overtaking Earth. Except for the Sun, all Solar System objects exhibit apparent retrograde motion. The magnitude of the effect diminishes with increasing distance from Earth. Retrograde motion is an illusion caused by the relative motion between Earth and the other planets.

Combining geometry with observations of the positions of the planets in the sky including their altitudes and the times they rise and set, Copernicus estimated the planet–Sun distances in terms of the Earth–Sun distance. These relative distances are remarkably close to those obtained by modern methods. From these

▶▶ **Nebraska Simulation:** Planetary Orbit Simulator

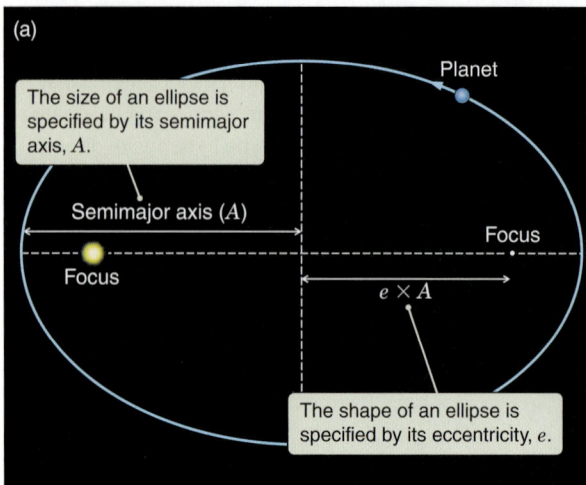

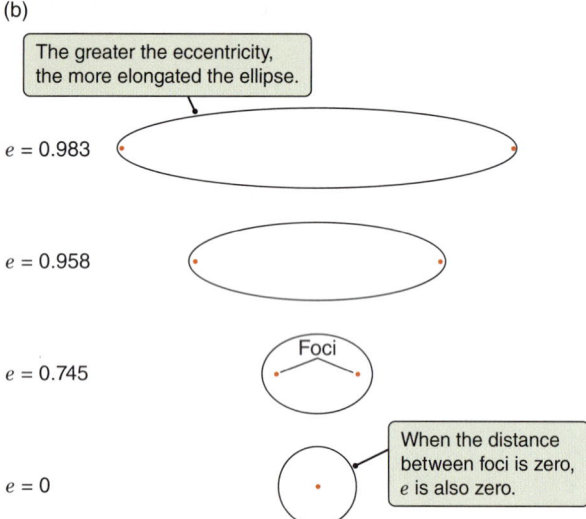

Figure 3.3 (a) Planets move in elliptical orbits with the Sun at one focus. The average radius of the orbit is equal to the semimajor axis, and the shape of the orbit is given by its eccentricity, e. (b) Ellipses range from circles to elongated eccentric shapes.

▶▶ **Nebraska Simulation:** Eccentricity Demonstrator

▶❙❙ **AstroTour:** Kepler's Laws

Vocabulary Alert

focus (plural *foci*): In common language, this word is used in several ways to indicate directed attention or the place where light is concentrated by a lens. In this context, it refers to a special point within an ellipse. An ellipse has two of these special points, and the sum of the distances from these points to any point on the ellipse is a constant.

observations, he also found when each of the planets aligns with Earth and the Sun. He used this information and geometry to figure out how long it took each planet to orbit the Sun. This model made testable predictions of the location of each planet on a given night.

Kepler's Laws

Tycho (conventionally referred to by his first name) was the last great observational astronomer before the invention of the telescope. Through careful measurements of the precise positions of planets in the sky, he developed the most comprehensive set of planetary data available at that time. When he died, his assistant, Kepler, received these records. Kepler used the data to deduce three rules that elegantly and accurately describe the motions of the planets. These three rules are now generally referred to as **Kepler's laws**. These laws are **empirical**: they use prior data to make predictions about future behavior but do not include an underlying theory of why the objects behave as they do.

Kepler's First Law When Kepler compared Tycho's observations with predictions from Copernicus's model, he expected the data to confirm circular orbits for planets orbiting the Sun. Instead, he found disagreements between his predictions and the observations. He was not the first to notice such discrepancies. Rather than simply discarding the model, Kepler adjusted Copernicus's idea until it matched the observations.

Kepler discovered that if he replaced circular orbits with *elliptical* orbits, the predictions fit the observations almost perfectly. An **ellipse** is a specific kind of oval. It is symmetric from right to left and from top to bottom. **Figure 3.3a** illustrates the vocabulary of ellipses. The dashed lines represent the two main axes of the ellipse. Half of the length of the long axis (the major axis) is called the **semimajor axis**, often denoted by the letter A. The semimajor axis of an orbit is equal to the average distance between the planet and the Sun. The shape of an ellipse is given by its **eccentricity (e)**, which in turn is determined by the distance between the **foci** of the ellipse. Foci are two mathematically important points along the major axis; the Sun is located at one focus of a planet's orbit, but there is nothing at the other. As the two foci approach each other, the figure becomes a circle with eccentricity 0, as shown in Figure 3.3b. Correspondingly, as the foci move farther apart, the ellipse becomes more elongated, and the eccentricity approaches 1.

Kepler's first law of planetary motion states that the orbit of a planet is an ellipse with the Sun at one focus. Most planetary objects in our Solar System have nearly circular orbits with eccentricities close to zero. As shown in **Figure 3.4a,** Earth's orbit is very nearly a circle centered on the Sun, with an eccentricity of 0.017. By contrast, Pluto's orbit, as shown in Figure 3.4b, has an eccentricity of 0.249. The orbit is noticeably elongated, with the Sun offset from center.

Kepler's Second Law From Tycho's observations of planetary motions, Kepler found that a planet moves fastest when it is closest to the Sun and slowest when it is farthest from the Sun. Kepler found an elegant way to describe the changing speed of a planet in its orbit around the Sun. **Figure 3.5** shows a planet at six different points in its orbit. The time elapsed between t_1 and t_2 is equal to the time elapsed between t_3 and t_4 and to the time elapsed between t_5 and t_6. The areas of the green, orange, and blue regions are also equal. To see how those areas are

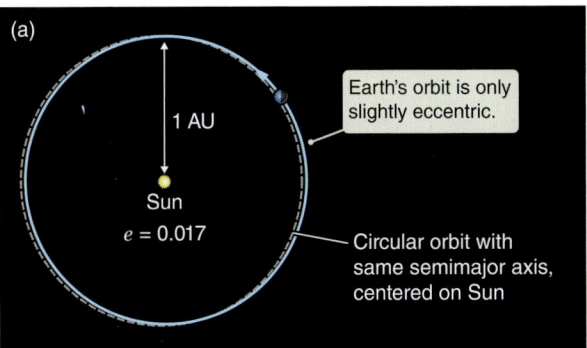
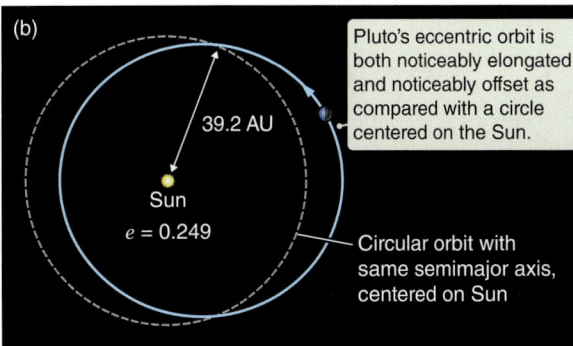

Figure 3.4 When the orbits of (a) Earth and (b) Pluto are compared with circles centered on the Sun, it becomes clear that they are elliptical. In the case of Pluto, you can also see that the Sun is not at the center of the orbit but instead lies at one focus of the ellipse.

constructed, imagine a straight line connecting the Sun with the planet. We can think of this line as "sweeping out" an area as the planet moves from one point to another. **Kepler's second law**, also called Kepler's **law of equal areas**, states that the area swept out by a planet during a specific time interval is always the same, regardless of the location of the planet in its orbit. In Figure 3.5, if the three time intervals are equal (that is, $t_1 \rightarrow t_2 = t_3 \rightarrow t_4 = t_5 \rightarrow t_6$), then the three areas A, B, and C will be equal as well.

Kepler's Third Law Planets close to the Sun travel on shorter orbits than planets that are far from the Sun. Jupiter's average distance from the Sun, for example, is 5.2 times larger than Earth's average distance from the Sun. Because an orbit's circumference is proportional to its radius, Jupiter must travel 5.2 times farther in its orbit around the Sun than Earth does in its orbit. If the two planets were traveling at the same speed, Jupiter would complete one orbit in 5.2 years. But Jupiter takes almost 12 years to complete one orbit. Jupiter not only has farther to go but also is moving more slowly than Earth. The farther a planet is from the Sun, the larger the circumference of its orbit and the lower its speed.

Kepler discovered a mathematical relationship between the **period** of a planet's orbit and its average distance from the Sun. **Kepler's third law** states that the period squared is equal to the distance cubed (**Working It Out 3.1**).

3.2 Galileo Was the First Modern Scientist

Galileo Galilei is one of the heroes of astronomy. He was the first to use a telescope to make significant discoveries about astronomical objects, and much has been written about the considerable danger that Galileo—as he is commonly known—faced as a result of his discoveries. Galileo's telescopes were relatively small, yet they were sufficient for him to observe spots on the Sun, the uneven surface and craters of the Moon, and the large numbers of stars in the band of light in the sky called the Milky Way.

Two other sets of observations made Galileo famous. When he turned his telescope to the planet Jupiter, he observed several "stars" in a line near Jupiter. Over time, he saw that there were actually four of these stars and that their positions changed from night to night. Galileo correctly reasoned that these were moons in orbit around Jupiter. These are the largest of Jupiter's many moons and are still referred to as the Galilean moons. This was the first observational evidence that some objects in the sky did not orbit Earth. He also observed that the planet Venus went through phases similar to the Moon's and that the phases

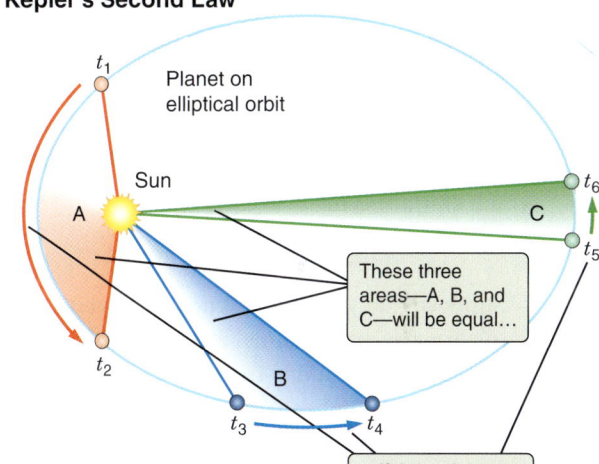

Figure 3.5 Kepler's second law states that an imaginary line between a planet and the Sun sweeps out an area as the planet orbits such that if the three time intervals shown are equal, then the areas A, B, and C will also be equal.

Vocabulary Alert

period In common language, the word *period* can mean how long a thing lasts. For example, we might talk about the "period of time spent at the grocery store." Astronomers use this word only when talking about repeating intervals, such as the time it takes for an object to complete one orbit.

Working It Out 3.1 | Kepler's Third Law

Just as *squaring* a number means that you multiply it by itself, as in $a^2 = a \times a$, *cubing* it means that you multiply it by itself again, as in $a^3 = a \times a \times a$. Kepler's third law states that the square of the period of a planet's orbit, P_{years}, measured in years, is equal to the cube of the semimajor axis of the planet's orbit, A_{AU}, measured in astronomical units (AU). Translated into math, the law says

$$(P_{years})^2 = (A_{AU})^3$$

Here, astronomers use nonstandard units as a matter of convenience. Years are handy units for measuring the periods of orbits, and astronomical units are handy units for measuring the sizes of orbits. When we use years and astronomical units as our units, we get the relationship shown in **Figure 3.6**, where the slope of the line is equal to 1. It is important to realize that our choice of units in no way changes the physical relationship we are studying. For example, if we instead chose seconds and meters as our units, this relationship would read

$$(3.2 \times 10^{-8} \text{ years/second} \times P_{seconds})^2$$
$$= (6.7 \times 10^{-12} \text{ AU/meter} \times A_{meters})^3$$

which simplifies to

$$(P_{seconds})^2 = 2.9 \times 10^{-19} \times (A_{meters})^3$$

Suppose that we want to know the average radius of Neptune's orbit in astronomical units. First, we need to find out how long Neptune's period is in Earth years, which can be determined by careful observation of Neptune's position relative to the fixed stars. Neptune's period is 165 years. Plugging this into Kepler's third law, we find that

$$(P_{years})^2 = (A_{AU})^3$$
$$(165)^2 = (A_{AU})^3$$

To solve this equation, we must first square the left side to get 27,225 and then take its cube root.

Calculator hint: A scientific calculator usually has a cube root function. It sometimes looks like $x^{1/y}$ and sometimes like $\sqrt[x]{y}$. You use it by typing the base number, hitting the button, and then typing the root you are interested in (2 for square root, 3 for cube root, and so on). Occasionally, a calculator will instead have a button that looks like x^y (or y^x). In this case, you need to enter the root as a decimal. For example, if you want to take the square root, you type 0.5 because the square root is denoted by ½. For the cube root, you type 0.333333333 (repeating) because the cube root is denoted by ⅓.

To find the length of the semimajor axis of Neptune's orbit, we might type 27,225 [$x^{1/y}$] 3. This gives

$$30.1 = A_{AU}$$

so the average distance between Neptune and the Sun is 30.1 AU.

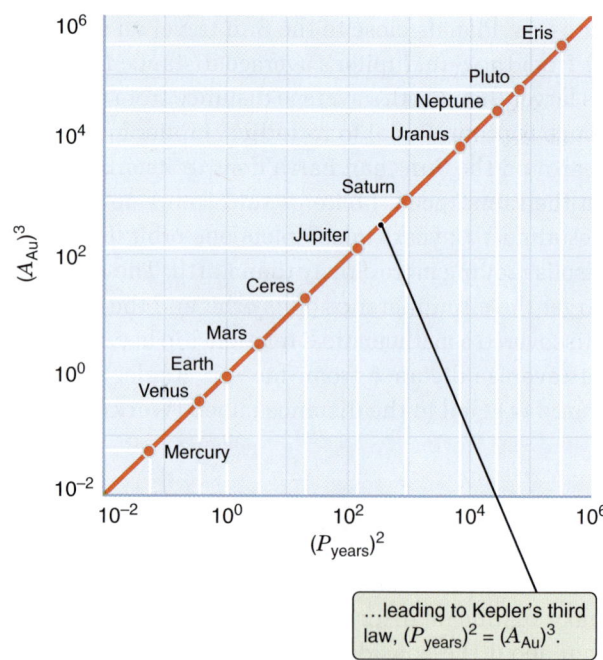

Figure 3.6 A plot of A^3 versus P^2 for the eight planets and three of the dwarf planets in our Solar System shows that they obey Kepler's third law. (Note that by plotting powers of 10 on each axis, we are able to fit both large and small values on the same plot. We will do this frequently.)

Figure 3.7 Modern photographs of the phases of Venus show that when we see Venus more illuminated, it also appears smaller, implying that Venus is farther away at that time.

correlated with the size of the image of Venus in his telescope (**Figure 3.7**). This is impossible to explain in a geocentric model, but it makes sense in a heliocentric model. These observations in particular convinced Galileo that Copernicus was correct to place the Sun at the center of our Solar System.

Galileo's public support for Copernican heliocentricity got him into trouble with the Catholic Church. In 1632, Galileo published his best-selling book, *Dialogo sopra i due massimi sistemi del mondo (Dialogue Concerning the Two Chief World Systems)*. In the *Dialogo*, the champion of the Copernican view of the universe, Salviati, is a brilliant philosopher. The defender of an Earth-centered universe, Simplicio, uses arguments made by the classical Greek philosophers and the pope, and he sounds silly and ignorant. Galileo had submitted drafts of his book to church censors, but the censors found the final version unacceptable. The perceived attack on the pope attracted the attention of the church, and Galileo was eventually placed under house arrest. The book was placed on the Index of Prohibited Works, along with Copernicus's *De Revolutionibus*; but it traveled across Europe, was translated into other languages, and was read by other scientists.

Galileo's work on the motion of objects was at least as fundamental a contribution as his astronomical observations. He conducted actual experiments with falling and rolling objects. In one famous experiment, he dropped two objects of different masses from the same very tall height and found that they landed at the same time. His work on falling objects demonstrated that gravity on Earth accelerates all objects at the same rate, independent of mass. Galileo's approach was different from that of prior natural philosophers who believed that one could understand the universe just by thinking about it—no experiments needed. Galileo's observations and experiments with many types of moving objects, such as carts and balls, led him to disagree with the philosophers about when and why objects continued to move or came to rest. Before Galileo, it was thought that the natural state of an object was to be at rest. But Galileo found that the natural state of an object is to keep doing what it was doing until a force acts on it. Put more precisely, Galileo found that an object in motion will continue moving along a straight line with a constant speed until a force acts on it to change its state of motion. This idea has implications for not only the motion of carts and balls but also the orbits of planets.

3.3 Newton's Laws Govern Motion

One of the earliest advances in theoretical science was also one of the greatest intellectual accomplishments. The work of Sir Isaac Newton on the nature of

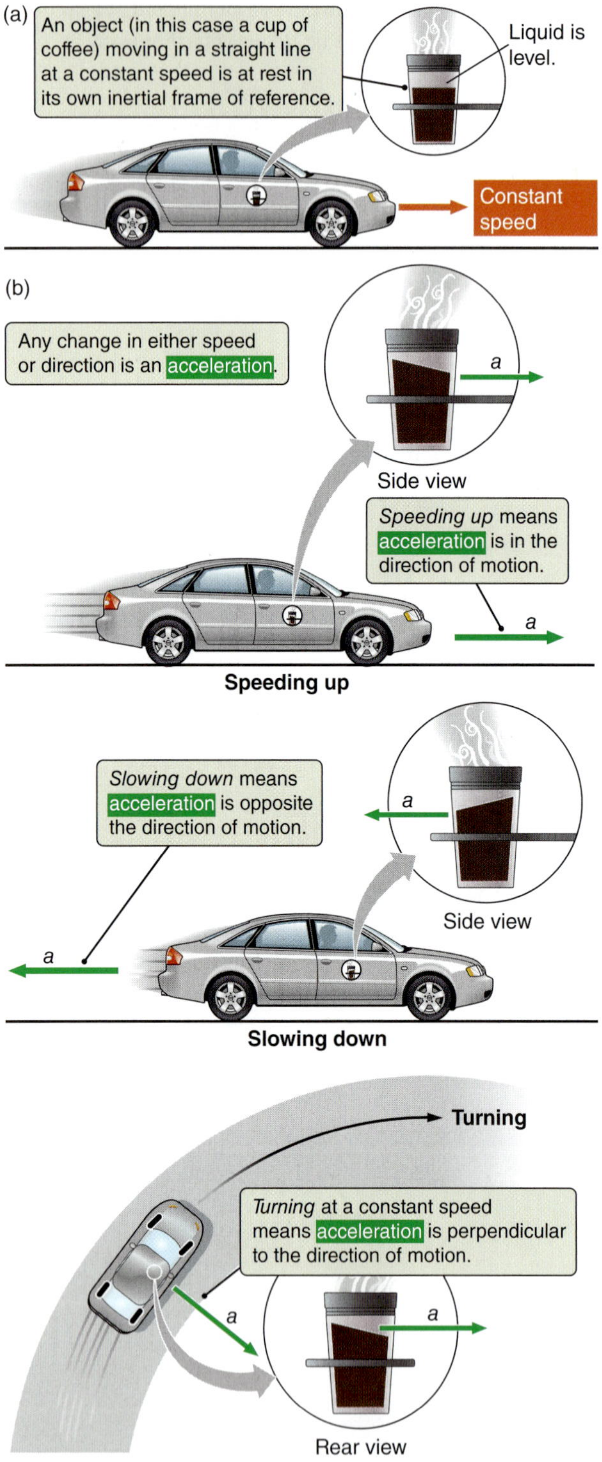

Figure 3.8 (a) An object moving in a straight line at a constant speed is not accelerating, so the coffee remains level in the cup. (b) When accelerating, the coffee in the cup sloshes forward or backward. (Throughout the text, velocity arrows are shown in red, and acceleration arrows are shown in green.) As shown in the bottom frame, even a change in the direction of motion with no change in speed indicates an acceleration that will produce sloshing of the coffee in the cup.

motion set the standard for what we now refer to as *scientific theory* and *physical law*. Newton proposed three elegant laws that govern the motions of all objects in the universe. They enabled Newton to connect phenomena on Earth to phenomena in the sky. Newton's laws are essential to our understanding of the motions of the planets and all other celestial bodies. In this section, we will look at each of the three laws in turn.

Newton's First Law: Objects at Rest Stay at Rest; Objects in Motion Stay in Motion

A **force** is a push or a pull. It is possible for two or more forces to oppose one another in such a way that they are perfectly balanced and cancel out. For example, gravity pulls down on you as you sit in your chair. But the chair pushes up on you with an exactly equal and opposite force. So you remain motionless. Forces that cancel out have no effect on an object's motion. When forces add together to produce an effect, we often use the term *net force*, or sometimes just *force*.

Imagine that you are driving a car, and your phone is on the seat next to you. A rabbit runs across the road in front of you, and you press the brakes hard. You feel the seat belt tighten to restrain you. At the same time, your phone flies off the seat and hits the dashboard. You have just experienced what Newton describes in his first law of motion. **Inertia** is the tendency of an object to maintain its state—either of motion or of rest—until it is pushed or pulled by a net force. In the case of the stopping car, you did not hit the dashboard because the force of the seat belt on you slowed you down. The phone did hit the dashboard because no such force acted upon it.

Newton's first law of motion describes inertia and states that an object in motion tends to stay in motion, in the same direction, until a net force acts upon it; and an object at rest tends to stay at rest until a net force acts upon it.

Recall from Section 3.1 the concept of a frame of reference. Within a frame of reference, only the relative motions between objects have any meaning. Without external clues, you cannot tell the difference between sitting still and traveling at constant speed in a straight line. For example, if you close your eyes while riding in the passenger seat of a quiet car on a smooth road, you would feel as though you were sitting still. Returning to the earlier example, your phone was "at rest" beside you on the front seat of your car, but a person standing by the side of the road would see the phone moving past at the same speed as the car. And people in a car approaching you would see the phone moving quite fast—at the speed they are traveling plus the speed you are traveling! All of these perspectives are equally valid, and all of these speeds of the phone are correct when measured in the appropriate reference frame.

A reference frame moving in a straight line at a constant speed is an **inertial frame of reference**. Any inertial frame of reference is as good as any other. As illustrated in **Figure 3.8a**, in the frame of reference of a cup of coffee, *it is at rest in its own frame* even if the car is speeding down the road. (Notice that we use particular colors throughout this text for different quantities. Here, green arrows are used for acceleration, and red arrows are used for speed or velocity, which we will describe later.)

Newton's Second Law: Motion Is Changed by Forces

What if a net force does act? In the previous example, you were traveling in the car, and your motion was slowed when the force of the seat belt acted upon you.

Forces change an object's motion—by changing either the speed or the direction. This reflects **Newton's second law of motion**: if a net force acts on an object, then the object's motion changes.

In the preceding paragraphs, we spoke of "changes in an object's motion," but what does that phrase really mean? When you are in the driver's seat of a car, a number of controls are at your disposal. On the floor are a gas pedal and a brake pedal. You use these to make the car speed up or slow down. A *change in speed* is one way the car's motion can change. But you also have the steering wheel in your hands. When you are moving down the road and you turn the wheel, your speed does not necessarily change, but the direction of your motion does. A *change in direction* is also a change in motion.

Together, the combined speed and direction of an object's motion is called an object's **velocity**. "Traveling at 50 kilometers per hour (km/h)" indicates speed; "traveling north at 50 km/h" indicates velocity. The rate at which the velocity of an object changes is called **acceleration**. Acceleration tells you how rapidly a change in velocity happens. For example, if you go from 0 to 100 km/h in 4 seconds, you feel a strong push from the seat back as it shoves your body forward, causing you to accelerate along with the car. However, if you take 2 minutes to go from 0 to 100 km/h, the acceleration is so slight that you hardly notice it.

Because the gas pedal on a car is often called the accelerator, some people think *acceleration* always means that an object is speeding up. But we need to stress that, as used in physics, any change in motion is an acceleration. Figure 3.8b illustrates the point by showing what happens to the coffee in a coffee cup as the car speeds up, slows down, or turns. Slamming on your brakes and going from 100 to 0 km/h in 4 seconds is just as much acceleration as going from 0 to 100 km/h in 4 seconds. Similarly, the acceleration you experience as you go through a fast, tight turn at a constant speed is every bit as real as the acceleration you feel when you slam your foot on the gas pedal or the brake pedal. Speeding up, slowing down, turning left, turning right—if you are not moving in a straight line at a constant speed, you are experiencing an acceleration.

Newton's second law of motion says that net forces cause accelerations. The acceleration an object experiences depends on two things. First, as shown in **Figure 3.9a**, the acceleration depends on the strength of the net force acting on the object to change its motion. Push three times as hard, and the object experiences three times the acceleration, shown in Figure 3.9b. The change in motion occurs

Vocabulary Alert

force In common language, the word *force* has many meanings. Astronomers specifically mean a push or a pull.

inertia In common language, we think of inertia as a tendency to remain motionless. Astronomers and physicists think more generally of inertia as the tendency of matter to resist a change in motion—of an object at rest to remain at rest, and of a moving object to remain in motion.

Figure 3.9 Newton's second law of motion says that the acceleration experienced by an object is determined by the force acting on the object, divided by the object's mass. (Throughout the text, force arrows are shown in blue.)

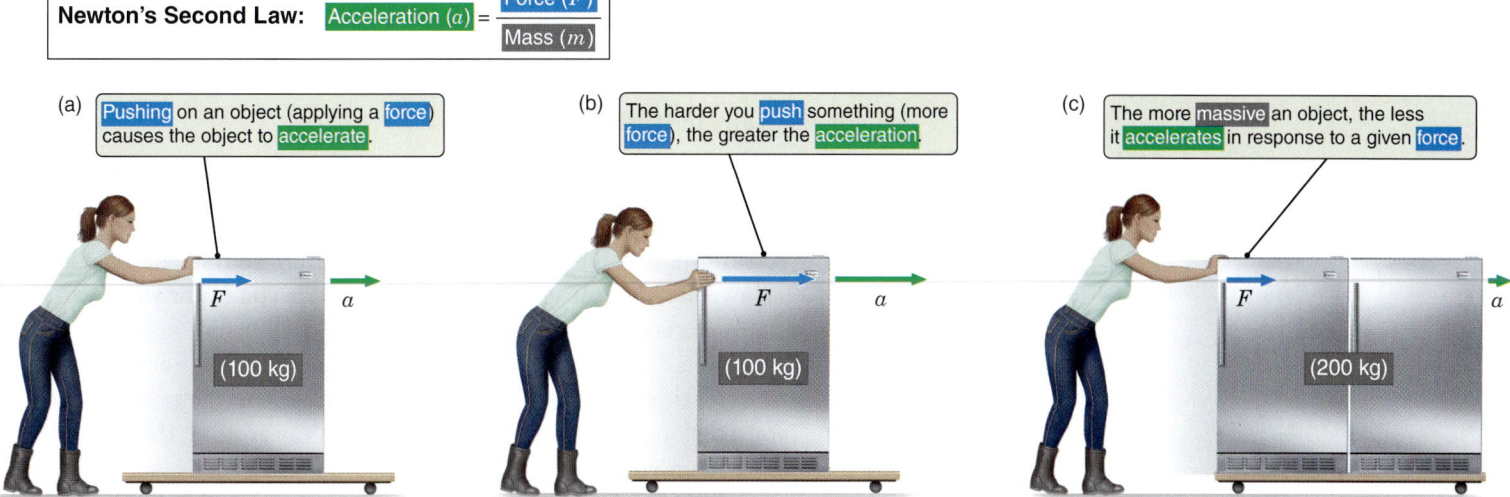

Newton's Second Law: Acceleration $(a) = \dfrac{\text{Force } (F)}{\text{Mass } (m)}$

(a) Pushing on an object (applying a force) causes the object to accelerate.

(b) The harder you push something (more force), the greater the acceleration.

(c) The more massive an object, the less it accelerates in response to a given force.

Working It Out 3.2 | Finding the Acceleration

The equation for acceleration states

$$\text{Acceleration} = \frac{\text{How much velocity changes}}{\text{How long the change takes to happen}}$$

We can write this more compactly by expressing the change in velocity as $v_2 - v_1$ and the change in time as $t_2 - t_1$. Acceleration is commonly abbreviated as a. So our equation can be translated to math as

$$a = \frac{v_2 - v_1}{t_2 - t_1}$$

For example, if an object's speed goes from 5 meters per second (m/s) to 15 m/s, then the final velocity is $v_2 = 15$ m/s. The initial velocity is $v_1 = 5$ m/s. So the change in velocity $(v_2 - v_1)$ is 10 m/s. If that change happens over the course of 2 seconds, then the change in time, $t_2 - t_1$, is 2 seconds. So the acceleration is 10 m/s divided by 2 seconds:

$$a = \frac{10 \text{ m/s}}{2 \text{ s}} = 5 \text{ m/s/s}$$

This is the same as saying "5 meters per second squared," which is written as 5 m/s^2, or 5 m/s^{-2}.

If we want to know how an object's motion is changing, we need to know two things: what net force is acting on the object, and what is the mass of the object? We can say this formula in words as "the acceleration is equal to the strength of the net force divided by the mass." We can translate it into math as

$$\text{Acceleration} = \frac{\text{Strength of net force}}{\text{Mass}}$$

This is a lot to write every time we want to talk about acceleration. Typically, we let a stand for acceleration, F for force, and m for mass:

$$a = \frac{F}{m}$$

Newton's second law is often written as $F = ma$ (to understand how we changed this equation, revisit Working It Out 2.1), giving force as units of mass multiplied by units of acceleration, or kilograms times meters per second squared (kg m/s^2). The units of force are named **newtons (N)**, so that 1 N = 1 kg m/s^2.

This is the mathematical statement of Newton's second law of motion. This equation says three things: (1) when you push on an object, that object accelerates in the direction you are pushing; (2) the harder you push on an object, the more it accelerates; and (3) the more massive the object is, the harder it is to accelerate it.

in the direction the net force points. Push an object away from you, and it will accelerate away from you.

The acceleration that an object experiences also depends on its inertia. Some objects—for example, the empty box that a refrigerator was delivered in—are easily shoved around by humans. However, a refrigerator, even though it is about the same size, is not easily shoved around. For our purposes, an object's **mass** is interchangeable with its inertia. The greater the mass, the greater the inertia, and thus less acceleration will occur in response to the same net force, as shown in Figure 3.9c. This relationship between acceleration, force, and mass is expressed mathematically in **Working It Out 3.2**.

Newton's Third Law: Whatever Is Pushed, Pushes Back

Imagine you are standing on a skateboard and pushing yourself along with your foot. Each shove of your foot against the ground sends you faster along your way. But why does this happen? Your muscles flex, and your foot exerts a force on the ground. (Earth does not noticeably accelerate because its great mass gives it great inertia.) Yet this does not explain why you experience an acceleration. The fact that you accelerate means that as you push on the ground, the ground must be pushing back on you.

Part of Newton's genius was his ability to see patterns in such everyday events. Newton realized that *every* time one object exerts a force on another, a matching force is exerted by the second object on the first. That second force is exactly as strong as the first force but is in exactly the *opposite* direction. When you are riding on the skateboard, you push backward on Earth, and Earth pushes you forward. As shown in **Figure 3.10**, a woman pulling a load on a cart pulls on the rope, and the rope pulls back. A car tire pushes back on the road, and the road pushes forward on the tire. Earth pulls on the Moon, and the Moon pulls on Earth. Turning back to the chapter-opening illustration, we see that a rocket engine pushes hot gases out of its nozzle, and those hot gases push back on the rocket, propelling it into space. Two equal and opposite pairs of forces are at work in this illustration; a pair of forces operates between the rocket and Earth, and a pair of forces operates between the rocket and the exhaust. Because the force of the exhaust on the rocket is larger than the force of gravity on the rocket, the rocket accelerates upward. The force pairs are examples of **Newton's third law of motion**, which says that forces always come in pairs, and the forces of a pair are always equal in strength but opposite in direction. The forces in these action-reaction pairs always act on two different objects. Your weight pushes down on the floor, and the floor pushes back on you with the same amount of force. For every force there is always an equal and opposite force. This is one of the few times when we can say "always" and really mean it.

Newton's Laws and Motion

To see how Newton's three laws of motion work together, study **Figure 3.11**. An astronaut is adrift in space, motionless with respect to the nearby space shuttle. According to Newton's first law, he's stuck! With no tether to pull on, how can the astronaut get back to the shuttle? Suppose the 100-kg astronaut throws a 1-kg wrench directly away from the shuttle at a speed of 10 m/s. Newton's second law says that to cause the motion of the wrench to change, the astronaut has to apply a force to it in the direction away from the shuttle. Newton's third law says that the wrench must therefore push back on the astronaut with just as much

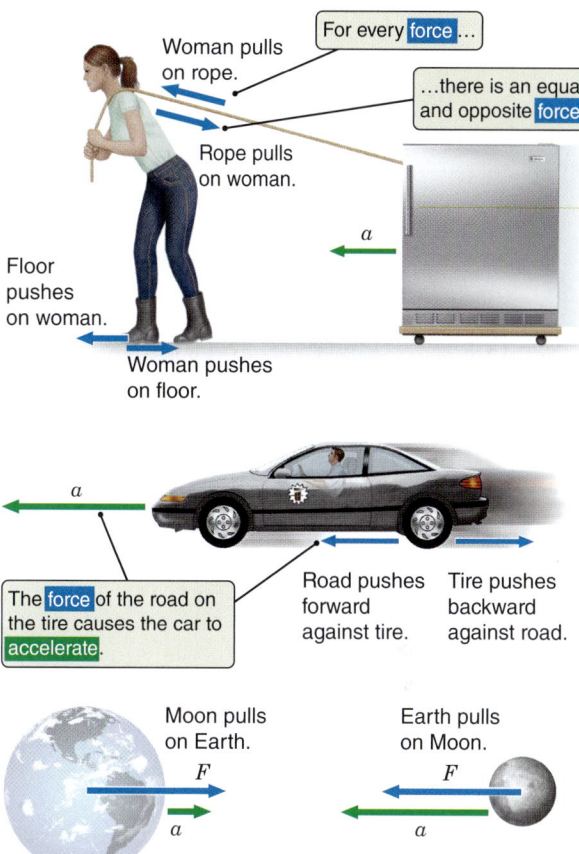

Figure 3.10 Newton's third law states that for every force there is always an equal and opposite force. These opposing forces (action-reaction pair) always act on two different objects.

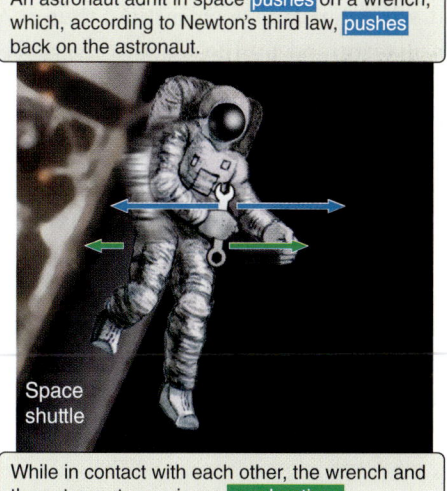

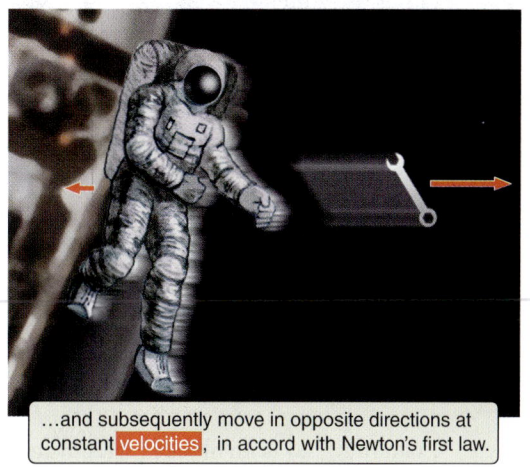

Figure 3.11 According to Newton's laws, if an astronaut adrift in space throws a wrench, the two will move in opposite directions. Their speeds will depend on their masses; the same force will produce a smaller acceleration of a more massive object than of a less massive object. (Acceleration and velocity arrows are not drawn to scale.)

force in the opposite direction. The force of the wrench on the astronaut causes the astronaut to begin drifting toward the shuttle. How fast will the astronaut move? Turn to Newton's second law again. Because the astronaut has more mass, he will accelerate less. The 100-kg astronaut will experience only 1/100 as much acceleration as the 1-kg wrench, and so he will have 1/100 the final velocity. The astronaut will drift toward the shuttle, but only at the leisurely rate of 0.1 m/s.

3.4 Gravity Is a Force between Any Two Massive Objects

▶|| **AstroTour:** Velocity, Acceleration, and Inertia

Newton's work on the motion of objects led to a great insight. Of course, gravity was known before Newton, so he did not "discover" gravity. But Newton realized that the force that was responsible for causing dropped keys to fall to the floor was also responsible for the orbits of the Moon around Earth and the planets around the Sun.

Gravity, Mass, and Weight

Consider what happens when you drop your keys to the floor. They begin at rest and then accelerate downward. This acceleration requires the action of a force, which is commonly known as the **gravitational force**: the mutually attractive force between two objects with mass. The gravitational force on an object attracted by a planet is also typically called its **weight**. Everyone already knew that dropped keys fall to the ground. Newton's genius was in connecting this everyday phenomenon to the motion of the planets around the Sun. Newton's theory of gravity united Kepler's empirical laws and his own laws of motion.

The acceleration due to the gravitational force—known as the gravitational acceleration—near the surface of Earth is usually written as g and has an average value across the surface of Earth of 9.8 m/s². Experiments show that all objects on Earth fall with this same acceleration. Whether you drop a marble or a cannonball, after 1 second it will be falling at a speed of 9.8 m/s, after 2 seconds at 19.6 m/s, and after 3 seconds at 29.4 m/s. (Note that air resistance becomes a factor at higher speeds, but it is negligible for dense, slow objects.)

Newton realized that if all objects fall with the same acceleration, then the gravitational force on an object must be determined by the object's mass. To see why, look back at Newton's second law—acceleration equals force divided by mass, or $a = F/m$. The only way gravitational acceleration can be identical for all objects is if the value of the force divided by the mass is the same for all objects. In other words, make an object twice as massive and you double the gravitational force acting on it. Make an object three times as massive and you triple the gravitational force acting on it.

On the surface of Earth, weight is just mass multiplied by Earth's gravitational acceleration, g. In everyday speech, people equate mass and weight. We often say that an object with a mass of 2 kg "weighs" 2 kg, but it is more correct to express a weight in terms of newtons (N), the metric unit of force:

$$F_{\text{weight}} = m \times g$$

where F_{weight} is an object's weight in newtons, m is the object's mass in kilograms, and g is Earth's gravitational acceleration in meters per second squared. On

Vocabulary Alert

weight In common language, we often use *weight* and *mass* interchangeably. For example, we often interchange kilograms (a unit of mass) and pounds (a unit of weight), but in scientific language this is incorrect. Astronomers use *mass* to refer to the amount of stuff in an object and *weight* to refer to the force exerted on that object by the planet's gravitational pull. Your weight changes with location, but your mass stays the same no matter what planet you are on

Earth, an object with a *mass* of 2 kg has a *weight* of 2 kg × 9.8 m/s², or 19.6 N. On the Moon, where the gravitational acceleration is 1.6 m/s², the 2-kg mass would have a weight of 2 kg × 1.6 m/s², or 3.2 N. Although your mass remains the same wherever you are, your weight varies. On the Moon, your weight would be about one-sixth of your weight on Earth.

Newton's Law of Gravity

Newton's next great insight came from applying his third law of motion to gravity. Newton's third law states that for every force there is an equal and opposite force. Therefore, if Earth exerts a force of 19.6 N on a 2-kg mass sitting on its surface, then that 2-kg mass *must* exert a force of 19.6 N on Earth as well. Drop a 10-kg frozen turkey and it falls toward Earth—but at the same time, Earth falls toward the 10-kg turkey. The reason we do not notice the motion of Earth is that Earth has a lot of inertia. In the time it takes a 10-kg turkey to fall to the ground from a height of 1 kilometer, Earth has "fallen" toward the turkey by a tiny fraction of the size of an atom.

Newton reasoned that this relationship should work with either object. If doubling the mass of an object doubles the gravitational force between the object and Earth, then doubling the mass of Earth ought to do the same thing. In short, the gravitational force between Earth and an object must be proportional to the product of the masses of Earth and the object:

Gravitational force = Something × Mass of Earth × Mass of object

If the mass of the object were two times greater, then the force of gravity would be two times greater. Likewise, if the mass of Earth were three times what it is, the force of gravity would have to be three times greater as well. If *both* the mass of Earth *and* the mass of the object were greater by these amounts, the gravitational force would increase by a factor of 2 × 3, or six times. Because objects fall toward the center of Earth, we know that gravity is an attractive force acting along a line between the two masses.

If gravity is a force that depends on mass, then there should be a gravitational force between *any* two masses. Suppose we have two masses—call them mass 1 and mass 2, or m_1 and m_2 for short. The gravitational force between them is something multiplied by the product of the masses:

Gravitational force between two objects = Something × m_1 × m_2

We have gotten this far just by combining Galileo's observations of falling objects with (1) Newton's laws of motion and (2) Newton's belief that Earth is a mass just like any other mass. But what about that "something" in the previous expression? Today we have sensitive instruments that allow scientists to put two masses close to each other in a laboratory, measure the force between them, and determine the value of that something directly. Yet Newton had no such instruments. He had to look elsewhere to continue his exploration of gravity.

Kepler had already thought about this question. He reasoned that because the Sun is the focal point for planetary orbits, the Sun must be responsible for exerting an influence over the motions of the planets. Kepler speculated that whatever this influence is, it must grow weaker with distance from the Sun. After all, it must surely require a stronger influence to keep the innermost planet, Mercury, whipping around in its tight, fast orbit than it does to keep

Working It Out 3.3 | Newton's Law of Gravity: Playing with Proportionality

One of the most useful ways to play with equations is to study the proportionalities. For example, in Newton's law of gravity, we can see that the force is proportional to each of the masses of the objects involved:

$$F = \frac{Gm_1m_2}{r^2}$$

What does that mean? It means that if the mass of one object is doubled, the force is as well. If the mass of one object is increased by a factor of 2.63147, so is the force. This provides a handy shorthand way of making calculations without actually having to plug G, r, and m_2 into the equation. Suppose that your mass, m_1, suddenly increased by a factor of 2 (doubled). How would this affect the force of gravity, F, on you; that is, your weight? F would also increase by a factor of 2, so that your weight would double. This is somewhat intuitive. If you suddenly had twice as much stuff in your insides, you should certainly weigh twice as much.

If you study the equation, you will see that we are just using a rule we learned in Working It Out 2.1: Whatever you do to one side of the equation, you also have to do to the other side. In this case, we multiplied one of the terms on the right by 2, so we also had to multiply the term on the left by 2.

In these examples, we have been focusing on terms that are directly proportional to each other. Inverse proportions are slightly more complicated: In inverse proportions, when one term is doubled, the other is halved. This happens when one term is in the denominator while the other is in the numerator. We saw something like this before when we learned about the relationship between distance, time, and speed:

$$s = \frac{d}{t}$$

Speed and distance are directly proportional to each other. If you traveled twice as far as your friend in the same amount of time, you were going twice as fast. But speed and time are inversely proportional. If it took you twice as long to travel that distance, you were going *half* as fast. Consider our rule about doing the same thing to both sides of the equation. On the right, you multiply the time by 2. But time is in the denominator, so you have effectively multiplied the right side by ½. You must also do this on the left and multiply s by ½. (Notice that two things that are inversely proportional never have the same units. This is often true of direct proportions as well.)

Let's return to Newton's law of gravity and ask what happens when we change r, the distance between the objects. We might ask, for example, "How would the gravitational force between Earth and the Moon change if the distance between them were doubled?" Newton's law of gravitation states that the force is inversely proportional to r^2:

$$F = \frac{Gm_1m_2}{r^2}$$

So if we double r, we expect F to decrease; so far so good. But what do we do about the square? If we put $(2r)$ in where (r) is in the equation, we see that we would have to write $(2r)^2$. This means that the r is squared *and* the 2 is squared. So $(2r)^2 = 4r^2$. We have effectively multiplied the right-hand side by ¼. We must also do that on the left, so the force is ¼ as strong as before:

$$\frac{F}{4} = \frac{Gm_1m_2}{r^2}$$

Doubling the distance reduces the force by a factor of 4. This $1/r^2$ proportionality occurs in many contexts in astronomy. Take a moment to calculate how the force would change if the distance increased by a factor of 3, 5, or 10, and if it decreased by a factor of ½, ¼, or ⅒. Once you can do this, you will have a tool you can use again and again in the remaining chapters.

---

the outer planets lumbering along their paths around the Sun. Kepler's speculation went even further. Although he did not know about forces or inertia or gravity, he did know quite a lot about geometry, and geometry alone suggested how this solar "influence" might change for planets progressively farther from the Sun.

Imagine that you have a certain amount of paint to spread over the surface of a sphere. If the sphere is small, you will get a thick coat of paint. But if the sphere is larger, the paint has to spread farther, and you get a thinner coat. The surface area of a sphere depends on the square of the sphere's radius. Double the radius of a sphere, and the sphere's surface becomes four times what it was. If you paint this new, larger sphere, the paint must cover four times as much area, and the thickness of the paint will be only one-fourth of what it was on the smaller sphere. Triple the radius of the sphere and the sphere's surface will be nine times as large, and the thickness of the coat of paint will be only one-ninth as thick.

The paint in this example describes how Kepler thought about the influence the Sun exerts over the planets. As the influence of the Sun extended farther and farther into space, it would spread out to cover the surface of a larger and larger imaginary sphere centered on the Sun. The influence of the Sun should diminish with the square of the distance from the Sun—a relationship known as an **inverse square law**.

Kepler had an interesting idea but no scientific hypothesis with testable predictions. He lacked an explanation for how the Sun influences the planets and the mathematical tools to calculate how an object would move. Newton had both. If gravity is a force between *any* two objects, then there should be a gravitational force between the Sun and each of the planets. Might this gravitational force be the same as Kepler's "influence"? If so, gravity might behave according to an inverse square law. Newton's expression for gravity came to look like this:

$$\text{Gravitational force between two objects} = \text{Something} \times \frac{m_1 \times m_2}{(\text{Distance between objects})^2}$$

There is still a "something" left in this expression, and that something is a constant: a number that does not change. This constant determines the strength of gravity between objects, and it is the same for all pairs of objects. Newton named it the **universal gravitational constant**, written as G. It was not until many years later that the actual value of G was first measured. Today, the value of G is known to be 6.673×10^{-11} m³/(kg s²).

Putting the Pieces Together: A Universal Law for Gravitation

Newton's **universal law of gravitation** states that gravity is a force between any two objects having mass and has these properties:

1. It is an attractive force, F, acting along a straight line between the two objects.
2. It is proportional to the mass of one object, m_1, multiplied by the mass of the other object, m_2. If we double m_1, F increases by a factor of 2. Likewise, if we double m_2, F increases by a factor of 2.
3. It is inversely proportional to the square of the distance r between the centers of the two objects. If we double r, F decreases by a factor of 4. If we triple r, F falls by a factor of 9 (**Working It Out 3.3**).

These properties are illustrated in **Figure 3.12**. Translated into mathematics, and including the constant of proportionality, G, the universal law of gravitation is

$$F = \frac{Gm_1m_2}{r^2}.$$

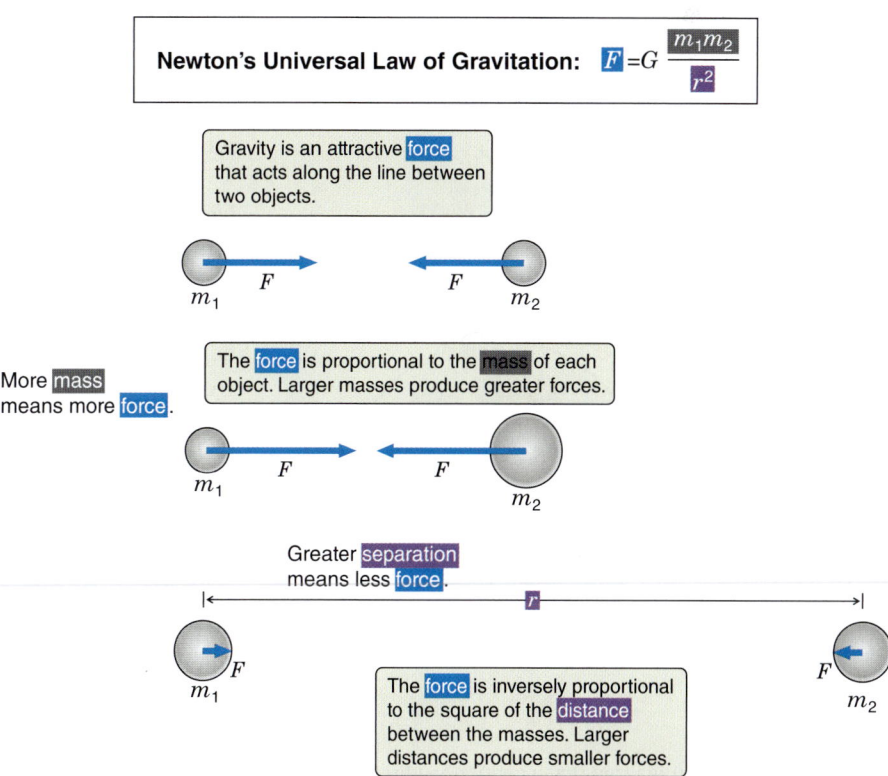

Figure 3.12 Gravity is an attractive force between two objects. The force of gravity depends on the masses of the objects, m_1 and m_2, and the distance, r, between them.

Newton's Universal Law of Gravitation: $F = G\dfrac{m_1m_2}{r^2}$

3.5 Orbits Are One Body "Falling Around" Another

Kepler's laws regarding the motions of planets allowed astronomers to predict the positions of the planets accurately, but these laws did not explain why the planets behaved as they did. Newton's work provided the answer to the question: Why do planets orbit the Sun?

Newton Explains Kepler

Newton used his laws of motion and his proposed law of gravity to calculate the paths that planets should follow as they move around the Sun. When he did so, his calculations predicted that:

- Planetary orbits should be ellipses with the Sun at one focus.
- Planets should travel faster when closer to the Sun.
- The square of the period of the orbit should equal the semimajor axis cubed (in appropriate units).

In short, Newton's universal law of gravitation *predicted* that planets should orbit the Sun in just the way that Kepler's empirical laws described. This was the moment when it all came together. By *explaining* Kepler's laws, Newton found important support for his law of gravitation. Newton argued that the same gravitational force governed the behavior of dropped keys and orbiting planets.

Gravity and Orbits

Newton's laws describe how an object's motion changes in response to forces and how objects interact with each other through gravity. To go from statements about how an object's motion is *changing* to more practical statements about where an object *is*, we carefully have to "add up" the object's motion over time. To see how we can do this, let's begin with a "thought experiment"—the same thought experiment that helped lead Newton to his understanding of how planets orbit the Sun.

Drop a cannonball and it falls directly to the ground, just as any mass does. However, if instead we fire the cannonball from a cannon that is level with the ground, as shown in **Figure 3.13a**, it behaves differently. The cannonball still falls to the ground in the same time as before, but while falling it is also traveling *over* the ground, following a curved path that carries it some horizontal distance before it finally lands. As you can see in Figure 3.13b, the faster the ball is fired from the cannon, the farther it will go before it hits the ground.

In the real world, this experiment reaches a natural limit. To travel through air, the cannonball must push the air out of its way—an effect we normally refer to as air resistance—which slows it down. But because this is only a thought experiment, we can ignore such real-world complications. Instead imagine that, having inertia, the cannonball continues along its course until it runs into something. As the cannonball is fired faster and faster, it goes farther and farther before hitting the ground. If the cannonball flies far enough, Earth's surface "curves out from under it," shown in Figure 3.13c. Eventually, the cannonball is flying so fast that the surface of Earth curves away from the cannonball at exactly the same rate that the cannonball is falling toward Earth. This is the case shown in Figure 3.13d. At this point the cannonball, which always falls *toward the center of Earth*, is literally "falling around the world."

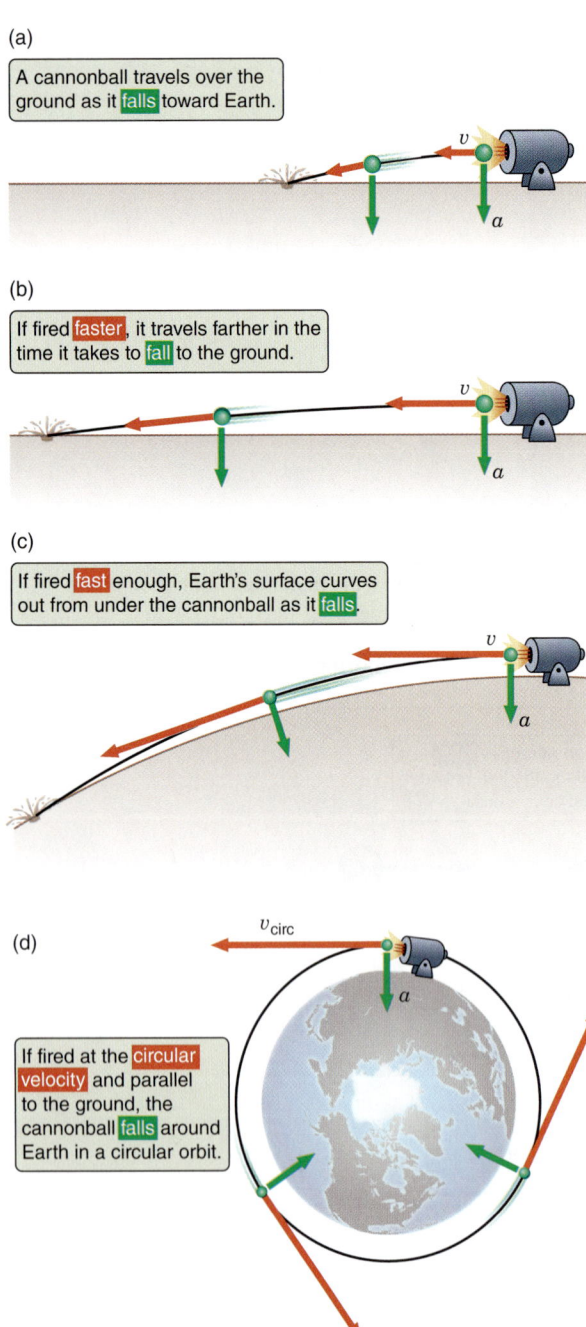

Figure 3.13 Newton realized that a cannonball fired at the right speed would fall around Earth in a circle.

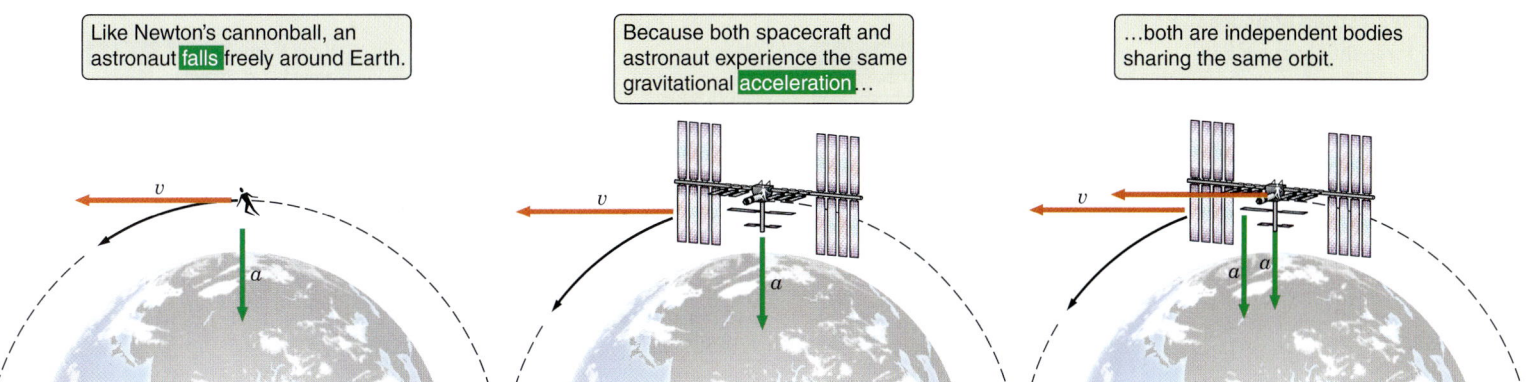

Figure 3.14 A "weightless" astronaut has not escaped Earth's gravity. Rather, an astronaut and a spacecraft share the same orbit as they fall around Earth together.

In 1957, the Soviet Union used a rocket to lift Sputnik 1, an object about the size of a basketball, high enough above Earth's upper atmosphere that air resistance ceased to be a concern. With this event, Newton's thought experiment became a matter of great practical importance; Sputnik 1 was moving so fast that it fell around Earth, just as the cannonball did in Newton's mind. Sputnik 1 was the first human-made object to orbit Earth. You now know enough about motions for a formal definition of the term *orbit*. An **orbit** is the path of one object that freely falls around another. "To orbit" is to fall freely around another object.

The concept of orbits also answers the question of why astronauts float freely about the cabin of a spacecraft. It is *not* because they have escaped Earth's gravity; it is Earth's gravity that holds them in their orbit. Instead the answer lies in Galileo's early observation that every object falls in just the same way, regardless of its mass. The astronauts and the spacecraft are both moving in the same direction, at the same speed, and are experiencing the same gravitational acceleration, so they fall around Earth together. **Figure 3.14** demonstrates this point. The astronaut is orbiting Earth just as the spacecraft is orbiting Earth. On the surface of Earth our bodies try to fall toward the center of Earth, but the ground gets in the way. We experience our weight when we are standing on Earth because the ground pushes on us hard enough to counteract the force of gravity, which is trying to pull us down. In the spacecraft, however, nothing interrupts the astronaut's fall because the spacecraft is falling around Earth in just the same orbit. The astronaut is in **free fall**: he is falling freely in Earth's gravity.

When one object is falling around another, much more massive object, we say that the less massive object is a **satellite** of the more massive object. Planets are satellites of the Sun, and moons are natural satellites of planets. Newton's imaginary cannonball is a satellite. The spacecraft and the astronauts are independent satellites of Earth that conveniently happen to share the same orbit.

▶❚❚ **AstroTour:** Newton's Laws and Universal Gravitation

Centripetal Force and Circular Velocity

If fired fast enough, Newton's cannonball falls around the world; but just how fast is "fast enough"? Newton's orbiting cannonball moves along a circular path at constant speed. This type of motion is referred to as **uniform circular motion**. You are probably familiar with other examples of uniform circular motion. For example, think about a ball whirling around your head on a string, as shown in

Figure 3.15 (a) A string provides the centripetal force that keeps a ball moving in a circle. (We are ignoring the smaller force of gravity that also acts on the ball.) (b) Similarly, gravity provides the centripetal force that holds a satellite in a circular orbit.

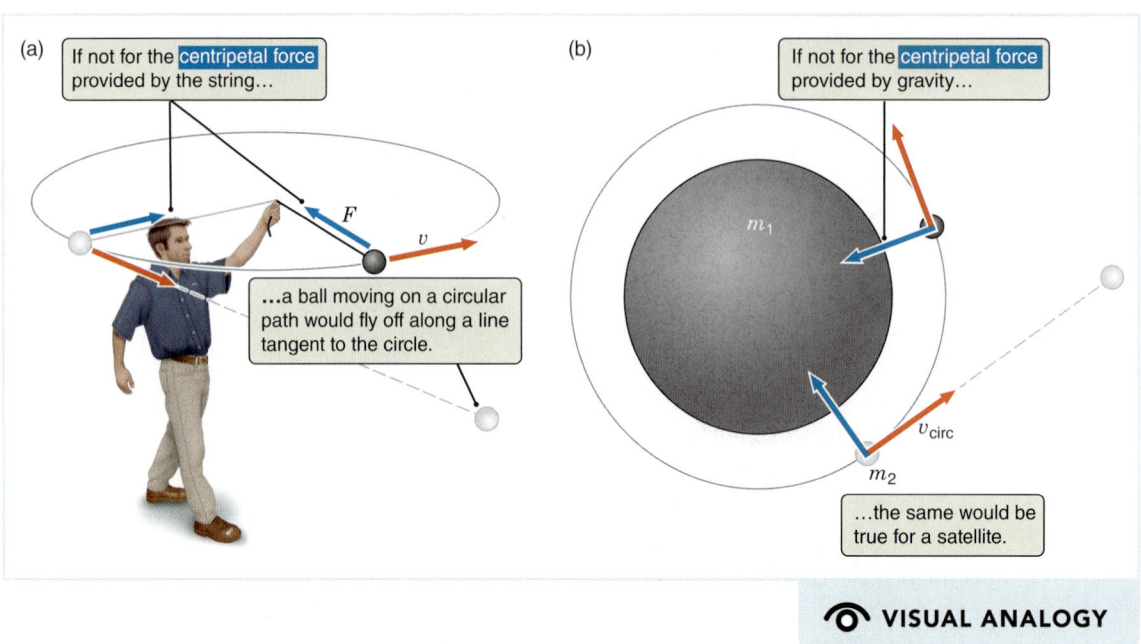

VISUAL ANALOGY

Figure 3.15a. If you were to let go of the string, the ball would fly off in a straight line in whatever direction it was traveling at the time. The string prevents the ball from flying off by constantly changing the direction the ball is traveling. This central force of the string on the ball is called a **centripetal force**: a force toward the center of a circle. Using a more massive ball, speeding up the ball's motion, or making the string shorter so that the turn is tighter all increase the force needed to keep a ball moving in a circle.

In the case of Newton's cannonball (or a satellite), there is no string to hold the ball in its circular motion. Instead, the force is provided by gravity, as illustrated in Figure 3.15b. For Newton's thought experiment to work, the force of gravity must be just enough to keep the satellite moving on its circular path. Because this force has a specific strength, it follows that the satellite must be moving at a particular speed, which we call its **circular velocity (v_{circ})**. If the satellite were moving at any other velocity, it would not be moving in a circular orbit. Remember the cannonball. If the cannonball were moving too slowly, it would drop below the circular path and hit the ground. Similarly, if the cannonball were moving too fast, its motion would carry it above the circular orbit. Only a cannonball moving at just the right velocity—the circular velocity—will fall around Earth on a circular path.

The Shape of Orbits

Some Earth satellites travel a circular path at constant speed. Just like the ball on a string, satellites traveling at the circular velocity remain the same distance from Earth at all times, neither speeding up nor slowing down in orbit. But what if the satellite were in the same place in its orbit and moving in the same direction, but traveling faster than the circular velocity? The pull of Earth is as strong as ever, but because the satellite has greater speed, its path is not bent by Earth's gravity sharply enough to hold it in a circle. So the satellite begins to climb above a circular orbit.

As the distance between the satellite and Earth increases, the satellite slows down. Think about a ball thrown into the air, as shown in **Figure 3.16a**. As the

AstroTour: Elliptical Orbits

ball climbs higher, Earth's gravity slows the ball down. The ball climbs more and more slowly until its vertical motion stops for an instant and then is reversed; the ball falls back toward Earth, speeding up along the way. The satellite does the same thing. As the satellite moves away from Earth, Earth's gravity slows the satellite down. The farther the satellite is from Earth, the more slowly the satellite moves—just like the ball thrown into the air. And just like the ball, the satellite reaches a maximum height on its curving path and then begins falling back toward Earth. As the satellite falls back toward Earth, Earth's gravity speeds it up.

This happens for any object in an elliptical orbit, including a planet orbiting the Sun. Kepler's second law says that a planet moves fastest when it is closest to the Sun and slowest when it is farthest from the Sun. Now we know why. As shown in Figure 3.16b, planets lose speed as they pull away from the Sun and then gain that speed back as they fall inward toward the Sun.

Newton's laws do more than explain Kepler's laws: they predict orbits beyond Kepler's empirical observations. **Figure 3.17** shows a series of satellite orbits, each with the same point of closest approach to Earth but with different velocities at that point. The greater the speed a satellite has at its closest approach to Earth, the farther the satellite is able to pull away from Earth, and the more eccentric its orbit becomes. As long as it remains elliptical, no matter how eccentric, the orbit is called a **bound orbit** because the satellite is gravitationally bound to the object it is orbiting.

In this sequence of faster and faster satellites there comes a point of no return—a point when the satellite is moving so fast that gravity is unable to reverse its outward motion, so the satellite coasts away from Earth, never to return. This indeed is possible. The lowest speed at which this happens is called the **escape velocity** from the orbit, v_{esc}. Once a satellite's velocity at closest approach equals or exceeds v_{esc}, it is in an **unbound orbit**. The object is no longer gravitationally bound to the body that it was orbiting. A comet traveling on an unbound orbit makes only a single pass around the Sun and then is back off into deep space, never to return.

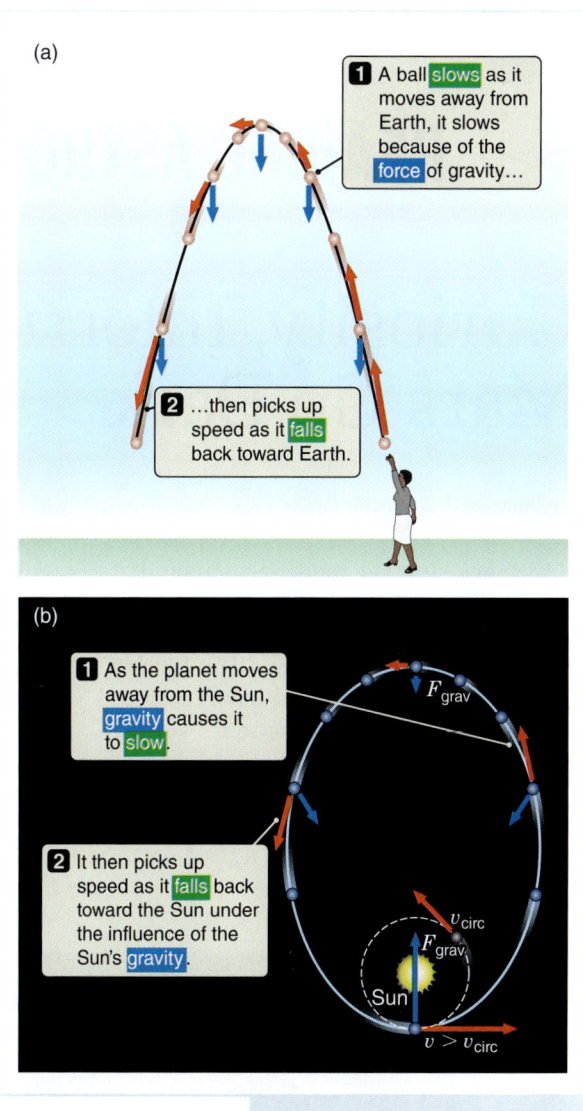

Figure 3.16 (a) A ball thrown into the air slows as it climbs away from Earth and then speeds up as it heads back toward Earth. (b) A planet on an elliptical orbit around the Sun does the same thing. (Although no planet has an orbit as eccentric as the one shown here, the orbits of comets can be far more eccentric.)

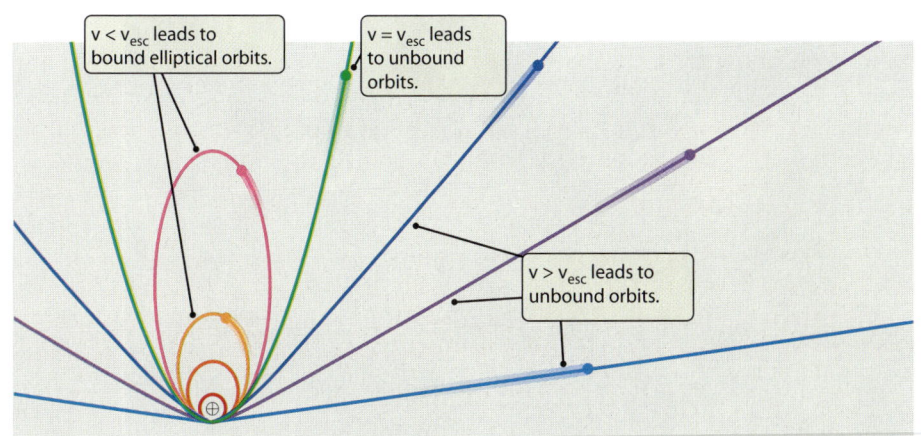

Figure 3.17 A range of different orbits that share the same point of closest approach but differ in velocity at that point. An object's velocity at closest approach determines the orbit shape and whether the orbit is bound or unbound. The dividing line between bound and unbound occurs when the object has a velocity equal to the escape velocity (green line). Objects with higher velocities have excess velocity, and their orbits are not bound (blue and purple lines)

READING ASTRONOMY News

As the planets orbit the Sun, their relative positions create challenges for the space program. For unmanned probes, it's a simple matter of having them become dormant and waiting until they emerge from eclipse. Thinking ahead to manned missions presents different problems.

Curiosity, Interrupted: Sun Makes Mars Go Dark

By **IAN O'NEILL**, *Discovery News*

NASA has stopped sending commands to Mars rover Curiosity and will soon follow suit for rover *Opportunity*, *Odyssey*, and the *Mars Reconnaissance Orbiter* (*MRO*). But don't worry, government cutbacks haven't severed interplanetary communications, you can blame the sun.

Once every 26 months the orbits of Earth and Mars align in such a way that our nearest star physically gets in the way of our line of sight. Known as a "Mars solar conjunction," from our point of view, the red planet almost passes directly behind the solar disk [**Figure 3.18**]. This means, inevitably, communications between the planets are severely disrupted.

Curiosity mission managers at NASA's Jet Propulsion Laboratory (JPL) in Pasadena, California, suspended communications with the 1-ton robot yesterday (April 4) so to avoid any corruption of data. They won't recommence transmissions until May 1.

The Sun's lower atmosphere (the corona) is buzzing with highly charged particles, causing interference with radio communications that pass through it. Also, each conjunction is different, depending on orbital inclination and solar activity. The Sun is currently undergoing "solar maximum"—peak activity for its approximate 11-year cycle. This conjunction will see Mars close in on the solar limb by only 0.4° on April 17—a maximum period for interference.

Although receiving data from the rover

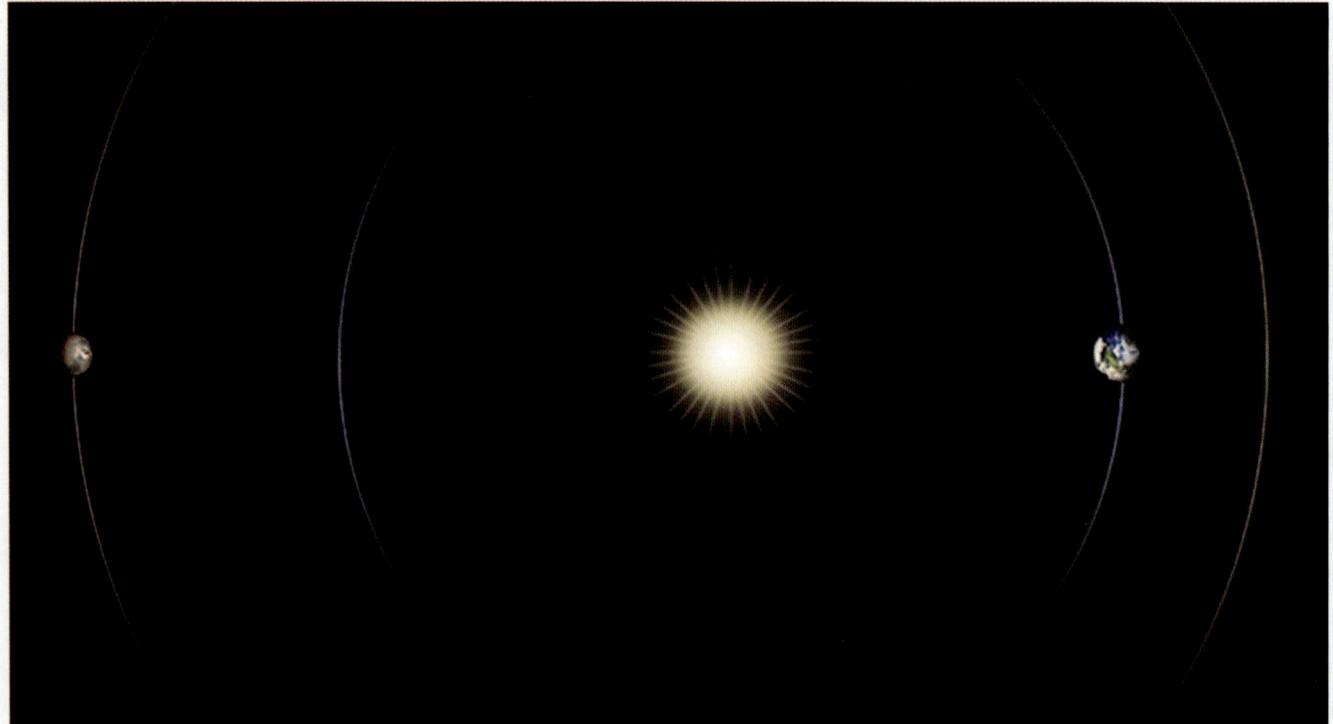

Figure 3.18 About every 2 years, Mars and Earth are on opposite sides of the Sun, making communication between the two planets impossible.

see Curiosity

CURIOSITY (CONT.)

isn't too much of a problem (blocks of missing data can be requested again later), sending commands, only for the data to become corrupted on the way to Mars, could be a serious issue for the health of the mission.

Commands to the NASA orbiters will also be suspended from April 9 to 26. The *MRO* has been switched to a "record only" mode whereas the veteran *Odyssey* will continue to transmit throughout the conjunction—although NASA is expecting the inevitable data dropouts particularly around April 17. Operations of the European *Mars Express* satellite, which has been orbiting Mars since 2003, will also be affected.

"For the entire conjunction period, we'll just be storing data on board (the *Mars Reconnaissance Orbiter*)," said JPL's Reid Thomas, *MRO* deputy mission manager. The orbiter is expected to accumulate 40 gigabits of science data from its own instruments and 12 gigabits of data from *Curiosity*, which will all be downloaded to Earth about May 1.

"This is our sixth conjunction for *Odyssey*," said Chris Potts, *Odyssey* mission manager at JPL. "We have plenty of useful experience dealing with them, though each conjunction is a little different." *Odyssey* has been orbiting Mars since 2001.

"The biggest difference for this 2013 conjunction is having *Curiosity* on Mars," added Potts.

Although NASA won't be transmitting, *Curiosity* will be sending regular "beeps" to Earth to reassure us that it's okay. "We will maintain visibility of rover status two ways," said Torsten Zorn, conjunction planning leader for *Curiosity*'s JPL engineering operations team. "First, *Curiosity* will be sending daily beeps directly to Earth. Our second line of visibility is in the *Odyssey* relays."

Curiosity landed inside Gale Crater on August 5, 2012, and has already made groundbreaking (literally) progress in the hunt of evidence for past (and present) habitability of the red planet. This communications blackout is the first time *Curiosity* will be alone.

Evaluating the News

1. Reading from context, what is the meaning of the word *conjunction*? Use a dictionary to find the correct astronomical meaning of this word. How close did you come to the actual definition when you reasoned from context?
2. Several different orbits are mentioned in this article: the orbit of Earth around the Sun, the orbit of Mars around the Sun, and the orbit of satellites around Mars. Rank these orbits in order from smallest to largest.
3. The period of Mars's orbit is 1.88 years, and the period of Earth's orbit is 1 year. How long is Mars's orbital period in months? How long is Earth's orbital period in months?
4. This solar conjunction occurs every 26 months. Draw a picture like the one in the article for one conjunction. Then add second locations for Mars and Earth that estimate their positions during the following conjunction. (Hint: See preceding question.)
5. If there were a manned mission on Mars today, this loss of contact would be very troubling. How might NASA plan to avoid losing contact with the astronauts on Mars during one of these conjunctions?

◆ SUMMARY

Early astronomers hypothesized that the Earth was stationary at the center of the Solar System. Later astronomers realized that a Sun-centered Solar System was much simpler. Kepler's laws describe the elliptical orbits of planets around the Sun, including details about how fast the planet travels at various points in its orbit. These laws helped Newton to advance science by developing his laws of motion, which govern the motion of all objects (not just orbiting ones). Kepler's laws are explained by Newton's theory of gravity, which describes how planets are bound to the Sun.

LG 1 Kepler's first law states that planets move in ellipses. Kepler's second law relates the speed of the planet to different places in its orbit (faster when closer to the Sun, slower when farther from the Sun), so that the planets sweep out equal areas in equal times. Kepler's third law gives an empirically determined relationship between the size of an orbit and the time it takes for the planet to orbit once (the period squared is proportional to the average distance cubed).

LG 2 Newton's three laws (in shorthand: inertia, $F = ma$, and "every action has an equal and opposite reaction") govern the motion of all objects. Net forces cause accelerations, or changes in motion. Mass is the property of matter that gives it inertia, or resistance to changes in motion. Gravity is one of the fundamental forces of nature and binds the universe together. Gravity is a force between any two objects due to their masses. The force of gravity is proportional to the product of the two masses and inversely proportional to the square of the distance between them.

LG 3 Planets orbit the Sun in bound, elliptical orbits because of gravity. Any circular orbit has a characteristic circular velocity, which is faster if the orbit is smaller because gravity is stronger when the objects are closer together. A planet on an elliptical orbit speeds up and slows down as it is closer and farther from the Sun, respectively. Orbits become "unbound" when the object reaches escape velocity. Knowing these properties of orbits allows astronomers to measure the masses of planets or stars.

LG 4 A reference frame is a system in which an observer makes measurements of both space and time. No reference frame is preferred over any other.

SUMMARY SELF-TEST

1. Complete the following descriptions of planetary orbits.
   a. A planet moves fastest when it is _____ to the Sun and slowest when it is _____ from the Sun.
   b. Each ellipse has two foci. At one focus is the _____. At the other focus is _____.
   c. A planet with a period of 84 Earth years has an orbit that is _____ than a planet with an orbit of 1 Earth year.

2. Newton's second law of motion states that
   a. objects have inertia.
   b. the acceleration of an object is proportional to the net force acting on it.
   c. every action has an equal and opposite reaction.
   d. the force of gravity is proportional to the masses of the two objects and the inverse of the square of the distance between them.

3. The connection between gravity and orbits enables astronomers to measure the _____ of stars and planets.
   a. distances
   b. sizes
   c. masses
   d. compositions

4. Suppose you are driving a car, and a coffee cup is on the seat beside you. Rank, in increasing order, the speed of the coffee cup in the reference frame of
   a. an astronaut on the International Space Station
   b. the cup itself
   c. an observer at the side of the road
   d. an observer in an oncoming car
   e. you, the driver

5. Place the following in order from largest to smallest semimajor axis:
   a. a planet with a period of 84 Earth days
   b. a planet with a period of 1 Earth year
   c. a planet with a period of 2 Earth years
   d. a planet with a period of 0.5 Earth years

6. What is the eccentricity of a circular orbit?

7. Suppose that you drop the following objects off of a tall tower. Rank the objects in terms of the gravitational force on them, from smallest to largest.
   a. an apple
   b. a decorative Styrofoam "apple"
   c. a solid gold "apple"

8. Imagine you are walking along a forest path. Which of the following is *not* an action-reaction pair in this situation?
   a. The gravitational force between you and Earth; the gravitational force between Earth and you.
   b. Your shoe pushes back on Earth; Earth pushes forward on your shoe.
   c. Your foot pushes back on the inside of your shoe; your shoe pushes forward on your foot.
   d. You push down on Earth; Earth pushes you forward.

9. A net force must be acting when (choose all that apply)
   a. an object accelerates.
   b. an object changes direction but not speed.
   c. an object changes speed but not direction.
   d. an object changes speed and direction.

10. Suppose you are transported to a planet with twice the mass of Earth, but the same radius of Earth. Your weight would _____ by a factor of _____.
    a. increase; 2   b. increase; 4
    c. decrease; 2   d. decrease; 4

QUESTIONS AND PROBLEMS

Multiple Choice and True/False

11. T/F: Kepler's three laws explain *why* the planets orbit the Sun as they do.

12. T/F: The natural state of objects is to be at rest. This is why a book, given a push across a table, will eventually slow to a stop.

13. T/F: To find the mass of a central object, such as the Sun, we only need to know the semimajor axis and period of an orbit, such as Earth's.

14. T/F: You are always at rest in your own reference frame.

15. T/F: A force is required to keep an object moving at the same velocity.

16. Kepler's second law says that
    a. planetary orbits are ellipses with the Sun at one focus.
    b. the square of a planet's orbital period equals the cube of its semimajor axis.
    c. for every action there is an equal and opposite reaction.
    d. planets move fastest when they are closest to the Sun.

17. Imagine that you are pulling a small child on a sled by means of a rope. Which of the following are action-reaction pairs?
    a. You pull forward on the rope; the rope pulls backward on you.
    b. The sled pushes down on the ground; the ground pushes up on the sled.
    c. The sled pushes forward on the child; the child pushes back on the sled.
    d. The rope pulls forward on the sled; the sled pulls backward on the rope.

18. Imagine a planet moving in a perfectly circular orbit around the Sun. Is this planet experiencing acceleration?
    a. Yes, because it is changing its speed all the time.
    b. Yes, because it is changing its direction of motion all the time.
    c. No, because its speed is not changing all the time.
    d. No, because planets do not experience accelerations.

19. Suppose you watch a person riding on the back of a truck that is traveling to your left. The person throws a ball off the back of the truck, in a direction opposite the motion of the truck. In your reference frame,
    a. the ball travels to the left.
    b. the ball travels to the right.
    c. the ball falls straight down.
    d. the behavior of the ball depends on the relative speeds of the ball and the truck.

20. Suppose you read in the newspaper that a new planet has been found. Its average speed in its orbit is 33 kilometers per second (km/s). When it is closest to its star, it moves at 31 km/s, and when it is farthest from its star, it moves at 35 km/s. This story is in error because
    a. the average speed is far too fast.
    b. Kepler's third law says the planet has to sweep out equal areas in equal times, so the speed of the planet cannot change.
    c. planets stay at a constant distance from their stars; they don't move closer or farther away.
    d. Kepler's second law says the planet must move fastest when it is closest, not when it is farthest away.
    e. using these numbers, the square of the orbital period will not be equal to the cube of the semimajor axis.

21. Suppose that you read about a new car that can go from 0 to 100 km/h in only 2.0 seconds. What is this car's acceleration?
    a. about 50 km/h
    b. about 14 m/s^2
    c. about 50 km/n/s
    d. about 200 km
    e. about 0.056 km/h^2

22. Imagine that you are standing at the top of a tall tower. You drop four objects, all the size of a bowling ball. Each is made of a different substance: Styrofoam, lead, Bubble Wrap, and pumpkin. Neglecting air resistance, in what order do they reach the ground?
    a. lead, pumpkin, Bubble Wrap, Styrofoam
    b. lead, Bubble Wrap, pumpkin, Styrofoam
    c. Styrofoam, lead, Bubble Wrap, pumpkin
    d. None of the above; they all reach the ground at the same time.

23. Two balls, one of gold with a mass of 100 kg and one of wood with a mass of 1 kg, are suspended 1 meter apart. How does the force of the gold ball on the wood ball compare to the force of the wood ball on the gold ball?
    a. The force of the gold ball on the wood ball is stronger than the force of the wood ball on the gold ball.
    b. The force of the wood ball on the gold ball is stronger than the force of the gold ball on the wood ball.
    c. Both forces are the same.
    d. Both forces are too small to measure, so the answer is not known.

24. *Weight* refers to the force of gravity acting on a mass. We often calculate the weight of an object by multiplying its mass by the local acceleration due to gravity. The value of gravitational acceleration on the surface of Mars is 0.377 times that on Earth. If your mass is 85 kg, your weight on Earth is 830 N ($m \times g = 85$ kg $\times$ 9.8 m/s^2 = 830 N). What would be your approximate mass and weight on Mars?
    a. mass 830 N; weight 8,300 N
    b. mass 85 kg; weight 830 N
    c. mass 85 kg; weight 31 kg
    d. mass 85 kg; weight 310 N

25. The average distance of Uranus from the Sun is about 19 times Earth's distance from the Sun. The Sun's gravitational force on Uranus is _____ than the Sun's gravitational force on Earth.
    a. 361 times weaker
    b. 19 times weaker
    c. 19 times stronger
    d. 361 times stronger

Conceptual Questions

26. Ellipses contain two axes, major and minor. Half the major axis is called the semimajor axis. What is especially important about the semimajor axis of a planetary orbit?

27. What is inertia? How is it related to mass?

28. Kepler's and Newton's laws all tell us something about the motion of the planets, but there is a fundamental difference between them. What is the difference?

29. Describe a situation in which you and a friend share the same reference frame. Describe a situation in which you are in two different reference frames.

30. The distance that Neptune has to travel in its orbit around the Sun is approximately 30 times greater than the distance that Earth must travel. Yet it takes nearly 165 years for Neptune to complete one trip around the Sun. Explain why.

31. What is the difference between speed and velocity?

32. When riding in a car, we can sense changes in speed or direction through the forces that the car applies on us. Do we wear seat belts in cars and airplanes to protect us from speed or from acceleration? Explain your answer.

33. An astronaut standing on Earth can easily lift a wrench having a mass of 1 kg, but not a scientific instrument with a mass of 100 kg. In the International Space Station, she is quite capable of manipulating both, although the scientific instrument responds much more slowly than the wrench. Explain why.

34. In 1920, a *New York Times* editor refused to publish an article based on rocket pioneer Robert Goddard's paper that predicted spaceflight, saying that "rockets could not work in outer space because they have nothing to push against" (a statement that the *New York Times* did not retract until July 20, 1969, the date of the *Apollo 11* Moon landing). You, of course, know better. What was wrong with the editor's logic?

35. Explain the difference between weight and mass.

36. On the Moon, your weight is different from your weight on Earth. Why?

37. Describe the difference between a bound orbit and an unbound orbit.

38. Two objects are leaving the vicinity of the Sun, one traveling in a bound orbit and the other in an unbound orbit. What can you say about the future of these two objects? Would you expect either of them eventually to return?

39. Suppose astronomers discovered an object approaching the Sun in an unbound orbit. What would that say about the origin of the object?

40. What is the advantage of launching satellites from spaceports located near the equator? Why are satellites never launched toward the west?

Problems

41. Suppose you discover a new dwarf planet in our Solar System with a semimajor axis of 46.4 AU. What is its period (in Earth years)?

42. Suppose you discover a planet around a Sun-like star. From careful observation over several decades, you find that its period is 12 Earth years. Find the semimajor axis cubed and then the semimajor axis.

43. Suppose you read in a tabloid newspaper that "experts have discovered a new planet with a distance from the Sun of 1 AU and a period of 3 years." Use Kepler's third law to argue that this is impossible nonsense.

44. A sports car accelerates from 0 km/h to 100 km/h in 4 seconds.
    a. What is its average acceleration?
    b. Suppose the car has a mass of 1,200 kg. How strong is the force on the car?
    c. What supplies the "push" that accelerates the car?

45. Flybynite Airlines takes 3 hours to fly from Baltimore to Denver at a speed of 800 km/h. To save fuel, management orders its pilots to reduce their speed to 600 km/h. How long will it now take passengers on this route to reach their destination?

46. Suppose that you are pushing a small refrigerator of mass 50 kg on wheels. You push with a force of 100 N.
    a. What is the refrigerator's acceleration?
    b. Assume the refrigerator starts at rest. How long will the refrigerator accelerate at this rate before it is moving as fast as you can run—of the order 10 m/s?

47. You are riding along on your bicycle at 20 km/h and eating an apple. You pass a bystander.
    a. How fast is the apple moving in your frame of reference?
    b. How fast is the apple moving in the bystander's frame of reference?
    c. Whose perspective is more valid?

48. Earth's mean radius and mass are 6,371 km and 5.97×10^{24} kg, respectively. Show that the acceleration of gravity at the surface of Earth is 9.80 m/s^2.

49. Suppose you go skydiving.
    a. Just as you fall out of the airplane, what is your gravitational acceleration?
    b. Would this acceleration be bigger, smaller, or the same if you were strapped to a flight instructor, and so had twice the mass?
    c. Just as you fall out of the airplane, what is the gravitational force on you (assume your mass is 70 kg)?
    d. Is the gravitational force bigger, smaller, or the same if you were strapped to a flight instructor and so had twice the mass?

50. Assume that a planet just like Earth is orbiting the bright star Vega at a distance of 1 astronomical unit (AU). The mass of Vega is twice that of the Sun.
    a. How long in Earth years will it take to complete one orbit around Vega?
    b. How fast is the Earth-like planet traveling in its orbit around Vega?

SMARTWORK

Norton's online homework system includes algorithmically generated versions of these questions, plus additional conceptual exercises. If your instructor assigns questions in SmartWork, log in at smartwork.wwnorton.com.

Exploration | Newtonian Features

wwnpag.es/uou2

In this Exploration, we will use the Planetary Orbit Simulator to explore the Newtonian features of Mercury's orbit. Visit the Student Site (wwnpag.es/uou2) and open the Planetary Orbit Simulator Nebraska Simulation in Chapter 3.

Accelerations

To begin exploring the simulation, set parameters for "Mercury" in the "Orbit Settings" panel and then click "OK." Click the Newtonian Features tab at the bottom of the control panel. Select "show solar system orbits" and "show grid" under Visualization Options. Change the animation rate to 0.01, and press the "start animation" button.

Examine the graph at the bottom of the panel.

1. Where is Mercury in its orbit when the acceleration is smallest?

2. Where is Mercury in its orbit when the acceleration is largest?

3. What are the values of the largest and smallest accelerations?

On the Newtonian Features graph, mark the boxes for "vector" and "line" that correspond to the acceleration. These will insert an arrow that shows the direction of the acceleration and a line that extends the arrow.

4. To what solar system object does the arrow point?

Think about Newton's second law.

5. In what direction is the force on the planet?

Velocities

Examine the graph at the bottom of the panel again.

6. Where is Mercury in its orbit when the velocity is smallest?

7. Where is Mercury in its orbit when the velocity is largest?

8. What are the values of the largest and smallest velocities?

Add the velocity vector and line to the simulation by clicking on the boxes in the graph window. Study the arrows carefully.

9. Are the velocity and the acceleration always perpendicular (is the angle between them always 90°)?

10. If the orbit were a perfect circle, what would the angle be between the velocity and the acceleration?

Hypothetical Planet

Use the Orbit Settings to change the semimajor axis to 0.8 AU.

11. How does this imaginary planet's orbital motion compare to Mercury's?

Now change the semimajor axis to 0.1 AU.

12. How does this planet's orbital period now compare to Mercury's?

13. Summarize your observations of the relationship between the speed of an orbiting object and the semimajor axis.

SMARTWORK • smartwork.wwnorton.com

4 Light and Telescopes

Nearly all of the objects in the sky are beyond the reach of direct investigation, even by robotic spacecraft. Our knowledge of the universe beyond Earth comes primarily from light given off or reflected by astronomical objects. Light from a star, for example, carries information about how hot it is, what it is made of, and how fast and in what direction it is traveling. It even tells us about the nature of the material between the object and the observer. In this chapter, you begin to learn about light and how astronomers use it to investigate the universe. In the illustration on the opposite page, a student has just acquired a new telescope and is eager to point it at the sky, but she has taken a moment to determine the path light takes as it comes through the telescope—an essential step toward figuring out how to use it.

◆ LEARNING GOALS

Astronomers try to learn the secrets of the universe from the light that reaches us from distant objects. This information must first be collected and processed before it can be analyzed and converted into useful knowledge. The student who has just gotten the new telescope is working out the path the light follows before it hits her eye. By the end of this chapter, you should be able to identify the type of telescope shown and evaluate the usefulness of such a telescope for astronomical observations. You should understand how observations of light give us information about the universe, and you should also be able to:

LG 1 Compare and contrast the properties of waves and particles and give examples of the wave and particle behavior of light.

LG 2 Describe the electromagnetic spectrum and the types of information that can be carried by light.

LG 3 Explain how light detectors have improved over time, and describe the advantages of modern detectors over historical ones.

LG 4 Understand how a telescope's aperture and focal length relate to resolution and image size.

APERTURE

EYEPIECE

SECONDARY MIRROR

PRIMARY MIRROR

Vocabulary Alert

vacuum In common language, we think of this as a completely empty space, without even an atom in it. Such a perfect vacuum does not exist in nature, so astronomers are comfortable using *vacuum* to mean "a place with hardly anything in it." For the purposes of determining what speed to use for light, the space between the galaxies is considered a vacuum, and so is the space between the stars.

4.1 What Is Light?

Understanding light and its interactions with matter has been a scientific quest at least since the time of the ancient Greeks. Throughout that time, light has been understood as a wave, as a particle, as a ray, and, finally, as an object that acts sometimes like a wave and sometimes like a particle. From this very statement, you can probably guess that light is complicated. In this section, we will examine the different properties of light, how light behaves, and the relationship between light and matter.

The Speed of Light

In the 1670s, Danish astronomer Ole Rømer (1644–1710) studied the movement of the moons of Jupiter. He measured the time at which each moon disappeared behind the planet. To his amazement, the observed times did not follow the regular schedule that he predicted. Sometimes the moons disappeared behind Jupiter sooner than expected, and at other times they disappeared behind Jupiter later than expected. Rømer realized that the difference depended on where Earth was in its orbit. If he began tracking the moons when Earth was closest to Jupiter, by the time Earth was farthest from Jupiter the moons were a bit more than 16½ minutes "late." But if he waited until Earth was once again closest to Jupiter, the moons again passed behind Jupiter at the predicted times.

Rømer's observations showed that light travels at a finite speed. As shown in **Figure 4.1**, the moons appeared "late" when Earth was farther from Jupiter because the light had to travel the extra distance between the two planets. The value of the speed of light that Rømer announced in 1676 was a bit on the low side—2.25×10^8 meters per second (m/s)—because the size of Earth's orbit was not well known. Modern measurements of the speed of light in a **vacuum** give a value of 2.99792458×10^8 m/s. In this book, we will round up to 3×10^8 m/s. The speed of light in a vacuum is a fundamental constant, c. Keep in mind, however, that light travels at this speed *only* in a vacuum. The speed of light through a medium such as air or glass is less than c.

The International Space Station moves around Earth at about 28,000 kilometers per hour (km/h), orbiting Earth in about 90 minutes. Light travels almost 40,000 times faster than this and could circle Earth in only $\frac{1}{7}$ of a second. Because light

Figure 4.1 Ole Rømer realized that apparent differences between the predicted and observed orbital motions of Jupiter's moons depend on the distance between Earth and Jupiter. He used these observations to measure the speed of light. (Recall that 1 AU is the average distance from Earth to the Sun, equal to 1.5×10^8 km.)

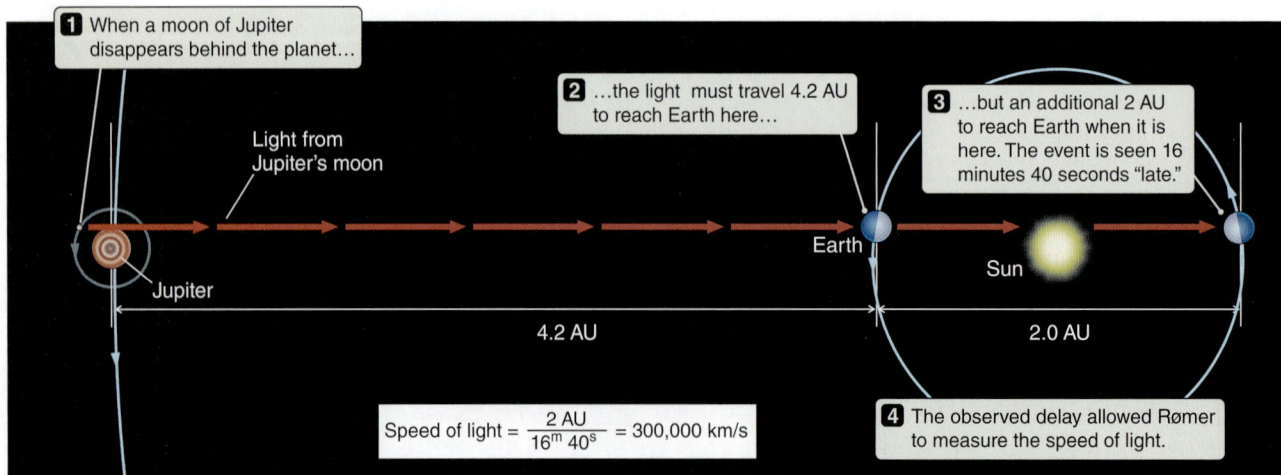

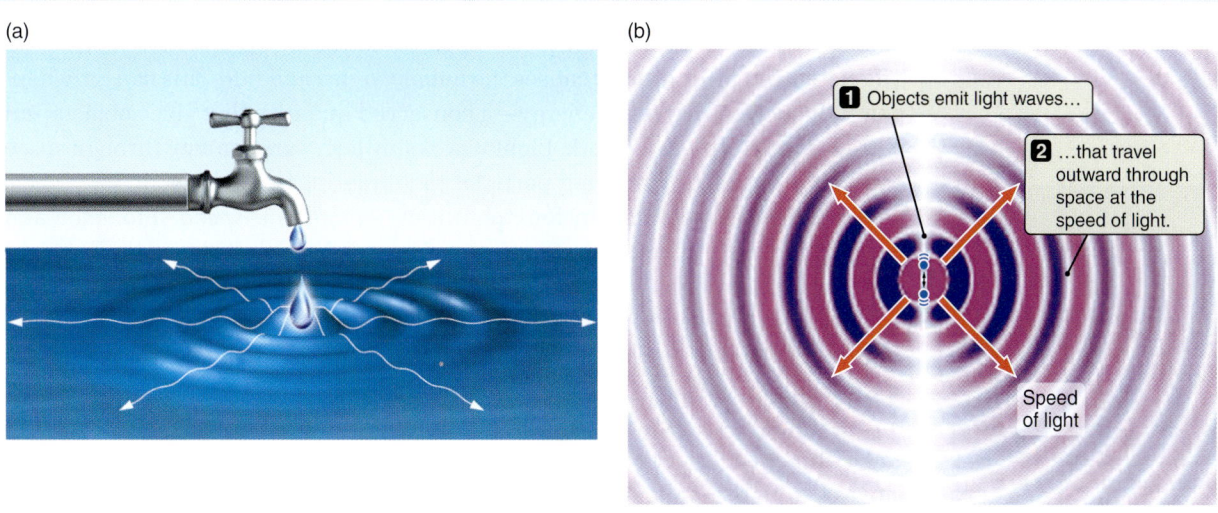

Figure 4.2 (a) A drop falling into water creates a wave that moves across the water's surface. (b) In similar fashion, an accelerating electric charge creates light waves that move away from the charge at the speed of light.

is so fast, the travel time of light is a convenient way of expressing cosmic distances, and the basic unit is the light-year—the distance light travels in 1 year. Pause for a moment to consider the light-year. Imagine traveling around Earth in $1/7$ of a second. Now try to imagine traveling at that speed for about 8 minutes, during which time you would cover the distance to the Sun. Now try to imagine traveling at that speed for an entire year. During that time, you would not travel even one-quarter of the way to the star closest to the Sun.

As light travels at this high speed, it carries energy from place to place. **Energy** is the ability to do work, and it comes in many forms. **Kinetic energy** is the energy of moving objects. **Thermal energy** is closely related to kinetic energy and is the sum of all the kinetic energy of the moving bits of matter inside a substance. The sum of the energies of all these random motions results in the object's temperature. For example, when light from the Sun strikes the pavement, the pavement heats up. That energy was carried from the Sun to the pavement by light. Rømer knew how long it took for light to travel a given distance, but it would take more than 200 years for physicists to figure out what light actually is.

Light as a Wave

Throughout the late 1800s and early 1900s, experimental results caused physicists some confusion. Sometimes light acted like a wave in water, and at other times it acted like a particle—an object that, for the moment, can be thought of as a very, very tiny baseball. Eventually, scientists came to understand the nature of light as an object that acts sometimes like a wave and at other times like a particle. In this subsection, we will discuss its wavelike properties. In the next subsection, we will discuss its particle properties.

When a drop of water falls from the faucet into a sink full of water, it causes a disturbance, or wave, like the one shown in **Figure 4.2a**. The wave moves outward as a ripple on the surface of the water. As you can see in Figure 4.2b, light moves out through space, away from its source in much the same way. However, the ripples in the sink are distortions of the water's surface, and they require a **medium**: a substance to travel through. Light waves do not require a medium—they can move through the vacuum of empty space.

76 CHAPTER 4 Light and Telescopes

Figure 4.3 (a) When waves moving across the surface of water reach a bubble, they cause the bubble to bob up and down. (b) Similarly, a passing light wave causes a charged particle to oscillate in response to the wave.

▶❙❙ **AstroTour: Light as a Wave, Light as a Photon**

Figure 4.4 (a) A wave is characterized by the distance from one peak to the next (wavelength, λ), the maximum height above the medium's undisturbed state (amplitude), and (b) the speed (v) at which the wave pattern travels. (c) A wave with a longer wavelength has a lower frequency. (d) Conversely, a wave with a shorter wavelength has a higher frequency.

Now imagine that a soap bubble is floating in the sink, as in **Figure 4.3a**. The bubble remains stationary until the ripple from the dripping faucet reaches it. The rising and falling water causes the bubble to rise and fall. This can only happen if the wave is carrying energy—a conserved quantity that gives objects and particles the ability to do work. Light waves similarly carry energy through space and cause electrically charged particles to vibrate, as in Figure 4.3b.

Waves are characterized by four quantities—*amplitude*, *speed*, *frequency*, and *wavelength*—as illustrated in **Figure 4.4**. The **amplitude** of a wave is the height of the wave above the undisturbed position (Figure 4.4a). For water waves, the amplitude is how far the water is lifted up by the wave. In the case of light, the amplitude of a light wave is related to the brightness of the light. A wave travels at a particular speed, v (Figure 4.4b), through the water. The water itself doesn't travel; it just moves up and down at the same location. For waves like those in water, this speed is variable and depends on the density of the substance the wave moves through, among other things. Light, in contrast, always moves through a vacuum at the same speed, c. The distance from one crest of a wave to the next is the **wavelength**, λ (this is the Greek letter *lambda*) (Figure 4.4c). The number of wave crests passing a point in space each second is the wave's **frequency**, f. Waves with longer wavelengths have lower frequencies, and waves with shorter wavelengths have higher frequencies. Frequency is measured in cycles per second, or **hertz** (Hz).

Waves travel a distance of one wavelength each cycle, so the speed of a wave can be found by multiplying the frequency and the wavelength. Translating this idea into math, we have $v = \lambda f$. The speed of light in a vacuum is always c, so once the wavelength of a wave of light is known, its frequency is known, and vice versa. Because light travels at constant speed, its wavelength and frequency are inversely proportional to each other: if the wavelength increases, the frequency decreases. **Working It Out 4.1** further explores this relationship.

Light as a Particle

Though the wave theory of light describes many observations, it does not provide a complete picture of the properties of light. Scientists working in the late 19th and early 20th centuries discovered that many of the puzzling aspects of light could be better understood by thinking of light as a particle. We now know that

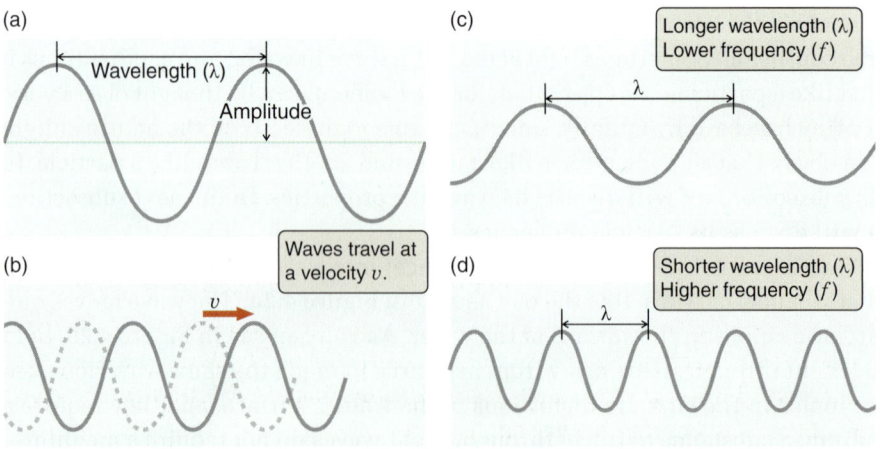

Working It Out 4.1 | Wavelength and Frequency

When you tune in to a radio station at 770 AM, you are receiving a signal that is broadcast at a frequency of 770 kilohertz (kHz), or 7.7×10^5 Hz. We can use the relationship between wavelength and frequency to calculate the wavelength of the AM signal:

$$\lambda = \frac{c}{f} = \frac{3 \times 10^8 \text{ m/s}}{7.7 \times 10^5 \text{/s}} = 390 \text{ m}$$

This AM wavelength is about four times the length of an American football field.

We can compare this wavelength with that of a typical FM signal, 99.5 FM, which is broadcast at a frequency of 99.5 megahertz (MHz), or 9.95×10^7 Hz:

$$\lambda = \frac{c}{f} = \frac{3 \times 10^8 \text{ m/s}}{9.95 \times 10^7 \text{/s}} = 3 \text{ m}$$

FM wavelengths are much shorter than AM wavelengths.

The human eye is most sensitive to green light, which has a wavelength of about 500–550 nanometers (nm)—roughly in the middle of the visible range. Green light with a wavelength of 520 nm has a frequency of

$$f = \frac{c}{\lambda} = \frac{3 \times 10^8 \text{ m/s}}{5.2 \times 10^{-7} \text{ m}} = \frac{5.8 \times 10^{14}}{s} = 5.8 \times 10^{14} \text{ Hz}$$

This frequency corresponds to 580 *trillion* (580 million million) wave crests passing by each second.

---

light sometimes acts like a wave and sometimes acts like a particle. In the particle model, light is made up of massless particles called **photons** (*phot-* means "light," as in *photograph*, and *-on* signifies a particle). Photons always travel at the speed of light, and they carry energy. When physicists speak of photons, they say that the light energy is *quantized*. The word **quantized** means that something is subdivided into individual units. A photon is a *quantum* of light.

The energy of a photon and the frequency of the electromagnetic wave are directly proportional to each other; the higher the frequency of the light wave, the greater the energy each photon carries. This relationship connects the particle and the wave concepts of light. For example, photons with higher frequencies carry more energy than do photons with lower frequencies. The constant of proportionality between the energy, E, and the frequency, f, is Planck's constant, h, which is equal to 6.63×10^{-34} joule-seconds (a joule is a unit of energy). Specifically, we write $E = hf$. For light that is visible to us, the energy of the light is related to the color: high-energy light is blue, and low-energy light is red.

Looking at **Figure 4.5**, we see that a beam of red light can carry just as much energy as a beam of blue light, but the red beam will have more photons than the blue beam. In this sense, light is similar to money. Ten dollars is 10 dollars, but it takes a lot more pennies (low-energy photons) than quarters (high-energy photons) to make up 10 dollars.

The Electromagnetic Spectrum

Light can have wavelengths that are much shorter or much longer than our eyes can perceive. Because light is related to electricity and magnetism, light waves are often referred to as **electromagnetic waves**. The whole range of different wavelengths of light is collectively referred to as the **electromagnetic spectrum**. Astronomers learn about distant objects by studying the electromagnetic spectrum of the light coming from them.

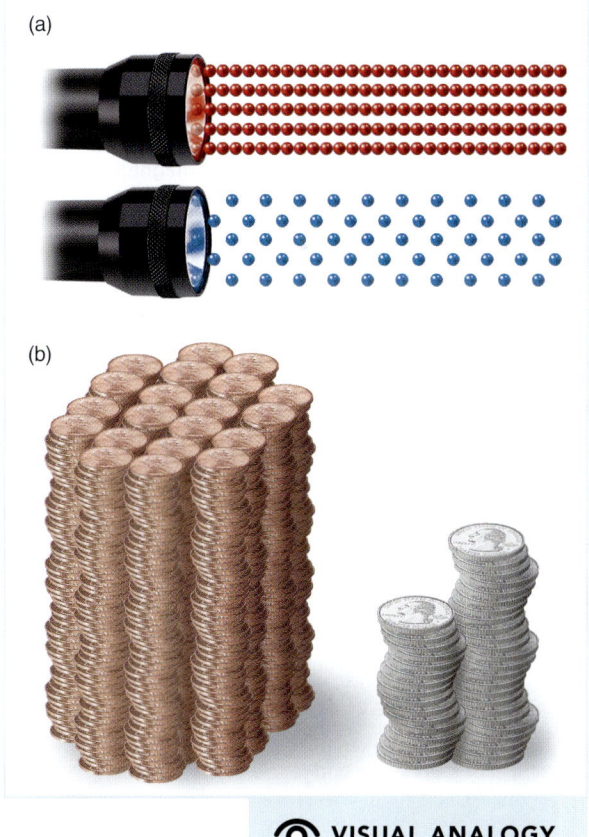

VISUAL ANALOGY

Figure 4.5 (a) Red light carries less energy than blue light, so it takes more red photons than blue photons to make a beam of a particular intensity. (b) Similarly, pennies are worth less than quarters, so it takes more pennies than quarters to add up to $10.

78 CHAPTER 4 *Light and Telescopes*

Figure 4.6 The visible part of the electromagnetic spectrum is laid out in the colors of this rainbow.

You have almost certainly seen a rainbow like the one in **Figure 4.6**. A rainbow is created when white light interacts with water droplets and is spread out by wavelength into its component colors. This spread of colors is called a **spectrum**. At the long-wavelength end of the visible spectrum is red light, between about 600 and 750 nm. At the other end of the visible spectrum (i.e., the short-wavelength end) is violet light, which is the bluest of blue light. Stretched out between the two, literally in a rainbow, is the rest of the visible spectrum, shown in **Figure 4.7**. The colors in the visible spectrum, in order of increasing wavelength, are

Violet Indigo Blue Green Yellow Orange Red

Recall that wavelength and frequency are inversely proportional and that frequency is directly proportional to energy. Short-wavelength light has large frequencies and high energies. Long-wavelength light has low frequencies and low energies.

Follow along in Figure 4.7 as we take a tour of the electromagnetic spectrum, beginning with the shortest wavelengths and working our way to the longest ones. The very shortest wavelengths of light are called **gamma rays**, or sometimes gamma radiation. Because this light has the shortest wavelengths, it is very high energy and penetrates matter easily. **X-rays**, which you may know about from visits to the dentist, have longer wavelengths than gamma rays. X-ray light has enough energy to penetrate through skin and muscle but is stopped by denser bone. **Ultraviolet (UV) radiation** has longer wavelengths than X-rays but shorter wavelengths than visible light. You are familiar with this type of light from sunburns: UV light has enough energy to penetrate into your skin, but not much farther.

The visible part of the spectrum—the part you can see with your eye—is a very small portion of the entire electromagnetic spectrum. The shortest wavelengths you can see are violet, at about 350 nm. The longest wavelengths you

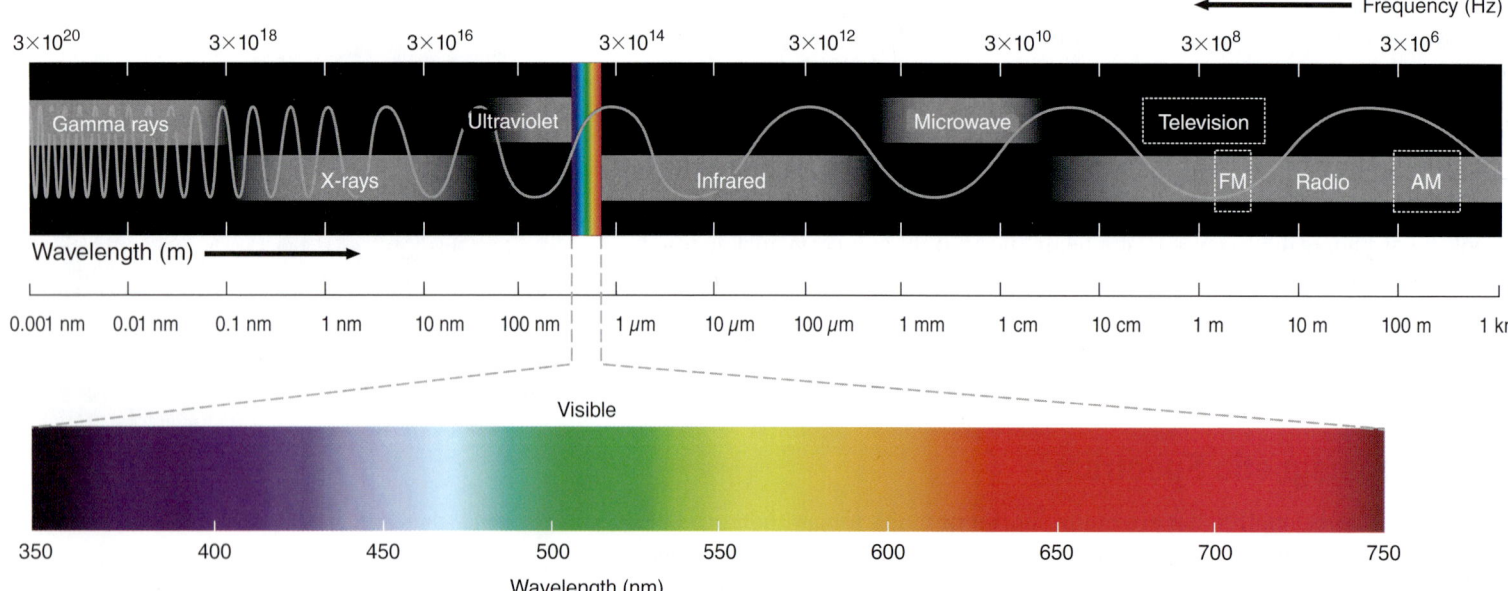

Figure 4.7 By convention, the electromagnetic spectrum is broken into loosely defined regions ranging from gamma rays to radio waves. Throughout this book, a labeled icon appears below individual astronomical images to identify what part of the spectrum was used to take the image: gamma rays (G), X-rays (X), ultraviolet (U), visible (V), infrared (I), or radio (R). If more than one region of the spectrum was used, multiple labels are highlighted in the icon.

can see are red, at about 750 nm. Most of the electromagnetic spectrum, and most of the information in the universe, is invisible to the human eye. To detect light outside the visible, we must use specialized detectors of various kinds that respond to photons with higher or lower energies than the ones that activate the cells in the eye.

Infrared (IR) radiation has longer wavelengths than the reddest wavelengths in the visible range. You are familiar with a small wavelength range of this kind of radiation because you often feel it as heat. When you hold your hand next to a hot stove, some of the heat you feel is carried to your hand by infrared radiation emitted from the stove. In this sense, you could think of your skin as being a giant infrared eyeball—it is sensitive to infrared wavelengths. Infrared radiation is also used in television remote controls, and night vision goggles detect infrared radiation from warm objects like animals. Beyond 500 millionths of a meter (5×10^{-4} meter), we start calling the light **microwave radiation**. The microwave in your kitchen heats the water in food using light of these wavelengths. The longest-wavelength light, with wavelengths longer than a few centimeters, is called **radio waves**. Light of these wavelengths in the form of FM, AM, television, and cell phone signals is used to transmit information all over the world.

Astronomers conventionally use nanometers when referring to wavelengths at visible and shorter wavelengths. A nanometer is one-billionth (10^{-9}) of a meter. One-millionth of a meter is a micrometer (μm), or "micron," useful for measurement of wavelengths in the infrared. Millimeters (mm), one-thousandth of a meter, centimeters (cm), one-hundredth of a meter, and meters (m) are used for wavelengths in the microwave and radio regions of the electromagnetic spectrum.

Light and Matter

Light and matter interact, and this interaction allows us to detect matter even at great distances in space. To understand this interaction, we must understand the building blocks of matter.

Matter is anything that occupies space and has mass. **Atoms**, such as the helium atom shown in **Figure 4.8a**, are the fundamental building blocks of matter, and all the objects around you are made of atoms. A block of pure carbon has only carbon atoms in it. **Molecules** are groups of atoms bound together by shared electrons in chemical bonds. Water (H_2O), shown in Figure 4.8b, has two hydrogen atoms and one oxygen atom.

While atoms are the fundamental building blocks of matter, an atom actually has several parts. **Protons** are positively charged particles, and **neutrons** are particles that have no charge. A massive nucleus of protons and neutrons sits at the center of the atom. A cloud of negatively charged **electrons** surrounds the nucleus. Atoms with the same number of protons are all of the same type, known as an **element**. For example, the helium atom in Figure 4.8a has two protons. Helium typically also has two neutrons in the nucleus and two electrons surrounding the nucleus. A rarer form has only one neutron.

Atoms and molecules both interact with light, absorbing or emitting it. It is also common for electrons to be torn loose from the atom and travel through space on their own. These electrons are negatively charged and leave behind a positively charged atom called an **ion**. Either of these objects, the ion or the electron, interact easily with light or other charged particles.

Recall the soap bubble floating in the sink. The bubble remains stationary until the ripple from the dripping faucet reaches it. The rising and falling water

Vocabulary Alert

radiation In common language, radiation is associated with emissions from nuclear bombs or radioactive substances. In some cases, this radiation actually is light in the form of gamma rays. In other cases, it is particles. Astronomers use the word to mean energy carried through space; that is, light. Astronomers often use the two words *light* and *radiation* interchangeably, especially when talking about wavelengths that are not in the visible range.

▶▶ **Nebraska Simulation:** EM Spectrum Module

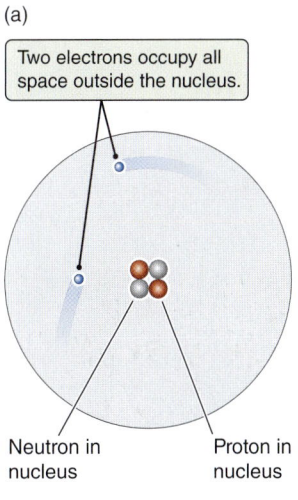

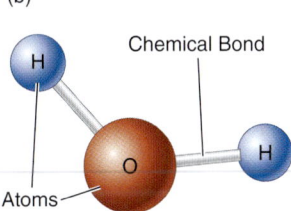

Figure 4.8 (a) The helium atom has two neutrons, two protons, and two electrons. (b) A molecule, such as a water molecule, consists of several atoms held together by chemical bonds.

causes the bubble to rise and fall. Similarly, light causes charged particles like electrons and protons to move about (Figure 4.3b). The reverse is also true: an oscillating electric charge (Figure 4.2b) causes a disturbance that moves outward through space as light.

It takes energy to produce light, and that energy is carried through space by the wave. Matter far from the source of the wave can absorb this energy; a process called **absorption**. Matter also emits energy; a process called **emission**. In this way, some of the energy lost, or *emitted*, by the particles that create the light is transferred to, or *absorbed* by, other charged particles. Emission and absorption of light is at the foundation of our understanding of the universe. We detect light by its interaction with matter.

4.2 Cameras and Spectrographs Record Astronomical Data

Astronomical observations began with the human eye—information about the overall colors of stars and their brightness in the night sky is apparent even to the "naked" eye, by which we mean unassisted by binoculars or telescopes or filters. Starting in the 1800s, the development of film photography and, later, digital photography revolutionized astronomy, allowing us to detect fainter and more distant objects than possible to detect with the eye alone.

The Eye

Human eyes respond to light with wavelengths ranging from about 350 nm (deep violet) to 750 nm (far red). **Figure 4.9** shows a simplified schematic of the human eye. The part of the human eye that detects light is called the retina, and the individual cells that respond to light falling on the retina are called rods and cones. Cones are located near the middle of the retina at the center of our vision, while rods are located away from the center of the retina and are responsible for our peripheral vision. As photons from a star enter the pupil of the eye, they strike cones at the center of vision. The cones then send a signal to the brain, which interprets this message as "I see a star." The limit of the faintest stars we can see with our unaided eyes is determined in part by two factors that are characteristic of all detectors of light: integration time and quantum efficiency. While the human eye is an imperfect analogy for astronomical detectors, it is the detector with which you have the most experience, and so we will use it to give you a sense of how astronomical detectors work.

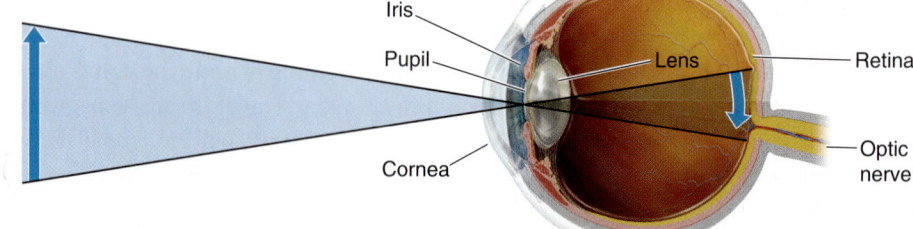

Figure 4.9 A sketch of the human eye shows how the image of the object is oriented upside-down on the retina.

Integration time is the time interval during which the eye can add up photons—this is analogous to leaving the shutter open on a camera. The brain "reads out" the information gathered by the eye about every 100 milliseconds (ms). Anything that happens faster than that appears to happen all at once. If two images flash on a computer screen 30 ms apart, you will see them as a single image because your eyes will add up (or integrate) whatever they see over an interval of 100 ms. However, if the images occur 200 ms apart, you will see them as separate images. This relatively brief integration time is the most important factor limiting our nighttime vision. Stars that produce too few photons for our eyes to detect in 100 ms are too faint for us to see.

Quantum efficiency determines how many responses occur for each photon received. For the human eye, 10 photons must strike a cone within 100 ms to activate a single response. So the quantum efficiency of our eyes is about 10 percent: for every 10 events, the eye sends one signal to the brain. Together, integration time and quantum efficiency determine the rate at which photons must arrive before the brain says, "Aha, I see something."

However, even when we receive a sufficient number of photons in a short-enough time span to see a star, our vision is further limited by the eye's **angular resolution**, which refers to how close two points of light can be to each other before we can no longer distinguish them. Unaided, the best human eyes can resolve objects separated by 1 arcminute ($1/60$ of a degree), an angular distance of about $1/30$ the diameter of the full Moon. This may seem small, but when we look at the sky, thousands of stars and galaxies may hide within a patch of sky with this diameter.

Until 1840, the retina of the human eye was the only astronomical detector. Permanent records of astronomical observations were limited to what an experienced observer could sketch on paper while working at the eyepiece of a telescope, as illustrated in **Figure 4.10a**.

Photographic Plates

In 1840, John W. Draper (1811–1882), a chemist, created the earliest known astronomical photograph, shown in Figure 4.10b. His subject was the Moon. Early photography was slow and very messy, and astronomers were reluctant to use it.

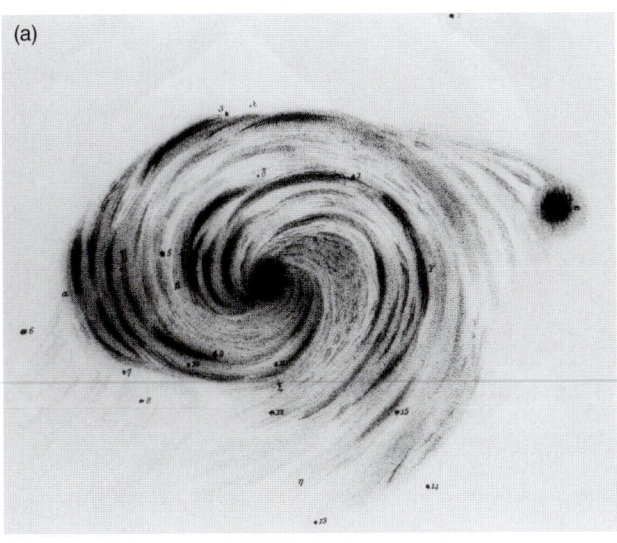

Figure 4.10 (a) William Parsons (Lord Rosse, 1800–1867) made this drawing of the galaxy M51 in 1845. (b) John W. Draper photographed the Moon in 1840.

CHAPTER 4 Light and Telescopes

In the late 1870s, a faster, simpler process was invented, and astrophotography took off. Astronomers could now create permanent images of planets, nebulae, and galaxies with ease.

The quantum efficiency of most photographic systems used in astronomy was very low—typically 1–3 percent, even poorer than that of the human eye. But unlike the eye, photography can overcome poor quantum efficiency by leaving the shutter open on the camera, increasing the integration time to many hours of exposure. Photography made it possible for astronomers to record and study objects that were previously invisible.

Photography is not without its problems. Very faint objects often require long exposures that can take up much of an observing night. By the middle of the 20th century, the search was on for electronic detectors that would overcome the deficiencies of this type of photography.

Charge-Coupled Devices

In 1969, scientists at Bell Laboratories invented a remarkable detector called a **charge-coupled device, or CCD**. Astronomers soon realized that this was the tool they had been looking for, and the CCD became the detector of choice in almost all astronomical imaging applications. The output from a CCD is a digital signal that can be sent directly from the telescope to image-processing software or stored electronically for later analysis.

A CCD is an ultrathin wafer of silicon (less than the thickness of a human hair) that is divided into a two-dimensional array of picture elements, or **pixels**, as seen in **Figure 4.11a**. When a photon strikes a pixel, it creates a small electric charge within the silicon. As each CCD pixel is read out, the digital signal that flows to the computer is almost precisely proportional to the accumulated charge. The first astronomical CCDs were small arrays containing a few hundred thousand pixels. The larger CCDs used in astronomy today—like the one seen in Figure 4.11b—may contain more than 100 million pixels. Still larger arrays are under development as ever-faster computing power keeps up with image-processing demands.

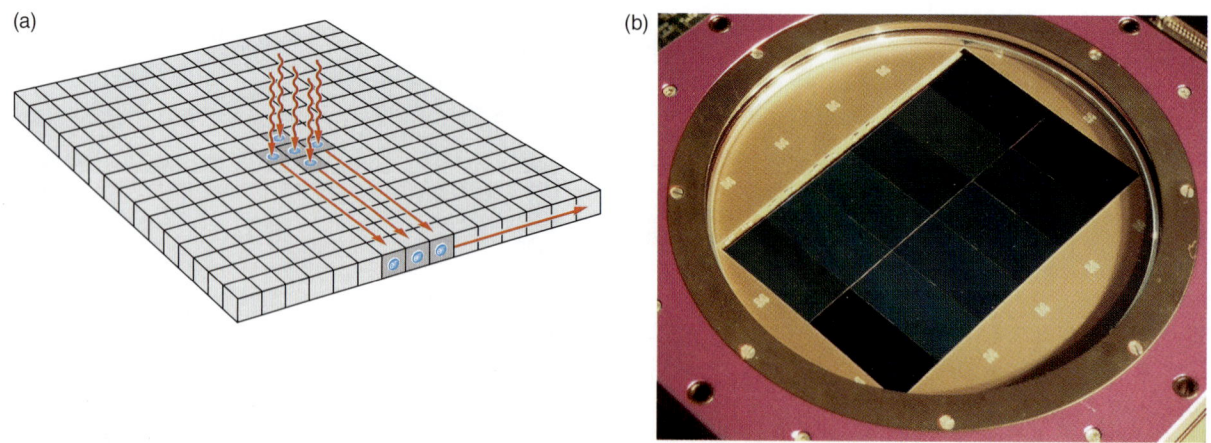

Figure 4.11 (a) A simplified diagram of a charge-coupled device (CCD) shows how light lands on pixels (gray squares) and produces free electrons within the silicon. These electrons are converted into an image by a computer. (b) This CCD has 12,288 × 8,192 pixels.

The quantum efficiency of CCDs is far superior to that of photography or even the eye. For some cameras, in the visible range of wavelengths, this quantum efficiency approaches 90 percent, with an electron produced for nearly every photon that strikes the CCD. This improvement dramatically increases our ability to view faint objects with short exposure times.

CCDs have found their way into many devices that we now take for granted, such as digital cameras, digital video cameras, and camera phones. Your cell phone takes color pictures by using a grid of CCD pixels arranged in groups of three. Each pixel in a group is constructed to respond only to a particular range of colors—only to red light, for example. This is also true for digital image displays. You can see this for yourself if you place a small drop of water on the screen of your smartphone or tablet and turn it on. The water magnifies the grid of pixels so that you can see them individually. This grid degrades the angular resolution of the camera because each spot in the final image requires three pixels of information. Astronomers choose instead to use all the pixels on the camera to measure the number of photons that fall on each pixel, without regard to color. They then put filters in front of the camera to allow only light of particular wavelengths to pass through. Color pictures are constructed by taking multiple pictures, coloring each one, and then carefully aligning and overlapping them to produce beautiful and informative images. Sometimes the colors are "true"; that is, they are close to the colors you would see if you were actually looking at the object with your eyes. At other times, the colors represent different portions of the spectrum and show the temperature or composition of different parts of the object. Using changeable filters instead of designated color pixels gives astronomers greater flexibility and greater angular resolution.

Spectra and Spectrographs

When astronomers want to know about an object in detail, they often take its spectrum by passing the light through a prism or grating to create an artificial rainbow. From the relative intensity of different colors in that rainbow, they can find the object's temperature: redder objects, which are dominated by lower-energy light, have lower temperatures; bluer objects, which are dominated by higher-energy light, have higher temperatures. For example, radio or infrared telescopes are used to study very cool, or even cold, dust. Gamma ray telescopes are used to study violently hot gas. In addition to the change in color, hotter objects emit more light of all colors, and so shine more brightly. You know this from the burners on an electric stove, which glow more brightly the hotter they are.

Still more information is embedded in the spectrum. Because atoms of different elements interact with different wavelengths of light, they amplify or reduce the spectrum at particular wavelengths. Atoms and molecules produce **emission lines** when they add light to the spectrum and **absorption lines** when they take light away. In either case, as we'll see in more detail in Chapter 10, each type of atom or molecule has a unique set of lines, which act like fingerprints telling astronomers what particular atoms and molecules the object is composed of.

Spectrographs—sometimes called **spectrometers**—are tools that take the spectrum of an object, split the light by wavelength, and then record it. **Spectroscopy** is the study of an object's light in terms of its component wavelengths, and we'll encounter its many applications throughout the chapters to come.

▶▶ **Nebraska Simulation:** Three Views Spectrum Demonstrator

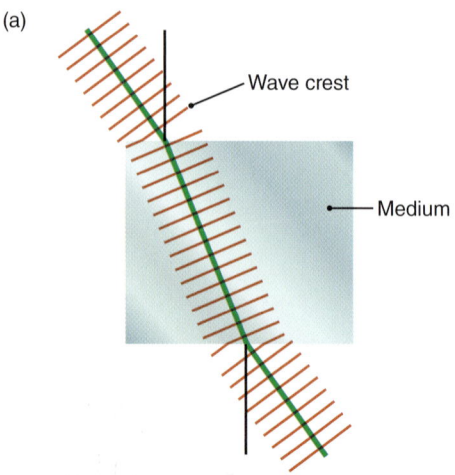
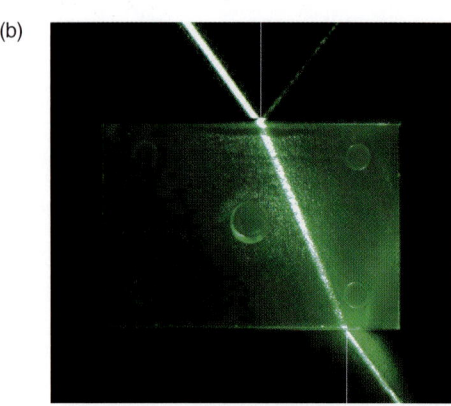
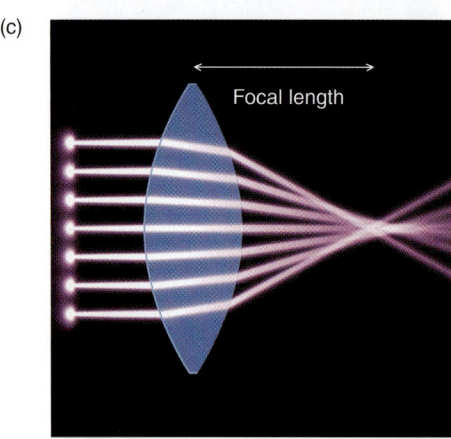

Figure 4.12 (a) When wavefronts enter a new medium, they bend in a new direction, relative to a line perpendicular to the surface (black lines). (b) An actual light ray entering and leaving a medium. Light waves are refracted (bent) when entering a medium with a higher index of refraction. They are refracted again as they reenter the medium with the lower index of refraction. (c) For a curved lens like the one shown, this phenomenon causes the light to focus to a point. This point is in a slightly different location for different wavelengths of light.

4.3 Telescopes Collect Light

A telescope is a device for collecting and focusing light. Different regions of the spectrum require not only different types of detectors but also different types of telescopes. Each region presents its own challenges. For example, because warm objects emit infrared light, an infrared telescope requires special shielding, even in space, to keep the entire instrument cool. Otherwise, the detector would receive so much light from the telescope itself that it would not be able to see any astronomical objects. Gamma and X-ray telescopes, because they collect light at the highest energies, require very different systems than those used for the rest of the electromagnetic spectrum—lenses and mirrors do not focus light of these wavelengths. For this discussion of telescopes, we confine ourselves to a consideration of lens and mirror systems.

For all telescopes, the "size" of the telescope refers to the diameter of the largest mirror or lens, which determines the light-collecting area. This diameter is called the **aperture**. A "4.5-meter telescope" has a primary mirror (or lens) that is 4.5 meters in diameter.

Refractors and Reflectors

Telescopes come in two primary types: **refracting telescopes**, which use lenses; and **reflecting telescopes**, which use mirrors. While the use of telescopes began with refractors, large modern telescopes are always reflectors. To understand why, you need to know something about how each type works.

Refraction is the basis for the refracting telescope. As light enters a new medium, its speed changes. If the light strikes the surface at an angle, some of the crest of the wave arrives at the surface earlier and some arrives later. You can see this in the schematic diagram of wave crests (red lines) striking a medium at an angle in **Figure 4.12a**. The light bends by an amount that depends on the materials involved and the angle at which the light strikes. Figure 4.12b shows an actual light ray passing into and out of a medium. Notice that the ray bends each time the medium changes. This bending of light as it enters a new medium is called **refraction**.

The amount of refraction is determined by the properties of the medium. The ratio of light's speed in a vacuum, c, to its speed in a medium, v, is the medium's **index of refraction**, n; that is, $n = c/v$. For example, n is approximately 1.5 for typical glass, so the speed of light in glass is only 200,000 kilometers per second (km/s). The amount of bend depends on the angle and on the index of refraction of the glass.

Because a telescope's glass lens is curved, light at the outer edges strikes the surface more obliquely than light near the center. Therefore, light at the outer edges is refracted more than light near the center. This concentrates the light rays entering the telescope, bringing them to a sharp focus at a distance referred to as the focal length, shown in Figure 4.12c. An image is created in the telescope's *focal plane*. **Figure 4.13a** shows the light from two stars passing through a lens and converging at the focal plane of the lens. Figure 4.13b shows the same situation for a lens with a longer focal length. Increasing the focal length decreases how sharply the lens must be curved. This in turn increases the size and separation of objects in the image, as shown by the red and blue light rays in Figure 4.13.

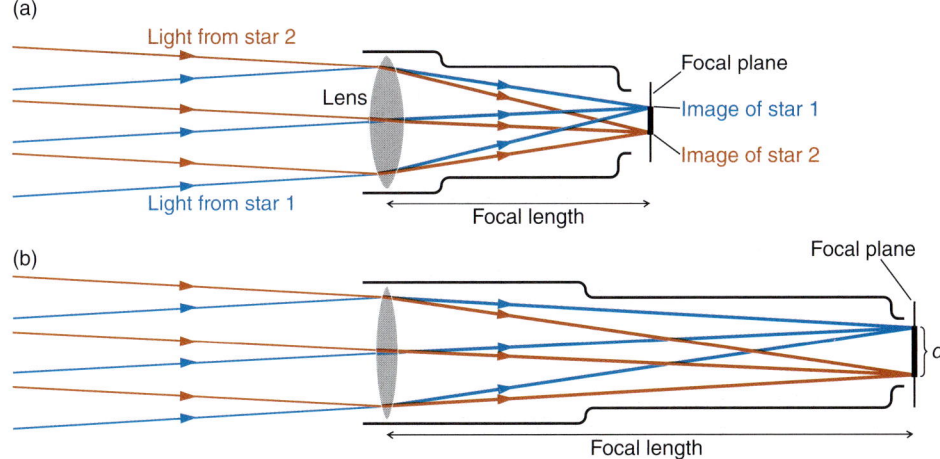

Figure 4.13 (a) A refracting telescope uses a lens to collect and focus light from two stars, forming images of the stars in its focal plane. (b) Telescopes with longer *focal length* spread the images of objects farther apart, producing larger, more widely separated images.

The larger the area of the lens, the more light-gathering power it has and the fainter the stars we can observe. However, there are physical limits on the size of refracting telescopes because large lenses are very heavy. There is another serious drawback to refracting telescopes: *chromatic aberration*. This is the effect that produces rainbows when sunlight passes through a piece of cut glass. Sunlight is made up of all the colors of the rainbow, and each color refracts at a slightly different angle because the index of refraction depends on the wavelength of the light. In astronomical applications, this chromatic aberration produces blurry images unless a very narrow filter is used to block all but a narrow range of wavelengths. As an example of this blurring, an image of a star will have a colored halo around the star's location; it will not appear as a crisp white star. Manufacturers of quality cameras and telescopes use a **compound lens**, composed of two types of glass, to correct for chromatic aberration.

Reflecting telescopes use reflection from mirrors rather than lenses to focus the light into an image. As you can see in **Figure 4.14**, light coming from a star first strikes the *primary mirror* and reflects off it back toward the sky. Typically, a *secondary mirror* then reflects the light back along the path from the primary mirror to the focal plane.

Reflecting telescopes have several important advantages over refracting telescopes. Because mirrors do not spread out white light into its component colors, chromatic aberration is no longer a problem. Primary mirrors can be supported from the back, and they can be made thinner and therefore weigh less than objective lenses. There are other advantages as well, and all large telescopes made today are of the reflector type. Look back at the chapter-opening illustration, which shows a reflecting telescope. In this case, the focal plane is located at the side of the telescope instead of at the bottom. This innovation makes the telescope easier to use because the telescope doesn't need to be lifted off the ground for the user to look through it. Innovations like this one are common, but the fundamental concept of using mirrors to focus the light to a focal plane is always the same.

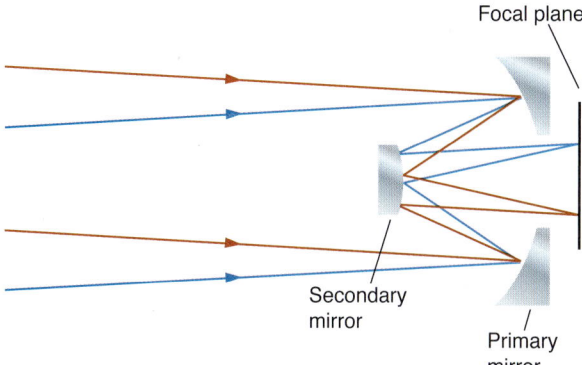

Figure 4.14 Reflecting telescopes use mirrors to collect and focus light. Large telescopes typically use a secondary mirror that directs the light back through a hole in the primary mirror to an accessible focal plane behind the primary mirror.

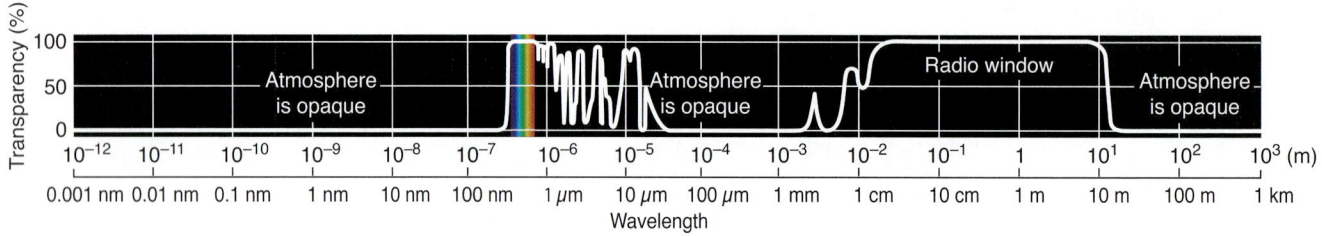

Figure 4.15 Earth's atmosphere is transparent in a few regions of the spectrum but blocks most electromagnetic radiation.

Radio Telescopes

Our atmosphere is transparent in several regions of the spectrum. The largest of these atmospheric "windows" is in the radio portion, and so we are able to build radio telescopes on the ground rather than in space (see **Figure 4.15**). Most **radio telescopes** are large, steerable dishes, typically tens of meters in diameter, such as the one shown in **Figure 4.16a**. The world's largest single-dish radio telescope is the 305-meter Arecibo dish, built into a natural bowl-shaped depression in Puerto Rico (**Figure 4.16b**). (China is building a larger one, scheduled for completion in late 2016.) This huge structure is too big to steer. Arecibo can only observe sources that pass within 20° of the zenith as Earth's rotation carries them overhead.

As large as radio telescopes are, they have relatively poor angular resolution compared to optical telescopes. A telescope's angular resolution is determined by the ratio λ/D, where λ is the wavelength of electromagnetic radiation and D is the telescope's aperture. A larger ratio means poorer resolution. Radio telescopes have diameters much larger than the apertures of most optical telescopes, and this is beneficial because a larger number in the denominator of the equation λ/D means a smaller (better) resolution. But the wavelengths of radio waves are much longer than the wavelengths of visible light. Because wavelength is in the top of the equation λ/D, a larger wavelength yields a larger (poorer) resolution. Radio telescopes are thus hampered by the very long wavelengths they are designed to receive. Consider the huge Arecibo dish. Its resolution is typically about 1 arcminute, little better than the unaided human eye.

Radio astronomers have had to develop ways to improve resolution. Mathematically combining the signals from two radio telescopes turns them into a telescope with a diameter equal to the separation between them. For example, if two 10-meter telescopes are located 1,000 meters apart, the D in λ/D is 1,000, not 10. This combination of two (or more) telescopes is called an **interferometer**, and it makes use of the wavelike properties of light. Usually several telescopes are employed in an **interferometric array**. Through the use of very large arrays, astronomers can attain and even exceed the angular resolution possible with optical telescopes.

One of the larger radio interferometric arrays is the Very Large Array (VLA) in New Mexico, shown in **Figure 4.17**. The VLA is made up of 27 individual movable dishes spread out in a Y-shaped configuration with a maximum antenna separation of 36 km. At a wavelength of 10 cm, this array can achieve resolutions of less than 1 arcsecond. The Very Long Baseline Array (VLBA) uses 10 radio telescopes spread out over more than 8,000 km from the Virgin Islands in the Caribbean to Hawaii in the Pacific. At a wavelength of 10 cm, this array

▶️ **AstroTour:** Geometric Optics and Lenses

▶▶ **Nebraska Simulation:** Telescope Simulator

Figure 4.16 (a) The Green Bank Telescope is the largest steerable telescope in the world. (b) The Arecibo radio telescope is the world's largest single-dish telescope. The steerable receiver suspended above the dish permits limited pointing toward celestial targets as they pass close to the zenith.

can reach resolutions better than 0.003 arcsecond. A radio telescope put into near-Earth orbit as part of a Space Very Long Baseline Interferometer (SVLBI) overcomes even this limit. Future SVLBI projects would extend the baseline to as much as 100,000 km, yielding resolutions far exceeding those of any existing optical telescope.

Optical telescopes can also be combined in an array to yield resolutions greater than those of single telescopes, although for technical reasons the individual units cannot be spread as far apart as radio telescopes. The Very Large Telescope (VLT), operated by the European Southern Observatory (ESO) in Chile, combines either four 8-meter telescopes or four movable 1.8-meter auxiliary telescopes (**Figure 4.18**). It has a baseline of up to 200 meters, yielding angular resolution in the milliarcsecond range.

Observing at Other Wavelengths

Earth's atmosphere distorts telescopic images, and molecules such as water vapor in Earth's atmosphere block large parts of the electromagnetic spectrum from getting through to the ground, so astronomers try to locate their instruments above as much of the atmosphere as possible. Most of the world's larger astronomical telescopes are located 2,000 meters or more above sea level. Mauna Kea, a dormant Hawaiian volcano and home of the Mauna Kea Observatories (MKO), rises 4,200 meters above the Pacific Ocean. At this altitude, the MKO telescopes sit above 40 percent of Earth's atmosphere; more important, 90 percent of Earth's atmospheric water vapor lies below. Still, for the infrared astronomer, the remaining 10 percent of water vapor is troublesome.

One way to solve the water vapor problem is to make use of high-flying aircraft. The Stratospheric Observatory for Infrared Astronomy (SOFIA), which began operations in 2011, carries a 2.5-meter telescope and works in the far-infrared region of the spectrum, from 1 to 650 µm. It flies in the stratosphere at an altitude of about 12 km, but still above 99 percent of the water vapor in Earth's lower atmosphere.

Airborne observatories overcome atmospheric absorption of infrared light by placing telescopes above most of the water vapor in the atmosphere. But gaining full access to the complete electromagnetic spectrum requires getting completely above Earth's atmosphere. The first astronomical satellite was the British Ariel 1, launched in 1962 to study solar ultraviolet and X-ray radiation and the energy spectrum of primary cosmic rays. Today a multitude of orbiting astronomical telescopes cover the electromagnetic spectrum from gamma rays to microwaves.

Observing in the X-ray or gamma ray regions as well as most of the infrared and ultraviolet regions of the electromagnetic spectrum must be done from space. These regions cannot be observed from the ground because the light from astronomical objects cannot penetrate through the atmosphere. Placing observatories in space enables astronomers to study objects that emit light of these wavelengths. Astronomers often find it helpful to combine multiple instruments into a single

Figure 4.17 The Very Large Array (VLA) in New Mexico combines signals from 27 different telescopes so that they act as one "very large" telescope.

Figure 4.18 The Very Large Telescope (VLT) operated by the European Southern Observatory in Chile operates as an optical interferometer.

88 CHAPTER 4 *Light and Telescopes*

Figure 4.19 The Hubble Space Telescope observes in the ultraviolet, visible, and infrared regions of the electromagnetic spectrum.

telescope, for example, optical instrumentation with ultraviolet or infrared, as was done with the Hubble Space Telescope (HST; **Figure 4.19**). Launched in 1990, HST has been collecting ultraviolet, visible, and infrared data for more than two decades. This telescope has been called a "discovery machine" because it has contributed so much useful data to so many subfields of astronomy. For other astronomical satellites, operating at other wavelengths, HST's altitude of 600 km is not nearly high enough.

The Chandra X-ray Observatory, NASA's X-ray telescope, cannot see through even the tiniest traces of atmosphere and therefore orbits more than 16,000 km above Earth's surface. And even this is not distant enough for some telescopes. NASA's Spitzer Space Telescope, an infrared telescope, is so sensitive that it needs to be completely free from Earth's own infrared radiation. The solution was to put it into a solar orbit, which means it trails tens of millions of kilometers behind Earth. Future space telescopes, including the James Webb Space Telescope—NASA's replacement for HST—will orbit at a point that is always farther from the Sun than Earth, along a line between Earth and the Sun.

You can now see the significance of the various types of telescopes that astronomers use. Each region presents different challenges that must be overcome if we are to see the entire universe. Space telescopes are required to observe many parts of the universe that cannot be seen from the ground. Even visible light observations are often improved when made from space.

Resolution and the Atmosphere

Figure 4.20 shows what happens as light waves pass through the aperture of a telescope; they spread out from the edges of the lens or mirror. The distortion that occurs as light passes the edge of an opaque object is called **diffraction**. Diffraction "diverts" some of the light from its path, slightly blurring the image made by the telescope. The degree of blurring depends on the wavelength of the light and the telescope's aperture. The larger the aperture, the smaller the problem posed by diffraction. The best resolution that a given telescope can achieve is known as the **diffraction limit** (see **Working It Out 4.2**). Larger telescopes have better resolution and can distinguish objects that appear closer together. Theo-

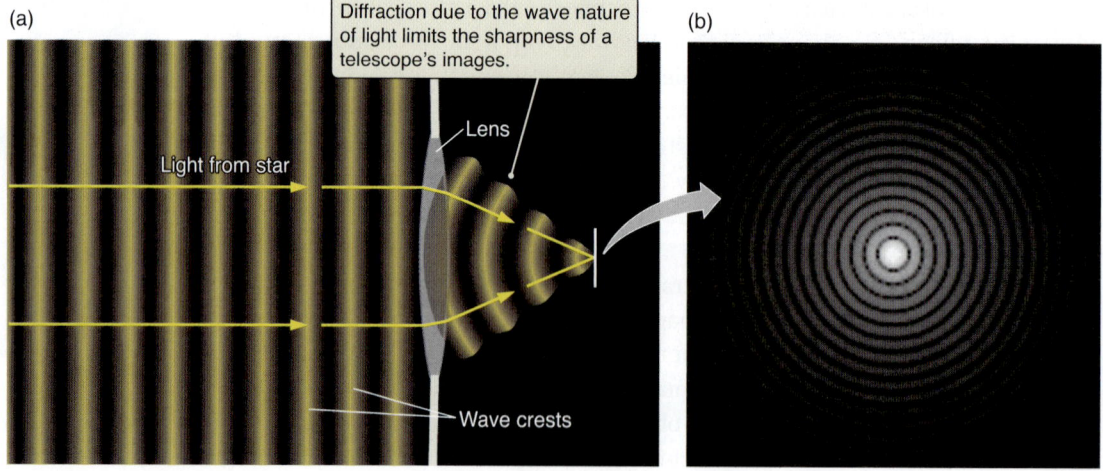

Figure 4.20 (a) Light waves from a star are diffracted by the edges of a telescope's lens or mirror. (b) This diffraction causes the stellar image to be blurred, limiting a telescope's ability to resolve objects.

Working It Out 4.2 | Diffraction Limit

The ultimate limit on the angular resolution, θ, of a lens—the diffraction limit—is determined by the ratio of the wavelength of light, λ, to the aperture, D:

$$\theta = 2.06 \times 10^5 \frac{\lambda}{D} \text{ arcseconds}$$

The constant, 2.06×10^5, changes the units to arcseconds. An arcsecond is a tiny angular measure found by first dividing a degree by 60 to get arcminutes, and then by 60 again to get arcseconds. To get an idea of the size of an arcsecond, imagine that you hand a tennis ball to your friend and ask her to run 8 miles (almost 13 km) away from you down a straight road and then hold up the tennis ball. The angle you perceive from one side of the tennis ball to the other is approximately 1 arcsecond.

Both λ and D must be expressed in the same units (usually meters). The smaller the ratio of λ/D, the better the resolution. The size of the human pupil can change from about 2 mm in bright light to 8 mm in the dark. A typical pupil size is about 4 mm, or 0.004 meter. Visible (green) light has a wavelength of 550 nm, or 5.5×10^{-7} meters. Using these values for the aperture and the wavelength gives

$$\theta = 2.06 \times 10^5 \left(\frac{5.5 \times 10^{-7} \text{ m}}{0.004 \text{ m}} \right) \text{ arcseconds}$$

$$= 28.3 \text{ arcseconds}$$

or about 0.5 arcminute. But, as we learned earlier, the best resolution that the human eye can achieve is about 1 arcminute, and 2 arcminutes is actually more typical. We do not achieve the theoretical resolution with our eyes because the optical properties of the lens and the physical properties of the retina are not perfect.

How does this compare to the resolution of a telescope? Consider the Hubble Space Telescope, operating in the visible part of the spectrum. Its primary mirror has a diameter of 2.4 meters. If we substitute this value for D and again assume visible light, we have

$$\theta = 2.06 \times 10^5 \left(\frac{5.5 \times 10^{-7} \text{ m}}{2.4 \text{ m}} \right) \text{ arcseconds}$$

$$= 0.047 \text{ arcseconds}$$

This is about 600 times better than the theoretical resolving power of the human eye.

---

retically, the 10-meter Keck telescopes located on a mountaintop in Hawaii have a diffraction-limited resolution of 0.0113 arcsecond in visible light, which would allow you to read newspaper headlines 60 km away.

For telescopes with apertures larger than about a meter, Earth's atmosphere stands in the way of better resolution. If you have ever looked out across a road on a summer day, you have seen the air shimmer as light is bent this way and that by turbulent bubbles of warm air rising off the hot pavement.

The problem of the shimmering atmosphere is less pronounced when we look overhead, but the twinkling of stars in the night sky is caused by the same phenomenon. As telescopes magnify the angular diameter of an object, they also magnify the effects of the atmosphere. The limit on the resolution of a telescope on the surface of Earth caused by this atmospheric distortion is called astronomical **seeing**. One advantage of launching telescopes such as the Hubble Space Telescope into orbit around Earth is that they are not hindered by astronomical seeing.

Modern technology has improved ground-based telescopes with computer-controlled **adaptive optics**, which compensate for much of the atmosphere's distortion. To understand better how adaptive optics work, we need to look more closely at how Earth's atmosphere smears out an otherwise perfect stellar image. Light from a distant star arrives at the top of Earth's atmosphere with flat, parallel wave crests (Figure 4.20). If Earth's atmosphere were perfectly uniform, the crests would remain flat as they reached the objective lens or primary mirror of a ground-based telescope. After making its way through the telescope's optical system, the crests would produce a tiny diffraction disk in the focal plane, as shown in Figure 4.20b. But Earth's atmosphere is not uniform. It is filled with small bubbles of air that have slightly different temperatures than their surroundings. Different temperatures mean different densities, and different densities mean each bubble bends light differently.

The small air bubbles act as weak lenses, and by the time the waves reach the telescope they are far from flat, as shown in **Figure 4.21**. Instead of a tiny diffraction disk, the image in the telescope's focal plane is distorted and swollen. Adaptive optics flattens out this distortion. First, an optical device within the telescope measures the wave crests. Then, before reaching the telescope's focal plane, the light is reflected off yet another mirror, which has a flexible surface. A computer analyzes the light and bends the flexible mirror so that it accurately corrects for the distortion caused by the air bubbles. Examples of images corrected by adaptive optics are shown in **Figure 4.22**. The widespread use of adaptive optics has made the image quality of ground-based telescopes competitive with that of the Hubble Space Telescope.

Image distortion is not the only problem caused by Earth's atmosphere. Nearly all of the X-ray, ultraviolet, and infrared light arriving at Earth fails to reach the ground because it is partially or completely absorbed by ozone, water vapor, carbon dioxide, and other atmospheric molecules.

Light is the primary messenger that astronomers have for exploring the universe, and the vast majority of what is known about the universe beyond Earth has come from studying its properties. Throughout the rest of this book, we will return again and again to the study of light and its detailed properties.

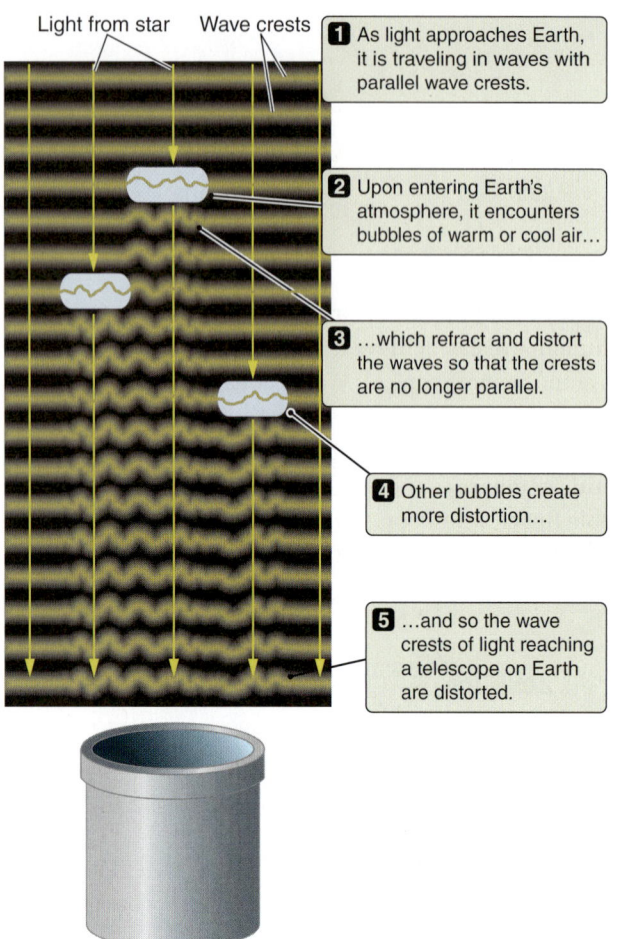

Figure 4.21 Bubbles of warmer or cooler air in Earth's atmosphere distort the wavefront of light from a distant object.

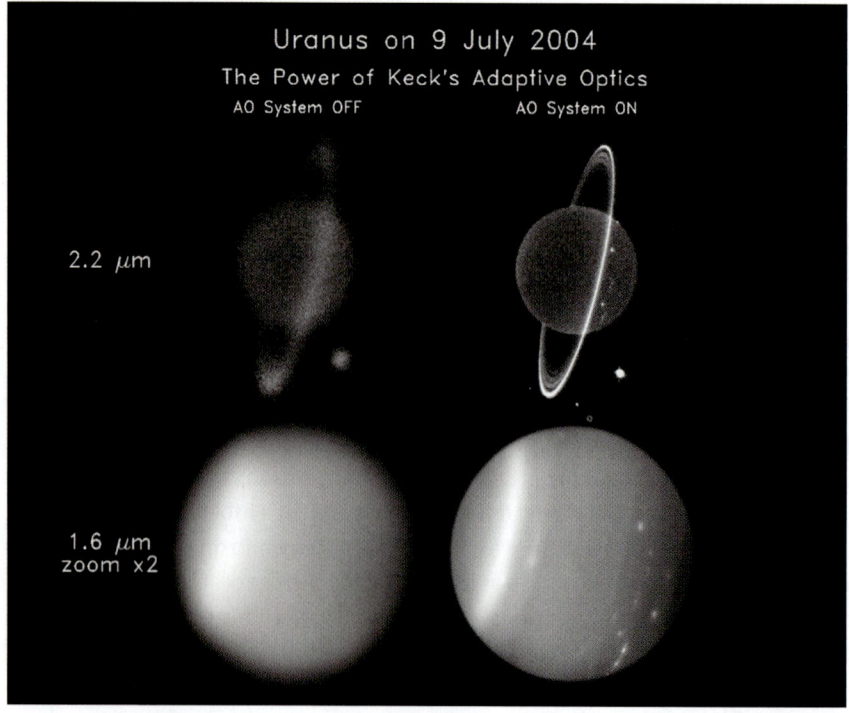

Figure 4.22 These near-infrared images of Uranus show the planet as seen without adaptive optics (left) and with the technology turned on (right).

READING ASTRONOMY News

The James Webb Space Telescope will replace the aging Hubble Space Telescope. This telescope will be much larger and much farther away, and it will be optimized in the infrared. Scientists are busy constructing both the telescope and the cameras that will be attached to it. This article describes progress on part of this telescope.

ATK Building World's Largest Space Telescope in Magna

By **NOAH BOND, ABC4.com**

MAGNA, Utah (ABC 4 News)—ATK is busy constructing the world's largest space telescope in Magna. The James Webb Space Telescope will be at least 100 times more powerful than the Hubble Space Telescope and open our view to never before seen planets and galaxies.

ATK engineers and technicians are constructing essentially the eye of the James Webb Space Telescope. It is the heart of the more than $8 billion NASA project.

Never in the history of the world has mankind come so close to seeing so far away. The James Webb Space Telescope under construction in Magna will dwarf the accomplishments of the Hubble Telescope.

Hubble opened our eyes to deep space galaxies. Scientists hope the James Webb will reveal the very edges of our Universe.

"Look back to the time of the Big Bang 13.8 billion years ago and better understand the formation of galaxies, stars, and planets," said the James Webb Space Telescope Program Manager, Bob Hellekson.

The Hubble is the size of a school bus with an 8 foot diameter mirror. The James Webb telescope is the size of a tennis court with a 21.3 foot diameter mirror.

The Hubble is 364 miles from Earth's surface, but the James Webb will remain a million miles above Earth, which is far above the outer reaches of our atmosphere.

"The telescope will be such a game change in the world of astronomy that it's going to rewrite textbooks over and above what the Hubble Space Telescope has already provided," said Hellekson.

For now, the future of our world's understanding of deep space sits in a room in Magna.

The eye of the James Webb Telescope will remain in Magna for about three more months before NASA ships it off to Huntsville, Alabama, for more testing.

NASA will launch it into space in 2018 from French New Guinea, which is on the northern tip of South America and close to the earth equator. The location will allow the telescope to more efficiently sling shot to its destination in Earth's shadow beyond our Moon.

Evaluating the News

1. It is common in news articles for journalists to use imprecise language. In the first paragraph, the James Webb Space Telescope (JWST) is described as being 100 times more "powerful" than the Hubble Space Telescope. Given what you've learned in this chapter, what could the author mean by "powerful"? What words might the author have used to be more precise?
2. In paragraph two, the author says the engineers are constructing the "eye" of the JWST. To what part of the telescope might he be referring? How could the author have been more precise? Why did he choose to call this the "eye"?
3. Consider the overall tone of the article. Would you describe it as objective? Why or why not?
4. Compare the diameters of James Webb Space Telescope and the Hubble Space Telescope. Are these primary mirror diameters or secondary mirror diameters? What is the advantage of the larger diameter of JWST?
5. What is the advantage in placing JWST so far from Earth? Are there any disadvantages to this placement of the telescope?

SUMMARY

Light is both a particle and a wave. Whether it is observed to have particle or wave properties depends on the type of observations being made. Visible light is only a tiny portion of the entire electromagnetic spectrum, and studying all the regions of the spectrum provides a range of information about the object being studied. Modern CCD cameras have improved quantum efficiency and longer integration times, which allow astronomers to study fainter and more distant objects than were visible with prior detectors. Telescopes are matched to the wavelengths of observation, with different technologies required for each region of the spectrum. The aperture of a telescope both determines its light-gathering power and limits its resolution; larger telescopes are better in both measures.

LG 1 Waves carry energy through space, and light waves do not require a medium. The speed of light in a vacuum is a fundamental constant of nature and is always the same, but when light enters a medium it slows. Photons are particles of light. Light sometimes behaves like a wave (for example, when it refracts) and sometimes behaves like a particle (for example, when it is absorbed by an atom).

LG 2 The electromagnetic spectrum spans a very large range of wavelengths. Short-wavelength light, like gamma rays, have very high energy. Long-wavelength light, like radio waves, is low in energy. Light carries information about the temperature and composition of objects.

LG 3 The original light detector was the eye. Through time, astronomers developed better instruments for collecting light from space, moving through film photography to digital photography using CCD cameras. Spectrographs are specialized instruments that take the spectrum of an object to reveal what the object is made of, along with many other physical properties. Modern detectors are both more responsive to individual photons and can integrate for a longer time than the human eye. In addition, detectors of various types can collect light in regions of the spectrum that cannot be seen by the eye.

LG 4 Telescopes of larger aperture focus more light on the detector. They therefore collect more light from very faint objects. Also, because the diffraction of the telescope is inversely proportional to the diameter, larger aperture telescopes have better resolution: objects can appear closer together and still be distinguished on the detector. Telescopes with long focal lengths produce larger images than telescopes with small focal lengths.

SUMMARY SELF-TEST

1. A light wave does *not* require
   a. a medium.
   b. a speed.
   c. a frequency.
   d. a wavelength.

2. An extremely hot object emits most of its light
   a. at very low energies.
   b. in the radio.
   c. in the visible.
   d. at very high energies.

3. CCD cameras are better astronomical detectors than the human eye because (choose all that apply)
   a. their quantum efficiency is higher.
   b. the integration time can be longer.
   c. they can observe at wavelengths beyond the visible.
   d. they turn photons into protons.

4. All large astronomical telescopes are reflectors because (choose all that apply)
   a. chromatic aberration is minimized.
   b. they are not as heavy.
   c. they can be shorter.
   d. they are not subject to diffraction.

5. Light acts like (choose all that apply)
   a. a wave, always.
   b. a particle, always.
   c. both a wave and a particle.
   d. neither a wave nor a particle.

6. When light enters a medium from space
   a. it slows down.
   b. it speeds up.
   c. it travels at the same speed.
   d. it changes frequency.

7. The amplitude of a light wave is related to
   a. its color.
   b. its speed.
   c. its frequency.
   d. its intensity.

8. Which of the following can be observed from the ground? (Choose all that apply.)
   a. radio waves
   b. gamma radiation
   c. ultraviolet light
   d. X-ray light
   e. visible light

9. Rank the following in order of decreasing wavelength.
   a. gamma rays
   b. visible light
   c. infrared radiation
   d. ultraviolet light
   e. radio waves

10. Match the following properties of telescopes (lettered) with their corresponding definitions (numbered).
    a. aperture
    b. resolution
    c. focal length
    d. chromatic aberration
    e. diffraction
    f. interferometer
    g. adaptive optics
    (1) several telescopes connected to act as one
    (2) distance from lens to focal plane
    (3) diameter
    (4) ability to distinguish objects that appear close together in the sky
    (5) computer-controlled active focusing
    (6) rainbow-making effect
    (7) smearing effect due to sharp edge

QUESTIONS AND PROBLEMS

Multiple Choice and True/False

11. **T/F:** Energy can travel faster than 3×10^8 m/s.
12. **T/F:** The frequency of a wave is related to the energy of the photon.
13. **T/F:** Blue light has a longer wavelength than red light.
14. **T/F:** Blue light has more energy than red light.
15. **T/F:** Visible light is an electromagnetic wave.
16. Which of the following is not a property of waves?
    a. speed
    b. wavelength
    c. frequency
    d. mass
17. Light at the lowest-energy end of the electromagnetic spectrum is in the _____ region.
    a. visible
    b. gamma ray
    c. ultraviolet
    d. radio
18. The advantage of an interferometer is that
    a. the resolution is dramatically increased.
    b. the focal length is dramatically increased.
    c. the light-gathering power is dramatically increased.
    d. diffraction effects are dramatically decreased.
    e. chromatic aberration is dramatically decreased.
19. Suppose that a telescope has a resolution of 1.5 arcseconds at a wavelength of 300 nm. What is its resolution at 600 nm?
    a. 3 arcseconds
    b. 1.5 arcseconds
    c. 0.75 arcseconds
    d. In order to know this, you need to know the diameter of the aperture. Not enough information is given.
20. If the wavelength of a beam of light were halved, how would that affect the frequency?
    a. The frequency would be four times larger.
    b. The frequency would be two times larger.
    c. The frequency would not change.
    d. The frequency would be two times smaller.
    e. The frequency would be four times smaller.
21. How does the speed of light in a medium compare to the speed in a vacuum?
    a. It's the same, as the speed of light is a constant.
    b. The speed in the medium is always faster than the speed in a vacuum.
    c. The speed in the medium is always slower than the speed in a vacuum.
    d. The speed in the medium may be faster or slower, depending on the medium.
22. Astronomers put telescopes in space to
    a. get closer to the stars.
    b. avoid atmospheric effects.
    c. look primarily at radio wavelengths.
    d. improve quantum efficiency.
23. The angular resolution of a ground-based telescope is usually determined by
    a. diffraction.
    b. refraction.
    c. the focal length.
    d. atmospheric seeing.
24. Gamma ray telescopes are placed in space because
    a. gamma rays are too fast to be detected by stationary telescopes.
    b. gamma rays do not penetrate the atmosphere.
    c. the atmosphere produces too many gamma rays and overwhelms the signal from space.
    d. gamma rays are too dangerous to collect in large numbers.
25. The mirror of the James Webb Space Telescope will have a diameter of about 21 feet. This is approximately how many meters?
    a. 63 meters
    b. 10 meters
    c. 7 meters
    d. 0.21 meters

Conceptual Questions

26. We know that the speed of light in a vacuum is 3×10^8 m/s. Is it possible for light to travel at a lower speed? Explain your answer.
27. Is a light-year a measure of time or distance, or both?
28. If photons of blue light have more energy than photons of red light, how can a beam of red light carry as much energy as a beam of blue light?
29. Galileo's telescope used simple lenses rather than compound lenses. What is the primary disadvantage of using a simple lens in a refracting telescope?
30. The largest astronomical refractor has an aperture of 1 meter. Why is it impractical to build a larger refractor with, say, twice the aperture?
31. Name and explain at least two advantages that reflecting telescopes have over refractors.
32. Your camera may have a zoom lens, ranging between wide angle (short focal length) and telephoto (long focal length). How would the size of an image in the camera's focal plane differ between wide angle and telephoto?
33. What causes refraction?
34. How do manufacturers of quality refracting telescopes and cameras correct for the problem of chromatic aberration?
35. Consider two optically perfect telescopes having different diameters but the same focal length. Is the image of a star larger or smaller in the focal plane of the larger telescope? Explain your answer.
36. Name two ways in which Earth's atmosphere interferes with astronomical observations.

37. Explain *why* stars twinkle.
38. Explain integration time and how it contributes to the detection of faint astronomical objects.
39. Explain quantum efficiency and how it contributes to the detection of faint astronomical objects.
40. Some people believe that we put astronomical telescopes in orbit because doing so gets them closer to the objects they are observing. Explain what is wrong with this common misconception.

Problems

41. The index of refraction, n, of a diamond is 2.4. What is the speed of light within a diamond?
42. You are tuned to 790 on AM radio. This station is broadcasting at a frequency of 790 kHz (7.90×10^5 Hz). What is the wavelength of the radio signal? You switch to 98.3 on FM radio. This station is broadcasting at a frequency of 98.3 MHz (9.83×10^7 Hz). What is the wavelength of this radio signal?
43. Many amateur astronomers start out with a 4-inch (aperture) telescope and then graduate to a 16-inch telescope. By what factor does the light-gathering power (which is proportional to the area) of the telescope increase?
44. Compare the light-gathering power of a large astronomical telescope (aperture 10 meters) with that of the dark-adapted human eye (aperture 8 mm).
45. Assume a telescope has an aperture of 1 meter. Calculate the telescope's resolution when observing in the near-infrared region of the spectrum ($\lambda = 1{,}000$ nm). Calculate the resolution in the violet region of the spectrum ($\lambda = 400$ nm). In which region does the telescope have better resolution?
46. The resolution of the human eye is about 1.5 arcminutes. What would the aperture of a radio telescope (observing at 21 cm) have to be to have this resolution? Even though the atmosphere is transparent at radio wavelengths, we do not see in the radio region of the electromagnetic spectrum. Using your calculations, explain why humans do not see in the radio region.
47. One of the earliest astronomical CCDs had 160,000 pixels, each recording 8 bits (256 levels of brightness). A new generation of astronomical CCDs may contain a billion pixels, each recording 15 bits (32,768 levels of brightness). Compare the number of bits of data that each CCD type produces in a single image.
48. The VLBA uses an array of radio telescopes ranging across 8,000 km of Earth's surface from the Virgin Islands to Hawaii.
    a. Calculate the angular resolution of the array when radio astronomers are observing interstellar water molecules at a microwave wavelength of 1.35 cm.
    b. How does this resolution compare with the angular resolution of two large optical telescopes separated by 100 meters and operating as an interferometer at a visible wavelength of 550 nm?
49. When operational, the SVLBI may have a baseline of 100,000 km. What will its angular resolution be when we are studying an interstellar molecule emitting at a wavelength of 17 mm from a distant galaxy?
50. The *Mars Reconnaissance Orbiter* (MRO) flies at an average altitude of 280 km above the martian surface. If its cameras have an angular resolution of 0.2 arcsecond, what is the size of the smallest objects that the MRO can detect on the martian surface?

SMARTWORK

Norton's online homework system includes algorithmically generated versions of these questions, plus additional conceptual exercises. If your instructor assigns questions in SmartWork, log in at smartwork.wwnorton.com.

Exploration | Light as a Wave

wwnpag.es/uou2

Visit the Student Site (wwnpag.es/uou2) and open the Light as a Wave, Light as a Photon AstroTour in Chapter 4. Watch the first section, and click through using the "Play" button until you reach section 2 of 3.

Here we will explore the following questions: How many properties does a wave have? Are any of these related to each other?

Work your way through to the experimental section, where you can adjust the properties of the wave. Watch the simulation for a moment to see how fast the frequency counter increases.

1. Increase the wavelength using the arrow key. What happens to the rate of the frequency counter?

2. Reset the simulation and then decrease the wavelength. What happens to the rate of the frequency counter?

3. How are the wavelength and frequency related to each other?

4. Imagine that you increase the frequency instead of the wavelength. How should the wavelength change when you increase the frequency?

5. Reset the simulation and increase the frequency. Did the wavelength change in the way you expected?

6. Reset the simulation and increase the amplitude. What happens to the wavelength and the frequency counter?

7. Decrease the amplitude. What happens to the wavelength and the frequency counter?

8. Is the amplitude related to the wavelength or frequency?

9. Why can't you change the speed of this wave?

5 The Formation of Stars and Planets

Stars and planets form from clouds of cool dust and gas. Hot gas fills the space between these clouds, pressing on them and helping to keep them together. In addition, each of the atoms and molecules in a cloud is gravitationally attracted to every other particle. This gravitational pull will cause some clouds to collapse. During the collapse, the clouds may fragment to form multiple stars and then further evolve to form planets. Only a few physical principles are required to describe the formation of stars and planets from dust and gas. The illustration on the opposite page shows a montage of photographs of various steps in the formation of stars and planets that a student is comparing to a sketch she is making of the structure of a proto–solar system.

In the past few decades, stellar astronomers and planetary scientists, working from different perspectives, have come to a similar understanding of our early Solar System. However, recent discoveries of exoplanets challenge this basic picture, furthering the development of a detailed theory of star and planet formation.

✧ LEARNING GOALS

The planets of our Solar System are by-products of the birth of the Sun. The discovery of planetary systems surrounding other stars has shown that our Solar System is not unique. By the end of this chapter, you should be able to look at an image, as the student is doing at right, and determine where it should be placed in the sequence of events that occur during star formation. You should also be able to:

LG 1 Understand the role that gravity and angular momentum play in the formation of stars and planets.

LG 2 Diagram the process by which dust grains in the disk around a young star stick together to form larger and larger solid objects.

LG 3 Know why planets orbit the Sun in a plane and why they revolve in the same direction that the Sun rotates.

LG 4 Explain how temperature at different locations in the disk affects the composition of planets, moons, and other bodies.

LG 5 Understand how astronomers find planets around other stars, and explain how we know that planetary systems around other stars are common.

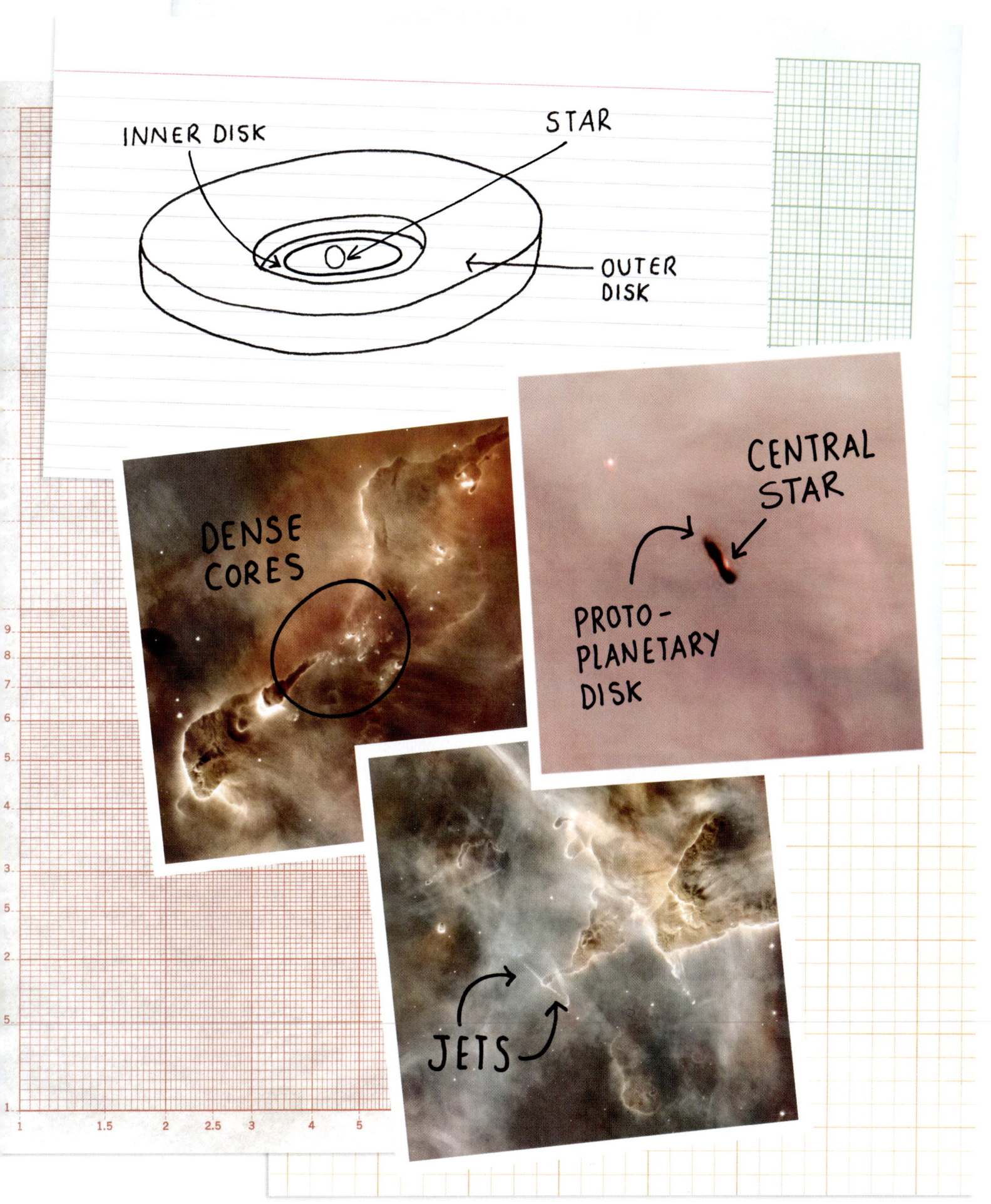

Vocabulary Alert

pressure In common language, we often use *pressure* interchangeably with the word *force*. Astronomers specifically use pressure to mean the force per unit area that atoms or molecules exert as they speed around and collide with each other and their surroundings.

dense In common language, we use this word in many ways, some of which are metaphorical and unkind, as in "You can't understand this? You are so dense!" Astronomers specifically use *density* to mean "the amount of mass packed into a volume"; denser material contains more mass in the same amount of space. In practical terms, you are familiar with density by how heavy an object feels for its size: a pool ball and a tennis ball are roughly the same size, but the pool ball has greater mass and therefore feels heavier because it is denser.

▶‖ **AstroTour:** Star Formation

5.1 Molecular Clouds Are the Cradles of Star Formation

A **star** is a dense cloud of gas that produces energy in its core by fusing light atoms into heavier ones. This energy production causes the outer parts of the star to shine because these parts have been heated to thousands of kelvins. Scientists measure the temperature of stars and other objects on the **Kelvin temperature scale**, which has increments of the same size as those of the more familiar Celsius scale but has a zero point at −273°C (that is, 0 K = −273°C, or *absolute zero*). Water freezes at 273 K and boils at 373 K. Stars are often accompanied by **planets**: large, round bodies that orbit the star in individual orbits. In general, a system of planets and other smaller objects surrounding a star is a **planetary system**, and there are many planetary systems in the Milky Way Galaxy. Our **Solar System** is the planetary system that surrounds the Sun. Stars and their associated planets share a common origin in a cloud of dust and gas. We begin our study of star formation by investigating the places where stars form.

Interstellar Clouds

Stars and planets form from large clouds of dust and gas. These clouds are held together partly by their own self-gravity and partly by the **pressure** of hot gas that occupies the space between the clouds. If parts of the cloud have a high enough **density**, these clouds will fragment and collapse to form stars and planets. The dust and gas in these clouds have usually been through several cycles of star formation and stellar death, and so the dust and gas have many different types of atoms that were formed within earlier generations of stars.

As shown in **Figure 5.1**, an **interstellar cloud**—the cloud of cool dust and gas in the space between stars—has *self-gravity*. **Self-gravity** is a gravitational attraction among all parts of the same object. In this case, each part of the cloud is gravitationally attracted to every other part of the cloud. The sum of all these forces (the *net* force) on each particle points toward the center of the cloud. In a stable, unchanging object like a planet or a star, the self-gravity is balanced in two ways. It can be balanced by structural strength; for example, the rocks that make up Earth. Self-gravity can also be balanced by the outward force resulting from gas pressure, such as pressure from hot gases within a star. When the forces on a cloud are balanced, the cloud is in **hydrostatic equilibrium**. However, the forces are not always in balance. If the outward force is weaker than self-gravity, the object is unstable and contracts. If the outward force is stronger than self-gravity, the object is unstable and expands. In most interstellar clouds, the internal gas pressure pushing out is much stronger than self-gravity, so the cloud should expand. But the much hotter gas surrounding the clouds exerts a pressure inward on the cloud that helps to hold the clouds together.

The densest, coolest interstellar clouds are called **molecular clouds** because they are primarily composed of hydrogen molecules, although they also contain dust and other gases. Some molecular clouds are massive enough, dense enough, and cool enough that their self-gravity overwhelms their internal pressure, and they collapse under their own weight. Other molecular clouds are pushed into a collapse by the explosion of nearby stars or by gravitational interactions with passing stars. We might expect that the collapse of a molecular cloud should

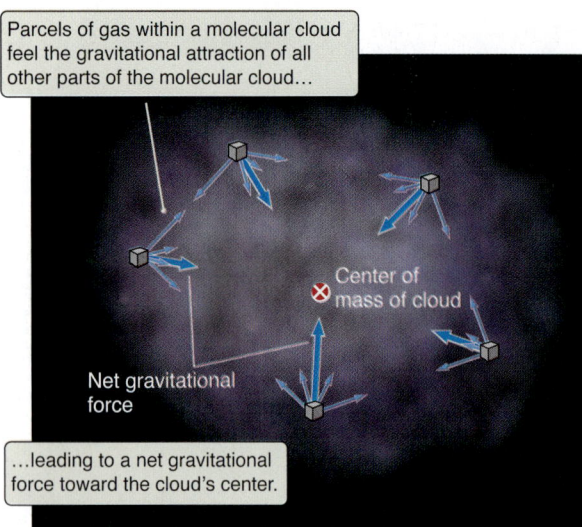

Figure 5.1 Self-gravity causes a molecular cloud to collapse, drawing parcels of gas toward a single point inside the cloud.

happen quickly. In practice, the process goes very slowly because several other effects slow the collapse. However, these effects are temporary, and gravity is weak but relentless. As the forces that oppose the cloud's self-gravity gradually fade away, the cloud slowly collapses.

Molecular-Cloud Fragmentation

Molecular clouds are never uniform. Some regions are denser and collapse more rapidly than surrounding regions. As these regions collapse, their self-gravity becomes stronger, so they collapse even faster. **Figure 5.2** shows the process of collapse in a molecular cloud. Because of slight variations in the density of the cloud, some regions of the cloud become very dense concentrations of gas. The result is that instead of collapsing into a single object, the cloud fragments into a number of very dense **molecular-cloud cores**. A single molecular cloud may form hundreds or thousands of molecular-cloud cores, each of which is typically a few light-months in size. Some of these cores will eventually form stars.

As a molecular-cloud core collapses, the gravitational forces grow stronger still, because the force of gravity is inversely proportional to the square of the radius. Suppose a core starts out being 4 light-years across. By the time the core has collapsed to 2 light-years across, the different parts of the cloud are, on average, only half as far apart as when the collapse started. As a result, the gravitational attraction they feel toward each other will be four times stronger. When the cloud is one-fourth as large as when the collapse began, the force of gravity will be 16 times as strong. As a core collapses, the inward force of gravity increases; as gravity increases, the collapse speeds up; as the collapse speeds up, the gravitational force increases even faster. Eventually, gravity is able to overwhelm all the opposing forces. This happens first near the center of the cloud core because that's where the cloud material is most strongly concentrated. The inner parts of the cloud core start to fall rapidly inward, "pulling the bottom out" from the more distant parts of the cloud. Without the support

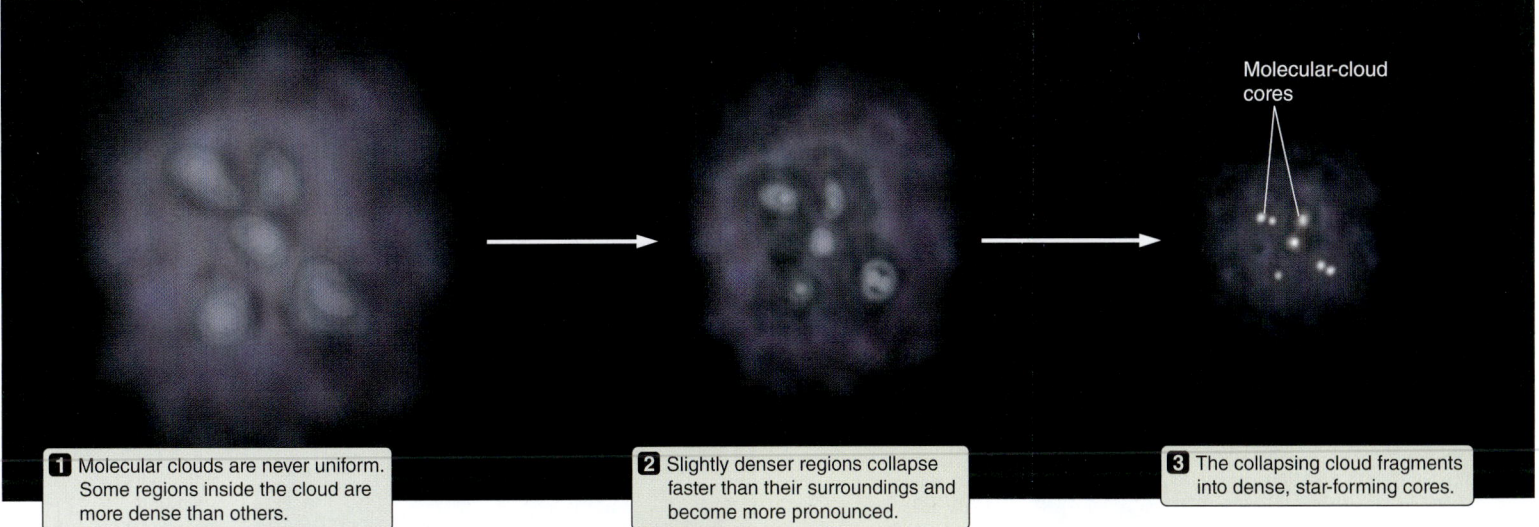

1 Molecular clouds are never uniform. Some regions inside the cloud are more dense than others.

2 Slightly denser regions collapse faster than their surroundings and become more pronounced.

3 The collapsing cloud fragments into dense, star-forming cores.

Figure 5.2 As a molecular cloud collapses, denser regions within the cloud collapse more rapidly than less dense regions. As this process continues, the cloud fragments into a number of very dense molecular-cloud cores that are embedded within the large cloud. These cloud cores may go on to form stars.

Vocabulary Alert

luminous In common language, *luminous* and *bright* are used interchangeably. Astronomers, however, observe objects at great distances. They use the word *bright* to describe how an object appears in our sky. They use the word *luminous* to describe how much light the object emits in all parts of the spectrum. A bright star might be luminous or it might just be very close to Earth. A faint star might be extremely luminous but very, very far away.

▶▶ **Nebraska Simulation: Blackbody Curves**

of that inner material, the more distant material begins to fall freely toward the center as well. The process continues: the cloud core collapses from the inside out, as shown in **Figure 5.3**. The whole structure comes crashing down.

5.2 The Protostar Becomes a Star

Once a molecular-cloud core begins to collapse, several things happen at once. The innermost core eventually becomes a star, while the outer parts may form planets. We will first follow what happens to that innermost core and then go back to find out what becomes of the rest of the dust and gas.

Stars and Protostars

The innermost part of a collapsing molecular-cloud core is called a **protostar**. As the cloud collapses, gravitational energy is converted to thermal energy, and the surface of the protostar is heated to a temperature of thousands of degrees, causing the protostar to shine. Particles are pulled toward the center by gravity. As they fall, they move faster and faster. As they become more densely packed, they begin to crash into each other, causing random motions and raising the temperature of the core. These random motions of particles are collectively known as the "thermal energy." When the particles are hotter, they are moving faster, and the thermal energy is higher. In the collapsing protostar, the thermal energy comes from the gravitational energy that was stored in the cloud when the particles were far apart. The collapse of the cloud converts gravitational energy to thermal energy.

Because it is hot, the surface of the protostar radiates away energy. The hotter it gets, the more energy it radiates, and the bluer that radiation becomes (see **Working It Out 5.1**). The surface of a protostar is tens of thousands of times larger than the surface of the Sun, and each square meter of that enormous surface radiates away energy. As a result, the protostar is thousands of times more **luminous** than our Sun.

Figure 5.3 When a molecular-cloud core gets very dense, it collapses from the inside out.

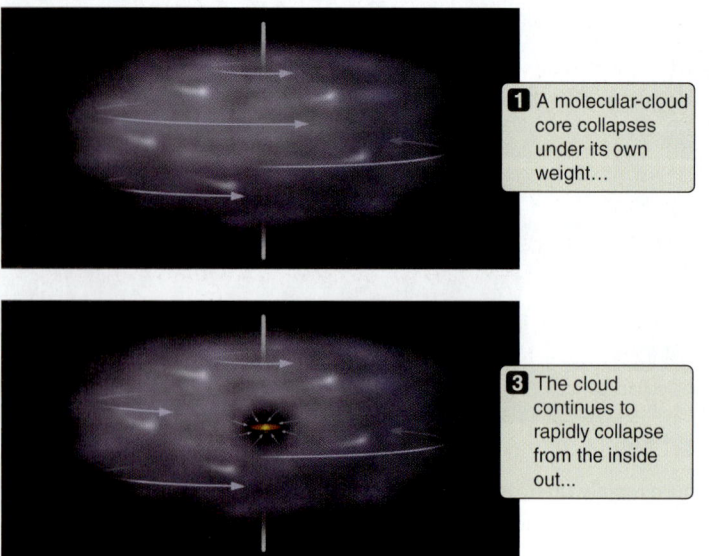

1 A molecular-cloud core collapses under its own weight...

3 The cloud continues to rapidly collapse from the inside out...

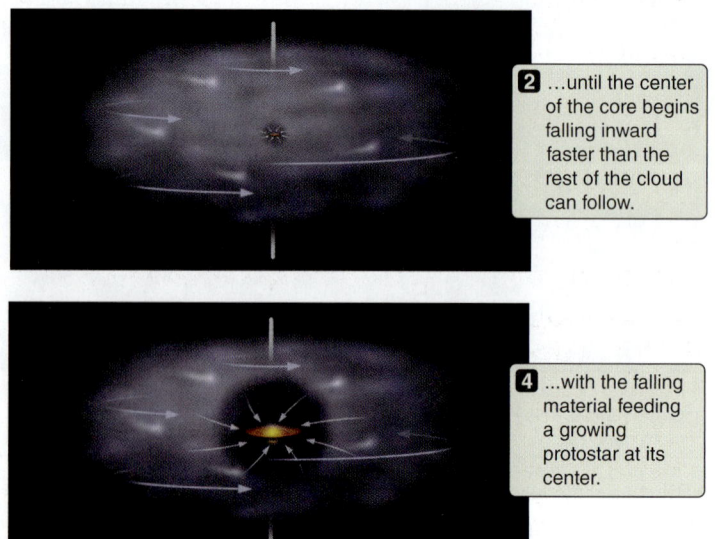

2 ...until the center of the core begins falling inward faster than the rest of the cloud can follow.

4 ...with the falling material feeding a growing protostar at its center.

Working It Out 5.1 | The Stefan-Boltzmann Law and Wien's Law

Figure 5.4 shows the spectra of a **blackbody**: a source that absorbs and emits all the electromagnetic energy it receives. If we graph the intensity (energy per unit area per second) of a blackbody's emitted radiation across all wavelengths, as in Figure 5.4, we obtain a characteristic curve called a **blackbody spectrum**. As the object's temperature increases, it emits more radiation at every wavelength, so the entire curve is higher. The **luminosity**, L, of the object is proportional to the *fourth power* of the temperature, T

$$L \propto T^4$$

This relationship between temperature and luminosity is known as the **Stefan-Boltzmann law** because it was discovered in the laboratory by physicist Josef Stefan (1835–1893) and derived by his student, Ludwig Boltzmann (1844–1906).

The amount of energy radiated by each square meter of the surface of an object each second is called the **flux**, $\mathcal{F}$. We can relate the flux to the temperature by using the **Stefan-Boltzmann constant**, σ (the Greek letter *sigma*). The value of σ is 5.67×10^{-8} W/(m² K⁴), where W stands for watts, a unit of power equal to 1 joule per second. Expressing all this in math, we find:

$$\mathcal{F} = \sigma T^4$$

Even modest changes in temperature can result in large changes in the amount of power radiated by an object. If the temperature triples, then the flux goes up by a factor of 3^4, or 81.

Suppose we want to find the flux and luminosity of Earth. Earth's average temperature is 288 K, so the flux from its surface is

$$\mathcal{F} = \sigma T^4$$
$$\mathcal{F} = [5.67 \times 10^{-8} \text{ W/(m}^2\text{K}^4)] \, (288 \text{ K})^4$$
$$\mathcal{F} = 390 \text{ W/m}^2$$

To find the total energy radiated every second, we must multiply by the surface area (A) of Earth, given by $4\pi R^2$. The radius of Earth is 6,378,000 meters, or 6.378×10^6 meters. So the luminosity is

$$L = \mathcal{F} \times A$$
$$L = \mathcal{F} \times 4\pi R^2$$
$$L = (390 \text{ W/m}^2) \, [4\pi \, (6.378 \times 10^6 \text{ m})^2]$$
$$L \approx 2 \times 10^{17} \text{ W}$$

Earth emits the equivalent of the energy used by 2,000,000,000,000,000 hundred-watt light bulbs.

Look again at Figure 5.4. Notice where the peak of each curve lines up along the horizontal axis. As the temperature, T, increases, the *peak* of the spectrum shifts toward shorter wavelengths. This is an inverse proportion. Translating this

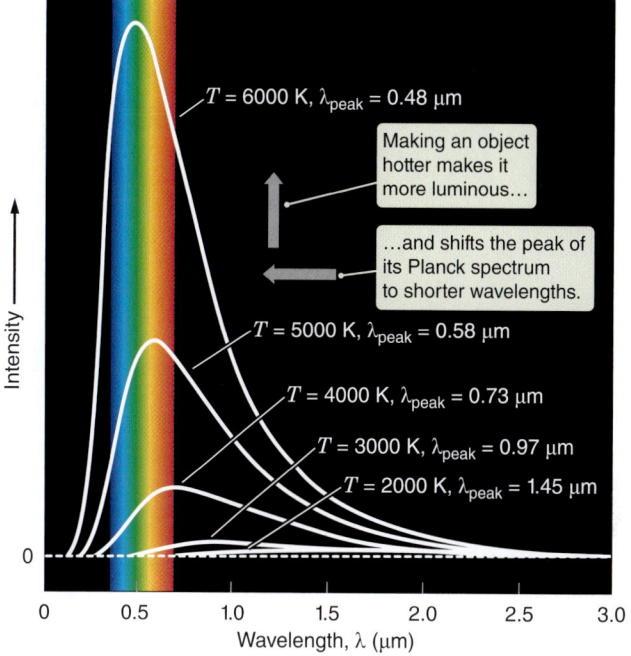

Figure 5.4 This graph shows blackbody spectra emitted by sources with temperatures ranging from 2000 K to 6000 K. At higher temperatures, the peak of the spectrum shifts toward shorter wavelengths, and the amount of energy radiated per second from each square meter of the source increases.

into math, and inserting the constant of 2.9×10^6 nm K (derived from measurements of this relationship) to fix the units, gives **Wien's law**:

$$\lambda_{\text{peak}} = \frac{2.9 \times 10^6 \text{ nm K}}{T}$$

Wien's law, pronounced "Veen's law," is named for physicist Wilhelm Wien (1864–1928), who discovered the relationship. In this equation, λ_{peak} (pronounced "lambda peak") is the wavelength where the electromagnetic radiation from an object is greatest.

If we insert Earth's average temperature of 288 K into Wien's law, we get

$$\lambda_{\text{peak}} = \frac{2.9 \times 10^6 \text{ nm K}}{T}$$

$$\lambda_{\text{peak}} = \frac{2.9 \times 10^6 \text{ nm K}}{288 \text{ K}}$$

$$\lambda_{\text{peak}} = 10{,}100 \text{ nm}$$

or slightly more than 10 μm. Earth's radiation peaks in the infrared region of the spectrum.

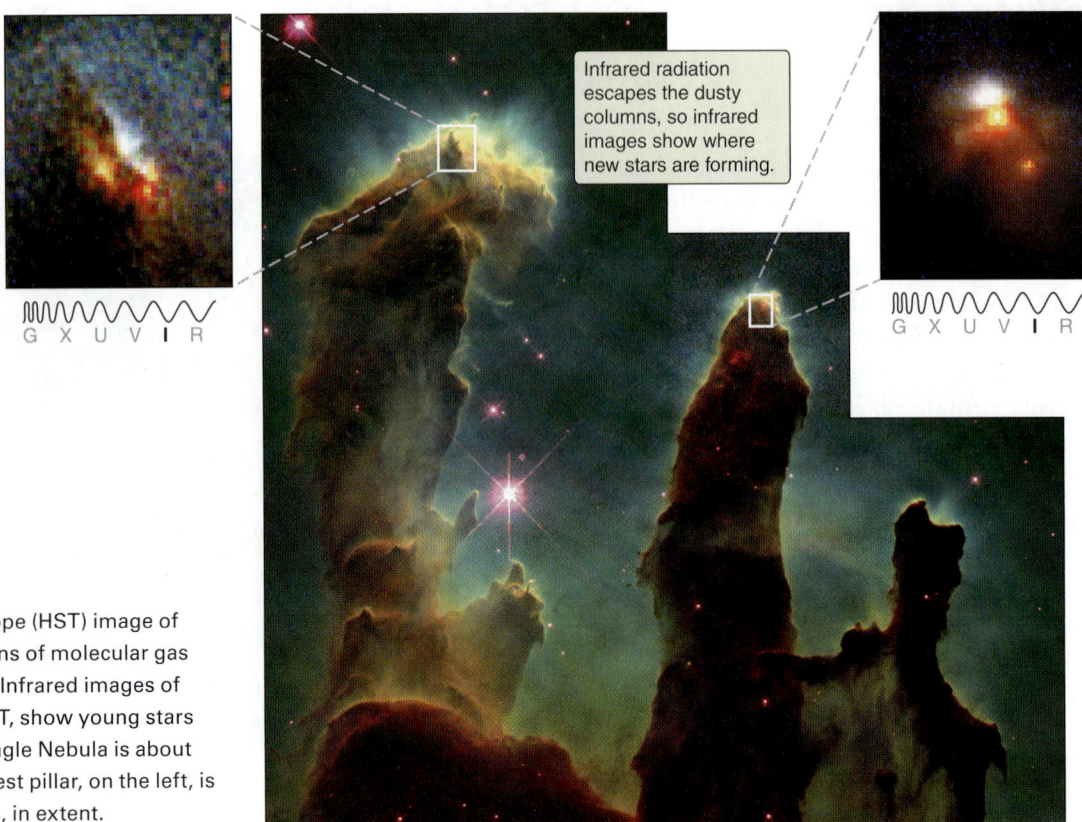

Figure 5.5 This Hubble Space Telescope (HST) image of the Eagle Nebula shows dense columns of molecular gas and dust illuminated by nearby stars. Infrared images of the same field, also taken with the HST, show young stars forming within these columns. The Eagle Nebula is about 7,000 light-years from Earth. The largest pillar, on the left, is about 4 light-years, or 24 trillion miles, in extent.

Although the protostar is emitting a lot of light, astronomers often cannot see it in visible light. There are two reasons for this. First, most of the protostar's radiation is in the infrared rather than the visible part of the spectrum. Second, the protostar is buried deep in the heart of a dense and dusty molecular cloud. Dust absorbs visible light. However, astronomers are able to view these objects in the infrared part of the spectrum because much of the longer-wavelength infrared light from the protostar *is* able to escape through the cloud. In addition, as the dust absorbs the visible light, it warms up, and this heated dust also glows in the infrared.

Sensitive infrared instruments developed in the past 30 years have revolutionized the study of protostars and other young stellar objects. Dark clouds have revealed themselves to be entire clusters of dense cloud cores, young stellar objects, and glowing dust when viewed in the infrared. **Figure 5.5** shows infrared pictures of stars forming within columns of gas and dust in the Eagle Nebula. **Nebula** is the most general term for an interstellar cloud of gas and dust.

A Shifting Balance: The Evolving Protostar

At any given moment, the protostar is in balance: the force from hot gas pushes outward and the force of gravity pulls inward, and these forces exactly oppose each other. However, this balance is constantly changing. How can an object be in perfect balance and yet be changing at the same time? **Figure 5.6a** shows a simple spring scale. If an object is placed on it, the spring compresses until the

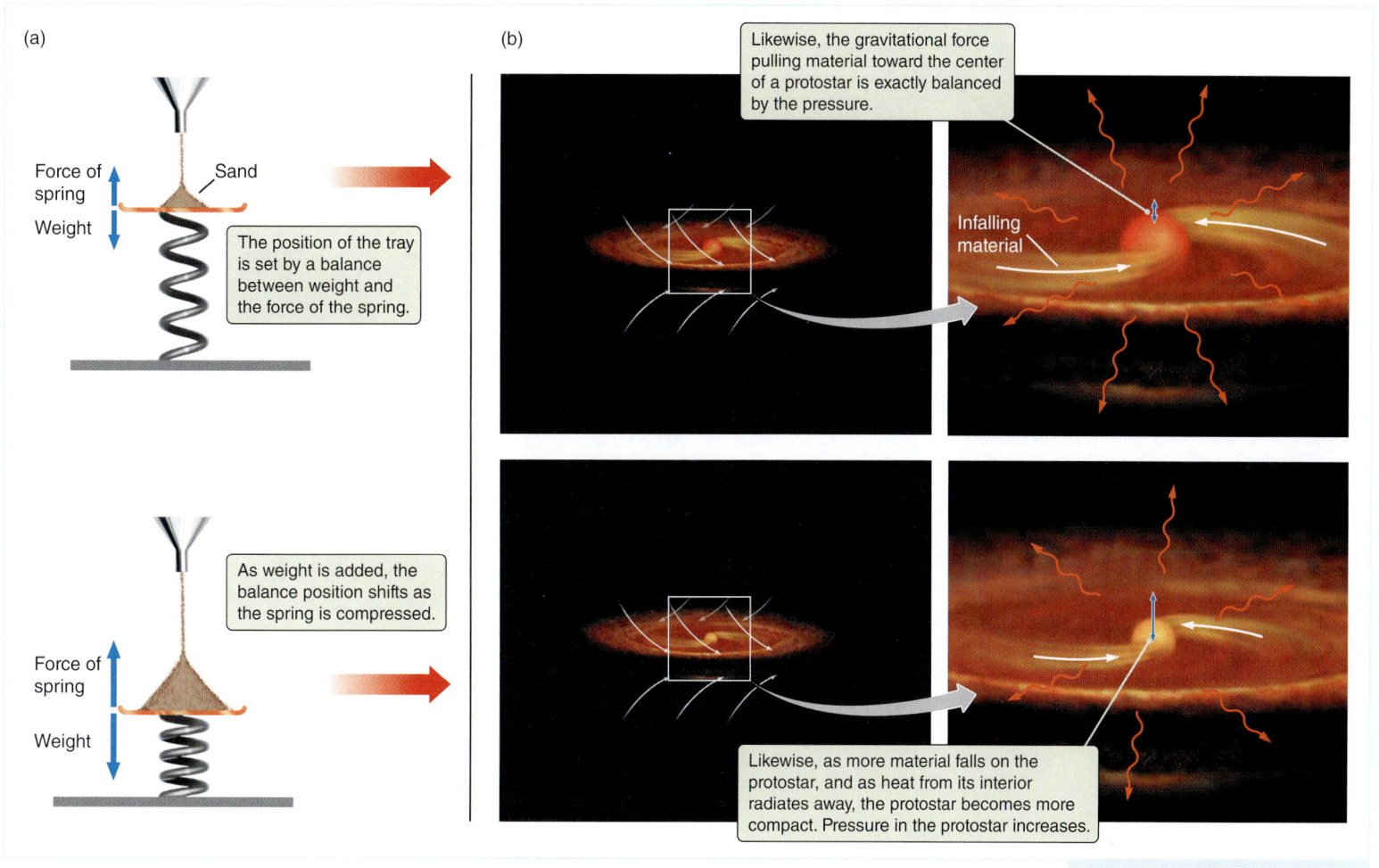

Figure 5.6 (a) A spring scale comes to rest at the point where the weight of the sand is matched by the upward force of the compressed spring. As sand is added, the location of this balance point shifts. (b) Similarly, the balance between pressure and gravity determines the structure of a protostar. Like the spring scale, the structure of the protostar constantly shifts as additional material falls onto the surface of the protostar and as the protostar radiates energy away.

downward force of the weight of the object is exactly balanced by the upward force of the spring. The more the spring is compressed, the harder the spring pushes back. We measure the weight of the object by determining the point at which the pull of gravity and the push of the spring are equal.

Let's now slowly pour sand onto our spring scale. At any time, the downward weight of the sand is balanced by the upward force of the spring. As the weight of the sand increases, the spring is slowly compressed. The spring and the weight of the sand are always in balance, but this balance is *changing with time* as more sand is added. The situation is analogous to our protostar (Figure 5.6b), in which the outward pressure of the gas behaves like the spring. Material falls onto the protostar, adding to its mass and gravitational pull. Additionally, the protostar slowly loses internal thermal energy by radiating it away. But the material that has fallen onto the protostar also compresses the protostar and heats it up. The interior becomes denser and hotter, and the pressure rises—just enough to balance the increased weight of the material above it. Dynamic balance is always maintained.

This dynamic balance persists as energy is radiated away and the protostar slowly contracts. Gravitational energy is converted to thermal energy, which heats the core, raising the pressure to oppose gravity. This process continues, with the protostar becoming smaller and smaller and its interior growing hotter and hotter, until the center of the protostar is finally hot enough to "ignite," or begin turning

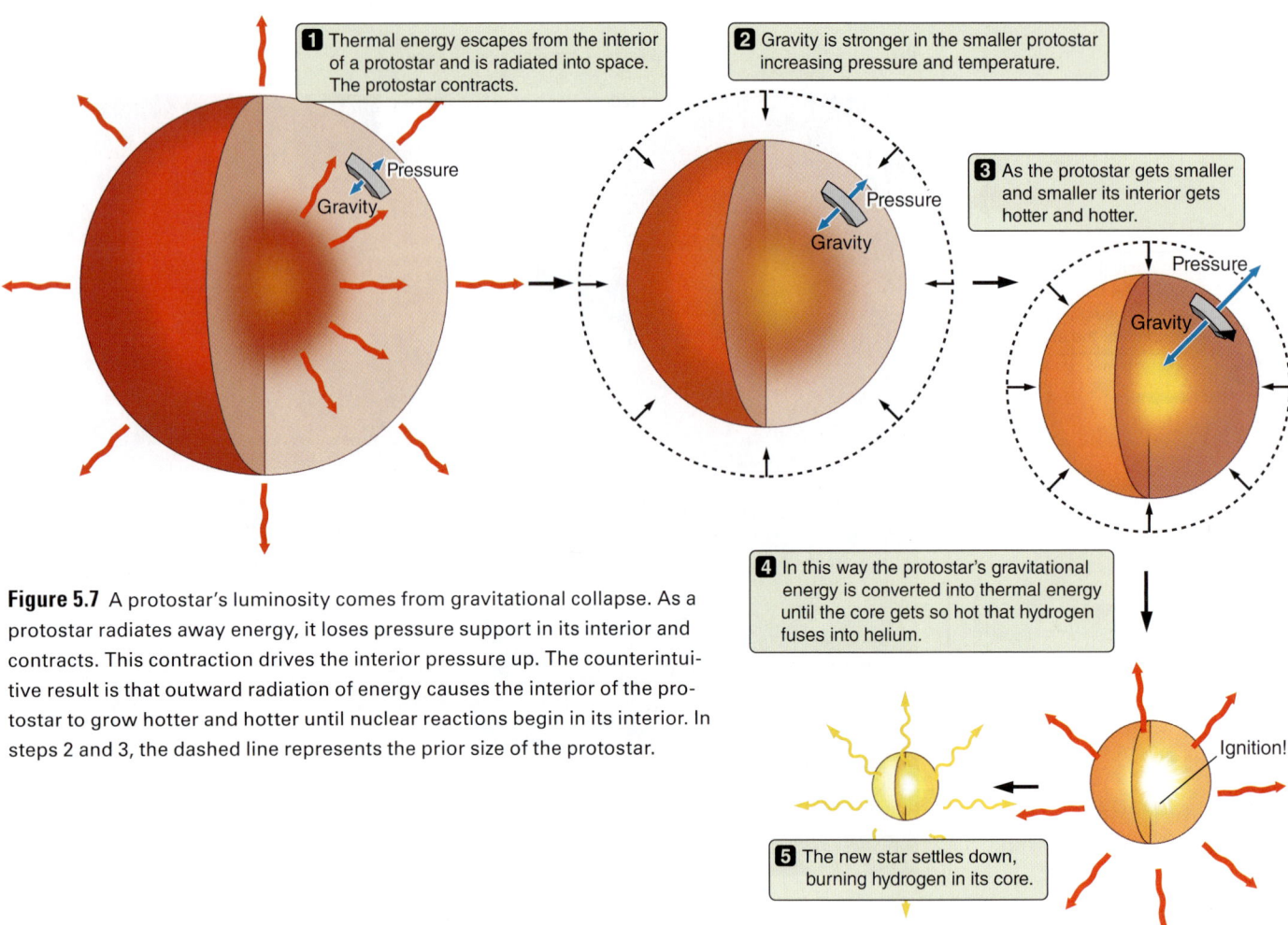

Figure 5.7 A protostar's luminosity comes from gravitational collapse. As a protostar radiates away energy, it loses pressure support in its interior and contracts. This contraction drives the interior pressure up. The counterintuitive result is that outward radiation of energy causes the interior of the protostar to grow hotter and hotter until nuclear reactions begin in its interior. In steps 2 and 3, the dashed line represents the prior size of the protostar.

hydrogen into helium. Once this ignition happens, the protostar becomes a star. The fusion of hydrogen into helium is a nuclear process that releases energy. This process will be discussed in more detail in Chapter 11. This released energy adds to the gas pressure and helps balance the gravitational force pulling inward on the star. This sequence of events is shown in **Figure 5.7**.

The protostar's mass determines whether it will actually become a star. As the protostar slowly collapses, the temperature at its center rises. If the protostar's mass is greater than about 0.08 times the mass of the Sun (0.08 $M_\odot$), the temperature in its core will eventually reach 10 million K, and the reaction that converts hydrogen into helium will begin. The newly born star will once again adjust its structure until it is radiating energy away at exactly the rate that energy is being released in its interior.

If the mass of the protostar is just under 0.08 $M_\odot$, it will never be hot enough to become a star. A **brown dwarf** is such a "failed" star: intermediate between a star and a planet, not quite massive enough to cause hydrogen fusion in its core. The energy emitted from a brown dwarf has the same source as the energy emitted from a protostar: gravitational collapse turns gravitational energy into thermal energy. As the years pass, a brown dwarf gets gradually smaller and fainter. About 2,000 brown dwarfs have been found since the first one was identified in the mid-1990s.

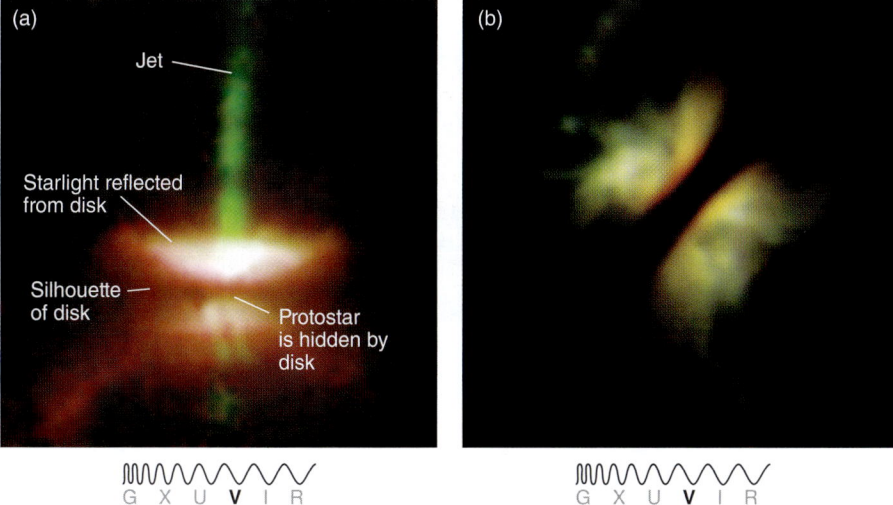

Figure 5.8 Hubble Space Telescope images show disks around newly formed stars. (a) The dark band is the silhouette of the disk seen more or less edge on. Bright regions are dust illuminated by starlight. The jets, shown in green, will be discussed shortly. (b) In this image, the disk is seen in silhouette. Planets may be forming or have already formed in this disk.

5.3 Planets Form in a Disk Around the Protostar

So far we have focused on the innermost portion of the collapsing core of a protostar. What happens to the rest of the dust and gas? Disks of gas and dust have been found surrounding young stellar objects like the ones shown in **Figure 5.8**. From this observational evidence, we know that the cloud collapses first into a disk. A piece of dust or molecule of gas in the disk eventually suffers one of three fates: it travels inward onto the protostar at its center, remains in the disk to form planets and other objects, or is thrown back into interstellar space. In this section, we focus on the formation of planets in the dusty disk.

▶❙❙ **AstroTour: Solar System Formation**

Convergence of Evidence

While astronomers were working to understand star formation by studying molecular clouds, other scientists—mainly geochemists and geologists—with very different backgrounds were piecing together the history of our Solar System. Planetary scientists looking at the current structure of the Solar System inferred what many of its early characteristics must have been. The orbits of all the planets in the Solar System lie very close to a single plane, so the early Solar System must have been flat. Additionally, all the planets orbit the Sun in the same direction, so the material from which the planets formed must have been orbiting in the same direction as well. To find out more, it would be helpful to have samples of the very early Solar System to study. Fortunately, rocks that fall to Earth from space, known as **meteorites**, include pieces of material that are left over from the Solar System's youth! Many meteorites, such as the one in **Figure 5.9**, resemble a piece of concrete in which pebbles and sand are mixed with a much finer filler, suggesting that the larger bodies in the Solar System must have grown from the aggregation of smaller bodies. Following this chain of thought back in time, we find an early Solar System in which the young Sun

Figure 5.9 Meteorites are fragments of the young Solar System that have landed on the surfaces of planets. It is clear from this cross section that this meteorite formed from many smaller components that stuck together.

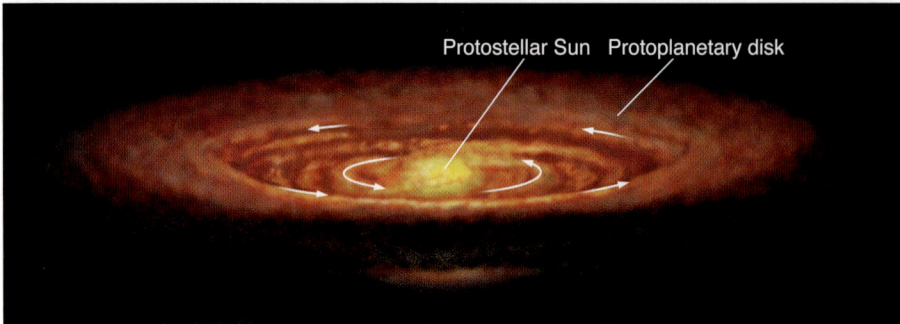

Figure 5.10 When you consider the young Sun, think of it as being surrounded by a flat, rotating disk of gas and dust that is flared at its outer edge.

was surrounded by a flattened disk of both gaseous and solid material. Our Solar System formed from this swirling disk of gas and dust.

As stellar astronomers and planetary scientists compared notes, they realized they had arrived at the same picture of the early Solar System from two completely different directions. The connection between the formation of stars and the origin and evolution of the Solar System is one of the cornerstones of both astronomy and planetary science—a central theme of our understanding of our Solar System.

Figure 5.10 shows the young Solar System as it appeared roughly 5 billion years ago. Surrounding the protostellar Sun was a flat, orbiting disk of gas and dust like those seen around protostars today. Each bit of the material in this thin disk orbited according to the same laws of motion and gravitation that govern the orbits of the planets. This disk is an example of an **accretion disk**: a disk that forms from the accretion of material around a massive object. The accretion disk around the forming Sun had only a fraction of the mass of the Sun, but this amount was more than enough to account for the bodies that make up the Solar System today.

The Collapsing Cloud and Angular Momentum

What is it about the process of star formation that leads not only to a star, but to a flat, orbiting collection of gas and dust as well? The answer to this question involves something called *angular momentum*. **Angular momentum** is a conserved quantity of a revolving or rotating system with a value that depends on both the velocity and distribution of the mass. The angular momentum of an isolated object is always conserved; that is, it remains unchanged unless acted on by an external force. You have likely seen a figure-skater spinning on the ice like the one shown in **Figure 5.11a**. Like any rotating object, the spinning skater has some amount of angular momentum. Unless an external force acts on her, such as the ice pushing on her skates, she will always have the same amount of angular momentum.

The amount of angular momentum depends on three things:

1. How fast the object is rotating. The faster an object is rotating, the more angular momentum it has.
2. The mass of the object. If a bowling ball and a basketball are spinning at the same speed, the bowling ball has more angular momentum because it has more mass.

3. How the mass of the object is distributed relative to the spin axis; that is, how far or spread out the object is. For an object of a given mass and rate of rotation, the more spread out it is, the more angular momentum it has. A spread-out object that is rotating slowly might have the same angular momentum as a compact object rotating rapidly.

Both an ice-skater and a collapsing interstellar cloud are affected by **conservation of angular momentum**: the angular momentum must remain the same in the absence of an external force. In order for angular momentum to be conserved, a change in one of the above quantities (rate of spin, mass, or distribution of mass) must be accompanied by a change of another quantity. For example, an ice-skater can control how rapidly she spins by pulling in or extending her arms. As she pulls in her arms to become more compact (decreasing her distribution of mass), she must spin faster to maintain the same angular momentum (increasing how fast she rotates). When her arms are held tightly in front of her, the skater's spin becomes a blur. She finishes with a flourish by throwing her arms and leg out—an action that abruptly slows her spin by spreading out her mass. Her angular momentum remains constant throughout the entire maneuver. Similarly—as shown in Figure 5.11b—the cloud that formed our Sun rotated faster and faster as it collapsed, just as the ice-skater speeds up when she pulls in her arms.

However, this description presents a puzzle. Suppose the Sun formed from a typical cloud—one that was about a light-year across and was rotating so slowly that it took a million years to complete one rotation. By the time such a cloud collapsed to the size of the Sun today, it would have been spinning so fast that one rotation would occur every 0.6 second. This is more than 3 million times faster than our Sun actually spins. At this rate of rotation, the Sun would tear itself apart. It appears that angular momentum was not conserved in the actual formation of the Sun—but that can't be right, because angular momentum must be conserved. We must be missing something. Where did the angular momentum go?

Figure 5.11 (a) A figure-skater relies on the principle of conservation of angular momentum to change the speed of her spin. (b) In the same way, a collapsing cloud spins faster as it becomes smaller. Angular momentum is conserved in a collapsing cloud.

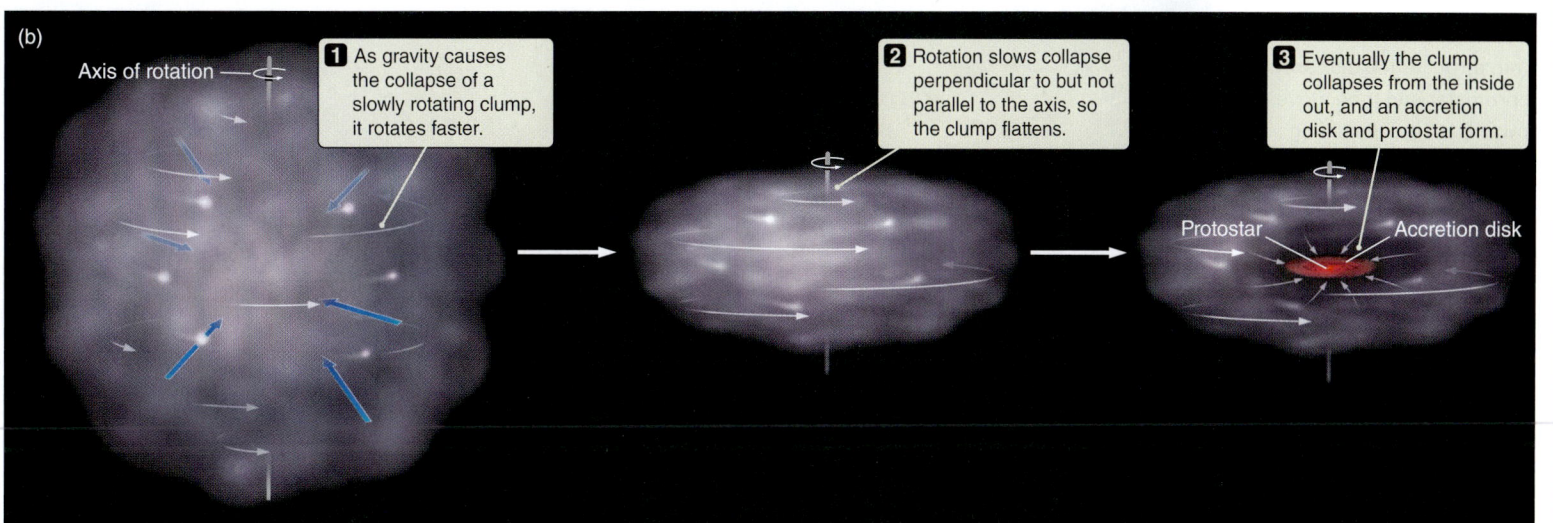

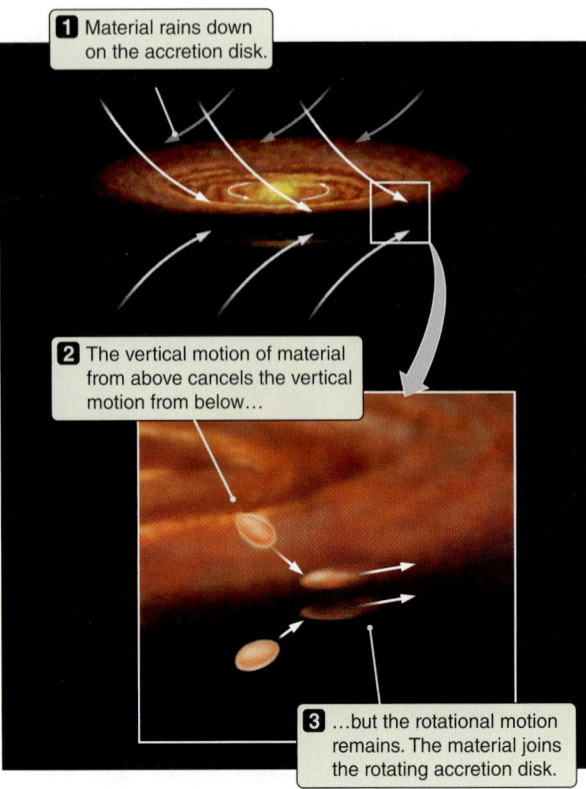

Figure 5.12 Gas from a rotating molecular cloud falls inward from opposite sides, piling up onto a rotating disk

1. Material rains down on the accretion disk.
2. The vertical motion of material from above cancels the vertical motion from below...
3. ...but the rotational motion remains. The material joins the rotating accretion disk.

The Formation of an Accretion Disk

To understand how angular momentum is conserved in disk formation, we must try to think in three dimensions. Imagine that the ice-skater bends her knees, compressing herself downward instead of bringing her arms toward her body. As she does this, she again makes herself less spread out, but her rate of spin does not change because no part of her body has become any closer to the axis of spin. As we saw in Figure 5.11b, a clump of a molecular cloud can flatten out without speeding up. As the clump collapses, its self-gravity increases, and the inner parts begin to fall freely inward, raining down on the growing object at the center. The outer portions of the clump lose the support of the collapsed inner portion, and they start falling inward, too. As this material makes its final inward plunge, it lands on a thin, rotating accretion disk.

The formation of accretion disks is common in astronomy, so it is worth taking a moment to examine this process, which is shown in **Figure 5.12**. As the material falls toward the protostar, it travels on curved, almost always elliptical paths. These paths are oriented randomly except for one key feature—either they all go clockwise or they all go counterclockwise, when viewed from a direction along the axis of rotation. This is what we mean when we say the cloud rotates. Imagine yourself in such a cloud, near the edge, looking toward the center. As you watch, all the material is orbiting from left to right, but some of it is traveling upward and some downward. Some of it is on steep orbits, traveling more vertically, and some is on shallow orbits, traveling mostly to the side. Now imagine that two pieces of material collide and stick together to form a larger piece. Both are on shallow orbits, but one is traveling up and

Figure 5.13 (a) Material falls onto an accretion disk around a protostar and then moves inward, eventually falling onto the star. In the process, some of this material is driven away in powerful jets that stream perpendicular to the disk. (b) This infrared Spitzer Space Telescope image shows jets streaming outward from a young, developing star. Note the nearly edge-on, dark accretion disk surrounding the young star.

(a) Material moves in toward the protostar in the accretion disk.

Bipolar jets and outflows flow away from the young star and disk.

Wind

Jets

This structure is seen in images of forming stars.

the other is traveling down. What will happen to the new, larger piece? It will still orbit from left to right, but the upward motion and the downward motion will cancel out, so the orbit will become shallower. Imagine this same scenario for two pieces on steep orbits. Again, the orbit will become less steep. In this way, the upward and downward motions of the material cancel each other out, and a disk is formed in which all the material has very shallow orbits, and all orbits still proceed in the same overall direction—either clockwise or counterclockwise. The angular momentum is unchanged because all the material has maintained its distance from the axis.

Now we can explain why the Sun does not have the same angular momentum that was present in the original clump of cloud. The radius of a rotating accretion disk is thousands of times greater than the radius of the star that will form at its center. Much of the angular momentum in the original interstellar clump is conserved in its accretion disk rather than in its central protostar.

Most of the matter that lands on the accretion disk either becomes part of the star or is ejected back into interstellar space, in the form of jets or other outflows, as seen in **Figure 5.13**. Material swirling in the bipolar jets carries angular momentum away from the accretion disk in the general direction of the poles of the rotation axis. However, a small amount of material is left behind in the disk. It is the objects in this leftover disk—the dregs of the process of star formation—that form planets and other objects that orbit the star.

Theoretical calculations by astronomers long predicted that accretion disks should be found around young stars. Look back at Figure 5.8, which shows Hubble Space Telescope images of edge-on accretion disks around young stars. The dark bands are the shadows of the edge-on disks, the top and the bottom of which are illuminated by light from the forming star. Our Sun and Solar System formed from a protostar and disk much like those in these pictures.

Creation of Large Objects

Random motions of material in the accretion disk push the smaller grains of solid material toward larger grains, as shown in **Figure 5.14**. As this happens, the smaller grains stick to the larger grains. The "sticking" process among smaller grains is due to the same static electricity that causes dust bunnies to grow under your bed. Starting out at only a few micrometers across, the slightly larger bits of dust grow to the size of pebbles and then to clumps the size of boulders, which are not as easily pushed around by gas. When clumps grow to about 100 meters across, the objects are so few and far between that they collide less frequently, and their growth rate slows down but does not stop.

For two large clumps to stick together rather than explode into many small pieces, they must bump into each other very gently; collision speeds must be about 0.1 m/s or less for colliding boulders to stick together. Your stride is probably about a meter, so to walk as slowly as the collision speed of 0.1 m/s, you would take one step every 10 seconds. The process is not a uniform movement toward larger and larger bodies. Violent collisions do occur in an accretion disk, and larger clumps break back into smaller pieces. But over a long period, large bodies do form.

Objects grow by "sweeping up" smaller objects that get in their way. These objects can eventually measure up to several hundred meters across. But as the clumps reach the size of about a kilometer, a different process becomes important. These kilometer-sized objects are massive enough that their gravity begins

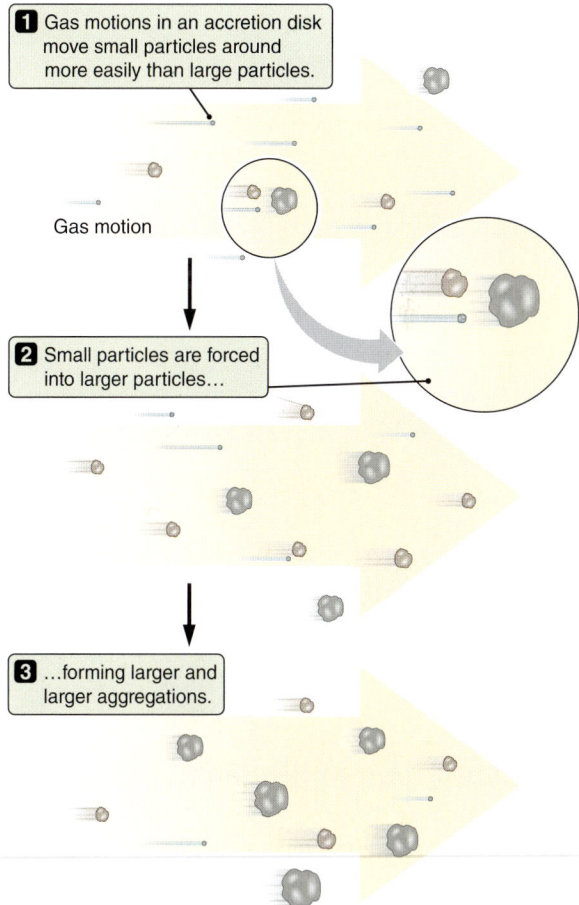

Figure 5.14 Motions of gas in an accretion disk force smaller particles of dust into larger particles, then make these large particles even larger. This process continues, eventually creating objects many meters in size.

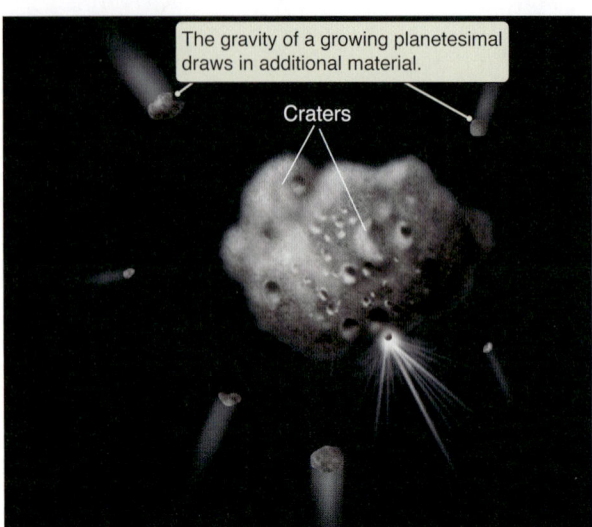

Figure 5.15 The gravity of a planetesimal is strong enough to attract surrounding material, which causes the planetesimal to grow more rapidly.

to pull on nearby bodies, as shown in **Figure 5.15**, at which point they are known as **planetesimals** (literally "tiny planets"). At this point, the planetesimal is not growing only by chance collisions with other objects; now it can pull in and capture small objects outside its direct path. The growth speeds up, and the larger planetesimals quickly consume most of the remaining bodies in the vicinity of their orbits. The final survivors of this process are large enough to be called planets. As with the major bodies in orbit around the Sun, some of the planets may be relatively small and others quite large.

Very large objects can acquire mini accretion disks as they capture gas and dust. Some of this material may grow into larger bodies in much the same way that material in the accretion disk formed planets. The result is a mini "solar system"—a group of moons that orbit about the planet.

5.4 The Inner and Outer Disk Have Different Compositions

In the Solar System, the inner planets are small and mostly rocky, while the outer planets are very large and mostly gaseous. This distinct difference between the inner and the outer Solar System must have some explanation in a model of the formation of stars and planets.

Rock, Metal, and Ice

On a hot summer day, ice melts and water quickly evaporates; but on a cold winter night, even the water in our breath freezes into tiny ice crystals before our eyes. Metals and rocky materials, such as iron, **silicates** (minerals containing silicon and oxygen), and carbon remain solid even at quite high temperatures. Substances that are capable of withstanding high temperatures without melting or being vaporized are referred to as **refractory materials**. Other materials, such as water, ammonia, and methane, remain in a solid form only if their temperature is very low. These materials, which become gases at moderate temperatures, are called **volatile materials**, or "volatiles."

Differences in temperature from place to place within the accretion disk significantly affect the makeup of the dust grains in the disk As **Figure 5.16** illustrates, in the hottest parts of the disk—the area closest to the protostar—only refractory substances exist. In the inner disk, dust grains are composed almost entirely of refractory materials. Somewhat farther out, some hardier volatiles,

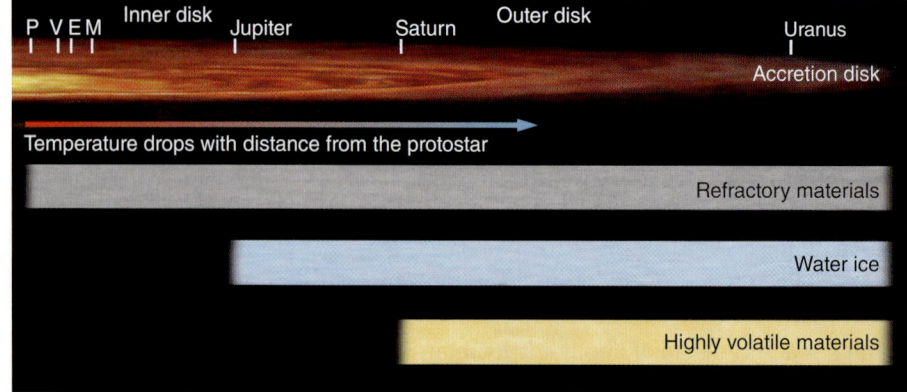

Figure 5.16 Differences in temperature within an accretion disk determine the composition of dust grains that then evolve into planetesimals and planets. The colored bars show where refractory materials, water ice, and volatiles can exist in the accretion disk. Shown here are the protostar (P) and the distances of Venus (V), Earth (E), Mars (M), Jupiter, Saturn, and Uranus.

such as water ice and certain organic substances, can survive in solid form, adding to the materials that make up dust grains. Highly volatile components such as methane, ammonia, and carbon monoxide ices and some simple organic molecules survive in solid form only in the coldest, outermost parts of the accretion disk, far from the central protostar. The differences in composition of dust grains within the disk are reflected in the composition of the planets formed from that dust. Planets closest to the central star are composed primarily of refractory materials such as rock and metals but are deficient in volatiles. Those that form farthest from the central star contain refractory materials, but they also contain large quantities of ices and organic materials.

Chaotic encounters can change this organization of planetary compositions. Through **planet migration**, gravitational scattering or interactions with gas in the protoplanetary disk can force some planets to end up far from where they formed. Uranus and Neptune originally may have formed near the orbits of Jupiter and Saturn but were then driven outward to their current locations by gravitational encounters with Jupiter and Saturn. A planet can also migrate when it gives up some of its orbital angular momentum to the disk material that surrounds it. Such a loss of angular momentum causes the planet to slowly spiral inward toward the central star. We will see examples from other planetary systems when we discuss *hot Jupiters* in Section 5.6.

Vocabulary Alert

organic *Organic* often means "pertaining to life"; or, in the case of food labeled "organic" in the United States, it means that a list of rules has been obeyed regarding the use of pesticides and herbicides. To scientists, *organic* means "carbon based," as in organic chemistry. Organic molecules are those with carbon in them. They need not be of biological origin.

ice In common language, *ice* refers specifically to the solid form of water, as we've used it so far in this chapter. The term *dry ice* is also common, referring to frozen carbon dioxide. To astronomers, ice refers to the solid form of any type of volatile material.

chaotic (or chaos) *Chaotic* is commonly used to mean "messy or disorganized." To scientists, a *chaotic* system is one that is very sensitive to tiny changes in the initial conditions: A small change in the initial state can lead to a large change in the final state of a system.

Atmospheres around Solid Planets

Once a solid planet has formed, it may continue to grow by capturing gas from the accretion disk. To do so, it must act quickly. Young stars and protostars emit fast-moving particles and intense radiation that can quickly disperse the remains of the accretion disk. Planets like Jupiter have about 10 million years to form and accumulate an atmosphere. Massive planets can capture more of the hydrogen and helium gas that makes up the bulk of the disk.

The gas—primarily hydrogen and helium—that is captured by a planet at the time of its formation is the planet's **primary atmosphere**. The primary atmosphere of a large planet can be more massive than the solid body, as in the case of giant planets such as Jupiter.

A less massive planet may also capture some gas from the accretion disk, only to lose it later. The gravity of small planets may be too weak to hold low-mass gases such as hydrogen and helium. Even if a small planet is able to gather some hydrogen and helium from its surroundings, this primary atmosphere will not last long. The atmosphere around a small planet like our Earth is a **secondary atmosphere**, which forms later in the life of a planet. Volcanism is one important source of a secondary atmosphere because it releases carbon dioxide, water vapor, and other gases from the planet's interior. In addition, volatile-rich comets that formed in the outer parts of the disk fall inward toward the new star and sometimes collide with planets. **Comets** are icy planetesimals that survive planetary accretion. They may provide a significant source of water, organic compounds, and other volatile materials on planets close to the central star.

5.5 A Case Study: The Solar System

The Solar System is a collection of planets, moons, and other smaller bodies surrounding an ordinary star that we call the Sun. Enough small bodies occupy par-

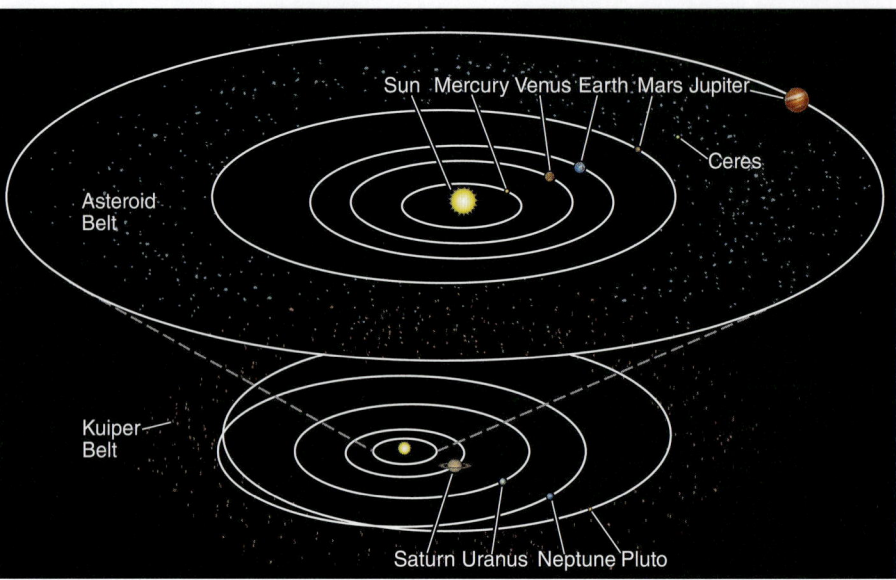

Figure 5.17 Our Solar System includes planets, moons, and other small bodies. Sizes and distances are not to scale in this sketch of the layout of the Solar System.

ticular regions of the Solar System to warrant separate names for these regions, such as asteroid belt and Kuiper Belt (see **Figure 5.17**). Do not confuse our Solar System with the universe. Our Solar System is a tiny part of our galaxy, which is a tiny part of the universe. You may wish to go back to Figure 1.2 to remind yourself of the size scales involved.

Nearly 5 billion years ago, our Sun was a protostar surrounded by an accretion disk of gas and dust. Over the course of a few hundred thousand years, much of the dust collected into planetesimals—clumps of rock and metal near the Sun and aggregates of rock, metal, ice, and organic materials farther from the Sun. Within the inner few astronomical units (AU) of the disk, several rock and metal planetesimals—probably less than a half dozen—quickly grew in size to become the dominant masses in their orbits. These few either captured most of the remaining planetesimals or ejected them from the inner part of the disk. These dominant planetesimals had now become planet-sized bodies with masses ranging between $1/20$ and 1 Earth mass ($M_\oplus$). They became the **terrestrial planets** (Earth-like, or rocky). Mercury, Venus, Earth, and Mars are the surviving terrestrial planets. Even though it is not a planet itself and formed in a different way, Earth's Moon is often grouped with these terrestrial planets because of its similar physical and geological properties. One or two others may have formed in the young Solar System but were later destroyed.

For several hundred million years after the formation of the four surviving terrestrial planets and Earth's Moon, leftover pieces of debris still in orbit around the Sun continued to rain down on their surfaces. Much of this barrage may have originated in the outer Solar System. Today, we can still see the scars of these early impacts on the cratered surfaces of some of the terrestrial planets, such as the surface of Mercury shown in **Figure 5.18**. This rain of debris continues today, but at a much lower rate.

Before the proto-Sun became a true star, gas in the inner part of the accretion disk was still plentiful. During this early period, Earth and Venus held on to weak primary atmospheres of hydrogen and helium, but these thin atmospheres were soon lost to space. The terrestrial planets did not develop thick atmospheres until the formation of the secondary atmospheres that now surround Venus, Earth, and Mars. Mercury's proximity to the Sun and the Moon's small mass prevented these bodies from retaining significant secondary atmospheres.

Beyond 5 AU from the Sun, planetesimals combined to form a number of bodies with masses about 10–20 times that of Earth. These planet-sized objects formed from planetesimals containing volatile ices and organic compounds in addition to rock and metal. Four such massive bodies later became the cores of the **giant planets**: Jupiter, Saturn, Uranus, and Neptune. These giant planets are many times the mass of any terrestrial planet and lack a solid surface. Mini accretion disks formed around these planetary cores, capturing large amounts of hydrogen and helium and funneling this material onto the planets' surfaces.

Jupiter's massive solid core captured and retained the most gas—roughly 300 times the mass of Earth. The other planetary cores captured lesser amounts of hydrogen and helium, perhaps because their cores were less massive or because less gas was available to them. Saturn has less than 100 Earth masses of gas, whereas Uranus and Neptune captured only a few Earth masses' worth of gas.

This description indicates that it could take up to 10 million years for a Jupiter-like planet to form. Some scientists do not think that our accretion disk could have survived long enough to form gas giants such as Jupiter through this process. All the gas may have dispersed in roughly half that time, cutting off Jupiter's supply of hydrogen and helium. An alternative explanation is that the accretion disk fragments into massive clumps, each of which is equivalent to that of a large planet. It is possible that both processes played a role in the formation of our own and other planetary systems. This part of the story is still incomplete.

During the formation of the planets, gravitational energy was converted into thermal energy as the individual atoms and molecules moved faster. This conversion warmed the gas surrounding the cores of the giant planets. Proto-Jupiter and proto-Saturn probably became so hot that they actually glowed a deep red color, similar to the heating element on an electric stove. Their internal temperatures may have reached as high as 50,000 K. However, they were never close to becoming stars. As we saw in Section 5.2, a ball of gas must have a mass at least 0.08 times the mass of the Sun for it to become a star. This is about 80 times the mass of Jupiter.

The composition of the moons of the giant planets followed the same trend as the planets that formed around the Sun: the innermost moons formed under the hottest conditions and therefore contained the smallest amounts of volatile material. When Jupiter's moon Io formed, Jupiter was glowing so intensely that it rivaled the distant Sun. The high temperatures created by the glowing planet evaporated most of the volatile substances nearby. Io today contains no water at all. However, water is probably plentiful on at least three of Jupiter's other large moons, Europa, Ganymede, and Callisto, because these moons formed farther from warm, glowing Jupiter.

Not all planetesimals in the disk went on to become planets. For example, **dwarf planets** orbit the Sun but have not cleared other, smaller bodies from their orbits. Many are not massive enough to be round. Ceres and Pluto, shown in Figure 5.17, are both dwarf planets. More dwarf planets, along with a large number of smaller bodies, are found in the **Kuiper Belt**, beyond Pluto's orbit. **Asteroids** are small bodies found interior to Jupiter's orbit; most are located in the main **asteroid belt** between Mars and Jupiter. Jupiter's gravity kept the region between Jupiter and Mars so stirred up that most planetesimals there never formed a large planet.

Planetesimals persist to this day in the outermost part of the Solar System as well. Formed in a deep freeze, these objects have retained most of the highly

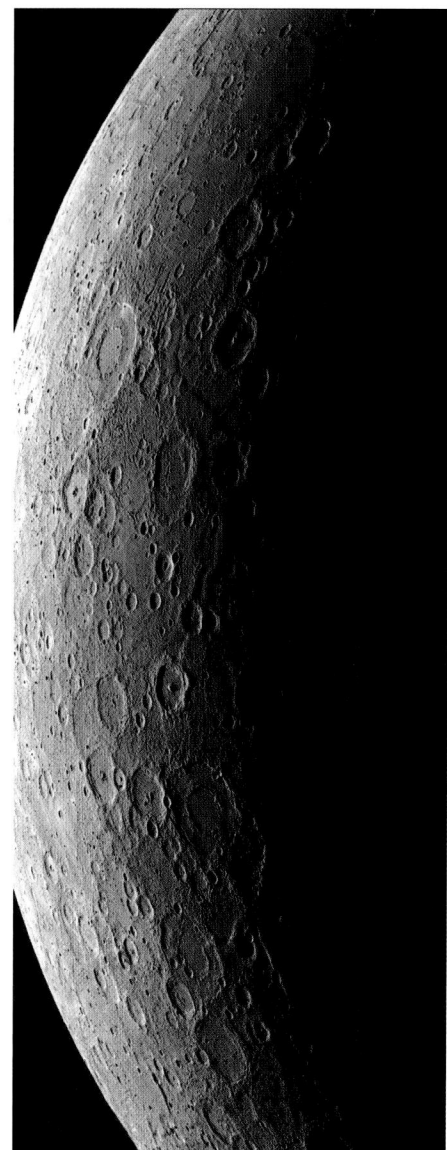

Figure 5.18 Large impact craters on Mercury (and on solid bodies throughout the Solar System) record the final days of the Solar System's youth, when smaller planetesimals rained down on their surfaces.

volatile materials found in the grains present at the formation of the accretion disk. Unlike the crowded inner part of the disk, the outermost parts of the disk had planetesimals that were too sparsely distributed for large planets to grow. Icy planetesimals in the outer Solar System remain today as **comet nuclei**—relatively pristine samples of the material from which our planetary system formed. The frozen, distant dwarf planets Pluto and Eris are especially large examples of these residents of the outer Solar System.

Even after the initial formation, the Solar System was a remarkably violent and chaotic place. Many objects in the Solar System show evidence of cataclysmic impacts that reshaped worlds. A dramatic difference between the terrains of the northern and southern hemispheres on Mars, for example, has been interpreted by some planetary scientists as the result of one or more colossal collisions. Mercury has a crater on its surface from an impact so devastating that it caused the crust to buckle on the opposite side of the planet. In the outer Solar System, one of Saturn's moons, Mimas, has a crater roughly one-third the diameter of the moon itself. Uranus suffered a collision that was violent enough to literally knock the planet on its side. Today, its axis of rotation is tilted to lie almost in its orbital plane.

Not even our own Earth escaped devastation by these cataclysmic events. The Moon itself may be the result of such a collision. According to the best current hypothesis for the formation of the Moon, the early Solar System included a protoplanet about the same size and mass as Mars. As the newly formed planets were settling into their present-day orbits, this fifth terrestrial planet suffered a grazing collision with Earth and was completely destroyed. The remains of the planet, together with material knocked from Earth's outer layers, formed a huge cloud of debris encircling Earth. For a brief period Earth may have displayed a magnificent group of rings like those of Saturn. In time, some of this debris coalesced into the single body we know as our Moon. This "impact formation" hypothesis is still an active area of research, as astronomers try better to explain all the observations about the Earth-Moon system.

Look back to the chapter-opening illustration, which shows images related to the formation of stars and planets. Now that we have explored both a generic example and our own Solar System, you can determine where these images fit in the story of the formation of stars and planets.

5.6 Planetary Systems Are Common

In 1995, astronomers announced the first confirmed **extrasolar planet**—a planet orbiting around a star other than the Sun. This planet orbits around a solar-type star and is a Jupiter-sized body orbiting surprisingly close to this star, 51 Pegasi. Today, the number of known extrasolar planets, sometimes called *exoplanets*, has grown to more than a thousand, and new discoveries are occurring almost daily.

The discovery of extrasolar planets raises the question of what we mean by the term *planet*. Within our own Solar System, we feel reasonably comfortable with our definition of a planet. But what about those extrasolar bodies? The International Astronomical Union defines an extrasolar planet as a body that orbits a star other than our Sun and has a mass less than 13 Jupiters. Objects more massive than this but less than $0.08\ M_\odot$ are brown dwarfs. Objects more massive than $0.08\ M_\odot$ are defined as stars.

The Search for Extrasolar Planets

Astronomers use several methods for finding extrasolar planets. The first planets were discovered indirectly, by observing their gravitational tug on the central star. As technology has improved, other methods have become more productive. Astronomers now have direct imagery of planets orbiting stars and have also been able to take the spectra of planets to observe the composition of their atmospheres. Almost certainly, between the time we write this and the time you read it, there will be new discoveries. The field is advancing extremely quickly. We will now look at each discovery method in turn.

▶❙ **AstroTour: The Doppler Effect**

The Radial Velocity Method As a planet orbits a star, the planet's gravity tugs the star around ever so slightly. We can sometimes detect this wobble and infer the properties of the planet—its mass and its distance from the star. To see how this works, we must understand the *Doppler effect* shown in **Figure 5.19**.

Have you ever listened to a fire truck speed by with sirens blaring? As the fire truck comes toward you, its siren has a certain high pitch. But as it passes by, the pitch of the siren drops noticeably. If you close your eyes and listen, you have no trouble knowing when the fire truck passed; the change in pitch of its siren indicates its position. You do not even need a fire engine to hear this effect. The sound of normal traffic behaves in the same way. As a car drives past, the pitch of the sound that it makes suddenly drops. A change in frequency due to motion is known as the **Doppler effect**.

The pitch of a sound is like the color of light: it is determined by the wavelength of the wave. What we perceive as higher pitch corresponds to sound waves with shorter wavelengths. Sounds that we perceive as lower in pitch are waves with longer wavelengths. When an object is moving toward us, the waves that it emits, whether light or sound or waves in the water, "crowd together" in front of the object. You can see how this works by looking at Figure 5.19, which shows the locations of successive wave crests emitted by a moving object.

The Doppler effect causes a shift in the light emitted from a moving object. If

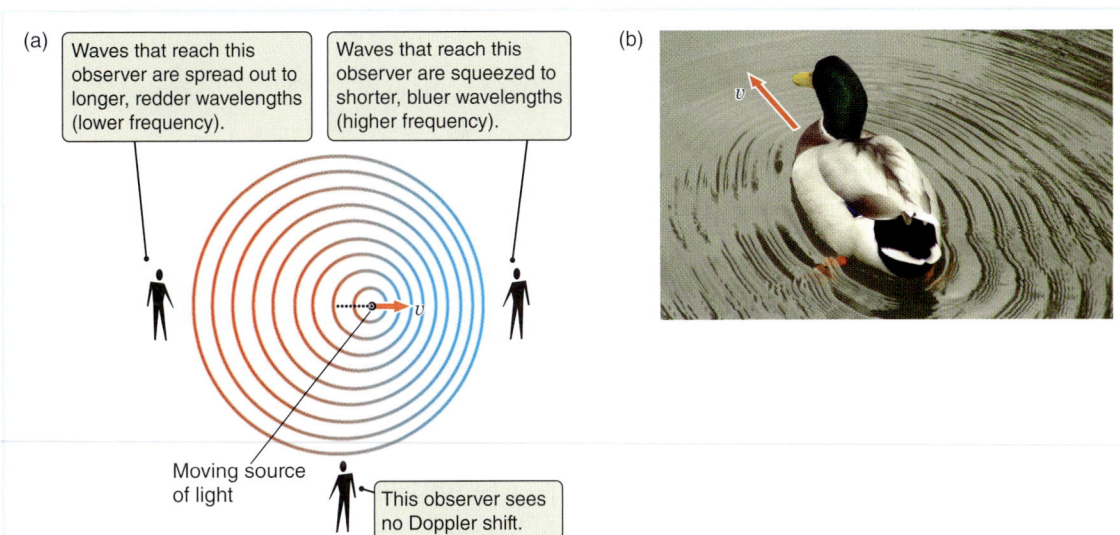

Figure 5.19 (a) Motion of a light or sound source relative to an observer causes waves to be spread out (*redshifted*, or made lower in pitch) or squeezed together (*blueshifted*, or made higher in pitch). Such a change in the wavelength of light or the frequency of sound is called a Doppler shift. (b) The phenomenon can be seen in waves of all kinds. The waves in front of the moving duck are compressed while the ones behind are stretched out.

👁 **VISUAL ANALOGY**

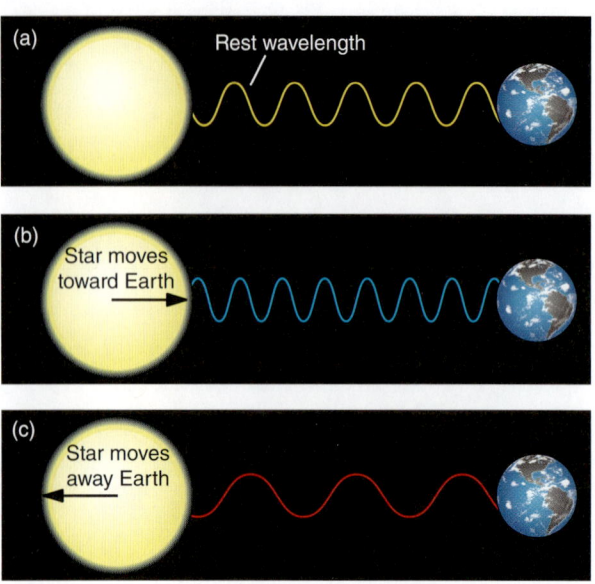

Figure 5.20 Light from an astronomical object will be observed at its rest wavelength (a), blueshifted (b), or redshifted (c), depending on whether the object is at rest (a), moving toward us (b), or moving away from us (c).

the object were at rest, it would emit light with the **rest wavelength**, as shown in **Figure 5.20a**. If an object such as a star is moving toward you, the light reaching you from the object has a shorter wavelength than its rest wavelength—the light is bluer than the rest wavelength, so it is **blueshifted** (Figure 5.20b). In contrast, light from a source that is moving away from you is shifted to longer wavelengths. The light that you see is redder than if the source were not moving away from you, so it is **redshifted**, as seen in Figure 5.20c. The faster the object is moving with respect to you, the larger the shift. The amount by which the wavelength of light is shifted by the Doppler effect is called the **Doppler shift** of the light, and it depends on the speed of the object emitting the light.

The Doppler shift provides information only about the **radial velocity** of the object: the part of the motion that is toward you or away from you. An object moving *across* the sky, for example, does not move toward or away from you, and so its light will not be Doppler shifted from your point of view. The technique of examining Doppler shifts in the light from stars to detect extrasolar planets is called the **spectroscopic radial velocity method**. In **Working It Out 5.2**, we explore how to use the Doppler shift to calculate the radial velocity of an astronomical object, or the amount by which its light has shifted from our standpoint.

We can see how the Doppler shift helps us find planets by using our own Solar System as an example. Imagine that an extrasolar astronomer points a spectro-

Working It Out 5.2 | Making Use of the Doppler Shift

We noted in Section 4.2 that atoms and molecules emit and absorb light only at certain wavelengths. The spectrum of an atom or molecule has absorption or emission lines that look something like a bar code instead of a rainbow, and each type of atom or molecule has a unique set of lines. These lines are called *spectral lines*. A prominent spectral line of hydrogen atoms has a rest wavelength, λ_{rest}, of 656.3 nanometers (nm). Suppose that using a telescope, you measure the wavelength of this line in the spectrum of a distant object and find that instead of seeing the line at 656.3 nm, you see it at an observed wavelength, λ_{obs}, of 659.0 nm. The mathematical form of the Doppler effect shows that the object is moving at a radial velocity (v_r) of

$$v_r = \frac{\lambda_{obs} - \lambda_{rest}}{\lambda_{rest}} \times c$$

$$v_r = \frac{659.0\,\text{nm} - 656.3\,\text{nm}}{656.3\,\text{nm}} \times (3 \times 10^8\,\text{m/s})$$

$$v_r = 1.2 \times 10^6\,\text{m/s}$$

The object is moving away from you (because the wavelength became longer and redder) with a speed of 1.2×10^6 m/s, or 1,200 kilometers per second (km/s).

Now consider our stellar neighbor, Alpha Centauri, which is moving toward us with a radial velocity of −21.6 km/s (−2.16 × 10^4 m/s). (Negative velocity means the object is moving toward us.) What is the observed wavelength, λ_{obs}, of a magnesium line in Alpha Centauri's spectrum having a rest wavelength, λ_{rest}, of 517.27 nm? First, we need to manipulate the Doppler equation to get λ_{obs} all by itself. Then we can plug in all the numbers.

$$v_r = \frac{\lambda_{obs} - \lambda_{rest}}{\lambda_{rest}} \times c$$

Solve this equation for λ_{obs} to get

$$\lambda_{obs} = \lambda_{rest} + \frac{v_r}{c}\lambda_{rest}$$

Both terms on the right contain λ_{rest}. Factor it out to make the equation a little more convenient:

$$\lambda_{obs} = \left(1 + \frac{v_r}{c}\right)\lambda_{rest}$$

We are ready to plug in some numbers to solve for the observed wavelength:

$$\lambda_{obs} = \left(1 + \frac{-2.16 \times 10^4\,\text{m/s}}{3 \times 10^8\,\text{m/s}}\right) \times 517.27\,\text{nm}$$

$$\lambda_{obs} = 517.23\,\text{nm}$$

Although the observed Doppler blueshift (517.23 − 517.27) is only −0.04 nm, it is easily measurable with modern instrumentation.

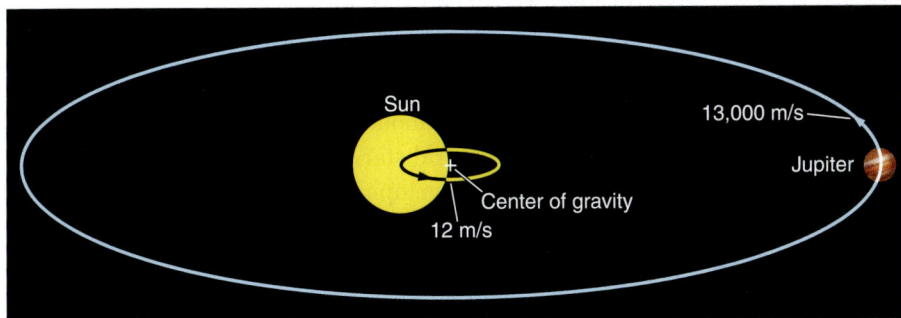

Figure 5.21 Both the Sun and Jupiter orbit around a common center of gravity, which lies just outside the Sun's surface. Spectroscopic measurements made by an extrasolar astronomer would reveal the Sun's radial velocity varying by ±12 m/s over an interval of 11.86 years, which is Jupiter's orbital period. Jupiter travels around its orbit at a speed of 13,000 m/s.

graph toward the Sun. Both the Sun and Jupiter orbit a common center of gravity (sometimes called center of mass; this is the location where the effect of one mass balances the other) that lies just outside the surface of the Sun, as shown in **Figure 5.21**. The astronomer would find that the Sun's radial velocity varies by ±12 m/s, with a period equal to Jupiter's orbital period of 11.86 years. From this information, the astronomer would rightly conclude that the Sun has at least one planet with a mass comparable to Jupiter's, but without greater precision, she would be unaware of the other, less massive major planets. But, spurred on by the excitement of the discovery of Jupiter, the astronomer would improve the sensitivity of her instruments. If the astronomer could measure radial velocities as small as 2.7 m/s, she would be able to detect Saturn, and if the precision extended to radial velocities as small as 0.09 m/s, she would be able to detect Earth.

Current technology limits the precision of our own radial velocity instruments to about 0.3 m/s, but to date it has been the most successful ground-based approach to finding extrasolar planets. This technique enables astronomers to detect giant planets around solar-type stars, but we are still far from being able to use it to reveal bodies with masses similar to those of our own terrestrial planets. Finding the signal of the Doppler shift in the noise of the observation requires the star to be quite bright in our sky. So this method is limited to nearby stars, within about 160 light-years from Earth. The next method, the transit method, does not have this limitation.

The Transit Method Another technique for finding extrasolar planets is the **transit method**, in which we observe the effect of a planet passing in front of its parent star. When this happens, the light from the star diminishes by a tiny amount, as illustrated in **Figure 5.22**. Try picturing the geometry required to observe a transit, however, and you may see a major limitation. For a planet to pass in front of a star from our perspective, Earth must lie nearly in the orbital plane of the planet. There is another important difference between the radial velocity and transit methods: Whereas the radial velocity method gives us the mass of the planet and its orbital distance from a star, the transit method provides the *size* of a planet. Current ground-based technology limits the sensitivity of the transit method to about 0.1 percent of a star's brightness.

Using the transit method, our aforementioned extrasolar astronomer could infer the existence of Earth only if the astronomer was located somewhere in the plane of Earth's orbit and could detect an 0.009 percent drop in the Sun's

▶▶ **Nebraska Simulation:** Exoplanet Radial Velocity Simulator

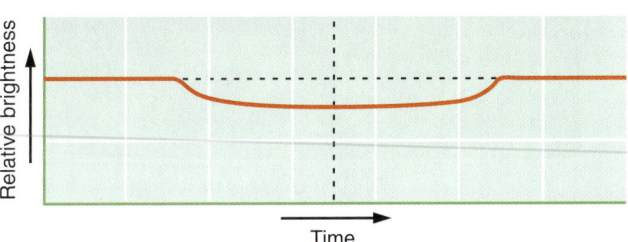

Figure 5.22 As a planet passes in front of a star, it blocks some of the light coming from the star's surface, causing the brightness of the star to decrease slightly. (The brightness decrease has been exaggerated in this illustration.)

▶▶ **Nebraska Simulation:** Exoplanet Transit Simulator

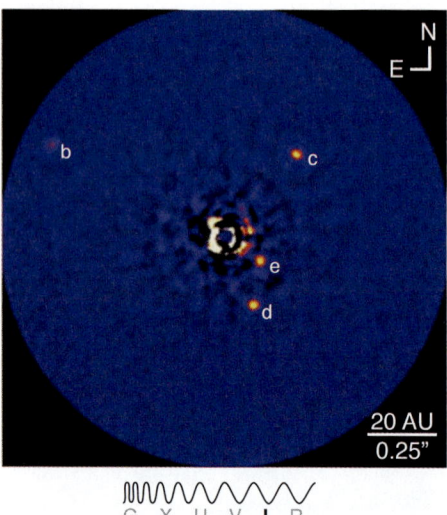

Figure 5.23 An infrared image shows four planets (labeled "b," "c," "d," and "e"), each with a mass several times that of Jupiter, orbiting the star HR 8799 (hidden behind a mask).

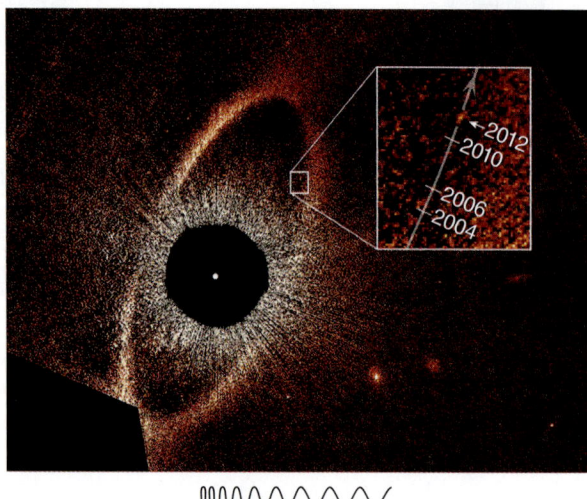

Figure 5.24 Fomalhaut b is seen here moving in its orbit around Fomalhaut, a nearby star easily visible to the naked eye. The parent star, hidden by an obscuring mask, is about a billion times brighter than the planet, which is located within a dusty debris ring that surrounds the star.

brightness. In 2009, NASA launched a solar-orbiting telescope called Kepler with instruments that are able to detect transits of Earth-sized planets. This telescope has found thousands of extrasolar planet candidates in just a few short years. Many of these are "Earth-like"—they are about Earth's size, with about Earth's mass, located at a distance from the central star that might allow for Earth-like temperatures. This is an exciting new development in the search for life in the universe. In 2013, the original Kepler mission came to an end due to an equipment failure, but discovery of new planets is expected to continue for several more years as planet candidates identified by Kepler are confirmed.

Gravitational Lensing The gravitational field of an unseen planet acts like a lens. If the planet passes in front of a background star, it causes the star to brighten temporarily while the planet is passing in front of it. Because the effect is small, it is usually called **microlensing**. Like the transit method used in space, lensing is also capable of detecting Earth-sized planets. This method provides an estimate of the mass of a planet.

Astrometry Planets may also be detected by **astrometry**—precisely measuring the position of a star in the sky. If the system is viewed from "above," the star moves in a mini-orbit as the planet pulls it around. This motion is generally tiny and therefore very difficult to measure. However, for systems viewed from above the plane of the planet's orbit, none of the prior methods will work because the planet neither passes in front of the star nor causes a shift in its speed along the line of sight. Several space missions, such as the *Gaia* spacecraft, which was launched in 2013 by the European Space Agency, will carry out observations of this kind.

Direct Imaging **Direct imaging** means taking a picture of the planet directly. This technique is conceptually straightforward but is technically difficult because it requires searching for a faint planet in the overpowering glare of a bright star—a challenge far more difficult than looking for a star in the dazzling brilliance of a clear daytime sky. Even when an object is detected by direct imaging, the astronomer must still determine whether the observed object is actually a planet. Suppose we detect a faint object near a bright star. Could it be a more distant star that just happens to be in the line of sight? Future observations could tell if the object shares the bright star's motion through space. But it could also be a brown dwarf rather than a true planet. The astronomer would need to make further observations to determine the object's mass.

Some planets have been discovered by this method using large, ground-based telescopes operating in the infrared region of the spectrum. **Figure 5.23** is an infrared image of four planets orbiting the star HR 8799. The first visible-light discovery was made from space while the Hubble Space Telescope was observing Fomalhaut, a bright naked-eye star only 25 light-years away. The planet Fomalhaut b is shown in **Figure 5.24**. It has a mass no more than three times that of Jupiter and orbits within a dusty debris ring some 17 billion km from the central star. A related form of direct observation involves separating the spectrum of a planet from the spectrum of its star to obtain information about the planet directly. Large ground-based telescopes have been able to obtain spectra of the atmospheres of extrasolar planets and have found, for example, carbon monoxide and water in these atmospheres.

The most exciting discoveries will probably come with future missions. Future observatories will not only detect Earth-like planets around nearby stars but also measure the planets' physical and chemical characteristics.

Other Planetary Systems

As we noted at the beginning of this section, the search for extrasolar planets has been remarkably successful. Since the first extrasolar planet around a solar-type star was confirmed in 1995, astronomers have found hundreds more stars with planets, many with multiple planets. **Figure 5.25** shows the distribution of planet candidates as of November 2013. Many of these candidates have yet to be confirmed as actual planets. Such confirmation requires follow-up observations that may take many years. The field is changing so fast that the most up-to-date information can only be found online or through mobile applications like the Kepler App.

The first discoveries included many **hot Jupiters**, which are Jupiter-type planets orbiting solar-type stars in tight circular orbits that are closer to their parent stars than Mercury is to our own Sun. These planets were among the first to be detected because they are relatively easy targets for the spectroscopic radial velocity method. The large mass of a nearby hot Jupiter tugs the star very hard, creating large radial velocity variations in the star. In addition, these large planets orbiting close to their parent stars are more likely to pass in front of the star periodically and reveal themselves with the transit method. Therefore, these hot Jupiter systems are easier to find than smaller, more distant planets. Scientists call this bias a *selection effect*.

Many astronomers were surprised by the hot Jupiters because, according to the formation theory they had at the time (based only on our Solar System), these giant, volatile-rich planets should not have been able to form so close to their parent stars. Jupiter-type planets should form in the more distant, cooler regions of the accretion disk, where the volatiles that make up much of their composition are able to survive. Perhaps hot Jupiters formed much farther away from their parent stars and subsequently migrated inward. The mechanism by which a planet could migrate involves an interaction with gas or planetesimals in which orbital angular momentum is transferred from the planet to its surroundings, allowing the planet to spiral inward.

Most of the new planets being discovered by Kepler are mini-Neptunes (gaseous planets in the range of roughly 2–10 Earth masses) or super-Earths (rocky planets more massive than Earth but less massive than Neptune), but improvements in technology will enable smaller planets to be found. Planets with longer orbital periods (and therefore larger orbits) can be discovered only when the observations have gone on long enough to observe more than one complete orbit. Some of the newly discovered planets have highly elliptical orbits, unlike planets in the Solar System. Some orbits of newly discovered planets are highly inclined compared to the plane of rotation of their stars. Yet multiple-planet systems so far discovered tend to reside in flat systems like our own, possibly supporting the accretion disk theory of planet formation. In addition, planets have been found wandering freely through the Milky Way, unattached to any star. These planets may have been ejected from forming systems.

Studies of planetary systems, many unlike our own, challenge some aspects of our understanding of planet formation. Nevertheless, the message conveyed by our discoveries is clear: The formation of planets frequently, and perhaps always, accompanies the formation of stars. The implications of this conclusion are profound. In a galaxy of more than 100 billion stars and a universe of hundreds of billions of galaxies, how many planets, or even moons, with Earth-like conditions might exist? And with all of these Earth-like worlds in the universe, how many might play host to the particular category of chemical reactions that we refer to as "life"? To answer these questions, we must become more familiar with our own Solar System, as we will do in the next few chapters.

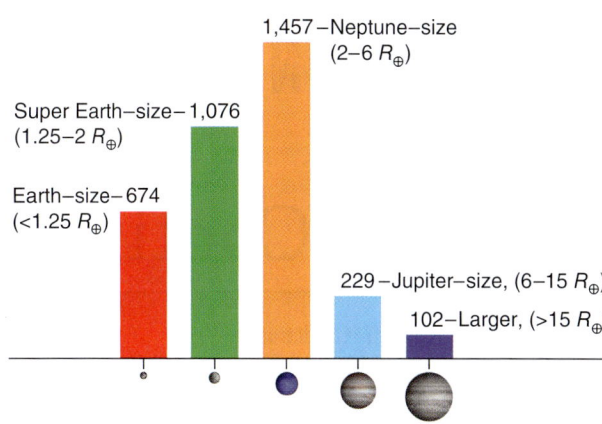

Figure 5.25 Planetary systems have been discovered around hundreds of stars other than the Sun, confirming what astronomers have long suspected—that planets are natural and common by-products of star formation. As time passes and technology improves, more smaller planets will be discovered. This graph shows the distribution of sizes (in Earth radii; $R_\oplus$) of planet candidates from the Kepler mission.

READING ASTRONOMY News

NASA's Kepler mission may be disabled, but researchers say the best results are yet to come!

Kepler's Continuing Mission

By **RACHEL COURTLAND, IEEE Spectrum**

In early August, the moment that Bill Borucki had been dreading finally arrived. As the principal investigator of NASA's Kepler space telescope, Borucki had been working with his colleagues to restore the spacecraft's ability precisely to point itself. The planet-hunting telescope has four reaction wheels—essentially, electrically driven flywheels—and at least three must be functional to maintain positioning. But in the past year, two of those wheels had been on the fritz. One went off line in July 2012 after showing elevated levels of friction, and a second followed suit in May 2013, effectively ending science operations. After a few months of recovery efforts, the telescope team was finally forced to call it quits, 6 months after the mission was originally scheduled to finish but years before they hoped it would.

The failures mark the end of an era for Kepler. With only two reaction wheels, the telescope can't steady itself well enough to ensure that light from each star hits the same fraction of a pixel on its charge-coupled devices for months on end without deviation. That's what Kepler needs in order to detect, with high precision, the transit of a planet: the slight dip in the brightness of a star that occurs when an orbiting planet crosses in front of it.

But the Kepler spacecraft might still have its uses, and the data it has already gathered almost certainly will. The telescope's managers are currently evaluating proposals for what might be done with a two-wheeled spacecraft. And the telescope's analysis team is gearing up for the rest of the science mission: a 2- to 3-year effort to crawl systematically through the 4 years of data that Kepler has collected since its launch in 2009.

That analysis effort, which will incorporate new machine-learning techniques and a bit of human experimentation, could yield a bounty of new potential planets on top of the 3,500 that Kepler has found so far. "We expect somewhere between several hundred more planets to maybe as many as a thousand," Borucki says. If all goes well, the revised hunt might even uncover the first handful of terrestrial twins—or at the very least, near cousins: roughly Earth-size planets on nearly yearlong orbits around Sun-like stars.

Uncovering those Earth analogues won't be easy. The orbits are slow and the planets themselves are small. "You're looking for a percent of a percent" dip in the brightness of a star, says Jon Jenkins, the telescope's analysis lead. "That's a very demanding and challenging measurement to make."

The task will be made even more difficult by an unexpected complication: Stars vary in brightness due to sunspots and flares, and Kepler's observations reveal that these variations are greater than scientists had previously estimated. Those fluctuations can hide the presence of a planet, reducing the telescope's sensitivity to terrestrial transits by 50 percent.

In April 2012, NASA granted Kepler a 4-year extension that would have compensated for the extra noise. But with the failure of the reaction wheels, Jenkins and his colleagues now must find a different way to uncover planetary signals.

Earlier this year, they moved the data processing from a set of computer clusters containing 700 microprocessors to the Pleiades supercomputer at the NASA Ames Research Center in Moffett Field, California, where they have the use of up to 15,000 of the machine's more than 160,000 cores. The team is also working on implementing a machine-learning process using an algorithm called the random forest, which will be trained with data already categorized by Kepler scientists. Once it's up and running, the software should be able to speedily differentiate false positives and data artifacts from promising candidates. Eventually, Jenkins says, the analysis team will insert fake data into the pipeline to test the performance of both the humans that ordinarily do the processing and the automated algorithms. "We need to know for every planet we detect how many we missed," Jenkins says.

No one can predict exactly how many planets Kepler will find. The telescope's main goal was to determine how common planets are in and around the habitable zones of stars—the areas around stars with the right temperature range for liquid water to be present. Such statistics could help astrophysicists decide how practical it would be to build a space telescope capable of directly detecting light from Earth-like planets, which is necessary to determine whether they have atmospheres that could support life.

For Earth-size planets in settings similar to our own, developing a good statistical estimate will be difficult. With small numbers, the uncertainty in the size of the overall population will be large. "The best-case scenario is that Kepler could still have, with very large error bars, a number for us at the end of the day," says Sara Seager, a professor of planetary

SEE KEPLER

KEPLER (CONT.)

science and physics at MIT and a participating scientist on the team.

Even if Kepler finds no Earth analogues, Seager says, the mission is a success. "Kepler revolutionized exoplanet science and, arguably, big-data astronomy," she says. "We'll see the data being mined for years to come."

Evaluating the News

1. What planet-finding method does Kepler use?
2. Recall the discussion of CCD cameras from Chapter 4. Kepler has 42 CCD chips, each of which measures 50 × 25 mm and has 2,200 × 1,024 pixels. The article states that before the failure, Kepler was steady enough to ensure that "light from each star hits the same fraction of a pixel on its charge-coupled devices for months on end without deviation." What is the approximate size of one of these pixels?
3. The article states that finding Earth analogues is difficult because "the orbits are slow." What do they mean by this, and why does this matter?
4. Even though we have not discussed sunspots yet, you can figure out what a sunspot does to the light of a star from the context of the article. Do sunspots tend to brighten or dim a star?
5. While it is conceptually straightforward to understand this method of detecting planets, the article implies that the actual process of doing so requires enormous computational resources. Why would this be?
6. As of the date of the article, scientists are looking for ways to repurpose the Kepler telescope. Use the Internet to find out if they have been successful and, if so, what they are using the telescope for now that its primary mission is over.

✧ SUMMARY

Stars and planets form from clouds of dust and gas. These clouds collapse under their own gravity, sometimes assisted by external events, such as nearby exploding protostars. As the clouds collapse, they fragment to form multiple protostars. Conservation of angular momentum produces an accretion disk around the protostars that often further fragments to form multiple planets, as well as smaller objects such as asteroids and dwarf planets, through the gradual accumulation of dust into larger and larger objects. There are multiple methods for finding planets around other stars, and these planets are now thought to be very common. This field of study is evolving very quickly as technology advances.

LG 1 Gravity pulls clumps of gas and dust together, causing them to shrink and heat up. Angular momentum must be conserved, leading to both a spinning central star and an accretion disk that rotates and revolves in the same direction as the central star.

LG 2 Dust grains in the accretion disk first stick together because of collisions and static electricity. As these objects grow, they eventually have enough mass to attract other objects gravitationally. Once this occurs, they begin emptying the space around them. Collisions of planetesimals lead to the formation of planets.

LG 3 As particles orbit the forming star, those on rising tracks impact those on falling tracks. The upward and downward motions cancel, and the cloud of dust and gas flattens into a plane. Conservation of angular momentum determines both the speed and the direction of the revolution of the objects in the forming system.

LG 4 Near the central protostar, the temperature is higher. This forces volatile elements, like water, to evaporate and leave the inner part of the disk. Planets in the inner part of the system will have fewer volatiles than those in the outer part of the disk.

LG 5 Astronomers find planets around other stars using a variety of methods: the radial velocity method, the transit method, gravitational lensing, astrometry, and direct imaging. As technology improves, the number and variety of known extrasolar planets has increased dramatically, with thousands of planets and planet candidates discovered in just the past few years. Although the first extrasolar planets around other stars were not discovered until the 1990s, since then detection rates have risen dramatically, and astronomers now know of planetary systems of all kinds: systems with many planets, systems that are much like the Solar System, and systems that are completely unlike the Solar System. This great diversity, and the remarkable rate of discovery, implies that planets are very common around stars.

SUMMARY SELF-TEST

1. Which of the following are reasons that gravity is important to star and planet formation? (Choose all that apply.)
   a. Gravity determines the direction in which the system rotates.
   b. Gravity causes the cloud to collapse.
   c. Without gravity, tiny dust particles would never come together to form larger particles.
   d. Gravity causes the collapsing cloud to form a disk because it acts downward more strongly than it acts inward.
   e. Once bodies are large enough, gravity pulls them together to make even larger bodies.
   f. Gravity causes atmospheres to form around planetesimals.

2. When dust grains first begin to grow into larger objects, this occurs because of
   a. gravity between dust grains.
   b. gravity from the central star.
   c. collisions between dust grains.
   d. gravity from large planetesimals.
   e. collisions of dust grains with large planetesimals.

3. The direction of revolution in the plane of the Solar System was determined by
   a. the plane of the galaxy in which the Solar System sits.
   b. the direction of the gravitational force within the original cloud.
   c. the direction of rotation of the original cloud.
   d. the amount of material in the original cloud.

4. The terrestrial planets are different from the giant planets because when they formed,
   a. the inner Solar System was richer in heavy elements.
   b. the inner Solar System was hotter than the outer Solar System.
   c. the outer Solar System took up a bigger volume, so there was more material to form planets.
   d. the inner Solar System was moving faster, so centrifugal force was more important.

5. The radial velocity method preferentially detects
   a. large planets close to the central star.
   b. small planets close to the central star.
   c. large planets far from the central star.
   d. small planets far from the central star.
   e. none of the above. (The method detects all of these equally well.)

6. Planetary systems are probably
   a. exceedingly common—nearly every star has planets.
   b. common—many stars in our galaxy have planets.
   c. rare—few stars have planets.
   d. exceedingly rare—only one star has planets.

7. Nuclear reactions require very high _____ and _____.
   a. temperature; density
   b. volume; density
   c. density; area
   d. mass; area
   e. temperature; mass

8. The transit method preferentially detects
   a. large planets close to the central star.
   b. small planets close to the central star.
   c. large planets far from the central star.
   d. small planets far from the central star.
   e. none of the above. (The method detects all of these equally well.)

9. Which of the following are true? An "Earth-like" planet
   a. has life on it.
   b. has water on it.
   c. has physical properties similar to Earth's.
   d. orbits a Sun-like star.

10. A planet in the "habitable zone"
    a. is close to the central star.
    b. is far from the central star.
    c. is the same distance from its star as Earth is from the Sun.
    d. is at a distance where liquid water can exist on the surface.
    e. is extremely rare—none have yet been found.

QUESTIONS AND PROBLEMS

Multiple Choice and True/False

11. **T/F:** All molecular clouds are held together solely by gravity.
12. **T/F:** Gravity and angular momentum are both important in the formation of planetary systems.
13. **T/F:** A protostar has nuclear reactions inside.
14. **T/F:** Volatile materials are solid only at low temperatures.
15. **T/F:** The Solar System formed from a giant cloud of dust and gas that collapsed under gravity.
16. Figure 5.4 shows a number of curves for objects of different temperatures. Suppose you observe a star with a temperature of 4500 K. What color would the peak wavelength be?
    a. red
    b. orange
    c. yellow
    d. green
    e. blue

17. Figure 5.4 shows a number of curves for objects of different temperatures. Suppose you observe a star with a temperature of 10,000 K. Where would its spectrum lie on this graph?
    a. below the $T = 2000$ K curve
    b. between the $T = 3000$ K curve and the $T = 4000$ K curve
    c. between the $T = 4000$ K curve and the $T = 6000$ K curve
    d. above the $T = 6000$ K curve

18. Figure 5.5 shows a molecular cloud in which stars are currently forming. Of the three visible pillars in this image, one is bright and two are dark. From front to back, how are these pillars and the light source arranged?
    a. colored pillar, light source, black pillars
    b. black pillars, light source, colored pillar
    c. light source, colored pillar, black pillars
    d. colored pillar, black pillars, light source

19. Molecular clouds collapse because of
    a. gravity.
    b. angular momentum.
    c. static electricity.
    d. nuclear reactions.

20. Because angular momentum is conserved, an ice-skater who throws her arms out will
    a. rotate more slowly.
    b. rotate more quickly.
    c. rotate at the same rate.
    d. stop rotating entirely.

21. Clumps grow into planetesimals by
    a. gravitationally pulling in other clumps.
    b. colliding with other clumps.
    c. attracting other clumps with opposite charge.
    d. both a and b.

22. The terrestrial planets and the giant planets have different compositions because
    a. the giant planets are much larger.
    b. the terrestrial planets are closer to the Sun.
    c. the giant planets are mostly made of solids.
    d. the terrestrial planets have few moons.

23. Of the following, which planet still has its primary atmosphere?
    a. Mercury   c. Mars
    b. Earth   d. Jupiter

24. Extrasolar planets have been detected by the
    a. spectroscopic radial velocity method.
    b. transit method.
    c. gravitational lensing method.
    d. direct imaging method.
    e. all of the above

25. Examine Figure 5.16. Why does Mars have no water ice at the distance of its orbit?
    a. because the inner disk was too hot
    b. because the inner disk was too dense
    c. because Earth is the only place where water (in any form) can be found in the Solar System
    d. because the outer planets are larger and have more gravity

Conceptual Questions

26. Compare the size of our Solar System with the size of the universe.

27. Describe the formation of the Solar System in a few sentences of your own.

28. Examine Figure 5.7. Suppose that the universe were different, and in step 2, the shrinking star caused the gravity to decrease, instead of increase, so that the pressure arrow became longer than the gravity arrow. What would this mean for the formation of stars?

29. Physicists describe certain properties, such as angular momentum and energy, as being "conserved." What does this mean? Do these conservation laws imply that an individual object can never lose or gain angular momentum or energy? Explain your reasoning.

30. How does the law of conservation of angular momentum control a figure-skater's rate of spin?

31. Look under your bed for "dust bunnies." If there aren't any, look under your roommate's bed, the refrigerator, or any similar place that might have some. Once you find them, blow one toward another. Watch carefully and describe what happens as they meet. What happens if you repeat this with another dust bunny? Will these dust bunnies ever have enough gravity to begin pulling themselves together? If they were in space, instead of on the floor, might that happen? What force prevents their mutual gravity from drawing them together into a "bunny-tesimal" under your bed?

32. Study the image of Mercury in Figure 5.18. How does this image support the idea that many, many planetesimals were once zooming around the early Solar System? Some of the large craters have smaller craters inside them. Which happened first, a larger crater or the smaller ones? How do you know? What does this tell you about their relative ages? This reasoning becomes extremely important in the next chapter.

33. There are two reasons why the inner part of an accretion disk is hotter than the outer part. What are they?

34. Why were the four giant planets able to collect massive gaseous atmospheres, whereas the terrestrial planets could not?

35. Explain the fate of the original atmospheres of the terrestrial planets.

36. What happened to all the leftover Solar System debris after the last of the planets formed?

37. Examine Figure 5.21. Redraw this figure looking straight down on the system. Now draw a series of pictures from that same orientation, showing one complete orbit of Jupiter around the Sun. Label the motions of the Sun and Jupiter (toward, away, neither) as they would be viewed by an observer off the page to the right. Are the Sun and Jupiter ever on the same side of the center of mass?

38. Examine Figure 5.22. Redraw this figure, paying close attention to where the line on the graph drops in brightness. Now, add three more graphs. In the first, show what happens to the light curve if the planet crosses much closer to the bottom of the star. In the second, show what happens to the light curve if the planet is much larger than the one in Figure 5.22. In the third, show what happens to the light curve if the planet crosses the precise middle of the star, but from top to bottom instead of side to side.

39. Many planets that astronomers have found orbiting other stars are giant planets with masses more like that of Jupiter than of Earth and with orbits located very close to their parent stars. How did this affect our understanding of the formation of planetary systems?

40. Step outside and look at the nighttime sky. Depending on the darkness of the sky, you may see dozens or hundreds of stars. Would you expect many or very few of those stars to be orbited by planets? Explain your answer.

Problems

41. Use information about the planets given in Appendix 2 to answer these questions:
    a. What is the total mass of all the planets in our Solar System, expressed in Earth masses ($M_\oplus$)?
    b. What fraction of this total planetary mass does Jupiter represent?
    c. What fraction does Earth represent?

42. Suppose a very young star has a peak wavelength of 0.97×10^{-6} m.
    a. In what region of the spectrum is this peak wavelength?
    b. What is its temperature?

43. The asteroid Vesta has a diameter of 530 km and a mass of about 3×10^{20} kg.
    a. Calculate the density of Vesta assuming it is spherical.
    b. The density of water is 1,000 kilograms per cubic meter (kg/m³) and that of rock is about 2,500 kg/m³. What does this difference tell you about the composition of this primitive body?

44. You observe a spectral line of hydrogen at a wavelength of 502.3 nm in a distant star. The rest wavelength of this line is 486.1 nm. What is the radial velocity of this star? Is it moving toward or away from Earth?

45. The best current technology can only measure radial velocities larger than about 0.3 m/s. Suppose that you are observing an iodine line with a wavelength of 575 nm. How large a shift in wavelength would a radial velocity of 1 m/s produce?

46. Earth tugs the Sun around as it orbits, but it causes a small effect on the radial velocity of the Sun, only 0.09 m/s. How large a shift in wavelength does this cause in the Sun's spectrum at 575 nm?

47. If an alien astronomer observed a plot of the light curve as Jupiter passed in front of the Sun, by how much would the Sun's brightness drop during the transit?

48. Suppose a planet has been found around a solar-mass star. It orbits the star in 200 days.
    a. What is the orbital radius of this extrasolar planet?
    b. Compare its orbit with that of Mercury around our own Sun. What environmental conditions must this planet experience?

49. The Kepler satellite has detected a planet with a diameter of 1.7 Earth diameters.
    a. How much larger is the volume of this planet than Earth's volume?
    b. Assume that the density of the planet is the same as Earth's density. How much more massive is this planet than Earth?

50. Suppose the planet in Problem 49 has a temperature of 400 K.
    a. What is the flux from a square meter of its surface?
    b. What is its total luminosity?
    c. What is the peak wavelength of its emission?

SMARTWORK

Norton's online homework system includes algorithmically generated versions of these questions, plus additional conceptual exercises. If your instructor assigns questions in SmartWork, log in at smartwork.wwnorton.com.

Exploration | Exploring Extrasolar Planets

wwnpag.es/uou2

Visit the Student Site (wwnpag.es/uou2) and open the Exoplanet Radial Velocity Simulator Nebraska Simulation in Chapter 5. This applet has a number of different panels that allow you to experiment with the variables that are important for measurement of radial velocities. First, in the window labeled "Visualization Controls," check the box to show multiple views. Compare the views shown in panels 1–3 with the colored arrows in the last panel to see where an observer would stand to see the view shown. Start the animation (in the "Animation Controls" panel), and allow it to run while you watch the planet orbit its star from each of the views shown. Stop the animation, and in the "Presets" panel, select "Option A" and then click "set."

1. Is Earth's view of this system most nearly like the "side view" or most nearly like the "orbit view"?

2. Is the orbit of this planet circular or elongated?

3. Study the radial velocity graph in the upper right panel. The blue curve shows the radial velocity of the star over a full period. What is the maximum radial velocity of the star?

4. The horizontal axis of the graph shows the "phase," or fraction of the period. A phase of 0.5 is halfway through a period. The vertical red line indicates the phase shown in views in the upper left panel. Start the animation to see how the red line sweeps across the graph as the planet orbits the star. The period of this planet is 365 days. How many days pass between the minimum radial velocity and the maximum radial velocity?

5. When the planet moves away from Earth, the star moves toward Earth. The sign of the radial velocity tells the direction of the motion (toward or away). Is the radial velocity of the star positive or negative at this time in the orbit? If you could graph the radial velocity of the planet at this point in the orbit, would it be positive or negative?

In the "Presets" window, select "Option B" and then click "set."

6. What has changed about the orbit of the planet as shown in the views in the upper left panel?

7. When is the planet moving fastest: when it is close to the star or when it is far from the star?

8. When is the star moving fastest: when the planet is close to it or when it is far away?

9. Explain how an astronomer would determine, from a radial velocity graph of the star's motion, whether the orbit of the planet was in a circular or elongated orbit.

10. Study the Earth view panel at the top of the window. Would this planet be a good candidate for a transit observation? Why or why not?

In the "System Orientation" panel, change the inclination to 0.0.

11. Now is Earth's view of this system most nearly like the "side view" or most nearly like the "orbit view"?

12. How does the radial velocity of the star change as the planet orbits?

13. Click the box that says "show simulated measurements," and change the "noise" to 1.0 m/s. The gray dots are simulated data, and the blue line is the theoretical curve. Use the slider bar to change the inclination. What happens to the radial velocity as the inclination increases? (Hint: Pay attention to the vertical axis as you move the slider, not just the blue line.)

14. What is the smallest inclination for which you would find the data convincing? That is, what is the smallest inclination for which the theoretical curve is in good agreement with the data?

6 Terrestrial Worlds in the Inner Solar System

The four innermost planets in our Solar System—Mercury, Venus, Earth, and Mars—are known collectively as the *terrestrial planets*. Although the Moon is not a planet but rather Earth's only natural satellite, we discuss it here because of its similarities to the terrestrial planets. The similarities and differences of these worlds highlight fundamental questions about them. For example, when we explain why the Moon is covered with craters, we must also explain why preserved craters on Earth are rare. An explanation for why Venus has a very dense atmosphere should also explain why Earth and Mars do not. Comparing worlds with one another teaches us what shapes a planet, both on the surface and in the interior.

The illustration on the opposite page shows a photograph of the Moon taken with a large telescope. A student is working out the relative ages of the craters and has identified a very young surface, with few craters on it. From images such as these, we can find the relative ages of different features, and comparisons to the ages of Moon rocks give the actual ages of the features.

✦ LEARNING GOALS

By the end of this chapter, you should be able to look at an image similar to that on the opposite page and identify which geological features occurred early in the history of that world and which occurred late. You should also be able to:

LG 1 Explain the four processes that shape a terrestrial planet's surface.

LG 2 Identify the relative ages of parts of a planet's surface from the concentration of craters, and explain how radiometric dating tells us the ages of rocks.

LG 3 Explain how scientists combine theory and observation to determine the structure of planetary interiors.

LG 4 Describe tectonism and volcanism and the forms they take on different planets.

LG 5 Describe the ways in which erosion modifies and wears down surface features.

6.1 Impacts Help Shape the Terrestrial Planets

Figure 6.1 In December 1968, *Apollo 8* astronauts photographed our planet in the sky above the Moon.

For most of human history, Earth has seemed vast and separate from the stars and planets. This view of our planet changed forever with a single snapshot taken in December 1968 by *Apollo 8* astronauts looking back at Earth while orbiting the Moon (**Figure 6.1**). This changed our perspective of Earth's place in the universe. We now know that Earth and the other terrestrial planets share a common origin and common physical processes.

When comparing planets, we first compare the basic physical characteristics, such as distance from the Sun, size and density, and gravitational pull at the surface. These characteristics reveal what a planet is made of, what its surface temperature is likely to be, and how well it can hold an atmosphere. **Table 6.1** compares the basic physical properties of the terrestrial planets and the Moon. These physical properties affect the geological properties of the planet. For

TABLE | 6.1

Comparison of Physical Properties of the Terrestrial Planets and the Moon

	MERCURY	VENUS	EARTH	MARS	MOON*
Orbital semimajor axis (AU)	0.387	0.723	1.000	1.524	384,000 km
Orbital period (years)[†]	0.241	0.615	1.000	1.881	27.32^d
Orbital velocity (km/s)	47.4	35.0	29.8	24.1	1.02
Mass ($M_\oplus = 1$)	0.055	0.815	1.000	0.107	0.012
Equatorial radius (km)	2,440	6,052	6,378	3,397	1,738
Equatorial radius ($R_\oplus = 1$)	0.383	0.949	1.000	0.533	0.272
Density (water = 1)	5.43	5.24	5.51	3.93	3.34
Rotation period[†]	58.65^d	243.02^d	23^h 56^m	24^h 37^m	27.32^d
Tilt of axis (degrees)[‡]	0.04	177.36	23.45	25.19	6.68
Surface gravity (m/s^2)	3.70	8.87	9.78	3.71	1.62
Escape velocity (km/s)	4.25	10.36	11.18	5.03	2.38

*The Moon's orbital radius and orbital period are given in kilometers and days, respectively. The Moon's orbital radius and velocity are given relative to the Earth.
[†]The superscript letters *d*, *h*, and *m* stand for days, hours, and minutes of time, respectively.
[‡]A tilt greater than 90° indicates that the planet rotates in a retrograde, or backward, direction.

example, planets without atmospheres are affected more by impacts than planets with an atmosphere.

Impacts and Craters

Although all terrestrial planets are subject to *tectonism, volcanism, impact cratering,* and *erosion,* the relative importance of each of these processes varies. **Impact cratering**—a process resulting from the collision of solid planetary bodies—leaves distinctive scars known as **impact craters** that tell of large collisions in the past. Of these processes, large impacts cause the most concentrated and sudden release of energy. Planets and other objects orbit the Sun at very high speeds; Earth, for example, orbits at an average speed of around 30 kilometers per second (km/s), equivalent to 67,000 miles per hour (mph). Collisions between orbiting bodies release huge amounts of energy. **Figure 6.2a** shows the process of impact cratering. When an object hits a planet, its energy heats and compresses the surface and throws material (labeled "ejected material" in the figure) far from the resulting impact crater. Sometimes this material falls back to the surface of the planet with enough energy to cause more scars known as **secondary craters**. The rebound of heated and compressed material can also lead to the formation of a central peak or a ring of mountains within the crater walls, as shown in the lunar crater in Figure 6.2b.

▶▶ **Nebraska Simulation:** Solar System Properties Explorer

▶‖ **AstroTour:** Processes That Shape the Planets

Figure 6.2 (a) Stages in the formation of an impact crater. (b) A lunar crater photographed by *Apollo* astronauts, showing the crater wall and central peak surrounded by ejected material, rays, and secondary craters—all typical features associated with impact craters.

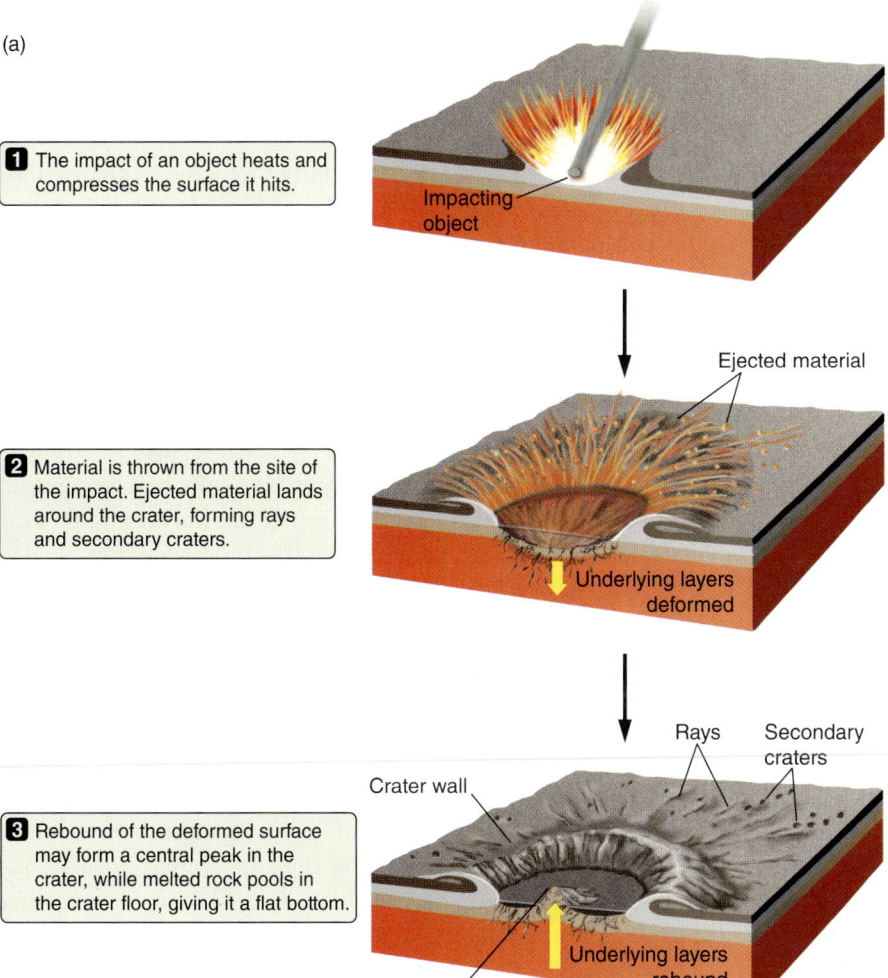

Figure 6.3 Meteor Crater (also known as Barringer Crater), located in northern Arizona, is an impact crater 1.2 km in diameter. It was formed some 50,000 years ago by the collision of a nickel-iron asteroid fragment with Earth.

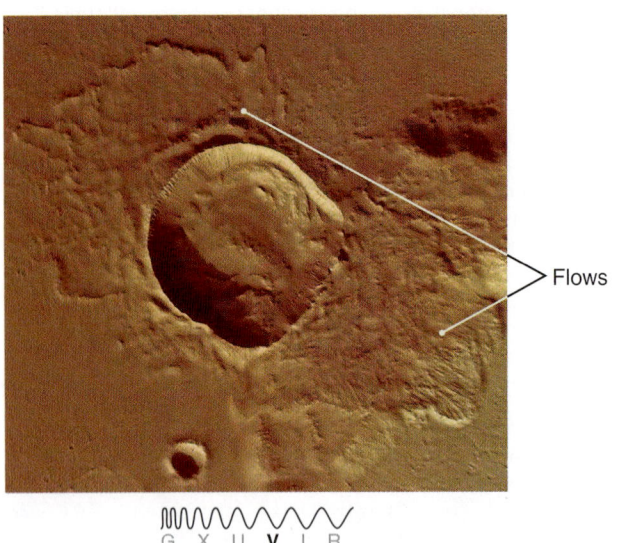

Figure 6.4 Some craters on Mars look like those formed by rocks thrown into mud, suggesting that material ejected from the crater contained large amounts of water. This crater is about 20 km across.

Vocabulary Alert

vaporize In common language, this often means "to destroy completely," with connotations that even the stuff the object was made of is not there anymore. Astronomers use this word specifically to mean "turn into vapor." Here, the rocks are being turned into gas by the energy of the collision.

The energy of an impact can be great enough to melt or even **vaporize** rock. The floors of some large craters are the cooled surfaces of melted rock that flowed as lava. The energy released in an impact can also lead to the formation of new minerals. Because some minerals form *only* during an impact, they are evidence of ancient impacts on Earth's surface. The space rocks that cause these impacts are referred to by three closely related terms: **Meteoroids** are small (less than 100 meters) cometary or asteroid fragments in space. A meteoroid that enters and burns up in a planetary atmosphere is called a **meteor**. Any meteoroids that survive to hit the ground are known as **meteorites**.

One of the best-preserved impact craters on our planet (**Figure 6.3**) is Meteor Crater in Arizona. This crater was caused by an impact about 50,000 years ago. From the crater's size and shape and from the remaining pieces of the impacting body, we know that the nickel-iron asteroid fragment was about 50 meters across, had a mass of about 300 million kilograms (kg), and traveled at 13 km/s when it hit the upper atmosphere. Approximately half of the original mass was vaporized in the atmosphere before the remainder hit the ground. This collision released about 300 times as much total energy as the first atomic bomb. Yet, at only 1.2 km in diameter, Meteor Crater is tiny compared with impact craters seen on the Moon and with more ancient impact scars on Earth.

Impact craters cover the surfaces of Mercury, Mars, and the Moon. On Earth and Venus, by comparison, most impact craters have been destroyed. Fewer than 200 impact craters have been identified on Earth, and about 1,000 have been found on Venus. Earth's crater shortage is primarily due to *plate tectonics* in Earth's ocean basins and erosion on land, while lava flows on Venus have destroyed its craters.

The surface of the Moon is directly exposed to bombardment from space. However, the surfaces of Earth and Venus are partially protected by their atmospheres. This is another reason for the shortage of craters on Earth and Venus. Rock samples from the Moon show craters smaller than a pinhead, formed by micrometeoroids. In contrast, most meteoroids smaller than 100 meters in diameter that enter Earth's atmosphere are either burned up or broken up by friction with the atmosphere before they reach Earth's surface. Small meteorites found on the ground are probably pieces of much larger bodies that broke up upon entering the atmosphere. With an atmosphere far thicker than that of Earth, Venus is even better protected.

We can tell a lot about the surface of a planet by studying its craters. Craters on the Moon's surface are often surrounded by strings of smaller secondary craters formed from material thrown out by the impact, like those shown in Figure 6.2b. Some craters on Mars have a very different appearance. These craters are surrounded by structures that look much like the pattern you might see if you threw a rock into mud (**Figure 6.4**). The flows appear to indicate that the martian surface rocks contained water or ice at the time of the impact. Not all martian craters have this feature, so the water or ice must have been concentrated in only some areas, and these icy locations might have changed with time.

At the time these craters formed, there may have been liquid water on the surface of Mars. Features resembling canyons and dry riverbeds are further evidence of this hypothesis. Another explanation for the appearance of these craters involves a change in surface temperature that might have occurred when the meteoroids hit causing liquid water to exist temporarily. Today, the surface of Mars is dry or frozen, which suggests that water that might once have been on the surface has soaked into the ground, much like water frozen in the ground in Earth's polar regions. The energy released by an impact could melt this ice, turning the surface material into slurry with a consistency much like wet concrete. When thrown from the crater by the force of the impact, this slurry would hit the surrounding terrain and slide across the surface, forming the mud-like craters we see today.

Calibrating a Cosmic Clock

Because many planetesimals were roaming around the early Solar System, every planet experienced a period of heavy bombardment early in its history. The number of *visible* craters on a planet is determined by the rate at which those craters are destroyed. Geological activity on planets such as Earth, Mars, and Venus erases most evidence of early impacts. By contrast, the Moon's surface still preserves the scars of craters dating from about 4 billion years ago. The lunar surface has remained essentially unchanged for more than a billion years because the Moon has no atmosphere or surface water and a cold, geologically dead interior. Mercury also has well-preserved craters, although recent evidence from the *Messenger* mission shows tilted crater floors that are higher on one side than the other—evidence that internal forces lifted the floors unevenly after the craters formed. Planetary scientists use the cratering record to estimate the **age** of planetary surfaces—extensive cratering signifies an older planetary surface. Look back at the chapter-opening illustration, which shows a student beginning to work out the relative ages of features by the number of craters and their relative positions. How does she know this is a young surface?

We can use the amount of cratering as a clock to measure the relative ages of surfaces. But to determine the exact age of a surface based on the number of craters, we need to know how fast the clock runs. In other words, we need to "calibrate the cratering clock."

Radioactive elements naturally decay into other elements. The relative amounts of these elements change over time. Scientists find the ages of different lunar regions by measuring these relative amounts—a process called **radiometric dating** (see **Working It Out 6.1**). Between 1969 and 1976, *Apollo* astronauts and Soviet unmanned probes visited the Moon and brought back samples from nine different locations on the lunar surface. The results of that work were surprising. Although smooth areas on the Moon were indeed younger than heavily cratered areas, they were still very old. The oldest, most heavily cratered regions on the Moon date back to about 4.4 billion years ago, whereas most of the smoother parts of the lunar surface are typically 3.1 billion to 3.9 billion years old. As you can see in **Figure 6.5**, almost all of the cratering in the Solar System took place within its first billion years.

Vocabulary Alert

age In common language, we might consider the age of a planet to be the length of time since it formed. But the age of the surface of a planet can be much younger than that, due to volcanism or tectonics moving material from the inside to the outside of the planet, for example. It is important to be aware of the distinction between the age of the planet and the age of its surface. This is not so strange. You are much older than the outermost layer of your skin, which is replaced about every month.

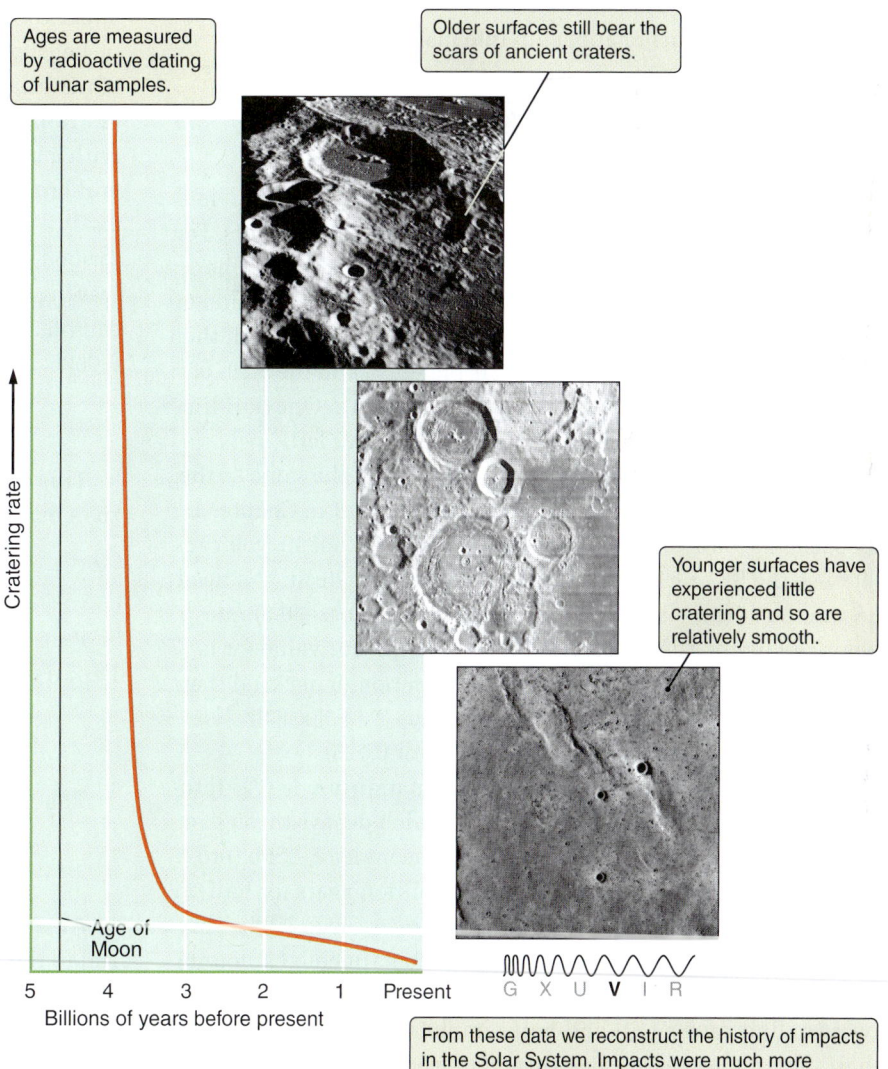

Figure 6.5 Radiometric dating of lunar samples returned from specific sites by *Apollo* astronauts is used to determine how the cratering rate has changed over time. Cratering records can then be used to tell us the age of other parts of the lunar surface.

Working It Out 6.1 | How to Read Cosmic Clocks

A geologist can find the age of a mineral by measuring the relative amounts of a radioactive element, known as a **radioisotope**, and the decay products it turns into. An isotope is an atom with the same number of protons but a different number of neutrons as other atoms of the same element. The radioactive element is known as the **parent element**, and the decay products are called **daughter products**. The time interval over which a radioisotope decays by half is called its **half-life**. With every half-life that passes, the remaining amount will decrease by a factor of 2. For example, over three half-lives, the final amount, P_F, of a parent radioisotope will be ½ × ½ × ½ = ⅛ of its original amount, P_O. If we express the number of half-lives more generally as n, then we can translate this relationship into math:

$$\frac{P_F}{P_O} = \left(\frac{1}{2}\right)^n$$

This equation has an exponential expression on the right-hand side. This means that we are raising one-half to the power of n. Try to remember some of the rules of exponents. For example, if $n = 0$, the right-hand side of this equation will be 1. That makes sense. If no half-lives have gone by, then the ratio of the final amount to the initial amount should be 1. If $n = 1$, then the right-hand side of the equation is one-half, and the ratio of the final amount to the initial amount is one-half. One half-life has gone by, and we have one-half of the material left.

Calculator hint: This kind of exponent is not the same as the "times 10 to the" operation that we learned about in Chapter 1. Because the base of the exponent in this case is not 10, you cannot use the EE or EXP key on your calculator. Instead, you need to use the x^y (sometimes y^x) key. First, type the base (0.5) into your calculator, then hit x^y, and then type in your value for n. For example, if you are calculating the fraction of material that is left after three half-lives, you would type [0][.][5][x^y][3][=] to find the answer (try it: you should get ⅛, or 0.125).

The most abundant form of the element uranium is ^{238}U (pronounced "uranium two-thirty-eight"), which decays through a series of reactions to an isotope of the element lead, ^{206}Pb (pronounced "lead two-oh-six"). It takes 4.5 billion years for half of a sample of ^{238}U to decay through these processes to ^{206}Pb. This means that a sample of pure ^{238}U would contain equal amounts of uranium and lead after 4.5 billion years had passed. If we were to find such a half-and-half sample, we would know that half the uranium atoms had turned to lead, so that

$$\frac{P_F}{P_O} = \frac{1}{2}$$

Compare this to the previous equation, and convince yourself that if $(½)^n = (½)$, then $n = 1$. So, we find that the mineral formed one half-life, or 4.5 billion years, ago.

When analyzing the chemical composition of rocks to find their age, you have the ratio on the left, not the number of half-lives. The number of half-lives is what you are looking for! Solving for this in the general case requires a nontrivial mathematical manipulation involving logarithms, so it may be easiest to try reasonable numbers in your calculator until you find approximately the right one.

Let's look at an example of finding the age from the fraction of material that has decayed, this time with a different form of uranium (^{235}U) that decays to a different form of lead (^{207}Pb) with a half-life of 700 million years. Suppose that a lunar mineral brought back by astronauts has 15 times as much ^{207}Pb as ^{235}U. Assuming that the sample was pure uranium when it formed, this means that ¹⁵⁄₁₆ of the ^{235}U has decayed to ^{207}Pb, leaving only ¹⁄₁₆ of the parent element remaining in the mineral sample. Now we know that the left side of the equation is ¹⁄₁₆, or 0.0625:

$$\frac{1}{16} = \left(\frac{1}{2}\right)^n$$

This is less than ½, so we know that n is bigger than 1 (more than one half-life has passed). Let's try 2:

$$\text{Does } \frac{1}{16} = \left(\frac{1}{2}\right)^2 \text{ ? No.}$$

Putting that into the calculator gives 0.25—that's too big. Let's try 5:

$$\text{Does } \frac{1}{16} = \left(\frac{1}{2}\right)^5 \text{ ? No.}$$

That gives 0.03125—too small. Try 4:

$$\text{Does } \frac{1}{16} = \left(\frac{1}{2}\right)^4 \text{ ? Yes!}$$

Aha! After $n = 4$ half-lives, the ratio of the remaining material to the original material is 0.0625, or ¹⁄₁₆. To find out how much time has passed, we multiply the number of half-lives by the length of a half-life: 4 × 700 million = 2.8 billion years old.

6.2 The Surfaces of Terrestrial Planets Are Affected by Processes in the Interior

While impact cratering is driven by forces external to the planet, two other important processes, tectonism and volcanism, are determined by conditions in the interior of the planet. To understand these processes, we must understand the structure and composition of the interiors. But how do we know what the interiors of planets are like? The deepest holes ever drilled are about 12 km deep; tiny when compared to Earth's radius of 6,378 km. Even so, scientists have determined a lot about the interior of Earth.

Interior Composition

The composition of Earth's interior can be determined in two different ways. In one approach, the mass of Earth can be found from the strength of Earth's gravity. Dividing the mass by the volume gives the average density: about 5,500 kilograms per cubic meter (kg/m^3) (see **Working It Out 6.2**). But rocky surface material averages only 2,900 kg/m^3. Because the density of the whole planet is greater than the density of the surface, the interior must contain material denser than surface rocks. The interior contains large amounts of iron, which has a density of nearly 8,000 kg/m^3. Another approach to determine the composition of Earth's interior is by considering meteorites. Because meteorites and Earth formed at the same time out of similar materials, we can reason that the overall composition of Earth should resemble the composition of meteorite material, which includes minerals with large amounts of iron. From these considerations, planetary scientists can determine the composition of Earth's interior.

Building a Model of Earth's Interior

It is impossible to drill down into Earth's core to observe Earth's interior structure directly. Instead, geologists use the laws of physics and the properties of materials and how they behave at different temperatures and pressures to model

---

Working It Out 6.2 | **The Density of Earth**

Density is defined as the mass divided by the volume. The mass of Earth can be determined from its gravitational pull to be 5.97×10^{24} kg. Earth's average radius is given in Table 6.1 as 6,378 km, or 6,378,000 meters. If we assume that Earth is a perfect sphere, then we can calculate the volume, V, from the radius, R, using the formula for the volume of a sphere:

$$V = \frac{4}{3} \pi R^3$$

$$V = \frac{4}{3} \pi (6,378,000 \text{ m})^3$$

$$V = 1.087 \times 10^{21} \text{ m}^3$$

To find the density, we must divide the mass by this volume:

$$\text{Density} = M \div V$$

$$\text{Density} = (5.97 \times 10^{24} \text{ kg}) \div (1.087 \times 10^{21} \text{ m}^3)$$

$$\text{Density} = 5,490 \text{ kg/m}^3$$

This is the average density of Earth. Because the rocky surface material has a much lower density than this average density, the interior density must be much higher to bring the density of the whole Earth up to roughly double the density of surface rocks.

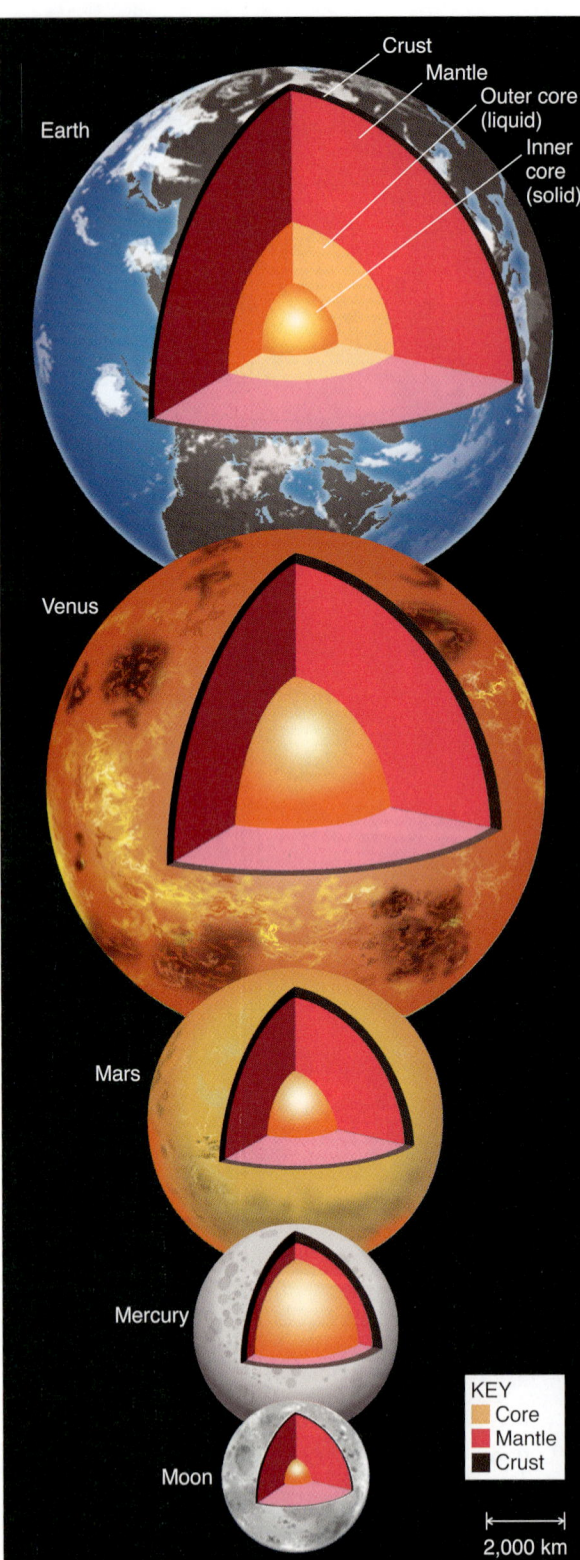

Figure 6.6 Comparing models of the interiors of the terrestrial planets and Earth's Moon, we see that the thicknesses of the components changes from one planet to another. Some fractions of the cores of Mercury, Venus, and Mars are probably liquid.

the structure of Earth's interior. The pressure at any point in Earth's interior must be just high enough that the outward forces balance the inward force of the weight of all the material above that point. From this information, a layered model of Earth's interior can be constructed. Waves produced by earthquakes, called **seismic waves**, can be used to test this model by comparing seismic-wave measurements at seismic stations around the globe with simulations of seismic waves traveling through the layered model. The extent to which the predictions agree with the observations points out both strengths and weaknesses of the model. Geologists adjust the model—always remaining consistent with the known physical properties of materials—until a good match is found between prediction and observation.

This is the method geologists used to arrive at our current picture of the interior of Earth. The innermost region of Earth's interior consists of a two-component **core**. Earth's core is primarily composed of iron, nickel, and other dense metals. Outside of the core lies a solid portion called the **mantle**, which is made of medium-density materials. Covering the mantle, the **crust** is a thin, hard layer of lower-density materials that is chemically distinct from the interior.

The cross sections in **Figure 6.6** show the structures of each of the terrestrial planets and the structure of Earth's Moon. As you can see, Earth's interior is not uniform. The materials have been separated by density, a process known as **differentiation**. When rocks of different types are mixed together, they tend to stay mixed. Once this rock melts, however, the denser materials sink to the center and the less dense materials float toward the surface. Today, little of Earth's interior is molten; but the differentiated structure shows that Earth was once much hotter, and its interior was liquid throughout. The cores of all the terrestrial planets and the core of the Moon were once molten.

The Moon's Structure and Formation

As Figure 6.6 shows, the Moon has only a tiny core. This core is composed of material similar to that of Earth's mantle. The best explanation of the Moon's composition and the size of its core hypothesizes a collision between Earth and a Mars-sized protoplanet when the Solar System was still full of planetesimals. The collision blasted off and vaporized some of Earth's crust and mantle. The debris from this collision formed the Moon. During the vaporization stage of the collision, most gases were lost to space. This explanation accounts for the similar composition of the Moon and Earth's mantle. It also explains the Moon's relative lack of water and other volatiles while Earth, Mars, and Venus are volatile rich. However, the origin of the Moon is still an active area of astronomical research, as this hypothesis does not yet explain every detail of the Moon's geology and composition.

The Evolution of Planetary Interiors

A number of factors influence how the interior of a planet evolves, which depends primarily upon its temperature. Factors that influence how the temperature changes over time include the size of the planet, the composition of the material, and heating from various sources, among others. The balance between energy received and energy produced and emitted governs the temperature within a planet. In particular, here we are concerned with thermal energy—the kinetic energy in a substance that determines the temperature. In general, the interior of a planet cools down over time as heat is emitted from the surface. Because it takes time for heat to travel through rock, the deeper we go within a planet, the higher the temperature climbs. This is similar to the effect of taking a hot pie out of the oven. Over time, the pie radiates heat from the surface and cools down, but the filling takes much longer to cool than the crust.

Cooling Planets lose thermal energy from their surfaces primarily through radiation. When you hold your hand over the stove to find out whether it is hot, you are detecting infrared radiation from the stove through your skin. All objects radiate energy, and the hotter they are, the more energy they radiate. The type of energy radiated (infrared, optical, ultraviolet, and so forth) also depends on the temperature of the object. Hotter objects emit more of their energy at shorter wavelengths.

The rate at which a planet cools depends on its size. A larger planet has a larger volume of matter and more thermal energy trapped inside. Thermal energy has to escape through the planet's surface, so the planet's surface area determines the rate at which energy is lost. Smaller planets have more surface area in comparison with their small volumes, so they cool off faster, whereas larger planets have a smaller surface area to volume ratio and cool off more slowly. Smaller objects like Mercury and the Moon are less geologically active than the terrestrial planets—Venus, Earth, and Mars—because their interiors are cooler.

Radioactive Heating Some of the thermal energy in the interior of Earth is left over from when Earth formed. The tremendous energy of collisions and the energy from short-lived radioactive elements melted the planet. As the surface of Earth radiated energy into space, it cooled rapidly. A solid crust formed above a molten interior. Because a solid crust does not transfer thermal energy well, it helped to retain the remaining heat. Over a long time, energy from the interior of the planet continued to leak through the crust and radiate into space. The interior of the planet slowly cooled, and the mantle and the inner core solidified.

Most of the rest of the thermal energy in Earth's interior comes from long-lived radioactive elements trapped in the mantle. As these radioactive elements decay, they release energy, which heats the planet's interior. Equilibrium between radioactive heating of the interior and the loss of energy to space determines Earth's interior temperature. As radioactive elements decay, the amount of thermal energy generated declines, and Earth's interior becomes cooler as it ages.

Tidal Heating If thermal energy were the only source of heating in Earth's interior, the outer core would have solidified completely. Given today's high internal temperatures, additional sources of energy must continue to heat the interior of Earth. One source of continued heating is friction from tidal effects of the Moon and Sun. Think about holding a ball of Silly Putty in your hand. If you pull it and squish it and pull it and squish it repeatedly, it will heat up. A similar process happens in the interiors of planets as they are pulled and squished by the gravity of other celestial objects.

Tidal effects are caused by the change in the strength of gravity across a solid object. **Figure 6.7** shows how the force of gravity on the part of Earth nearest the Moon is stronger than the force of gravity on the part of Earth farthest from the Moon. This stretches Earth and causes a tidal bulge. While the oceans respond dramatically to the tidal stresses from the Moon and Sun, these tidal stresses also affect the *solid* body of Earth. Earth is somewhat elastic (like a rubber ball), and tidal stresses cause the ground to rise under you by about 30 centimeters (cm) twice each day. This takes a lot of energy, which ends up converted to heat. On

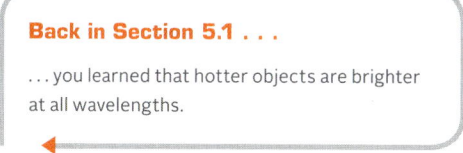

Back in Section 5.1 . . .
. . . you learned that hotter objects are brighter at all wavelengths.

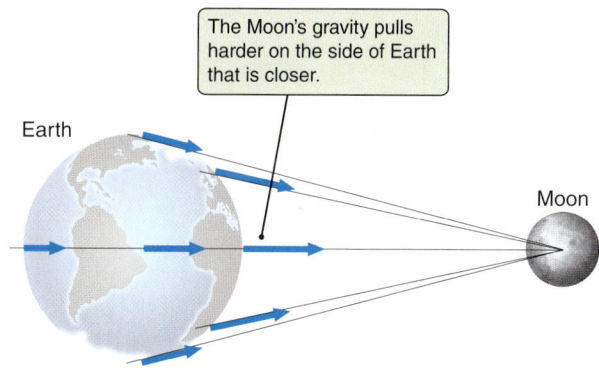

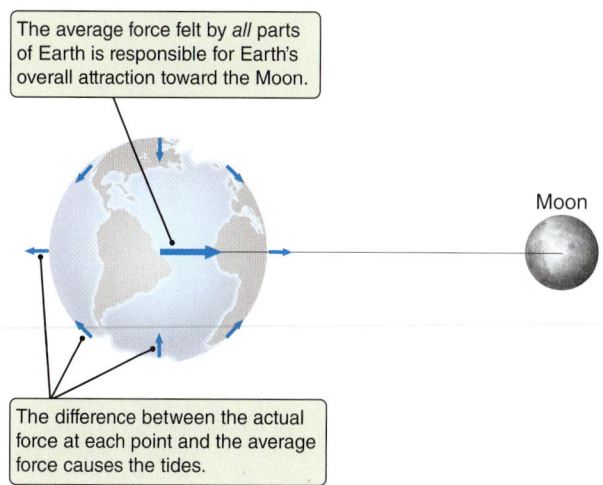

Figure 6.7 Tides stretch Earth along the line between Earth and the Moon but compress Earth perpendicular to this line.

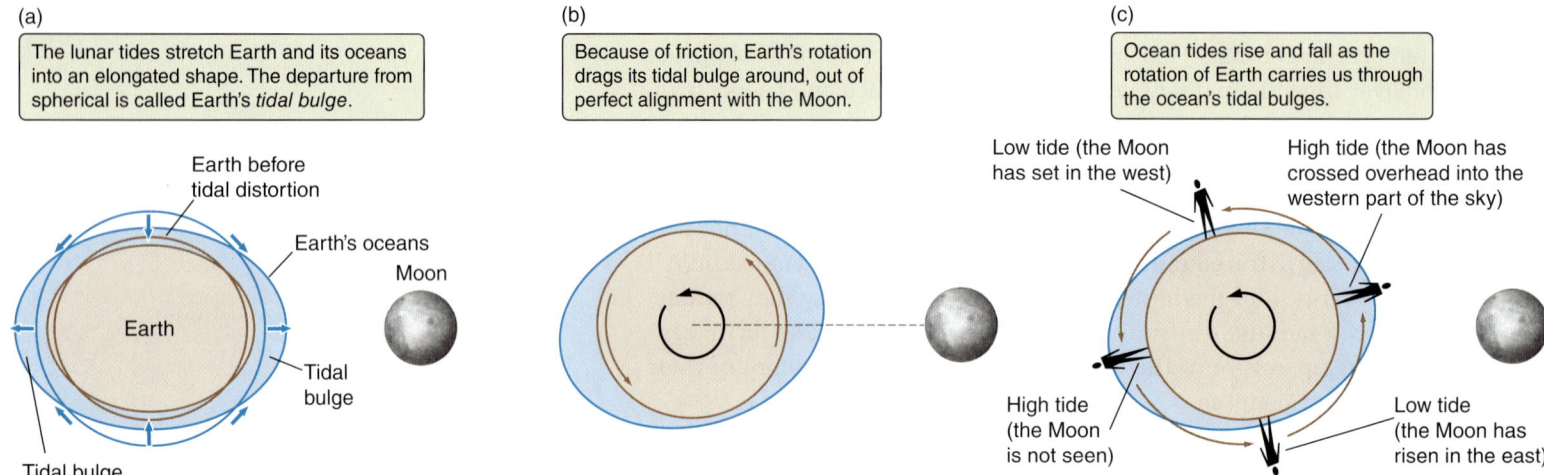

Figure 6.8 (a) Tidal effects pull Earth and its oceans into a tidal bulge. (b) Earth's rotation pulls its tidal bulge slightly out of alignment with the Moon. (c) As Earth's rotation carries us through these bulges, we experience the ocean tides. The magnitude of the tides has been exaggerated in these diagrams for clarity. In the diagrams, the observer is looking down from above Earth's North Pole. Sizes and distances are not to scale.

Earth, friction from these tidal effects of the Moon and Sun contribute a small amount—about 6 percent—to the heating in the interior.

This stretching of the mass of one body due to the gravitational pull of another is called a **tide**. On Earth, this stretching is most noticeable in the rise and fall of the oceans, as Earth rotates through a bulge caused by the gravity of the Moon and of the Sun. The rise and fall of the oceans are also called tides. The gravitational pull of the Moon causes Earth to stretch along a line pointing approximately in the direction of the Moon. The resulting tides are called **lunar tides**. The gravitational pull of the Sun causes Earth to stretch along a line pointing approximately in the direction of the Sun. The resulting tides, which are about half as strong as those caused by the Moon, are called **solar tides**. **Figure 6.8** illustrates how the rotation of Earth drags the tides slightly ahead of the position of the Moon in the sky. As you can see in **Figure 6.9**, when the Moon, Earth, and Sun are all in a line—at new and full Moon—the lunar and solar tides overlap. This creates

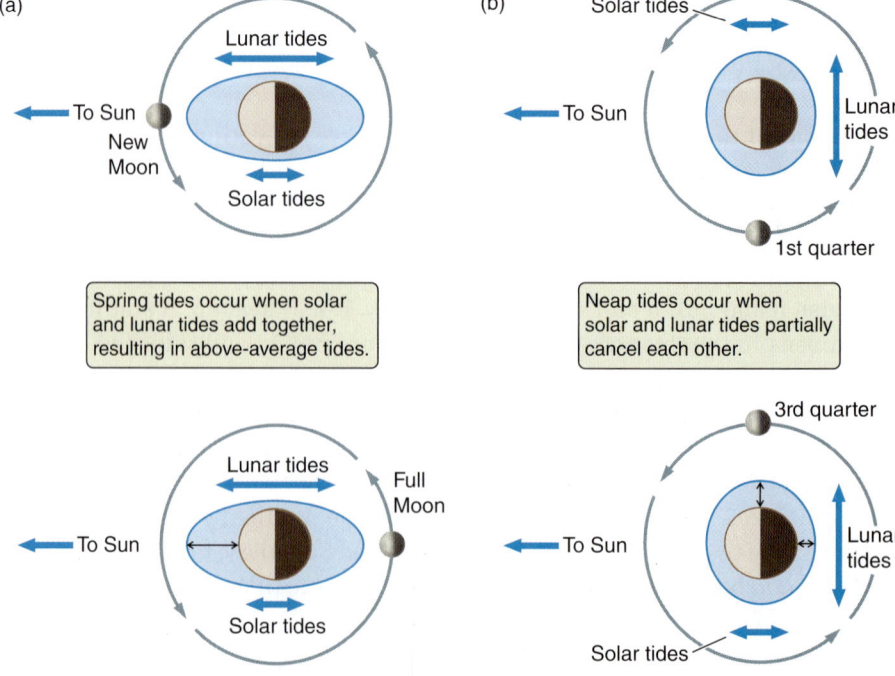

Figure 6.9 Solar tides are about half as strong as lunar tides. The interactions of solar and lunar tides result in either (a) spring tides when they are added together or (b) neap tides when they partially cancel each other.

more extreme tides ranging from extra-high high tides to extra-low low tides. An extreme tide created by this effect is called **spring tide**—not because of the season, but because the water appears to spring out of the sea. When the Moon, Earth, and Sun make a right angle, at the Moon's first and third quarters, the lunar and solar tides stretch Earth in different directions, creating less extreme tides. This is called **neap tide** from the Saxon word *neafte*, which means "scarcity"; at these times of the month, shellfish and other food gathered in the tidal region are less accessible because the low tide is higher than at other times.

▶︎ll **AstroTour:** Tides and the Moon

The Effect of Temperature and Pressure on Material Temperature plays an important role in a planet's interior structure, but it's not the only influence—whether a material is solid or liquid depends on pressure as well. Higher pressure forces atoms and molecules closer together and makes the material more likely to become a solid. Toward the center of Earth, the effects of temperature and pressure oppose each other: The higher temperatures make it more likely that material will melt, but the higher pressures favor a solid form. In the outer core of Earth the high temperature wins, allowing the material to exist in a molten state. At the center of Earth, even though the temperature is higher, the pressure is so great that the inner core of Earth is solid.

Magnetic Fields

Planetary magnetic fields are created when the right conditions exist within the planet's interior. A **magnetic field** is created by moving charges and exerts a force on magnetically reactive objects, such as iron. A compass is a familiar example on Earth. A compass needle lines up with Earth's magnetic field and points "north" and "south," as shown in **Figure 6.10a**. The north-pointing end does not point at the *geographic* North Pole (about which Earth spins), but rather at a location in the Arctic Ocean off the coast of northern Canada. This is Earth's north *magnetic* pole. Earth's south magnetic pole is off the coast of Antarctica, 2,800 km from Earth's geographic South Pole. Earth behaves as if it contained a giant bar mag-

Figure 6.10 (a) Earth's magnetic field can be visualized as though it were a giant bar magnet tilted relative to Earth's axis of rotation. Compass needles line up along magnetic field lines and point toward Earth's north magnetic pole. (b) Iron filings sprinkled around a bar magnet help us visualize such a magnetic field.

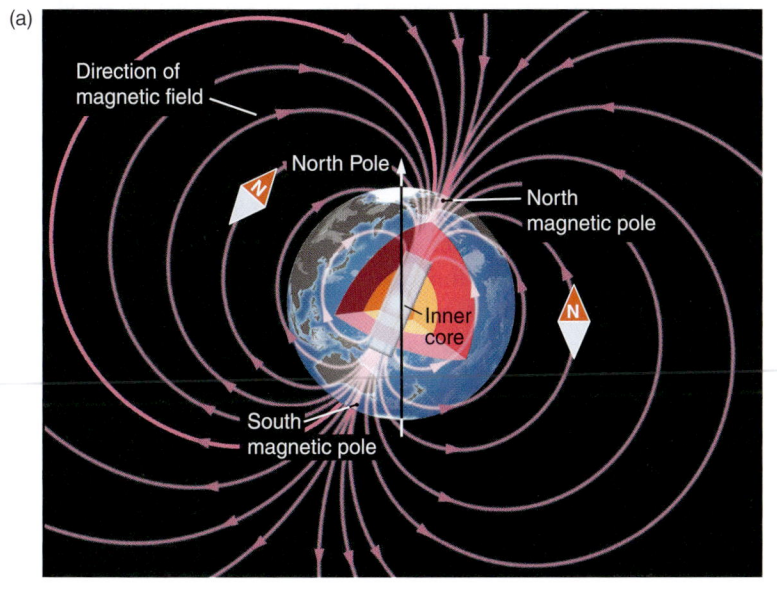
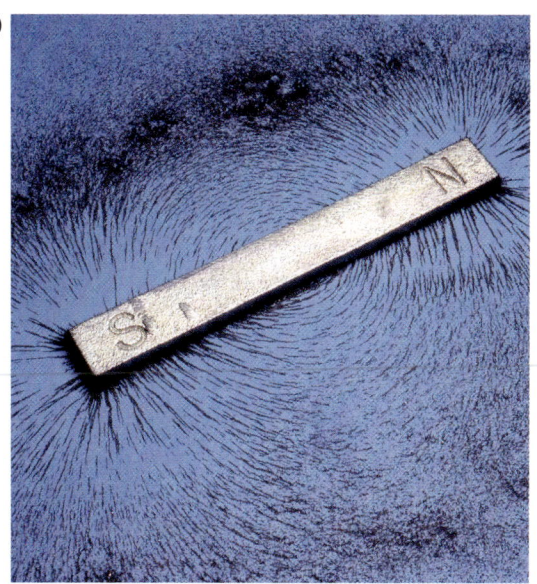

VISUAL ANALOGY

net that was slightly tilted with respect to Earth's rotation axis and had its two endpoints near the two magnetic poles, as shown in Figure 6.10b.

Earth's magnetic field is not actually due to a bar magnet buried within the planet. A magnetic field is the result of moving electric charges. Earth's magnetic field is created by the combination of Earth's rotation about its axis and a liquid, electrically conducting, circulating outer core. From this combination, Earth converts mechanical energy into magnetic energy. The magnetic field of a planet is an important probe into its internal structure.

Earth's magnetic field is constantly changing. At the moment, the north magnetic pole is traveling several tens of kilometers per year toward the northwest. If this rate and direction continue, the north magnetic pole could be in Siberia before the end of the century. The magnetic pole tends to wander, constantly changing direction as a result of changes in the core.

The geological record shows that much more dramatic changes in the magnetic field have occurred over the history of our planet. When a magnet made of material such as iron gets hot enough, it loses its magnetization. As the material cools, it again becomes magnetized by any magnetic field surrounding it. Thus, iron-bearing minerals record the Earth's magnetic field at the time that they cooled. From these minerals, geologists learn how Earth's magnetic field has changed over time. Although Earth's magnetic field has existed for billions of years, the north and south magnetic poles switch from time to time. On average, these reversals in Earth's magnetic field take place about every half-million years.

During the *Apollo* program, astronauts measured the Moon's local magnetic field, and small satellites have searched for global magnetism. The Moon has a very weak field, possibly none at all, because the Moon is very small and therefore has a solid (not liquid and rotating) inner core. The Moon also has a very small core. However, remnant magnetism is preserved in lunar rocks from an earlier time when the Moon likely had a molten core and a magnetic field.

Other than Earth, Mercury is the only terrestrial planet with a significant magnetic field today. Rotation and a large iron core, parts of which are molten and circulating, cause Mercury's magnetic field. Planetary scientists expected that Venus would have a magnetic field because its mass and distance from the Sun imply an iron-rich core and partly molten interior like Earth's. Its lack of a magnetic field might be attributed to its extremely slow rotation (once every 243.0 Earth days), but this explanation is still uncertain.

Mars has a weak magnetic field, presumably frozen in place early in its history. The magnetic signature occurs only in the ancient crustal rocks. Geologically younger rocks lack this residual magnetism, so the planet's magnetic field has long since disappeared. The lack of a strong magnetic field today on Mars might be the result of its small core. However, given that Mars is expected to have a partly molten interior and has a rotation rate similar to that of Earth, the lack of a field is still not yet fully understood.

6.3 Planetary Surfaces Evolve through Tectonism

Now that we have looked at planetary interiors, we can connect the interior conditions to the processes that shape the surface. The crust and part of the upper mantle form the **lithosphere** of a planet. **Tectonism** modifies the lithosphere, warping, twisting, and shifting it to form visible surface features. If you have been on a drive through mountainous or hilly terrain, you may have noticed places like the one shown in **Figure 6.11**, where the roadway has been cut through rock. The exposed layers tell the story of Earth through the vast expanse of geologic time. In this section, we will look at tectonic processes that create these layers and play an important part in shaping the surface of a planet.

The Theory of Plate Tectonics

Early in the 20th century, it became apparent that parts of Earth's continents could be fitted together like pieces of a giant jigsaw puzzle. In addition, the layers in the rock and the fossil record on the east coast of South America match those on the west coast of Africa. Based on evidence like this, Alfred Wegener (1880–1930) proposed a hypothesis that the continents were originally joined in one large landmass that broke apart as the continents began to "drift" away from each other over millions of years. This hypothesis was further developed into the theory known today as **plate tectonics**.

This idea was met with great skepticism because it was difficult to imagine a mechanism that could move such huge landmasses. In the 1960s, however, studies of the ocean floor provided compelling evidence for plate tectonics. These surveys showed surprising characteristics in bands of *basalt*—a type of rock formed from cooled lava—that were found on both sides of ocean rifts. As **Figure 6.12** shows, hot material in these rifts rises toward Earth's surface, creating new ocean floor. When this hot material cools, it becomes magnetized along the

Figure 6.11 Tectonic processes fold and warp Earth's crust, as seen in the rocks along this roadside cut.

Figure 6.12 (a) As new seafloor is formed at a spreading center, the cooling rock becomes magnetized. The magnetized rock is then carried away by tectonic motions. (b) Maps like this one of banded magnetic structures in the seafloor near Iceland provide support for the theory of plate tectonics.

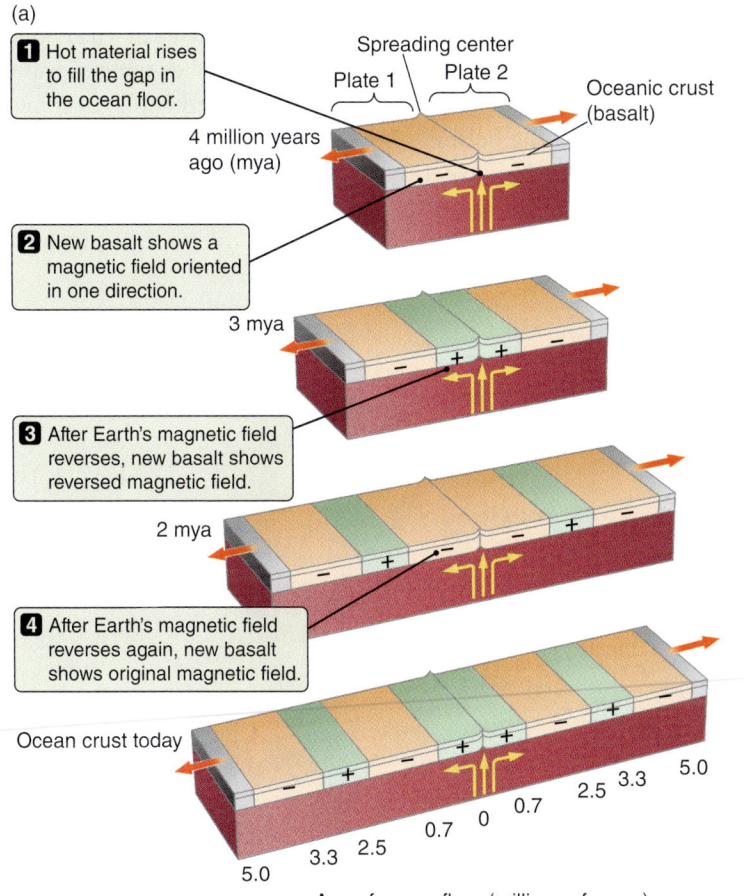

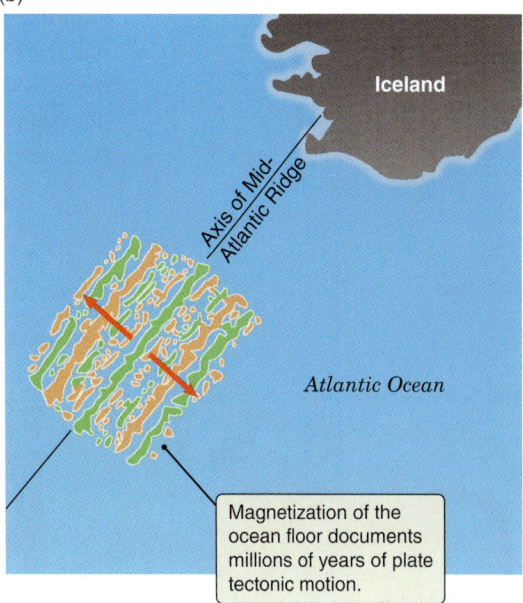

direction of Earth's magnetic field, thus recording the direction of Earth's magnetic field at that time. Greater distance from the rift indicates an older ocean floor and an earlier time. Combined with radiometric dates for the rocks, this magnetic record showed that the continental plates have moved over long geologic time spans.

More recently, precise surveying techniques and global positioning system (GPS) methods have confirmed these results more directly. Some areas are being pulled apart by about the length of a pencil each year. Over millions of years of geologic time, such motions add up. Over 10 million years—a short time by geological standards—15 cm/year becomes 1,500 km, and maps definitely need to be redrawn.

Today, geologists recognize that Earth's crust is composed of a number of relatively brittle segments, or **lithospheric plates**, and that motion of these plates is constantly changing the surface of Earth. The theory of plate tectonics is perhaps the greatest advance in 20th-century geology. Plate tectonics is responsible for a wide variety of geological features on our planet, including the continental drift that Wegener hypothesized.

The Role of Convection

Movement of lithospheric plates requires immense forces. These forces are the result of thermal energy escaping from the interior of Earth. The transport of thermal energy by the movement of packets of gas or liquid is known as **convection**. **Figure 6.13a** illustrates the process of convection, which should be familiar to you if you have ever watched water in a heated pot on a stovetop. Thermal energy from the stove warms water at the bottom of the pot. The warm water expands slightly, becoming less dense than the cooler water above it. The cooler water sinks, displacing the warmer water upward. When the lower-density water reaches the surface, it gives up part of its energy to the air and cools, becomes denser, and sinks back toward the bottom of the pot. Water rises in some locations and sinks in others, forming convection "cells."

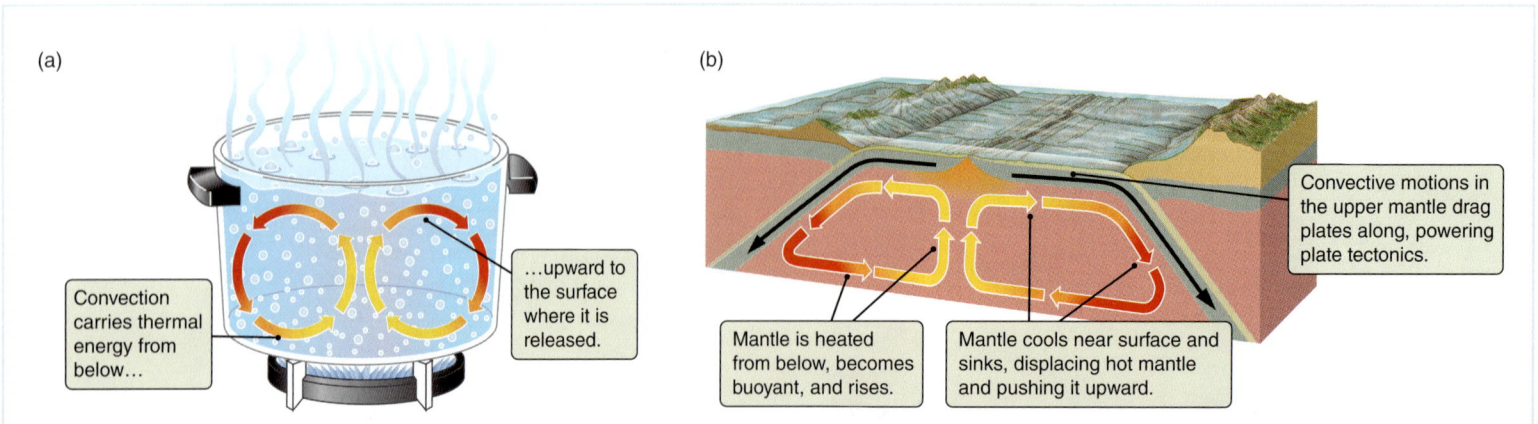

Figure 6.13 (a) Convection occurs when a fluid is heated from below. (b) Similarly, convection in Earth's mantle drives plate tectonics, although the timescale and velocities involved are very different from those in a pot boiling on your stovetop.

VISUAL ANALOGY

Figure 6.13b shows how convection works in Earth's mantle. Radioactive decay provides the heat source to drive convection in the mantle. Earth's mantle is not molten, but it is somewhat mobile and so allows convection to take place very slowly. Earth's crust is divided into seven major plates and a half-dozen smaller plates floating on top of the mantle. Convection cells in Earth's mantle drive the plates, carrying both continents and ocean crust along with them. Convection also creates new crust along rift zones in the ocean basins, where mantle material rises up, cools, and slowly spreads out.

Figure 6.14 illustrates plate tectonics and some of its consequences. If material rises and spreads in one location, then it must converge and sink in another. In sinking regions, one plate slides beneath the other, and convection drags the crust material down into the mantle. The Mariana Trench—the deepest part of Earth's ocean floor—is such a place. Much of the ocean floor lies between rising and sinking zones, and so the ocean floor is the youngest portion of Earth's crust. In fact, the oldest seafloor rocks are less than 200 million years old.

In some places, the plates are not sinking but colliding with each other and being shoved upward. The highest mountains on Earth, the Himalayas, grow 0.5 meter per century as the Indo-Australian Plate collides with the Eurasian Plate. In still other places, plates meet at oblique angles and slide along past each other. One such place is the San Andreas Fault in California, where the Pacific Plate slides past the North American Plate. A **fault** is a fracture in a planet's crust along which material can slide.

Locations where plates meet tend to be very active geologically. One of the best ways to see the outline of Earth's plates is to look at a map of where earth-

▶ll **AstroTour:** Continental Drift

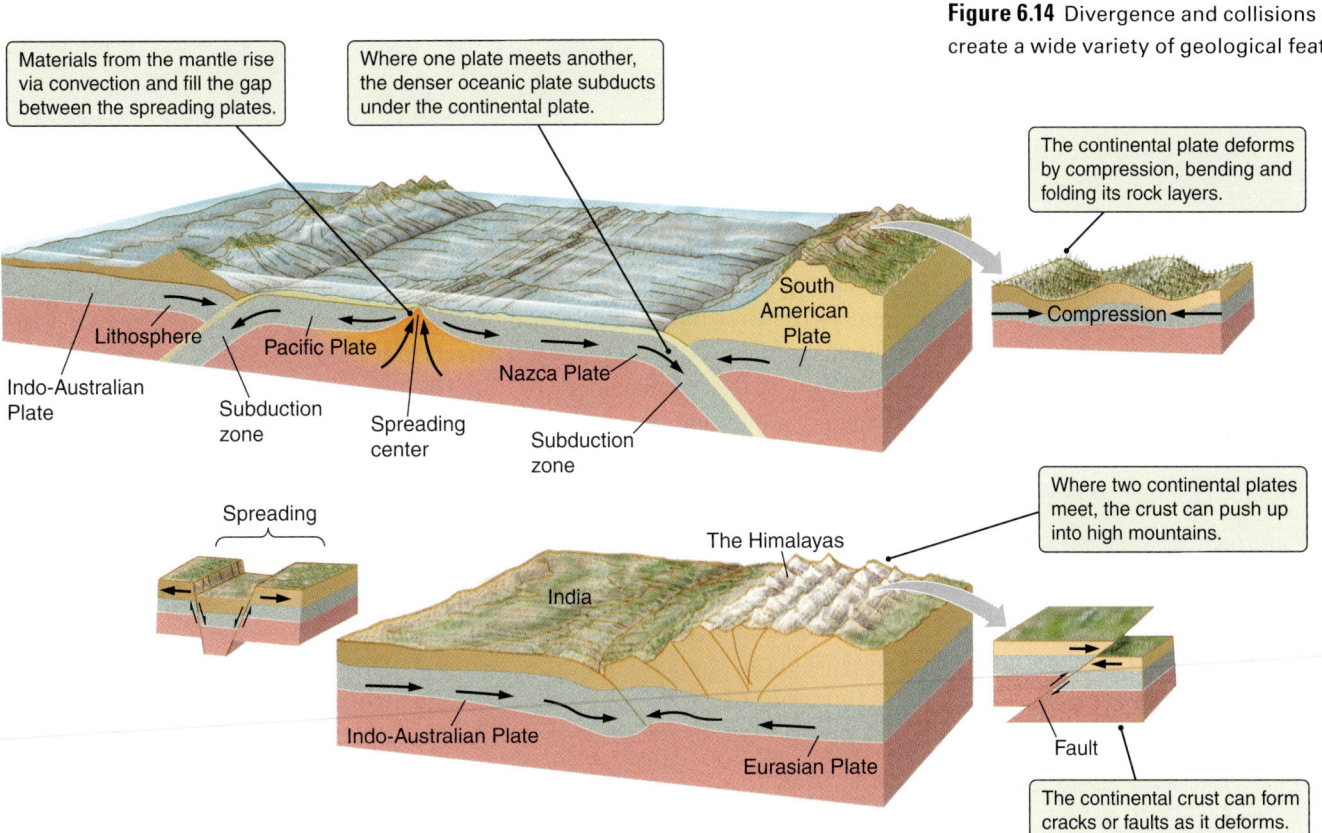

Figure 6.14 Divergence and collisions of tectonic plates create a wide variety of geological features.

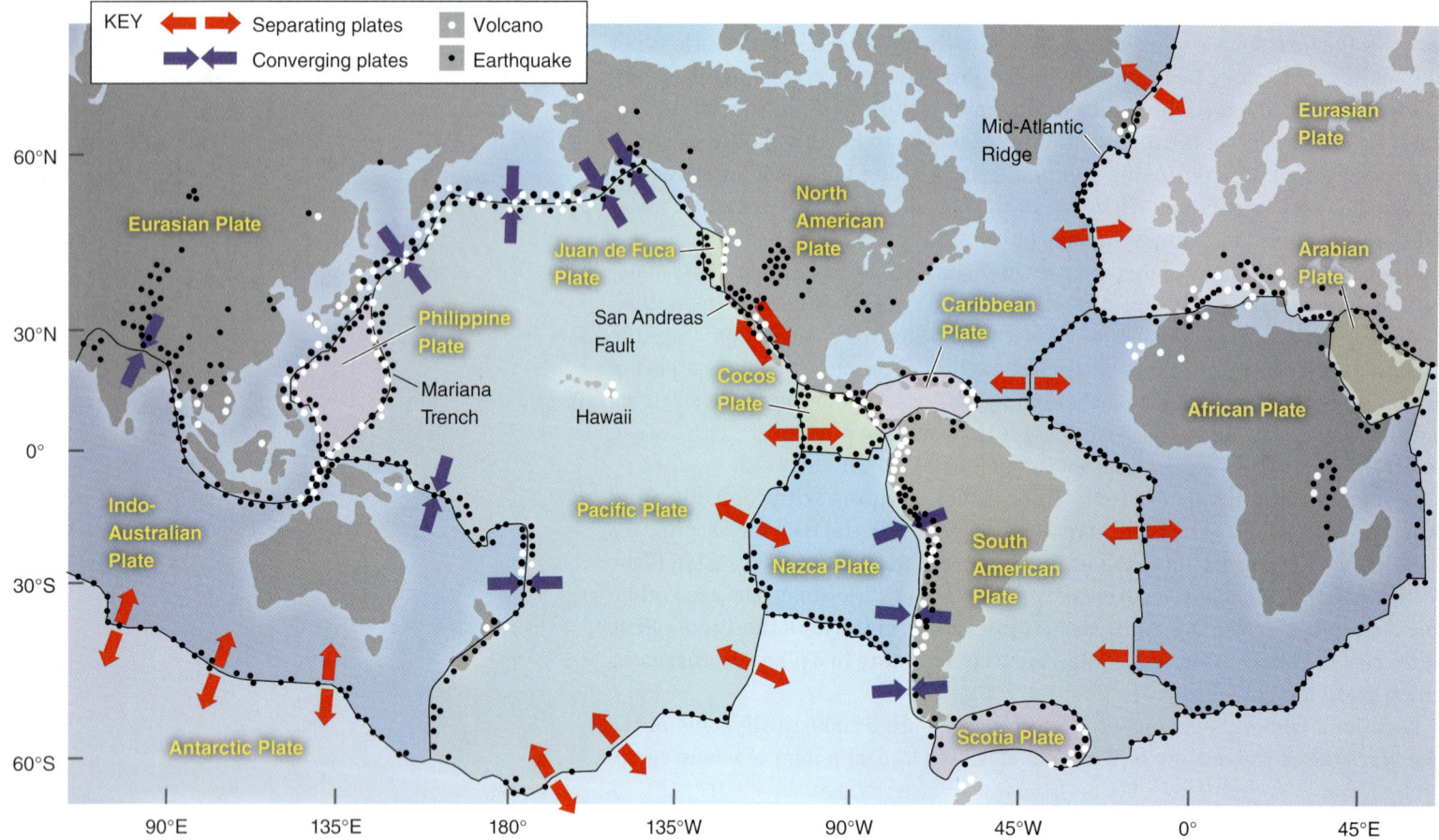

Figure 6.15 Major earthquakes and volcanic activity are often concentrated along the boundaries of Earth's principal tectonic plates.

quakes and volcanism occur, like the map in **Figure 6.15**. Where plates run into each other, enormous stresses build up. Earthquakes occur when a portion of the boundary between two plates finally and suddenly slips, relieving the stress. Volcanoes are created when friction between plates melts rock, which is then pushed up through cracks to the surface. Earth also has numerous **hot spots**, such as the Hawaiian Islands, where hot deep-mantle material rises, releasing thermal energy. As plates shift, some parts move more rapidly than others, causing the plates to stretch, buckle, or fracture. These effects are seen on the surface as folded and faulted rocks. Mountain chains also are common near converging plate boundaries, where plates buckle and break.

Tectonism on Other Planets

We have observed plate tectonics only on Earth. However, all of the terrestrial planets and some moons show evidence of tectonic disruptions. Fractures have cut the crust of the Moon in many areas, leaving fault valleys. Most of these features are the result of large impacts that have cracked and distorted the lunar crust.

Mercury also has fractures and faults. In addition, numerous cliffs on Mercury are hundreds of kilometers long. Like the other terrestrial planets, Mercury was once molten. After the surface of the planet cooled and its crust formed, the interior of the planet continued to cool and shrink. As the planet shrank, Mercury's

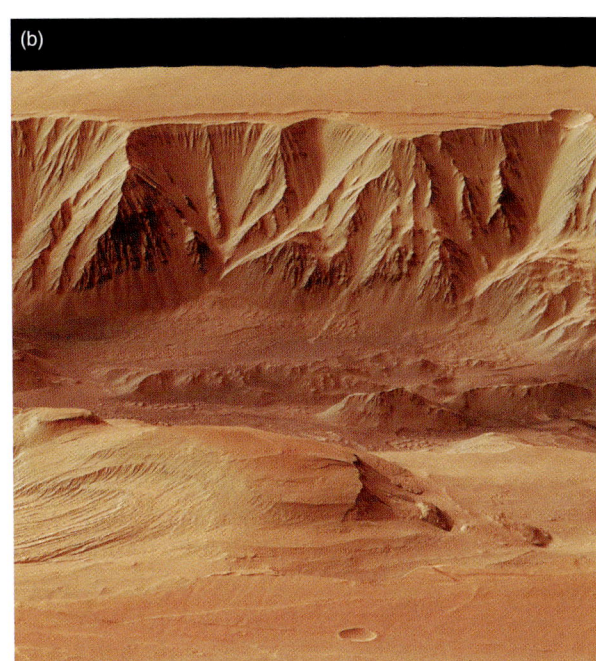

Figure 6.16 (a) A mosaic of *Viking Orbiter* images shows Valles Marineris, the major tectonic feature on Mars, stretching across the center of the image from left to right. This canyon system is more than 4,000 km long. The dark spots on the left are huge shield volcanoes in the Tharsis region. (b) This close-up perspective view of the canyon wall was photographed by the European Space Agency's *Mars Express* spacecraft.

crust cracked and buckled in much the same way that a grape skin wrinkles as it shrinks to become a raisin. To explain the faults seen on the planet's surface, the volume of Mercury must have shrunk by about 5 percent after the planet's crust formed.

Possibly the most impressive tectonic feature in the Solar System is the Valles Marineris on Mars (**Figure 6.16**). Stretching nearly 4,000 km and nearly four times as deep as the Grand Canyon, this chasm, if located on Earth, would link San Francisco with New York. Valles Marineris includes a series of massive cracks in the crust of Mars that are thought to have formed as local forces, perhaps related to mantle convection, pushed the crust upward from below. The surface could not be equally supported by the interior everywhere, and unsupported segments fell in. Once formed, the cracks were eroded by wind, water, and landslides, resulting in the structure we see today. Other parts of Mars have faults similar to those on the Moon, but cliffs like those on Mercury are absent.

The *Magellan* spacecraft mapped about 98 percent of the surface of Venus, providing the first high-resolution views of the surface of the planet. *Magellan*'s view of one face of Venus is shown in **Figure 6.17**. Although Venus has a wealth of volcanic features and tectonic fractures, there is no evidence of lithospheric plates or plate motion of the sort seen on Earth. Yet the relative scarcity of impact craters on Venus suggests that most of its surface is less than 1 billion years old.

The mass of Venus is only 20 percent less than that of Earth, and its radius is only 5 percent smaller than the radius of Earth. Because of the similarities between the two planets, the interior of Venus should be very much like the interior of Earth, and convection should be occurring in its mantle. On Earth, mantle convection and plate tectonics release thermal energy from the interior.

Figure 6.17 The atmosphere of Venus blocks our view of the surface in visible light. This false-color view of Venus is a radar image made by the *Magellan* spacecraft. Bright yellow and white areas are mostly fractures and ridges in the crust. Some circular features seen in the image may be hot spots—regions of mantle upwelling. Most of the surface is formed by lava flows, shown in orange.

On Venus, hot spots may be the principal way that thermal energy escapes from the planet's interior. Circular fractures called *coronae* on the surface of Venus, ranging from a few hundred kilometers to more than 2,500 km across, may be the result of upwelling plumes of hot mantle that have fractured Venus's lithosphere. Alternatively, energy may build up in the interior until large chunks of the lithosphere melt and overturn, releasing an enormous amount of energy. Then the surface cools and solidifies. Our understanding of why Venus and Earth are so different with regard to plate tectonics is still very uncertain.

6.4 Volcanism Reveals a Geologically Active Planet

You are probably familiar with the image of a volcano spewing molten rock onto the surface of Earth. This molten rock, known as **magma**, originates deep in the crust and in the upper mantle, where sources of thermal energy combine. These sources include rising convection cells in the mantle, heating by friction from movement in the crust, and concentrations of radioactive elements. In this section, we will look at the occurrence of volcanic activity on a planet or moon, which is called **volcanism**. Volcanism not only shapes planetary surfaces but also is a key indicator of a geologically active planet.

Terrestrial Volcanism Is Related to Tectonism

Because Earth's thermal energy sources are not uniformly distributed, volcanoes are usually located along plate boundaries and over hot spots. Maps such as the one shown in Figure 6.15 leave little doubt that most terrestrial volcanism is linked to the same forces responsible for plate motions. A tremendous amount of friction is generated as plates slide under each other. This friction raises the temperature of rock toward its melting point.

Material at the base of a lithospheric plate is under a great deal of pressure because of the weight of the plate pushing down on it. This pressure increases the melting point of the material, forcing it to remain solid even at high temperature. As this material is forced up through the crust, the pressure drops; as the pressure drops, so does the material's melting point. Material that was solid at the base of a plate may melt as it nears the surface. Places where convection carries hot mantle material toward the surface are frequent sites of eruptions. Iceland, which is one of the most volcanically active regions in the world, sits astride one such place—the Mid-Atlantic Ridge (see Figure 6.15).

Once lava reaches the surface of Earth, it can form many types of structures. Flows often form vast sheets, especially if the eruptions come from long fractures. If very **fluid** lava flows from a single "point source," it can spread out over the surrounding terrain or ocean floor, forming a **shield volcano,** shown in **Figure 6.18a**. A **composite volcano** forms when thick lava flows alternating with explosively generated rock deposits build a steep-sided structure, shown in Figure 6.18b.

Terrestrial volcanism also occurs where convective plumes rise toward the surface in the interiors of lithospheric plates, creating local hot spots like the Hawaiian islands, shown in Figure 6.18c. Volcanism over hot spots works much like volcanism elsewhere, except that the convective upwelling occurs at a single spot rather than in a line along the edge of a plate. These hot spots

▶|| **AstroTour:** Hot Spot Creating a Chain of Islands

Vocabulary Alert

fluid In common language, this is used mainly as a noun. Here, it is used as an adjective, describing how easily a substance flows. If it is very fluid, it flows like water. If it is not very fluid, it flows like molasses or tar. Another adjective often used in this context is *viscous*. A viscous substance does not flow easily.

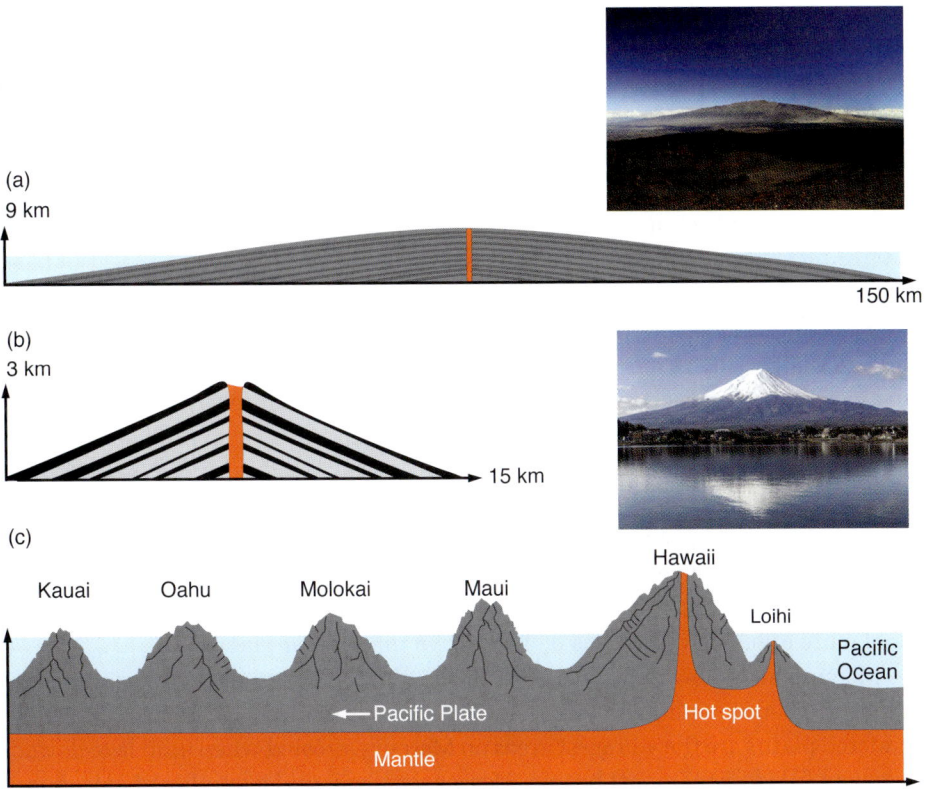

Figure 6.18 Magma reaching Earth's surface commonly forms (a) shield volcanoes, such as Mauna Loa, which have gently sloped sides built up by fluid lava flows; and (b) composite volcanoes, such as Mt. Fuji, which have steeply symmetrical sides built up by viscous lava flows. (c) Hot spots form a series of volcanoes as the plate above them slides by.

force mantle and lithospheric material toward the surface, where it emerges as liquid lava.

Earth has numerous hot spots, including the regions around Yellowstone Park and the Hawaiian Islands. The Hawaiian Islands are a chain of shield volcanoes that formed as their lithospheric plate moved across a hot spot. The island ceases to grow as the plate motion carries the island away from the hot spot. Erosion, occurring since the island's inception, continues to wear the island away. Meanwhile, a new island grows over the hot spot. Today the Hawaiian hot spot is located off the southeast coast of the Big Island of Hawaii, where it continues to power the active volcanoes. On top of the hot spot, the newest Hawaiian island, Loihi, is forming. Loihi is already a massive shield volcano, rising more than 3 km above the ocean floor. Loihi will eventually break the surface of the ocean and merge with the Big Island of Hawaii—but not for another 100,000 years.

Volcanism in the Solar System

Even before the *Apollo* astronauts brought back rock samples that confirmed volcanism, photographs showed flowlike features in the dark regions of the Moon. Some of the first observers to use telescopes thought that these dark areas looked like bodies of water—thus the name **maria** (singular: *mare*), Latin for "seas." The maria are actually vast hardened lava flows, similar to volcanic rocks known

Figure 6.19 This rock sample from the Moon, collected by the *Apollo 15* astronauts from a lunar lava flow, shows gas bubbles typical of gas-rich volcanic materials. This rock is about 6 × 12 cm.

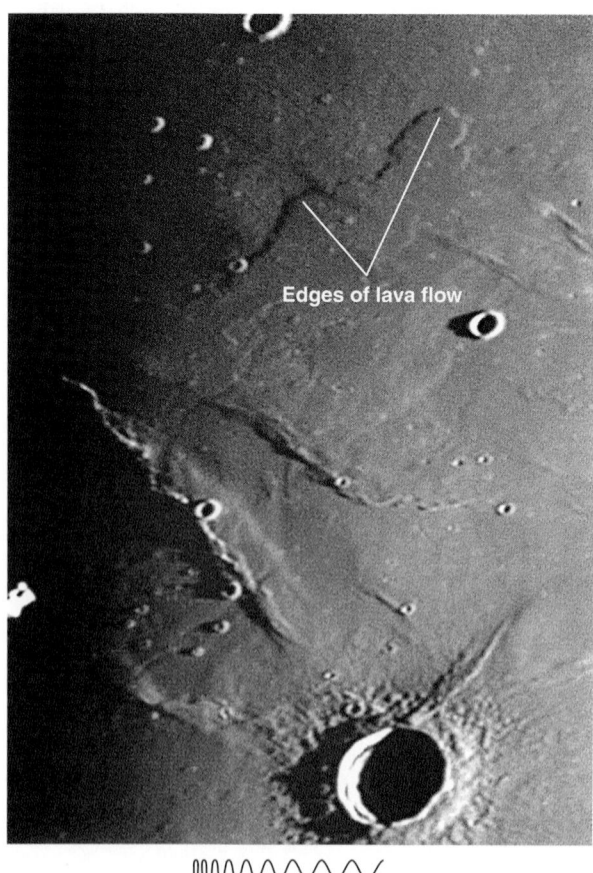

Figure 6.20 Lava flowing across the surface of Mare Imbrium on the Moon must have been relatively fluid to have spread out for hundreds of kilometers in sheets that are only tens of meters thick.

as basalts on Earth. Because the maria contain relatively few craters, these volcanic flows must have occurred after the period of heavy bombardment ceased.

When the *Apollo* astronauts returned rock samples from the lunar maria, many of them were found to contain gas bubbles typical of volcanic materials (**Figure 6.19**). The lava that flowed across the lunar surface must have been relatively fluid (**Figure 6.20**). The fluidity of the lava, due partly to its chemical composition, explains why lunar basalts form vast sheets that fill low-lying areas such as impact basins. It also explains the Moon's lack of classic volcanoes: the lava was too fluid to pile up.

The samples also showed that most of the lunar lava flows are older than 3 billion years. Only in a few limited areas of the Moon are younger lavas thought to exist; most of these have not been sampled directly. Samples from the heavily cratered terrain of the Moon also originated from magma, so the young Moon must have gone through a molten stage. These rocks cooled from a "magma ocean" and are more than 4 billion years old, preserving the early history of the Solar System. Most of the sources of heating and volcanic activity on the Moon shut down some 3 billion years ago—unlike on Earth, where volcanism continues. This conclusion is consistent with our earlier argument that smaller planets should cool more efficiently and thus be less active than larger planets.

Mercury also shows evidence of past volcanism. *Mariner 10* and *Messenger* missions revealed smooth plains similar in appearance to lunar maria. These sparsely cratered plains are the youngest areas on Mercury, created when fluid lava flowed into and filled huge impact basins. *Messenger* also found a number of volcanoes.

Venus has more volcanoes than the other terrestrial planets. Radar images reveal a wide variety of volcanic landforms. These include highly fluid flood lavas covering thousands of square kilometers, enormous shield volcanoes, dome volcanoes, and lava channels thousands of kilometers long. These lavas must have been extremely hot and fluid to flow for such long distances.

Mars has also been volcanically active. More than half the surface of Mars is covered with volcanic rocks. Lava covered huge regions of Mars, flooding the older, cratered terrain. Most of the vents or long cracks that created these flows are buried under the lava that poured forth from them. Among the more impressive features on Mars are its enormous shield volcanoes. These volcanoes are the largest mountains in the Solar System. Olympus Mons, standing 27 km high at its peak and 550 km wide at its base (**Figure 6.21**), would tower over Earth's largest mountains. Despite the difference in size, most of the very large volcanoes of Mars are shield volcanoes, just like their Hawaiian counterparts. Olympus Mons and its neighbors grew as the result of hundreds of thousands of individual eruptions. These volcanoes have remained over their hot spots for billions of years, growing ever taller and broader with each successive eruption. This is evidence that Mars lacks plate tectonics, which would have moved the volcanoes over time, creating chains of volcanoes like the Hawaiian Islands.

Lava flows and other volcanic landforms span nearly the entire history of Mars, estimated to extend from the formation of crust some 4.4 billion years ago to geologically recent times, and to cover more than half of the red planet's surface. "Recent" in this sense could still be more than 100 million years ago. Although some "fresh-appearing" lava flows have been identified on Mars, until rock samples are radiometrically dated we will not know the age of these latest eruptions. Mars could, in principle, experience eruptions today.

Figure 6.21 The largest known volcano in the Solar System, Olympus Mons on Mars is a 27-km-high shield-type volcano, similar to but much larger than Hawaii's Mauna Loa. This oblique view was created from an overhead *Viking* image and topographic data provided by the *Mars Orbiter* laser altimeter.

6.5 Wind and Water Modify Surfaces

Tectonism, volcanism, and impact cratering affect Earth's surface by creating variations in the height of the surface. **Erosion** is the wearing away of a planet's surface by mechanical action. The term *erosion* covers a wide variety of processes. Erosion by running water, wind, and the actions of living organisms wears down hills, mountains, and craters; the resulting debris fills in valleys, lakes, and canyons. If erosion were the only geological process operating, it would eventually smooth out the surface of the planet completely. Because Earth is a geologically and biologically active world, however, its surface is an ever-changing battleground between processes that build up topography and those that tear it down.

Weathering

Weathering is the first step in the process of erosion. During weathering, rocks are broken into smaller pieces and may be chemically altered. For example, rocks on Earth are physically weathered along shorelines, where the pounding waves break them into beach sand. Other weathering processes include chemical reactions, such as when oxygen in the air combines with iron in rocks to form a type of rust. One of the most efficient forms of weathering is caused by water; liquid water runs into crevices and then freezes. As the water freezes it expands and shatters the rock.

After weathering, the resulting debris can be carried away by flowing water, glacial ice, or blowing wind and deposited in other areas as sediment. Where material is eroded, we can see features such as river valleys, wind-sculpted hills, or mountains carved by glaciers. Where eroded material is deposited, we see features such as river deltas, sand dunes, or piles of rock at the bases of mountains and cliffs. Erosion is most efficient on planets with water and wind. On Earth, where water and wind are so dominant, most impact craters on Earth's continents have been worn down and filled in.

Even though the Moon and Mercury have almost no atmosphere and no running water, a type of erosion is still at work. Radiation from the Sun and from deep space very slowly decomposes some types of minerals, effectively weathering the rock. Such effects are only a few millimeters deep at most. Impacts of micrometeoroids also chip away at rocks. In addition, landslides can occur wherever gravity and differences in elevation are present. Although water enhances landslide activity, landslides are also seen on Mercury and the Moon.

Wind Erosion

Earth, Mars, and Venus all show the effects of windstorms. Images of Mars and Venus returned by spacecraft landers show surfaces that have been subjected to the forces of wind. Sand dunes are common on Earth and Mars, and some have been identified on Venus. Orbiting spacecraft have also found wind-eroded hills and surface patterns called wind streaks. These surface patterns appear, disappear, and change in response to winds blowing sediments around hills, craters, and cliffs. They serve as local "wind vanes," telling planetary scientists about the direction of local prevailing surface winds. Planet-encompassing dust storms occur on Mars.

Water Erosion

Today, Earth is the only planet where the temperature and atmospheric conditions allow extensive liquid surface water to exist. Water is an extremely powerful agent of erosion and dominates erosion on Earth. Every year, rivers and streams on Earth deliver about 10 billion metric tons of sediment into the oceans. Even though today there is no liquid water on the surface of Mars, at one time water likely flowed across its surface in vast quantities. Features resembling water-carved channels on Earth such as those shown in **Figure 6.22** suggest the past presence of liquid water on Mars. In addition, many regions on Mars show small networks of valleys that are thought to have been carved by flowing water. Some parts of Mars may once have contained even oceans and glaciers.

The Search for Water in the Solar System

Life, as we know it on Earth, requires water as a solvent and as a delivery mechanism for essential chemistry. Because of this, the search for water is central to the search for life in the Solar System and also to considerations of future human space travel.

In 2004, NASA sent two instrument-equipped roving vehicles, *Opportunity* and *Spirit*, to search for evidence of water on Mars. *Opportunity* landed inside a small crater. For the first time, martian rocks were available for study in the original order in which they were laid down. Previously, the only rocks that landers and rovers had studied were those that had been dislodged from their original settings by either impacts or river floods.

The layered rocks at the *Opportunity* site revealed that they had once been soaked in or transported by water. The form of the layers was typical of layered sandy deposits laid down by gentle currents of water. Rover instruments found a mineral so rich in sulfur that it had almost certainly formed by precipitation from water. Magnified images of the rocks showed "blueberries," small spherical grains a few millimeters across that probably formed in place among the layered rocks. These are similar to terrestrial features that form by the percolation of water through sediments. Analysis of these blueberries revealed abundant hematite, an iron-rich mineral that forms in the presence of water. Further observations by the European Space Agency's *Mars Express* and NASA's *Mars Odyssey* orbiters have shown the hematite signature and the presence of sulfur-rich compounds in a

Figure 6.22 An image of gully channels in a crater on Mars taken by the *Mars Reconnaissance Orbiter*. The gullies coming from the rocky cliffs near the crater's rim (out of the image, to the upper left) show meandering and braided patterns similar to those of water-carved channels on Earth.

Figure 6.23 This image compares a photograph taken by NASA's *Curiosity* rover (left) with a photograph of a streambed on Earth (right). The Mars image shows water-worn gravel embedded in sand, sure evidence of an ancient streambed.

vast area surrounding the *Opportunity* landing site. These observations suggest an ancient martian sea larger than the combined area of Earth's Great Lakes and as much as 500 meters deep.

    Spirit landed in Gusev, a 170-km-wide impact crater. This site was chosen because it showed signs of ancient flooding by a now-dry river. Scientists hoped that surface deposits would provide further evidence of past liquid water. Surprise, and perhaps some disappointment, followed when *Spirit* revealed that the flat floor of Gusev consisted primarily of basaltic rock. Only when the rover ventured cross-country to some low hills located 2.5 km from the landing site did it find basaltic rocks showing clear signs of having been chemically altered by liquid water.

    In August 2012, the Mars rover *Curiosity* landed in Gale Crater, a large (150 km) crater just south of Mars's equator. *Curiosity* found evidence of a stream that flowed at a rate of about 1 meter per second and was as much as 2 feet deep. The streambed is identified by water-worn gravel, shown in **Figure 6.23**. The rover, which is about the size of a car, has instrumentation that includes not only cameras but also a drill and an instrument to measure chemical composition. When the rover drilled into a rock, it found sulfur, nitrogen, hydrogen, oxygen, phosphorus, and carbon, together with clay minerals that formed in a water-rich environment that was not very salty. Taken together, these pieces of evidence indicate that Mars may have had conditions suitable to support Earth-like microbial life in the distant past.

    Where did the water go? Some escaped into the thin atmosphere of Mars, and at least some of it is locked up as ice in the polar regions, just as the ice caps on Earth hold much of our planet's water. Unlike our own polar caps, those on Mars are a mixture of frozen carbon dioxide and frozen water. But water must be hid-

Figure 6.24 Water ice appears a few centimeters below the surface of Mars in this trench dug by a robotic arm on the *Phoenix* lander. The trench measures about 20 × 30 cm.

ing elsewhere on Mars. Small amounts of water are found on the surface. Water ice crystals have been found mixed in with the martian soil, and NASA's *Phoenix* lander found water ice just a centimeter or so beneath surface soils at high northern latitudes (**Figure 6.24**). However, most of the water on Mars appears to be trapped well below the surface of the planet. Radar imaging by NASA's *Mars Reconnaissance Orbiter* (*MRO*) indicates huge quantities of subsurface water ice, not only in the polar areas as expected but also at lower latitudes. In addition, some recent *MRO* images suggest that there might be seasonal saltwater flows on the surface far from the poles. Saltwater freezes at a lower temperature, so some sites could be warm enough to have temporary liquid saltwater.

Evidence for water on Venus comes primarily from water vapor in its atmosphere, but there are some geological indications of past water, such as color differences between highland and lowland regions. This may indicate the presence of granite, which forms in the presence of water.

Although Earth and Mars are the only terrestrial planets that show strong evidence for liquid water at any time in their histories, water ice exists on the Moon and could exist on Mercury today. Some deep craters in the polar regions of both Mercury and the Moon have floors in perpetual shadow. Temperatures in these permanently shadowed areas remain below 180 K. For many years, planetary scientists speculated that ice—perhaps from comets—could be found in these craters. In the early 1990s, radar measurements of Mercury's north pole and infrared measurements of the Moon's polar areas seemed to support this possibility. The *Messenger* spacecraft has reported the strong possibility of ice at both of Mercury's poles.

In 2009, NASA's *Lunar Reconnaissance Orbiter* (*LRO*) and a companion satellite known as the *Lunar Crater Observation and Sensing Satellite* (*LCROSS*) were put into lunar orbit to continue the search for possible sources of subsurface water ice. *LCROSS* intentionally crashed the second stage of its Centaur rocket into a lunar polar-region crater, Cabeus. *LRO* collected data from the resulting plume, finding that more than 150 kg of water ice and water vapor were blown out of the crater by the impact. These data revealed the presence of large amounts of water buried beneath the crater's floor. Other evidence of water on the Moon was found by later spacecraft observations. These results are important to scientists who are thinking about future missions to the Moon.

READING ASTRONOMY News

> Nearly all of what people know about the terrestrial planets (except Earth) and our Moon comes from observations of the surface properties of these worlds.

Moon Is Wetter, Chemically More Complex than Thought, NASA Says

By **AMINA KHAN**, *Los Angeles Times*

The Moon is a much wetter—and more chemically complicated—place than scientists had believed, according to new data released Thursday by NASA.

Last year, after the space agency hurtled a rocket into a frozen crater at the Moon's south pole and measured the stuff kicked up by the crash, scientists calculated that the plume contained about 25 gallons of water. But further analysis over the last 11 months indicates that the amount of water vapor and ice was more like 41 gallons—an increase of 64 percent.

"It's twice as wet as the Sahara desert," said Anthony Colaprete, the lead scientist for the *Lunar Crater Observation and Sensing Satellite*, or *LCROSS*, mission at NASA Ames Research Center in Northern California.

The instruments aboard the satellite, including near-infrared and visible light spectrometers, scanned the lunar debris cloud and identified the compounds it contained. They determined that about 5.6 percent of the plume was made of water, give or take 2.9 percent. It also included a surprising array of chemicals, including mercury, methane, silver, calcium, magnesium, pure hydrogen, and carbon monoxide.

The findings were reported in six related papers published online Thursday by the journal *Science*.

"The lunar closet is really at the poles, and I think there's a lot of stuff crammed into the closet that we really haven't investigated yet," said Peter Schultz, a planetary geologist at Brown University in Providence, Rhode Island, and one of the *LCROSS* team members.

The new measurements allowed Colaprete to estimate that the entire Cabeus Crater could hold as much as 1 billion gallons of water.

Potentially, that would be really handy for future space explorers who might use the Moon as an interplanetary way station. Along with providing water to drink, it could be mined for breathable oxygen and used to make hydrogen fuel for long-distance spacecraft.

"You can't take a lot of big things to the Moon and you can't take much water, so we're learning to live off the land," said Lawrence Taylor, a planetary geochemist at the University of Tennessee in Knoxville, who was not involved in the studies.

However, plans to develop a space colony on the Moon were put on hold by the Obama administration this year.

Evaluating the News

1. In early analysis of the data, scientists discovered 25 gallons of water. In this article, that number has been revised upward to 41 gallons. What astronomical observation methods did they use to discover this?
2. The article states that 5.6 percent of the plume was made of water, give or take 2.9 percent. What is the smallest percentage of the plume that could be water? What is the largest? Is this a large range?
3. Average daily urban water use in the United States is roughly 100 gallons per household. If Cabeus Crater does have 1 billion gallons of water, how many American households could this crater support for 1 day? For 1 month? For 1 year? For 10 years?
4. What do your answers to question 3 imply for the long-term survival of a colony on the Moon? How does this prediction change if intense methods of water reuse and recycling are implemented in the colony? How does this prediction change if the water is also being used to make oxygen to breathe and hydrogen for fuel?

SUMMARY

The terrestrial planets in the Solar System include Mercury, Venus, Earth, and Mars. Because Earth's Moon is similar in many ways to these terrestrial planets, it is included in the list of terrestrial worlds. The interiors of planets are extremely hot at formation. Tectonics and volcanism are a result of these hot interiors. Over time, the interiors cool, and tectonics and volcanism weaken. On Earth, radioactive decay and tidal effects from the Moon contribute to heat in the interior. Impacts deliver energy and some materials to the surface of a planet, tectonism and volcanism deform the surface, and erosion gradually scrubs away the evidence of impacts, tectonism, and volcanism. Surface features on the terrestrial planets, such as tectonic plates, volcanoes, mountain ranges, or canyons, are the result of the interplay between these four processes.

LG 1 Four processes shape the surfaces of the terrestrial planets: impact cratering, volcanism, tectonics, and erosion. Impact cratering is the result of a direct interaction of an astronomical object with the surface of the planet. Active volcanism and tectonics are the results of a "living" planetary interior: one that is still hot inside. Erosion is a surface phenomenon that results from weathering by wind or water.

LG 2 The relative position of craters gives their relative ages, with more recent craters found superimposed on older ones. Crater densities can be used to find the relative ages of regions on a surface, with more heavily cratered regions being older than less cratered ones. Finding the absolute age of a region requires a calibration using radioactive isotopes and their products.

LG 3 Earth's interior has been mapped using a combination of theory and observation. Models are used to predict how seismic waves should propagate through the interior, and these predictions are compared to actual observations of seismic waves. The interiors of other planets are modeled upon basic physical principles and verified by observations of magnetic fields.

LG 4 Tectonism folds, twists, and cracks the outer surface of a planet. Plate tectonics is unique to Earth, although other types of tectonic disruptions are observed on the other terrestrial planets, such as cracking and buckling on the surface.

LG 5 Weathering by wind and water causes erosion on planetary surfaces. In general, the consequence of erosion is to erase features that are the result of impact cratering, tectonics, and volcanism. Geological evidence of water erosion is one factor in the search for water in the Solar System. This search is important to both the search for extraterrestrial life and the possibilities of human colonization of space.

SUMMARY SELF-TEST

1. _____, _____, and _____ build up structures on the terrestrial planets, while in general, _____ tears them down.
   a. Impacts, erosion, volcanism; tectonism
   b. Impacts, tectonism, volcanism; erosion
   c. Tectonism, volcanism, erosion; impacts
   d. Tectonism, impacts, erosion; volcanism

2. If crater A is inside crater B, we know that
   a. crater A was formed before crater B.
   b. crater B was formed before crater A.
   c. both craters were formed at about the same time.
   d. crater B formed crater A.
   e. crater A formed crater B.

3. Scientists learn about the interior structure of planets by using (select all that apply)
   a. ground-penetrating radar.
   b. deep mine shafts.
   c. observations of seismic waves.
   d. models of Earth's interior.
   e. observations of magnetic fields.
   f. X-ray observations from satellites.

4. Lava flows on the Moon and Mercury created large, smooth plains. We don't see similar features on Earth because
   a. Earth has less lava.
   b. Earth had fewer large impacts in the past.
   c. Earth has plate tectonics and erosion that modify the surface.
   d. Earth is large compared to the size of these plains, so they are not as noticeable.
   e. Earth's rotation rate is much faster than that of either of these other worlds.

5. Erosion is most efficient on planets with
   a. tectonism.
   b. volcanoes.
   c. wind and water.
   d. large masses.

6. If a radioactive element A decays into radioactive element B with a half-life of 20 seconds, then after 40 seconds,
   a. none of element A will remain.
   b. none of element B will remain.
   c. half of element A will remain.
   d. one-quarter of element A will remain.

7. On which of the following worlds is wind erosion negligible? (Select all that apply.)
   a. Mercury   b. Venus
   c. Earth   d. Moon
   e. Mars

8. On which of the following does plate tectonics occur? (Select all that apply.)
   a. Mercury   b. Venus
   c. Earth   d. Moon
   e. Mars

9. On which of the following has erosion by wind and water occurred? (Select all that apply.)
   a. Mercury
   b. Venus
   c. Earth
   d. Moon
   e. Mars

10. Which of the following worlds shows evidence of water? (Select all that apply.)
    a. Mercury
    b. Venus
    c. Earth
    d. Moon
    e. Mars

QUESTIONS AND PROBLEMS

Multiple Choice and True/False

11. T/F: Volcanism has occurred on all the terrestrial planets.
12. T/F: The propagation of seismic waves reveals the structure of the interior of Earth.
13. T/F: Large worlds remain geologically active longer than small ones.
14. T/F: Mercury has no volcanoes.
15. T/F: Wind erosion is an important process on Venus.
16. Of the four processes that shape the surface of a terrestrial world, the one with the greatest potential for future catastrophic rearrangement is
    a. impacts.
    b. volcanism.
    c. tectonism.
    d. erosion.
17. Geologists can find the relative age of impact craters on a world because
    a. the ones on top must be older.
    b. the ones on top must be younger.
    c. the larger ones must be older.
    d. the larger ones must be younger.
    e. all the features we can see are the same age.
18. Geologists can find the actual age of features on a world by
    a. radioactive dating of rocks retrieved from the world.
    b. comparing cratering rates on one world to those on another.
    c. assuming that all features on a planetary surface are the same age.
    d. both a and b
    e. both b and c
19. Impacts on the terrestrial worlds
    a. are more common than they used to be.
    b. have occurred at approximately the same rate since the formation of the Solar System.
    c. are less common than they used to be.
    d. periodically become more common and then are less common for a while.
    e. never occur any more.
20. Earth has fewer craters than Venus. Why?
    a. Earth's atmosphere provides better protection than Venus's.
    b. Earth is a smaller target than Venus.
    c. Earth is closer to the asteroid belt.
    d. Earth's surface experiences more erosion.
21. The terrestrial worlds that may still be geologically active are
    a. Earth, Moon, and Mercury.
    b. Earth, Mars, and Venus.
    c. Earth, Venus, and Mercury.
    d. Earth only.
    e. Earth and Venus only.
22. Spring tides occur only when
    a. the Sun is near the vernal equinox in the sky.
    b. the Moon is in first or third quarter.
    c. the Moon, Earth, and Sun form a right angle.
    d. the Moon is in new or full phase.
    e. the Sun is in full phase.
23. On Earth, one high tide each day is caused by the Moon pulling on that side of Earth. The other is caused by
    a. the Sun pulling on the opposite side of Earth.
    b. the Earth rotating around so that the opposite side is under the Moon.
    c. the Moon pulling the center of Earth away from the opposite side, leaving a tidal bulge behind.
    d. the resonance between the rotation and revolution of the Moon.
24. Scientists know the history of Earth's magnetic field because
    a. the magnetic field hasn't changed since the formation of Earth.
    b. they see how it's changing today and project that back in time.
    c. the magnetic field gets frozen into rocks, and plate tectonics spreads them out.
    d. they compare the magnetic fields on other planets to Earth's.
    e. there are written documents of magnetic field measurements since the beginning of Earth.
25. Water erosion is an important ongoing process on
    a. all the terrestrial worlds.
    b. Earth only.
    c. Earth and Mars only.
    d. Earth, Mars, and the Moon only.
    e. Earth, Mars, and Venus only.

Conceptual Questions

26. List evidence of the four geological processes that shape the terrestrial planets: tectonism, volcanism, impact cratering, and erosion.
27. In discussing the terrestrial planets, why do we include our Moon?
28. Explain how scientists know that rock layers at the bottom of Arizona's Grand Canyon are older than those found on the rim.
29. One region on the Moon is covered with craters, while another is a smooth volcanic plain. Which is older? How do we know?

30. Suppose that you have two rocks, each containing a radioactive isotope and its decay products. In rock A, there is an equal amount of the parent and the daughter. In rock B, there is twice as much daughter as parent. Which rock is older? Explain how you know this from the information given.

31. Describe the sources of heating that are responsible for the generation of Earth's magma.

32. Explain why the Moon's core is cooler than Earth's.

33. Compare and contrast tectonism on Venus, Earth, and Mercury.

34. Volcanoes have been found on all of the terrestrial planets. Where are the largest volcanoes in the inner Solar System?

35. Describe the collision theory of the formation of the Moon.

36. Examine Figure 6.7a. This figure shows two tidal bulges on Earth, both caused by the Moon. Compare this figure to Figure 6.9. In your own words, describe why the Moon's gravity causes two tides on Earth—one on the side closest to the Moon, and one on the side farthest away.

37. Describe and explain the evidence for reversals in the polarity of Earth's magnetic field.

38. Why do earthquakes and volcanoes tend to occur near plate boundaries?

39. Does the age of a planetary surface tell you the age of the planet? Why or why not?

40. Explain some of the evidence that Mars once had liquid water on its surface. Why is there no liquid water on Mars today?

Problems

41. Use the mass and radius of Venus given in Table 6.1 to find the average density of Venus. From this density, determine whether Venus has an iron core. (Assume that Venus's surface is rocky, like Earth's.)

42. Earth has a radius of 6,378 km.
    a. What is its volume? (Hint: The volume of a sphere of radius R is $[4/3]\pi R^3$.)
    b. What is its surface area? (Hint: The surface area of a sphere of radius R is $4\pi R^2$.)
    c. Suppose Earth's radius suddenly became twice as big. By what factor would the volume change? By what factor would the surface area change?

43. Suppose you find a piece of ancient pottery and take it to the laboratory of a physicist friend. He finds that the glaze contains radium, a radioactive element that decays to radon and has a half-life of 1,620 years. He tells you that the glaze couldn't have contained any radon when the pottery was being fired but that it now contains three atoms of radon for each atom of radium. How old is the pottery?

44. Archaeological samples are often dated by radiocarbon dating. The half-life of carbon-14 is 5,700 years.
    a. After how many half-lives will the sample have only $1/64$ as much carbon-14 as it originally contained?
    b. How much time will have passed?
    c. Suppose that the daughter of carbon-14 is present in the sample when it forms (even before any radioactive decay happens). Then you cannot assume that every daughter you see is the result of carbon-14 decay. If you did make this assumption, would you overestimate or underestimate the age of a sample?

45. Different radioisotopes have different half-lives. For example, the half-life of carbon-14 is 5,700 years, the half-life of uranium-235 is 704 million years, the half-life of potassium-40 is 1.3 billion years, and the half-life of rubidium-87 is 49 billion years.
    a. Explain why you would not use an isotope with a half-life similar to that of carbon-14 to determine the age of the Solar System.
    b. The age of the universe is approximately 14 billion years. Does the long half-life mean that no rubidium-87 has decayed yet?

46. Recall Working It Out 5.1. The average temperature of Mars is about 210 K. What is the blackbody flux from a square meter of Mars?

47. Recall Working It Out 5.2. The average temperature of Venus is about 773 K. What is the peak wavelength of radiation from a blackbody the same temperature as Venus?

48. Suppose that the flux from a planet is measured to be 350 W/m². What is its average temperature?

49. The object that created Arizona's Meteor Crater was estimated to have a radius of 25 meters and a mass of 300 million kg. Follow Working It Out 6.2 to calculate the density of the impacting object, and explain what that may tell you about its composition.

50. Assume that the east coast of South America and the west coast of Africa are separated by an average distance of 4,500 km. Assume also that GPS measurements indicate that these continents are now moving apart at a rate of 3.75 cm/year. If this rate has been constant over geologic time, how long ago were these two continents joined together as part of a supercontinent?

SMARTWORK

Norton's online homework system includes algorithmically generated versions of these questions, plus additional conceptual exercises. If your instructor assigns questions in SmartWork, log in at smartwork.wwnorton.com.

Exploration | Earth's Tides

wwnpag.es/uou2

Visit the Student Site (wwnpag.es/uou2) and open the Tidal Bulge Simulator Nebraska Simulation in Chapter 6.

Before you start the simulation, examine the setup. You are looking down on Earth from the North Pole.

1. Are the sizes of Earth and the Moon approximately to scale in this image?

2. Is the distance between them to scale in this image?

3. Are the tides shown to scale?

4. Explain why the authors of the simulation made the scaling choices they did.

5. In this position, is the east coast of North America experiencing high or low tide?

Recall from Chapter 2 that Earth rotates counterclockwise when viewed from this vantage point. Click the box that says "Include Effects of Earth's Rotation."

6. What happened to the tidal bulges?

7. Why does Earth's rotation have this effect?

Click the box that says "Run."

8. Over the course of 1 day, how many high tides does the east coast of North America experience?

9. How many low tides does it experience in 1 day?

10. Why are there two high tides? That is, what causes the tide on the side of Earth away from the Moon?

11. Do the tides change as the Moon orbits?

Once the Moon has orbited back to the right side of the window, stop the simulation by unchecking "Run." Click the box that says "Include Sun."

12. What happened to the tides when you added the Sun to the simulation?

Now, run the simulation again. Stop the simulation when the Moon is at first quarter. Remove the check mark from the "Include Sun" box.

13. What changed about the tides when you removed the Sun?

Run the simulation until the Moon is full. Stop the simulation by unchecking "Run." Click the box that says "Include Sun."

14. What changed about the tides when you added the Sun back in?

15. Which astronomical body dominates the tides on Earth, the Moon or the Sun?

SMARTWORK • smartwork.wwnorton.com

7 Atmospheres of Venus, Earth, and Mars

Earth's atmosphere is responsible for the weather on Earth. The atmosphere lifts water from lakes and oceans and deposits it on mountaintops and plains to make streams and rivers. Without an atmosphere, Earth would look something like the Moon, and life would not exist on our planet. Earth's atmosphere also interacts with material that flows outward from the Sun. The illustration on the opposite page shows a photograph of an aurora, which is caused by the interaction of charged particles from the Sun with Earth's atmosphere, as indicated in the sketch a student is making as he associates the Sun's radiation with Earth-bound observations.

Among the five terrestrial bodies that we discussed in Chapter 6, only Venus and Earth have dense atmospheres. Mars has a very-low-density atmosphere, and the atmospheres of Mercury and the Moon are so sparse that they can hardly be detected. To understand the origins of the atmospheres of Venus, Earth, and Mars, how they have changed over time, how they compare to one another, and how they are likely to evolve in the future requires us to look back nearly 5 billion years to a time when the planets were just completing their growth.

✦ LEARNING GOALS

A thick blanket of atmosphere warms and sustains Earth's climate. Venus and Earth have dense atmospheres. The thinner atmosphere on Mars is a useful point of comparison. By the end of this chapter, you should be able to explain the origin of the aurora borealis and explain why it is so colorful. You should also be able to:

LG 1 Explain the origins and differences of primary and secondary atmospheres.

LG 2 Define the layers of atmospheres for Earth, Venus, and Mars, and explain the existence of these layers.

LG 3 Compare the strength of the greenhouse effect and differences in the atmospheres of Earth, Venus, and Mars.

LG 4 Describe how Earth's atmospheric composition has been reshaped by life.

LG 5 Describe the evidence that shows Earth's climate is changing.

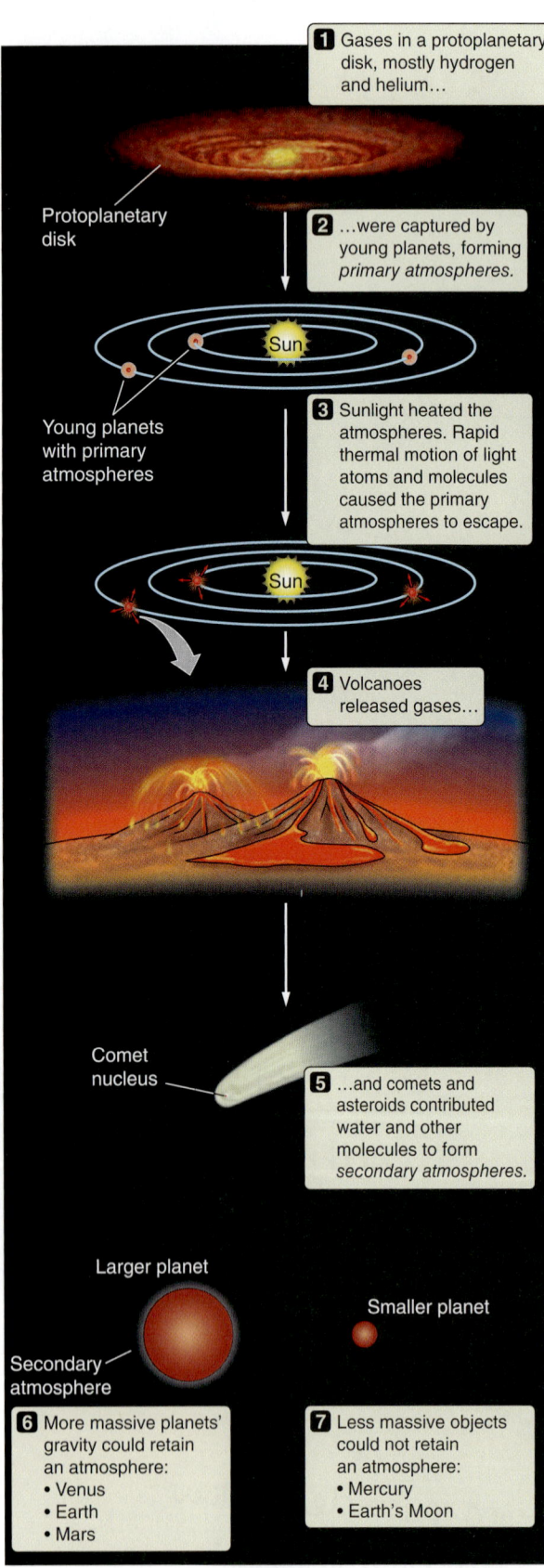

Figure 7.1 Planetary atmospheres form and evolve in phases.

7.1 Atmospheres Change over Time

An atmosphere is a layer of gas that sits above the surface of a terrestrial planet. It has currents and eddies much like an ocean. An atmosphere, even Earth's atmosphere, is a very thin layer compared to the radius of a planet. The atmospheres of the terrestrial planets formed in phases. **Figure 7.1** sketches the entire history of planetary atmospheres, from planet formation to modern times. The evolution of planetary atmospheres involves both a primary and a secondary atmosphere as well as several processes, including volcanism and impacts. Differences in mass, composition, and distance from the Sun affect the evolution of atmospheres. Venus and Earth are similar in both mass and composition, and their orbits are less than 0.3 astronomical units (AU) apart—about one-third of Earth's average distance from the Sun. Mars is also similar in composition, but its mass is only about one-tenth that of Earth.

Formation and Loss of Primary Atmospheres

Young planets captured some of the residual hydrogen and helium that filled the protoplanetary disk surrounding the Sun (Figure 7.1). Gas capture continued until the supply of gas ran out. The gaseous atmosphere collected by a newly formed planet is called its primary atmosphere. This primary atmosphere was lost from the terrestrial planets as these lightweight atoms and molecules escaped from the planet's gravity. To understand this, we must study more closely how particles move within a planetary atmosphere.

Imagine a large box that contains air. In thermal equilibrium, each type of molecule in the box, from the lightest to the most massive, will have the same average kinetic energy. Because the kinetic energy of a molecule is determined by its mass and its speed, if each type has the same average energy, then the lightest molecules must be moving faster than the more massive ones. For example, in a mixture of hydrogen and oxygen at room temperature, hydrogen molecules will be rushing around the box at about 2,000 meters per second (m/s) on average, while the much more massive oxygen molecules are moving at only 500 m/s. Remember, though, that these are the *average* speeds. A few of the molecules will always be moving much faster or slower than average.

In an atmosphere, fast molecules near the ground almost certainly collide with other molecules before the fast molecules have a chance to escape. Higher regions of the atmosphere contain fewer molecules. Therefore, fast molecules in the upper atmosphere are less likely to collide with other molecules and have a better chance of escaping as long as they are heading more or less upward. At a given temperature, lighter molecules such as hydrogen and helium move faster and are more quickly lost to space than more massive molecules such as nitrogen or carbon dioxide.

In the early Solar System, the terrestrial planets were heated by the Sun, and so the molecules were moving swiftly. In addition, small planets, like terrestrial planets, have only a weak gravitational grasp. These conditions caused the terrestrial planets to lose the hydrogen and helium they had acquired as a primary atmosphere. This process was likely assisted by collisions with other planetesimals. Because the giant planets were farther from the Sun, they were far more massive and also cooler; stronger gravity and lower temperatures enabled them to retain nearly all of their massive primary atmospheres.

The Formation of Secondary Atmospheres

Although Earth's primary atmosphere was lost, we do have an atmosphere today, known as a secondary atmosphere. Where did this secondary atmosphere come from? Accretion, volcanism, and impacts are responsible for Earth's atmosphere today. During the planetary accretion process, minerals containing water, carbon dioxide, and other volatile matter collected in Earth's interior. Later, as the interior heated up, these gases were released from the minerals that had held them. Volcanism then brought the gases to the surface, where they accumulated and created our secondary atmosphere, as shown in step 4 of Figure 7.1.

Impacts by huge numbers of comets and asteroids were another important source of gases. As the giant planets of the outer Solar System grew, they **perturbed** the orbits of comets and asteroids, scattering some of them into the inner Solar System. Upon impact with the terrestrial planets, these objects brought water, carbon monoxide, methane, and ammonia. On Earth, and perhaps on Mars as well, most of the water vapor then condensed as rain and flowed into the lower areas to form the earliest oceans.

Sunlight also influenced the composition of secondary atmospheres. Ultraviolet (UV) light from the Sun easily fragments molecules such as ammonia and methane. Ammonia, for example, is broken down into hydrogen and nitrogen. When this happens, the lighter hydrogen atoms quickly escape to space, leaving behind the much heavier nitrogen atoms. Pairs of nitrogen atoms then combine to form more massive nitrogen molecules (N_2), and these molecules are even less likely to escape into space. Decomposition of ammonia by sunlight became the primary source of molecular nitrogen in the atmospheres of Venus, Earth, and Mars.

Mercury's relatively small mass and its proximity to the Sun caused it to lose nearly all of its secondary atmosphere to space, just as it had previously lost its primary atmosphere. Even molecules as massive as carbon dioxide can escape from a small planet if the temperature is high enough, as it is on Mercury's sunlit side. Furthermore, intense UV radiation from the Sun can break molecules into less massive fragments, which are lost to space even more quickly. Because the distance from the Sun to the Moon is much farther than the distance from the Sun to Mercury, the Moon is much cooler than Mercury. But the Moon's mass is so small that molecules can easily escape, even at low temperatures. Both the Moon and Mercury have virtually no atmosphere today because of their small masses and relative proximity to the Sun.

▶▶ **Nebraska Simulation:** Gas Retention Simulator

▶ǁ **AstroTour:** Atmospheres: Formation and Escape

7.2 Secondary Atmospheres Evolve

Although Venus, Earth, and Mars most likely started out with atmospheres of similar composition and comparable quantity, they ended up being very different from one another. All three are volcanically active today or have been volcanically active in their geological past. In addition, they must have shared the intense cometary showers that took place in the early Solar System. Their similar geological histories suggest that their early secondary atmospheres might also have been quite similar. However, Earth's secondary atmosphere has changed significantly since it formed; the development of life increased the amount of oxygen. Earth's atmosphere is made up primarily of nitrogen and oxygen, with only a trace of carbon dioxide. In contrast, the atmospheres

Vocabulary Alert

perturb This word is uncommon in everyday usage, although the form *perturbed* is sometimes used to indicate that someone is upset. Astronomers use it to mean changes to an object's orbit, often caused by gravitational interactions.

of Venus and Mars today are nearly identical in composition—mostly carbon dioxide, with much smaller amounts of nitrogen. The atmospheres of these planets differ for two reasons we will explore: mass and the greenhouse effect.

The Effect of Planetary Mass on a Planet's Atmosphere

Despite the similarities in atmospheric composition, Mars and Venus have vastly different *amounts* of atmosphere. The atmospheric pressure on the surface of Venus is nearly a hundred times greater than that of Earth. By contrast, the average surface pressure on Mars is less than a hundredth of that on our own planet. Venus is nearly eight times as massive as Mars, so we can assume that it probably had about eight times as much carbon within its interior to produce carbon dioxide, the principal secondary-atmosphere component of both planets. Even allowing for the differences in planetary mass, however, Venus today has greater than 2,500 times more atmospheric mass than Mars. Why such a large difference? We can find the answer by considering the relative strengths of their surface gravity, which involves both the mass and radius of a planet. Venus has the gravitational pull necessary to hang onto its atmosphere; Mars has less gravitational attraction to keep its atmosphere. Furthermore, when a planet such as Mars loses so much of its atmosphere to space, the process begins to take on a "runaway" behavior. With a thinner atmosphere, there are fewer slow molecules to keep fast molecules from escaping, and the rate of escape increases. This process in turn leads to even less atmosphere and still faster escape rates.

The Atmospheric Greenhouse Effect

Differences in the present-day masses of the atmospheres of Venus, Earth, and Mars have a large effect on their surface temperatures. The temperature of a planet is determined by a balance between the amount of sunlight being absorbed and the amount of energy being radiated back into space. If the planet's temperature remains constant, then it must radiate as much energy as it absorbs, just as the bucket in **Figure 7.2** fills and empties at the same rate, keeping the water level constant. In **Working It Out 7.1**, we calculate the temperature of a planet by finding the equilibrium between the amount of energy it receives and the amount of energy it radiates. This calculation gives a good result for planets without atmospheres; but for Earth, and especially Venus, the calculations are far from actual measured values. The cause is the **atmospheric greenhouse effect**, which traps solar radiation.

The atmospheric greenhouse effect in planetary atmospheres and the **conventional greenhouse effect** operate in different ways, although the end results are much the same. Planetary atmospheres and the interiors of greenhouses are both heated by trapping the Sun's energy, but here the similarities end. The conventional greenhouse effect is what happens in a car on a sunny day when you leave the windows closed or what allows plants to grow in the winter in a greenhouse. Sunlight pours through the glass, heating the interior and raising the internal air temperature. With all the windows closed, hot air is trapped and temperatures can climb as high as 180°F (about 80°C).

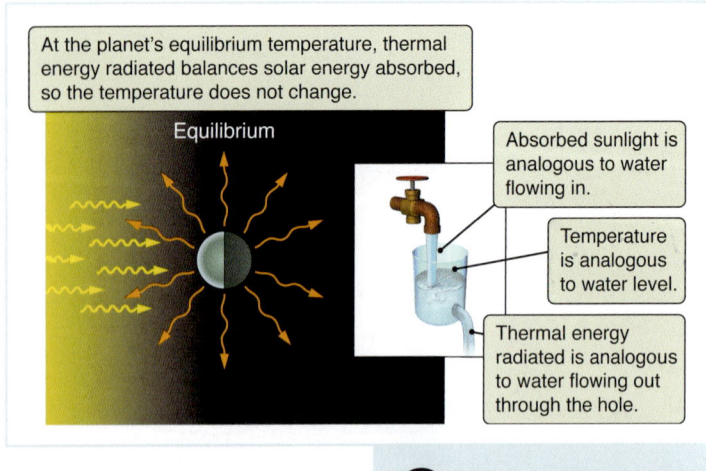

Figure 7.2 Planets are heated in part by absorbing sunlight and cooled by emitting thermal radiation into space. If there are no other sources of heating or means of cooling, then the equilibrium between these two processes determines the temperature of the planet.

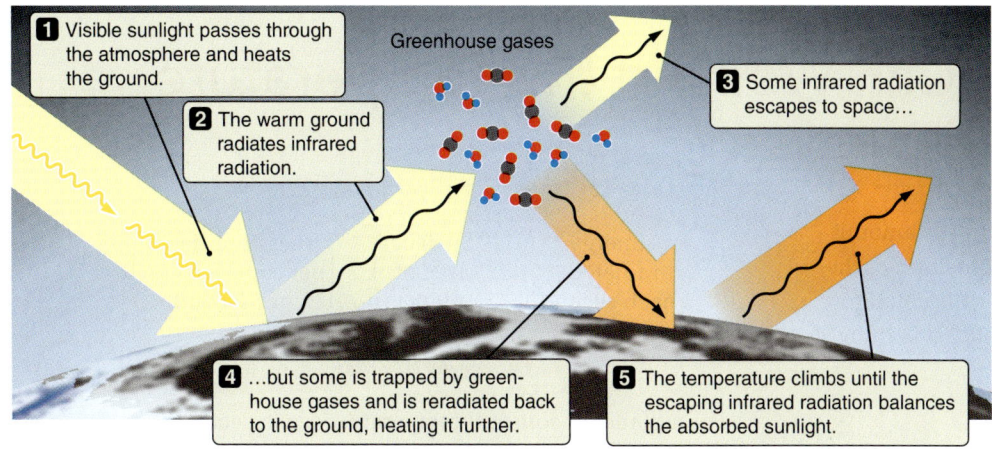

Figure 7.3 In the atmospheric greenhouse effect, greenhouse gases such as water vapor and carbon dioxide absorb infrared radiation and reradiate it in all directions, slowing the transport of energy out of the atmosphere. This causes the temperature of the planet to rise to a new equilibrium.

The atmospheric greenhouse effect is illustrated in **Figure 7.3**. Atmospheric gases freely transmit visible light, allowing the Sun to warm the planet's surface. The warmed surface radiates the excess energy in the infrared region of the spectrum. Some atmospheric gases strongly absorb infrared radiation and convert it to thermal energy, which is released in random directions. Some of the thermal energy continues into space, but much of it goes back to the ground, which causes a planet's surface temperature to rise. As a result of this radiation, the planet receives thermal energy from both the Sun and the atmosphere. Gases that transmit visible radiation but absorb infrared radiation are known as **greenhouse gases**. In addition to carbon dioxide and water vapor, greenhouse gases found in Earth's atmosphere include methane, nitrous oxide, and chlorofluorocarbons [CFCs]. The presence of greenhouse gases increases the temperature of a planet.

This rise in temperature continues until the surface becomes sufficiently hot—and therefore radiates enough energy—that the fraction of infrared radiation leaking out through the atmosphere is just enough to balance the absorbed sunlight. Convection also helps maintain equilibrium by transporting thermal energy to the top of the atmosphere, where it can be more easily radiated to space. In short, the temperature rises until equilibrium between absorbed sunlight and thermal energy radiated away by the planet is reached.

▶❙❙ **AstroTour: Greenhouse Effect**

Similarities and Differences among the Terrestrial Planets

Let's look more closely at how the atmospheric greenhouse effect operates on Venus, Earth, and Mars. What really matters is the actual *number* of greenhouse molecules in a planet's atmosphere, not the *fraction* they represent. For example, even though the atmosphere of Mars is composed almost entirely of carbon dioxide—an effective greenhouse molecule—the atmosphere is very thin and contains relatively few greenhouse molecules compared to the atmospheres of Venus or Earth. As a result, the atmospheric greenhouse effect is relatively weak on Mars and raises the mean surface temperature by only about 5 K. At the other extreme, Venus's massive atmosphere of carbon dioxide and sulfur compounds raises its mean surface temperature by more than 400 K, to about 737 K. At such high temperatures, any water and most carbon dioxide locked up in surface rocks are driven into the atmosphere, further enhancing the atmospheric greenhouse effect.

The atmospheric greenhouse effect on Earth is not as severe as it is on Venus—

Working It Out 7.1 | How Can We Find the Temperature of a Planet?

To predict the temperature of a planet, begin with the amount of sunlight being absorbed. As viewed from the position of the Sun, a planet looks like a circular disk with a radius equal to the radius of the planet, R. The area of the planet that is lit by the Sun is

$$(\text{Absorbing area of planet}) = \pi R^2$$

The amount of energy striking a planet also depends on the intensity of sunlight at the distance at which the planet orbits. Imagine a light bulb on a stage in a large theater. If you are standing on the stage with the light bulb, the light is intense—every part of your body receives a lot of energy each second from the bulb. At the back row of the theater, the light is less intense: at this greater distance, the bulb's energy is spread out over a much larger area. The intensity falls off as the area of a sphere, $4\pi d^2$, where d is the distance from you to the bulb. Similarly, the intensity of sunlight at a planet's orbit is the luminosity of the Sun, L, divided by $4\pi d^2$:

$$(\text{Intensity of sunlight}) = \frac{L}{4\pi d^2}$$

A planet does not absorb all the sunlight that falls on it. **Albedo**, a, is the fraction of light that *reflects* from a planet. The fraction of the sunlight that is *absorbed* by the planet is 1 minus the albedo. A planet covered entirely in snow would have a high albedo (close to 1), while a planet covered entirely by black rocks would have a low albedo (close to zero).

$$(\text{Fraction of sunlight absorbed}) = 1 - a$$

We can now calculate the energy absorbed by the planet each second. Writing this relationship as an equation, we say that

$$\begin{pmatrix}\text{Energy} \\ \text{absorbed} \\ \text{each second}\end{pmatrix} = \begin{pmatrix}\text{Absorbing} \\ \text{area of} \\ \text{planet}\end{pmatrix} \times \begin{pmatrix}\text{Intensity} \\ \text{of} \\ \text{sunlight}\end{pmatrix} \times \begin{pmatrix}\text{Fraction of} \\ \text{sunlight} \\ \text{absorbed}\end{pmatrix}$$

$$= \pi R^2 \times \frac{L}{4\pi d^2} \times (1-a)$$

The amount of energy that the planet radiates away depends on the surface area of the planet, $4\pi R^2$. The Stefan-Boltzmann law tells us that each square meter radiates energy equal to σT^4 every second. Putting these two pieces together gives

$$\begin{pmatrix}\text{Energy} \\ \text{radiated} \\ \text{each second}\end{pmatrix} = \begin{pmatrix}\text{Surface} \\ \text{area of} \\ \text{planet}\end{pmatrix} \times \begin{pmatrix}\text{Energy radiated} \\ \text{per square meter} \\ \text{per second}\end{pmatrix}$$

$$= 4\pi R^2 \times \sigma T^4$$

For a planet at constant temperature, each second the "Energy radiated" must be equal to "Energy absorbed." When we set these two quantities equal to each other, we arrive at the expression

$$\begin{pmatrix}\text{Energy radiated} \\ \text{each second}\end{pmatrix} = \begin{pmatrix}\text{Energy absorbed} \\ \text{each second}\end{pmatrix}$$

$$4\pi R^2 \, \sigma T^4 = \pi R^2 \frac{L}{4\pi d^2}(1-a)$$

or

$$T^4 = \frac{L}{16\sigma \pi d^2}(1-a)$$

This gives us a T^4, but we want just T. If we take the fourth root of each side, we get

$$T = \left[\frac{L(1-a)}{16\sigma\pi d^2}\right]^{1/4}$$

Fortunately, L, σ, and π are known. If we express the distance from the Sun in astronomical units, the equation simplifies to:

$$T = 279 \text{ K} \times \left[\frac{(1-a)}{d^2_{\text{AU}}}\right]^{1/4}$$

For a blackbody ($a = 0$) at 1 AU from the Sun, the temperature is 279 K. For Earth, with an albedo of 0.306 and a distance from the Sun of 1 AU, the temperature is

$$T = 279 \text{ K} \times \left[\frac{(1-0.306)}{1^2}\right]^{1/4}$$

$$T = 255 \text{ K}$$

Calculator hint: To take the fourth root of a number, you can either use the x^y (or sometimes y^x) button on your calculator and put in 0.25 for the exponent or you may have a button labeled $x^{1/y}$, which allows you to put in 4 for the y. For example, if you are calculating $3^{1/4}$, you can either type [3][x^y][0][.][2][5] or [3][$x^{1/y}$][4].

Figure 7.4 plots the predicted and actual temperatures of the planets of the Solar System. The vertical bars show the range of temperatures measured for the surface on each terrestrial planet and at the tops of the clouds above each giant planet. The large black dots show our predictions using the final equation above to calculate temperature. Overall, there is fairly good agreement, so our basic understanding of *why* planets have the temperatures they do is probably close to the mark. The data and predictions for Mercury and Mars agree particularly well.

For Earth and the giant planets, however, the actual temperatures are a bit higher than predicted. For Venus, the actual temperature is much higher than our prediction. As we built our mathematical model for the equilibrium temperatures of planets, we made a number of assumptions. For example, we assumed that the temperature of a planet is the same everywhere. This is clearly not true; for example, we might expect planets to be hotter on the day side than on the night side. We also assumed that a planet's only source of energy is sunlight. Finally, we assumed that a planet is able to radiate energy into space freely as a blackbody.

The discrepancies between our model and the measured temperatures tell us that for some of these planets, some or all of these assumptions must be incorrect. The question of why these planets are hotter than predicted by this model leads us to a number of new and interesting insights.

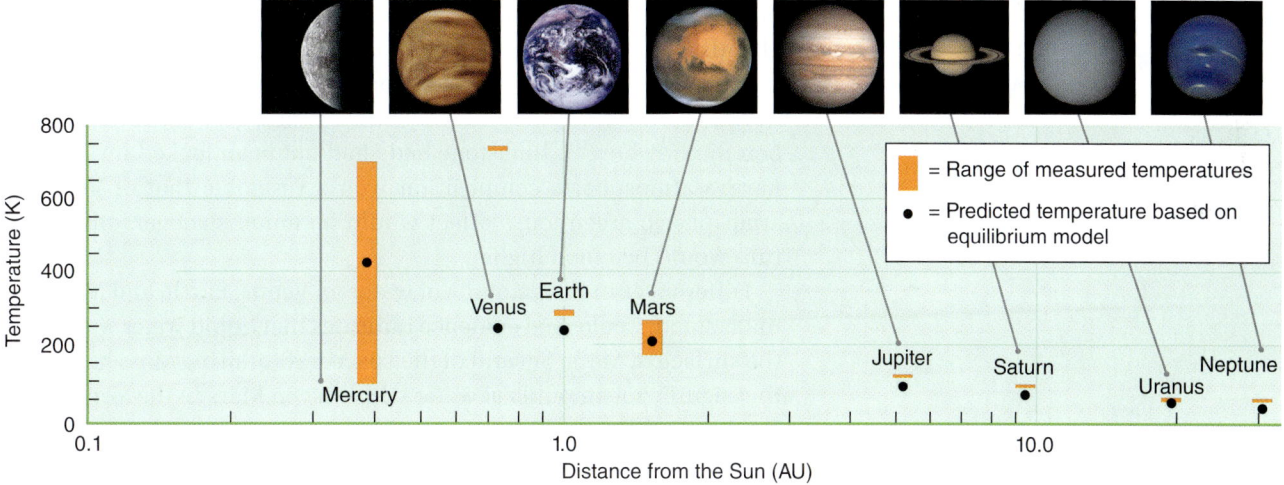

Figure 7.4 Predicted planetary temperatures, based on the equilibrium between absorbed sunlight and thermal radiation into space, are compared with ranges of observed surface temperatures.

the average global temperature near Earth's surface is about 288 K (15°C). Temperatures on Earth's surface are about 33 K warmer than they would be in the absence of an atmospheric greenhouse effect, mainly because of water vapor and carbon dioxide. Yet this comparatively small difference has been crucial in shaping the Earth we know. Without the greenhouse effect, Earth's average global temperature would be –18°C, well below the freezing point of water, leaving us with a world of frozen oceans and ice-covered continents.

How has the atmospheric greenhouse effect made the composition of Earth's atmosphere so different from the high-carbon-dioxide atmospheres of Venus and Mars? We may find the answer in Earth's particular location in the Solar System. Consider early Earth and early Venus: each had about the same mass, but Venus orbits somewhat closer than Earth to the Sun. Volcanism poured out large amounts of carbon dioxide and water vapor to form early secondary atmospheres on both planets. Most of Earth's water quickly rained out of the atmosphere to fill vast ocean basins. But because Venus was closer to the Sun, its surface temperatures were higher than those of Earth. Most of the rainwater on Venus immediately re-evaporated, much as water does in Earth's desert regions. Venus was left with a surface that contained very little liquid water and an atmosphere filled with water vapor. The continuing buildup of both water vapor and carbon dioxide in the atmosphere of Venus led to a runaway atmospheric greenhouse effect that drove up the surface temperature of the planet. Ultimately, the surface of Venus became so hot that no liquid water could exist on it.

The early difference between a watery Earth and an arid Venus changed the ways that their atmospheres and surfaces evolved. On Earth, water erosion caused by rain and rivers continually exposed fresh minerals, which then reacted chemically with atmospheric carbon dioxide to form solid carbonates. This reaction removed some of the atmospheric carbon dioxide, burying it within Earth's crust as a component of a rock called limestone. Later, the development of life in Earth's oceans accelerated the removal of atmospheric carbon dioxide. Tiny sea creatures built their protective shells of carbonates, and as they died they built up massive beds of limestone on the ocean floors. Water erosion and the chemistry of life tied up all but a trace of Earth's total inventory of carbon dioxide in limestone beds. Earth's particular location in the Solar System seems to have spared it from the runaway atmospheric greenhouse effect. If all the carbon dioxide now in limestone beds had not been locked up by these reactions, Earth's atmospheric composition would resemble that of Venus or Mars; the atmospheric greenhouse effect would be much stronger, and Earth's temperature would be much higher.

Differences in the amounts of water on Venus, Earth, and Mars are not so well understood. Geological evidence indicates that liquid water was once plentiful on the surface of Mars. Several of the spacecraft orbiting Mars have found evidence that significant amounts of water still exist on Mars in the form of subsurface ice. Earth's liquid and solid water supply is even greater: about 10^{21} kilograms (kg), or 0.02 percent of its total mass. More than 97 percent of Earth's water is in the oceans, which have an average depth of about 4 kilometers (km). Earth today has 100,000 times more water than Venus. What happened to all the water on Venus? One possibility is that water molecules high in its atmosphere were broken apart into hydrogen and oxygen by solar UV radiation. The low-mass hydrogen atoms were quickly lost to space. Oxygen, however, migrated downward to the planet's surface, where it was removed from the atmosphere by bonding with surface minerals. The *Venus Express* spacecraft has measured hydrogen and some oxygen escaping from the upper levels of Venus's atmosphere.

7.3 Earth's Atmosphere Has Detailed Structure

Now that we understand some of the overall processes that have influenced the evolution of the terrestrial planet atmospheres, we will look in depth at each of them. We begin with the composition, structure, and weather of Earth's atmosphere, not only because we know it best but also because it will help us better understand the atmospheres of other worlds.

Life and the Composition of Earth's Atmosphere

Earth's atmosphere is relatively uniform on a global scale. Two principal gases make up our atmosphere: about four-fifths of it is nitrogen, and one-fifth is oxygen. (Here, we are talking about oxygen molecules, O_2.) There are also many impor-

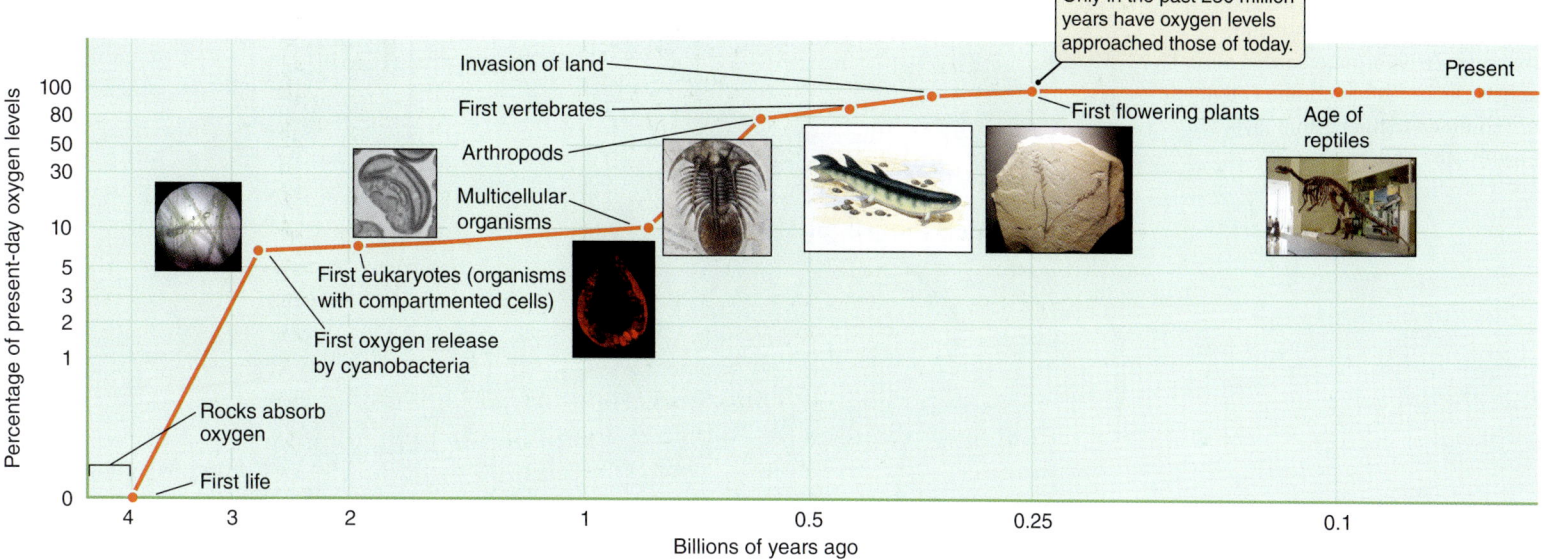

Figure 7.5 The amount of oxygen in Earth's atmosphere has built up over time as a result of plant life on the planet.

tant minor constituents such as water vapor and carbon dioxide, the amounts of which vary depending on global location and season.

Earth's atmosphere contains abundant amounts of oxygen, while the atmospheres of other planets do not. Why should this be so? Oxygen is a highly reactive gas. It chemically combines with, or oxidizes, almost any material it touches. The rust (iron oxide) that forms on steel is an example. A planet that has significant amounts of oxygen in its atmosphere requires a means of replacing what is lost through oxidation. On Earth, plants perform this task.

The oxygen concentration in Earth's atmosphere has changed over the history of the planet, as shown in **Figure 7.5**. When Earth's secondary atmosphere first appeared about 4 billion years ago, it had very little oxygen because free oxygen is not found in volcanic gases or comets. Studies of ancient sediments show that about 2.8 billion years ago, an ancestral form of cyanobacteria (single-celled organisms that contain chlorophyll, which enables them to obtain energy from sunlight) began releasing oxygen into Earth's atmosphere as a waste product. At first, this oxygen combined readily with exposed metals and minerals in surface rocks and soils and so was removed from the atmosphere as quickly as it formed. Ultimately, the explosive growth of plant life accelerated the production of oxygen, building up atmospheric concentrations that approached today's levels only about 250 million years ago. All true plants, from tiny green algae to giant redwoods, use the energy of sunlight to build carbon compounds out of carbon dioxide and produce oxygen as a waste product in the process called photosynthesis. Earth's atmospheric oxygen content is held in a delicate balance primarily by plants. If plant life disappeared, so would nearly all of Earth's atmospheric oxygen, and therefore all animal life—including humans.

The Layers of Earth's Atmosphere

Earth's atmosphere is a blanket of gas several hundred kilometers deep. It has a total mass of approximately 5×10^{18} kg (about 5,000 trillion metric tons), which is less than one-millionth of Earth's total mass. The weight of Earth's atmosphere presses with 100,000 newtons on each square meter of our planet's surface, equiva-

Figure 7.6 (a) Temperature and (b) pressure determine Earth's atmospheric layers. Most human activities are confined to the bottom layers of Earth's atmosphere.

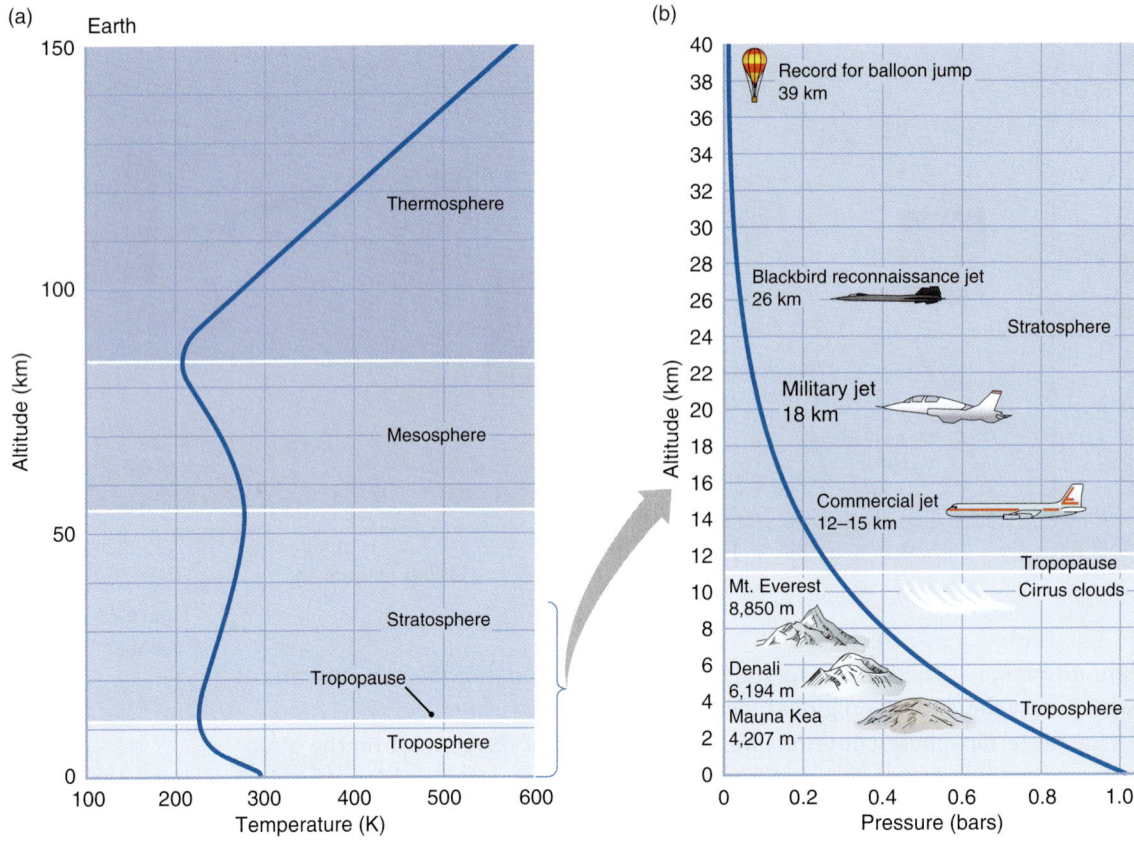

Vocabulary Alert

bar This common word has many everyday meanings, but scientists often use the bar as a unit of pressure equal to the average atmospheric pressure at sea level on Earth. A millibar (mb) is one-thousandth of 1 bar and is more commonly used in meteorology and in weather reports.

lent to about 14.7 pounds pressing on every square inch. This amount of pressure is called a **bar**. We are largely unaware of Earth's atmospheric pressure because the same pressure exists both inside and outside our bodies, so the force pushing out precisely balances the force pushing in. As you learned in Chapter 5, the pressure at any point within a star must be great enough to balance the weight of the overlying layers. The same principle holds true in a planetary atmosphere: The atmospheric pressure on a planet's surface must be great enough to hold up the weight of the overlying atmosphere.

This blanket of gas is made up of several distinct layers, shown in **Figure 7.6a**. These layers are distinguished by the changes in temperature and pressure through the atmosphere. Earth's lowermost atmospheric layer, the one in which we live and breathe, is called the **troposphere**. It contains 90 percent of Earth's atmospheric mass and is the source of all our weather. Within the troposphere, atmospheric pressure, density, and temperature all decrease as altitude increases. Convection circulates air between the lower and upper levels, and such circulation tends to diminish the temperature extremes caused by heating at the bottom and cooling at the top.

Convection also affects the vertical distribution of atmospheric water vapor. The ability of air to hold water in the form of vapor depends very strongly on the air temperature: the warmer the air, the more water vapor it can hold. As air is convected upward, it cools. When the air temperature decreases to the point at which the air can no longer hold all of its water vapor, water begins to condense to tiny droplets or ice crystals. In large numbers these become visible to us as clouds. When these droplets combine to form large drops, they fall as rain or snow. For this reason, most of the water vapor in Earth's atmosphere

Figure 7.7 Observatories are usually located high on mountaintops, to rise above most of Earth's water vapor to improve access to the infrared portion of the spectrum. Visitors to the Mauna Kea Observatories in Hawaii have a view of the tops of the clouds, stretching away from the mountaintop.

stays within 2 km of the surface. At an altitude of 4 km, the Mauna Kea Observatories (Figure 7.6b) are higher than approximately one-third of Earth's atmosphere, but they lie above nine-tenths of the atmospheric water vapor. This is important for astronomers who observe the sky in the infrared region of the spectrum, because water vapor strongly absorbs infrared light. The water in the atmosphere is more often visible as condensed water in the form of clouds and ice, as is shown in **Figure 7.7**.

Above the troposphere is the **stratosphere** (see Figure 7.6). The boundary between these two regions is called the **tropopause**. In the stratosphere, the temperature-altitude relationship reverses, and the temperature begins to *increase* with altitude. Therefore, convection does not take place. This temperature reversal is caused by the ozone layer. In this layer, oxygen molecules (O_2) absorb UV radiation from the Sun and are split into individual oxygen atoms. These atoms then combine with other oxygen molecules to form ozone (O_3). Because light is absorbed in this process, it warms the stratosphere. Ozone absorbs high-energy UV photons—harmful to terrestrial life—preventing them from reaching the ground.

The region above the stratosphere is called the **mesosphere**. It extends from an altitude of 50 km to about 90 km. In the mesosphere there is no ozone to absorb sunlight, so temperatures once again decrease with altitude. The base of the stratosphere and the upper boundary of the mesosphere are the two coldest levels in Earth's atmosphere.

Higher in Earth's atmosphere, interactions with space begin to be important. The **solar wind** is a flow of high-energy particles that stream continuously from the Sun. At altitudes above 90 km, solar UV radiation and the high-energy particles of the **solar wind** ionize atoms and molecules in the atmosphere, once again causing the temperature to increase with altitude. This region is called the **thermosphere**, and it is the hottest part of the atmosphere. Near the top of the thermosphere, at an altitude of 600 km, the temperature can reach 1000 K. The gases within and beyond the thermosphere are ionized by UV photons and high-energy particles from the Sun to form a **plasma** layer known as the **ionosphere**. This layer overlaps the thermosphere but also extends farther into space.

Even farther out is Earth's **magnetosphere**. It has a radius approximately 10 times the radius of Earth and a volume greater than 1,000 times the volume of

Vocabulary Alert

plasma In common usage, this word almost always refers to a component of blood (or a type of television). But it has another scientific meaning, referring to a gas in which most of the atoms and molecules have lost some electrons, leaving them with a net positive charge. Because these particles are charged, they interact with electric and magnetic fields.

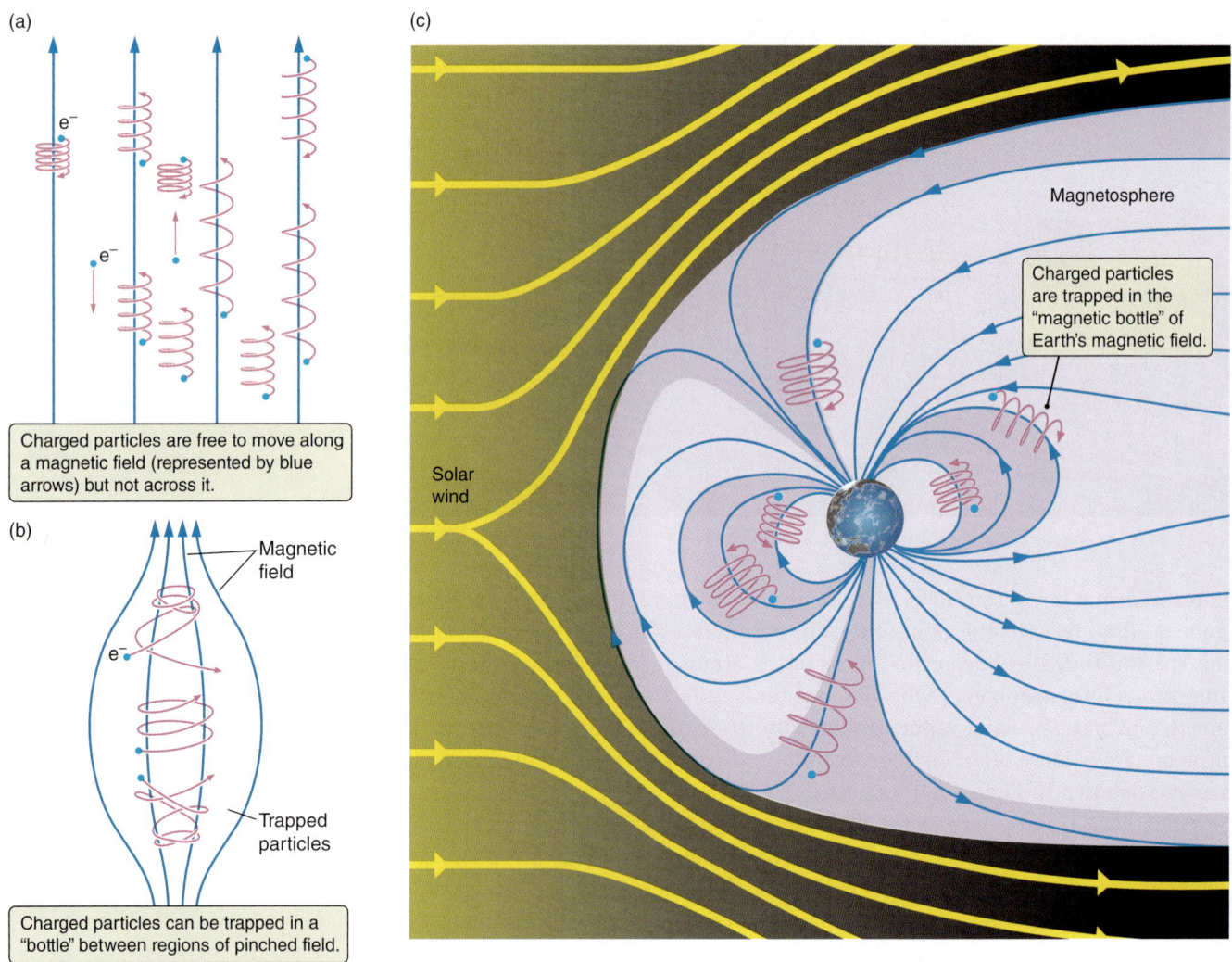

Figure 7.8 (a) Charged particles, in this case electrons, spiral in a uniform magnetic field. (b) When the field is pinched, charged particles can be trapped in a "magnetic bottle." (c) Earth's magnetic field acts like a bundle of magnetic bottles, trapping particles in Earth's magnetosphere. In all these images, the radius of the helix the charged particle follows is greatly exaggerated.

the planet. This large region is filled with charged particles from the Sun that have been captured by Earth's magnetic field. Magnetic fields only affect *moving* charges. Charged particles move freely *along* the direction of the magnetic field but cannot cross magnetic field lines. If they try to move *across* the direction of the field, they experience a force that causes them to loop around the direction of the magnetic field, as illustrated in **Figure 7.8a**. This force is perpendicular to the motion of the particle and perpendicular to the direction of the magnetic field.

If the field is pinched together at some point, particles moving into the pinch will feel a magnetic force that reflects them back along the direction they came from, creating a sort of "magnetic bottle" that contains the charged particles. If charged particles are located in a region where the field is pinched on both ends, as shown in Figure 7.8b, then they may bounce back and forth many times.

Earth and its magnetic field are immersed in the solar wind. When these charged particles first encounter Earth's magnetic field, the smooth flow is

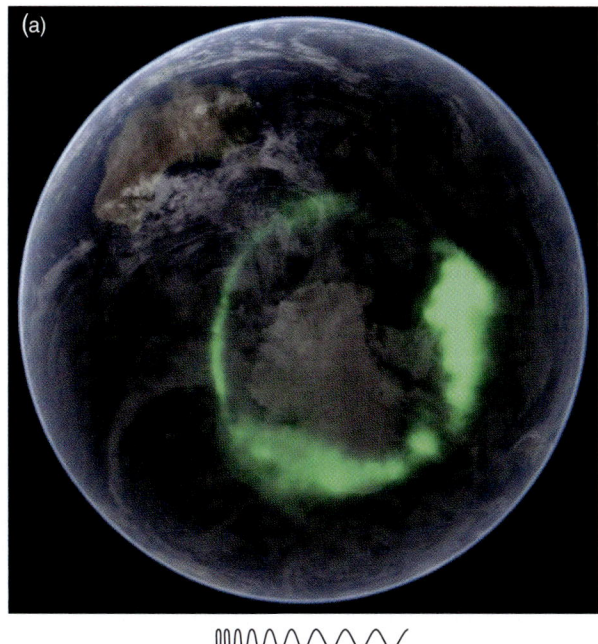

Figure 7.9 Auroras result when particles trapped in Earth's magnetosphere collide with molecules in the upper atmosphere. (a) An auroral ring around Earth's south magnetic pole, as seen from space. (b) Aurora borealis—the "northern lights"—viewed from the ground.

interrupted and their speed suddenly drops—they are diverted by Earth's magnetic field like a river is diverted around a boulder. As they flow past, some of these charged particles become trapped by Earth's magnetic field, where they bounce back and forth between Earth's magnetic poles as illustrated in Figure 7.8c.

Some regions in the magnetosphere contain especially strong concentrations of energetic charged particles, which can be very damaging to both electronic equipment and humans. Disturbances in Earth's magnetosphere caused by changes in the solar wind can affect Earth's magnetic field enough to trip power grids, cause blackouts, and disrupt communications.

Earth's magnetic field also funnels energetic charged particles down into the ionosphere in two rings located around the magnetic poles. These charged particles (mostly electrons) collide with atoms such as oxygen, nitrogen, and hydrogen in the upper atmosphere, causing them to glow like the gas in a neon sign. Interactions with different atoms cause different colors. These glowing rings, called **auroras**, can be seen from space (**Figure 7.9a**). When viewed from the ground (Figure 7.9b), they look like eerie, shifting curtains of multicolored light. People living far from the equator view the aurora borealis ("northern lights") in the Northern Hemisphere or the aurora australis in the Southern Hemisphere. When the solar wind is particularly strong, auroras can even be seen at lower latitudes. Look back at the chapter-opening illustration, which shows both a photograph of an aurora and a handmade sketch. You should now be able to describe how this aurora formed and sketch the interaction of the particles and the magnetic field that gave rise to them. Auroras are not unique to Earth. They have also been observed on Venus, Mars, all of the giant planets, and some moons.

The general structure we have described here is not unique to Earth's atmosphere. The major components—troposphere, stratosphere, and ionosphere—also exist in the atmospheres of Venus and Mars as well as in the atmospheres of Saturn's moon Titan and the giant planets. The magnetospheres of the giant planets are among the largest structures in the Solar System.

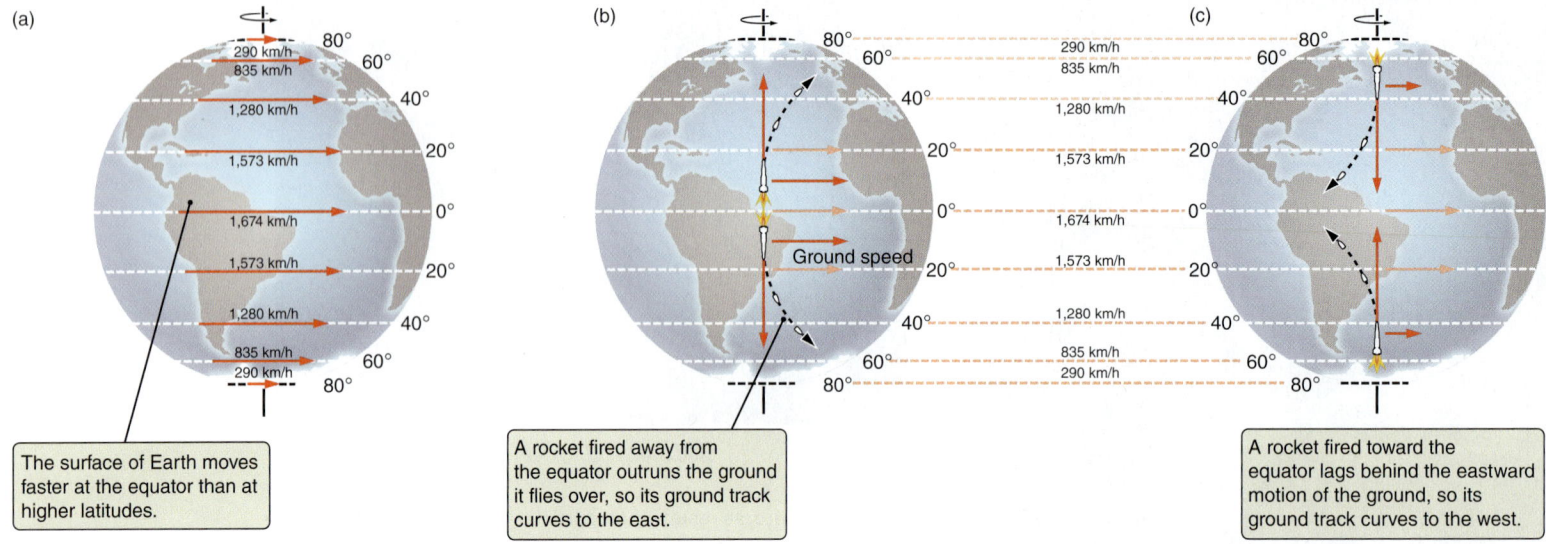

Figure 7.10 The Coriolis effect causes objects to appear to be deflected as they move across the surface of Earth.

Wind

Heating a gas increases its pressure, which causes it to push into its surroundings. These pressure differences cause winds. Winds are the natural movement of air, both locally and on a global scale, in response to variations in temperature from place to place. The air is usually warmer in the daytime than at night, warmer in the summer than in winter, and warmer at the equator than near the poles. Large bodies of water, such as oceans, also affect atmospheric temperatures. The strength of the winds is governed by the magnitude of the temperature difference.

The effect of Earth's rotation on winds—and on the motion of any object—is called the **Coriolis effect**. This effect results from the rotation of Earth and the inertia of moving objects. Objects near the equator move in the direction of rotation (toward the east) faster than objects closer to the poles do, as shown in **Figure 7.10a**. Now imagine that a rocket is launched directly north from a point in the Northern Hemisphere, as shown in Figure 7.10b; its path would appear to curve to the east because the rocket is already traveling to the east faster than other objects farther north—including the ground. Note that if the rocket travels from a northern point toward the equator (Figure 7.10c), the path would appear to curve toward the west, because it is traveling east more slowly than the ground at the equator.

As air in Earth's equatorial regions is heated by the warm surface, convection causes it to rise. The warmed surface air displaces the air above it, which then has no place to go but toward the poles. This air becomes cooler and denser as it moves toward the poles, and so it sinks back down through the atmosphere. It displaces the surface polar air, which is forced back toward the equator, completing the circulation. As a result, the equatorial regions remain cooler and the polar regions remain warmer than they otherwise would be. Air moves between the equator and poles of a planet in a pattern known as **Hadley circulation,** which is shown in **Figure 7.11a**.

On Earth, other factors break up the planet-wide flow into a series of smaller Hadley cells. Most planets and their atmospheres rotate rapidly, and the Coriolis effect strongly interferes with Hadley circulation by redirecting the horizontal

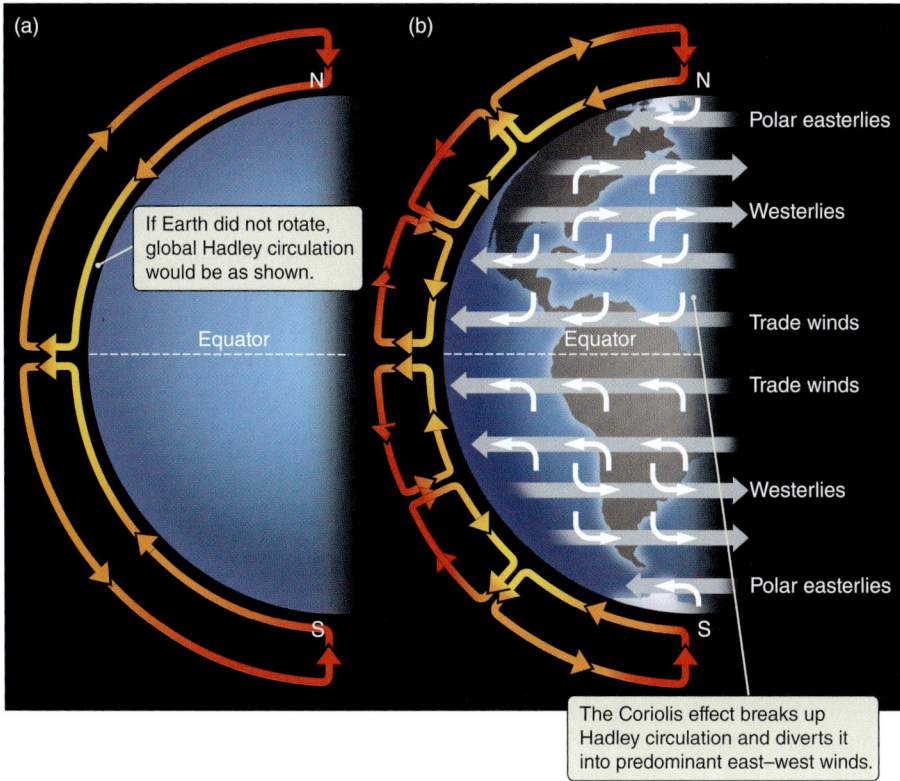

Figure 7.11 (a) The Hadley circulation covers an entire hemisphere. (b) On Earth, Hadley circulation breaks up into smaller circulation cells. The Coriolis effect diverts north–south flow into east–west flow.

flow, shown in Figure 7.11b. The Coriolis effect creates winds that blow predominantly in an east–west direction and are often confined to relatively narrow bands of latitude. Meteorologists call these **zonal winds**. More rapidly rotating planets have a stronger Coriolis effect and stronger zonal winds. Between the equator and the poles in most planetary atmospheres, the zonal winds alternate between winds blowing from the east toward the west and winds blowing from the west toward the east.

In Earth's atmosphere, several bands of alternating zonal winds lie between the equator and each hemisphere's pole. This zonal pattern is called Earth's **global circulation**. The best-known zonal currents are the subtropical trade winds—more or less easterly winds that once carried sailing ships from Europe westward to the Americas—and the midlatitude prevailing westerlies that carried them home again (see Figure 7.11b).

Embedded within Earth's global circulation pattern are systems of winds associated with large high- and low-pressure regions. A combination of a low-pressure region and the Coriolis effect produces a circulating pattern called **cyclonic motion**. Cyclonic motion is associated with stormy weather, including hurricanes. Similarly, high-pressure systems are localized regions where the air pressure is higher than average. We can think of these regions of greater-than-average air concentration as "mountains" of air. Owing to the Coriolis effect, high-pressure regions rotate in a direction opposite to that of low-pressure regions. These high-pressure circulating systems experience **anticyclonic motion** and are generally associated with fair weather.

172 CHAPTER 7 Atmospheres of Venus, Earth, and Mars

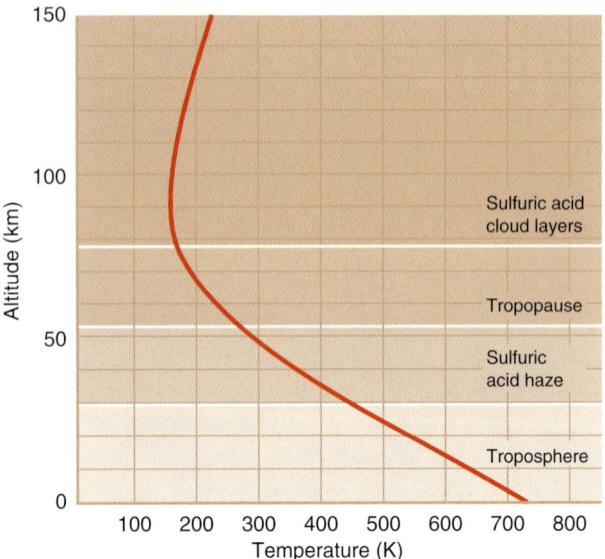

Figure 7.12 The temperature of the atmosphere on Venus primarily drops with increase in altitude, unlike temperatures in Earth's atmosphere, which fall and rise and fall again through the troposphere, stratosphere, and mesosphere, respectively.

Figure 7.13 Thick clouds obscure our view of the surface of Venus.

7.4 The Atmospheres of Venus and Mars Differ from Earth's

The atmospheres of Venus, Earth, and Mars are very different. The atmosphere of Venus is very hot and dense compared to that of Earth, while the atmosphere of Mars is very cold and thin. Whereas Earth is a lush paradise, the greenhouse effect has turned Venus into a convincing likeness of hell—an analogy made complete by the presence of choking amounts of sulfurous gases. Compared to Venus, the surface of Mars is almost hospitable. For this reason, Mars is the planet of choice as humans consider the colonization of other planets. Understanding why and how these atmospheres are so different helps us understand how Earth's atmosphere may evolve in the future.

Venus

Venus and Earth are similar in many ways—so similar that they might be thought of as sister planets. When we used the laws of radiation to predict temperatures for the two planets, we concluded that they should be very similar. But that was before we considered the greenhouse effect and the role of carbon dioxide in trapping the infrared radiation that a planetary surface typically emits. Venus's atmosphere is 96 percent carbon dioxide, with a small amount (3.5 percent) of nitrogen and still less amounts of other gases. This thick blanket of carbon dioxide traps the infrared radiation from Venus, driving the temperature at the surface of the planet to a sizzling 737 K, which is hot enough to melt lead.

Large variations in the observed amounts of sulfurous compounds in the high atmosphere of Venus suggest that the source of sulfur may be sporadic episodes of volcanic activity. Additionally, the atmospheric pressure at the surface of Venus is 92 times greater than the pressure on Earth's surface; the atmospheric pressure at the surface of Venus is equal to the water pressure at an ocean depth of 900 meters on Earth. This is enough pressure to crush the hull of a modern nuclear submarine.

As you can see in **Figure 7.12,** the atmospheric temperature of Venus decreases continuously with increase in altitude throughout the planet's troposphere—similar to Earth. The atmospheric temperature of Venus drops to a low of about 160 K at the tropopause. At an altitude of approximately 50 km, Venus's atmosphere has an average temperature and pressure similar to our own atmosphere at sea level. At altitudes between 50 and 80 km, the atmosphere is cool enough for sulfurous oxide vapors to react with water vapor to form clouds of concentrated sulfuric acid droplets (H_2SO_4). These dense clouds completely block our view of the surface of Venus, as **Figure 7.13** shows. Throughout the 1960s, radio telescopes and spacecraft with cloud-penetrating radar provided low-resolution views of the surface of Venus. But it was not until 1975, when the Soviet Union succeeded in landing cameras there, that people got a clear picture of the surface. Radar images taken by the *Magellan* spacecraft in the early 1990s (see Figure 6.17) produced a global map of the surface of Venus.

Unlike the other planets, Venus rotates on its axis in a direction opposite to its motion around the Sun. This is called *retrograde rotation.* (Recall that *retrograde motion* is the movement of a planet in a direction opposite from its normal motion.) Relative to the stars, Venus rotates on its axis once every 243 Earth days. However, a solar day on Venus—the time it takes for the Sun to return to the same place in the sky—is only 117 Earth days. Venus has no seasons because its axis of rotation is nearly perpendicular to the orbital plane. Its extremely slow rotation means that the Coriolis effect is small, so the global circulation is quite close to a classic Hadley pattern shown in Figure 7.11a. Venus is the only planet in the Solar System known to behave in this way.

Imagine yourself standing on the surface of Venus. Because sunlight cannot easily penetrate the dense clouds above you, noontime on the surface of Venus is no brighter than a very cloudy day on Earth. The Sun is setting, which takes a long time because the planet rotates so slowly. Just when you think you'll get some relief from the scorching heat of the setting Sun, you find that the thick atmosphere efficiently retains heat throughout the night. The temperature does not drop at night, nor does it drop near the poles. Such small temperature variations also mean there is almost no wind at the surface, so wind erosion is weak compared to wind erosion on Earth and Mars. High in the atmosphere, however, you would find winds of 110 m/s. High temperatures and very light winds keep the lower atmosphere of Venus free of clouds and hazes, and strong scattering of light by the dense atmosphere turns any view you might have of distant scenes hazy and bluish. We see the same effect, but to a lesser extent, in our own atmosphere.

Mars

Mars is a stark landscape, colored reddish by the oxidation of iron-bearing surface minerals (**Figure 7.14**). The sky is sometimes a dark blue, but more often it has a pinkish color caused by windblown dust. The lower density of the Mars atmosphere makes it more responsive than Earth's atmosphere to heating and cooling, so its temperature extremes are greater. The surface near the equator at noontime is a comfortable 20°C—a cool room temperature. At night, however, temperatures typically drop to a frigid −100°C, and during the polar night the air temperature can reach −150°C, cold enough to freeze carbon dioxide out of the air in the form of dry-ice frost.

The average atmospheric surface pressure on Mars is equivalent to the pressure at an altitude of 35 km above sea level on Earth, far higher than our highest mountain. Pressures range from a high of 11.5 millibars in the lowest impact

Figure 7.14 This true-color image of the surface of Mars was taken by the rover *Spirit*. In the absence of dust, the sky's thin atmosphere would appear deep blue. In this image, wind-blown dust turns the sky pinkish in color.

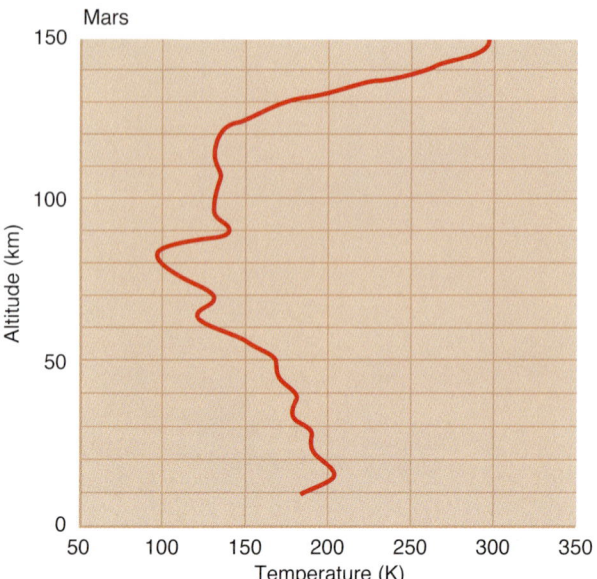

Figure 7.15 The temperature profile of the atmosphere of Mars. Note the differences in temperature and structure between this profile and the profile of the atmosphere of Venus (see Figure 7.12).

basins of Mars to a mere 0.3 millibars at the summit of Olympus Mons. Recall that Earth's pressure at sea level is about 1 bar, so the highest pressure on Mars is only 1.1 percent of Earth's pressure at sea level. The temperature profile of the atmosphere of Mars is more similar to Earth's than to that of Venus, with a range of only 100 degrees up to about 125 km, above which the temperature rises because of absorption of sunlight in the upper atmosphere (**Figure 7.15**).

The inclination of Mars' equator to its orbital plane is similar to that of Earth, so both planets have similar seasons. But the effects on Mars are larger for two reasons: Mars varies more in its annual orbital distance from the Sun than does Earth, and the low density of the Mars atmosphere makes it more responsive to seasonal change. The large daily, seasonal, and latitudinal surface temperature differences on Mars often create locally strong winds; some are estimated to be higher than 100 m/s. High winds can stir up huge quantities of dust and distribute it around the planet's surface (**Figure 7.16**). For more than a century, astronomers have watched the seasonal development of springtime dust storms on Mars. The stronger storms spread quickly and can envelop the entire planet in a shroud of dust within a few weeks (**Figure 7.17**). Such large amounts of windblown dust can take many months to settle out of the atmosphere. Seasonal movement of dust from one area to another alternately exposes and covers large areas of dark, rocky surface.

7.5 Greenhouse Gases Affect Global Climates

Observations of the temperature and composition of Earth's atmosphere show that Earth's climate is changing. To understand observations of climate change, we must first clearly understand the distinction between *climate* and *weather*. The state of Earth's atmosphere at any given time and place is **weather**. Weather is small scale and short term. **Climate** is the term used to define the *average* state

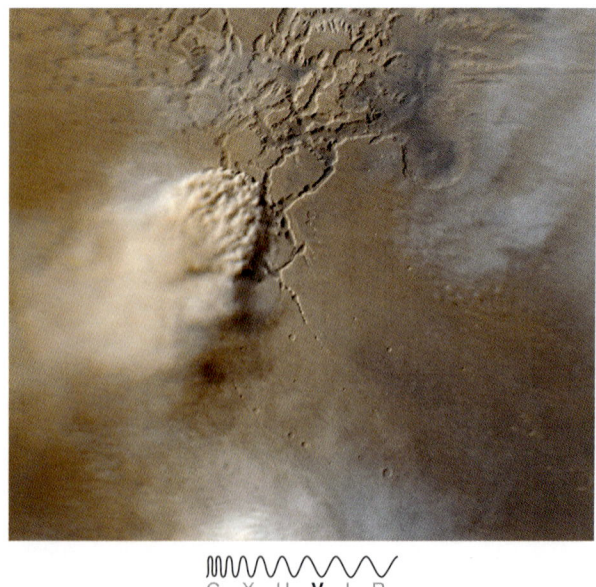

Figure 7.16 This image shows a dust storm raging in the canyon lands of Mars.

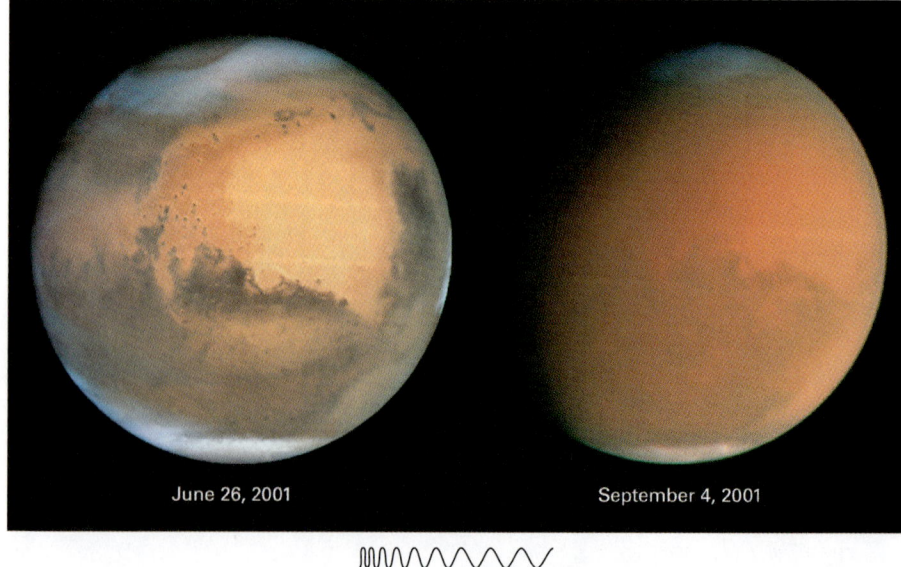

Figure 7.17 Hubble Space Telescope images show the development of a global dust storm that enshrouded Mars in September 2001. The same region of the planet is shown in both images.

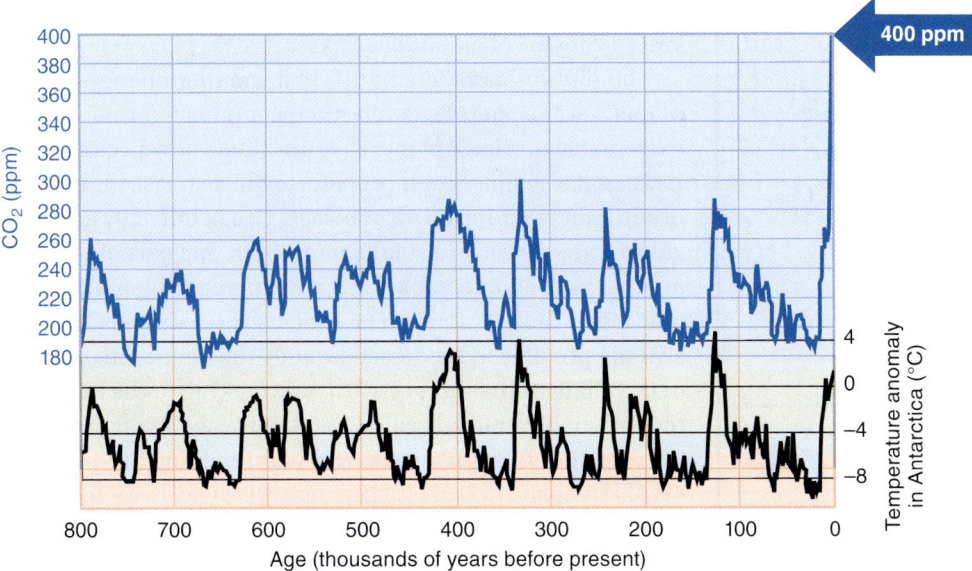

Figure 7.18 Temperature anomaly and carbon dioxide level over the past 800,000 years of Earth's history have been plotted on the same graph to make the similarities and differences easier to see. Notice the two *y*-axes on this graph. The right *y*-axis plots temperature data, and the left *y*-axis plots the CO_2 data. The temperature and the CO_2 levels are very tightly correlated. The modern carbon dioxide level is indicated by the large blue arrow. This level is much higher than at any time in the past 800,000 years.

of Earth's atmosphere, including temperature, humidity, winds, and so on. Climate describes the planet as a whole over longer timescales.

As you can see in **Figure 7.18,** Earth's climate has lengthy temperature cycles, which can last hundreds of thousands of years and occasionally produce shorter cold periods called *ice ages*. These data were obtained from Antarctic ice cores. The temperature data are given as the *anomaly*: the difference of the temperature at a given time from a long-term time average over the past few thousand years. A positive anomaly means the global temperature was warmer than the average. A negative anomaly means the temperature was cooler than the average. These changes in the mean (average) global temperature are far smaller than typical geographic or seasonal temperature changes, but Earth's atmosphere is so sensitive to mean global temperature that a drop of only a few degrees can plunge our climate into an ice age. Scientists still do not understand all the mechanisms controlling these climate-changing temperature swings. An external influence, such as small changes in the Sun's energy output, may be one of the causes. Some researchers have also suggested that changes in Earth's orbit or the inclination of its rotation axis have affected the climate. Temperature changes may be triggered internally by volcanic eruptions, which can produce global sunlight-blocking clouds or hazes. Another factor may be long-term interactions between Earth's oceans and its atmosphere.

As mentioned in Section 7.2, the atmospheric greenhouse effect clearly is responsible for warming Earth and Venus above the temperatures that would be expected due to their distance from the Sun. Though Mars has a thin atmosphere, it is warmer than it would be without any atmosphere. In our own Solar System, we have astronomical evidence of how the atmospheric greenhouse effect can influence planetary atmospheres, including Earth's. Astronomers view these

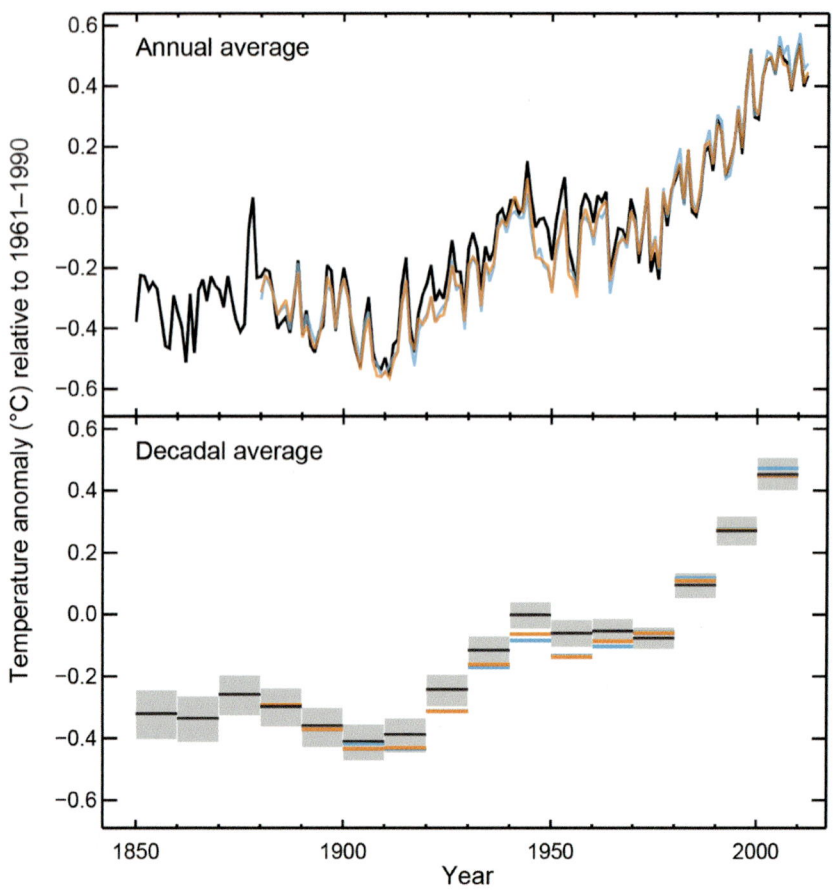

Figure 7.19 Global temperature variations of Earth since 1850 show that the temperature has been rising over this time period, whether considered year by year (top) or decade by decade (bottom). Different data sets are represented in different colors and yield the same result. The zero-point of the anomaly in this graph is slightly different than in the previous graph because a different time span was used to calculate the average. This does not, however, affect the result that the average global temperature has increased significantly over this time period.

different planets as a cautionary tale, showing the results of varied "doses" of greenhouse gases.

The physics associated with real, smaller changes to our planet's atmosphere is much more complicated than in this extreme scenario. An increase in cloud cover, caused by increased water in the atmosphere, might decrease the amount of sunlight reaching Earth's surface. Ocean currents are critical in transporting thermal energy from one part of Earth to another, but no one yet knows how increased temperatures may affect those systems. The process is so complex that it is still not possible to predict accurately the long-term outcome of the changes that humans are now making to the composition of Earth's atmosphere. In a real sense, we are *experimenting* with the atmosphere of Earth. We are asking the question, "What happens to Earth's climate if we steadily increase the number of greenhouse molecules in its atmosphere?" We do not yet know the answer, but we are already seeing some results. Figure 7.18 shows the carbon dioxide levels and temperature of Earth's atmosphere over the past 800,000 years. Notice that temperature and carbon dioxide are clearly correlated. When one rises, so does the other. Since the industrial revolution, the carbon dioxide level has risen higher than ever before. At the scale of this illustration, it is difficult to see what's happening to temperature during the past few hundred years. We need to "zoom in" to see this more clearly.

Figure 7.19 plots temperature measurements over the past 160 years. The data show a steady increase in the mean global temperature. Most computer models suggest that this trend represents the beginning of a longer-term change caused by the buildup of human-made greenhouse gases. Earth's climate is a complex, chaotic system within which tiny changes can produce enormous and often unexpected results. To add to the complexity, Earth's climate is intimately tied to ocean temperatures and currents. We see examples of this connection in the periodic El Niño and La Niña conditions, small shifts in ocean temperature that cause much larger global changes in air temperature and rainfall. A further complication is that the ocean acts as a heat reservoir. While atmospheric warming has remained flat in recent years, the temperature of the oceans continues to rise. Studies suggest that changes in the flow of the Gulf Stream current in the North Atlantic have very large effects on the climates of North America and northern Europe—and that these changes may take place not over centuries but within a matter of decades. The best models show that a small addition of greenhouse gases to the atmosphere is extremely likely to have large effects over the relatively short period of about 50–100 years.

READING ASTRONOMY News

Scientists use many different techniques to reconstruct past events. Here, they use isotopes to investigate the history of Mars' atmosphere.

Curiosity Rover Sees Signs of Vanishing Martian Atmosphere

By **Amina Khan**, *Los Angeles Times*

Before going incommunicado behind the Sun for a month, NASA's Mars *Curiosity* rover sent Earth evidence that the red planet has lost much of its original atmosphere.

The findings, announced by Jet Propulsion Laboratory scientists at the European Geosciences Union meeting in Vienna, bolster the idea that the martian atmosphere was once much thicker than it is today—and come less than a month after the rover drilled its first rock and found signs that Mars was once hospitable to life.

Curiosity's Sample Analysis at Mars instrument sniffed the martian atmosphere and counted up the isotopes of argon in the air. Isotopes are heavier and lighter versions of the same element, and when a planet starts to lose its atmosphere, the lighter isotopes in the upper layers are the first to go. So if scientists see fewer of the lighter isotopes than expected, it might mean that there was once much more air there.

The researchers looked at two isotopes of argon: the heavier argon-38 and the lighter argon-36. They found that there was four times as much argon-36 as argon-38—a lower share of the lighter argon than expected based on data from other parts of the Solar System.

Some of that missing argon-36 must have escaped because the top of the atmosphere had started to blow away, they surmised. This could mean Mars's atmosphere was much thicker in the past than what *Curiosity* picks up today.

The scientists recently used this technique with carbon isotopes, finding similar results. But the new argon measurements provide much firmer evidence than older, less certain data on argon isotopes from the *Viking* mission in 1976, as well as from martian meteorites.

Curiosity is experiencing what scientists call "solar conjunction," when the Sun comes between Earth and Mars, blocking communications until May. But *Curiosity* isn't taking a vacation. The rover has to take weather and radiation measurements every day—part of a detailed list of chores sent by the Mars Science Laboratory team at JPL.

Evaluating the News

1. Is the atmosphere being discussed in this article the primary atmosphere or the secondary atmosphere of Mars? How do you know?
2. Explain why the lighter isotopes disappear from the atmosphere first. Where do those isotopes go?
3. The argument is made here that there "should be" more argon-36 than was found. How do the scientists know how much argon-36 to expect to find?
4. The article states that the scientists had carried out this experiment with carbon isotopes, finding similar results. Some might argue that this argon experiment was therefore unnecessary and inefficient. Using what you learned in Chapter 1 about the scientific method, argue that the argon experiment should have been done.
5. Sketch the Earth-Mars-Sun system as viewed from above during the "solar conjunction" referenced in the article. Why is *Curiosity* "incommunicado" during the time of the article?

SUMMARY

Earth, Venus, and Mars all have significant atmospheres that are different from the atmospheres they captured when they formed. These atmospheres are complex, both in chemical composition and in physical characteristics such as temperature and pressure. The climates of Earth, Venus, and Mars are all modified by their individual atmospheres. Venus, Earth, and Mars are warmer than they would be from solar illumination alone. The atmospheres of Earth, Venus, and Mars have different chemical compositions. These, in turn, lead to dramatic differences in temperature and pressure. Life has altered Earth's atmosphere several times, most notably in the distant past from an increase in the amount of oxygen in the atmosphere and in modern times from an increase in greenhouse gases.

LG 1 The primary atmospheres of the terrestrial planets were lost early in the history of the Solar System. Secondary atmospheres arose from volcanism and impacts, and they differ in chemical composition from the primary atmospheres. These secondary atmospheres continue to exist today.

LG 2 The atmospheres of Venus, Earth, and Mars have different temperatures, pressure, and composition. Earth's atmosphere, in particular, has many layers. The layers are determined by the variations in temperature and absorption of solar radiation vertically throughout the atmosphere.

LG 3 Naturally occurring greenhouse gases exist on Venus, Earth, and Mars, which increase the average surface temperature of each planet. The amount by which these greenhouse gases raise the temperature of a planet depends on the number of greenhouse gas molecules in the atmosphere. The difference in global temperatures among these planets can be explained in part by their distance from the Sun. However, the different compositions and densities of their atmospheres is a far more significant factor in global temperature.

LG 4 The oxygen levels in Earth's atmosphere have been enhanced through photosynthesis. Increased oxygen led to the development of other forms of life that continue to modify the atmosphere today.

LG 5 Records of Earth's global temperature go back hundreds of thousands of years. Large variations in global temperature correlate strongly with the number of greenhouse molecules in the atmosphere. The current level of greenhouse gases in Earth's atmosphere is higher than any seen in the past 800,000 years and correlates with a subsequent rise in temperature.

SUMMARY SELF-TEST

1. Place in chronological order the following steps in the formation and evolution of Earth's atmosphere.
    a. Plant life converts CO_2 to oxygen.
    b. Hydrogen and helium are lost from the atmosphere.
    c. Volcanoes, comets, and asteroids increase the inventory of volatile matter.
    d. Hydrogen and helium are captured from the protoplanetary disk.
    e. Oxygen enables the growth of new life-forms.
    f. Life releases CO_2 from the subsurface into the atmosphere.

2. Place in order, from lowest to highest, the layers of Earth's atmosphere.
    a. magnetosphere   b. mesosphere
    c. thermosphere   d. troposphere
    e. stratosphere

3. The _____ of greenhouse gas molecules affects the temperature of an atmosphere.
    a. percentage   b. fraction
    c. number   d. mass

4. The oxygen molecules in Earth's atmosphere
    a. were part of the primary atmosphere.
    b. arose when the secondary atmosphere formed.
    c. are the result of life.
    d. are being rapidly depleted by the burning of fossil fuels.

5. Studying climate on other planets is important to understanding climate on Earth because (select all that apply)
    a. underlying physical processes are the same on every planet.
    b. other planets offer a range of extremes to which Earth can be compared.
    c. comparing climates on other planets helps scientists understand which factors are important.
    d. other planets can be used to test atmospheric models.

6. The Coriolis effect
    a. causes winds to circulate north to south.
    b. causes winds to flow only eastward.
    c. causes winds to flow only westward.
    d. causes winds to circulate east and west.

7. The difference in climate between Venus, Earth, and Mars is primarily caused by
    a. the compositions of their atmospheres.
    b. their relative distances from the Sun.
    c. the thickness of their atmospheres.
    d. the time at which their atmospheres formed.

8. Which of the following statements are true? (Choose all that apply.)
    a. Earth's magnetosphere shields us from the solar wind.
    b. Earth's magnetosphere is essential to the formation of auroras.
    c. Earth's magnetosphere extends far beyond Earth's atmosphere.
    d. Earth's magnetosphere is weaker than Mercury's.

9. On which planet(s) is the greenhouse effect present? (Choose all that apply.)
   a. Mercury
   b. Venus
   c. Earth
   d. Mars

10. The words *weather* and *climate*
    a. mean essentially the same thing.
    b. refer to different size scales.
    c. refer to different timescales.
    d. both b and c

QUESTIONS AND PROBLEMS

Multiple Choice and True/False

11. T/F: The current atmospheres of the terrestrial planets were formed when the planets formed.

12. T/F: Comets and asteroids may be the source of most of the water on Earth.

13. T/F: The stratosphere is where most weather happens.

14. T/F: All other things being equal, a planet with a high albedo will have a lower temperature than a planet with a low albedo.

15. T/F: The temperature of Earth has measurably risen over the past 100 years.

16. Venus is hot and Mars is cold primarily because
    a. Venus is closer to the Sun.
    b. Venus has a much thicker atmosphere.
    c. the atmosphere of Venus is dominated by CO_2, but the atmosphere of Mars is not.
    d. Venus has stronger winds.

17. The atmosphere of Mars is often pinkish because
    a. the atmosphere is dominated by carbon dioxide.
    b. the Sun is at a low angle in the sky.
    c. Mars has no oceans to reflect blue light to the sky.
    d. winds lift dust into the atmosphere.

18. Convection in the ____ causes weather on Earth.
    a. stratosphere
    b. mesosphere
    c. troposphere
    d. ionosphere

19. Auroras are the result of
    a. the interaction of particles from the Sun and Earth's atmosphere.
    b. upper-atmosphere lightning strikes.
    c. destruction of the upper atmosphere, which leaves a hole.
    d. the interaction of Earth's magnetic field with Earth's atmosphere.

20. The ozone layer protects life on Earth from
    a. high-energy particles from the solar wind.
    b. micrometeorites.
    c. ultraviolet radiation.
    d. charged particles trapped in Earth's magnetic field.

21. Hadley circulation is broken into zonal winds by
    a. convection from solar heating.
    b. hurricanes and other storms.
    c. interactions with the solar wind.
    d. the planet's rapid rotation.

22. Earth experiences long-term climate cycles spanning _____ of years.
    a. hundreds of millions
    b. hundreds of thousands
    c. thousands
    d. hundreds

23. Over the past 800,000 years, Earth's temperature has closely tracked
    a. solar luminosity.
    b. oxygen levels in the atmosphere.
    c. nitrogen levels in the atmosphere.
    d. carbon dioxide levels in the atmosphere.

24. Uncertainties in climate science are dominated by
    a. uncertainty about physical causes.
    b. uncertainty about current effects.
    c. uncertainty about past effects.
    d. uncertainty about future effects.

25. The atmospheric greenhouse effect
    a. occurs only on Earth.
    b. lets through more solar radiation.
    c. traps visible light in the atmosphere.
    d. traps infrared light in the atmosphere.

Conceptual Questions

26. Primary atmospheres of the terrestrial planets were composed almost entirely of hydrogen and helium. Explain why they contained these gases and not others.

27. How were the secondary atmospheres of the terrestrial planets created?

28. Some of Earth's water was released above ground by volcanism. What is another likely source of Earth's water?

29. The force of gravity holds objects tightly to the surfaces of the terrestrial planets. Yet atmospheric molecules are constantly escaping into space. Explain how these molecules are able to overcome gravity's grip. How does the mass of a molecule affect its ability to break free?

30. Examine Figure 7.3. Why is the range of temperatures on Mercury so much larger than for any other planet?

31. Given that the atmospheres of both Venus and Mars are dominated by carbon dioxide, an effective "greenhouse" gas, why is Venus very hot and Mars very cold?

32. In what ways does plant life affect the composition of Earth's atmosphere?

33. You check the barometric pressure and find that it is reading only 920 millibars. Two possible effects could be responsible for this lower-than-average reading. What are they?

34. What is the principal cause of winds on the terrestrial planets?
35. How does the solar wind affect Earth's upper atmosphere, and what problems can this create?
36. Why are humans able to get a clear view of the surface of Mars but not Venus?
37. Explain why surface temperatures on Venus vary only slightly between day and night and between the equator and the poles.
38. In 1975, the Soviet Union landed two camera-equipped spacecraft on Venus, giving planetary scientists the only close-up views ever seen of the planet's surface. Both cameras ceased to function after an hour. What environmental conditions probably caused them to stop working?
39. Explain the difference between climate and weather.
40. Describe the overall trend in Figure 7.19.

Problems

41. Study Figure 7.18. What is the earliest data? What does this graph imply about the relationship between CO_2 concentrations and temperature?
42. Figure 7.5 shows the development of oxygen in Earth's atmosphere over the history of the planet. This graph is logarithmic on both axes. Scientists use logarithmic graphs to study exponential behavior—a type of process that goes increasingly faster. Study the y-axis to see how the numbers increase with height above the x-axis. Is the distance on the page between 0 and 1 the same as the distance between 1 and 2? How about the distance between 0 and 10 versus the distance between 90 and 100? Redraw the y-axis of this graph on a linear scale by making a graph with evenly spaced intervals every 10 percent. Compare your new graph with the one in the text. How are they different?
43. Exponentials can be a difficult mathematical concept to master, but they are critical in modeling much of the behavior of the natural world. For example, the runaway atmospheres discussed in this chapter are exponential behaviors because of the feedback loops involved. To help you understand exponentials, think about a checkerboard. A checkerboard has 8 rows of 8 squares, so it has 64 squares on it.
    a. Imagine that you put one penny on the first square, two pennies on the second, three pennies on the third, four on the fourth, and so forth. This is linear behavior: each step is the same size; each time you go to a new space, you add the same number of pennies—one. How many pennies will be on the 64th square?
    b. Now imagine that you do this in an exponential fashion, so that you put one penny on the first space, two pennies on the second, *four* pennies on the third, *eight* pennies on the fourth, and so on. Predict how many pennies will be on the 64th square. Now, calculate this the long way on your calculator, multiplying by 2 for each square, all the way to the last square. How many pennies will be on the 64th square? Were you close with your prediction? You may notice two things. (1) It was easier to visualize and to see the pattern for the linear case. (2) The exponential case seems to increase by reasonable amounts, until the numbers suddenly become enormous.

44. The total mass of Earth's atmosphere is 5×10^{18} kg. Carbon dioxide (CO_2) makes up about 0.06 percent of Earth's atmospheric mass.
    a. What is the mass of CO_2 (in kilograms) in Earth's atmosphere?
    b. The annual global production of CO_2 is now estimated to be nearly 3×10^{13} kg. What annual fractional increase does this represent?
45. Figure 7.18 shows two sets of data on the same graph. The x-axis is time, which is the same for both data sets. The y-axis on the left is the amount of CO_2 in the atmosphere. The y-axis on the right is the temperature anomaly, the difference between the temperature that year and the average temperature over the past 10,000 years. Do both these lines show the same trend? Does this *necessarily* mean that one causes the other? What other information do we have that leads us to believe the increasing CO_2 levels are driving the increasing temperatures?
46. The atmospheric pressure at the surface of Venus is 92 bars; equal to the water pressure at an ocean depth of 900 meters on Earth. How many pounds of force would press on a square inch of your body at that pressure?
47. A planet with no atmosphere at 1 AU from the Sun would have an average blackbody surface temperature of 279 K if it absorbed all the Sun's electromagnetic energy falling on it (albedo = 0).
    a. What would be the average temperature on this planet if its albedo were 0.1, typical of a rock-covered surface?
    b. What would be the average temperature if its albedo were 0.9, typical of a snow-covered surface?
48. Consider a hypothetical planet named Vulcan. If such a planet were in an orbit one-fourth the size of Mercury's and had the same albedo as Mercury, what would be the average temperature on Vulcan's surface? Assume that the average temperature on Mercury's surface is 450 K.
49. The moon's average albedo is approximately 0.12.
    a. Calculate the moon's average temperature in degrees Celsius.
    b. Does this temperature meet your expectations? Explain why or why not.
50. The orbit of Eris, a dwarf planet, carries it out to a maximum distance of 97.7 AU from the Sun. Assuming an albedo of 0.8, what is the average temperature of Eris when it is farthest from the Sun?

SMARTWORK

Norton's online homework system includes algorithmically generated versions of these questions, plus additional conceptual exercises. If your instructor assigns questions in SmartWork, log in at smartwork.wwnorton.com.

Exploration | Climate Change

One prediction about climate change is that as the planet warms, ice in the polar caps and in glaciers will melt. This certainly seems to be occurring in the vast majority of glaciers and ice sheets around the planet. It is reasonable to ask whether this actually matters, and why. Here you will explore several consequences of the melting ice on Earth.

Experiment 1: Floating Ice

For this experiment, you will need a permanent marker, a translucent plastic cup, water, and ice. Place a few ice cubes in the cup, and add water until the ice cubes float (that is, they don't touch the bottom). Mark the water level on the outside of the cup with the marker, and label this mark so that later you will know it was the initial water level.

1. As the ice melts, what do you expect to happen to the water level in the cup?

Wait for the ice to melt completely; then mark the cup again.

2. What happened to the water level in the cup when the ice melted?

3. Given the results of your experiment, what can you predict will happen to global sea levels when the Arctic ice sheet, which floats on the ocean, melts?

Experiment 2: Ice on Land

For this experiment, you will need the same materials as in experiment 1, plus a paper or plastic bowl. Fill the cup about halfway with water and then mark the water level, labeling it so you know as the initial level. Poke a hole in the bottom of the bowl, and set the bowl over the cup. Add some ice cubes to the bowl.

4. As the ice melts, what do you expect to happen to the water level in the cup?

Wait for the ice to melt completely; then mark the cup again.

5. What happened to the water level in the cup when the ice melted?

6. In this experiment, the water in the cup is analogous to the ocean, and the ice in the bowl is analogous to ice on land. Given the results of your experiment, what can you predict will happen to global sea levels when the Antarctic ice sheet, which sits on land, melts?

Experiment 3: Why Does It Matter?

Search online using the phrase "Earth at night" to find a satellite picture of Earth taken at night. The bright spots on the image trace out population centers. In general, the brighter they are, the more populous the area (although there is a confounding factor relating to technological advancement).

7. Where do humans tend to live—near coasts or inland?

Coastal regions are, by definition, near sea level. If both the Arctic and Antarctic ice sheets melted completely, sea levels would rise by 80 meters.

8. How would a sea-level rise of about 2 meters (in the range of reasonable predictions) over the next few decades affect the global population? (To help you think about this, remember that a meter is about 3 feet, and one story of a building is about 10 feet.)

SMARTWORK • smartwork.wwnorton.com

8 The Giant Planets

Although they are very far away, the giant planets—Jupiter, Saturn, Uranus, and Neptune—are so large that with the unaided eye, we can see all but Neptune. The brilliance of Jupiter and Saturn is comparable to that of the brightest stars. Uranus is only slightly brighter than the faintest stars visible on a dark night. Neptune is the outermost planet in our Solar System. It cannot be seen without the aid of binoculars.

Because of their visibility, Jupiter and Saturn have been known since antiquity. In 1781, William Herschel was producing a catalog of the sky when he noticed a tiny disk that he thought was a comet. The object's slow nightly motion soon convinced him that it was a new planet—Uranus. In the 19th century, astronomers found that Uranus strayed from its predicted path in the sky and hypothesized that it might be subject to the gravitational pull of an unknown planet. In 1846, Johann Gottfried Galle found Neptune in the position predicted by mathematicians.

All four giant planets have rings and moons. All four have dense cores surrounded by very large atmospheres. All four giant planets rotate more rapidly than Earth. Although they are large, they have rotation periods of less than one Earth day. The illustration on the opposite page shows two observations of the Great Red Spot on Jupiter; a student has used this information to estimate Jupiter's rotation period.

✦ LEARNING GOALS

Unlike the solid, rocky planets of the inner Solar System, four worlds in the outer Solar System were able to capture and retain gases and volatile materials from the Sun's protoplanetary disk and swell to enormous size and mass. By the end of this chapter, you should be able to use two observations of a feature on a planet to determine the rotation period, as the student has done at right. You should also be able to:

LG 1 Differentiate the giant planets from each other and from the terrestrial planets.

LG 2 Describe why the giant planets look the way they do.

LG 3 Describe how gravitational energy turns into thermal energy and how that affects the atmospheres of the giant planets.

LG 4 Explain the extreme conditions deep within the interiors of the giant planets.

LG 5 Describe the origin and general structure of the rings of the giant planets.

~Pat/Desktop/jupiter.fits
Diameter = 315px
Circumference = 990px
x_0 = 70px
t = 23:07

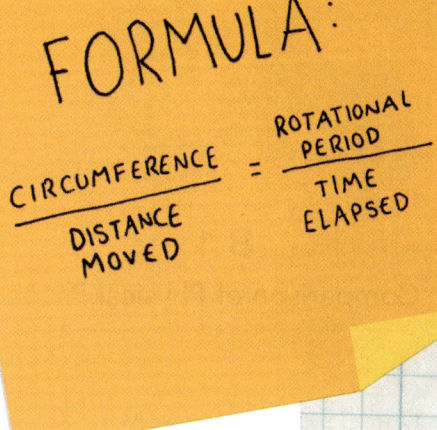

FORMULA:

$$\frac{\text{CIRCUMFERENCE}}{\text{DISTANCE MOVED}} = \frac{\text{ROTATIONAL PERIOD}}{\text{TIME ELAPSED}}$$

~Pat/Desktop/jupiter_1.fits
Diameter = 315px
Circumference = 990px
x_0 = 70px
x_1 = 240px
Δx = 170px
t = 00:47
Δt = 1.67h

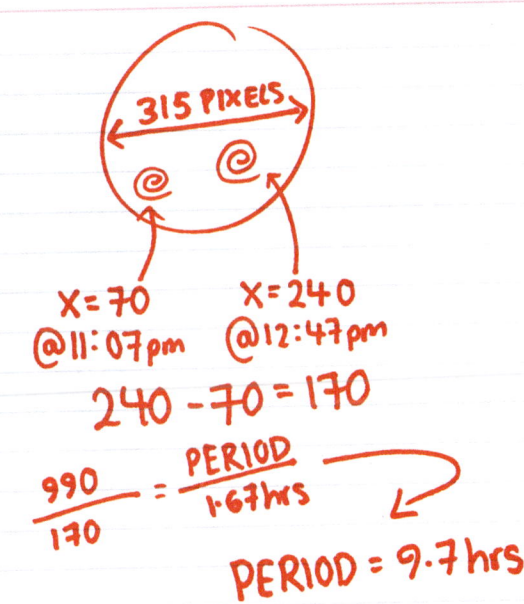

315 PIXELS

X=70 @ 11:07pm   X=240 @ 12:47pm

240 - 70 = 170

$$\frac{990}{170} = \frac{\text{PERIOD}}{1.67 \text{hrs}}$$

PERIOD = 9.7 hrs

8.1 Giant Planets Are Large, Cold, and Massive

As with the terrestrial planets, we learn much about the giant planets by comparing them to each other. We begin our comparison of the giant planets by investigating their physical properties and their compositions, shown in **Table 8.1**. Comparative planetology is useful both within planetary groups and between groups. Throughout most of the chapter, we will be comparing giant planets with giant planets. But it is useful to fix in your mind a comparison of at least one giant planet with Earth as a reference point. For example, to understand the size of the giant planets, it is helpful to know that Jupiter is 11 times larger than Earth, while its mass is 318 times as large. In this section, we examine the physical properties of the giant planets.

TABLE 8.1
Comparison of Physical Properties of the Giant Planets

PROPERTY	JUPITER	SATURN	URANUS	NEPTUNE
Orbital semimajor axis (AU)	5.20	9.6	19.2	30
Orbital period (Earth years)	11.9	29.5	84.0	164.8
Orbital velocity (km/s)	13.1	9.7	6.8	5.4
Mass ($M_\oplus = 1$)	317.8	95	14.5	17.1
Equatorial radius (km)	71,490	60,270	25,560	24,300
Equatorial radius ($R_\oplus = 1$)	11.2	9.5	4.0	3.8
Density (water = 1)	1.33	0.69	1.27	1.64
Rotation period*	9^h56^m	10^h39^m	17^h14^m	16^h6^m
Tilt of the axis (degrees)†	3.13	26.7	97.8	28.3
Surface gravity (m/s²)	24.8	10.4	8.7	11.2
Escape velocity (km/s)	59.5	35.5	21.3	23.5

*The superscript letters h and m stand for hours and minutes of time, respectively.
†An axial tilt greater than 90° indicates that the planet rotates in a retrograde, or backward, direction.

Characteristics of the Giant Planets

Jupiter, the closest giant planet, is more than 5 astronomical units (AU) from the Sun. The Sun shines only dimly and provides very little warmth in this remote part of the Solar System. From Jupiter, the Sun appears as a tiny disk, just 1/27 as bright as it appears from Earth. At the distance of Neptune, the Sun no longer looks like a disk at all; it appears as a brilliant star about 500 times brighter than the full Moon in our own sky. Daytime on Neptune is equivalent to twilight here on our own planet. With so little sunlight available for warmth, daytime temperatures hover around 123 K at the cloud tops on Jupiter. They can dip to just 37 K on Neptune's moon Triton.

Jupiter, Saturn, Uranus, and Neptune are enormous compared to their rocky terrestrial counterparts. Jupiter is the largest of the eight planets; it is more than one-tenth the diameter of the Sun itself. Saturn is only slightly smaller than Jupiter, with a diameter of 9.5 Earths. Uranus and Neptune are each about 4 Earth diameters across; Neptune is the slightly smaller of the two. **Figure 8.1** shows the most accurate method for finding the diameter of a planet. As the planet passes in front of a star, the star is eclipsed. This is known as a **stellar occultation**. Because we know the relative speeds of Earth and the giant planets, we can calculate the size of the eclipsing giant planet from the length of time the star is eclipsed. An example of this type of calculation is given in **Working It Out 8.1**.

The giant planets contain 99.5 percent of all the mass in the Solar System, not counting the mass of the Sun. All other Solar System objects—terrestrial planets, dwarf planets, moons, asteroids, and comets—are included in the remaining 0.5 percent. Jupiter is more than twice as massive as all the other planets in the Solar System combined: it is 3.5 times as massive as Saturn, its closest rival. Even so, its mass is only about a thousandth that of the Sun. Uranus and Neptune are the lightweights among the giant planets, but each is still more than 15 times as massive as Earth.

Before the space age, scientists measured a planet's mass by observing the motions of its moons. Chapter 3 demonstrated how to use Newton's law of gravitation and Kepler's third law to predict the motion of a moon using the planet's mass. The reverse is also true: it is possible to find a planet's mass if you know the motion of its moon. However, this technique only works for planets with moons, and the accuracy of those early calculations was limited by how precisely astronomers could measure the positions of the moons with ground-based telescopes. Planetary spacecraft now make it possible to measure the masses of planets much more accurately. As a spacecraft flies by, the planet's gravity deflects it. By tracking and comparing the spacecraft's radio signals using several antennae here on Earth, astronomers can detect tiny changes in the spacecraft's path and accurately measure the planet's mass.

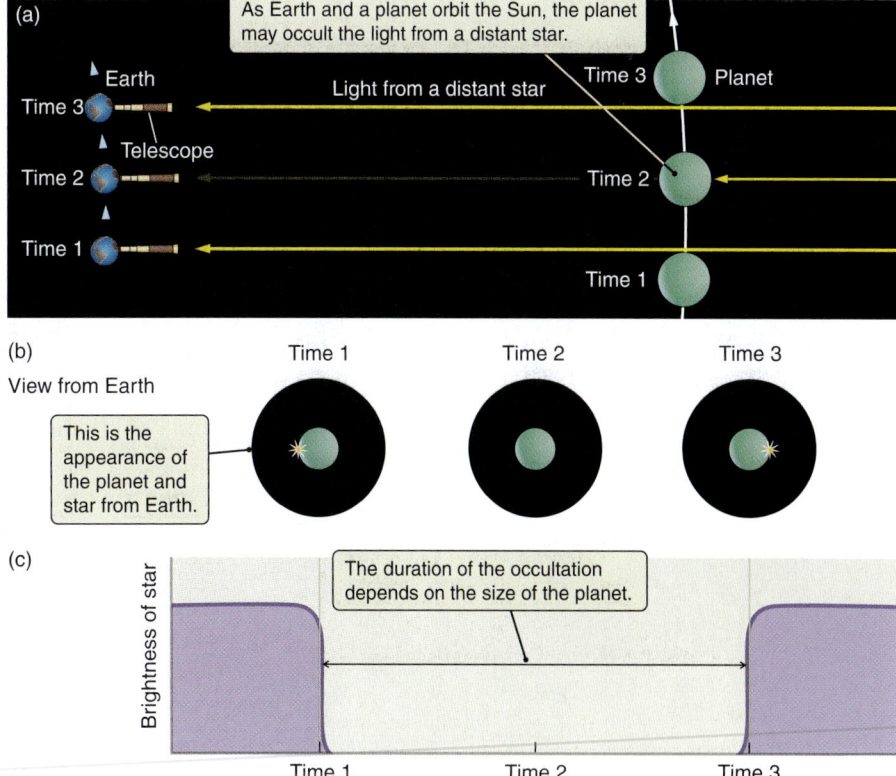

Figure 8.1 (a) Occultations occur when a planet, moon, or ring passes in front of a star. (b) As the planet moves (from right to left as seen from Earth), the starlight is blocked. (c) The amount of time that the star is hidden, combined with information about how fast the planet is moving, gives the size of the planet.

Working It Out 8.1 | Finding the Diameter of a Giant Planet

If the distance to a planet is known, the orbital speed can be found from Kepler's and Newton's laws (Chapter 3). Suppose a particular planet is moving at a speed of precisely 25 kilometers per second (km/s) relative to Earth's own motion. As the planet moves across the sky toward a background star, you begin to make careful observations. You note the moment when the planet first eclipses the star, and you wait while the star passes behind the center of the planet and emerges from the opposite side. You find that this event takes exactly 2,000 seconds. The planet has traveled a specific distance (its diameter), at a specific speed, during this time. We can use the equation from Working It Out 2.1 that relates speed, distance, and time to find this distance:

$$\text{Speed} = \frac{\text{Distance}}{\text{Time}}$$

We can rearrange this equation to solve for distance, first multiplying both sides by time and then exchanging the right and left sides of the equation:

$$\text{Distance} = \text{Speed} \times \text{Time}$$

Plugging in our speed of 25 km/s and our time of 2,000 seconds, we find the distance the planet traveled:

$$\text{Distance} = 25 \text{ km/s} \times 2{,}000 \text{ s}$$
$$\text{Distance} = 50{,}000 \text{ km}$$

The planet's diameter is equal to the distance it traveled during the 2,000 seconds, or 50,000 km.

The exact center of a planet rarely passes directly in front of a star, but observations of occultations from several widely separated observatories can provide the geometry necessary to calculate both the planet's size and its shape. Occultations of radio signals transmitted from orbiting spacecraft have also provided accurate measures of the sizes and shapes of planets and their moons, as have images taken by spacecraft cameras.

Composition of the Giant Planets

The giant planets are made up primarily of gases and liquids. Jupiter and Saturn are composed of hydrogen and helium and are therefore known as **gas giants**. Uranus and Neptune are known as **ice giants** because they both contain much larger amounts of water and other ices than Jupiter and Saturn. On a giant planet, a relatively shallow atmosphere merges seamlessly into a deep liquid **ocean**, which in turn merges smoothly into a denser liquid or solid core. Although the atmospheres of the giant planets are shallow compared with the depth of the liquid layers below, they are still much thicker than those of the terrestrial planets—thousands of kilometers rather than hundreds. Only the very highest levels of these atmospheres are visible to us. In the case of Jupiter or Saturn, we see the tops of a layer of thick **clouds**, the highest of many other layers that lie below. Although a few thin clouds are visible on Uranus, we mostly find ourselves looking into a clear, seemingly bottomless atmosphere. Atmospheric models tell us that thick cloud layers must lie below, but strong scattering of sunlight by molecules (see Chapter 7) in the clear part of the atmosphere prevents us from seeing these lower cloud layers. Neptune displays a few high clouds with a deep, clear atmosphere showing between them.

In Chapter 6, you learned that the terrestrial planets are composed mostly of rocky minerals, such as silicates, along with various amounts of iron and other metals. While the atmospheres of the terrestrial planets contain lighter materials, the masses of these atmospheres—and even of Earth's oceans—are insignificant compared with the total planetary masses. The terrestrial planets are the densest objects in the Solar System, ranging from 3.9 (Mars) to 5.5 (Earth) times the density of water. The giant planets have lower densities because they are composed mainly of

Vocabulary Alert

ocean Here on Earth, this word commonly and specifically means a vast expanse of salty liquid water. Astronomers have expanded the definition of *ocean* to mean a vast expanse of any liquid—not necessarily water.

clouds Just as oceans on other worlds are not necessarily bodies of water, the *clouds* that an astronomer discusses are not necessarily clouds of water vapor—remember the clouds of sulfuric acid we saw on Venus.

lighter materials such as hydrogen and helium. Among giant planets, Neptune has the highest density, about 1.6 times that of water. Saturn has the lowest density, only 0.7 times the density of water. This means that Saturn would actually float in water with 70 percent of its volume submerged—if you could find a large-enough lake. Jupiter and Uranus have densities between those of Neptune and Saturn.

The chemical compositions of the giant planets are not all the same. Astronomers use the relative amounts of the elements in the Sun as a standard reference, termed solar **abundance**. As illustrated in **Figure 8.2**, hydrogen (H) is the most abundant element, followed by helium (He). Jupiter's chemical composition is quite similar to that of the Sun. Jupiter has about a dozen hydrogen atoms for every atom of helium, which is typical of the Sun and the universe as a whole. Only 2 percent of its mass is made up of **heavy elements**, which astronomers define as all elements more massive than helium. Atoms of carbon (C), nitrogen (N), sulfur (S), and oxygen (O) have combined with hydrogen to form molecules of methane (CH_4), ammonia (NH_3), hydrogen sulfide (H_2S), and water (H_2O), respectively. More complex combinations are also common. Helium and certain other gases, such as neon and argon, are **inert** gases that do not combine with other elements or with themselves to make molecules in the atmosphere. Jupiter's liquid core, which contains most of the planet's iron and silicate and much of its water, is left over from the original rocky planetesimal around which Jupiter grew.

Vocabulary Alert

abundance In common language, this word means a more than adequate supply, or simply "enough." At times, it carries the connotation of having *much* more than is needed. Astronomers use this word very specifically to refer to percentages of chemical composition. If you take a sample of an object, and a certain fraction of it is hydrogen, then that fraction is the *abundance* of hydrogen in your sample. This means that if one star is more abundant in hydrogen than another, it must be less abundant in everything else, even if it has more atoms overall. This is because all the fractions must necessarily add together to equal 1.

inert In common language, this word can mean "immobile" or "sluggish." Scientists use *inert* in its chemical sense to describe an element that generally doesn't form molecules with other elements—that is, it does not react chemically.

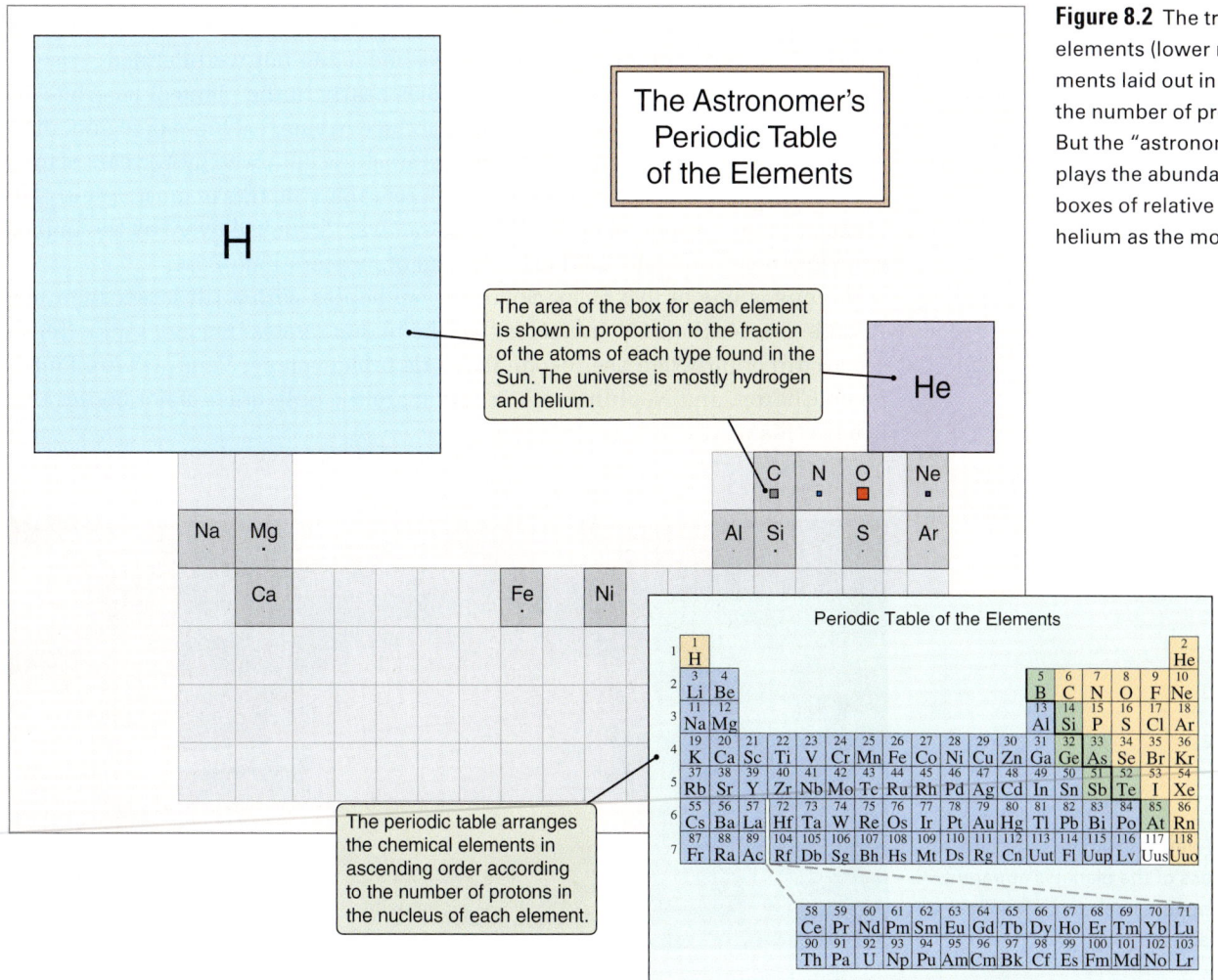

Figure 8.2 The traditional periodic table of the elements (lower right) shows the chemical elements laid out in ascending order according to the number of protons in the nucleus of each. But the "astronomer's periodic table" displays the abundances of the Sun's elements in boxes of relative size, showing hydrogen and helium as the most abundant.

The principal compositional differences among the four giant planets lie in their abundances of heavy elements. Because of its larger mass, Jupiter accumulated more hydrogen and helium when it formed than the other planets did. Saturn is more abundant in heavy elements than Jupiter and therefore less abundant in hydrogen and helium. In Uranus and Neptune, heavy elements are so abundant that they are major components of these two planets. These compositional differences support the model of the formation of the Solar System discussed in Chapter 5.

Rotation of the Giant Planets

Giant planets rotate rapidly, so their days are short. A day on Jupiter is just under 10 hours long, and a day on Saturn is only a little longer. Neptune and Uranus have rotation periods of 16 and 17 hours, respectively, so the length of their days is between those of Jupiter and Earth.

If they did not rotate, the giant planets would be perfectly spherical. Instead, the rapidly rotating giant planets are **oblate**—they bulge at their equators because the inertia of the rotating material near the equator acts to counter gravity. Saturn is very oblate; its equatorial diameter is almost 10 percent greater than its polar diameter (**Figure 8.3**). In comparison, the oblateness of Earth is only 0.3 percent.

The intensity of a planet's seasons is determined primarily by the tilt of the planet's axis. Earth's tilt of 23.5° causes our distinct seasons. With a tilt of only 3°, Jupiter has almost no seasons at all. The tilts of Saturn and Neptune are slightly greater than those of Earth, which causes moderate but well-defined seasons. Curiously, Uranus spins on an axis that lies nearly in the plane of its orbit—its tilt is about 98°. This causes its seasons to be extreme; each polar region alternately experiences 42 years of continuous sunshine, followed by 42 years of total darkness. Why is the tilt of Uranus so different than the tilts of most of the other planets? Many astronomers think the planet was "knocked over" by the impact of an Earth-size planetesimal near the end of its accretion phase.

Uranus is one of five major Solar System bodies with a tilt larger than 90°. A value greater than 90° indicates that the planet rotates in a clockwise (retrograde) direction when seen from above its orbital plane. Venus, Pluto, Pluto's moon Charon, and Neptune's moon Triton are the only other major bodies that behave this way.

Figure 8.3 This Hubble Space Telescope image of Saturn was taken in 1999. The oblateness of the planet is apparent. The large orange moon Titan appears near the top of the disk of Saturn, along with its black shadow.

Figure 8.4 Jupiter was imaged by the *Cassini* spacecraft while it was on its way to Saturn.

8.2 The Giant Planets Have Clouds and Weather

When we observe the giant planets through a telescope or in visible images from a spacecraft, we are seeing only the top layers of the atmosphere. In some cases, we can see a bit deeper into the clouds, but in essence, we are seeing a two-dimensional view of the cloud tops. The deeper cloud layers on these giant planets are inferred from physical models of the temperature as a function of depth. In this section, we explore the atmospheres of the giant planets.

Viewing the Cloud Tops

Jupiter is the most colorful of all the giant planets (**Figure 8.4**). Parallel bands, ranging in hue from bluish gray to various shades of orange, reddish brown, and pink, stretch out across its large, pale yellow disk. The darker bands are called belts, and the lighter ones are called zones. Many small clouds appear along the edges of, or within, the belts. The most prominent feature is a large, often brick-red oval in Jupiter's southern hemisphere known as the **Great Red Spot**, seen at the lower right in Figure 8.4. With a length of 25,000 km and a width of 12,000 km, the Great Red Spot could comfortably hold two Earths side by side within its boundaries, as shown in **Figure 8.5**.

The Great Red Spot has been circulating in Jupiter's atmosphere for at least 300 years. It was first seen shortly after the invention of the telescope, and it has since varied unpredictably in size, shape, color, and motion as it drifts among Jupiter's clouds. Observations of small clouds circling the perimeter of the Great Red Spot show that it is an enormous atmospheric whirlpool, swirling in a counterclockwise

VISUAL ANALOGY

Figure 8.5 The Great Red Spot on Jupiter is twice the size of Earth.

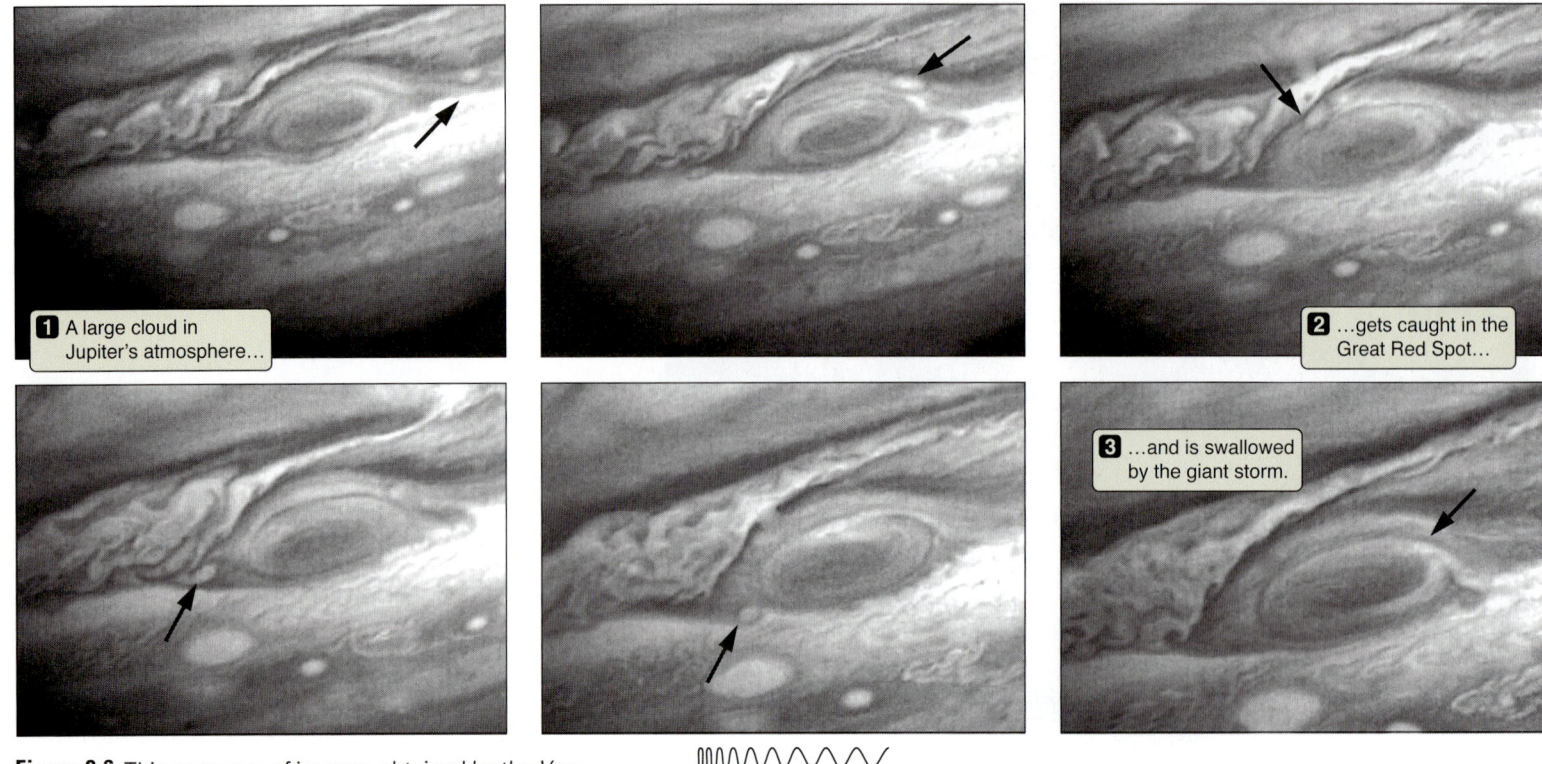

Figure 8.6 This sequence of images, obtained by the *Voyager 2* spacecraft during its encounter with Jupiter, shows the swirling, anticyclonic motion of Jupiter's Great Red Spot. These images span a time frame of about 16 Earth days, during which a large cloud was captured by the Great Red Spot.

1 A large cloud in Jupiter's atmosphere…
2 …gets caught in the Great Red Spot…
3 …and is swallowed by the giant storm.

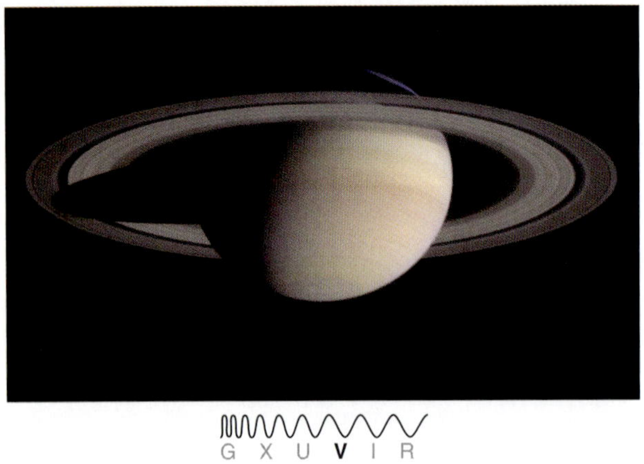

Figure 8.7 This *Cassini* image of Saturn was taken in 2004 as the spacecraft approached the planet. This was the last time the whole planet and its rings fit in the camera's field of view. Sunlight is scattered by the cloud-free upper atmosphere, creating a sliver of light in the northern hemisphere.

direction with a period of about a week. Its cloud pattern looks a lot like that of a terrestrial hurricane, but it rotates in the opposite direction, exhibiting anticyclonic rather than cyclonic flow (recall from the discussion of cyclonic motion in Chapter 7 that anticyclonic flow indicates a high-pressure system). Comparable whirlpool-like behavior is observed in many of the smaller oval-shaped clouds found elsewhere in Jupiter's atmosphere and in similar clouds observed in the atmospheres of Saturn and Neptune. Whirlpool-like, swirling features are known as **vortices** (the singular is *vortex*). These vortices are familiar to us on Earth as high- and low-pressure systems, hurricanes, and supercell thunderstorms.

Jupiter is a turbulent, swirling giant with atmospheric currents and vortices so complex that scientists still do not fully understand the details of how they interact with one another, even after decades of analysis. The Great Red Spot alone displays more structure than was visible over all of Jupiter before the space age. Dynamically, it also reveals some rather bizarre behavior, such as "cloud cannibalism." In a series of time-lapse images, the *Voyager 2* spacecraft observed a number of Alaska-sized clouds being swept into the Great Red Spot. Some of these clouds were carried around the vortex a few times and then ejected, while others were swallowed up and never seen again (**Figure 8.6**). Other smaller clouds with structure and behavior similar to that of the Great Red Spot are seen in Jupiter's middle latitudes.

Saturn is both farther away than Jupiter and somewhat smaller in radius, so from Earth it appears less than half as large as Jupiter. Like Jupiter, Saturn displays atmospheric bands, but they tend to be wider and their colors and contrasts much more subdued than those on Jupiter (**Figure 8.7**). A relatively narrow, meandering band in the mid-northern latitudes encircles the planet in a manner similar to that of our own terrestrial jet stream. Individual clouds on Saturn have been

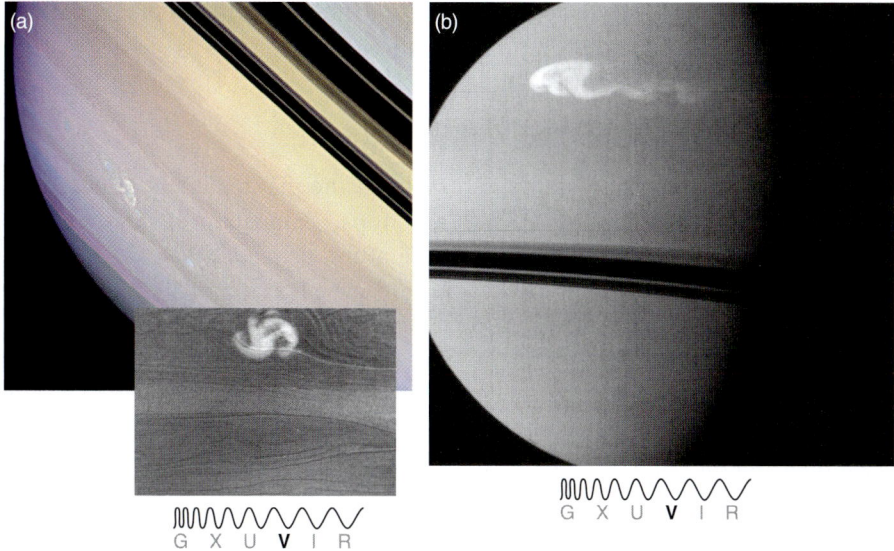

Figure 8.8 Violent storms are known to erupt on Saturn. (a) An enhanced *Cassini* image of an intense lightning-producing storm (left of center) located in Saturn's "storm alley." The inset shows a similar storm on Saturn's night side, illuminated by sunlight reflecting off Saturn's rings. (b) In December 2010, an enormous storm in Saturn's northern hemisphere was discovered by amateur astronomers and subsequently imaged by *Cassini*, as shown here.

seen only rarely from Earth. On these infrequent occasions, large, white, cloud-like features suddenly erupt in the tropics, spread out in longitude, and then fade away over a period of a few months. The largest clouds are larger than the continental United States, but many that we see are smaller than terrestrial hurricanes. Close-up views from the *Cassini* spacecraft (**Figure 8.8**) show immense lightning-producing storms in a region of Saturn's southern hemisphere known to mission scientists as "storm alley."

From Earth, even through the largest telescopes, Uranus and Neptune look like tiny, featureless, pale bluish-green disks. Infrared imaging reveals individual clouds and belts, giving these distant planets considerably more character (**Figure 8.9**). The strong absorption of reflected sunlight by methane causes the atmospheres of Uranus and Neptune to appear dark in the near infrared, allowing the highest clouds and bands to stand out in contrast against the dark background.

A large, dark, oval feature in Neptune's southern hemisphere, first observed in images taken by *Voyager 2* in 1989, reminded astronomers of Jupiter's Great Red Spot, so they called it the Great Dark Spot. However, the Neptune feature was gray rather than red, and it changed its length and shape more rapidly than the Great

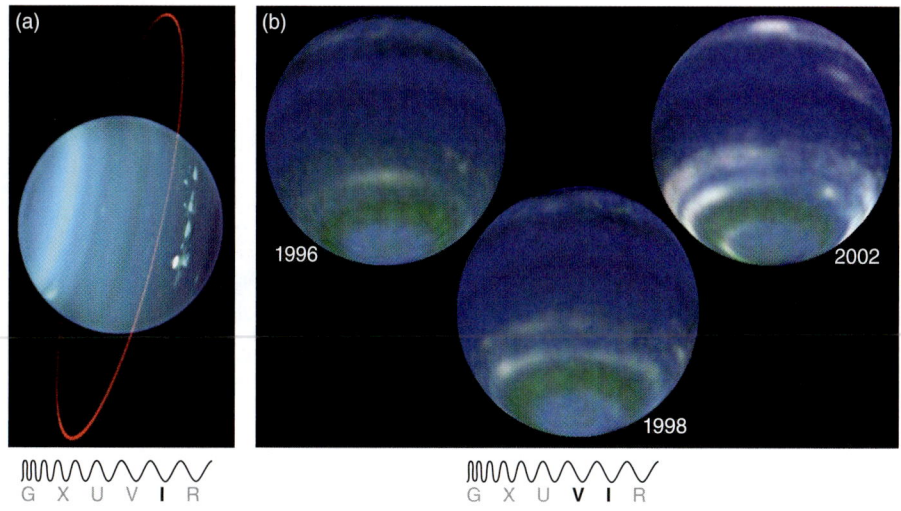

Figure 8.9 The ground-based Keck telescope image of Uranus (a) and the Hubble Space Telescope images of Neptune (b) were taken at a wavelength of light that is strongly absorbed by methane. The visible clouds are high in the atmosphere. The rings of Uranus show prominently because the brightness of the planet has been subdued by methane absorption. Seasonal changes in cloud formation on Neptune are evident over a 6-year interval.

192 CHAPTER 8 *The Giant Planets*

Vocabulary Alert

limb In common language, we use this word to refer to something that sticks out—an arm, a leg, or a tree branch. Astronomers use it to refer to the edge of the visible disk of a planet, moon, or the Sun.

Red Spot. By 1994, it had disappeared—but a different dark spot of comparable size had appeared briefly in Neptune's northern hemisphere.

The Structure Below the Cloud Tops

We have noted that our visual impression of the giant planets is based on a two-dimensional view of their cloud tops. Atmospheres, though, are three-dimensional structures whose temperature, density, pressure, and even chemical composition vary with height and over horizontal distances. As a rule, atmospheric temperature, density, and pressure all increase with decreasing altitude, although temperature is sometimes higher at very high altitudes, as in Earth's thermosphere. A thin haze above the cloud tops is visible in profile above the **limbs** of the planets. The composition of the haze particles remains unknown, but they may be smog-like products created when ultraviolet sunlight acts on hydrocarbon gases such as methane.

Water is the only substance in Earth's lower atmosphere that can condense into clouds, but the atmospheres of the giant planets contain a variety of volatile materials that can form clouds. **Figure 8.10** shows how the ice layers are stacked in the tropospheres of the giant planets. Because each kind of volatile condenses at a particular temperature and pressure, each forms clouds at a different altitude. Convection carries volatile materials upward along with all other atmospheric gases, and when a particular volatile reaches an altitude with its condensation

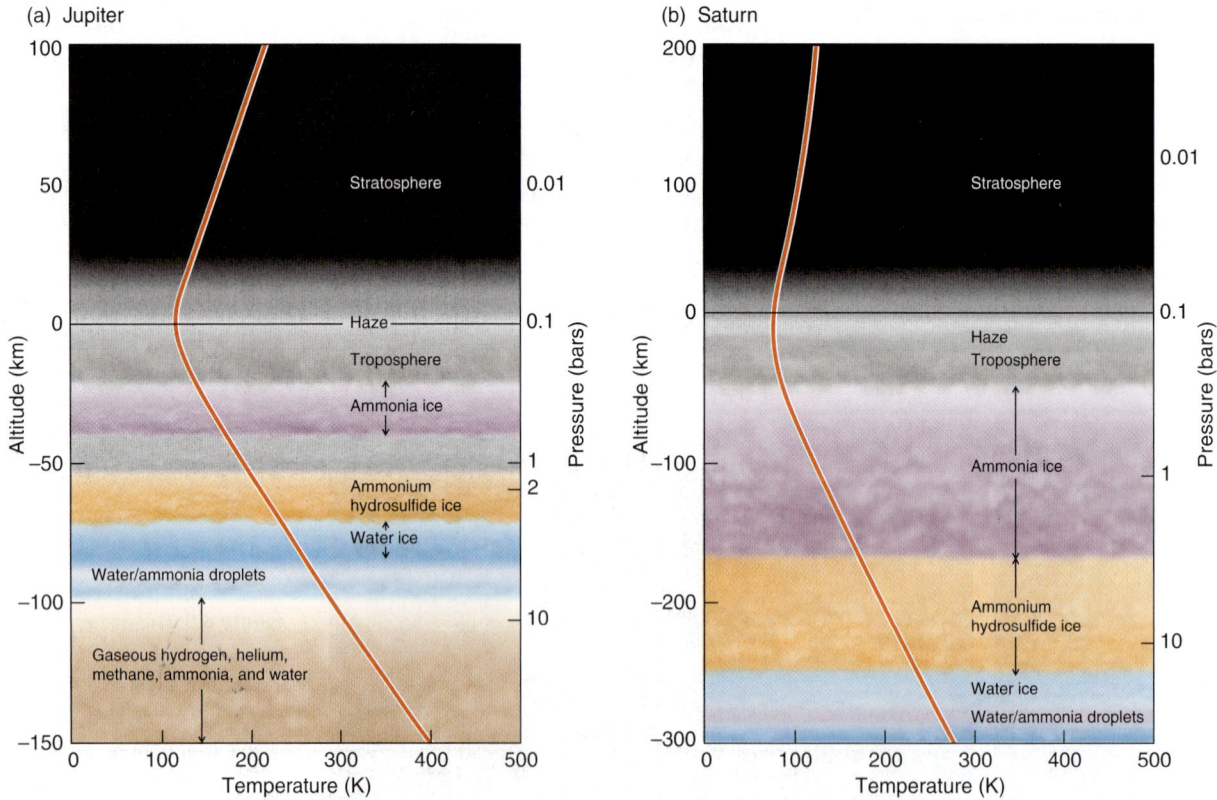

Figure 8.10 Volatile materials condense at different levels in the atmospheres of the giant planets, leading to chemically different types of clouds at different depths in the atmosphere. The red line in each diagram shows how atmospheric temperature changes with height. The arbitrary zero points of altitude are at 0.1 bar for Jupiter (a) and Saturn (b) and 1.0 bar for

temperature, most of the volatile condenses and separates from the other gases, so very little of it is carried higher aloft. These volatiles form dense layers of cloud separated by regions of relatively clear atmosphere.

The farther a planet is from the Sun, the colder its troposphere will be. Distance from the Sun thus determines the altitude at which a particular volatile, such as ammonia or water, will condense to form a cloud layer on each of the planets (see Figure 8.10). If temperatures are too high, some volatiles may not condense at all. The highest clouds in the frigid atmospheres of Uranus and Neptune are crystals of methane ice. The highest clouds on Jupiter and Saturn are made up of ammonia ice. Methane never freezes to ice in the warmer atmospheres of Jupiter and Saturn.

Why are some clouds so colorful, especially Jupiter's? These tints and hues must come from impurities in the ice crystals, similar to the way that syrups color snow cones. These impurities are probably elemental sulfur and phosphorus, as well as various organic materials produced when ultraviolet sunlight breaks up hydrocarbons and the fragments recombine to form complex organic compounds.

Methane gas is much more abundant in the atmospheres of Uranus and Neptune than in the atmospheres of Jupiter and Saturn. Like water, methane gas absorbs the longer wavelengths of light—yellow, orange, and red. Absorption of the longer wavelengths leaves only the shorter wavelengths—green and blue—to be scattered from the relatively cloud-free atmospheres of Uranus and Neptune.

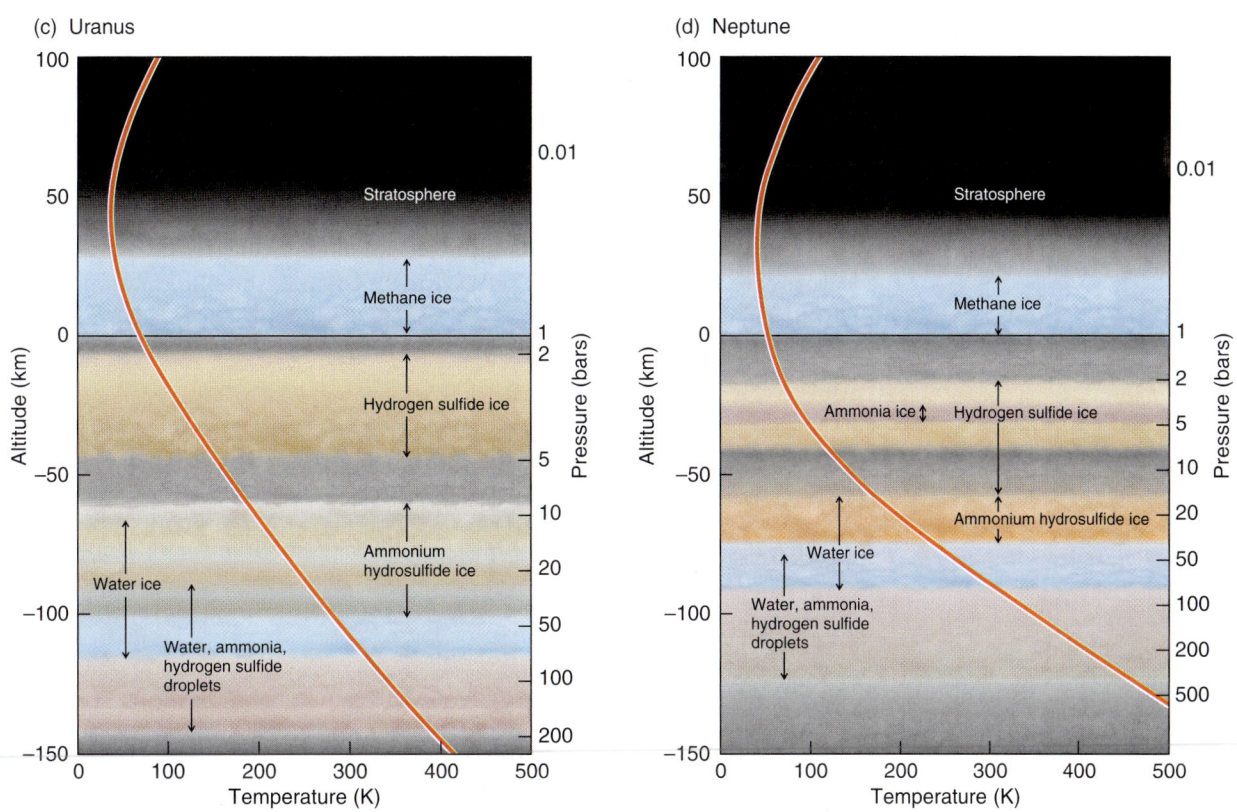

Uranus (c) and Neptune (d). Recall that 1.0 bar corresponds approximately to the atmospheric pressure at sea level on Earth. Note that Saturn's altitude scale is compressed to show the layered structure better.

Convection and Weather

The giant planets have much stronger zonal winds than the terrestrial planets. Because they are farther from the Sun, less thermal energy is available. However, they rotate rapidly, which makes the Coriolis effect very strong. In fact, the Coriolis effect is more important than atmospheric temperature patterns in determining the structure of the global winds.

On the giant planets, the thermal energy that drives convection comes in part from the Sun but primarily from the hot interiors of the planets themselves. The Coriolis effect shapes that convection into atmospheric vortices. Convective vortices are visible as isolated circular or oval cloud structures, such as the Great Red Spot on Jupiter and the Great Dark Spot on Neptune. As the atmosphere ascends near the centers of the vortices, it expands and cools. Cooling condenses certain volatile materials into liquid droplets, which then fall as rain. As they fall, the raindrops collide with surrounding air molecules, stripping electrons from the molecules and thereby developing tiny electric charges in the air. The cumulative effect of countless falling raindrops can generate an electric charge and a resulting electric field so great that it produces a surge of current and a flash of lightning. A single observation of Jupiter's night side by *Voyager 1* revealed several dozen lightning bolts occurring within an interval of 3 minutes. The strength of these bolts is estimated to be equal to or greater than the "superbolts" that occur in the tops of high convective clouds in Earth's tropics. *Cassini* has also imaged lightning flashes in Saturn's atmosphere, and radio receivers on *Voyager 2* picked up lightning static in the atmospheres of both Uranus and Neptune.

Winds on Jupiter and Saturn

Look back at the chapter-opening illustration: A student has used two images of Jupiter to work out the rotation period of Jupiter, assuming the whole planet rotates at the same speed that the Great Red Spot travels. This turns out to be a good assumption. If we know the radius of the planet, we can find out how fast the features are moving, as shown in **Working It Out 8.2**, and find rotation speeds and wind speeds.

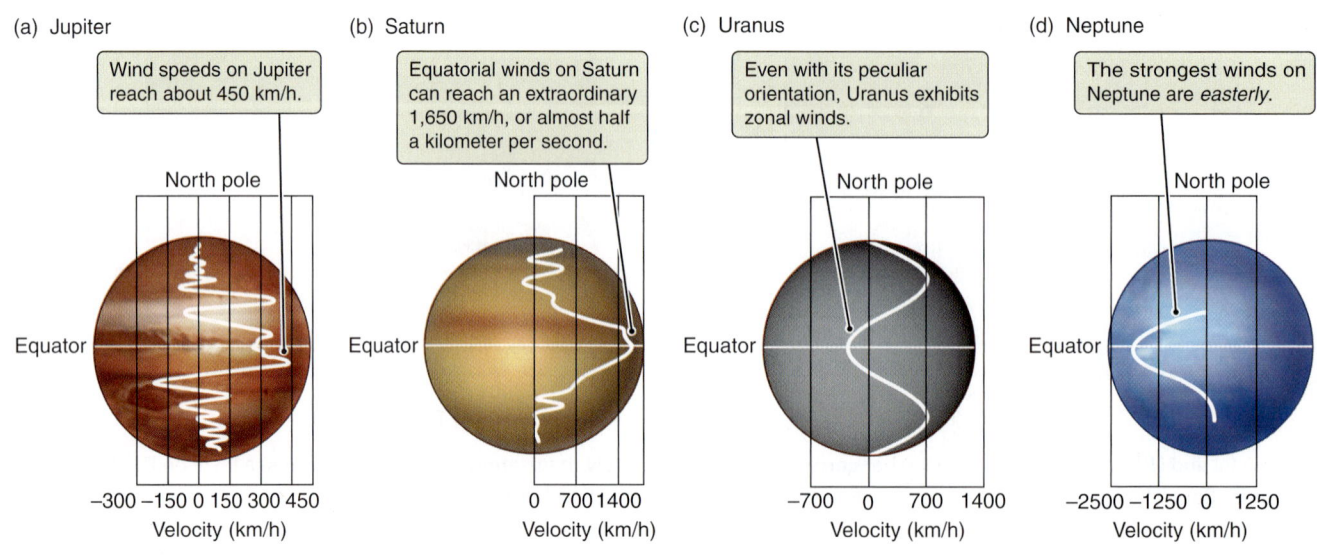

Figure 8.11 Strong winds blow in the atmospheres of the giant planets, driven by powerful convection and the Coriolis effect on these rapidly rotating worlds. The speed of the wind at various latitudes is shown by the white line. When the speed is positive (to the right of zero), the wind is westerly. When the speed is negative (to the left of zero), the wind is easterly.

Working It Out 8.2 | Measuring Wind Speeds on Distant Planets

How do astronomers measure wind speeds on planets that are so far away? If we can see individual clouds in their atmospheres, we can measure their winds. As on Earth, clouds are carried by the local winds. By measuring the positions of individual clouds and noting how much they move during an interval of a day or so, we can calculate the local wind speed. However, we also need to know how fast the planet is rotating so we can measure the speed of the winds with respect to the planet's rotating surface.

In the case of the giant planets, of course, there is no solid surface against which to measure the winds. We must instead assume a hypothetical surface—one that rotates as though it were somehow "connected" to the planet's deep interior. As we will see near the end of this chapter, periodic bursts of radio energy caused by the rotation of a planet's magnetic field tell us how fast the interior of the planet is rotating. Notably, radio bursts from Saturn have, at different times, implied different rotation periods. This is puzzling because there can be only one true rotation period. Because this phenomenon is not yet understood, astronomers have adopted an average value.

In the chapter-opening illustration, the student worked out the rotation period without knowing the actual radius of Jupiter (in kilometers). Instead, she was able to take a ratio of sizes in the image. That's helpful in many situations. However, if we do know the radius, we can find the speed of the features.

Let's see an example of how this works using a small white cloud in Neptune's atmosphere. The cloud, on Neptune's equator, is observed to be at longitude 73.0° west on a given day. The spot is then seen at longitude 153.0° west exactly 24 hours later. Neptune's equatorial winds have carried the white spot 153.0° − 73.0° = 80.0° in longitude in 24 hours.

The circumference, C, of a planet is given by $2\pi r$, where r is the equatorial radius. The equatorial radius of Neptune is 24,760 km. So the circumference is

$$C = 2\pi r$$
$$C = 2\pi (24{,}760 \text{ km})$$
$$C = 155{,}600 \text{ km}$$

There are 360° of longitude in the full circle represented by the circumference, so each degree of longitude stretches across 432 km (155,600 kilometers ÷ 360 degrees = 432 kilometers/degree). Because Neptune's equatorial winds have carried the spot 80°, we conclude that it has traveled 432 kilometers/degree × 80 degrees = 34,560 kilometers in 1 day. This means that the speed is

$$\text{Speed} = \frac{\text{Distance}}{\text{Time}}$$
$$\text{Speed} = \frac{34{,}560 \text{ km}}{1 \text{ day}}$$

Converting days to hours gives more familiar units:

$$\text{Speed} = \frac{34{,}560 \text{ km}}{1 \text{ day}} \times \frac{1 \text{ day}}{24 \text{ hours}}$$
$$\text{Speed} = 1{,}440 \text{ km/h}$$

The wind speed is 1,440 km/h.

On Jupiter, the strongest winds are equatorial westerlies, which have speeds of 550 kilometers per hour (km/h), as seen in **Figure 8.11a**. (Remember from Chapter 7 that westerly winds are those that blow *from*, not toward, the west.) At higher latitudes, the winds alternate between easterly and westerly in a pattern that seems to be related to Jupiter's banded structure. Near a latitude of 20° south, the Great Red Spot vortex lies between a pair of easterly and westerly currents with opposing speeds of more than 200 km/h. If you think this might imply something about the relationship between zonal flow and vortices, you are right.

The equatorial winds on Saturn are also westerly, but they are stronger than those on Jupiter. In the early 1980s, the two *Voyager* spacecraft measured speeds as high as 1,650 km/h. Later, the Hubble Space Telescope (HST) recorded maximum speeds of 990 km/h, and more recently *Cassini* has found them to be intermediate between the *Voyager* and HST measurements. What can be happening here? Saturn's winds appear to decrease with height in the

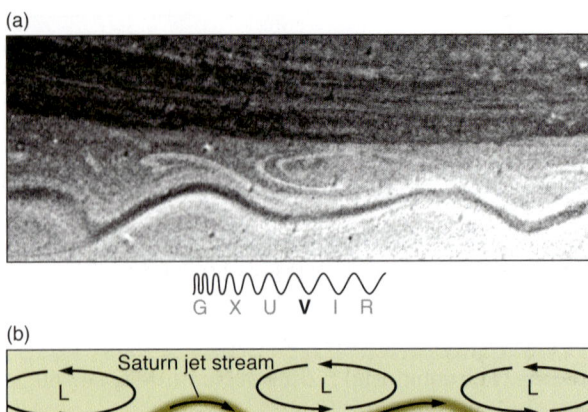

Figure 8.12 (a) This jet stream in Saturn's northern hemisphere seen in this *Voyager* image is similar to jet streams in our terrestrial atmosphere. (b) The jet stream dips equatorward around regions of low pressure and moves poleward around regions of high pressure.

atmosphere, so the *apparent* time variability of Saturn's equatorial winds may be nothing more than changes in the height of the cloud tops. Alternating easterly and westerly winds also occur at higher latitudes, but unlike Jupiter's case, this alternation seems to bear no clear association with Saturn's atmospheric bands (Figure 8.11b). This is one example of the many unexplained differences among the giant planets.

Saturn's jet stream, at latitude 45° north (**Figure 8.12a**), is a narrow, meandering river of atmosphere with alternating crests and troughs. Figure 8.12b shows how the jet stream curves around regions of high and low pressure to create the wave-like structure. It is similar to Earth's jet streams, where high-speed winds blow generally from west to east but wander toward and away from the poles. Nested within the crests and troughs of Saturn's jet stream are anticyclonic and cyclonic vortices. These are similar in both form and size to terrestrial high- and low-pressure systems, which bring us alternating periods of fair and stormy weather.

Winds on Uranus and Neptune

Our knowledge of global winds on Uranus is poorer than that of global winds on the other giant planets (see Figure 8.11c). When *Voyager 2* flew by Uranus in 1986, the few visible clouds were in its southern hemisphere because its northern hemisphere was in complete darkness at the time. The strongest winds observed were 650-km/h westerlies in the middle to high southern latitudes, and no easterly winds were detected. Because Uranus's peculiar orientation makes its poles warmer than its equator, some astronomers had predicted that the global wind system of Uranus might be very different from that of the other giant planets. But *Voyager 2* observed that the Coriolis forces dominate on Uranus as they do on other planets, so the dominant winds on Uranus are zonal, just as they are on the other giant planets.

As Uranus has continued in its orbit, previously hidden regions have become visible (**Figure 8.13**). Observations by HST and ground-based telescopes have shown bright cloud bands in the far north extending more than 18,000 km in length and have revealed wind speeds of up to 900 km/h. As Uranus's long year passes, we will continue to learn much more about the northern hemisphere of the planet.

Each season on Neptune lasts 40 Earth years. On Neptune, the southern hemisphere's summer solstice occurred in 2005, so much of the north is still in darkness. We'll have to wait a while before we can get a good look at its northern hemisphere. As on Jupiter and Saturn, the strongest winds on Neptune occur in the tropics (see Figure 8.11d). The surprise is that they are easterly rather than westerly, with speeds in excess of 2,000 km/h. Westerly winds with speeds higher than 900 km/h have been seen in Neptune's south polar regions. With wind speeds five times greater than those of the fiercest hurricanes on Earth, Neptune and Saturn are the windiest planets known.

Figure 8.13 Uranus is approaching equinox in this 2006 HST image. Much of its northern hemisphere is becoming visible. The dark spot in the northern hemisphere (to the right) is similar to but smaller than the Great Dark Spot seen on Neptune in 1989.

8.3 The Interiors of the Giant Planets Are Hot and Dense

At the center of the giant planets is a dense, liquid core consisting of a very hot mixture of heavy materials such as water, rock, and metals. **Figure 8.14** shows the structures of the giant planets. The overlying layers compress the liquid core, raising the temperature. For example, the pressure at Jupiter's core is about 45 million bars, and this high pressure heats the fluid to 35,000 K. Central temperatures and pressures of the other, less massive giant planets are correspondingly lower than those of Jupiter. It may seem strange that water is still liquid at temperatures of tens of thousands of degrees. Like an enormous version of a pressure cooker, the extremely high pressures at the centers of the giant planets prevent water from turning to steam.

The thermal energy from the core drives convection in the atmosphere and eventually escapes to space as radiation (**Working It Out 8.3**). With energy continually escaping from the interiors of the giant planets, we might wonder how they have maintained their high internal temperatures over the past 4.5 billion years. The answer is that they have been and are still shrinking in size, thereby converting gravitational energy into thermal energy. This continual production of thermal energy replaces the energy that is escaping from their interiors. This is the primary energy source for replacing the internal energy that leaks out of the interior of Jupiter, and it is probably an important source for the other giant planets as well. Jupiter is contracting by only 1 millimeter (mm) or so per year. If it were to continue at this rate, in a billion years, Jupiter would shrink by only 1,000 km, a little more than 1 percent of its radius.

The pressure within the atmospheres of Jupiter and Saturn increases with depth because overlying layers of atmosphere press down on lower layers, compressing the gas. At depths of a few thousand kilometers, the atmospheric gases of Jupiter and Saturn are so compressed by the weight of the overlying atmosphere that they turn to liquid. This roughly marks the lower boundary of the atmosphere.

Figure 8.14 The interiors of the giant planets have central cores and outer liquid shells. Only Jupiter and Saturn have significant amounts of the molecular and metallic forms of liquid hydrogen surrounding their cores.

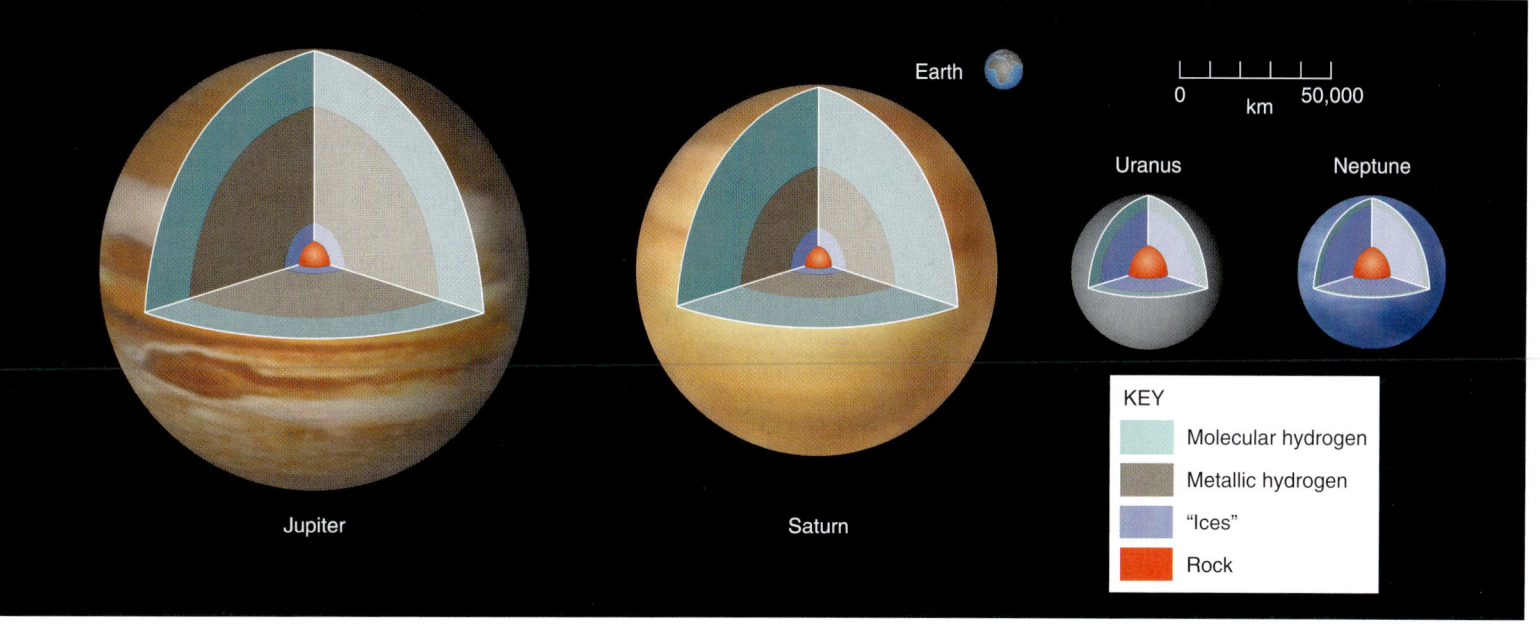

Working It Out 8.3 | Internal Thermal Energy Heats the Giant Planets

In Chapter 7, you learned about the equilibrium between the absorption of sunlight and the radiation of infrared light into space, and we saw how the resulting equilibrium temperature is modified by the greenhouse effect on Venus, Earth, and Mars. Yet when we calculate this equilibrium for three of the giant planets, we find that something seems amiss. According to these calculations, the equilibrium temperature for Jupiter should be 109 K; but when it is measured, we find instead an average temperature of about 124 K. A difference of 15 K might not seem like much, but remember that according to the Stefan-Boltzmann law, the energy radiated by an object depends on its temperature raised to the fourth power. Applying this relationship to Jupiter, we find

$$\left(\frac{T_{\text{actual}}}{T_{\text{expected}}}\right)^4 = \left(\frac{124}{109}\right)^4 = 1.67$$

This result is somewhat startling: Jupiter is radiating roughly two-thirds more energy into space than it absorbs in the form of sunlight. Similarly, the energy escaping from Saturn is about 1.8 times greater than the sunlight that it absorbs. Neptune emits 2.6 times as much energy as it absorbs from the Sun. Strangely, whatever internal energy may be escaping from Uranus is negligible compared with the absorbed solar energy.

The difference between a liquid and a highly compressed, very dense gas is subtle, so on Jupiter and Saturn there is no clear boundary between the atmosphere and the ocean of liquid that lies below. Jupiter's atmosphere is about 20,000 km deep, and Saturn's atmosphere is about 30,000 km deep; at these depths the pressure climbs to 2 million bars and the temperature reaches 10,000 K. Under these conditions, hydrogen molecules are battered so violently that their electrons are stripped free, and the hydrogen acts like a liquid metal. In this state, it is called *metallic hydrogen*. These oceans of hydrogen and helium are tens of thousands of kilometers deep. Uranus and Neptune are less massive than Jupiter and Saturn, have lower interior pressures, and contain a smaller fraction of hydrogen—their interiors probably contain only a small amount of liquid hydrogen, with little or none of it in a metallic state.

Differentiation has occurred and is still occurring in Saturn, and perhaps in Jupiter too. On Saturn, helium condenses out of the hydrogen-helium oceans. Helium can also be compressed to a metal, but it does not reach this metallic state under the physical conditions existing in the interiors of the giant planets. Because these droplets of helium are denser than the hydrogen-helium liquid in which they condense, they sink toward the center of the planet, converting gravitational energy to thermal energy. This process heats the planet and enriches helium in the core while depleting it in the upper layers. In Jupiter's hotter interior, by contrast, the liquid helium is mostly dissolved along with the liquid hydrogen.

The heavy-element components of the cores of Jupiter and Saturn have masses of about 10–20 Earth masses. Jupiter and Saturn have total masses of 318 and 95 Earth masses, respectively. The heavy materials in their cores contribute little to their average chemical composition. This means we can think of both Jupiter and Saturn as having approximately the same composition as the Sun and the rest of the universe: about 98 percent hydrogen and helium, leaving only 2 percent for everything else.

Uranus and Neptune are about twice as dense as Saturn, so they must be made of denser material than Saturn and Jupiter. Neptune, the densest of the giant planets, is about 1.5 times denser than uncompressed water and only about half as dense as uncompressed rock. Uranus is less dense than Neptune. These observations tell us that water and other low-density ices, such as ammonia and methane, must be the major compositional components of Uranus and Neptune, along with lesser amounts of silicates and metals. The total amount of hydrogen and helium in these planets is probably limited to no more than 1 or 2 Earth masses, and most of these gases reside in the relatively shallow atmospheres of the planets.

Why do Jupiter and Saturn have so much hydrogen and helium compared with Uranus and Neptune? Why is Jupiter so much more massive than Saturn? The answers may lie both in the time that it took for these planets to form and in the distribution of material from which they formed. The cores of Uranus and Neptune were smaller and formed much later than those of Jupiter and Saturn, at a time when most of the gas in the protoplanetary disk had been blown away by the emerging Sun. Why did the cores of Uranus and Neptune form so late? Probably because the icy planetesimals from which they formed were more widely dispersed at their greater distances from the Sun. With more space between planetesimals, their cores would have taken longer to build up. Saturn may have captured less gas than Jupiter, both because its core formed somewhat later and because less gas was available at its greater distance from the Sun.

8.4 The Giant Planets Are Magnetic Powerhouses

All of the giant planets have magnetic fields that are much stronger than Earth's: their field strengths range from 50 to 20,000 times stronger. However, because field strength falls off with distance, fields at the cloud tops of Saturn, Uranus, and Neptune are comparable in strength to Earth's surface field. Even in the case of Jupiter's exceptionally strong field, the field strength at the cloud tops is only about 15 times that of Earth's surface field. In Jupiter and Saturn, magnetic fields are generated by circulating currents within deep layers of metallic hydrogen. In Uranus and Neptune, magnetic fields arise within deep oceans of liquid water and ammonia made electrically conductive by dissolved salts. The magnetospheres of the giant planets are very large and interact with both the solar wind (as Earth's does) and the rings and moons that orbit the giant planets.

The Size and Shape of the Magnetospheres

We can illustrate the geometry of these magnetic fields as if they came from bar magnets, as shown in **Figure 8.15**. The orientations of the magnetic field axes provide a mystery. Jupiter's magnetic axis is inclined 10° to its rotation axis—an orientation similar to Earth's—but it is offset about a tenth of a radius from the planet's center (Figure 8.15a). Saturn's magnetic axis is located almost precisely at the planet's center and is almost perfectly aligned with the rotation axis (Figure 8.15b). *Voyager 2* measured the magnetic field around Uranus and found that the magnetic axis of Uranus is inclined nearly 60° to its rotation axis and is offset by a third of a radius from the planet's center (Figure 8.15c). The orientation of Neptune's rotation axis is similar to that of Earth, Mars, and Saturn. But Neptune's magnetic axis is inclined 47° to its rotation axis, and the center of this magnetic field is displaced from the planet's center by more than half the radius—an offset even greater than that of Uranus (Figure 8.15d). The displacement of the field is primarily toward Neptune's southern hemisphere, thereby creating a field 20

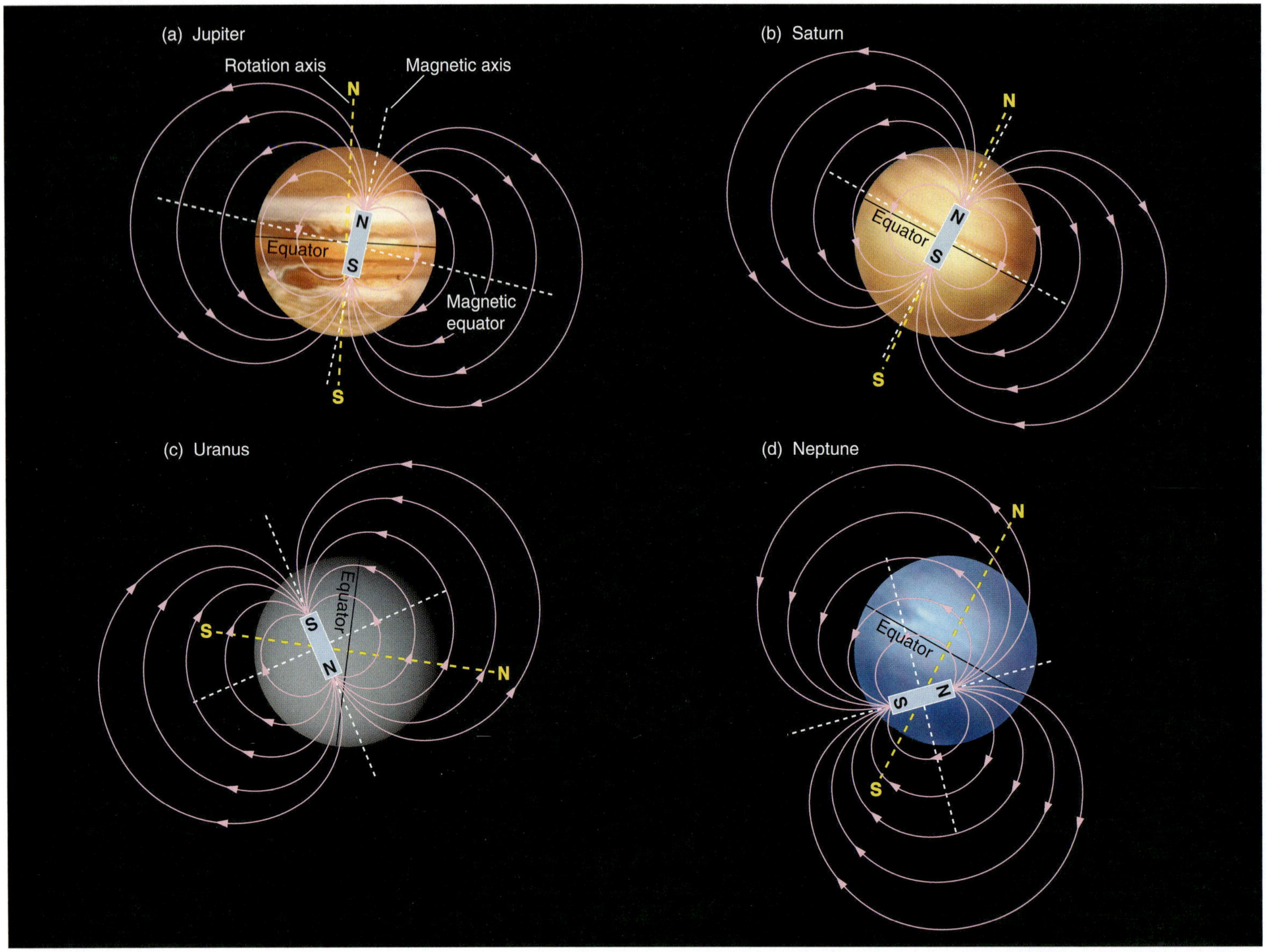

Figure 8.15 The magnetic fields of the giant planets can be approximated by the fields from bar magnets offset and tilted with respect to the rotation axes of the planets. Compare these with Earth's magnetic field, shown in Figure 6.10.

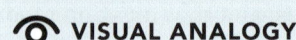 **VISUAL ANALOGY**

times stronger at the southern cloud tops than at the northern cloud tops. The reason for the unusual geometry of the magnetic fields of Uranus and Neptune remains unknown, but it is not related to the orientations of their rotation axes.

Just as Earth's magnetic field traps energetic charged particles to form Earth's magnetosphere, the magnetic fields of the giant planets also trap energetic particles to form magnetospheres of their own. Our magnetosphere is tiny in comparison with those of the giant planets. By far the most colossal of these is Jupiter's magnetosphere. Its radius is 100 times that of the planet itself, roughly 10 times the radius of the Sun. Even the relatively weak magnetic fields of Uranus and Neptune form magnetospheres that are comparable in size to the Sun.

The magnetosphere is also influenced by solar wind. In Chapter 7, we saw that the solar wind supplies some of the particles for a magnetosphere. In addition,

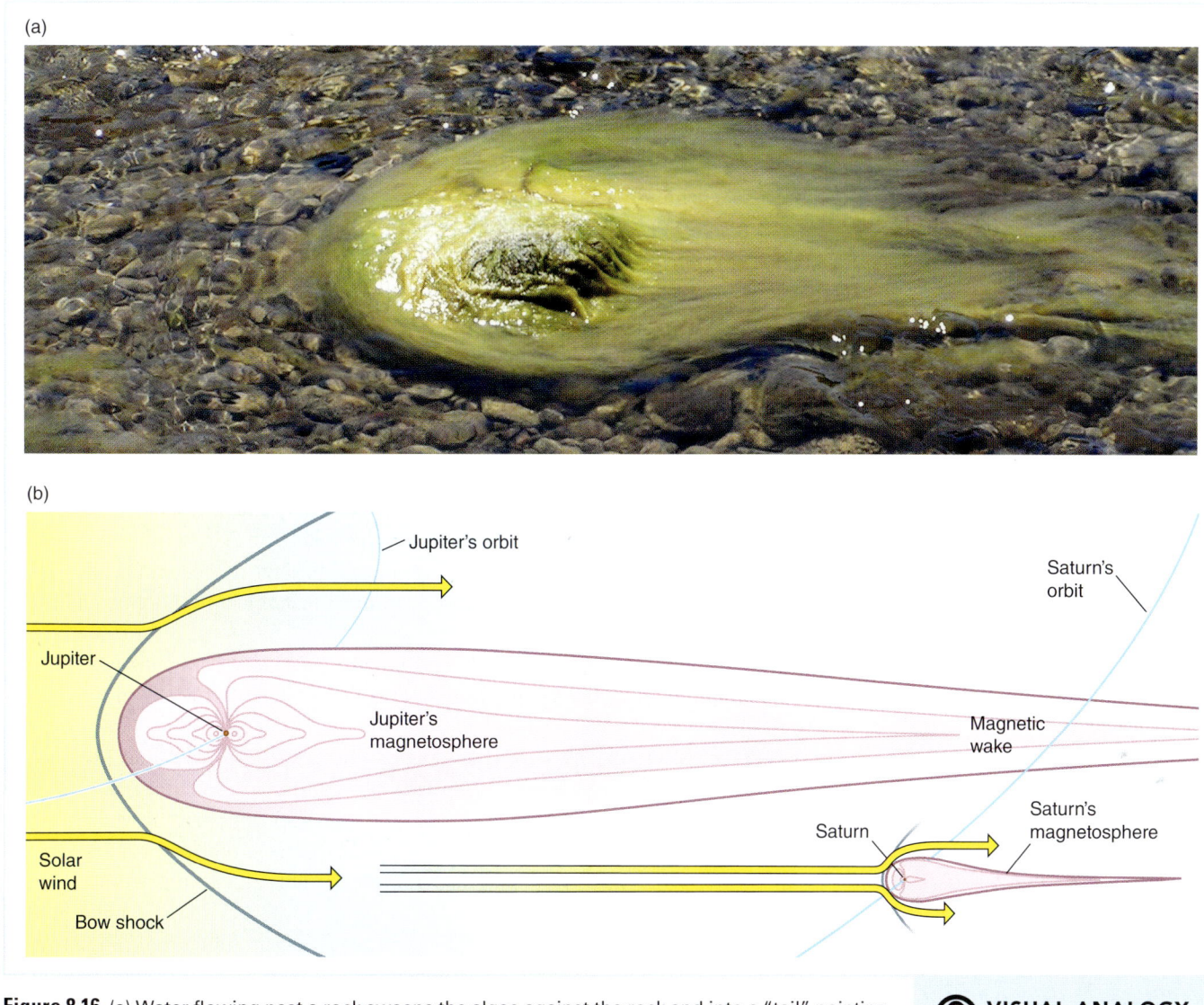

Figure 8.16 (a) Water flowing past a rock sweeps the algae against the rock and into a "tail" pointing in the direction of the water's flow. (b) The solar wind compresses Jupiter's (or any other) magnetosphere on the side toward the Sun and draws it out into a magnetic tail away from the Sun. Jupiter's tail stretches beyond the orbit of Saturn. Note that this drawing is not to scale.

👁 **VISUAL ANALOGY**

the pressure of the solar wind also pushes on and compresses a magnetosphere, so the size and shape of a planet's magnetosphere depends on how the solar wind is blowing at any particular time. As illustrated in **Figure 8.16**, the tail of Jupiter's magnetosphere extends well past the orbit of Saturn. The magnetic tails of Uranus and Neptune have a curious structure. For both Uranus and Neptune, the tilt and the large displacement of the magnetic field from the center of the planet cause the magnetosphere to wobble as the planet rotates. This wobble causes the tail of the magnetosphere to twist like a corkscrew as it stretches away.

Charged Particles and Auroras of the Giant Planets

As Jupiter rotates, it drags its magnetosphere around with it, and charged particles are swept around at high speeds. These fast-moving charged particles slam into neutral atoms, and the energy released in the resulting high-speed collisions heats the plasma to extreme temperatures. In 1979, while passing through Jupiter's magnetosphere, *Voyager 1* encountered a region of tenuous plasma with a temperature of more than 300 million K; this is 20 times the temperature at the center of the Sun. *Voyager 1* did not melt when passing through this region because the plasma is so tenuous that the plasma's particles were very far apart in space. Although each particle was extraordinarily energetic, there were so few of them that the probe passed unscathed through the plasma.

Charged particles are trapped in planetary magnetospheres. The locations where these particles are trapped are referred to as *radiation belts*. Although Earth's radiation belts are severe enough to worry astronauts, the radiation belts that surround Jupiter are searing in comparison. In 1974, the *Pioneer 11* spacecraft passed through the radiation belts of Jupiter. During its brief encounter, *Pioneer 11* picked up a radiation dose of 400,000 rads, or about 1,000 times the lethal dose for humans. Several of the onboard instruments were permanently damaged as a result, and the spacecraft itself barely survived to continue its journey to Saturn.

When fast-moving charged particles interact with magnetic fields, they emit radiation primarily in the radio portion of the spectrum. This is called **synchrotron radiation**. Jupiter has so many trapped high-energy electrons that within this radio part of the spectrum, the second-brightest object in our sky is Jupiter's magnetosphere—only the Sun is brighter.

Although Jupiter's magnetosphere is much farther away, it appears much larger than the Sun in the sky. Saturn's magnetosphere is also large but much fainter than Jupiter's. Saturn does have a strong magnetic field, but pieces of rock, ice, and dust in Saturn's rings absorb magnetospheric particles. With far fewer magnetospheric electrons, there is much less radio emission from Saturn.

Except in the case of Saturn, precise measurement of periodic variations in the radio signals from the giant planets tells us the true rotation periods of these planets (see Working It Out 8.2). The magnetic field of each planet is locked to the conducting liquid layers deep within the planet's interior, so the magnetic field rotates with exactly the same period as that of the deep interior of the planet.

In addition to protons and electrons from the solar wind, the magnetospheres of the giant planets contain large amounts of various elements, including sodium, sulfur, oxygen, nitrogen, and carbon from several sources, which include the planets' extended atmospheres and the moons that orbit within them. The most intense radiation belt in the Solar System is a **torus** (a doughnut-shaped ring) associated with Io, the innermost of Jupiter's four Galilean moons. As we shall see in Chapter 9, Io has low surface gravity and violent volcanic activity. Some of the gases erupting from Io's interior escape and become part of Jupiter's radiation belt. As charged particles are slammed into the moon by the rotation of Jupiter's magnetosphere, even more material is knocked free of Io's surface and ejected into space. Images of the region around Jupiter, taken in the light of emission lines from atoms of sulfur or sodium, show a faintly glowing torus of material supplied by Io (**Figure 8.17**).

Other moons also influence the magnetospheres of the planets they orbit. The atmosphere of Saturn's largest moon, Titan, is rich in nitrogen. Leakage of this gas into space is the major source of a torus that forms in Titan's wake. The density of this rather remote radiation belt is highly variable because the orbit of Titan is sometimes inside and sometimes outside Saturn's magnetosphere, depending on the strength of the solar wind. When Titan is outside Saturn's magnetosphere, any nitrogen molecules lost from Titan's atmosphere are carried away by the solar wind.

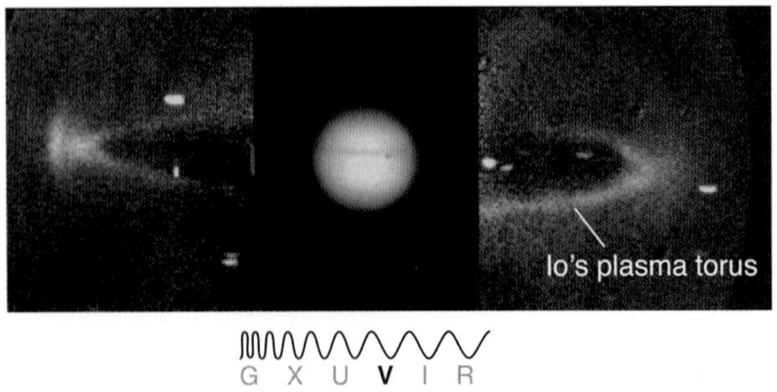

Figure 8.17 A faintly glowing torus of plasma surrounds Jupiter (center). The torus is made up of atoms knocked free from the surface of Io by charged particles. Io is the innermost of Jupiter's Galilean moons. (A semitransparent mask was placed over the disk of Jupiter to prevent its light from overwhelming the much fainter torus.)

Figure 8.18 The Hubble Space Telescope took images of auroral rings around the poles of Jupiter and Saturn. The auroral images were taken in ultraviolet light and then superimposed on visible-light images. (High-level haze obscures the ultraviolet views of the underlying cloud layers, as the insets show.)

Charged particles spiral along the magnetic-field lines of the giant planets, bouncing back and forth between the two magnetic poles just like they do around Earth. The results are bright auroral rings, shown in **Figure 8.18**. These auroral rings surround the magnetic poles of the giant planets, just as the aurora borealis and aurora australis ring the north and south magnetic poles of Earth.

8.5 Rings Surround the Giant Planets

A planetary ring is a collection of particles—varying in size from tiny grains to house-sized boulders—that orbit individually around the planet, forming a flat disk. Ring systems, which do not occur in the terrestrial planets, are found around all of the giant planets. **Figure 8.19** shows how the ring system of each giant planet varies in size and complexity, with some systems extending for hundreds of thousands of kilometers and some systems having detailed structure that includes numerous small rings. In this section, we discuss ring formation, composition, and evolution.

Orbits of Ring Particles

Kepler's laws dictate that the speeds and orbital periods of all ring particles must vary with their distance from the planet, and that the closest particles move the fastest and have the shortest orbital periods. The orbital periods of particles in Saturn's three bright rings, for example, range from 5 hours 45 minutes at their inner edge to 14 hours 20 minutes at their outer one. In the densely packed rings of Saturn, collisions between these particles circularize their orbits and also force them into the same plane. Ring particles have low speeds relative to one another because they are all orbiting in the same direction. A particle moving on an upward trajectory will bump into another particle on a downward trajec-

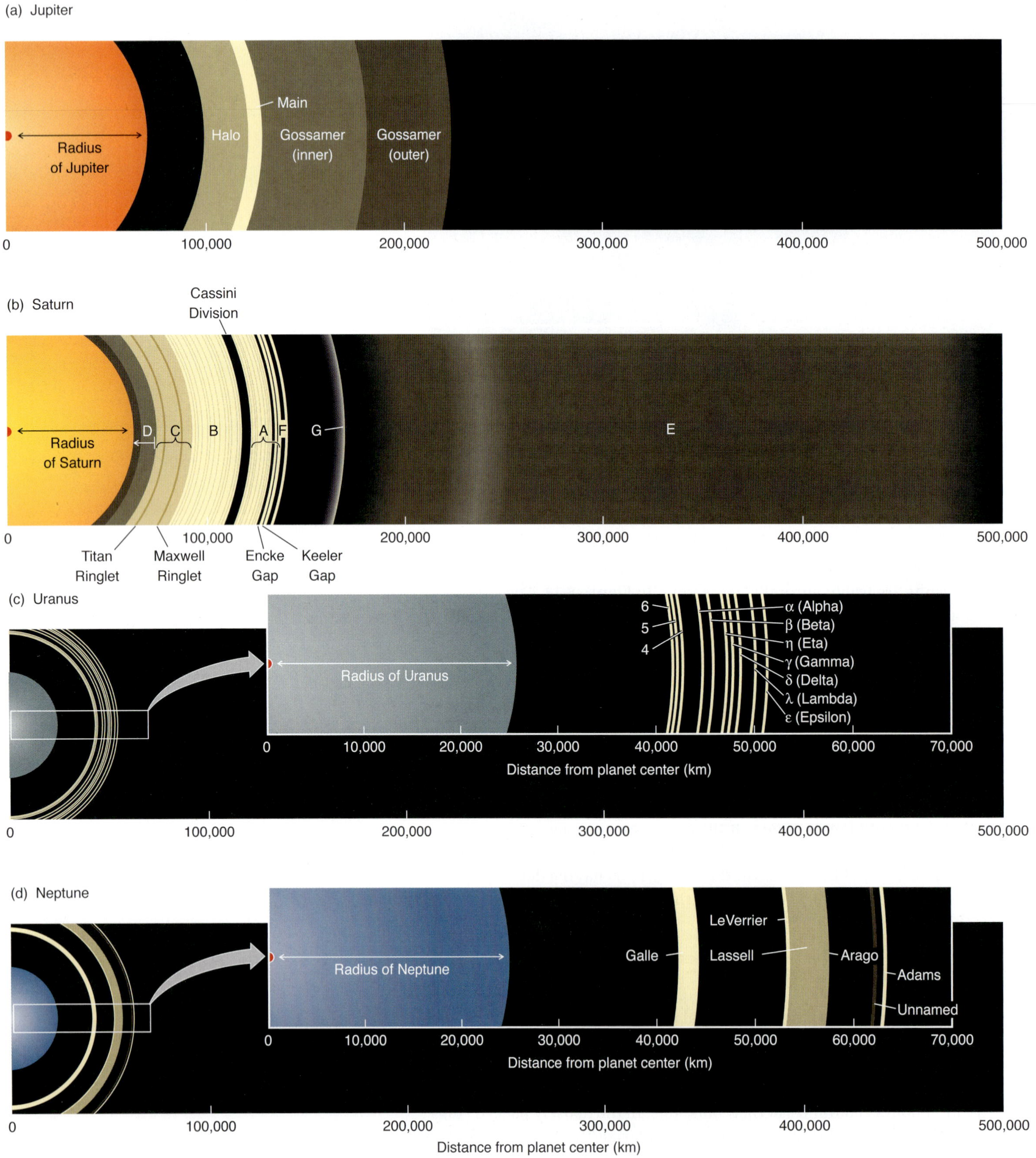

204 CHAPTER 8 *The Giant Planets*

Figure 8.19 The ring systems of the four giant planets vary in size and complexity. Saturn's system, with its broad E-Ring, is by far the largest and has the most complex structure in the inner rings.

tory and the upward and downward motion will cancel, leaving the particles moving in the same plane. A similar process occurs for particles moving inward and outward, leaving the particles moving at a constant radius.

The orbits of ring particles can also be influenced by their planet's larger moons. If the moon is massive enough, it exerts a significant gravitational tug on the ring particles as it passes by. If this happens over and over through many orbits, the particles are pulled out of the area, leaving a lower-density gap. Such is the case with Saturn's moon Mimas, which causes the famous gap in the rings around Saturn called the Cassini Division (Figure 8.19b). Other effects are also possible. For example, most narrow rings are caught up in a tug-of-war with nearby moons, known as **shepherd moons** because of the way they "herd" the flock of ring particles. A shepherd moon just outside a ring robs orbital energy from any particles that drift outward beyond the edge of the ring, causing them to move back inward. A shepherd moon just inside a ring gives up orbital energy to a particle that has drifted too far in, nudging it back outward.

Ring Formation and Evolution

Tidal stresses are thought to be responsible for much of the material found in planetary rings. If a moon (or other planetesimal) orbits a large planet, the force of gravity will be stronger on the side of the moon closest to the planet and weaker on the side farther away. This stretches out the moon, as we saw in the discussion of tidal forces in Chapter 6. If the tidal stresses are greater than the self-gravity that holds the moon together, the moon will be torn apart. The distance at which the tidal stresses exactly equal the self-gravity is known as the **Roche limit**. The Roche limit does not apply to objects that are held together by other forces—objects like people or bowling balls. It only applies to objects that are held together by their own gravity. If a moon or planetesimal comes within the Roche limit of a giant planet, it is pulled apart by tidal stresses, leaving many small particles to orbit the planet. These particles gradually spread out. Collisions circularize and flatten out the orbits, as described earlier, and rings are formed.

Planetary rings do not have the long-term stability of most Solar System objects. Ring particles are constantly colliding with one another in their tightly packed environment, either gaining or losing orbital energy. This redistribution of orbital energy can cause particles at the ring edges to leave the rings and drift away, aided by nongravitational influences such as the pressure of sunlight. Although moons may help guide the orbits of ring particles and delay the dissipation of the rings themselves, at best this can be only a temporary holding action. Most planetary rings eventually dissipate. At least one ring, however, seems immune from this eventual demise: Because the volcanic emissions from Saturn's moon Enceladus are constantly supplying icy particles to Saturn's E Ring, replacing those that drift away, the E Ring will survive for as long as Enceladus remains geologically active. Other rings may be maintained by collisions between objects in their vicinity.

Many of the planetary rings we see today have probably not existed in their current form since the Solar System's beginning. Astronomers think that ring systems have come and gone over the history of the Solar System. Even our own planet has probably had several short-lived rings at various times during its long history. Any number of comets or asteroids must have passed close enough to Earth to disintegrate into a swarm of small fragments to create a temporary ring. However, unlike the giant planets, Earth lacks shepherd moons to provide orbital stability.

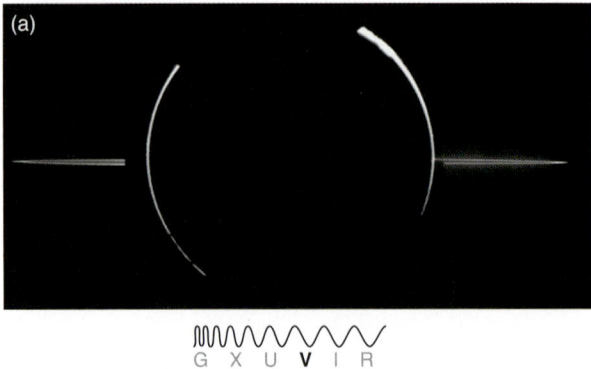

Figure 8.20 (a) This backlit *Galileo* view of Jupiter's rings also shows the forward scattering of sunlight by tiny particles in the upper layers of Jupiter's atmosphere. (b) A diagram of Jupiter's ring system and the small moons that form the rings.

The Composition of Rings

Because much of the material in the rings of the giant planets comes from their moons, the composition of the rings is similar to the composition of the moons. Saturn's bright rings probably formed when a moon or planetesimal came within the Roche limit of Saturn. These rings reflect about 60 percent of the sunlight falling on them. They are made of water ice, though a slight reddish tint tells us they are not made of pure ice but must contain small amounts of other materials, such as silicates. The icy moons around Saturn or the frozen comets of the outer Solar System could easily provide this material.

Saturn's rings are the brightest in the Solar System and are the only ones that we know are composed of water ice. In stark contrast, the rings of Uranus and Neptune are among the darkest objects known in our Solar System. Only 2 percent of the sunlight falling on them is reflected back into space; thus the ring particles are blacker than coal or soot. No silicates or similar rocky materials are this dark, so these rings are likely composed of organic materials and ices that have been radiation-darkened by high-energy charged particles from the planets' magnetospheres. (Radiation blackens organic ices such as methane by releasing carbon from the molecules of the ice.) Jupiter's rings are of intermediate brightness, suggesting that they may be rich in dark silicate materials, like the innermost of Jupiter's small moons.

Moons can contribute material to rings in other ways as well, as we see in the case of Jupiter's ring system, shown in **Figure 8.20**. The brightest of Jupiter's rings is a relatively narrow strand only 6,500 km across, consisting of material from the moons Metis and Adrastea. These two moons orbit in Jupiter's equatorial plane, and the ring they form is narrow. Beyond this main ring, however, are the very different **gossamer rings**, so called because they are extremely tenuous. The gossamer rings are supplied with dust by the moons Amalthea and Thebe. The innermost ring in Jupiter's system, called the halo ring, consists mostly of material from the main ring. As the dust particles in the main ring drift slowly inward toward the planet, they pick up electric charges and are pulled into this rather thick torus by electromagnetic forces associated with Jupiter's powerful magnetic field.

Finally, moons may contribute ring material through volcanism. Volcanoes on Jupiter's moon Io continually eject sulfur particles into space, many of which are pushed inward by sunlight and find their way into a ring. The particles in Saturn's E Ring are ice crystals ejected from icy geysers on the moon Enceladus, which is located in the very densest part of the E Ring.

Structure of Ring Systems

Saturn is adorned by a magnificent and complex system of rings, unmatched by any other planet in the Solar System. Figure 8.19b shows the individual components of Saturn's ring system and its major divisions and gaps. The most conspicuous features are its expansive bright rings, which dominate all photographs of Saturn. Among the four giant planets, only Saturn has rings so wide and so bright. The outermost bright ring, the A Ring, is the narrowest of the three bright rings. It has a sharp outer edge and contains several narrow gaps. On the A Ring's inner edge, the conspicuous Cassini Division is so wide (4,700 km) that the planet Mercury would almost fit within it. Astronomers once thought that it was completely empty, but images taken by *Voyager 1* show the Cassini Division is filled with material, although it is less dense than the other material in bright rings.

The B Ring, whose width is roughly twice Earth's diameter, is the brightest of Saturn's rings and has no internal gaps on the scale of those seen in the other bright rings. The C Ring is so much fainter than neighboring rings that it often fails to show up in normally exposed photographs. Through the eyepiece of a telescope this ring appears like delicate gauze. There is no known gap between the C Ring and either of the adjacent rings; only an abrupt change in brightness marks the boundary between them. The cause of this sharp change in the amount of ring material remains an unanswered question. Too dim to be seen next to Saturn's bright disk, the D Ring is a fourth wide ring that was unknown until imaged by *Voyager 1*. It shows less structure than any of the bright rings, and it does not appear to have a definable inner edge. The D Ring may extend all the way down to the top of Saturn's atmosphere, where its ring particles would burn up as meteors.

Saturn's bright rings are not uniform. The A and C rings contain hundreds, and the B Ring thousands, of individual **ringlets**, some only a few kilometers wide (**Figure 8.21**). Each of these ringlets is a narrowly confined concentration of ring particles bounded on both sides by regions of relatively little material.

About every 15 years, the plane of Saturn's rings lines up with Earth, and we view them edge on. The rings are so thin that they all but vanish for a day or so in even the largest telescopes. While the glare of the rings is absent, astronomers search for undiscovered moons or other faint objects close to Saturn. In 1966, an astronomer was looking for moons when he found weak but compelling evidence for a faint ring near the orbit of Saturn's moon Enceladus. In 1980, *Voyager 1* confirmed the existence of this faint ring—named the E Ring—and found another even closer, named the G Ring. The E and G rings are examples of what astronomers call a *diffuse ring*. In a **diffuse ring**, particles are far apart, and rare collisions between them can cause their individual orbits to become eccentric, inclined, or both. Because collisions are rare, the particles tend to remain in these disturbed orbits. Diffuse rings spread out horizontally and thicken vertically, sometimes without any obvious boundaries. Diffuse rings are difficult to detect except when lit from behind, when they appear bright because of the strong forward scattering of sunlight by very small ring particles (Figure 8.20a). In 2009, astronomers using the Spitzer Space Telescope discovered another diffuse ring. This dusty ring is thicker than other rings, about 20 times larger than Saturn from top to bottom, and is tilted 27° with respect to the plane of the rest of the rings.

Although Saturn's bright rings are very wide—more than 62,000 km from the inner edge of the C Ring to the outer edge of the A Ring—they are extremely thin. Saturn's bright rings are no more than a hundred meters and probably only a few tens of meters from their lower to upper surfaces. The diameter of Saturn's bright ring system is 10 million times the thickness of the rings. If the bright rings of Saturn were the thickness of a page in a book, six football fields laid end to end would stretch across them.

Figure 8.21 This *Cassini* image of the rings of Saturn shows so many ringlets and minigaps that it looks like a close-up of a phonograph record. The cause of most of this structure has yet to be explained in detail.

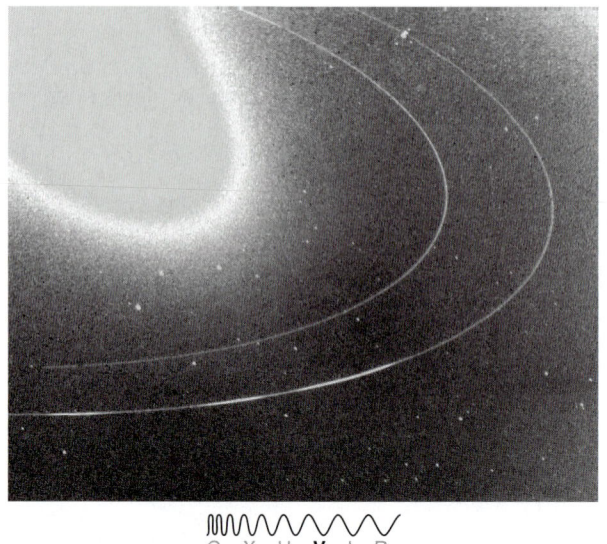

Figure 8.22 Neptune itself is very much overexposed in this Voyager 2 image of the three brightest arcs in Neptune's Adams Ring.

Ring structure among the other giant planets is not as diverse as that of Saturn. Turning back to Figure 8.19, you can see that rings other than Saturn's are quite narrow, although a few are diffuse. Jupiter's rings are made of fine dust dislodged by meteoritic impacts on the surfaces of Jupiter's small inner moons, which orbit among the rings. Nine of the 13 rings of Uranus are very narrow and widely spaced relative to their widths. Most are only a few kilometers wide, but they are many hundreds of kilometers apart. The space between these rings is filled with dust, which appears dark from our perspective. The dust is the right size to scatter sunlight forward instead of reflecting it backward. *Voyager 2* imaged this forward scattering after it passed Uranus and showed that the gaps were not empty. Four of Neptune's six rings are very narrow. In one of these rings, the Adams Ring, the material is clumped together in several **ring arcs**, which are high-density segments of the narrow ring (**Figure 8.22**). When first discovered, these ring arcs were a puzzle because collisions should have spread them uniformly around their orbit. Most astronomers now attribute this clumping to gravitational interactions with the moon Galatea. Some parts of Neptune's rings may be unstable: Ground-based images taken 13 years after *Voyager 2* show decay in the ring arcs. One of the ring arcs may disappear entirely before the end of the century.

> While the "island" of scientific knowledge is vast, the "ocean" of ignorance is larger still. Even the moons and rings in our own Solar System constantly surprise us.

READING ASTRONOMY News

Giant Propeller Structures Seen in Saturn's Rings

By **Denise Chow, MSNBC**

Giant propeller-shaped structures have been discovered in the rings of Saturn and appear to be created by a new class of hidden moons, NASA announced Thursday.

NASA's *Cassini* spacecraft spotted the distinctive structures inside some of Saturn's rings, marking the first time scientists have managed to track the orbits of individual objects from within a debris disk like the one that makes up Saturn's complicated ring system.

"Observing the motions of these disk-embedded objects provides a rare opportunity to gauge how the planets grew from, and interacted with, the disk of material surrounding the early Sun," said the study's co-author Carolyn Porco, one of the lead researchers on the Cassini imaging team based at the Space Science Institute in Boulder, Colorado. "It allows us a glimpse into how the solar system ended up looking the way it does."

Photos of the propellers taken by Cassini show them to be huge structures several thousands of miles long [**Figure 8.23**]. By understanding how they form, astronomers hope to glean insight into the debris disks around other stars as well, researchers said.

The results of the study are detailed in the July 8 issue of the journal *Astrophysical Journal Letters*.

Cassini scientists have seen double-armed propeller structures in Saturn's rings before, but on a smaller scale than the larger, new-found features. They were first spotted in 2006 in an area now known as the "propeller belt," which is located in the middle of Saturn's outermost dense ring, the A Ring.

SEE G*IANT* P*ROPELLER*

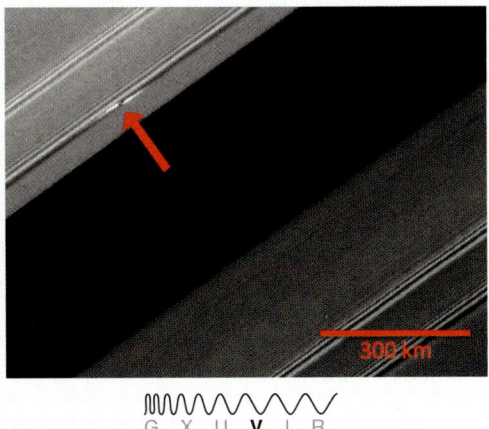

Figure 8.23 *Cassini* took this close-up image of a propeller-shaped structure in the rings of Saturn.

The propellers are actually gaps in the ring material [that] were created by a new class of objects, called moonlets, that are smaller than known moons but larger than the particles making up Saturn's rings. It is estimated that these moonlets could number in the millions, according to *Cassini* scientists.

The moonlets clear the space immediately around them to generate the propeller-like features, but are not large enough to sweep clear their entire orbit around Saturn, as seen with the moons Pan and Daphnis.

But in the new study, researchers [saw] a new legion of larger and rarer moons in a separate part of the A Ring, farther out from Saturn. These much larger moons create propellers that are hundreds of times larger than those previously described, and these objects have been tracked for about 4 years.

The study was led by *Cassini* imaging team associate Matthew Tiscareno at Cornell University in Ithaca, New York.

The propeller features for these larger moons are up to thousands of miles long and several miles wide. The moons embedded in Saturn's rings appear to kick up ring material as high as 1,600 feet above and below the ring plane.

This is much greater than the typical ring thickness of about 30 feet, researchers said.

Still, the *Cassini* spacecraft is too far away to see the moons amid the swirling ring material that surrounds them. Yet, scientists estimate that the moons measure approximately half a mile in diameter, based on the size of the propellers.

According to their research, Tiscareno and his colleagues estimate that there are dozens of these giant propellers. In fact, 11 of them were imaged multiple times between 2005 and 2009.

One such propeller, nicknamed Bleriot after the famous aviator Louis Bleriot, has shown up in more than 100 separate *Cassini* images and one ultraviolet imaging spectrograph observation during this time.

"Scientists have never tracked disk-embedded objects anywhere in the universe before now," said Tiscareno. "All the moons and planets we knew about before orbit in empty space. In the propeller belts, we saw a swarm in one image and then had no idea later on if we were seeing the same individual objects. With this new discovery, we can now track disk-embedded moons individually over many years."

Over their 4 years of observation, the researchers noticed shifts in the orbits of the giant propellers as they travel around Saturn, but the cause of these disturbances has not yet been determined.

The shifting orbits could be caused by collisions with other smaller ring particles or could be responses to these particles' gravity, the researchers said. The orbital paths of these moonlets could also be altered due to the gravitational attraction of large moons outside of Saturn's rings.

Scientists will continue to monitor the moons to see if the disk itself is driving the changes, similar to the interactions that occur in young solar systems. If so, Tiscareno said, this would be the first time such a measurement has been made directly.

"Propellers give us unexpected insight into the larger objects in the rings," said Linda Spilker, *Cassini* project scientist based at NASA's Jet Propulsion Laboratory in Pasadena, California. "Over the next 7 years, *Cassini* will have the opportunity to watch the evolution of these objects and to figure out why their orbits are changing."

NASA launched the *Cassini* probe in 1997 and it arrived at Saturn in 2004, where it dropped the European *Huygens* probe on the cloudy surface of Titan, Saturn's largest moon. *Cassini* was slated to be decommissioned in September of this year but received a life extension that now runs through 2017.

Evaluating the News

1. What new type of structure was observed in Saturn's rings?
2. Why had these structures never been observed before?
3. In the story, Carolyn Porco states: "Observing the motions of these disk-embedded objects provides a rare opportunity to gauge how the planets grew from, and interacted with, the disk of material surrounding the early Sun." Explain how the rings of Saturn might tell us something about the protoplanetary disk.
4. What is "new" about the "new class of moons" that are driving the formation of propeller shapes?
5. Put this story in context of all the information that you know about Saturn's rings. How does this discovery demonstrate the ongoing nature of the scientific process and the provisional nature of scientific knowledge?

SUMMARY

The giant planets are much larger and less dense than the terrestrial planets, consist primarily of light elements rather than rock, and are much colder. Because of their rapid rotation and the Coriolis effect, zonal winds are very strong on these planets. Volatiles become ices at various heights in these atmospheres, leading to a layered cloud structure. Jupiter, Saturn, and Neptune are still shrinking, and their gravitational energy is being converted to thermal energy, heating both the cores and the atmospheres from the inside. Uranus does not seem to have as large a heat source inside. Temperatures and pressures in the cores of the giant planets are very high, leading to novel states of matter, such as metallic hydrogen. All four giant planets have ring systems, which are transitory, created from and maintained by moons also in orbit around these planets.

LG 1 Jupiter and Saturn are made up mostly of hydrogen and helium—a composition similar to that of the Sun. Uranus and Neptune contain larger amounts of "ices" such as water, ammonia, and methane than Jupiter and Saturn. These compositions set them apart from the terrestrial planets. In addition, all four giant planets are much larger than Earth.

LG 2 We see only atmospheres on the giant planets because solid or liquid surfaces, if they exist, are deep below the cloud layers. Clouds on Jupiter and Saturn are composed of various kinds of ice crystals colored by impurities. Uranus and Neptune have relatively few clouds, and so their atmospheres appear more uniform.

LG 3 The ongoing collapse of the giant planets converts gravitational energy to thermal energy. This process heats most of the giant planets from within, producing convection. Powerful convection and the Coriolis effect drive high-speed winds in the upper atmospheres of all of the giant planets.

LG 4 The interiors of the giant planets are very hot and very dense because of the high pressures exerted by the overlying atmosphere.

LG 5 All four giant planets are surrounded by rings, many of which have their origin in the collisions of moons. Planetary rings form a series of concentric bands that lie in a flat disk.

SUMMARY SELF-TEST

1. Jupiter and Saturn have compositions similar to the Sun and are called _____ giants. Uranus and Neptune, conversely, are called _____ giants.

2. The differing colors of the giant planets are due primarily to
   a. differences in temperature.
   b. differences in composition.
   c. the Doppler effect from the varying wind speeds.
   d. differences in atmospheric depth.

3. The interiors of the giant planets are heated by gravitational contraction. We know this because
   a. the cores are very hot.
   b. the giant planets radiate more energy than they receive from the Sun.
   c. the giant planets have strong magnetic fields.
   d. the giant planets are mostly atmosphere.

4. As depth increases, _____ and _____ increase, which causes changes in the chemical composition of clouds in giant planet atmospheres.

5. Saturn's bright rings are located within the Roche limit of Saturn. This supports the theory that these rings (choose all that apply)
   a. are formed of moons torn apart by tidal stresses.
   b. formed at the same time that Saturn formed.
   c. are relatively recent.
   d. are temporary.

6. Zonal winds on the giant planets are stronger than those on the terrestrial planets because
   a. the giant planets have more thermal energy.
   b. the giant planets rotate faster.
   c. the moons of the giant planets provide additional pull.
   d. the moons of the giant planets feed energy to their planet through the magnetosphere.

7. Individual cloud layers in the giant planets have different compositions. This happens because
   a. the winds are all in the outermost layer.
   b. the Coriolis effect only occurs close to the "surface" of the inner core.
   c. there is no convection on the giant planets.
   d. different volatiles freeze out at different temperatures.

8. Deep in the interiors of the giant planets, water is still a liquid even though the temperatures are tens of thousands of degrees above the boiling point of water. This can happen because
   a. the density inside the giant planets is so high.
   b. the pressure inside the giant planets is so high.
   c. the outer Solar System is so cold.
   d. space has very low pressure.

9. Volcanos on Enceladus affect the E Ring of Saturn by
   a. pushing the ring around.
   b. stirring the ring particles.
   c. supplying ring particles.
   d. dissipating the ring.

10. Imagine a giant planet, very similar to Jupiter, that was ejected from its solar system at formation. (These objects exist, and are probably numerous, although their total number is still uncertain.) This planet would almost certainly still have (choose all that apply)
    a. a magnetosphere.   b. thermal energy.
    c. auroras.   d. rings.

QUESTIONS AND PROBLEMS

Multiple Choice and True/False

11. **T/F:** Uranus has extreme seasons because its poles are nearly in the plane of the Solar System.

12. **T/F:** All the giant planets have clouds and belts.

13. **T/F:** Water never forms visible clouds in the atmospheres of giant planets.

14. **T/F:** Storms on giant planets last much longer than storms on Earth.

15. **T/F:** The cores of the giant planets are all similar.

16. Stellar occultations are the most accurate way to measure the _____ of a Solar System object.
    a. mass
    b. density
    c. temperature
    d. diameter

17. The chemical compositions of Jupiter and Saturn are most similar to those of
    a. Uranus and Neptune.
    b. the terrestrial planets.
    c. their moons.
    d. the Sun.

18. Uranus and Neptune are different from Jupiter and Saturn in that
    a. Uranus and Neptune have a higher percentage of ices in their interiors.
    b. Uranus and Neptune have no rings.
    c. Uranus and Neptune have no magnetic field.
    d. Uranus and Neptune are closer to the Sun.

19. The Great Red Spot on Jupiter is
    a. a surface feature.
    b. a "storm" that has been raging for more than 300 years.
    c. caused by the interaction between the magnetosphere and Io.
    d. about the size of North America.

20. The different colors of clouds on Jupiter are primarily a result of
    a. temperature.
    b. composition.
    c. motion of the clouds toward or away from us.
    d. all of the above

21. Consider a northward-traveling particle of gas in Jupiter's northern hemisphere. According to the Coriolis effect, which way will this particle of gas turn?
    a. in the direction of rotation
    b. opposite the direction of rotation
    c. north
    d. south

22. Metallic hydrogen is *not*
    a. a metal that acts like hydrogen.
    b. hydrogen that acts like a metal.
    c. common in the cores of giant planets.
    d. a result of high temperatures and pressures.

23. Which of the giant planets has the most extreme seasons?
    a. Jupiter
    b. Saturn
    c. Uranus
    d. Neptune

24. The magnetic fields of the giant planets
    a. align closely with the rotation axis.
    b. extend far into space.
    c. are thousands of times stronger at the cloud tops than at Earth's surface field.
    d. have an axis that passes through the planet's center.

25. The rings of Saturn periodically disappear and reappear when
    a. observed in the direction of the Sun.
    b. the Sun has set on Saturn.
    c. the rings dissipate.
    d. viewed edge-on.

Conceptual Questions

26. Jupiter's chemical composition is more like that of the Sun than Earth's is. Yet both planets formed from the same protoplanetary disk. Explain why they are different today.

27. Astronomers take the unusual position of lumping together all atomic elements other than hydrogen and helium into a single category, which they call heavy elements. Why is this a reasonable position for astronomers to take?

28. None of the giant planets are truly round. Explain why they have a flattened appearance.

29. The Great Red Spot is a long-lasting atmospheric vortex in Jupiter's southern hemisphere. Winds rotate counterclockwise around its center. Is the Great Red Spot cyclonic or anticyclonic? Is it a region of high or low pressure? Explain.

30. What is the source of color in Jupiter's clouds? Uranus and Neptune, when viewed through a telescope, appear distinctly bluish green in color. What are the two reasons for their striking appearance?

31. Lightning has been detected in the atmospheres of all the giant planets. How is it detected?

32. What drives the zonal winds in the atmospheres of the giant planets?

33. Jupiter, Saturn, and Neptune all radiate more energy into space than they receive from the Sun. Does this violate the law of conservation of energy? What is the source of the additional energy?

34. What creates metallic hydrogen in the interiors of Jupiter and Saturn, and why do we call it metallic?

35. Jupiter's core is thought to consist of rocky material and ices, all in a liquid state at a temperature of 35,000 K. How can materials such as water be liquid at such high temperatures?

36. Saturn has a source of internal thermal energy that Jupiter may not have. What is it, and how does it work?

37. What creates auroras in the polar regions of Jupiter and Saturn?

38. Will the particles in Saturn's bright rings eventually stick together to form one solid moon orbiting at the average distance of all the ring particles? Explain your answer.

39. Explain the mechanism that creates gaps in Saturn's bright ring system.

40. Astronomers believe that most planetary rings eventually dissipate. Explain why the rings do not last forever. Name one ring that might continue to exist indefinitely, and explain why it could survive when others might not.

Problems

41. Using the radius and mass information in Table 8.1, calculate the density of Saturn. How does this density compare to the density of water? Would Saturn sink or float if you could find a big enough bathtub to put it in?

42. The Sun appears 400,000 times brighter than the full Moon in our sky. How far from the Sun (in astronomical units) would you have to go for the Sun to appear only as bright as the full Moon appears in our nighttime sky? Compare your answer with the semimajor axis of Neptune's orbit.

43. Uranus occults a star at a time when the relative motion between Uranus and Earth is 23.0 km/s. An observer on Earth sees the star disappear for 37 minutes 2 seconds and notes that the center of Uranus passes directly in front of the star.
    a. Based on these observations, what value would the observer calculate for the diameter of Uranus?
    b. What could you conclude about the planet's diameter if its center did not pass directly in front of the star?

44. Jupiter is an oblate planet with an average radius of 69,900 km compared to Earth's average radius of 6,370 km.
    a. Remembering that volume is proportional to the cube of the radius, how many Earth volumes could fit inside Jupiter?
    b. Jupiter is 318 times as massive as Earth. Show that Jupiter's average density is about one-fourth that of Earth's.

45. The tilt of the axis of Uranus is 98°. If you were located at one of the planet's poles, how far from the zenith would the Sun appear at the time of summer solstice?

46. Compare the graphs in Figure 8.10a and b. Does atmospheric pressure increase more rapidly with depth on Jupiter or on Saturn? Compare the graphs in Figure 8.10c and d. Does pressure increase more rapidly with depth on Uranus or on Neptune? Of the four giant planets, which has the fastest pressure rise with depth? Which has the slowest?

47. Using Figure 8.10, find the altitude at which water ice forms in the atmospheres of each of the four giant planets.

48. A small cloud in Jupiter's equatorial region is observed to be at longitude 122.0° west in a coordinate system that rotates at the same rate as the deep interior of the planet. (West longitude is measured along a planet's equator toward the west.) Another observation made exactly 10 Earth hours later finds the cloud at longitude 118.0° west. Jupiter's equatorial radius is 71,500 km. What is the observed equatorial wind speed in kilometers per hour? Is this an easterly or a westerly wind?

49. The equilibrium temperature for Saturn should be 82 K, but instead we find an average temperature of 95 K. How much more energy is Saturn radiating into space than it absorbs from the Sun?

50. Neptune radiates into space 2.6 times as much energy as it absorbs from the Sun. Its equilibrium temperature is 47 K. What is its true temperature?

SMARTWORK

Norton's online homework system includes algorithmically generated versions of these questions, plus additional conceptual exercises. If your instructor assigns questions in SmartWork, log in at smartwork.wwnorton.com.

Exploration | Estimating Rotation Periods of Giant Planets

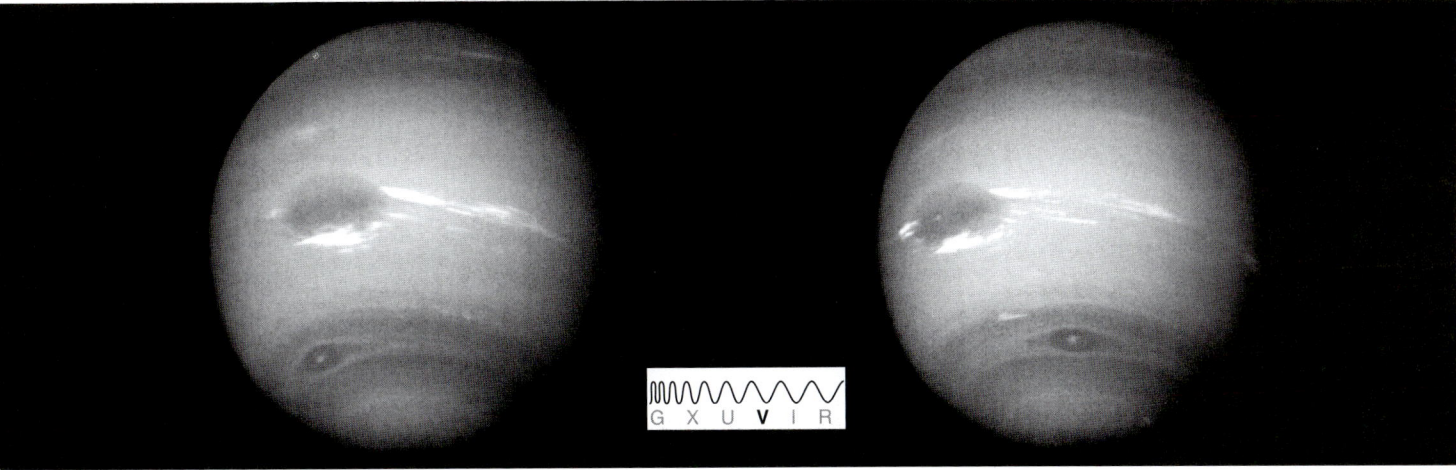

Figure 8.24 Two images of Neptune, taken 17.6 hours apart by *Voyager 2*.

Study the two images of Neptune in **Figure 8.24**. The image on the left was taken first, and the image on the right was taken 17.6 hours later; during this time, the Great Dark Spot completed nearly one full rotation. The small storm at the bottom of the image of the planet completed slightly more than one rotation. You would be very surprised to see this result for locations on Earth.

We can find the rotation period of the smaller storm in Figure 8.24 by equating two ratios. First, use a ruler to find the distance (in millimeters) from the left limb of the planet to the small storm (as shown in **Figure 8.25a**). Do this for both images in Figure 8.24. Estimate the radius of the circle traveled by the storm by measuring from the limb to a line through the center of the planet (as shown in **Figure 8.25b**).

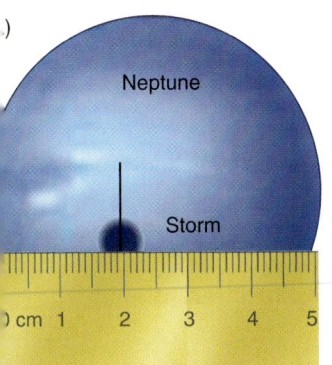
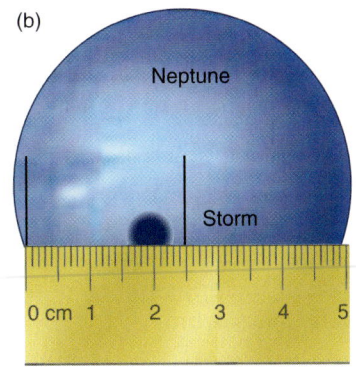

Figure 8.25 (a) How to measure the position of the storm. (b) How to measure the radius of the circle the storm traveled.

1. **Estimate the radius of the circle the small storm makes around Neptune (in millimeters) by making the appropriate measurements on Figure 8.24. Then calculate the circumference of the circle (in millimeters).**

Because the small storm has rotated *more than* one time, the total distance it has traveled is the circumference of the circle plus the distance between its location in the two images, which can be measured directly.

2. **Add those two numbers together to get the total distance traveled (in millimeters) between these images.**

Now we are ready to find the rotation period. The ratio of the rotation period, T, to the time elapsed, t, must be equal to the ratio of the circumference of the circle around which it travels, C (in millimeters), to the total distance traveled, D (in millimeters).

$$\frac{T}{t} = \frac{C}{D}$$

3. **What is the small storm's rotation period? (To check your work, note that your answer should be less than 17.6 hours. Why?)**

SMARTWORK • smartwork.wwnorton.com

9 Small Bodies of the Solar System

We began our discussion of the Solar System with its origins. You learned that very early on—at the same time our Sun was becoming a star—tiny grains of primitive material stuck together to produce swarms of small bodies called planetesimals. Those that formed in the hotter, inner part of the Solar System were composed mostly of rock and metal, while those in the colder, outer parts were made up of ice, organic compounds, and rock. Some of these objects collided to become planets and moons, and others were ejected from the Solar System by gravitational encounters with larger bodies. However, many of these bodies are still around. In the illustration on the opposite page, a student has found an asteroid while studying the stars in this field, and she then confirmed her finding using a catalog of asteroids. On the index card, she has determined the path of the moving asteroid. Asteroids like this one, as well as other small bodies, represent a small but scientifically important component of our present-day Solar System.

In this chapter, we explore these small bodies: dwarf planets, moons, asteroids, and comets. These remaining planetesimals, and the fragments that some of them continually create, provide planetary scientists with the opportunity to look back to the physical and chemical conditions of the earliest moments in the history of the Solar System. The study of these small bodies as a group also informs our understanding of the history of the Solar System.

✧ LEARNING GOALS

Dwarf planets are among the smaller worlds in the Solar System. Asteroids and comets are even smaller, yet these objects—and their fragments that survive the atmosphere to fall to Earth as meteorites—have told us much of what we know about the early history of the Solar System. By the end of this chapter, you should understand how observations like the ones shown in the illustration at right are especially interesting for a class of asteroids known as near-Earth objects. You should also be able to:

LG 1 List the categories of small bodies.

LG 2 Classify moons by their geological activity.

LG 3 Explain why some asteroids differentiated while others did not.

LG 4 Describe the origin of different types of meteorites.

LG 5 Sketch the appearance of a comet at different locations in its orbit.

LG 6 Explain the importance of meteorites, asteroids, and comets to the history of the Solar System.

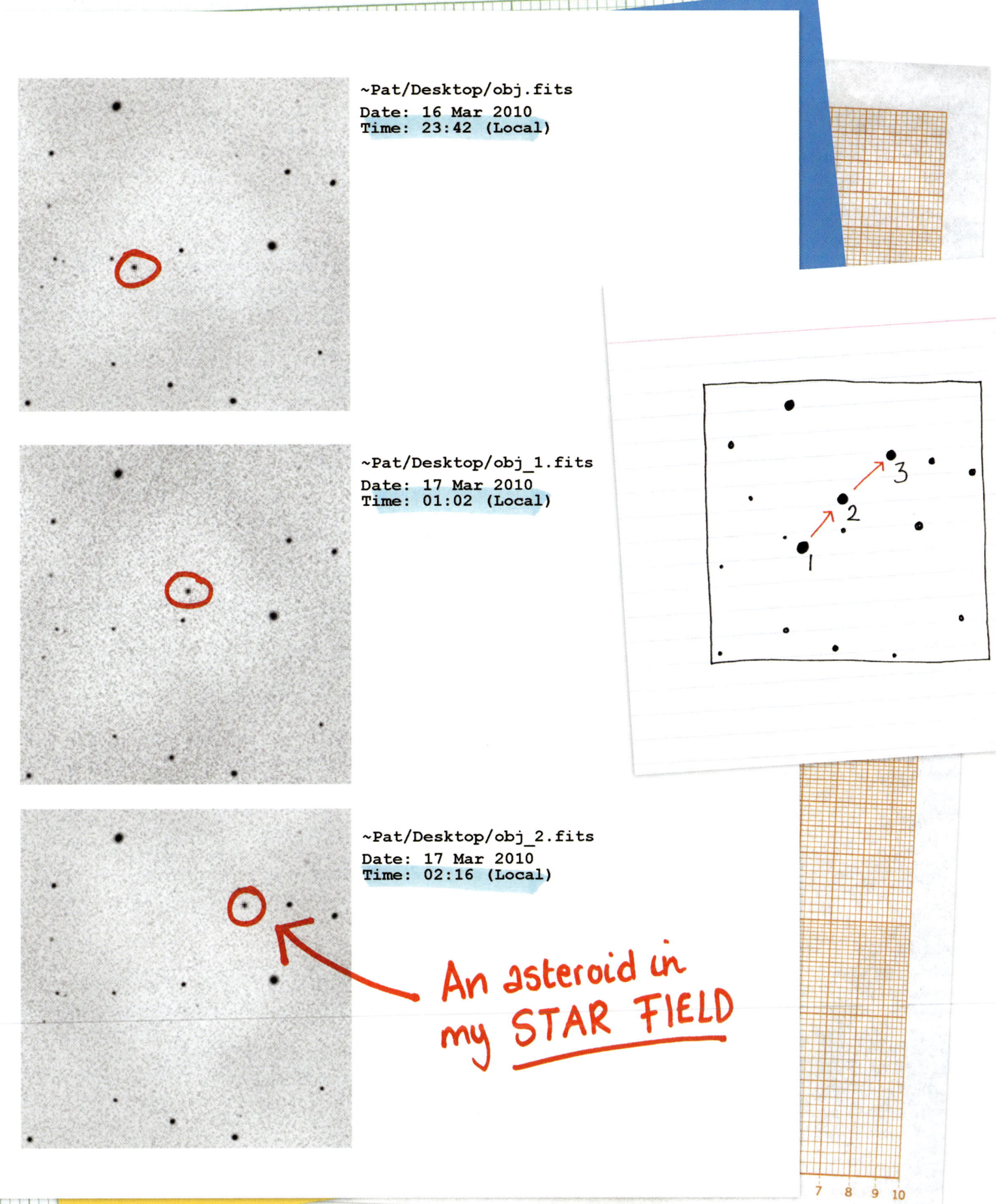

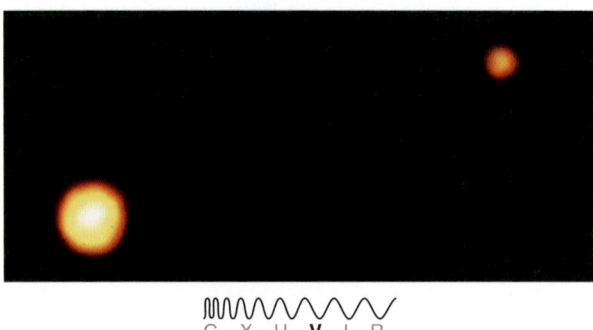

Figure 9.1 The Hubble Space Telescope imaged the dwarf planet Pluto and its largest moon, Charon. Pluto and Charon are so similar in size that they could be considered a "double dwarf planet."

9.1 Dwarf Planets May Outnumber Planets

There are five officially recognized dwarf planets in the Solar System: Pluto, Eris, Haumea, Makemake, and Ceres. Of these, all but Ceres are located in the outer Solar System. The four distant dwarf planets represent just the beginning of our knowledge about these objects in the Solar System. Recall from Chapter 5 that a *planet* is defined as an object that orbits the Sun and is large enough to (1) pull itself into a round shape and (2) clear the area around its orbit, so that there are no other comparable objects sharing the orbit. Dwarf planets are objects that meet criterion 1 by being round, but because they have not cleared the region near their orbits, they fail to meet criterion 2. Among the known objects in our Solar System, at least 100 could be dwarf planets, but their shapes have not yet been measured well enough for certain classification. All of these are located beyond the orbit of Pluto. A few astronomers estimate that there may be as many as 10,000 dwarf planets orbiting the Sun.

Pluto's orbit is 248 Earth years long and is quite elliptical, periodically bringing the dwarf planet inside Neptune's nearly circular orbit—from 1979 to 1999, Pluto was closer to the Sun than Neptune. Pluto has five known moons, the largest of which is Charon, about half the size of Pluto (**Figure 9.1**), which in turn is only two-thirds as large as our Moon. The total mass of the Pluto-Charon system is 1/400 that of Earth. Pluto and Charon each have a rocky core surrounded by a water-ice mantle. Pluto's surface is an icy mixture of frozen water, carbon dioxide, nitrogen, methane, and carbon monoxide, but Charon's surface is primarily water ice. Pluto has a thin atmosphere of nitrogen, methane, and carbon monoxide; these gases freeze out of the atmosphere when Pluto is more distant from the Sun and therefore colder.

Pluto and Charon are a tidally locked pair: each has one hemisphere that always faces the other. We know little else about the surface features of these worlds and nothing about their geological history. We hope to know more soon—the NASA spacecraft *New Horizons* will reach Pluto and Charon in 2015.

Eris (**Figure 9.2**) is about the same size as Pluto but is more massive. Eris also has a relatively large moon, called Dysnomia. The highly eccentric orbit of Eris carries it from 37.8 astronomical units (AU) out to 97.6 AU, with an orbital period of 562 years. By chance, Eris was found near the most distant point in its orbit, making it currently the most remote known object in the Solar System. (The eccentric orbits of some other Solar System bodies will eventually carry them farther from the Sun.) Eris is highly reflective, so it must have a coating of pristine ice. Spectra indicate this ice is probably mostly methane, which will form an atmosphere when Eris comes closest to the Sun in the year 2257.

Haumea and Makemake are both smaller and have slightly larger orbits than that of Pluto. Haumea has two known moons—Hi'iaka and Namaka. Haumea spins so rapidly on its axis that its shape is distorted into a flattened ellipsoid with an equatorial radius that is approximately twice its polar radius. This is the most distorted shape of any known planet, dwarf or otherwise. No moons have been discovered orbiting around Makemake, and we know less about this dwarf planet than about Pluto, Haumea, or Eris.

With a diameter of about 975 kilometers (km), Ceres is larger than most moons but smaller than any planet. It contains about a third of the total mass in the main

Figure 9.2 (a) The distant dwarf planet Eris (shown within the white circle) is about the same size as Pluto and has similar physical characteristics. Small bodies of the Solar System are typically indistinguishable from stars in individual images and are found by comparing images of the same field over time. Solar System objects change position from one image to the next, but the stars remain in the same place. (b) The orbit of Eris is both highly eccentric and highly inclined to the rest of the Solar System.

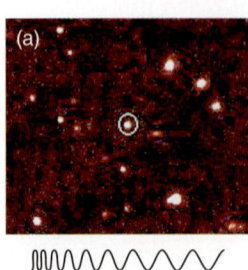
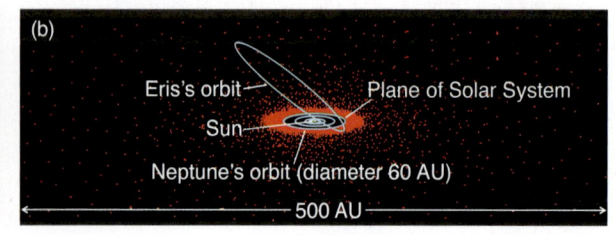

asteroid belt—the region between Mars and Jupiter that contains most of the Solar System's asteroids. Ceres rotates on its axis with a period of about 9 hours, typical of many asteroids (**Figure 9.3**). As a large planetesimal, Ceres seems to have survived mostly intact, although astronomers have found indications that it underwent differentiation at some point in its early history. Spectra show that the surface of Ceres contains hydrated minerals such as clays and carbonates, indicating the presence of significant amounts of water in its interior. Perhaps as much as a quarter of its mass exists in the form of a water-ice mantle that surrounds a rocky inner core. A recent discovery of water vapor coming from two locations on Ceres indicates that there is water ice in specific locations on the surface. This is the first direct detection of water in the asteroid belt. NASA's *Dawn* mission completed a 1-year exploration of asteroid Vesta in 2011, and the spacecraft is now traveling on to reach Ceres in 2015. Ceres is a dwarf planet because although it is round, it has not cleared its surroundings.

9.2 Moons as Small Worlds

The giant planets have many moons in orbit around them, ranging in size from small boulders to planet-sized objects. The major moons of the Solar System are shown in **Figure 9.4**. We will organize our tour of the moons of the Solar Sys-

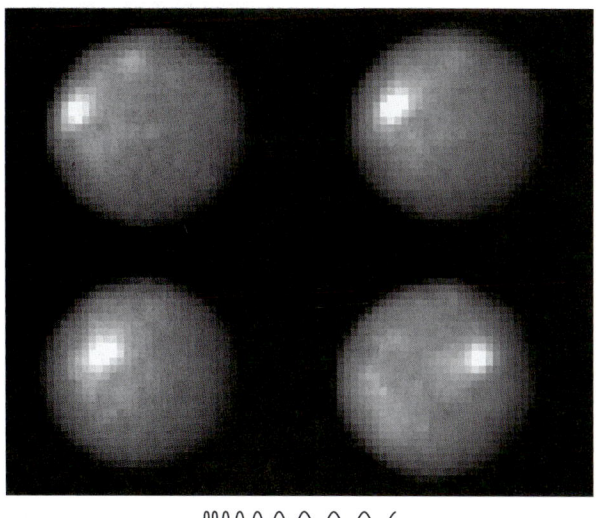

Figure 9.3 Ceres is spherical, one of the criteria giving Ceres its dwarf planet status. The nature of the bright spot is unknown, but it reveals the rotation of Ceres, as seen in this set of four images taken by the Hubble Space Telescope.

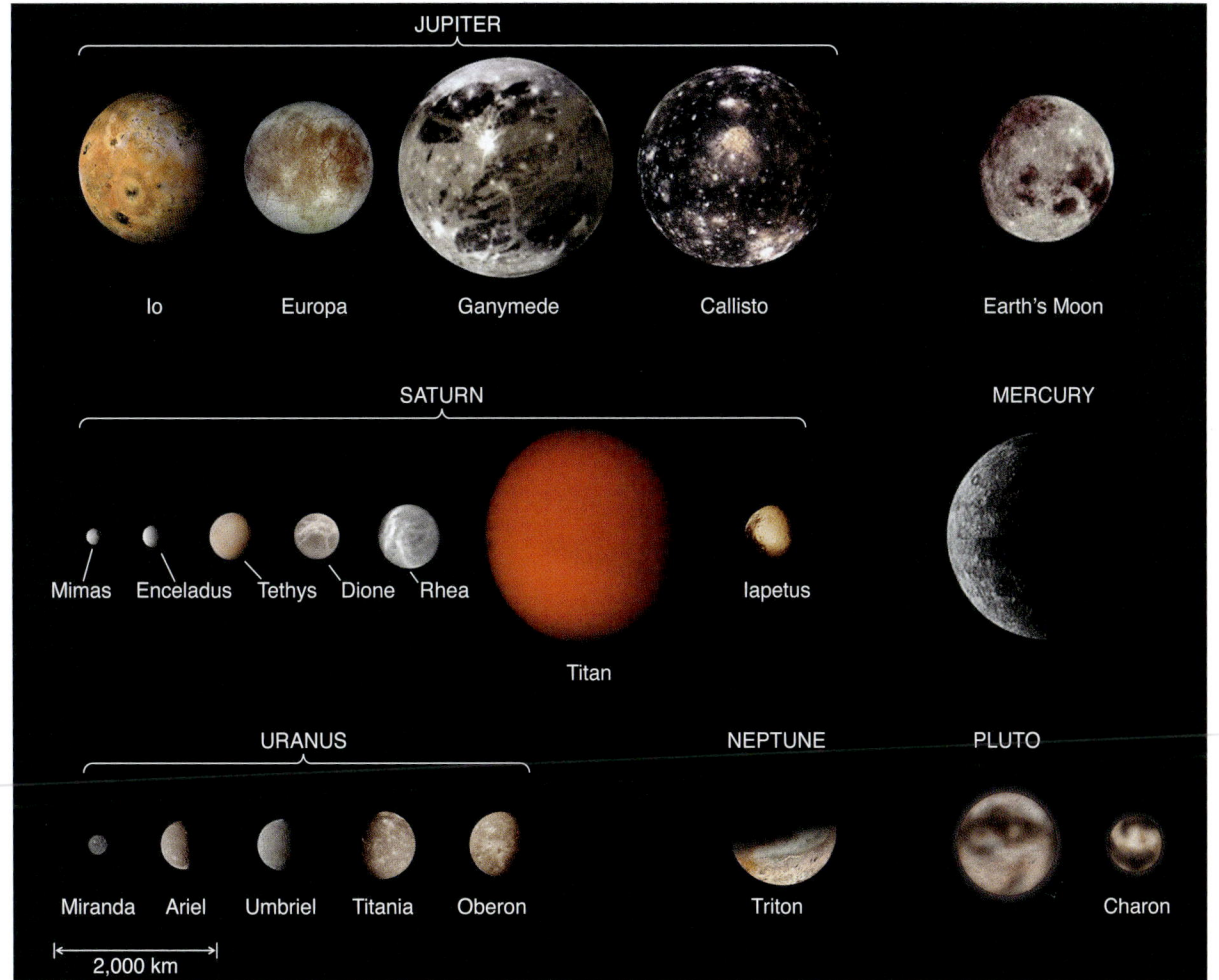

Figure 9.4 The major moons of the Solar System are shown to scale, along with Mercury and Pluto for comparison. Both Ganymede and Titan are larger than Mercury.

tem by their geological histories as shown by surface features. Comparing these worlds in this way allows us to draw conclusions about how they formed and the physical or geological principles governing their evolution. Some moons have been frozen in time since their formation during the early history of the Solar System, while others are even more geologically active than Earth. We identify three categories of geological activity that are the most instructive for our comparative approach: (1) definitely active today, (2) probably or possibly active today, (3) active in the past but not today. Moons in a fourth category—those apparently not active at any time since their formation—have heavily cratered surfaces and show no modifications other than those caused by a long history of impacts. This category includes Jupiter's Callisto, Uranus's Umbriel, and a large assortment of irregular moons. (Irregular moons have orbits that are highly inclined, very elliptical, or revolve in the opposite direction from the planet's spin.)

Geologically Active Moons

Volcanic features abound on Io, a moon of Jupiter (**Figure 9.5**). Because Io's orbit is elliptical, Jupiter's gravitational pull flexes Io's crust and generates enough thermal energy to melt parts of it, powering the most active volcanism known in the Solar System. Lava flows and volcanic ash bury impact craters as quickly as they form, so no impact craters have been observed on the surface. Explosive eruptions from more than 150 active volcanoes send debris hundreds of kilometers above Io's surface. The moon is so active that several huge eruptions are often occurring at the same time.

Mixtures of sulfur, sulfur dioxide frost, and sulfurous salts of sodium and potassium likely cause the wide variety of colors on Io's surface. Bright patches may be fields of sulfur dioxide snow. Images taken by *Voyager* and *Galileo* reveal plains, irregular craters, and flows, all related to eruption of mostly silicate mag-

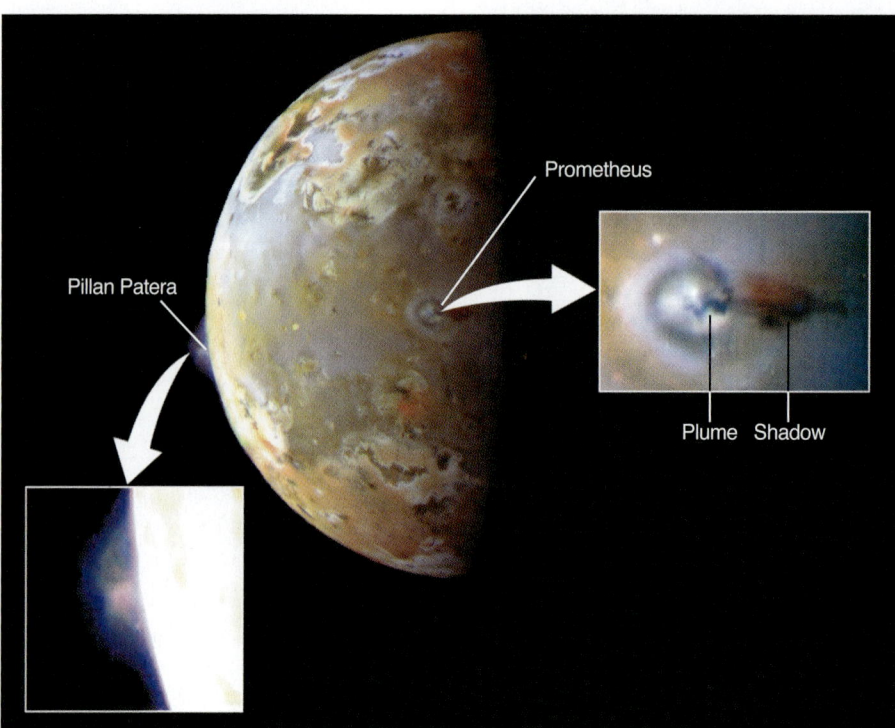

Figure 9.5 When this image of Io was obtained by *Galileo*, two volcanic eruptions could be seen at once. The plume of Pillan Patera rises 140 km above the limb of the moon on the left, while the shadow of a 75-km-high plume can be seen to the right of the vent of Prometheus, near the moon's center.

mas onto the surface of the moon. They also show tall mountains, some nearly twice the height of Mount Everest. Huge structures have multiple summit craters, showing a long history of repeated eruptions followed by collapse of the partially emptied magma chambers. Volcanoes on Io are spread around the moon much more randomly than those on Earth, implying a lack of plate tectonics.

Because of its active volcanism, Io's mantle has turned inside-out more than once in the past. Volatiles such as water and carbon dioxide escaped into space long ago. Heavier materials sank to the interior to form a core. Sulfur and various sulfur compounds, as well as silicate magmas, are constantly being recycled to form the complex surface we see today.

Farther out in the Solar System, we find volcanism of a different type. **Cryovolcanism** is similar to terrestrial volcanism but is driven by subsurface low-temperature volatiles such as water and nitrogen rather than molten rock. Enceladus, one of Saturn's icy moons, shows a wide variety of ridges, faults, and smooth plains. This evidence of tectonic processes is unexpected for a small (500-km) body. Some impact craters appear softened, perhaps by the viscous flow of ice, like the flow that occurs in the bottom layers of glaciers on Earth. Parts of the moon have no craters, indicating recent resurfacing.

Terrain near the south pole of Enceladus is cracked and twisted (**Figure 9.6a**). Enceladus has a liquid ocean, buried beneath 30–40 km of ice crust, and 10 km deep. The cracks are warmer than their surroundings, implying that tidal heating and radioactive decay within the moon's rocky core heat ice and drive it to the surface. Active cryovolcanic plumes (Figure 9.6b) are energetic enough to overcome the moon's low gravity, sending tiny ice crystals into space to replace particles continually lost from Saturn's E Ring. It is a mystery that Enceladus is so active while Mimas, a neighboring moon of about the same size and also subject to tidal heating, appears to be completely dead.

Cryovolcanism also occurs on Neptune's largest moon, Triton. Triton orbits Neptune in the opposite direction from Neptune's spin, implying that Triton must have been captured by Neptune after the planet's formation. Because angular momentum is conserved, moons that formed with their planets orbit in the same direction in which the planet rotates. As Triton achieved its current circular, synchronous orbit, it experienced extreme tidal stresses from Neptune, generating large amounts of thermal energy. The interior may have melted, allowing Triton to become chemically differentiated.

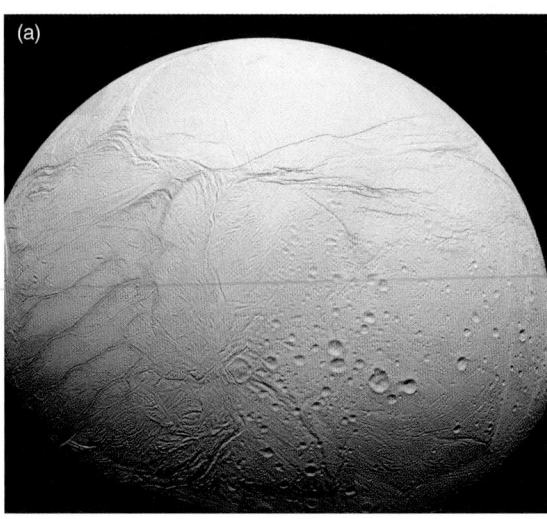

Figure 9.6 *Cassini* images of Enceladus show evidence of cryovolcanism. (a) The twisted and folded surface of deformed ice cracks near the moon's south pole (shown blue in false color) is warmer than the surrounding terrain and has been found to be the source of cryovolcanism. (b) Cryovolcanic plumes in the south polar region spew ice particles into space. In this image, there are two light sources. The surface of the moon is illuminated by sunlight reflected from Saturn while the plumes are backlit by the Sun itself, located almost directly behind Enceladus.

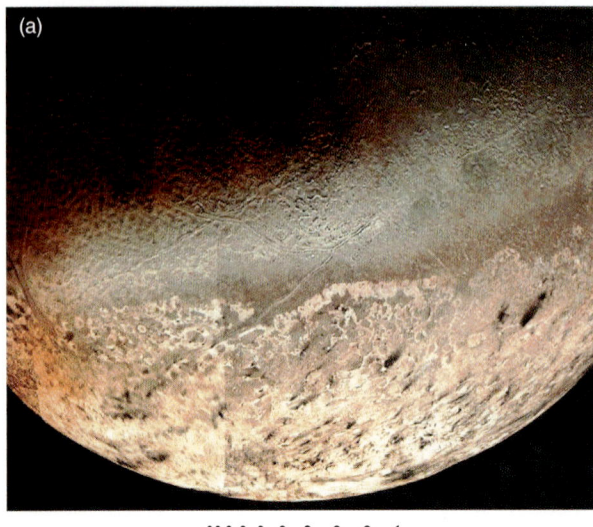

Figure 9.7 *Voyager 2* images of Triton show evidence of cryovolcanism. (a) This mosaic shows various terrains on the Neptune-facing hemisphere of Triton. "Cantaloupe terrain" is visible at the top; its lack of impact craters indicates that this area is geologically younger than the bright, cratered terrain at the bottom. (b) This irregular basin on Triton has been partly filled with frozen water, forming a relatively smooth ice surface. The state of New Jersey could just fit within the basin's boundary.

Triton has a thin atmosphere and a surface composed mostly of ices and frosts of methane and nitrogen at a temperature of about 38 K. A relative absence of craters tells us the surface is geologically young. Part of Triton is covered with terrain that looks like the skin of a cantaloupe (**Figure 9.7a**), with irregular pits and hills that may be caused by slushy ice emerging onto the surface from the interior. Vein-like features include grooves and ridges that could result from ice oozing out along fractures. The rest of Triton is covered with smooth volcanic plains. Irregularly shaped depressions as wide as 200 km (Figure 9.7b) formed when mixtures of water, methane, and nitrogen ice melted in the interior of Triton and erupted onto the surface, much as rocky magmas erupted onto the lunar surface and filled impact basins on Earth's Moon.

Clear nitrogen ice creates a localized greenhouse effect, in which solar energy trapped beneath the ice raises the temperature at the base of the ice layer. A temperature increase of only 4°C vaporizes the nitrogen ice. As this gas is formed, the expanding vapor exerts very high pressures beneath the ice cap. Eventually, the ice ruptures and vents the gas explosively into the low-density atmosphere. *Voyager 2* found four of these active cryovolcanoes on Triton. Each consisted of a plume of gas and dust as much as 1 km wide rising 8 km above the surface, where the plume was caught by upper atmospheric winds and carried for hundreds of kilometers downwind. Dark material, perhaps silicate dust or radiation-darkened methane ice grains, is carried along with the expanding vapor into the atmosphere, from which it subsequently settles to the surface and forms dark patches streaked out by local winds, as seen near the bottom of Figure 9.7a.

Possibly Active Moons

Jupiter's moon Europa is slightly smaller than our Moon but has an outer shell of water ice. Like Io, Europa experiences continually changing tidal stresses from Jupiter that generate heat and possibly volcanism. Regions of chaotic terrain, as shown in **Figure 9.8**, are places where the icy crust has been broken into slabs

Figure 9.8 A high-resolution *Galileo* image of Jupiter's moon Europa shows where the icy crust has been broken into slabs that, in turn, have been rafted into new positions. These areas of chaotic terrain are characteristic of a thin, brittle crust of ice floating atop a liquid or slushy ocean. The area shown in this image measures approximately 25 × 50 km.

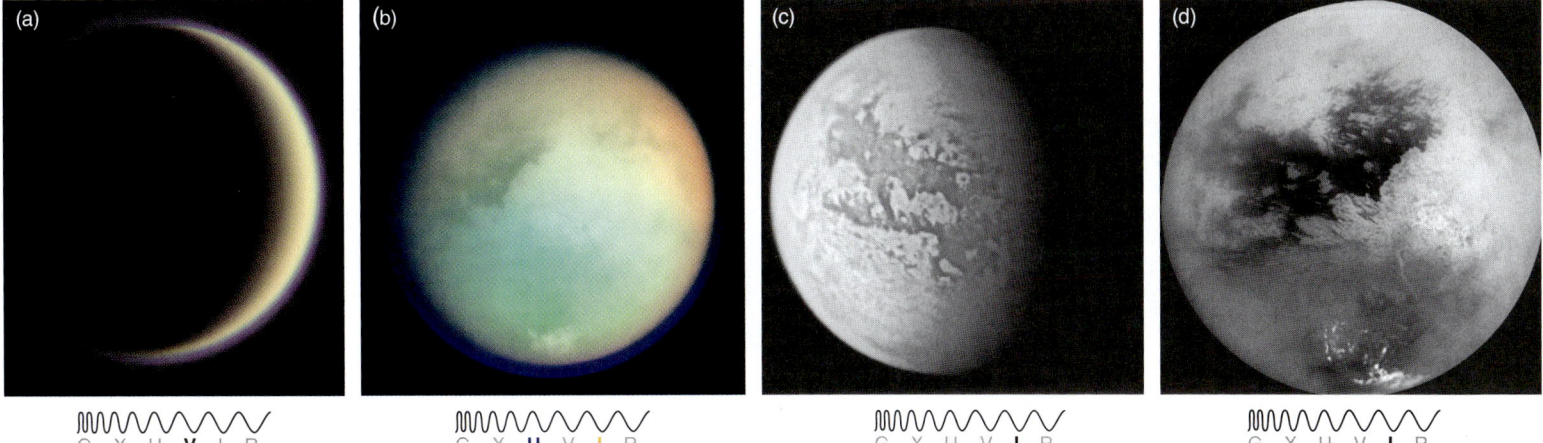

Figure 9.9 *Cassini* took images of Saturn's largest moon, Titan, in several regions of the electromagnetic spectrum. (a) Visible-light imaging shows Titan's orange atmosphere, which is caused by the presence of organic smog-like particles. (b) This infrared- and ultraviolet-light combined image shows surface features and a bluish atmospheric haze caused by scattering of ultraviolet sunlight by small atmospheric particles. Bright clouds in Titan's lower atmosphere are seen near the moon's south pole. (c) Infrared imaging penetrates Titan's smoggy atmosphere and reveals surface features. (d) This infrared image covers the same general area of Titan as seen in (b). The large dark area, called Xanadu, is about the size of the contiguous United States. Methane ice clouds are visible at the bottom of the image, near Titan's south pole.

that have shifted into new positions. In other areas the crust has split apart, and the gaps have filled in with new dark material rising from the interior.

When these features were formed, Europa's crust consisted of a thin, brittle shell overlying either liquid water or slushy ice. The lightly cratered and thus geologically young surface indicates that Europa may still possess a global ocean 100 km deep that contains more water than all of Earth's oceans. Such an ocean would be salty with dissolved minerals. This reasoning is supported by *Galileo*'s magnetometer data, which show that Europa's magnetic field is variable, indicating an internal electrically conducting fluid. Spectra from the Hubble Space Telescope further support this conclusion by identifying a water geyser erupting from the icy surface. Water erupting from the geyser falls back to the surface as frost after reaching an altitude of about 200 km above the surface. Europa's ocean may also contain an abundance of organic material. In fact, Europa is not so different from some places on Earth that support life. The conditions that create and support life—liquid water, heat, and organic material—could all be present in Europa's oceans, making Europa a high-priority target in the search for extraterrestrial life.

Saturn's largest moon, Titan, is bigger than Mercury and has a composition of about 45 percent water ice and 55 percent rocky material. What makes Titan especially remarkable is its thick atmosphere. Titan's mass and distance from the Sun have allowed it to retain an atmosphere that is 30 percent denser than that of Earth. Titan's atmosphere, like Earth's, is mostly nitrogen. As Titan differentiated, various ices, including methane (CH_4) and ammonia (NH_3), emerged from the interior to form an early atmosphere. Ultraviolet photons from the Sun have enough energy to break apart ammonia and methane molecules—a process called **photodissociation**. Photodissociation of ammonia is the likely source of Titan's atmospheric nitrogen. Methane breaks into fragments that recombine to form organic compounds including complex hydrocarbons such as ethane. These compounds tend to cluster in tiny particles, creating organic smog much like the air over Los Angeles on a bad day; this gives Titan's atmosphere its characteristic orange hue (**Figure 9.9a**).

The *Cassini* spacecraft, which began orbiting Saturn in 2004, gave astronomers the first close-up views of Titan's surface (Figure 9.9b). Haze-penetrating infrared imaging shows broad regions of dark and bright terrain (Figure 9.9c and d). Radar imaging shows irregularly shaped features in Titan's northern hemi-

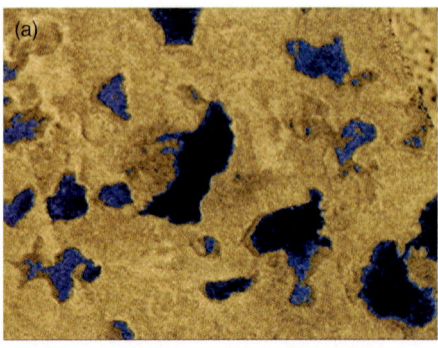

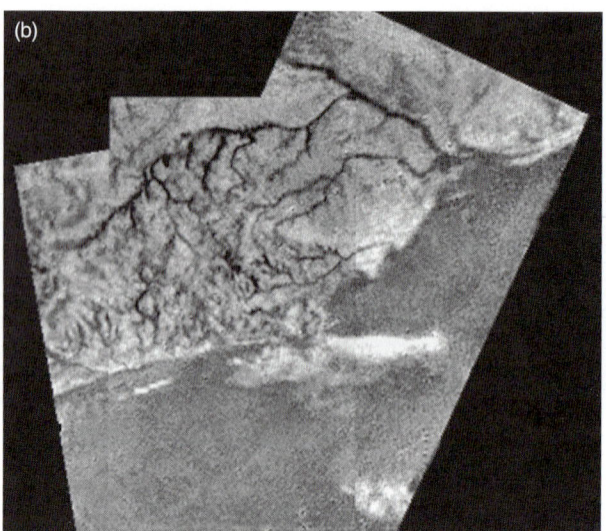

Figure 9.10 (a) Features commonly associated with terrestrial lakes, such as islands, bays, and inlets, are clearly visible in this false color radar image of Titan. (b) The surface of Titan viewed from the *Huygens* probe during its descent to the surface. The dark drainage patterns resemble river systems on Earth.

sphere that appear to be widespread lakes of methane, ethane, and other hydrocarbons (**Figure 9.10a**). Photodissociation should have destroyed all atmospheric methane within about 50 million years, so there must be some process for renewing the methane that is being destroyed by solar radiation. Heat supplied by radioactive decay could cause cryovolcanism that releases "new" methane from underground. To date, there is no direct evidence of active cryovolcanism on Titan, but the presence of abundant atmospheric methane and of methane lakes strongly suggests that Titan is indeed geologically active.

Titan has terrains reminiscent of those on Earth (Figure 9.10b), with networks of channels, ridges, hills, and flat areas that may be dry lake basins. The near absence of impact craters on the surface indicates recent erosion. These features suggest a cycle in which methane rain falls to the surface, washes the ridges free of dark hydrocarbons, and then collects in drainage systems that empty into low-lying liquid methane pools. An infrared camera photographed a reflection of the Sun from such a lake surface. The type of reflection observed proves that the lake contains a liquid and is not frozen or dry.

Cassini carried a probe, *Huygens*, that plunged through Titan's atmosphere measuring composition, temperature, pressure, and winds and taking pictures as it descended. *Huygens* confirmed the presence of nitrogen-bearing organic compounds in the clouds. These are key components in the production of proteins found in terrestrial life. Once on Titan's surface, *Huygens* took pictures and made physical and compositional measurements. The surface was wet with liquid methane, which evaporated as the probe—heated to 2000 K during its passage through the atmosphere—landed in the frigid soil. The surface was also rich with other organic compounds, such as cyanogen and ethane. As shown in **Figure 9.11**, the surface around the landing site is relatively flat and littered with rounded "rocks" of water ice. The dark "soil" is probably a mixture of water and hydrocarbon ices.

In many ways, Titan resembles a primordial Earth, albeit at much lower temperatures. The presence of organic compounds that could be biological precursors in the right environment makes Titan another high-priority target for continued exploration.

Formerly Active Moons

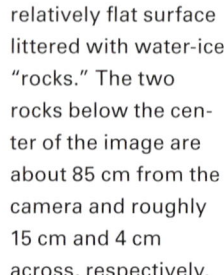

Figure 9.11 This view of the surface of Titan, obtained from *Huygens*, shows a relatively flat surface littered with water-ice "rocks." The two rocks below the center of the image are about 85 cm from the camera and roughly 15 cm and 4 cm across, respectively. The dark "soil" is probably made up of hydrocarbon ices.

Some moons show clear evidence of past ice volcanism and tectonic deformation but no current geological activity. For example, Jupiter's moon Ganymede is the largest moon in the Solar System, larger than the planet Mercury. The surface is composed of two prominent terrains: a dark, heavily cratered (and therefore ancient) terrain and a bright terrain characterized by ridges and grooves. The most extensive region of ancient dark terrain includes a semicircular area about the size of the contiguous United States on the leading hemisphere. Parallel ripples occurring in many dark areas are among Ganymede's oldest surface features. They may represent surface deformation from internal processes or they may be relics of impact cratering.

Impact craters on Ganymede range up to hundreds of kilometers in diameter, and the larger craters are proportionately shallower. The icy crater rims slowly slump, like a lump of soft clay. They are seen as bright, flat, circular patches found principally in the moon's dark terrain (**Figure 9.12**) and are thought to be scars left by early impacts onto a thin, icy crust overlying water or slush (**Figure 9.13**).

In Chapter 6, we discussed how planetary surfaces can be fractured by faults or folded by compression resulting from movements initiated in the mantle. On Ganymede, fracturing and faulting may have completely deformed the icy crust, destroying all signs of older features such as impact craters and creating the bright terrain. The energy that powered Ganymede's early activity was liberated during a period of differentiation when the moon was very young. Once the differentiation process was complete, that source of internal energy ran out and geological activity ceased.

Ganymede is not the only formerly active moon. Several other large moons in the outer Solar System have also been active in the past, including Saturn's moons Tethys and Dione and Uranus's moons Miranda and Ariel.

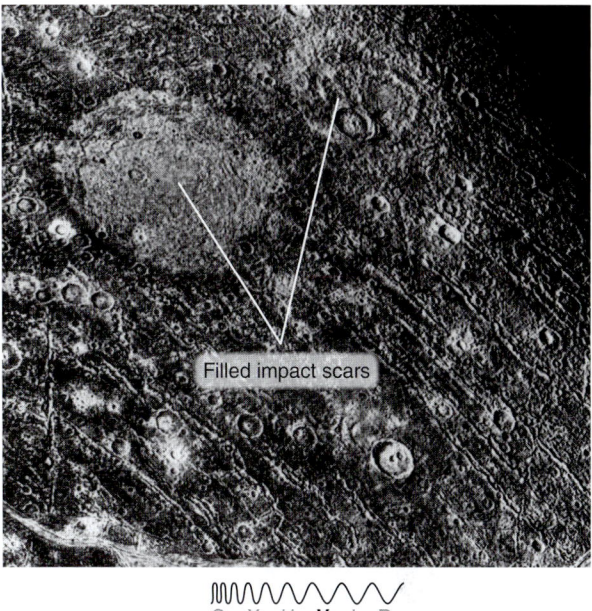

Figure 9.12 This *Voyager* image shows filled impact scars on Jupiter's moon Ganymede.

9.3 Asteroids Are Pieces of the Past

Recall from Chapter 5 that an asteroid is a primitive planetesimal that did not become part of the accretion process that formed planets. They are typically irregular in shape, somewhat like potatoes. The planetesimals that formed our Solar System's planets and moons have been so severely modified by planetary processes that nearly all information about their original physical condition and chemical composition has been lost. By contrast, asteroids constitute an ancient and far more pristine record of what the early Solar System was like. Asteroids continue to collide with one another today and produce small fragments of rock and metal, some of which crash to Earth's surface as meteorites. If you visit a planetarium or science museum, you may find on display a meteorite that you can touch. The meteorite may be older than Earth itself and might even contain tiny grains of material that predate the formation of our Solar System.

Asteroid Groups

Asteroids are found throughout the Solar System. However, most are main-belt asteroids, located in the **main asteroid belt** between the orbits of Mars and Jupiter. There are several other groups of asteroids, which are divided according to their orbital characteristics. Trojan asteroids (Trojans) share Jupiter's orbit and are held in place by interactions with Jupiter's gravitational field. Three other groups are defined by their relationship to the orbits of Earth and Mars: Apollos cross the orbits of Earth and Mars, Atens cross Earth's orbit but not that of Mars, and Amors cross the orbit of Mars but not that of Earth

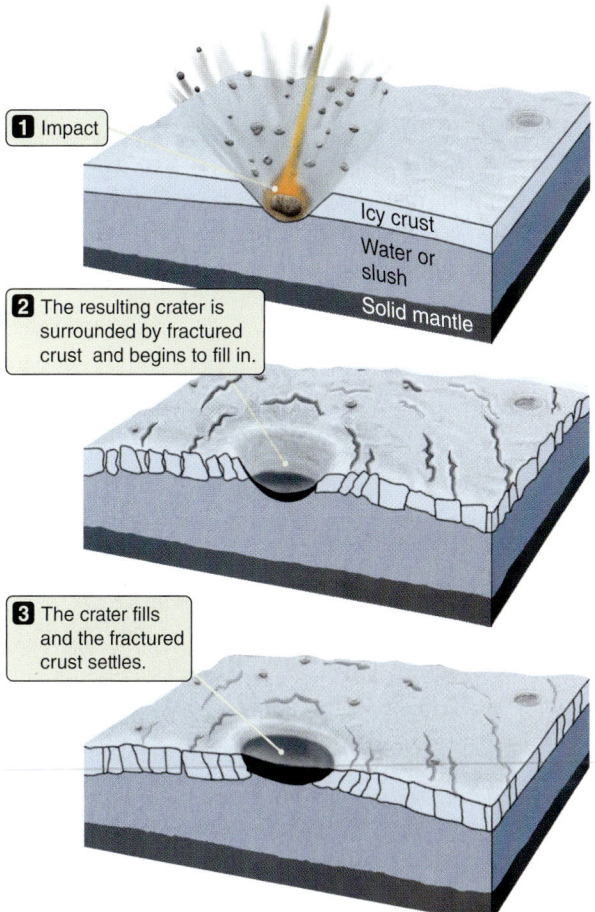

Figure 9.13 Filled impact scars form as viscous flow smooths out structures left by impacts on icy surfaces.

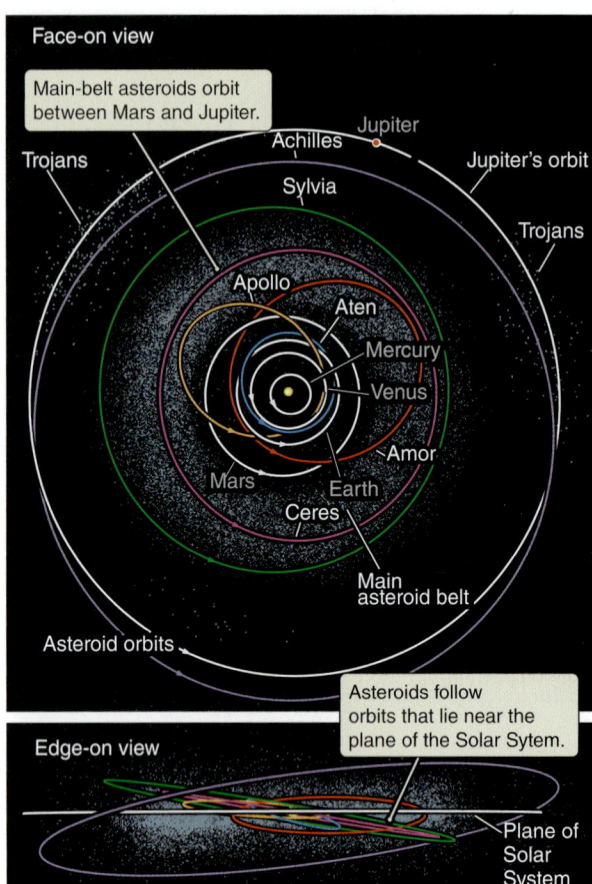

Figure 9.14 Blue dots show the locations of known asteroids at a single point in time. The orbits of Aten, Amor, and Apollo (prototype members of some groups of asteroids) are shown.

(**Figure 9.14**). All three of these groups are named for a prototype that is representative of the group.

Asteroids whose orbits bring them within 1.3 AU of the Sun are called **near-Earth asteroids**. These asteroids, along with a few comet nuclei, are known collectively as **near-Earth objects** (**NEOs**) and occasionally collide with Earth or the Moon. Astronomers estimate that between 500 and 1,000 NEOs have diameters larger than 1 km. Collisions with these large objects are geologically important and have dramatically altered life on Earth, as explained in Chapter 6.

Look back at the chapter-opening illustration, in which a student has identified an asteroid. Study its motion through the star field. Because asteroids are in the Solar System, many of them move quickly enough across the sky that their motion is noticeable over a few hours. Detecting an asteroid, particularly if you weren't looking for it, is exciting and often inspires amateur astronomers to continue their observations. The work of amateur astronomers has added important information to our body of knowledge about asteroids, including orbital parameters, rotation periods, and colors.

Asteroid Composition

Most asteroids are relics of rocky or metallic planetesimals that formed in the region between Mars and Jupiter. Although early collisions between these planetesimals created several bodies large enough to differentiate, Jupiter's tidal disruption prevented them from forming a single planet.

There are many more small asteroids than large ones—indeed, most are too small for their self-gravity to have pulled them into a spherical shape. Some are very elongated, suggesting they are either fragments of larger bodies or the result of haphazard collisions between smaller bodies. Asteroids account for only a tiny fraction of the mass in the Solar System. If all of the asteroids were combined into a single body, it would be only about one-third the mass of Earth's Moon.

Astronomers have measured the masses of some asteroids by noting the effect of their gravity on spacecraft passing nearby. Knowing the mass and the size of asteroids enables us to determine their densities, which range between 1.3 and 3.5 times the density of water. Those at the lower end of this range are "rubble piles" with large gaps between the fragments, and they are considerably less dense than the meteorite fragments they create. This is what we would expect of objects that were assembled from smaller objects and then suffered a history of violent collisions.

Visits to Asteroids

In 1991, the *Galileo* spacecraft passed by the asteroid Gaspra and obtained images of its surface. Gaspra is cratered and irregular in shape, about 9 × 10 × 20 km in size. Faint, groovelike patterns may be fractures from the impact that chipped Gaspra from a larger planetesimal. Distinctive colors imply that Gaspra is covered with a variety of rock types.

Later in its mission, *Galileo* returned to the main belt, passing close to the asteroid Ida (**Figure 9.15**). *Galileo* flew so close to Ida that its cameras could see details as small as 10 meters across—about the size of a small house. Ida is 54 km long, ranging from 15 to 24 km in diameter. Ida's craters indicate that the surface is about a billion years old, at least twice the age estimated for Gaspra. Like Gaspra, Ida is fractured. The fractures indicate that these asteroids must be made of

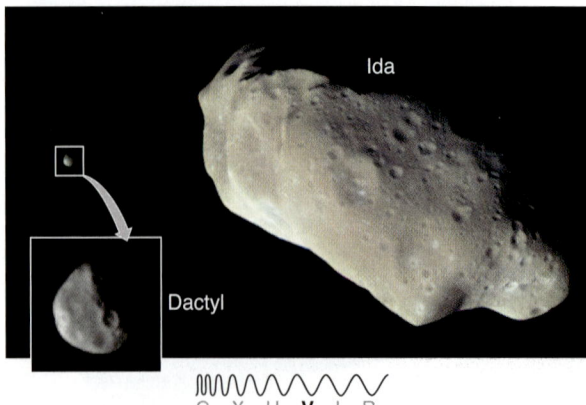

Figure 9.15 The *Galileo* spacecraft imaged the asteroid Ida with its tiny moon, Dactyl (also shown enlarged in the inset).

relatively solid rock. (You can't "crack" a loose pile of rubble.) This supports the idea that some asteroids are chips from larger, solid objects.

The *Galileo* images also revealed a tiny moon orbiting the asteroid. Ida's moon, called Dactyl, is only 1.4 km across and is cratered from impacts. Moons have now been found around nearly 200 asteroids, and at least five asteroids are known to have two moons. At least one asteroid has a ring around it.

In early 2000, the *NEAR Shoemaker* spacecraft entered orbit around the asteroid Eros to begin long-term observations (**Figure 9.16**). Eros is one of more than 5,000 known objects whose orbits bring them within 1.3 AU of the Sun. Eros is roughly a cylinder 34 km long and 11 km in diameter. Like Gaspra and Ida, Eros shows a surface with grooves, rubble, and impact craters, including a crater 8.5 km across. The scarcity of smaller craters suggests that its surface is younger than Ida's. After a year of observing, the spacecraft landed on the asteroid's surface. Chemical analyses confirmed that the composition of Eros is like that of primitive meteorites.

On its way to Eros, *NEAR Shoemaker* flew past the asteroid Mathilde. Mathilde is about 50 km in size and has a surface about as reflective as charcoal. The overall density of Mathilde is about 1,300 kilograms per cubic meter (kg/m^3; this is 1.3 times the density of water), which implies that Mathilde is a rubble pile, composed of chunks of rocky material with open spaces between them. Mathilde is covered with craters, the largest being more than 33 km across. The great number of craters suggests that Mathilde likely dates back to the very early history of the Solar System.

In November 2005, the Japanese spacecraft *Hayabusa* made contact with asteroid Itokawa. It collected samples and then returned to Earth in 2010. Later that year, the Japan Aerospace Exploration Agency confirmed that the *Hayabusa* spacecraft had completed the first sample-return mission from an asteroid—it had brought back particles from Itokawa. In addition to the significant engineering accomplishment, this mission was important because the dust was a pristine sample from a specific, well-characterized asteroid. Usually, scientists cannot match a meteorite sample to a specific asteroid.

Vesta provides a counterexample to this general rule. In 2011, NASA's *Dawn* spacecraft visited Vesta and found that it is a leftover intact protoplanet that formed within the first 2 million years of the condensation of the first solid bodies in the Solar System. It has an iron core and is differentiated, so it is more like the planets than like other asteroids. Vesta's spectrum matches the reflection spectrum of a peculiar group of meteorites. A collision that created one of the two large impact basins in the south polar region of Vesta blasted material into space that then landed on Earth as these meteorites. This transfer of material from one object to another is a useful way to gather detailed information about objects in the Solar System.

The spectra of the tiny moons of Mars, Deimos and Phobos, and their appearance (**Figure 9.17**) are similar to those of some asteroids. Many scientists think these moons may be asteroids that have been captured by the gravity of Mars. But are Deimos and Phobos really asteroids? Controversy about this question has never been put to rest. Some scientists argue that it is unlikely Mars could have captured two asteroids, proposing that Deimos and Phobos must somehow have evolved together with Mars. Another possibility is that Mars and its moons were all parts of a much larger body that was fragmented by a collision early in the planet's history. A visit to these moons is required to sort out these possibilities.

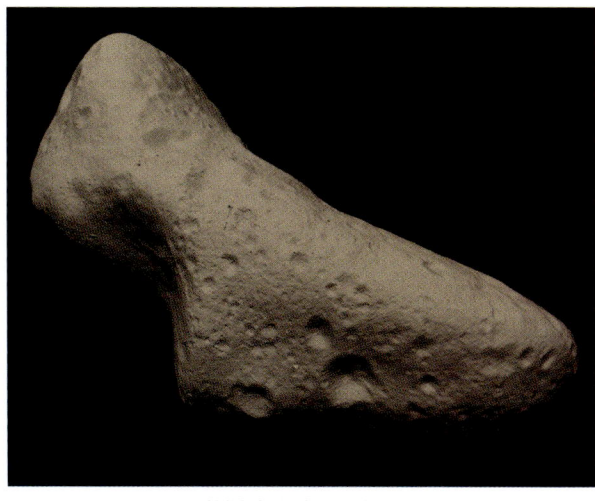

Figure 9.16 This image of the asteroid Eros was produced via high-resolution scanning of the asteroid's surface by the *NEAR Shoemaker* spacecraft's laser range finder. *NEAR Shoemaker* became the first spacecraft to land on an asteroid when it was gently crashed onto the surface of Eros.

Figure 9.17 *Mars Reconnaissance Orbiter* and *Mars Express* imaged the two tiny moons of Mars: (a) Deimos and (b) Phobos.

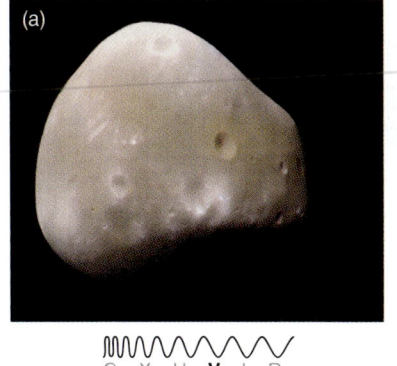
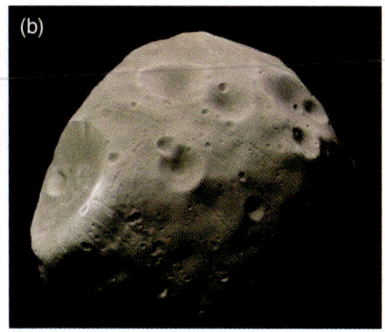

9.4 Comets Are Clumps of Ice

Comets are icy planetesimals that formed from primordial material. They spend most of their lives adrift in the frigid outer reaches of the Solar System. Most are much too small and far away to be seen, so no one really knows how many there are. Estimates for our Solar System range as high as a trillion (10^{12}) comets—more than all the stars in the Milky Way Galaxy. They put on a show only when they come deep enough into the inner Solar System to suffer destructive heating from the Sun. When they are close enough to show the effects of solar heating, they are called **active comets**.

The Homes of the Comets

We know where comets come from by observing their orbits as they pass through the inner Solar System. Comets fall into two distinct groups named for scientists Gerard Kuiper (1905–1973) and Jan Oort (1900–1992).

The Kuiper Belt is a disk-shaped population of comet nuclei that begins at about 30 AU from the Sun, near the orbit of Neptune, and extends outward to about 55 AU (**Figure 9.18**). Comets from the Kuiper Belt orbit the Sun in a disk-shaped region aligned with the Solar System. The innermost part of the Kuiper Belt (between about 40 and 48 AU) contains tens of thousands of icy planetesimals known as **Kuiper Belt objects** (**KBOs**). The largest KBOs are similar in size to Pluto. With a few exceptions, the sizes of KBOs are difficult to determine. Although brightness and approximate distance are known, their albedos are uncertain. Reasonable limits for the albedos can set maximum and minimum values for their size. Some KBOs have moons, and at least one has three moons. More than 1,000 KBOs have been discovered, and astronomers suspect many smaller ones also exist in the region. We know very little of the chemical and physical properties of KBOs because of their great distance. This lack of information may change soon, however: following its encounter with Pluto in 2015, the *New Horizons* spacecraft will continue outward to the Kuiper Belt, where it will attempt to fly close to one or more KBOs.

The **Oort Cloud** is a spherical distribution of planetesimals that are too distant to be seen by even the most powerful telescopes. We know the size and shape of the Oort Cloud from the orbits of this region's comets, which approach the Sun from all directions and from as far as 100,000 AU away—nearly halfway to the nearest stars.

The Kuiper Belt and Oort Cloud are both enormous reservoirs of icy planetesimals that now and then fall into the inner Solar System. Gravitational perturbations by objects beyond the Solar System may kick an Oort Cloud planetesimal in toward the Sun. About every 5 million to 10 million years, a star passes within about 100,000 AU of the Sun, perturbing the orbits of these planetesimals. The gravitational attraction from huge clouds of dense interstellar gas can also stir up the Oort Cloud.

Unlike the icy planetesimals in the Oort Cloud, those in the Kuiper Belt are packed closely enough to interact gravitationally from time to time. In such events, one object gains energy while the other loses it. The "winner" may gain enough energy to be sent into an orbit that reaches far beyond the boundary of the Kuiper Belt. The "loser" falls inward toward the Sun.

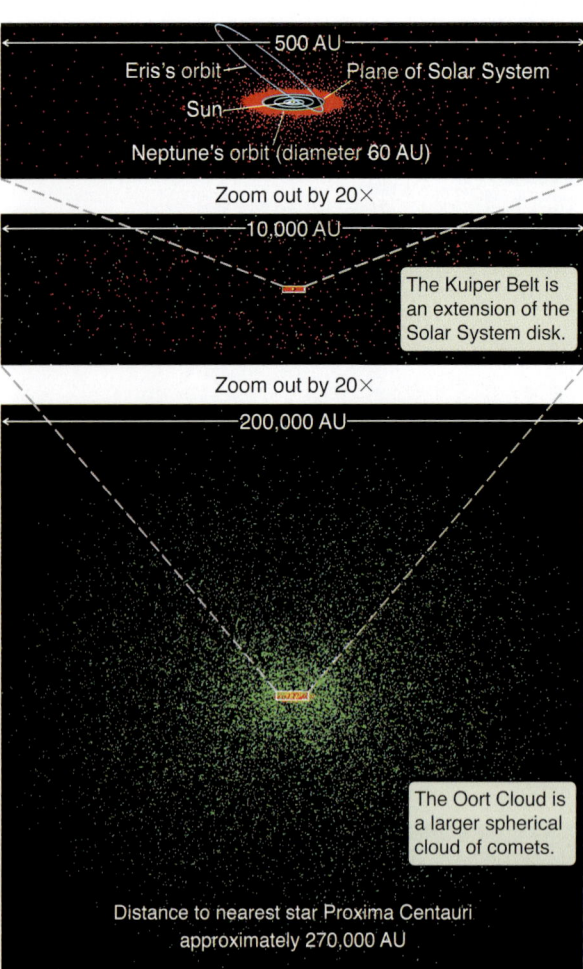

Figure 9.18 Most comets near the inner Solar System (shown in red) populate an extension to the disk of the Solar System called the Kuiper Belt. The spherical Oort Cloud is far larger and contains many more comets (shown in green).

The Orbits of Comets

The lifetime of a comet nucleus depends on how frequently it passes by the Sun and how close it comes. There are about 400 known **short-period comets**, which by definition have periods less than 200 years. Additionally, each year astronomers discover about six new **long-period comets**, whose orbital periods are longer than 200 years. The total number of long-period comets observed to date is about 3,000.

Figure 9.19 shows the orbits of a number of comets. Long-period comets were scattered to the outer solar system by gravitational interactions, so their orbits are random. They come into the inner Solar System from all directions, some orbiting the Sun in the same direction that the planets orbit (prograde) and some orbiting in the opposite direction (retrograde). Short-period comets tend to be prograde and to have orbits in the ecliptic plane, and they frequently pass close enough to a planet for its gravity to change the comet's orbit around the Sun. Short-period comets presumably originated in the Kuiper Belt, but as they fell in toward the Sun, they were forced into their current short-period orbits by gravitational encounters with planets.

More than 600 long-period comets have well-determined orbits. Some have orbital periods of hundreds of thousands or even millions of years. Almost all their time is spent in the Oort Cloud in the frigid, outermost regions of the Solar System. Because of their very long orbital periods, these comets have made at most one appearance throughout the course of recorded history.

▶❚❚ **AstroTour: Cometary Orbits**

Figure 9.19 Orbits of a number of comets (colored lines) are shown in face-on and edge-on views of the Solar System. Populations of (a) short-period comets and (b) long-period comets have very different orbital properties. Comet Halley, which appears in both diagrams for comparison, is a short-period comet.

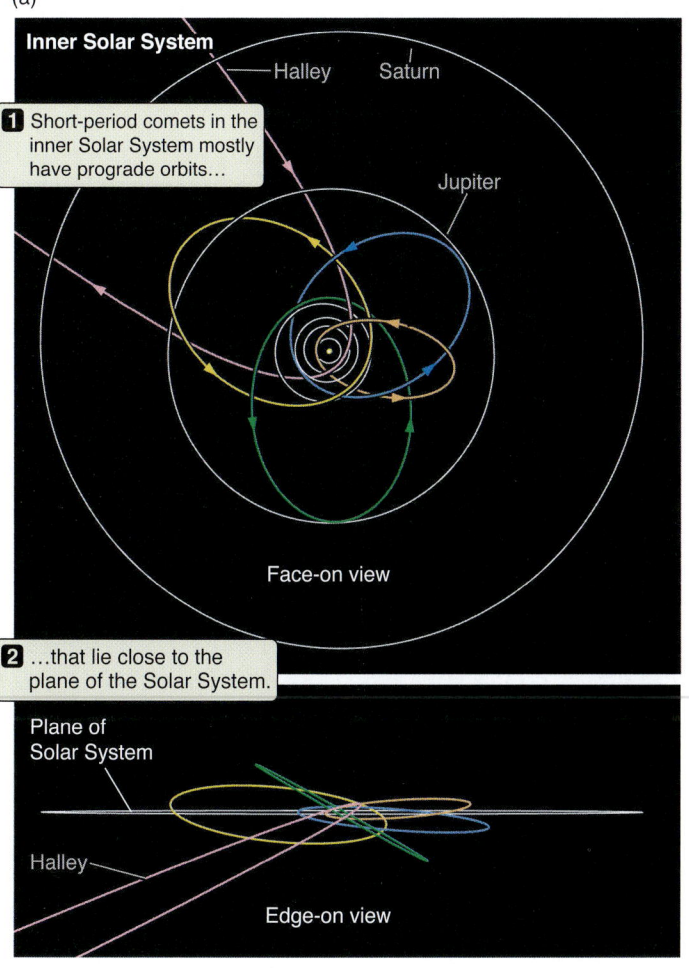

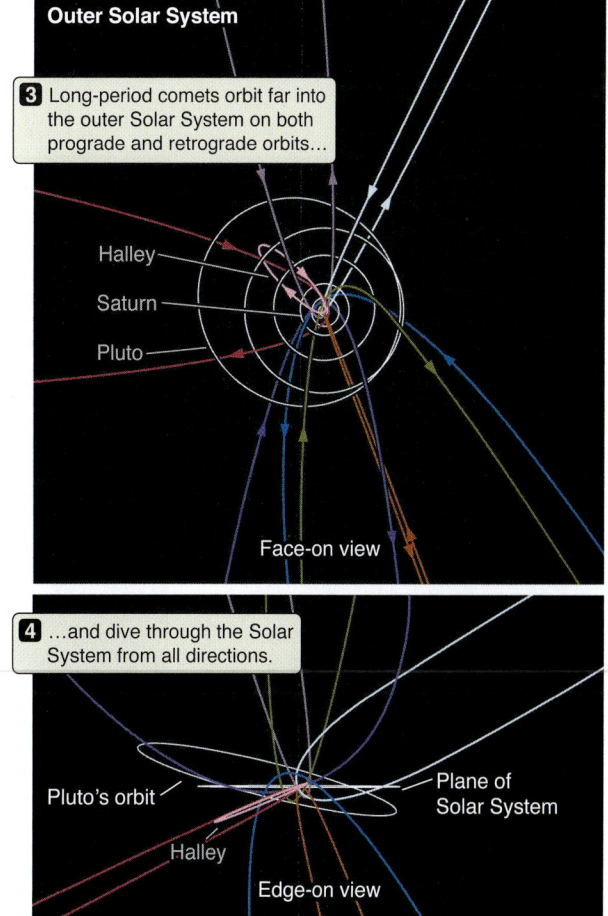

Anatomy of an Active Comet

The small object at the center of the comet—the icy planetesimal itself—is the comet **nucleus**. This is by far the smallest component of a comet, but it is the source of all the material that we see stretched across the sky as a comet nears the Sun (**Figure 9.20**). Comet nuclei range in size from a few dozen meters to several hundred kilometers across. These "dirty snowballs" are composed of ice, organic compounds, and dust grains.

As a comet nucleus nears the Sun, sunlight heats its surface, vaporizing ices that stream away from the nucleus, and these gases carry dust particles along with them. The gases and dust driven from the nucleus of an active comet form a nearly spherical atmospheric cloud around the nucleus called the **coma**. The nucleus and the inner part of the coma are sometimes referred to collectively as the comet's **head**. Pointing from the head of the comet in a direction more or less away from the Sun are long streamers of dust, gas, and ions called the **tails**.

The tails are the largest and most spectacular part of a comet. Active comets have two different types of tails, as shown in Figure 9.20. One is the **ion tail**. Many of the atoms and molecules that make up a comet's coma are ions. Because they are electrically charged, ions in the coma feel the effect of the solar wind, the stream of charged particles that blows continually away from the Sun. The solar wind pushes on these ions, rapidly accelerating them to speeds of more than 100 kilometers per second (km/s)—far greater than the orbital velocity of the comet itself—and sweeps them out into a long, wispy structure. Because the particles that make up the ion tail are so quickly picked up by the solar wind, ion tails are usually very straight and point from the head of the comet directly away from the Sun.

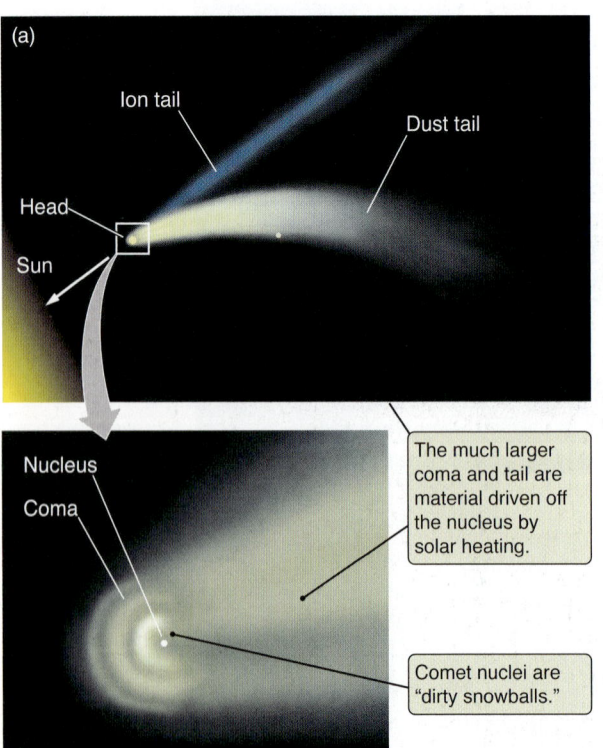

Figure 9.20 (a) The principal components of a fully developed active comet are the nucleus, the coma, and two types of tails called the dust tail and the ion tail. Together, the nucleus and the coma are called the head. (b) Comet Hale-Bopp, the great comet of 1997. The ion tail is blue in this image, and the dust tail is white.

Dust particles also have a net electric charge and feel the force of the solar wind. In addition, sunlight itself exerts a force on cometary dust. But dust particles are much more massive than individual ions, so they are accelerated more gently and do not reach such high relative speeds as the ions do. As a result, the **dust tail** often curves gently away from the head of the comet as the dust particles are gradually pushed from the comet's orbit in the direction away from the Sun.

Figure 9.21 shows the tails of a comet at various points in its orbit. Remember that both types of tails always point *away from the Sun*, regardless of which direction the comet is moving. As the comet approaches the Sun, its two tails trail behind its nucleus. But the tails extend *ahead* of the nucleus as the comet moves outward from the Sun.

Tails vary greatly from one comet to another. Some comets display both types of tails simultaneously; others, for reasons that we do not understand, produce no tails at all. A tail often forms as a comet crosses the orbit of Mars, where the increase in solar heating drives gas and dust away from the nucleus.

The gas in a comet's tail is even more tenuous than the gas in its coma; its density is no more than a few hundred molecules per cubic centimeter. This is much, much less than the density of Earth's atmosphere, which at sea level contains more than 10^{19} molecules per cubic centimeter. Dust particles in the tail are typically about 1 micron (μm) in diameter, roughly the size of smoke particles.

Most naked-eye comets develop first a coma and then an extended tail as they approach the inner Solar System. Comet McNaught in 2007 was such a comet (**Figure 9.22a**). But there are exceptions. Comet Holmes was a very faint telescopic object when it reached its closest point to the Sun just beyond the orbit of Mars. Then, several months later, as it was well on its way outward toward Jupiter's orbit, the comet suddenly became a half-million times brighter in just 42 hours. Comet Holmes became a bright naked-eye comet that graced Northern Hemisphere skies for several months (Figure 9.22b). Astronomers remain puzzled over the cause of this dramatic eruption. Explanations range from a meteoroid impact to a sudden (but unexplained) buildup of subsurface gas.

Generally, long-period comets provide a more spectacular display than short-period comets. The nuclei of short-period comets are eroded from repeated heating by the Sun, and as the volatile ices are driven away, some of the dust and organic compounds are left behind on the surface of the nucleus. The buildup of

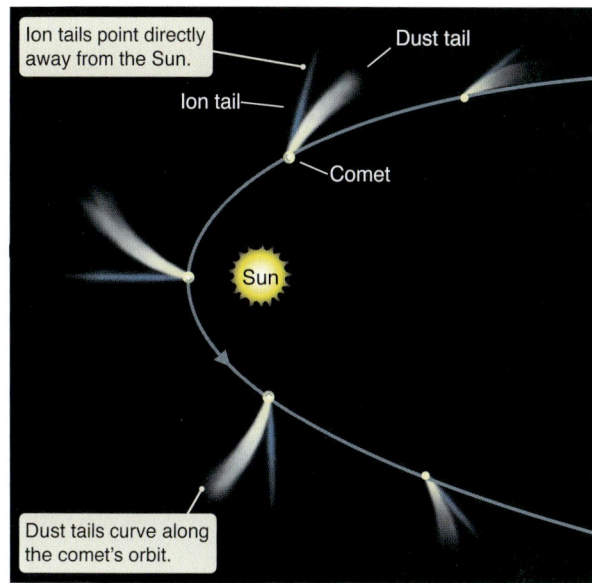

Figure 9.21 The orientation of the dust and ion tails changes as the comet orbits. The ion tail points directly away from the Sun while the dust tail curves along the comet's orbit.

Figure 9.22 (a) Comet McNaught, in 2007, was the brightest comet to appear in decades, but its true splendor was visible only to observers in the Southern Hemisphere. (b) Months after its closest approach to the Sun in 2007, Comet Holmes suddenly became a half-million times brighter within hours, turning into a naked-eye comet with an angular diameter larger than that of the full Moon. Comet Holmes favored Northern Hemisphere observers.

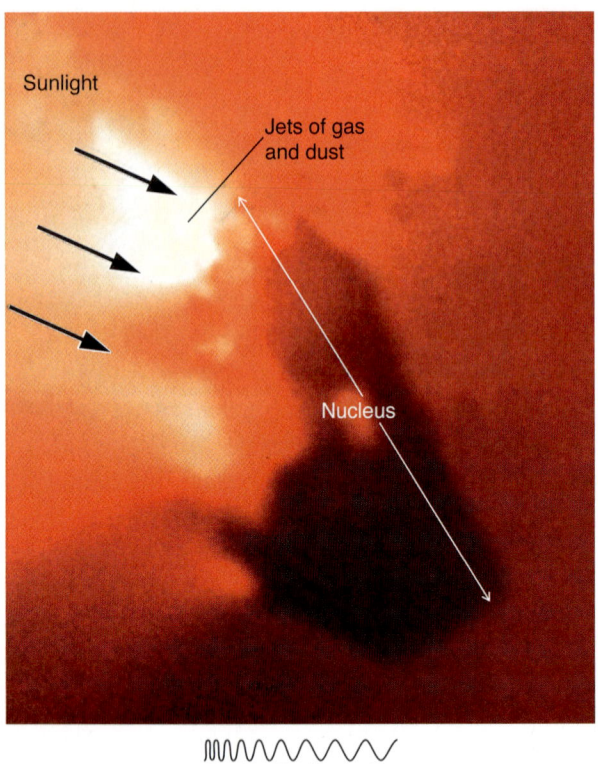

Figure 9.23 The nucleus of Comet Halley was imaged by the *Giotto* spacecraft in 1986. Here it is shown in false color to emphasize details.

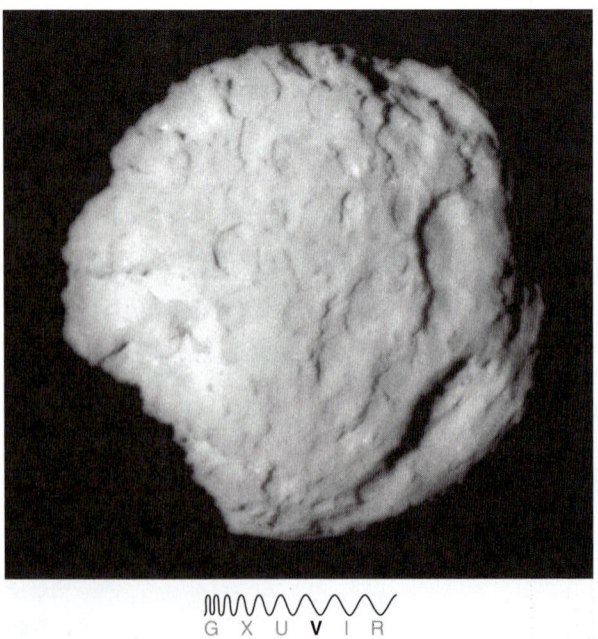

Figure 9.24 The nucleus of Comet Wild 2 was imaged by the *Stardust* spacecraft, which also sampled its tail.

this covering slows down cometary activity and makes their solar approaches less spectacular. In contrast, long-period comets are relatively pristine. More of their supply of volatile ices remains close to the surface of the nucleus, and they can produce a truly magnificent show. However, most pass too far from Earth or the Sun and never become bright enough to attract much public attention.

When will the next bright comet come along? On average, a spectacular comet appears about once per decade, but it is all a matter of chance. It might be many years from now—or it could happen tomorrow.

Visits to Comets

Comets provide an engineering challenge for spacecraft designers. We seldom have enough advance knowledge of a comet's visit or its orbit to mount a successful mission to intercept it. The closing speed between an Earth-launched spacecraft and a comet can be extremely high. Observations must be made very quickly, and there is a danger of high-speed collisions with debris from the nucleus. Still, nearly a dozen spacecraft have been sent to rendezvous with comets, including a five-spacecraft armada sent to Comet Halley by the Soviet, European, and Japanese space agencies in 1986. Much of what we know about comet nuclei and the innermost parts of the coma comes from data sent back by these missions.

The Soviet *Vega 1* and *Vega 2* and the European *Giotto* spacecraft entered the coma of Comet Halley when they were still nearly 300,000 km from its nucleus. We learned that the dust from Comet Halley was a mixture of light organic substances and heavier rocky material, and the gas was about 80 percent water and 10 percent carbon monoxide with smaller amounts of other organic molecules. The surface of the comet's nucleus is among the darkest-known objects in the Solar System, which means that it is rich in complex organic matter that must have been present as dust in the disk around the young Sun—perhaps even in the interstellar cloud from which the Solar System formed. As the three spacecraft passed close by the nucleus (**Figure 9.23**), they observed jets of gas and dust moving away from its surface at speeds of up to 1 km/s, far above the escape velocity. Material was coming from several small fissures that covered only about a tenth of the surface.

In 2001, NASA's *Deep Space 1* spacecraft flew within 2,200 km of Comet Borrelly's nucleus. Its tar-black surface is also among the darkest seen on any Solar System object and showed no evidence of water ice.

In 2004, NASA's *Stardust* spacecraft flew within 235 km of the nucleus of Comet Wild 2 (pronounced "vilt 2"). This comet is a newcomer to the inner Solar System. A close encounter with Jupiter in 1974 perturbed its orbit, bringing this relatively pristine body in from its orbit between Jupiter and Uranus. The nearly spherical nucleus of Wild 2 is about 5 km across (**Figure 9.24**). At least 10 gas jets were active, some of which carried large chunks of surface material. (A few particles as large as a bullet penetrated the outer layer of the spacecraft's protective shield.) The surface of Wild 2 is covered with features that may be impact craters modified by ice sublimation, small landslides, and erosion by jetting gas. Some show flat floors, suggesting a relatively solid interior beneath a porous surface layer.

In 2005, NASA's *Deep Impact* spacecraft launched a 370-kg impacting projectile into the nucleus of Comet Tempel 1 at a speed of more than 10 km/s. The

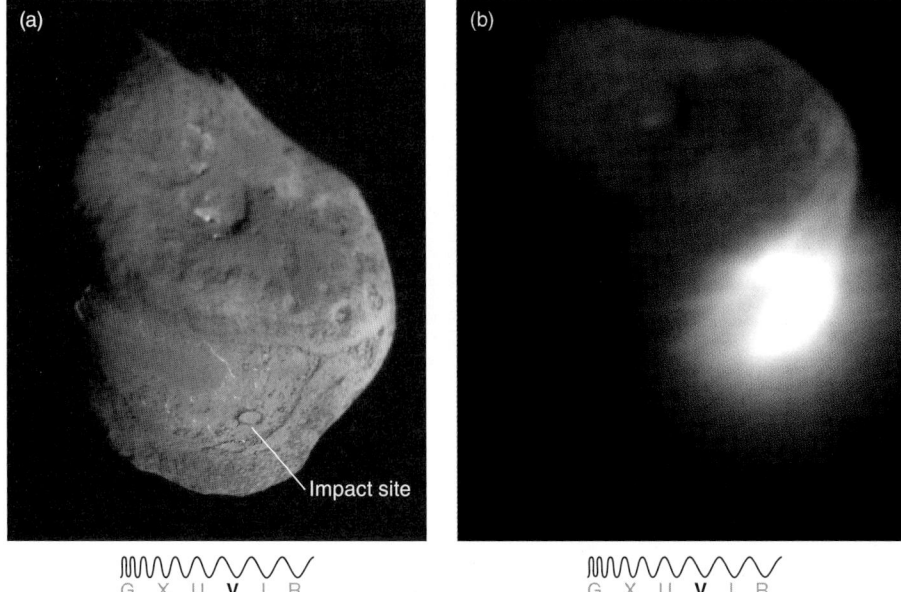

Figure 9.25 (a) The surface of the nucleus of Comet Tempel 1 was imaged just before impact by the *Deep Impact* projectile. The impact occurred between the two 370-meter-diameter craters located near the bottom of the image. The smallest features appearing in this image are about 5 meters across. (b) Sixteen seconds after the impactor struck the comet, the parent spacecraft took this image of the ejected gas and dust.

impact sent 10,000 tons of water and dust flying off into space at speeds of 50 m/s (**Figure 9.25**). A camera mounted on the projectile snapped photos of its target until it was vaporized by the impact. Observations of the event were made both locally by *Deep Impact* and back on Earth by a multitude of orbiting and ground-based telescopes.

Water, carbon dioxide, hydrogen cyanide, iron-bearing minerals, and a host of complex organic molecules were identified. The comet's outer layer is composed of fine dust with a consistency of talcum powder. Beneath the dust are layers made up of water ice and organic materials. One surprise for scientists was the presence of well-formed impact craters, which had been absent in close-up images of Comets Borrelly and Wild 2. Why some comet nuclei have fresh impact craters and others none remains a question.

In late 2010, the *EPOXI* spacecraft flew past Comet Hartley 2 (**Figure 9.26**), imaging not only jets of dust and gas, indicating a remarkably active surface, but also an unusual separation of rough and smooth areas that have very different natures. The "waist"—the narrow part at the middle—is a smooth inactive area where ejected material has fallen back onto the cometary nucleus. Carbon dioxide jets shoot out from the rough areas. It is unclear whether this unusual shape is a result of how Comet Hartley 2 formed 4.5 billion years ago or is due to more recent evolution of the comet. Further observations by the Herschel Space Observatory showed that the water on this comet has the same ratio of hydrogen isotopes as the water in Earth's oceans. This is strong evidence that some of Earth's water came from the Kuiper Belt. Comets from the Oort Cloud have a different ratio, and so they have been ruled out as the source of Earth's water.

Figure 9.26 In this image of Comet Hartley 2 taken by the *EPOXI* spacecraft, we see two distinct surface types.

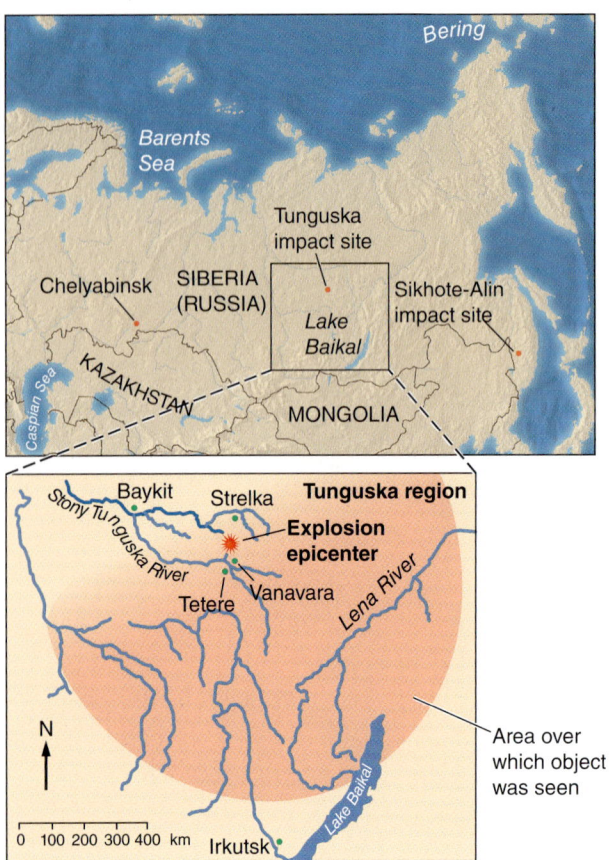

Figure 9.27 A large region of forest near the Tunguska River in Siberia was flattened in 1908 by an atmospheric explosion caused by the impact of a small asteroid or comet.

9.5 Comet Collisions Still Happen Today

Almost all hard-surfaced objects in the Solar System still bear the scars of a time when tremendous impact events were common. Although such impacts are far less frequent today than they once were, they still happen.

The Tunguska River flows through a remote region of western Siberia. In summer 1908, the region was blasted with the energy equivalent of 2,000 times the atomic bomb dropped over Hiroshima. **Figure 9.27** shows a map of the region, along with a photograph of the devastation caused by the blast. Eyewitness accounts detailed the destruction of dwellings, the incineration of reindeer (including one herd of 700), and the deaths of at least five people. Although trees were burned or flattened for more than 2,150 square kilometers—an area greater than metropolitan New York City—no crater was left behind! The Tunguska event was the result of a tremendous high-altitude explosion that occurred when a small body hit Earth's atmosphere, ripped apart, and formed a fireball before reaching Earth's surface. Recent expeditions to the Tunguska area have recovered resin from the trees blasted by the event. Chemical traces in the resin suggest that the impacting object may have been a stony asteroid.

On February 12, 1947, yet another planetesimal struck Earth, this time in the Sikhote-Alin region of eastern Siberia. Composed mostly of iron, the object had an estimated diameter of about 100 meters and broke into a number of fragments before hitting the ground, leaving a cluster of craters and widespread devastation. Witnesses reported a fireball brighter than the Sun and sound that was heard 300 km away.

In February 2013, a known near-Earth object about half the size of an American football field passed so close to Earth that it came within the orbit of man-made satellites. This near miss was uneventful, and the object simply continued on its way. However, in an unrelated event on the same day, a previously unknown

Figure 9.28 (a) In February 2013, a meteoroid entered the atmosphere over Russia, creating a fireball that eyewitnesses said was brighter than the Sun. (b) The shock wave from the meteor damaged buildings, and more than 1,000 people were injured by flying glass and other debris.

meteoroid estimated to have a radius of about 10 meters (**Working It Out 9.1**) exploded over Chelyabinsk, Russia (**Figure 9.28a**), damaging thousands of buildings in six cities (Figure 9.28b) and injuring more than 1,000 people. Several rock fragments of meteoritic composition have been recovered, and detailed study of this object and its origins is under way.

These impacts are sobering events. It is highly improbable that a populated area on Earth will experience a collision with a large asteroid within our lifetimes. Comets and smaller asteroids, however, are less predictable. There may be as many as 10 million asteroids larger than a kilometer across, but only about 130,000 have well-known orbits, and most of the unknowns are too small to see until they come very close to Earth. The U.S. government—along with the governments of several other nations—is aware of the risk posed by NEOs. Although the probability of a collision between a small asteroid and Earth is quite small, the consequences could be catastrophic, so NASA has been given a congressional mandate to catalog all NEOs and to scan the skies for those that remain undiscovered.

Comets present a more serious problem. Half a dozen unknown long-period comets enter the inner Solar System each year. If one happens to be on a collision course with Earth, we might not notice it until just a few weeks or months before impact. Although this has become a favorite theme of science-fiction disaster stories, Earth's geological and historical record suggests that impacts by large bodies are infrequent events.

Collisions also still occur on other planets. Comet Shoemaker-Levy 9 passed so close to Jupiter in 1992 that tidal stresses broke it into two-dozen major fragments, which subsequently spread out along its orbit. The fragments took one more orbit around the planet, and throughout a week in 1994, the string of fragments crashed into Jupiter. The impacts occurred just behind the limb of the planet, where they could not be observed from Earth, but the *Galileo* spacecraft was able to image some of the impacts. Immense plumes rose from the impacts to heights of more than 3,000 km above the cloud tops. Sulfur and carbon compounds released by the impacts formed Earth-sized scars in the atmosphere that persisted for months.

Working It Out 9.1 | Finding the Radius of a Meteoroid

How do astronomers know how big the Chelyabinsk meteoroid was? The energy released by the meteor has been estimated from the strength of the shock wave to be the equivalent of about 3 months of worldwide energy usage. Converting that number to more ordinary units gives about 1.8×10^{15} joules (J).

The speed at which the object was traveling can be found from video of the object's motion through the sky: 18.6 km/s. The energy of motion (the kinetic energy) of a moving object is related to both its mass and its speed. The kinetic energy is given by

$$E_K = \frac{1}{2} mv^2$$

where E_K is the kinetic energy in Joules, m is the mass in kilograms, and v is the magnitude of the velocity (the speed) in meters per second. So now we have a relationship between the estimated energy (which we know), and the speed (which we know), and the mass (which we do not know, but which is related to the size). Solving this equation for mass requires multiplying both sides by 2, and dividing both sides by v^2:

$$E_K = \frac{1}{2} mv^2$$

$$2E_K = mv^2$$

$$\frac{2E_K}{v^2} = m$$

$$m = \frac{2E_K}{v^2}$$

In the last step, we flipped the equation left to right. Plugging in the above values for E_K and v gives

$$m = \frac{2E_K}{v^2}$$

$$m = \frac{2\,(1.8 \times 10^{15})}{(18.6 \times 10^3)^2}$$

$$m = 10{,}000{,}000 \text{ kg}$$

To find the radius, we must use the fact that the density of an object is equal to the mass divided by the volume. Therefore, the volume of an object can be found by dividing the mass by the density. In this case, we assume the meteoroid was spherical, so it has a volume of $4/3\pi r^3$, where r is the radius:

$$\text{Volume} = \frac{\text{Mass}}{\text{Density}}$$

$$\frac{4}{3}\pi r^3 = \frac{m}{\text{Density}}$$

Because this meteorite was made of rock, we know the density was about 2,500 kg/m³. Solving for r gives

$$\frac{4}{3}\pi r^3 = \frac{m}{\text{Density}}$$

$$r^3 = \frac{3m}{(4\pi) \times \text{Density}}$$

$$r = \sqrt[3]{\frac{3m}{(4\pi) \times \text{Density}}}$$

$$r = \sqrt[3]{\frac{3 \times 10{,}000{,}000}{(4\pi) \times 2{,}500}}$$

$$r = \sqrt[3]{950}$$

$$r = 9.8 \text{ meters}$$

This is precisely how scientists estimated the radius of the Chelyabinsk meteoroid.

9.6 Meteorites Are Remnants of the Early Solar System

Recall from Chapter 6 that when a meteoroid enters Earth's atmosphere, it produces an atmospheric phenomenon called a meteor, and any fragment of the meteoroid that survives to land on a planet's surface is called a meteorite. In this section, we will look more closely at meteorites and what can be learned from them about the early Solar System.

Origins of Meteorites

Comet nuclei that enter the inner Solar System disintegrate within a few hundred thousand years as a result of their repeated trips near the Sun. Asteroids have much longer lives but still are slowly broken into pieces from occasional collisions with each other. Disintegration of comet nuclei and asteroid collisions create most of the debris that fills the inner part of the Solar System.

This cometary and asteroidal debris is the source of most of the meteoroids that Earth encounters. About 100,000 kg of meteoritic debris is swept up by Earth every day, and particles smaller than 100 µm eventually settle to the ground as fine dust.

Meteor showers are larger than normal displays of meteors, which happen when Earth's orbit crosses the orbit of a comet. Bits of dust and other debris from a comet nucleus remain in orbits near the orbit of the original comet nucleus. When Earth passes through this concentration of cometary debris, the result is a meteor shower. More than a dozen comets have orbits that come close enough to Earth's orbit to produce annual meteor showers. Because the meteoroids in a shower are all in similar orbits, they all enter our atmosphere moving in the same direction—the paths through the sky are parallel to one another. Therefore, all the meteors appear to originate from the same point in the sky (**Figure 9.29a**), just as the parallel rails of a railroad track appear to vanish to a single point in the distance (Figure 9.29b). This point is the shower's **radiant**.

Fragments of asteroids are much denser than cometary meteoroids. If an asteroid fragment is large enough—about the size of your fist—it can become a mete-

Figure 9.29 (a) Meteors appear to stream away from the radiant of the Leonid meteor shower. (b) Such streaks are actually parallel paths that appear to emerge from a vanishing point, as in our view of these railroad tracks.

VISUAL ANALOGY

orite. The fall of a 10-kg meteoroid can produce a fireball so bright that it lights up the night sky more brilliantly than the full Moon. Such a large meteoroid may create a sonic boom heard hundreds of kilometers away or explode into multiple fragments as it nears the end of its flight. Some glow with a brilliant green color caused by metals in the original meteoroid.

Types of Meteorites

Thousands of meteorites reach the surface of Earth every day, but only a tiny fraction of these are ever found and identified. Antarctica offers the best meteorite hunting in the world. Meteorites are no more likely to fall in Antarctica than anywhere else, but in Antarctica they are far easier to distinguish from their surroundings because in many places the *only* stones to be found on the ice are meteorites. Because Antarctica is actually very dry, Antarctic meteorites also tend to show little weathering or contamination from terrestrial dust or organic compounds, making them excellent specimens for study. Scientists compare meteorites to rocks found on Earth and the Moon and contrast their structure and chemical makeup with those of rocks studied by spacecraft that have landed on Mars and Venus. Meteorites are also compared with asteroids and other objects based on the colors of sunlight they reflect and absorb.

Meteorites are grouped into three categories according to their materials and the degree of differentiation they experienced within their parent bodies. More than 90 percent of meteorites are included in the first category, **stony meteorites** (**Figure 9.30**), which are similar to terrestrial silicate rocks. A stony meteorite is

Figure 9.30 Cross sections of several kinds of meteorites: (a) a chondrite (a stony meteorite with chondrules), (b) an achondrite (a stony meteorite without chondrules), (c) an iron meteorite, and (d) a stony-iron meteorite.

Figure 9.31 A basketball-sized iron meteorite was found lying on the surface of Mars by the Mars exploration rover *Opportunity*.

characterized by the thin coating of melted rock that forms as it passes through the atmosphere. Many stony meteorites contain **chondrules** (see Figure 9.30a), once-molten droplets that rapidly cooled to form crystallized spheres ranging in size from sand grains to marbles. Stony meteorites containing chondrules are called **chondrites**. Conversely, stony meteorites without chondrules are called **achondrites** (see Figure 9.30b). **Carbonaceous chondrites**, chondrites that are rich in carbon, are thought to be the building blocks of the Solar System. Indirect measurements suggest that these meteorites are about 4.56 billion years old—consistent with all other measurements of the time that has passed since the Solar System was formed.

The second category of meteorites, **iron meteorites** (see Figure 9.30c), is the easiest to recognize. The surface of an iron meteorite has a melted and pitted appearance generated by frictional heating as it streaked through the atmosphere. Even so, many are never found, either because they land in water or simply because no one happens to recognize them. Mars rover *Opportunity* has discovered a handful of iron meteorites on the surface of Mars (**Figure 9.31**). Both their appearance—typical of iron meteorites found on Earth—and their position on the smooth, featureless plains made them instantly recognizable.

The third category of meteorites is the **stony-iron meteorites** (see Figure 9.30d), which consist of a mixture of rocky material and iron-nickel alloys. Stony-iron meteorites are relatively rare.

Meteorites and the History of the Solar System

Meteorites come from asteroids, which in turn come from stony-iron planetesimals. During the growth of the terrestrial planets, large amounts of thermal energy were released as larger planetesimals accreted smaller objects, and these bodies were heated further as radioactive elements inside them decayed. Despite this heating, some planetesimals never reached the high temperatures needed to melt their interiors—instead, they simply cooled off, and they have since remained pretty much as they were when they formed. These planetesimals are known as **C-type asteroids**, which are composed of primitive material that astronomers believe is essentially unmodified since the origin of the Solar System almost 4.6 billion years ago.

Some planetesimals, however, did melt and differentiate, with denser matter such as iron sinking to their center. Lower-density material—such as compounds of calcium, silicon, and oxygen—floated toward the surface and combined to form a mantle and crust of silicate rock. **S-type asteroids** are pieces of the mantles and crusts of such differentiated planetesimals. They are chemically more similar to igneous rocks found on Earth than to C-type asteroids, because they were hot enough at some point to lose their carbon compounds and other volatile materials to space. Similarly, **M-type asteroids** (from which iron meteorites come) are fragments of the iron- and nickel-rich cores of one or more differentiated planetesimals that shattered into small pieces during collisions with other planetesimals. Slabs cut from iron meteorites show large, interlocking crystals characteristic of iron that cooled very slowly from molten metal. Rare stony-iron meteorites may come from the transition zone between the stony mantle and the metallic core of such a planetesimal.

Some types of meteorites fail to follow the patterns just discussed. Whereas most achondrites have ages in the range of 4.5 billion to 4.6 billion years, some members of one group are less than 1.3 billion years old and are chemically and physically similar to the soil and atmospheric gases that NASA's lander instruments have measured on Mars. The similarities are so strong that most planetary scientists believe these meteorites are pieces of Mars that were knocked into space by asteroidal impacts. This means researchers have pieces of another planet that they can study in laboratories here on Earth. Sometimes the findings are controversial. In 1996, a NASA research team announced that the meteorite ALH84001 showed possible physical and chemical evidence of past life on Mars. The team's extraordinary conclusions have been challenged, and the debate continues even today.

If pieces of Mars have reached Earth from its orbit almost 80 million km beyond our own orbit, we might expect that pieces of our companion Moon would have found their way to Earth as well. Indeed, meteorites of another group bear striking similarities to rock samples returned from the Moon. Like the meteorites from Mars, these are chunks of the Moon that were blasted into space by impacts and later fell to Earth.

The story of how planetesimals, asteroids, and meteorites are related is one of the great successes of planetary science. Scientists have assembled a wealth of information about this diverse collection of objects to piece together a picture of how planetesimals grow, differentiate, and then shatter in subsequent collisions.

READING ASTRONOMY News

Before science "news" gets to be "news," often someone in the scientific community must recognize it as interesting and issue a press release. These are typically very short pieces to let members of the press know something is happening, and if they want more information, they should call and ask about it. NASA issued one such press release about a spacecraft in development in 2013.

NASA's Asteroid Sample-Return Mission Moves into Development

NASA's first mission to sample an asteroid is moving ahead into development and testing in preparation for its launch in 2016. The *Origins-Spectral Interpretation Resource Identification Security Regolith Explorer* (*OSIRIS-REx*) passed a confirmation review Wednesday called Key Decision Point (KDP)-C. NASA officials reviewed a series of detailed project assessments and authorized the spacecraft's continuation into the development phase.

OSIRIS-REx will rendezvous with the asteroid Bennu in 2018 and return a sample of it to Earth in 2023.

"Successfully passing KDP-C is a major milestone for the project," said Mike Donnelly, *OSIRIS-REx* project manager at NASA's Goddard Space Flight Center in Greenbelt, Maryland. "This means NASA believes we have an executable plan to return a sample from Bennu. It now falls on the project and its development team members to execute that plan."

Bennu could hold clues to the origin of the Solar System. *OSIRIS-REx* will map the asteroid's global properties, measure nongravitational forces, and provide observations that can be compared with data obtained by telescope observations from Earth. *OSIRIS-REx* will collect a minimum of 2 ounces (60 grams) of surface material.

"The entire *OSIRIS-REx* team has worked very hard to get to this point," said Dante Lauretta, *OSIRIS-REx* principal investigator at the University of Arizona in Tucson. "We have a long way to go before we arrive at Bennu, but I have every confidence when we do, we will have built a supremely capable system to return a sample of this primitive asteroid."

The mission will be a vital part of NASA's plans to find, study, capture, and relocate an asteroid for exploration by astronauts. NASA recently announced an asteroid initiative proposing a strategy to leverage human and robotic activities for the first human mission to an asteroid while also accelerating efforts to improve detection and characterization of asteroids.

NASA's Goddard Space Flight Center in Greenbelt, Maryland, will provide overall mission management, systems engineering, and safety and mission assurance. The University of Arizona in Tucson is the principal investigator institution. Lockheed Martin Space Systems of Denver will build the spacecraft. *OSIRIS-REx* is the third mission in NASA's New Frontiers Program. NASA's Marshall Space Flight Center in Huntsville, Alabama, manages New Frontiers for NASA's Science Mission Directorate in Washington, D.C.

Evaluating the News

1. What is the primary mission of *OSIRIS-REx*?
2. The press release states that "Bennu could hold clues to the origin of the Solar System." Why are the spacecraft developers able to make this claim?
3. This mission is part of a larger plan for a manned mission to an asteroid. What are the advantages of landing on (and taking off from) an asteroid compared to a planet? What challenges can you see for that manned mission team that make landing on an asteroid more difficult than landing on a planet?
4. Consider the last paragraph, which lists the institutions involved in this project. How are the efforts divided among government agencies, public institutions, and private industry?
5. Consider the length of time involved in this project. According to its website, planning for the mission began in 2011. This Key Decision Point occurred in 2013. The mission is planned to launch in 2016 and will return samples in 2023. The team is planning on 2 years of sample analysis once the samples are returned. All together, about how much time does it take to plan such a mission, carry it out, and analyze the results?

SUMMARY

Pluto, Eris, Haumea, Makemake, and Ceres are classified as dwarf planets, rather than planets, because they have not cleared their orbits. The moons of the outer Solar System are composed of rock and ice. A few moons are geologically active, but most are dead. Asteroids are small Solar System bodies made of rock and metal. Some asteroids cross Earth's orbit and are potentially dangerous. Comets are small, icy planetesimals that reside in the frigid regions beyond the planets. Comets that venture into the inner Solar System are warmed by the Sun, often producing an atmospheric coma and a long tail. Very large asteroids or comets striking Earth create enormous explosions that can dramatically affect terrestrial life. Meteoroids are small fragments of asteroids and comets. When a meteoroid enters Earth's atmosphere, frictional heat causes the air to glow, producing a phenomenon called a meteor. A meteoroid that survives to a planet's surface is called a meteorite.

LG 1 Small bodies in the Solar System include dwarf planets, moons, asteroids, comets, Kuiper Belt objects, and meteoroids.

LG 2 The moons of the outer planets can be classified as geologically active, possibly geologically active, formerly active, and never active. Surface features provide evidence of geological activity, either current or in the past.

LG 3 The collisions that created massive asteroids also provided sufficient heat to make them molten. This allowed the materials in these asteroids to differentiate. Less massive asteroids did not become molten and remain undifferentiated.

LG 4 Meteorites are fragments of asteroids that did not burn up in the atmosphere and thus survived to reach the ground. Different types of meteorites come from different types of asteroids or from different parts of differentiated asteroids. Some meteorites are fragments of larger objects (for example, Mars) that were blasted into space during an impact.

LG 5 As a comet orbits the Sun, its appearance changes. Both the size and the orientation of its two tails vary with location in the orbit.

LG 6 Meteorites, asteroids, and comets provide samples of material from the entire history of the Solar System.

SUMMARY SELF-TEST

1. _____, _____, _____, and _____ are four categories of small bodies in the Solar System.

2. Classifying moons according to their geology allows us to (choose all that apply)
   a. compare their features.
   b. explain the formation mechanisms.
   c. determine their magnetic field strength.
   d. identify physical mechanisms responsible for their evolution.

3. A differentiated asteroid was once _____ enough to be molten.

4. The three types of meteorites come from different parts of their parent bodies. Stony-iron meteorites are rare because
   a. they are hard to find.
   b. only a small amount of a parent body has *both* stone and iron.
   c. there is very little iron in the Solar System.
   d. the magnetic field of the Sun attracts the iron.

5. As a comet leaves the inner Solar System, its ion tail always points
   a. back along the orbit.
   b. forward along the orbit.
   c. toward the Sun.
   d. away from the Sun.

6. Meteorites can tell us about (select all that apply)
   a. the early composition of the Solar System.
   b. the composition of asteroids.
   c. other planets.
   d. the Oort Cloud.

7. From the following, select the ways in which Titan resembles early Earth. (Choose all that apply.)
   a. It has a thick atmosphere.
   b. Its atmosphere is mostly nitrogen.
   c. It has liquid water on the surface.
   d. It has terrain similar to Earth's.
   e. It is rich in organic compounds.

8. Pluto differs significantly from the eight Solar System planets in that (choose all that apply)
   a. it is farther from the Sun than any classical planet.
   b. it has a different composition than any classical planet.
   c. its orbit is chaotic.
   d. it is not round.
   e. it has not cleared its orbit.

9. If an asteroid is not spherical, what does that tell you?
   a. It is made of iron.
   b. Its mass is low.
   c. It is an M-type.
   d. It is very young.

10. From the following list of terms, select the three terms that are parts of a comet, and place them in order from smallest to largest.
    radiant
    Oort Cloud
    coma
    Kuiper Belt
    nucleus
    chondrule
    tail

QUESTIONS AND PROBLEMS

Multiple Choice and True/False

11. **T/F** There are no dwarf planets interior to Jupiter's orbit.

12. **T/F** Most large moons of the outer Solar System are, or once were, geologically active.

13. **T/F** Asteroids are mostly rock and metal; comets are mostly ice.

14. **T/F** Comet tails always point directly away from the Sun.

15. **T/F** Major impacts on Earth don't happen any more.

16. Pluto is classified as a dwarf planet because
    a. it is not round.
    b. it orbits the Sun too far away to be considered a planet.
    c. it has company in its orbit.
    d. it is made mostly of ice.

17. Io is an example of a moon that
    a. is definitely active today.
    b. is probably or possibly active today.
    c. was active in the past but not today.
    d. was not active at any time since its formation.

18. Asteroids are
    a. small rock and metal objects orbiting the Sun.
    b. small icy objects orbiting the Sun.
    c. small rock and metal objects found only between Mars and Jupiter.
    d. small icy bodies found only in the outer Solar System.

19. Aside from their periods, short- and long-period comets differ because
    a. short-period comets orbit prograde, while long-period comets have either prograde or retrograde orbits.
    b. short-period comets formed with less ice, while long-period comets contained more.
    c. short-period comets do not develop ion tails, while long-period comets do.
    d. short-period comets come closer to the Sun at closest approach than long-period comets.

20. Congress tasked NASA with searching for near-Earth objects because
    a. they might impact Earth, as others have in the past.
    b. they are close by and easy to study.
    c. they are moving fast.
    d. they are scientifically interesting.

21. Large impacts of asteroids and comets with Earth are
    a. impossible.
    b. infrequent.
    c. unimportant.
    d. unknown.

22. What is the source of meteors we see during a meteor shower?
    a. near-Earth asteroids
    b. dust left over from the original disk of dust and gas around the Sun
    c. comet debris
    d. dust left over from the formation of Earth's moon

23. A meteoroid is found _____, a meteor is found _____, and a meteorite is found _____.
    a. in space; in the atmosphere; on the ground
    b. on the ground; in space; between Mars and Jupiter
    c. between Mars and Jupiter; in the atmosphere; on the ground
    d. between Mars and Jupiter; in the atmosphere; elsewhere in the Solar System

24. On average, a bright comet appears about once each decade. Statistically, this means that
    a. one will definitely be observed every tenth year.
    b. one will definitely be observed in each 10-year period.
    c. exactly 10 comets will be observed each century.
    d. about 10 comets will be observed each century.

25. During a meteor shower, all meteors trace back to a single region in the sky known as the radiant. This happens because
    a. the meteors originate from the same point in the atmosphere.
    b. all the meteors are traveling the same direction, relative to Earth.
    c. all the meteors burn up at the same altitude.
    d. all the meteors come from the direction of the Sun.

Conceptual Questions

26. Most asteroids are found between the orbits of Mars and Jupiter, but astronomers are especially interested in the relative few whose orbits cross that of Earth. Why?

27. Explain the process that drives volcanism on Jupiter's moon Io.

28. Describe cryovolcanism, and explain its similarities and differences with respect to terrestrial volcanism.

29. Europa and Titan may both be geologically active. What is the evidence for this?

30. Discuss evidence supporting the idea that Europa might have a subsurface ocean of liquid water.

31. Titan contains abundant amounts of methane. Why does this require an explanation? What process destroys methane in this moon's atmosphere?

32. Some moons display signs of past geological activity. Identify some of the evidence for past activity.

33. Explain why it is unlikely that the main-belt asteroids will coalesce to form a planet.

34. Describe differences between the Kuiper Belt and the Oort Cloud as sources of comets.

35. Kuiper Belt objects (KBOs) are actually comet nuclei. Why do they not display tails?

36. Sketch the appearance of a long-period comet at different locations in its orbit.

37. Explain the importance of meteorites, asteroids, and comets to the study of the history of the Solar System.

38. If collisions of comet nuclei and asteroids with Earth are rare events, why should we be concerned about the possibility of such a collision?

39. Make a table of the types of meteorites, a distinguishing characteristic of each type, and the origin of each type.

40. What are the differences between a comet and a meteor in terms of their size, distance, and how long they remain visible?

Problems

41. Io has a mass, M, of 8.9×10^{22} kg and a radius, R, of 1,820 km.
    a. The speed needed to escape from the surface of a world is called the escape velocity, v_{esc}. It is given by
    $$v_{esc} = \sqrt{\frac{2GM}{R}}$$
    Calculate the escape velocity at Io's surface.
    b. How does Io's escape velocity compare with the vent velocities of 1 km/s from this moon's volcanoes?

42. Planetary scientists have estimated that Io's extensive volcanism could be covering this moon's surface with lava and ash to an average depth of up to 3 millimeters (mm) per year.
    a. Io's radius is 1,820 km. If we model Io as a sphere, what are its surface area and volume?
    b. What is the volume of volcanic material deposited on Io's surface each year?
    c. How many years would it take for volcanism to perform the equivalent of depositing Io's entire volume on its surface?
    d. How many times might Io have "turned inside-out" during the life of the Solar System?

43. Electra is a 182-km-diameter asteroid accompanied by a small moon orbiting at a distance of 1,325 km in a circular orbit with a period of 5.25 days. Refer to Chapter 3 to answer the following questions.
    a. What is the mass of Electra?
    b. What is Electra's density?

44. At the time that *Giotto*, *Vega 1*, and *Vega 2* visited Comet Halley, the nucleus was losing 20,000 kg of gas and 10,000 kg of dust each second. The period of this comet is 76 years.
    a. If the comet constantly lost mass at this rate, how much mass would it lose in 1 year?
    b. What percentage is this of its total mass (2.2×10^{14} kg)?
    c. How many trips like this around the Sun would Halley make before it was completely destroyed?
    d. How long would that take?

45. The orbital periods of Comets Encke, Halley, and Hale-Bopp are 3.3 years, 76 years, and 2,530 years, respectively.
    a. What are the semimajor axes (in astronomical units) of the orbits of these comets?
    b. Assuming that the distance from the comet to the Sun at closest approach is negligible, what are the maximum distances from the Sun (in astronomical units) reached by Comets Halley and Hale-Bopp in their respective orbits?
    c. Which would you guess is the most pristine comet among the three? Which is the least? Explain your reasoning.

46. Comet Halley has a mass of approximately 2.2×10^{14} kg. It loses about 3×10^{11} kg each time it passes the Sun.
    a. The first confirmed observation of the comet was made in 230 BCE. Assuming a constant period of 76.4 years, how many times has it reappeared since that early sighting?
    b. How much mass has the comet lost since 230 BCE?
    c. What percentage of its total mass does this amount represent?

47. A cubic centimeter of the air you breathe contains about 10^{19} molecules. A cubic centimeter of a comet's tail may typically contain 10 molecules. Calculate the size of a cubic volume of comet tail material that would hold 10^{19} molecules.

48. A 1-megaton hydrogen bomb releases 4.2×10^{15} J of energy. Compare this amount of energy with that released by a 10-km-diameter comet nucleus ($m = 5 \times 10^{14}$ kg) hitting Earth at a speed of 20 km/s. You will need to use the fact that $E_K = \frac{1}{2}mv^2$ (where E_K is the kinetic energy in joules, m is the mass in kilograms, and v is the magnitude of the velocity [the speed] in meters per second).

49. One recent estimate concludes that nearly 800 meteorites with mass greater than 100 grams (massive enough to cause personal injury) strike the surface of Earth each day. Assuming that you present a target of 0.25 square meters (m²) to a falling meteorite, what is the probability that you will be struck by a meteorite during your 100-year lifetime? (Note that the surface area of Earth is approximately 5×10^{14} m².)

50. Assuming that 800 meteorites with mass greater than 100 grams strike the surface of Earth each day, how much material from meteorites of this size falls on Earth each day? How much each year? If this were the only transfer of mass onto Earth, how long would it take for the mass of Earth to double?

SMARTWORK

Norton's online homework system includes algorithmically generated versions of these questions, plus additional conceptual exercises. If your instructor assigns questions in SmartWork, log in at smartwork.wwnorton.com.

Exploration | Comparative Dwarf Planetology

Much of astronomy consists of gathering information about individual objects and then comparing and contrasting them to find similarities and differences. We used this approach for the planets of the Solar System, in which case it is called *comparative planetology*. The first step is always to gather together the fundamental information about the objects in one place, so that we can compare objects with one another and find patterns (see Tables 6.1 and 8.1).

Make a table of information (similar to the tables in Chapters 6 and 8) about the dwarf planets discussed in this chapter. In some cases, all the information is not yet available for some of these dwarf planets. In such a case, mark the relevant cell in the table with an N/A for "not available." You might wish to add rows to the table that are of interest to you, such as the number of moons, for example, or the year of discovery. Check online to see if any new dwarf planets have been discovered, and if so, include data about those dwarf planets as well.

To help you get started, the partial table in **Figure 9.32** shows how you might lay out the rows and columns. Be sure to include units of measurement for all of the data. Then answer the following questions.

1. Compare the masses of these dwarf planets with the mass of the Moon, listed in Table 6.1. Are these dwarf planets more or less massive than the Moon?

2. Compare the densities of these dwarf planets with the densities of the terrestrial planets (Table 6.1) and the giant planets (Table 8.1). From this comparison, what can you determine about the compositions of these dwarf planets?

3. Compare the surface gravities of these dwarf planets with the surface gravity of Earth. About how much less would you weigh on Eris than you do on Earth?

4. Compare Pluto's axial tilt to the axial tilts of the terrestrial planets (Table 6.1) and the giant planets (Table 8.1). Which planet will have seasons most similar to Pluto?

5. Compare the orbital velocity of these dwarf planets. How does the orbital velocity change as the orbital radius grows larger? Use concepts from Chapter 2 (that is, Kepler's laws) to explain this trend.

6. Add a row to your table for the orbital inclination of these dwarf planets. How does orbital inclination change as the orbital radius grows larger? Use concepts from Chapter 5 (The Formation of Stars and Planets) to explain this trend.

Comparison of the Properties of the Dwarf Planets

Property	Pluto	Ceres	Eris
Semimajor axis (AU)			
Orbital period (Earth years)			
Orbital velocity (km/s)			
Mass (Earth masses)			

Figure 9.32 A table of information about the dwarf planets is useful for comparison. This example shows how the rows and columns might be laid out.

SMARTWORK • smartwork.wwnorton.com

10 Measuring the Stars

Humans, by nature, are curious. How far away is that star? How big is it? How luminous is it? These are questions that people have wondered about from the beginning of human history. Asking such questions is also an important aspect of the process of science.

Unlike our exploration of the Solar System, we cannot send space probes to a star to take close-up pictures or land on its surface. We study the stars by observing their light, by applying our current understanding of the laws of physics, and by finding patterns in subgroups of stars that enable us to extrapolate to other stars. One of these patterns is the Hertzsprung-Russell diagram (H-R diagram), which we will focus on in this chapter. In the illustration on the opposite page, a student is using the H-R diagram to learn more about Vega, the bright star in her photograph of the constellation Lyra. She has used the data in Appendix 3 to plot Vega and the Sun and compare their relative locations on the chart. Using this same technique, astronomers are able to compare individual stars to each other. They combine this analysis with knowledge of geometry, radiation, and orbits to answer humanity's age-old questions about the stars and our relationship to them.

✦ LEARNING GOALS

To all but the most powerful of telescopes, a star is just a point of light in the night sky. But by applying our understanding of light, matter, and motion to what we see, we are able to build a remarkably detailed picture of the physical properties of stars. By the end of this chapter, you should be able to place any star in the correct location on the H-R diagram, as the student is doing with Vega in the illustration at right. You should also be able to:

LG 1 Use the brightness of nearby stars and their distances from Earth to discover how luminous they are.

LG 2 Infer the temperatures and sizes of stars from their colors.

LG 3 Determine the composition and mass of stars.

LG 4 Classify stars, and organize this information on a Hertzsprung-Russell diagram.

LG 5 Explain why the mass and composition of a main-sequence star determine its luminosity, temperature, and size.

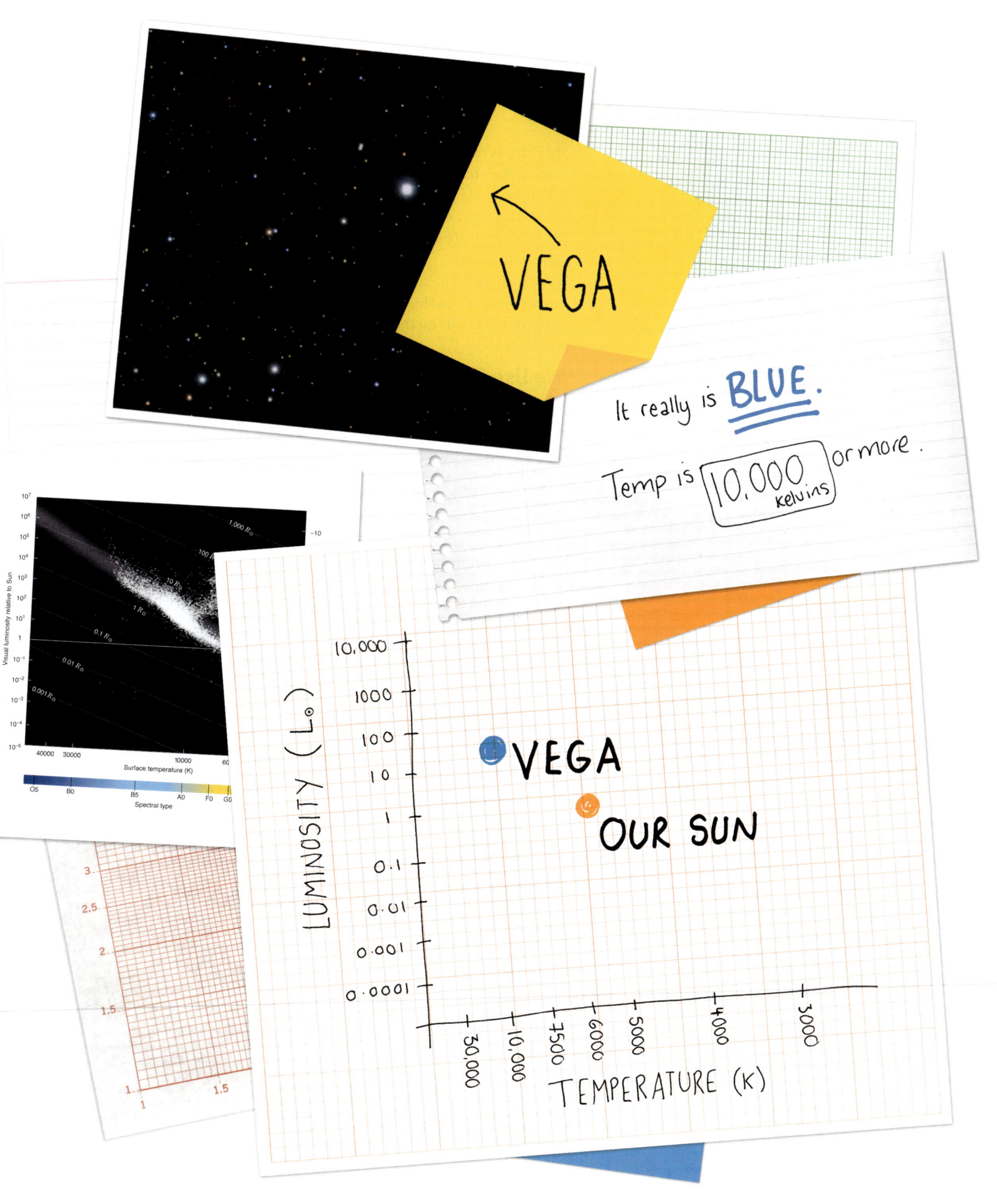

10.1 The Luminosity of a Star Can Be Found from the Brightness and the Distance

Brightness refers to how bright an individual star appears in our sky, while *luminosity* refers to how much light the star actually emits. The distance to a star must be known before it is possible to determine if it appears faint because it emits very little light or because it is very far away. Determining the brightness of a star is a conceptually straightforward task that usually involves comparing it to other nearby stars whose brightness is known. Finding the distance is somewhat more difficult. The method used depends on whether the star is relatively near or relatively far. In this section, you will learn how to find the distance to nearby stars and to combine that with the brightness to find the luminosity.

We Use Parallax to Measure Distances to Nearby Stars

Hold up your finger in front of you, quite close to your nose. View it with your right eye only and then with your left eye only. As you can see, your two eyes have different views of the world. Each eye sends a slightly different image to your brain, and so your finger *appears* to move back and forth relative to the background behind it. Now hold up your finger at arm's length, and blink your right eye, then your left eye. Your finger appears to move much less. The way your brain combines the different information from your eyes to perceive the distances to objects around you is called **stereoscopic vision**.

Stereoscopic vision allows you to judge the distances of objects as far away as 10 meters, but beyond that it is of little use. Your two eyes have identical views of a mountain several kilometers away—all you can determine is that the mountain is too far away for you to judge its distance stereoscopically. The distance over which our stereoscopic vision works is limited by the separation between our two eyes, about 6 centimeters (cm). If you could separate your eyes by several meters, you could judge the distances to objects that were about half a kilometer away.

Although we cannot literally take our eyes out of our heads and hold them apart at arm's length, we can compare pictures taken from two widely separated locations. The greatest separation we can get without leaving Earth is to let Earth's orbital motion carry us from one side of the Sun to the other. If we take a picture of the sky tonight and then wait 6 months and take another picture, the distance between the two locations is the diameter of Earth's orbit (2 astronomical units [AU]), which gives us very powerful stereoscopic vision. **Figure 10.1a** shows an overhead view of the experiment you just performed with your finger. The left eye sees the blue pencil nearly directly between the green balls on the bookshelf. But the right eye sees the blue pencil to the left of both balls. Similarly, the position of the pink pencil varies. Because the pink pencil is closer to the observer, its position changes more than the position of the blue pencil—it moves from the right of the blue pencil to the left of the blue pencil, so it must have shifted farther. Figure 10.1b shows how astronomers apply this concept to measure the distance to stars. This illustration shows Earth's orbit as viewed from far above the Solar System. The change in position of Earth over 6 months is like the distance between the right eye and the left eye in Figure 10.1a. The nearby (pink and blue) stars are like the pink and blue pencils, while the distant yellow stars

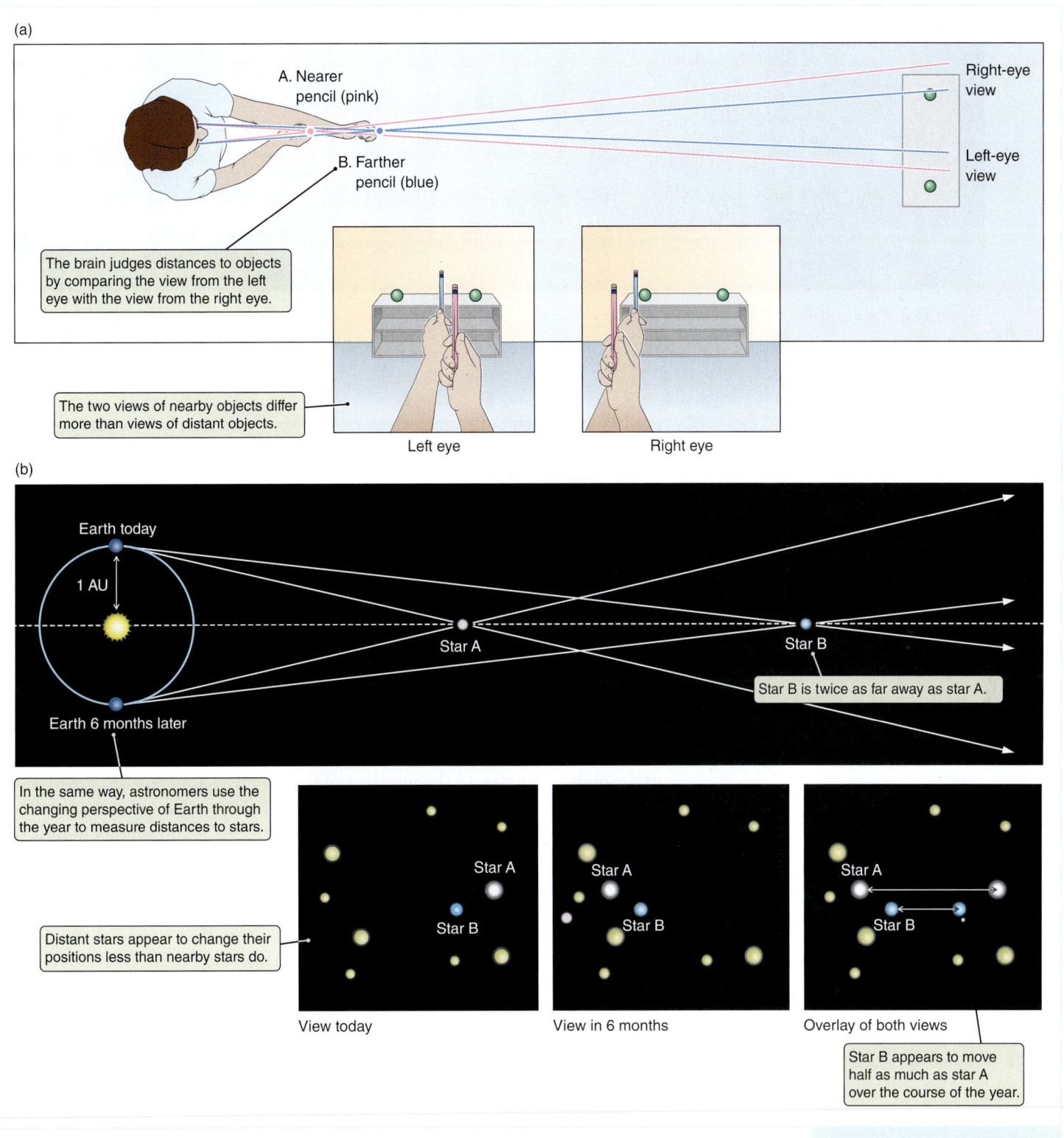

Figure 10.1 (a) Stereoscopic vision allows you to judge the distance to an object by comparing the view from each eye. (b) Similarly, comparing views from different places in Earth's orbit allows us to determine the distance to stars. As Earth moves around the Sun, the apparent positions of nearby stars change more than the apparent positions of more distant stars. (The diagram is not to scale.) This is the starting point for measuring the distances to nearby stars.

VISUAL ANALOGY

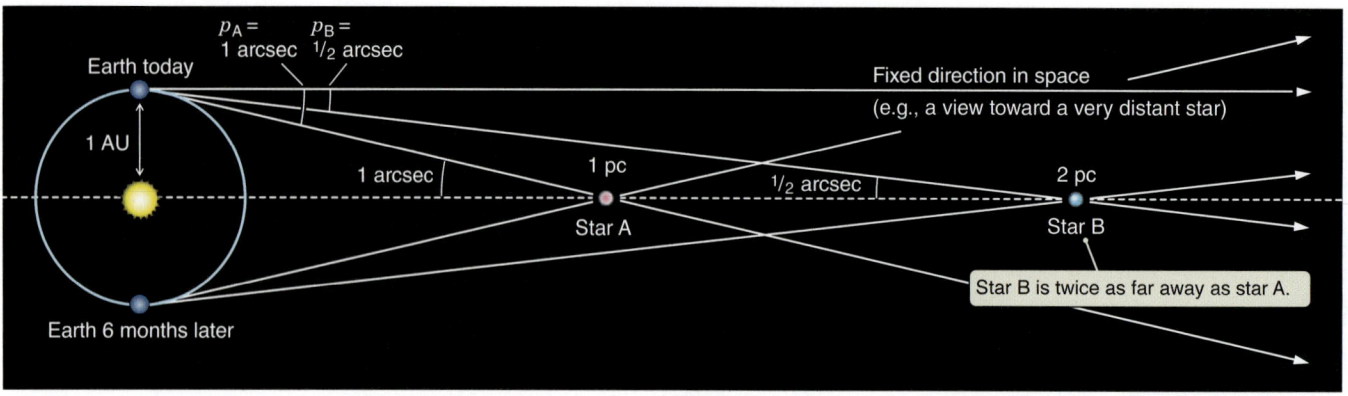

Figure 10.2 The parallax (*p*) of a star is inversely proportional to its distance. More distant stars have smaller parallaxes. (The diagram is not to scale.)

are like the green balls on the bookcase. Because of the shift in perspective as Earth orbits the Sun, nearby stars appear to shift their positions. The pink star, which is closer, appears to move farther than the more distant blue star. Over the course of 1 full year, the nearby star appears to move and then move back again with respect to distant background stars, returning to its original position 1 year later. We can determine the distance to the star using the amount of this apparent shift and geometry.

Figure 10.2 shows the same configuration as Figure 10.1b. Look first at star A, the closer star. When Earth is at the top of the figure, it forms a right triangle with the Sun and star A at the other corners. (Remember that a right triangle is one with a 90° angle in it.) The short leg of the triangle is the distance from Earth to the Sun, which is 1 AU. The long leg of the triangle is the distance from the Sun to star A. The small angle at the star A corner of the triangle measures the change in the apparent position of the star. This change in position, measured as an angle, is known as the **parallax** of the star. The apparent motion of a star across the sky, described earlier for Figure 10.1b, is equal to twice the parallax.

More distant stars make longer and skinnier triangles and smaller parallaxes. Star B is twice as far away as star A, and its parallax is half the parallax of star A. The parallax of a star is inversely proportional to its distance: when one goes up, the other goes down.

The parallaxes of real stars are tiny. Astronomers have a special set of units to talk about such distances. Just as an hour on the clock is divided into minutes and seconds, a degree can be divided into *arcminutes* and *arcseconds*. An **arcminute** (abbreviated **arcmin**) is 1/60 of a degree, and an **arcsecond** (abbreviated **arcsec**) is 1/60 of an arcminute. An arcsecond is about equal to the angle formed by the diameter of a golf ball at a distance of 5.5 miles.

The distances to real stars are large, and in this book we normally use units of light-years to describe them. As we saw in Chapter 4, 1 light-year is the distance that light travels in 1 year—about 9.5 trillion kilometers. We use this unit because it is the unit you are most likely to see in a newspaper article or a popular book about astronomy. When astronomers discuss distances to stars and galaxies, however, the unit they generally use is the **parsec** (which is short for *parallax second* and is abbreviated **pc**—a star at a distance of 1 parsec has a parallax of 1 arcsecond); 1 parsec is equal to 3.26 light-years.

▶▶ **Nebraska Simulation: Parallax Explorer**

Working It Out 10.1 | Parallax and Distance

Recall from earlier chapters that "inversely proportional" means that on one side of the equation, a variable is in the numerator, while on the other side, a different variable is in the denominator. The relationship between distance (d) and parallax (p) is an inverse proportion:

$$p \propto \frac{1}{d} \text{ or } d \propto \frac{1}{p}$$

The parsec has been adopted by astronomers because it makes the relationship between distance and parallax easier than using light-years:

$$\begin{pmatrix}\text{Distance measured} \\ \text{in parsecs}\end{pmatrix} = \frac{1}{\begin{pmatrix}\text{Parallax measured} \\ \text{in arseconds}\end{pmatrix}}$$

or

$$d \text{ (pc)} = \frac{1}{p \text{ (arcsec)}}$$

Notice that the proportionality sign has turned into an equals sign: You don't have to remember any constants if the distance is in parsecs and the parallax is in arcseconds.

Suppose that the parallax of a star is measured to be 0.5 arcsec. To find the distance to the star, we substitute that into the parallax equation for p:

$$d = \frac{1}{0.5} = \frac{1}{1/2} = 2 \text{ pc}$$

Suppose that the parallax of a star is measured to be 0.01 arcsec. What is its distance in light-years? First, we find its distance in parsecs:

$$d = \frac{1}{0.01} = \frac{1}{1/100} = 100 \text{ pc}$$

Then, we convert to light-years by remembering that a parsec is 3.26 light-years:

$$d = 100 \text{ pc} \times \frac{3.26 \text{ light-years}}{1 \text{ pc}}$$

$$d = 100 \times 3.26 \text{ light-years}$$

$$d = 326 \text{ light-years}$$

The star closest to us after the Sun is Proxima Centauri. Located at a distance of 4.22 light-years, Proxima Centauri is a faint member of a system of three stars called Alpha Centauri. What is this star's parallax? First, we must convert from light-years back to parsecs:

$$d = 4.22 \text{ light-years} \times \frac{1 \text{ pc}}{3.26 \text{ light-years}}$$

$$d = 1.29 \text{ pc}$$

Then we find the parallax from the distance:

$$d = \frac{1}{p}$$

Solve for p to get:

$$p = \frac{1}{d}$$

Then insert our value for the distance in parsecs:

$$p = \frac{1}{1.29}$$

$$p = 0.77 \text{ arcsec}$$

This star has a parallax of only 0.77 arcsec.

When astronomers began to apply parallax to stars in the sky, they discovered that stars are very distant objects (see the example in **Working It Out 10.1**). The first successful parallax measurement was made by F. W. Bessel (1784–1846), who in 1838 reported a parallax of 0.314 arcsec for the star 61 Cygni. This finding implied that 61 Cygni was 3.2 pc away, or 660,000 times as far away as the Sun. With this one measurement, Bessel increased the known size of the universe by a factor of 10,000. Today, only about 60 stars are known within 15 light-years of the Sun. In the neighborhood of the Sun, each star (or star system) has about 235 cubic light-years of space all to itself.

Astronomers worked hard to find the parallax of known stars for more than 100 years. But most stars were so far away that this motion relative to background stars was too small to measure using ground-based telescopes. Knowledge of our stellar neighborhood took a tremendous step forward during the 1990s, when the *Hipparcos* satellite measured the positions and parallaxes of 120,000 stars. Even this catalog has its limits. The accuracy of any given *Hipparcos* parallax measurement is about ±0.001 arcsec. Because of this observational **uncertainty**, our measurements of the distances to stars are not perfect. For example, a star with a *Hipparcos* parallax of 0.004 ± 0.001 arcsec really has a parallax between 0.003 arcsec and 0.005 arcsec. This gives a corresponding distance range of 200 to 330 pc from Earth. As an analogy, consider your speed while driving down the road. If your digital speedometer says 10 kilometers per hour (km/h), you might actually be traveling 10.4 km/h or 9.6 km/h. The precision of your speedometer is limited to the nearest 1 km/h, but that doesn't mean you don't have any idea about your speed. You are certainly not traveling 100 km/h, for example. With current technology, we cannot reliably measure stellar distances of more than a few hundred parsecs using parallax. Other methods—to be discussed later—are used for more distant stars.

The Brightness of a Star

Two thousand years ago, the Greek astronomer Hipparchus classified the brightest stars he could see as being "of the first magnitude" and the faintest as being "of the sixth magnitude." **Magnitude** has come to mean a measure of a star's brightness in the sky. Note that this means a brighter object has a *smaller* magnitude. It is also an example of logarithmic behavior, so that an object of magnitude 2 is much more than twice as bright as an object of magnitude 4. Each increase of 2.5 magnitudes corresponds to an increase of a factor of 10 in brightness. Hipparchus himself must have had typical eyesight, as an average person under dark skies can see stars only as faint as 6th magnitude. Modern telescopes can see much "**deeper**" than this. The Hubble Space Telescope can detect stars as faint as 30th magnitude—4 billion times fainter than what the naked eye can see.

Objects that are brighter than 1st magnitude in this system have magnitudes of less than one, and the magnitude can even be negative. For example, Sirius, the brightest star in the sky in the visible wavelengths, has a magnitude of −1.46. Venus can be bright enough to cast shadows, at magnitude −4.4. The magnitude of the full Moon is −12.7 and that of the Sun is −26.7. Thus, the Sun is 14.0 magnitudes (about 400,000 times) brighter than the full Moon.

The magnitude of a star, as we have discussed it, is called the star's **apparent magnitude** because it is the brightness of the star as it *appears* to us in our sky. Stars are found at different distances from us, so a star's apparent magnitude does not tell us how much light it actually emits. To find out how much light it emits—its luminosity—we must know the distance. Then we can put it on a scale with all other stars of known distances, and we can calculate how bright they *would* be if they were all located 10 pc from us. This is called the **absolute magnitude**: the brightness of each star if it were located at a distance of 10.0 pc (32.6 light-years).

The brightness of astronomical objects generally varies with wavelength region (color), so astronomers use special symbols to represent magnitudes at certain colors. For example, they use *V* and *B*, respectively, to represent magnitudes in the visual (yellow-green) and blue regions of the spectrum. The term *visual* is used because yellow-green light roughly corresponds to the range of wavelengths to which our eyes are most sensitive.

Vocabulary Alert

uncertainty An uncertain distance does not mean that the distance is completely unknown. Uncertainty is simply a way of expressing how accurately the distance is known. For scientists, the uncertainty is sometimes the most important number, and they often get very excited when a new experiment reduces the uncertainty, even when it doesn't change the value. When astronomers discovered that the age of the universe was 13.8 billion years, with an uncertainty of 0.1 billion years, many astronomers were more excited about the 0.1 than the 13.8. This is because the measurement was so much more accurate than any that came before it.

deep In common language, this word has many meanings; for instance, referring to depth in the ocean or to a profound idea. Astronomers use this word to refer to an object's distance—how *deep* it is in space. Because distant objects are typically fainter, *deep* and *faint* are closely related and sometimes used somewhat interchangeably.

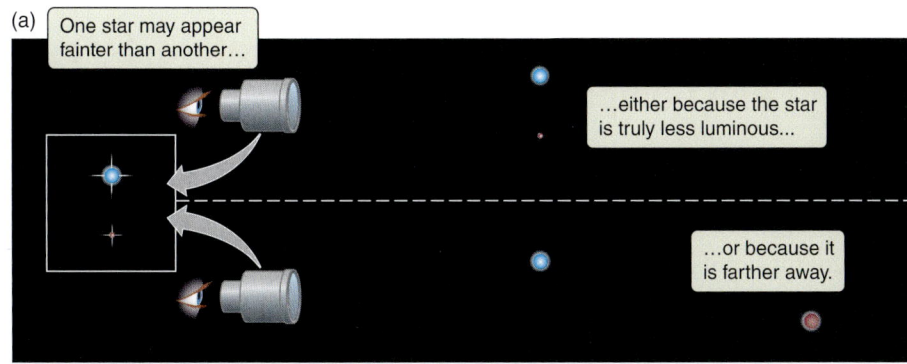

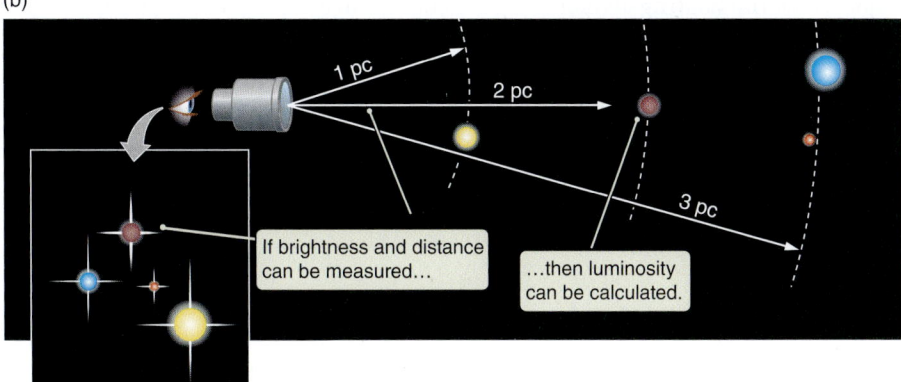

Figure 10.3 The brightness of a visible star in our sky depends on both its luminosity—how much light it emits—and its distance. When brightness and distance are measured, luminosity can be calculated.

Finding Luminosity

Although the brightness of a star is directly measurable, it does not immediately tell us much about the star itself. As illustrated in **Figure 10.3**, an apparently bright star in the night sky may in fact be intrinsically dim but close by. Conversely, a faint star may actually be very luminous, but because it is very far away, it appears faint to us. We can specifically say that its apparent brightness, which measures the starlight that reaches us, is inversely proportional to the square of our distance from the star. If we know this distance, we can then use our measurement of the star's apparent brightness to find its luminosity.

The range of possible luminosities for stars is very large. The Sun provides a convenient yardstick for measuring the properties of stars, including their luminosity. (We compare stars to the Sun so often that properties of the Sun have a special subscript, $_\odot$, which always means "Sun," so that $L_\odot$ is the luminosity of the Sun.) The most luminous stars are more than a million times the luminosity of the Sun. The least luminous stars have luminosities less than 1/10,000 $L_\odot$. The most luminous stars are therefore more than 10 billion (10^{10}) times more luminous than the least luminous stars. Very few stars are near the upper end of this range of luminosities, and the vast majority of stars are far less luminous than our Sun. **Figure 10.4** shows the relative number of stars compared to their luminosity in solar units.

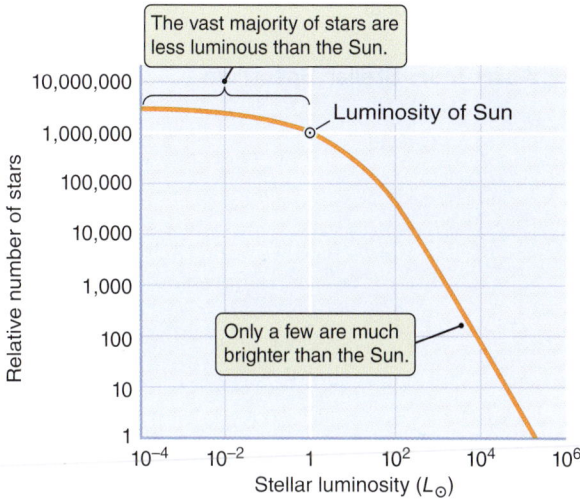

Figure 10.4 The distribution of the luminosities of stars is plotted here logarithmically, so that increments are in powers of 10. For example, for every million stars with a luminosity equal to the Sun, there are only about 100 with a luminosity 10,000 times greater.

10.2 Radiation Tells Us the Temperature, Size, and Composition of Stars

Two everyday concepts—stereoscopic vision and the fact that objects appear brighter when closer—have given us the tools we need to measure the distance and luminosity of the closest stars. In these two steps, stars have gone from being merely faint points of light in the night sky to being extraordinarily powerful beacons located at great distances.

Stars are gaseous, but they are fairly dense—dense enough that the radiation from a star comes close to obeying the same laws as the radiation from solid objects like the heating element on an electric stove. That means we can use our understanding of blackbody radiation to understand the radiation from stars. Recall the Stefan-Boltzmann law, which states that among same-sized objects, the hotter objects are more luminous; and Wien's law, which states that hotter objects are bluer. In this section, we will use these two laws to measure the temperatures and sizes of stars. We will also develop a more detailed understanding of the line emission mentioned in Chapters 4 and 5 to obtain information about the composition of stars.

Wien's Law Revisited: The Color and Surface Temperature of Stars

Wien's law (see Working It Out 5.1) shows that the temperature of an object determines the peak wavelength of its spectrum. Hotter objects emit bluer light. Stars with especially hot surfaces are blue, stars with especially cool surfaces are red, and our Sun is yellow. If you obtain a spectrum of a star and measure the wavelength at which the spectrum peaks, then Wien's law will tell you the temperature of the star's **surface**. The color of a star tells us only about the temperature at the surface, because this layer is giving off most of the radiation that we see. Stellar interiors are far hotter than this, as we will discover.

Look back at the chapter-opening illustration. The student identified that Vega is blue in her image and immediately determined that it must be about 10,000 K or hotter. Other stars in the image have other colors. From these colors alone, you should be able to identify several stars that are definitely cooler than Vega.

In practice, it is usually not necessary to obtain a complete spectrum of a star to determine its temperature. Instead, astronomers often measure the colors of stars by comparing the brightness at two different, specific wavelengths. The brightness of a star is often measured through an optical **filter**—sometimes just a piece of colored glass—that lets through only a small range of wavelengths. Two of the most common filters are a blue filter and a "visual" (again, yellow-green) filter. From a pair of pictures of a group of stars, each taken through a different filter, we can find an approximate value of the surface temperature of every star in the picture—perhaps hundreds or even thousands—all at once. When we do, we find there are many more cool stars than hot stars. We also discover that most stars have surface temperatures lower than that of the Sun.

Atomic Energy Levels

So far, we have concentrated on what we can learn about stars by applying our understanding of thermal radiation. However, the spectra of stars are not smooth,

---

Back in Section 5.1 . . .

. . . you learned that hotter objects are bluer and also brighter at all wavelengths.

▶▶ **Nebraska Simulation:** Blackbody Curves and Filters Explorer

▶❙ **AstroTour:** Stellar Spectrum

Vocabulary Alert

surface In common language, we don't use the word *surface* to refer to a layer within a gaseous body. But here, astronomers mean the part of the star that gives off most of the radiation that we see. A star's surface is not solid like the surface of a terrestrial planet, and stars usually have more layers outside of this "surface."

continuous blackbody spectra. Instead, when we pass the light of stars through a prism, we see dark and bright lines at specific wavelengths in their spectra. To understand these lines (which tell us much of what we know about the universe), we must know how light interacts with matter. You learned a little about light and matter in Chapter 4. Now we will explore more about how they interact.

In Chapter 4, we saw that the electrons in an atom are much less massive than protons or neutrons. As illustrated in **Figure 10.5**, most of the space in an atom is occupied by electrons, although almost all the mass of an atom is found in its nucleus. Just as waves of light have particle-like properties, particles of matter have wavelike properties.

Because of these wavelike properties, electrons in an atom can take on only certain specific energies that depend on the energy states of the atom. We can imagine these energy states as a set of shelves in a bookcase, as shown in **Figure 10.6a**. The energy of an atom might correspond to the energy of one shelf or to the energy of the next shelf, but the energy of the atom will never be found between the two states, as a book will never be found floating between two shelves. Astronomers keep track of the allowed states of an atom using energy level diagrams, as shown in Figure 10.6b, where each energy level is like a shelf on the bookcase, and the ground state is at the bottom. Both of these metaphors (the bookcase and the energy level diagram) are simplifications of the possible energies of a three-dimensional system.

The lowest possible energy state for a system (or part of a system) such as an atom is called the **ground state**. When the atom is in this state, the electron has

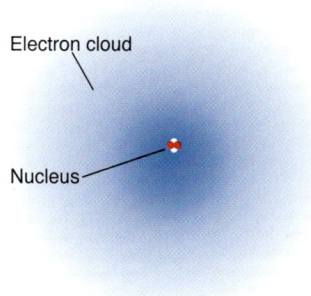

Figure 10.5 Electrons form a cloud around the nucleus of an atom.

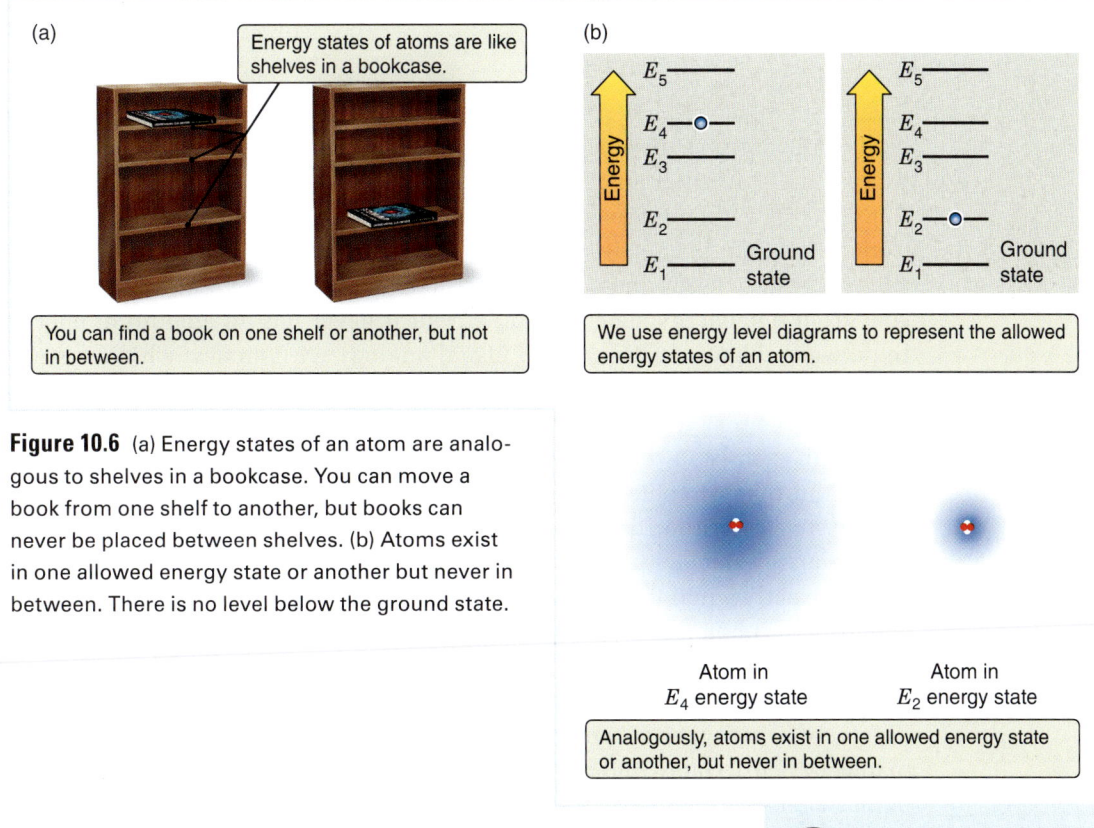

Figure 10.6 (a) Energy states of an atom are analogous to shelves in a bookcase. You can move a book from one shelf to another, but books can never be placed between shelves. (b) Atoms exist in one allowed energy state or another but never in between. There is no level below the ground state.

VISUAL ANALOGY

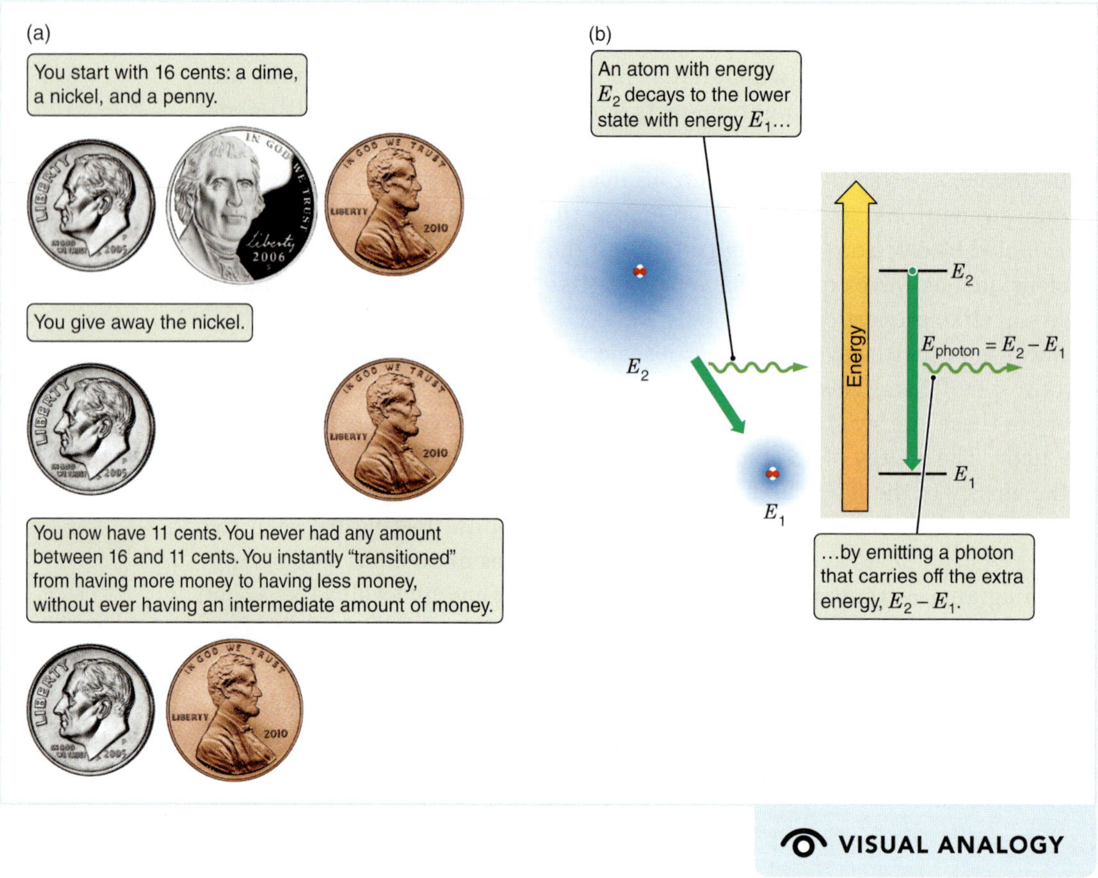

Figure 10.7 (a) The energy associated with transitions between energy states is analogous to individual coins in a handful. If you begin with a dime, a nickel, and a penny, you might give someone the nickel, leaving you with 11 cents. But you never had exactly 13 cents. First you had 16 cents, then you had 11 cents. (b) Similarly, an atom can give up photons with only specific energies. A photon with energy $E_2 - E_1$ is emitted when an atom in the higher-energy state decays to the lower-energy state.

(a) You start with 16 cents: a dime, a nickel, and a penny.

You give away the nickel.

You now have 11 cents. You never had any amount between 16 and 11 cents. You instantly "transitioned" from having more money to having less money, without ever having an intermediate amount of money.

(b) An atom with energy E_2 decays to the lower state with energy E_1…

$E_{\text{photon}} = E_2 - E_1$

…by emitting a photon that carries off the extra energy, $E_2 - E_1$.

VISUAL ANALOGY

▶‖ **AstroTour:** Atomic Energy Levels and the Bohr Model

its minimum energy. It can't give up any more energy to move to a lower state, because there isn't a lower state. An atom will remain in its ground state forever unless it gets energy from outside.

Energy levels above the ground state are called **excited states**. An atom in an excited state might transition to the ground state by getting rid of the "extra" energy all at once. It does this when the electron emits a photon. The atom goes from one energy state to another, but it never has an amount of energy in between. **Figure 10.7a** makes a visual analogy with money, which is similarly *quantized*. If you have a penny, a nickel, and a dime, you have 16 cents. Now imagine that you give away the nickel. You are left with 11 cents. But you never had exactly 13 cents, and certainly you never had 13.6 cents. You had 16 cents, and then 11 cents. Atoms don't accept and give away money to change energy states, but they do accept and give away photons. Atoms falling from a higher-energy state with energy E_2 to a lower-energy state, E_1, lose an amount of energy exactly equal to the difference in energy levels, $E_2 - E_1$. Therefore, the energy of the photon emitted must be $E_{\text{photon}} = E_2 - E_1$. This change is illustrated in Figure 10.7b, where the downward arrow indicates that the atom went from the higher state to the lower state. Electron transitions between these states lead to two different types of spectra: **emission spectra**, in which atoms are falling to lower-energy states, and **absorption spectra**, in which atoms are jumping to higher-energy states.

The energy level structure of an atom determines the wavelengths of the pho-

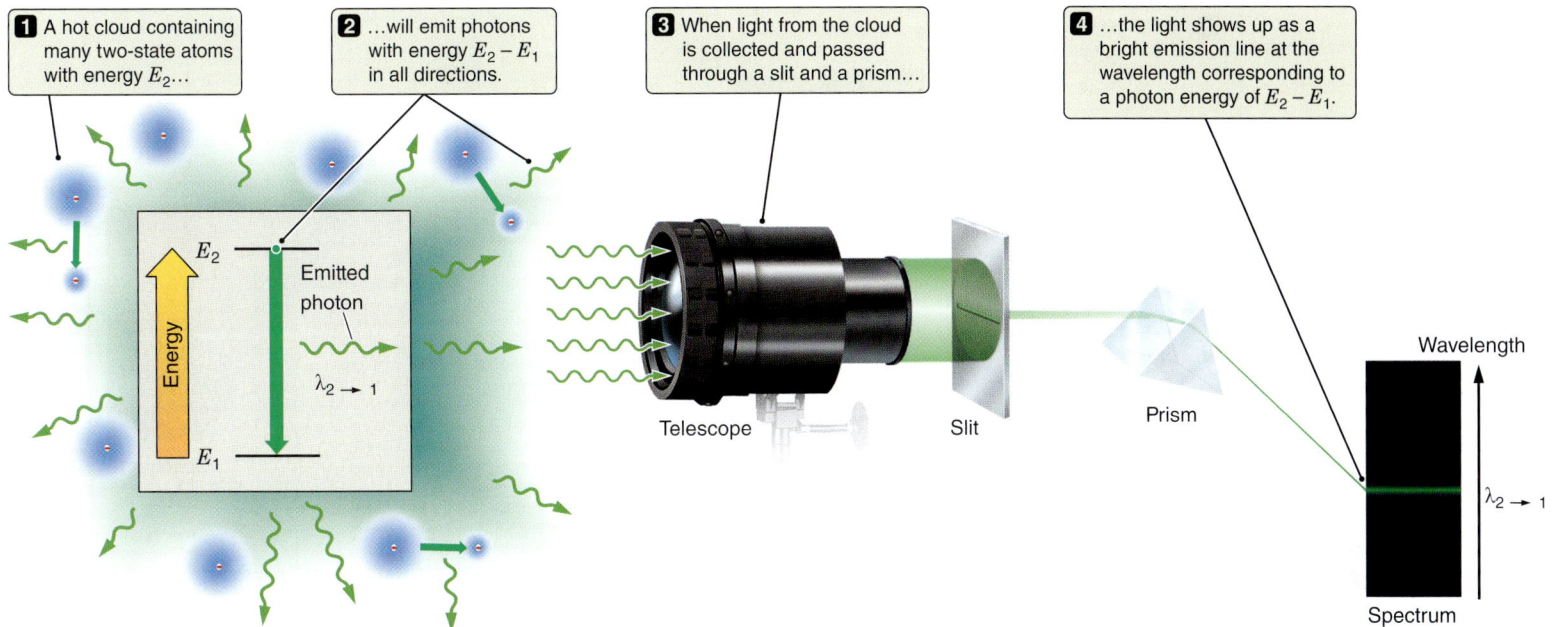

Figure 10.8 A hot cloud of gas containing atoms with two energy states, E_1 and E_2 (left), emits photons with an energy $E_2 - E_1$. When these photons pass through an astronomical instrument (middle), they appear in the spectrum (right) as a single *emission line*.

tons it emits—the color of the light that the atom gives off. An atom can emit photons with energies corresponding *only* to the difference between two of its allowed energy states. Recall from Chapter 4 that the energy, E, wavelength, λ, and frequency, f, of photons are all related ($E = hf$ and $\lambda = c/f$, where h and c are constants). A photon of energy $E_{\text{photon}} = E_2 - E_1$ has a specific wavelength $\lambda_{2 \rightarrow 1}$ and a specific frequency $f_{2 \rightarrow 1}$. Therefore, these emitted photons have a very specific color, and every photon emitted in any transition from E_2 to E_1 will have this same color.

Why was the atom in the excited state E_2 in the first place? An atom sitting in its ground state will remain there forever unless it absorbs just the right amount of energy to kick it up to an excited state. In general, the atom either absorbs the energy of a photon or it collides with another atom, or perhaps an unattached electron, and absorbs some of the other particle's energy. Atoms moving from a lower-energy state E_1 to a higher-energy state E_2 can *absorb* only energy $E_2 - E_1$, whether it comes in the form of photons or collisions.

Imagine a cloud of hot gas consisting of atoms with only two energy states, E_2 and E_1, as shown in **Figure 10.8**. Because the gas is hot, the atoms are continually zooming around and colliding, thus getting kicked up from the ground state, E_1, into the higher-energy state, E_2. Any atom in the higher-energy state quickly decays and emits a photon in a random direction. This emitted light contains only photons with the specific energy $E_2 - E_1$. In other words, all of the light coming from the cloud is the same color. If passed through a slit and a prism, it forms a single bright line of one color, called an *emission line*. This is how some neon signs work: Each color in a "neon" sign comes from a different gas (not necessarily neon) trapped inside the glass tubes. A spectrum like the one shown to the right in Figure 10.8 is an emission spectrum, identifiable because it is dominated by emission lines.

▶▶ **Nebraska Simulation: Hydrogen Atom Simulator**

256 CHAPTER 10 *Measuring the Stars*

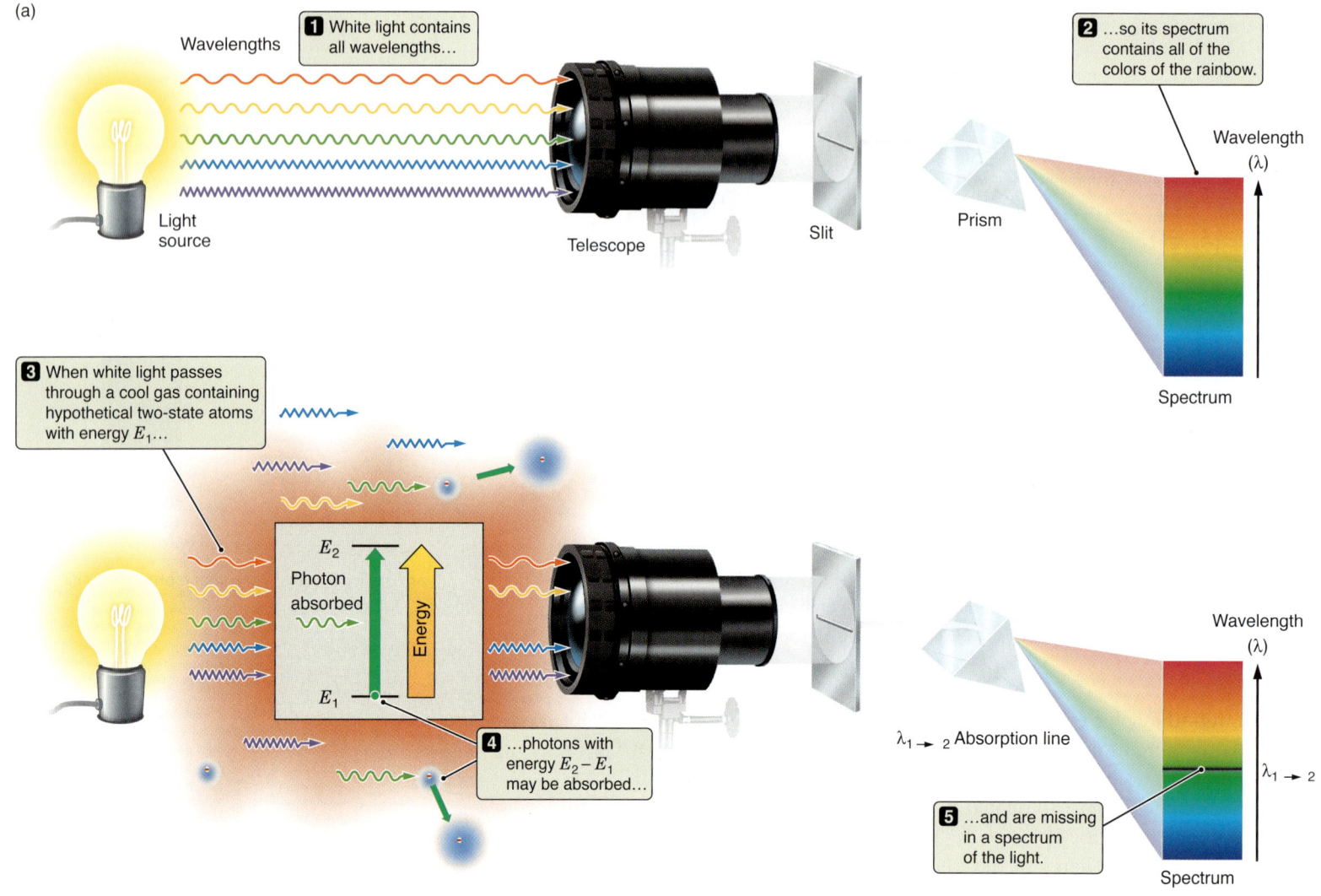

In **Figure 10.9a**, we imagine viewing a white light (one with all wavelengths of photons in it) through a spectrometer. All the photons pass through and create a rainbow (a spectrum) on a detector. However, when the white light passes through a cool cloud of gas, some photons will be absorbed. Almost all of the photons will pass through the cloud of gas unaffected because they do not have the right amount of energy ($E_2 - E_1$) to be absorbed by atoms of the gas. However, photons with just the right amount of energy will be absorbed; as a result, they will be *missing* from the spectrum. We will see a sharp, dark line at the wavelength corresponding to this energy. This process is called *absorption*, and the dark line is called an *absorption line*. Figures 10.9b and 10.9c show such absorption lines in the spectrum of a star. The spectrum is shown in two different ways here: as a rainbow with light missing, and then again as a graph of the brightness at every wavelength. Comparing the top and bottom versions of the spectrum, you can see that where there are dark lines, the brightness drops abruptly at a particular wavelength. Places between the lines are brighter and therefore higher on the graph than the absorption lines.

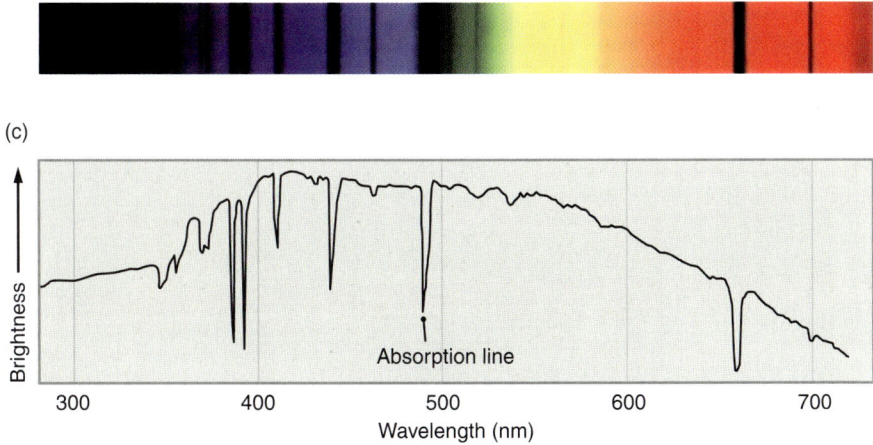

Figure 10.9 (a) When passed through a prism, white light produces a spectrum containing all colors. If light of all colors passed through a cloud of atoms with only two possible states, photons with energy $E_2 - E_1$ would be absorbed, leading to the dark absorption line in the spectrum. Absorption lines in the spectrum of a star may be viewed two ways: (b) The camera attached to the telescope captures an image of the "rainbow" with the dark lines where absorption has occurred. (c) Astronomers measure the brightness at each wavelength and make a graph that shows the shape of the absorption line in more detail.

For any element, the absorption lines occur at exactly the same wavelength as the emission lines. The energy difference between the two levels is the same whether the electron in the atom is emitting a photon or absorbing one, so the energy of the photon involved will be the same in either case. The spectrum shown in Figure 10.9b is typical of an absorption spectrum: The blackbody spectrum of the object is bright, with dark lines superimposed on it where light has been absorbed by atoms.

When an atom absorbs a photon, it may quickly return to its previous energy state, emitting a photon with the same energy as the photon it just absorbed. If the atom emits a photon just like the one it absorbed, you might reasonably ask why the absorption matters at all, as the photon taken out of the spectrum was replaced by an identical one. The photon was replaced, it's true, but all of the absorbed photons were originally traveling in the *same direction*, whereas the emitted photons are now traveling in *random directions*. In other words, most of the photons have been diverted from their original paths. If you look at a white light *through* the cloud, you will observe an absorption line at a wavelength of $\lambda_{1 \to 2}$, but if you look at the cloud from another direction, you will observe an emission line at this same wavelength.

▶ll **AstroTour: Atomic Energy Levels and Light Emission and Absorption**

Emission and Absorption Lines Are the Spectral Fingerprints of Atoms

Real atoms can occupy many more than just two possible energy states, so an atom of a given element is capable of emitting and absorbing photons at many different wavelengths. In an atom with three energy states, for example, the electron might jump from state 3 to state 2, or from state 3 to state 1, or from state 2 to state 1. Its spectrum will have three distinct emission lines.

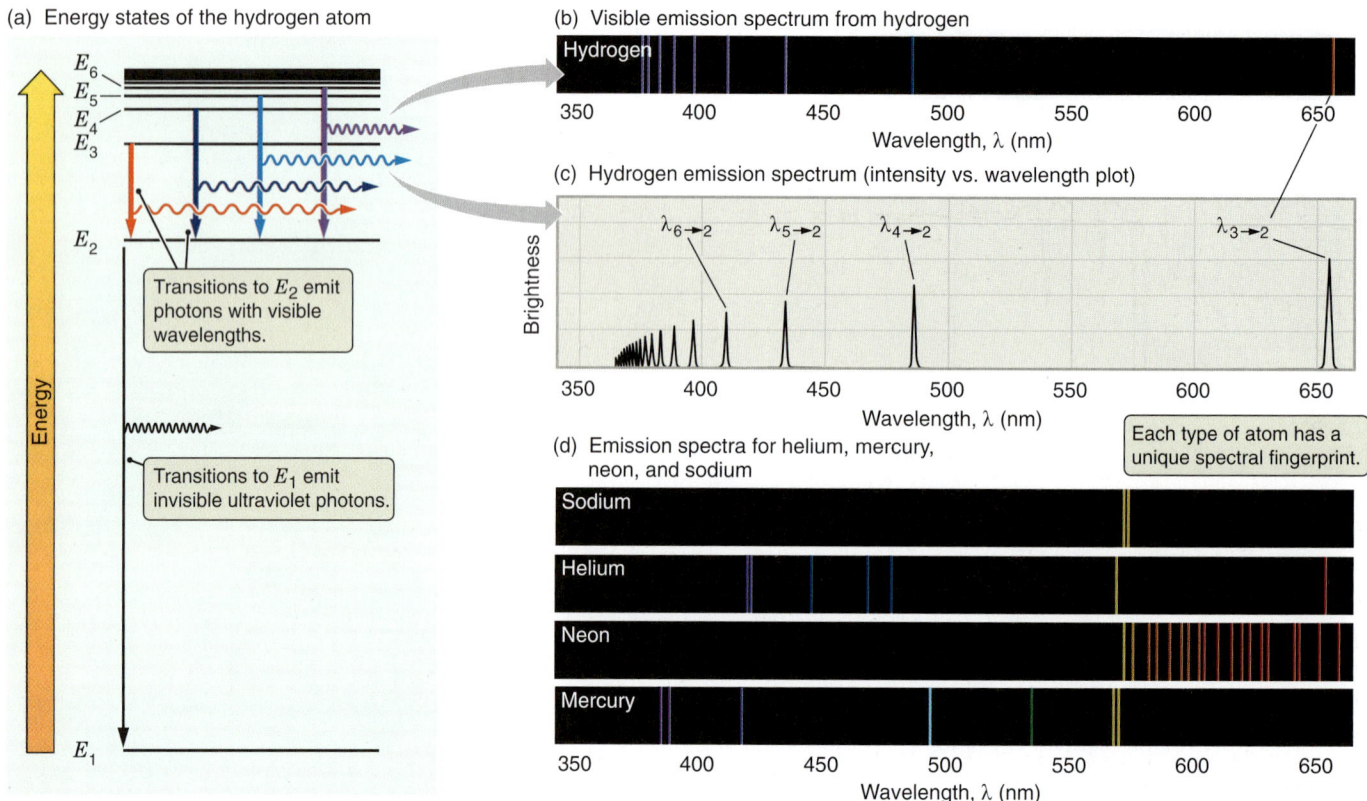

Figure 10.10 (a) Electrons make transitions between the energy states of the hydrogen atom. Transitions from higher levels to level E_2 emit photons in the visible part of the spectrum. (b) The light from a hydrogen lamp produces an emission spectrum. This is the image a camera would produce if you took a picture of the spectrum. (c) The brightness at every wavelength can be measured to produce a graph of the brightness of spectral lines versus their wavelength. (d) Emission spectra from several other types of gases.

Every hydrogen atom has the same energy states available to it, and all hydrogen atoms have the same emission and absorption lines. **Figure 10.10a** shows the energy level diagram of hydrogen. Figure 10.10b shows the spectrum of emission lines for hydrogen in the visible part of the spectrum. Figure 10.10c displays this same information as a graph. Each different element has a unique set of available energy states and therefore a unique set of wavelengths at which it can emit or absorb radiation. Figure 10.10d shows the set of emission lines from different kinds of atoms. These unique sets of wavelengths serve as unmistakable spectral "fingerprints" for each element.

If we see the spectral lines of hydrogen, helium, carbon, oxygen, or any other element in the light from a distant object, then we know that element is present in that object. The **strength** of a line is determined in part by how many atoms of that type are present in the source. By measuring the strength of the lines from an element, astronomers can often infer the abundance of the element in the object—and sometimes the temperature, density, and pressure of the material as well.

Vocabulary Alert

strength In this context, *strength* means how bright the emission line is or how faint the absorption line is. More atoms either add more light (in the case of emission) or remove more light (in the case of absorption), making a stronger line. A strong line may be deep, wide, or both.

Classification of Stars

Although the hot "surface" of a star emits radiation with a spectrum very close to a smooth blackbody curve, this light must then escape through the outer layers of the star's atmosphere. The atoms and molecules in the cooler layers of the

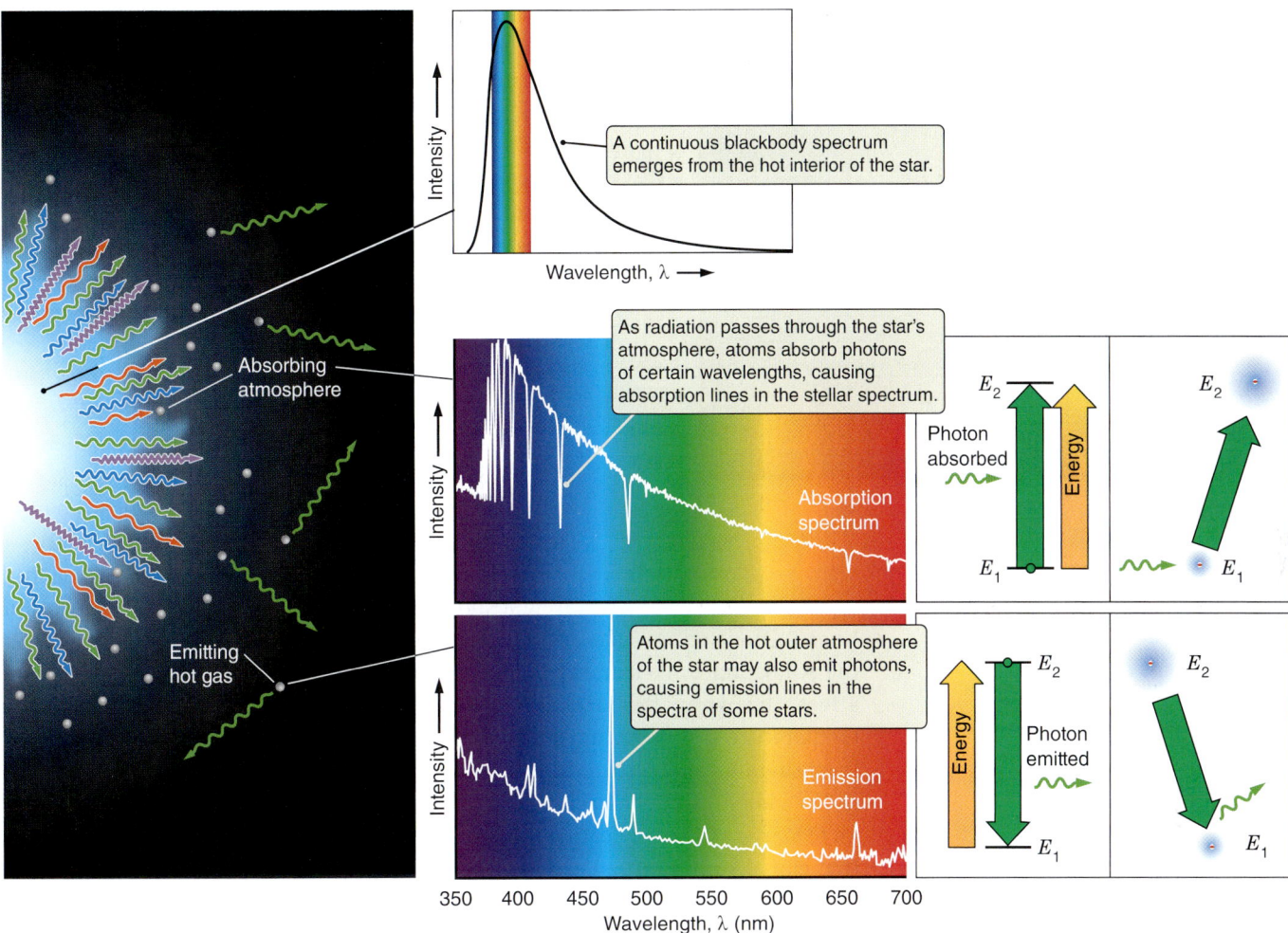

Figure 10.11 Absorption and emission lines both appear in the spectra of stars. The blackbody spectrum is the light emitted from a hot object, just because it is hot (recall Chapter 5). As that light passes through a gas, some of it is absorbed, producing an absorption spectrum. Hot gas also emits light and produces emission lines in the spectra of some stars.

star's atmosphere leave their absorption line fingerprints in this light, as shown in **Figure 10.11**. These atoms and molecules, along with any gas that might be found in the vicinity of the star, can also produce emission lines in stellar spectra. Although absorption and emission lines complicate how we use the laws of blackbody radiation to interpret light from stars, spectral lines more than make up for this trouble by providing a wealth of information about the state of the gas in a star's atmosphere.

The spectra of stars were first classified during the late 1800s, long before stars, atoms, or radiation were well understood. Stars with the strongest hydrogen lines were labeled "A stars," stars with somewhat weaker hydrogen lines were labeled "B stars," and so on. The classification we use today is based on the prominence of particular absorption lines seen in the spectra.

Annie Jump Cannon (1863–1941) led an effort at the Harvard College Observatory to examine and classify systematically the spectra of hundreds of thousands of stars. She dropped many of the earlier spectral types, keeping only seven that were subsequently reordered based on surface temperatures. Spectra of stars of

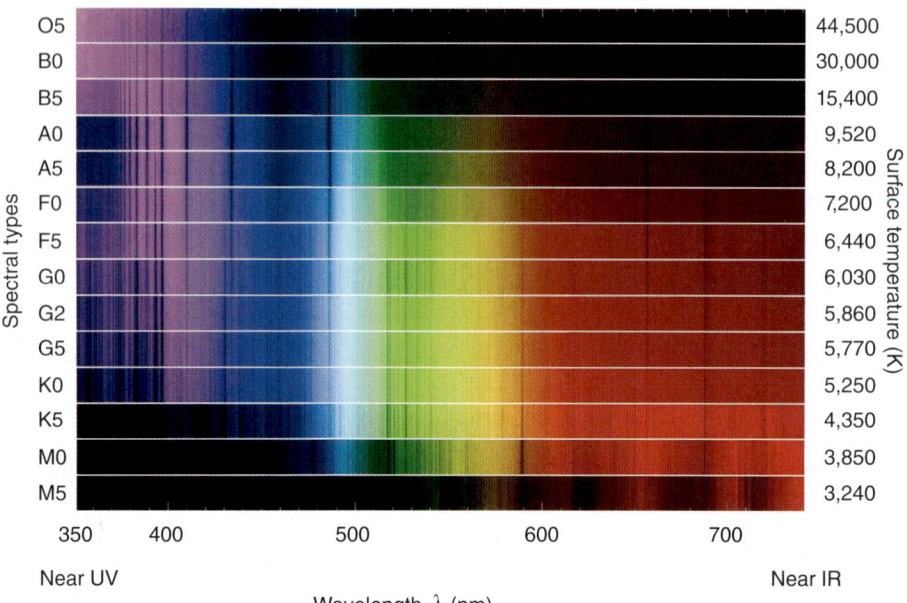

Figure 10.12 Spectra of stars with different spectral types are shown, ranging from hot blue O stars to cool red M stars. Hotter stars are more luminous at shorter wavelengths. The dark lines are absorption lines.

different types are shown in **Figure 10.12**. The hottest stars, with surface temperatures above 30,000 K, are labeled "O stars." O stars have only weak absorption lines from hydrogen and helium. The coolest stars—"M stars"—have temperatures as low as 2800 K. M stars show myriad lines from many different types of atoms and molecules. The complete sequence of **spectral types** of stars, from hottest to coolest, is O, B, A, F, G, K, M. The boundaries between spectral types are not precise. A hotter-than-average G star is very similar to a cooler-than-average F star.

Astronomers divide the main spectral types into subclasses by adding numbers to the letter designations. For example, the hottest B stars are called B0 stars, slightly cooler B stars are called B1 stars, and so on. The coolest B stars are B9 stars, which are only slightly hotter than A0 stars. The Sun is a G2 star.

Returning to Figure 10.12, we can see that not only are hot stars bluer than cool stars, but the absorption lines in their spectra are quite different as well. The temperature of the gas in the atmosphere of a star affects the state of the atoms in that gas, which in turn affects the energy level transitions available to absorb radiation. In O stars, the temperature is so high that most atoms have had one or more electrons stripped from them by energetic collisions within the gas. Few transitions are available in the visible part of the electromagnetic spectrum, so the visible spectrum of an O star is relatively featureless. At lower temperatures, there are more atoms that can absorb light in the visible part of the spectrum, so the visible spectra of cooler stars are more complex than the spectra of O stars.

Most absorption lines have a temperature at which they are strongest. For example, absorption lines from hydrogen are most prominent at temperatures of about 10,000 K, which is the surface temperature of an A star. (Spectral-type A stars are so named because they are the stars with the strongest lines of hydrogen in their spectra.)

At the very lowest stellar temperatures, atoms in the atmosphere of the star form molecules. Molecules such as titanium oxide (TiO) are responsible for much of the absorption in the atmospheres of cool M stars.

Because different spectral lines are formed at different temperatures, we can use these absorption lines to measure a star's temperature directly. The temperatures of stars measured in this way agree extremely well with the temperatures of stars measured using Wien's law, again confirming that the physical laws that apply on Earth apply to stars also.

Information from Spectral Lines

Because each type of atom has different energy levels, each type of atom has different spectral lines. For example, if a star has absorption lines that correspond to the energy difference between two levels in the calcium atom, then we know that calcium is present in the atmosphere of the star. Consider this carefully. We can determine what a star is made of simply by looking at its light! In fact, this is a common way to find out the composition of various compounds on

Earth—chemists will heat them to very high temperatures and then study their spectra.

The strengths of various absorption lines tell us not only what kinds of atoms are present in the gas but also the abundance of each. However, we must take great care in interpreting spectra to account properly for the temperature and density of the gas in the atmosphere of a star. Typically, more than 90 percent of the atoms in the atmosphere of a star are identified as hydrogen, while helium accounts for most of what remains. All of the other elements are present only in very small amounts.

The spectral lines also tell us about other physical properties of stars, such as pressure and magnetic-field strength. In addition, by making use of the Doppler shift, we can measure rotation rates, motions within the atmosphere, expansion and contraction, "winds" driven away from stars, and other dynamic properties of stars.

The Stefan-Boltzmann Law and Finding the Sizes of Stars

Stars are so far away that the vast majority of them cannot be imaged as more than point sources. Finding the size of a star, then, involves putting together other measurements that are observable: the temperature and the luminosity.

The temperature of a star can be found directly, either from Wien's law, illustrated in **Figure 10.13a**, or from the strength of its spectral lines. The temperature of a star is one factor that influences its luminosity. If a large star and a small star are the same temperature, they will emit the same energy from every patch of surface, but the large star has more patches, so it is more luminous altogether. Conversely, if two stars are the same size, the hotter one will be more luminous than the cooler one. This is an application of the Stefan-Boltzmann law, shown in Figure 10.13b.

The luminosity of a star can be found from its brightness and its distance. Because the luminosity depends on both the temperature and the size of the star, combining the luminosity with the temperature allows us to determine the radius of the star, as shown in Figure 10.13c. Recall that we carried out a calculation relating these three quantities in Working It Out 5.1. In that case, Earth was the example, but the laws of physics remain the same for planets and for stars.

The luminosity-temperature-radius relationship has been used to estimate the radii of thousands of stars. When we talk about the sizes of stars, we again use our Sun as a yardstick. The radius of the Sun, written as $R_\odot$, is about 700,000 km. When we look at stars around us, we find that the smallest stars we see, called white dwarfs, have radii that are only about 1 percent of the radius of the Sun ($R = 0.01\ R_\odot$). The largest stars that we see, called red supergiants, can have radii more than 1,000 times that of the Sun. There are many more stars toward the small end of this range, smaller than our Sun, than there are giant stars.

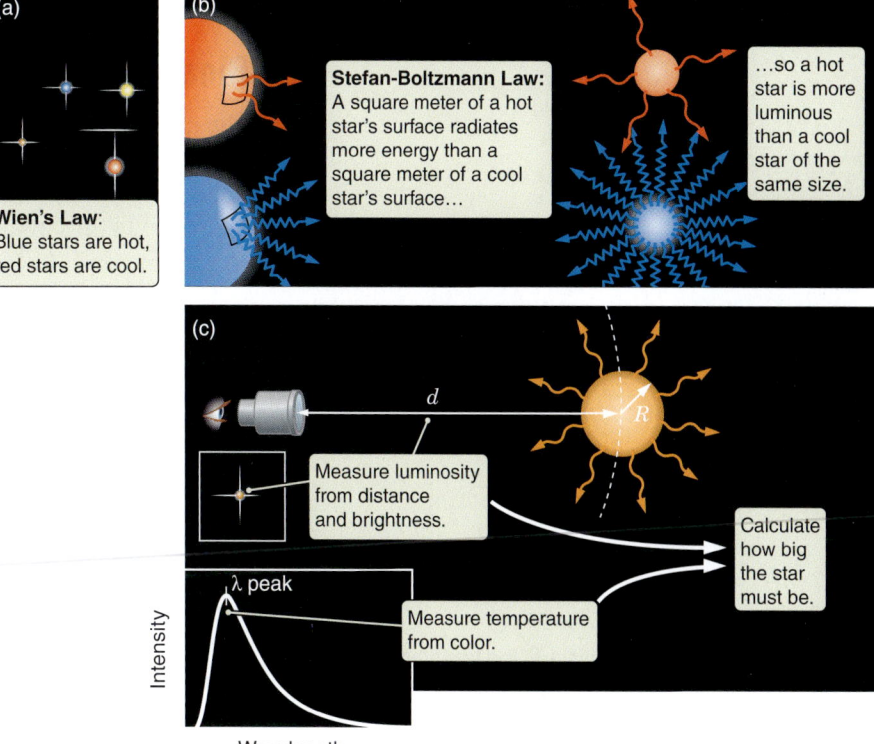

Figure 10.13 (a) The temperature of a star can be found from its color through Wien's law. (b) The luminosity depends on both the temperature and the size of the star. (c) Once the temperature and the luminosity are known, the size of the star can be calculated.

Figure 10.14 The center of mass of two objects is the "balance" point on a line joining the centers of the two masses.

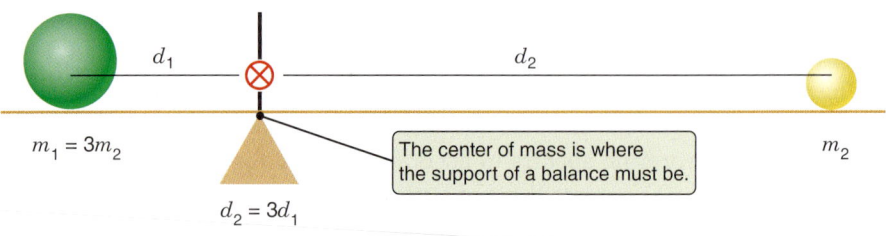

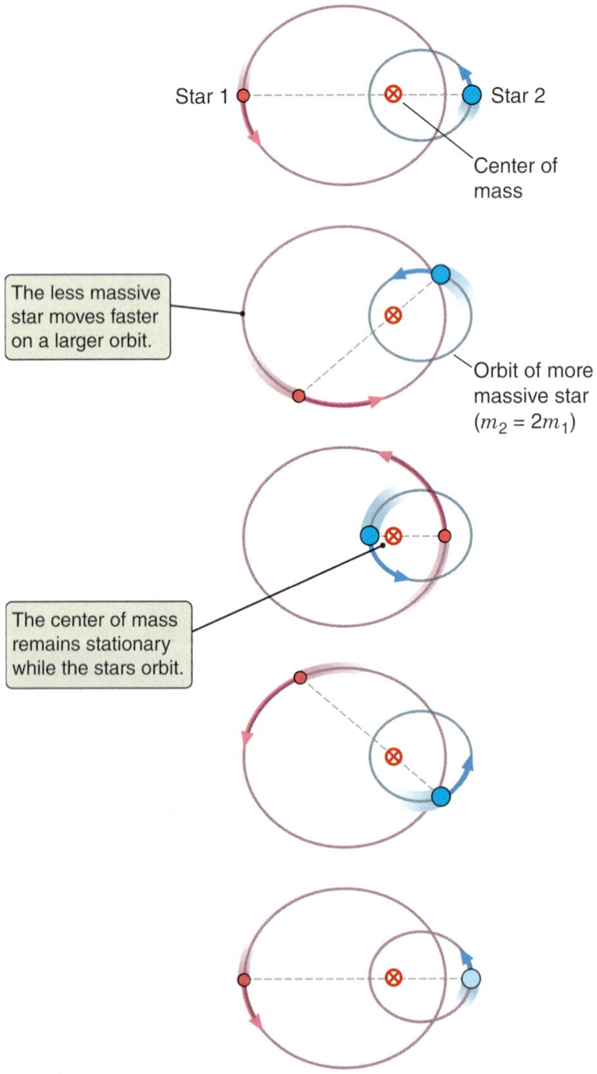

10.3 The Mass of a Star Can Be Determined in Some Binary Systems

Determining the mass of an object can be tricky. We certainly cannot rely on the amount of light from an object or the object's size as a measure of its mass. Massive objects can be large or small, faint or luminous. However, more massive objects *always* have stronger gravity. When astronomers are trying to determine the masses of astronomical objects, they almost always wind up looking for the effects of gravity.

In Chapter 3, we found that Kepler's laws of planetary motion are the result of gravity, and we showed that the orbit of a planet can be used to measure the mass of the Sun. We can also study two *stars* that orbit around each other. About half of the higher-mass stars in the sky are actually systems consisting of several stars moving around under the influence of their mutual gravity. Most of these are **binary stars** in which two stars orbit each other as predicted by Newton's version of Kepler's laws. This version of Kepler's laws can be used to find the mass of a star, as we will show in this section. However, most stars are single, and their mass cannot be found this way.

Binary Stars Orbit a Common Center of Mass

The **center of mass** is the balance point of a system. If the two objects were sitting on a seesaw in a gravitational field, the support of the seesaw would have to be directly under the center of mass for the objects to balance, as shown in **Figure 10.14**. In a binary system, the two stars orbit the center of mass, a point in space that is seldom located inside either star but usually somewhere in between.

When Newton applied his laws of motion to the problem of orbits, he found that two objects must move in elliptical orbits around each other, and that their common center of mass lies at one focus shared by both of the ellipses, as shown in **Figure 10.15**. The center of mass, which lies along the line between the two objects, remains stationary. The two objects will always be found on exactly opposite sides of the center of mass.

Because the orbit of the less massive star is larger than the more massive star's orbit, the less massive star has farther to go than the more massive star. But it must cover that distance in the same amount of time, so the less massive star must be moving *faster* than the more massive star. The velocity of a star in a binary system is inversely proportional to its mass.

Imagine that you are watching a binary star as shown in **Figure 10.16a**. As seen from above, two stars orbit the common center of mass. The less massive star (star 2) must complete its orbit in the same time that the more massive star does. Because the less massive star has farther to go around the center of mass,

Figure 10.15 In a binary star system, the two stars orbit on elliptical paths about their common center of mass. In this case, star 2 has twice the mass of star 1. The eccentricity of the orbits is 0.5. There are equal time steps between the frames.

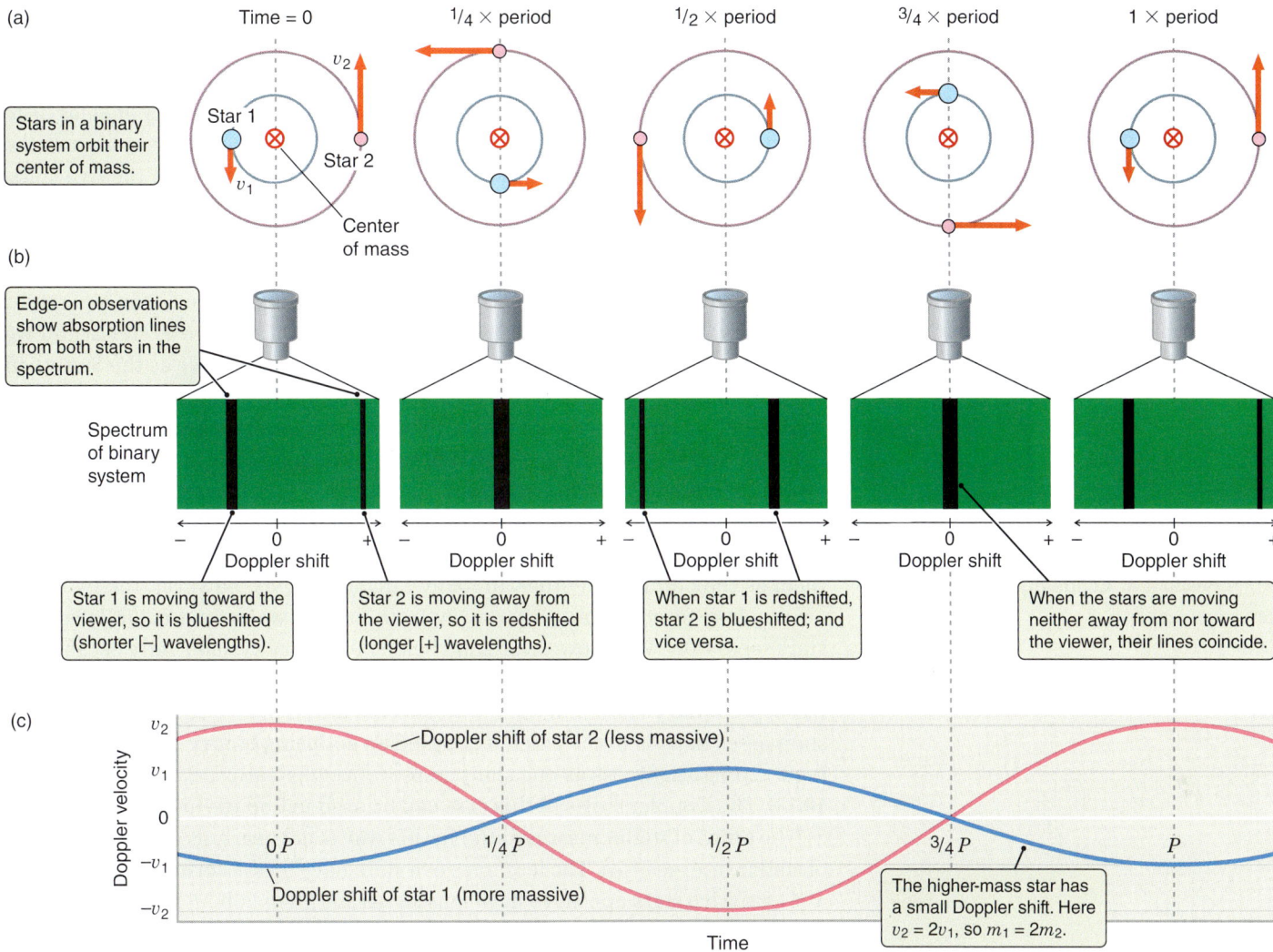

Figure 10.16 (a) The view from "above" the binary system shows that both stars orbit a common center of mass. (b) The spectrum of the combined system (seen edge-on) shows the spectral lines of each star shift back and forth. (c) Graphing the Doppler shift of star 1 with star 2 versus time reveals that star 1 has half the maximum Doppler shift, so star 1 is twice as massive as star 2. *P* is the period of the orbit.

it must also move more quickly. In this view, no determination of the Doppler shift (recall Chapter 5) can be made because all the motion is in the plane of the sky, and none is toward or away from the observer.

When the system is edge-on to the observer, however, the observer can take advantage of the Doppler shift to find out about the motion. Observations of the spectrum of the combined system (Figure 10.16b) show that the spectral lines of the stars shift back and forth as they move toward and away from the observer. When star 1 approaches, star 2 recedes. The light coming from star 1 will be shifted to shorter wavelengths by the Doppler effect as it approaches, so the light will be blueshifted, and the light coming from star 2 will be shifted to longer wavelengths as it recedes, so the light will be redshifted. Half an orbital period later, the situation would be reversed: lines from star 1 would be redshifted, and lines from star 2 would be blueshifted.

The less massive star has a larger orbit—and consequently moves more quickly—than the more massive star. Comparing the maximum Doppler shift for star 1 with the maximum Doppler shift for star 2 (Figure 10.16c) gives the *ratio* of the masses of the two stars. That is, we can find that star 1 is two times as massive as star 2. But we can't find the actual mass of either star from these observations alone.

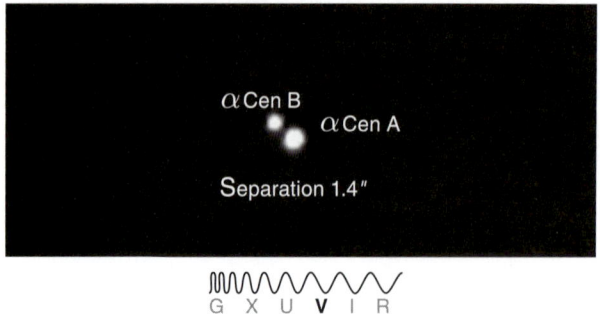

Figure 10.17 The two stars of this visual binary are resolved. These stars are two components of Alpha Centauri, the nearest star system to the Sun.

▶▶ **Nebraska Simulation: Eclipsing Binary Simulator**

▶▶ **Nebraska Simulation: Center of Mass Simulator**

Kepler's Third Law and Total Mass of a Binary System

In Chapter 3, we ignored the complexity of the motion of two objects around their common center of mass. Now, however, this very complexity enables us to measure the masses of the two stars in a binary system. If we can measure the period of the binary system and the average separation between the two stars, then Kepler's third law gives us the total mass in the system: the sum of the two masses. Because the analysis in the previous subsection gives us the ratio of the two masses, we now have two different relationships between two different unknowns. We have all we need to determine the mass of each star separately. In other words, if we know that star 1 is two times as massive as star 2, and we know that star 1 and star 2 together are three times as massive as the Sun, then we can calculate separate values for the masses of star 1 and star 2.

Depending on the type of system, there are two ways to measure the average distance and the period. In a **visual binary** system, shown in **Figure 10.17**, the system is close enough to Earth, and the stars are far enough from each other, that we can take pictures that show the two stars separately. We can directly measure the shapes and period of the orbits of the two stars, just by watching them as they orbit each other. In many binary systems, however, the two stars are so close together and so far away from us that we cannot actually see the stars separately. We know these stars belong to binary systems only because we see the Doppler shift in the spectral lines of the two stars; these are called **spectroscopic binary** stars. If a binary system is viewed nearly edge-on so that one star passes in front of the other, it is called an **eclipsing binary**. An observer will see a dip in brightness as one star passes in front of (eclipses) the other (**Figure 10.18**). The Doppler shifts in this case can be used to find the masses of the stars.

The range of stellar masses found in this way is not nearly as great as the range of stellar luminosities. The least massive stars have masses of about 0.08 $M_\odot$; the most massive stars appear to have masses greater than 200 $M_\odot$. You might wonder why the mass of a star should have any limits. These limits are determined solely by the physical processes that go on deep in the interior of the star. You will learn in the chapters ahead that a minimum stellar mass is necessary to ignite the nuclear furnace that keeps a star shining, but the furnace can run out of control if the stellar mass is too great. Thus, although the most luminous stars are 10^{10}, or 10 billion, times more luminous than the least luminous ones, the most massive stars are only about 10^3, or a thousand, times more massive than the least massive stars.

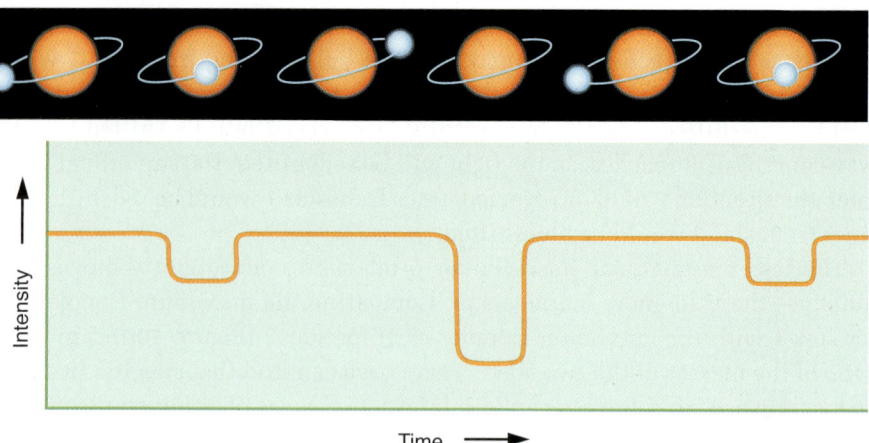

Figure 10.18 In an eclipsing binary system, the system is viewed nearly edge-on, so that the stars repeatedly pass behind one another, blocking some of the light. Even though the blue star here is smaller, it is significantly more luminous because its temperature is higher. When the blue star passes in front of the larger, cooler star, less light is blocked than when it passes behind the red star. The shape of the dips in the light curve of an eclipsing binary can reveal information about the relative size and surface brightness of the two stars.

10.4 The H-R Diagram Is the Key to Understanding Stars

We have come a long way in our effort to measure the physical properties of stars. However, just knowing some of the basic properties of stars does not mean that we understand stars. The next step involves looking for patterns in the properties we have determined. The first astronomers to take this step were Ejnar Hertzsprung (1873–1967) and Henry Norris Russell (1877–1957). In the early part of the 20th century, Hertzsprung and Russell studied the properties of stars independently; each plotted the luminosities of stars versus their surface temperatures. The resulting plot is referred to as the Hertzsprung-Russell diagram, or simply the **H-R diagram** (Figure 10.19). The H-R diagram is one of the most used and useful diagrams in astronomy. Many stages in the evolution of stars are shown on the H-R diagram. In this section, we take a first look at this important diagram and the way stars are organized within it.

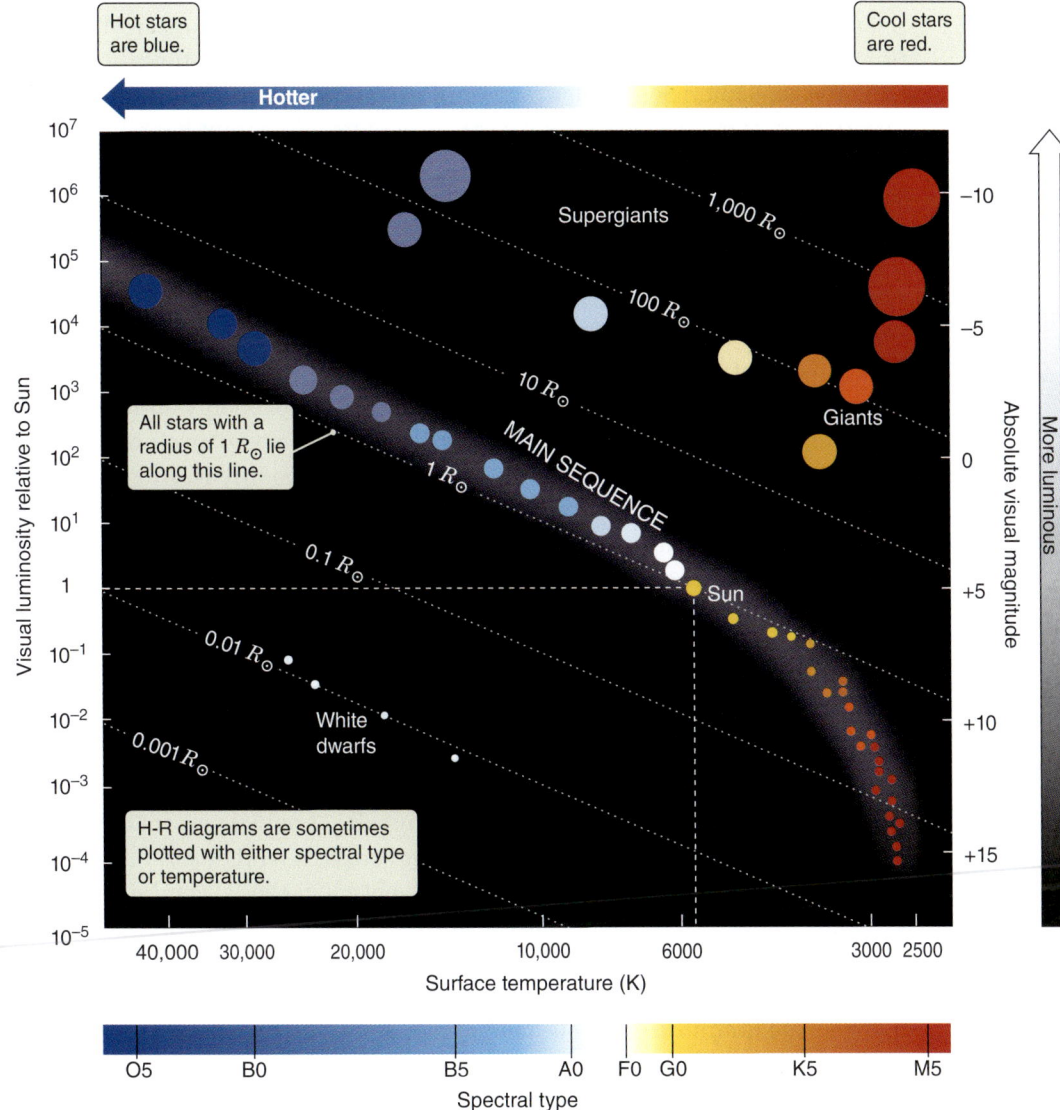

Figure 10.19 The Hertzsprung-Russell, or H-R, diagram is used to plot the properties of stars. More luminous stars are at the top of the diagram. Hotter stars are on the left. Stars of the same radius (R) lie along the dotted lines moving from upper left to lower right.

▶▶ **Nebraska Simulation:** Hertzsprung-Russell Diagram Explorer

▶❙❙ **AstroTour:** H-R Diagram

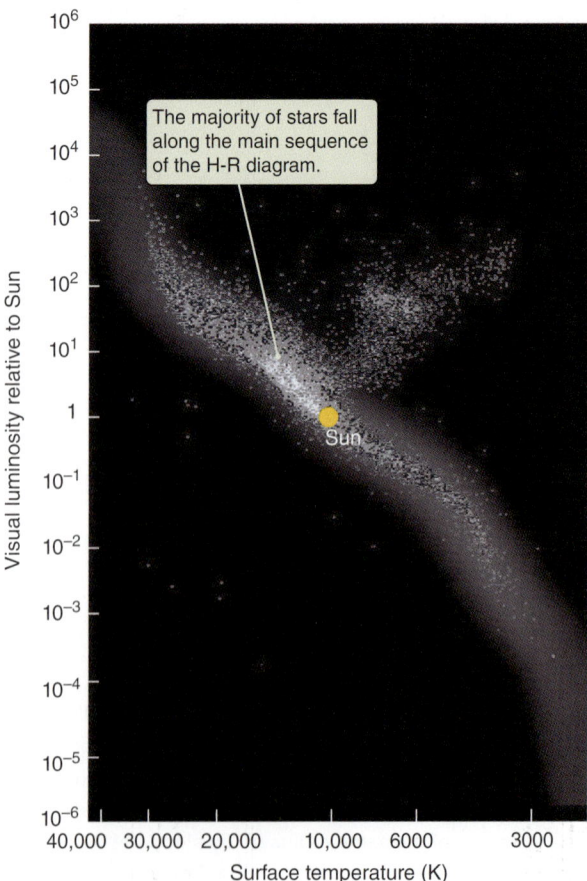

Figure 10.20 An H-R diagram for 16,600 stars plotted from data obtained by the *Hipparcos* satellite clearly shows the main sequence. Most of the stars lie along this band running from the upper left of the diagram toward the lower right.

The H-R Diagram

We begin with the layout of the H-R diagram itself, shown in Figure 10.19. The surface temperature is plotted on the horizontal axis (the *x*-axis), but it is plotted backward: temperature is high on the left and low on the right. Hot blue stars are on the left side of the H-R diagram; cool red stars are on the right. Temperature is plotted logarithmically, which means that the size of an interval along the axis from a point representing a star with a surface temperature of 40,000 K to one with a surface temperature of 20,000 K—a temperature change by a factor of 2—is the same as the size of an interval between points representing a star with a temperature of 10,000 K and a star with a temperature of 5000 K, which is also a temperature change by a factor of 2. The temperature axis is sometimes labeled with another characteristic that corresponds to temperature, such as the spectral type, as shown at the bottom of the graph.

Along the vertical axis (the *y*-axis), we plot the luminosity of stars—the total amount of energy that a star radiates each second. More luminous stars are toward the top of the diagram, and less luminous stars are toward the bottom. Luminosities are plotted logarithmically, in this case with each step along the left-hand *y*-axis corresponding to a multiplicative factor of 10. To understand why the plotting is done this way, recall that the most luminous stars are 10 billion times more luminous than the least luminous stars, yet all of these stars must fit on the same plot. Sometimes the luminosity axis is labeled with the absolute visual magnitude instead of luminosity, as shown on the right-hand *y*-axis.

Because each point on the H-R diagram is specified by a surface temperature and a luminosity, we can use the luminosity-temperature-radius relationship to find the radius of a star at that point as well. A star in the upper right corner of the H-R diagram is very cool, so each square meter of its surface radiates a small amount of energy. But this star is also extremely luminous. It must be huge to account for its high luminosity despite the feeble radiation coming from each square meter of its surface. Conversely, a star in the lower left corner of the H-R diagram is very hot, which means that a large amount of energy is coming from each square meter of its surface. However, this star has a very low luminosity, so it must be very small. Stars in the lower left corner of the H-R diagram are small. Moving up and to the right takes you to larger and larger stars. Moving down and to the left takes you to smaller and smaller stars. All stars of the same radius lie along slanted lines across the H-R diagram.

The Main Sequence

Figure 10.20 shows 16,600 nearby stars plotted on an H-R diagram. The data are based on observations obtained by the Hipparcos satellite. A quick look at this diagram immediately reveals a remarkable fact, one that was first discovered in the original diagrams of Hertzsprung and Russell. About 90 percent of the stars in the sky lie in a well-defined region running across the H-R diagram from lower right to upper left, known as the **main sequence**. On the left end of the main sequence are the O stars: hotter, larger, and more luminous than the Sun. On the right end of the main sequence are the M stars: cooler, smaller, and fainter than the Sun. If you know where a star lies on the main sequence, then you know its approximate luminosity, surface temperature, and size.

The H-R diagram supplies a useful method for finding the distance to main-sequence stars. We can determine whether a star is a main-sequence star by looking at the absorption lines in its spectrum. We can also determine its temperature from these spectral lines. The star must lie on the main sequence above that specific temperature. We can then read across to the y-axis to find the star's luminosity. Recall that the luminosity, brightness, and distance are all connected. This method of determining distances to main-sequence stars from the spectra, luminosity, and the brightness is called **spectroscopic parallax**. Despite the similarity between the names, this method is very different from the *parallax* method using geometry, discussed earlier in the chapter. Spectroscopic parallax is useful to much larger distances than the geometric method of parallax, although it is less precise.

From a combination of observations of binary star masses, parallax, luminosity measurements, and mathematical models, astronomers have determined that stars of different masses lie on different parts of the main sequence. If a main-sequence star is less massive than the Sun, it will be smaller, cooler, redder, and less luminous than the Sun; it will be located to the lower right of the Sun on the main sequence. Conversely, if a main-sequence star is more massive than the Sun, it will be larger, hotter, bluer, and more luminous than the Sun; it will be located to the upper left of the Sun on the main sequence. The mass of a star determines where on the main sequence the star will lie. In the chapter-opening illustration, the student used this feature of the H-R diagram to find the luminosity and temperature of Vega.

For stars of similar chemical composition, the mass of a main-sequence star alone determines all of its other characteristics. Knowing the mass and chemical composition of a main-sequence star tells us how large it is, what its surface temperature is, how luminous it is, what its internal structure is, how long it will live, how it will evolve, and what its final fate will be. Therefore, its mass alone determines its position on the H-R diagram, as shown in **Figure 10.21**.

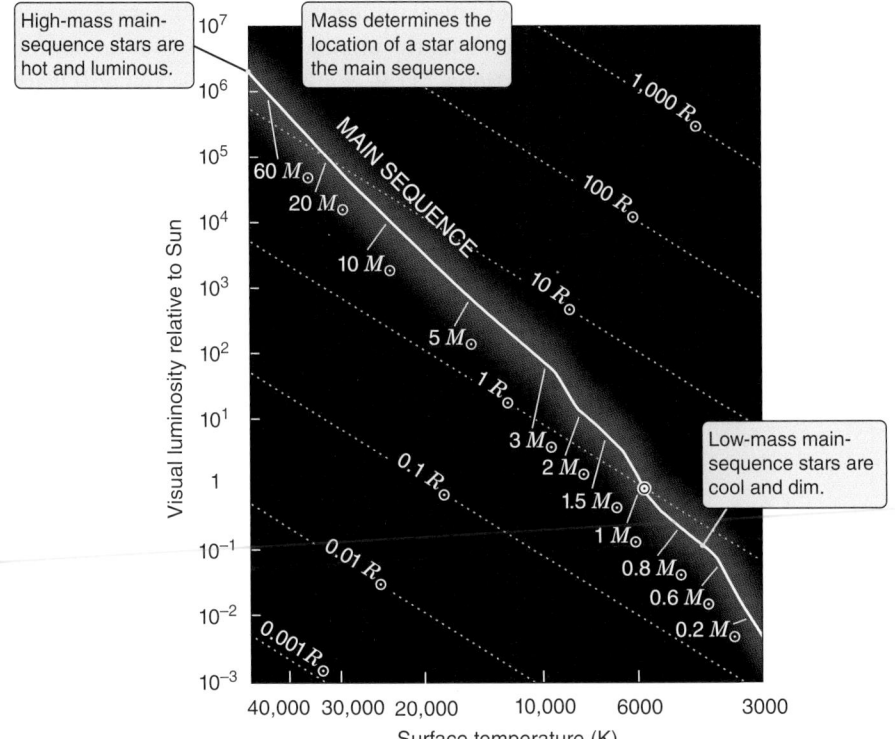

Figure 10.21 The main sequence of the H-R diagram is a sequence of masses.

▶▶ **Nebraska Simulation:** Spectroscopic Parallax Explorer

Stars Not on the Main Sequence

Although 90 percent of stars are main-sequence stars, some stars are found in the upper right portion of the H-R diagram, well above the main sequence. So they must be large, cool giants with radii hundreds or thousands of times the radius of the Sun. At the other extreme are stars found in the far lower left corner of the H-R diagram. These stars are tiny, comparable to the size of Earth. Their small surface areas explain why they have such low luminosities despite having such high temperatures.

Stars that lie off the main sequence on the H-R diagram can be identified by their luminosities (determined by their distance) or by slight differences in their spectral lines. The density and temperature of gas in a star's atmosphere affect the width of a star's spectral lines. In general, hotter stars have broader lines. Puffed-up stars above the main sequence have lower densities and lower temperatures compared to main-sequence stars. When using the H-R diagram to estimate the distance to a star by the spectroscopic parallax method, astronomers must know whether the star is on, above, or below the main sequence in order to find the star's luminosity. Stars both on and off the main sequence have a property called **luminosity class**, which tells us the *size* of the star. Supergiant stars, which are the largest stars that we see, are luminosity class I, bright giants are class II, giants are class III, subgiants are class IV, main-sequence stars are class V, and white dwarfs are class WD. Luminosity classes I through IV lie above the main sequence, while class WD falls below and to the left of the main sequence. The labels in Figure 10.19 show the approximate locations of these classes of stars on the H-R diagram. Thus, the complete spectral classification of a star includes both its spectral type, which tells us temperature and color; and its luminosity class, which indicates size.

The existence of the main sequence, together with the fact that the mass of a main-sequence star determines where on the main sequence it will lie, is a grand pattern that points to the possibility of a deep understanding of what stars are and what makes them tick. By the same token, the existence of stars that do *not* follow this pattern raises yet more questions. In the coming chapters, you will learn that the main sequence tells us what stars are and how they work, and that stars off the main sequence tell us how stars form, how they evolve, and how they die.

READING ASTRONOMY News

Astronomers are still finding out about stars. Sometimes we are lucky enough to get new data that help settle a long-standing issue.

Mystery of Nearby SS Cygni Star System Finally Resolved

By **JOHN P. MILLIS**, for Your Universe Online

In 1990, the Hubble Space Telescope measured the distance to a nearby star system known as SS Cygni, composed of a low-mass main-sequence star and a compact object known as a white dwarf—a stellar remnant about the mass of our Sun, but compressed to the size of Earth.

The distance measured by Hubble puzzled scientists, as the measured brightness of the system was considerably higher than expected. If correct, it would call into question the mechanisms by which a white dwarf interacts with a nearby companion.

"If SS Cygni was actually as far away as Hubble measured, then it was far too bright to be what we thought it was, and we would have had to rethink the physics of how systems like this worked," noted James Miller-Jones from the Curtin University campus of the International Centre for Radio Astronomy Research (ICRAR).

Miller-Jones and other astronomers have used two of the most powerful radio telescope networks in the world—the VLBA [Very Long Baseline Array] and EVLA [Expanded Very Large Array]—to measure the distance to SS Cygni, attempting to resolve the dilemma created by the Hubble result. The team used a method known as parallax, whereby the system is observed at various points during Earth's orbit around the Sun, and then the position of the system is measured against the fixed, distant background.

"If you hold your finger out at arm's length and move your head from side to side, you should see your finger appear to wobble against the background. If you move your finger closer to your head, you'll see it starts to wobble more. We did the exact same thing with SS Cygni—we measured how far it moved against some very distant galaxies as Earth moved around the Sun," noted Miller-Jones.

"The wobble we were detecting is the equivalent of trying to see someone stand up in New York from as far as away as Sydney."

The team found that SS Cygni is about 372 light-years from Earth, considerably closer than previous measurements made using the Hubble Space Telescope.

"The pull of gas off a nearby star onto the white dwarf in SS Cygni is the same process that happens when neutron stars and black holes are orbiting with a nearby companion, so a lot of effort has gone in to understanding how this works," explained Miller-Jones.

"Our new distance measurement has solved the puzzle of SS Cygni's brightness, it fits our theories after all."

Evaluating the News

1. In the second paragraph, the article states that "the distance measured" was a puzzle, because "the measured brightness of the system was considerably higher than expected." Did the author really mean "brightness" or "luminosity" here? How can you tell?
2. Evaluate the author's description of parallax. Is his explanation accessible and correct?
3. Given the distance of 372 light-years, calculate the measured parallax. (This is the opposite calculation of the one the astronomers did.) Is the analogy of the person in New York as viewed from Sydney approximately correct?
4. Sketch an H-R diagram. Label the locations of the two stars of SS Cygni.
5. Think back to Chapter 1. How do the studies described in this article reflect what you learned about the scientific method in that chapter?

SUMMARY

Finding the distances to stars is a difficult but important task for astronomers. Parallax and spectroscopic parallax are two of the methods that astronomers use to determine distances to stars. Combining the brightness with the distance yields the luminosity. Careful study of the light from a star, including its spectral lines, gives the temperature, size, and composition. Study of binary systems gives the mass of stars of various spectral types, which we can extend to all stars of the same spectral type. The H-R diagram shows the relationship among the various physical properties of stars. A star's luminosity and temperature can be combined to find the radius. The major determining factor in all the properties of a star is the mass.

LG 1 The luminosity of a star is found by combining stellar distances with the brightness of the star in the sky. The luminosity is the energy emitted from the star.

LG 2 The temperature of a star is determined by its color, with blue stars being hotter and red stars being cooler. Combining this temperature information with the luminosity of the star gives the radius.

LG 3 Spectral lines carry a great deal of information about a star. Perhaps most important, they show what elements and molecules are present in the star. The masses of stars can be obtained by observing spectral lines in binary systems.

LG 4 The H-R diagram is a key to understanding stellar properties. Temperature increases to the left, so that hotter stars lie on the left side of the diagram, while cooler stars lie on the right. Brightness or luminosity increase vertically, so that the brightest stars lie near the top of the diagram. Ninety percent of stars lie along the main sequence.

LG 5 The mass of a main-sequence star is the fundamental determining factor of all of the star's other characteristics: luminosity, temperature, and size. The main sequence on the H-R diagram is a sequence of masses.

SUMMARY SELF-TEST

For the following three questions, star A is blue and star B is red. They have equal luminosities, but star A is twice as far away as star B.

1. Which star appears brighter in the night sky?
   a. star A   b. star B

2. Which star is hotter?
   a. star A   b. star B

3. Which one is larger?
   a. star A   b. star B

4. If a star has very strong hydrogen absorption lines, which of the following are true?
   a. The temperature is right for hydrogen to make lots of transitions.
   b. Hydrogen is abundant in the star because hydrogen is absorbing at these wavelengths.
   c. Hydrogen is depleted in the star because hydrogen is not emitting at these wavelengths.

5. To find the masses of both stars in a binary system, you must find the _____ of each star, the _____ of the orbit, and the average _____ between the stars.

6. Suppose you are studying a star with a luminosity of 100 $L_\odot$ and a surface temperature of 4000 K. According to the H-R diagram, this star is a
   a. main-sequence star.
   b. giant red star.
   c. white dwarf.
   d. giant blue star.

7. If a star has very weak hydrogen lines and is blue, what does that most likely mean?
   a. The star is too hot for hydrogen lines to form.
   b. The star has no hydrogen.
   c. The star is too cold for hydrogen lines to form.
   d. The star is moving too fast to measure the lines.

8. Which of the following can be determined from the location of a main-sequence star on the H-R diagram? (Select all that apply.)
   a. temperature
   b. luminosity
   c. radius
   d. mass

9. If a star is found *directly* to the right of the Sun on the H-R diagram, what can you conclude about its temperature?
   a. It is hotter than the Sun.
   b. It is cooler than the Sun.
   c. It is exactly the same temperature as the Sun.
   d. You would need to know if it was on the main sequence to answer this question.

10. If a star has the same mass as the Sun, what can you conclude about its temperature?
    a. It is hotter than the Sun.
    b. It is cooler than the Sun.
    c. It is the same temperature as the Sun.
    d. You would need to know if it was on the main sequence to answer this question.

QUESTIONS AND PROBLEMS

Multiple Choice and True/False

11. **T/F:** Red stars have cooler surfaces than blue stars.
12. **T/F:** The brightness of a star in the sky tells you its luminosity.
13. **T/F:** An atom can emit or absorb a photon of any wavelength.
14. **T/F:** *Visual binary* is just another term for eclipsing binaries.
15. **T/F:** The mass of an isolated star must be inferred from the star's position on the H-R diagram.
16. Two stars have equal luminosities, but star A has a much larger radius than star B. What can you say about these stars?
    a. Star A is hotter than star B.
    b. Star A is cooler than star B.
    c. Star A is farther away than star B.
    d. Star A is brighter in our sky than star B.
17. Most stars are
    a. cool and low-luminosity.
    b. hot and low-luminosity.
    c. cool and high-luminosity.
    d. hot and high-luminosity.
18. Suppose an atom has energy levels at 1, 3, 4, and 4.3 (in arbitrary units). Which of the following is not a possible energy for an emitted photon?
    a. 2   b. 1
    c. 1.3   d. 2.3
19. An eclipsing binary system has a primary eclipse (star A is eclipsed by star B) that is deeper (more light is removed from the light curve) than the secondary eclipse (star B is eclipsed by star A). What does this tell you about stars A and B?
    a. Star A is hotter than star B.
    b. Star B is hotter than star A.
    c. Star B is larger than star A.
    d. Star B is moving faster than star A.

For the following three questions, suppose you are studying a main-sequence star of 10 solar masses. Consult the H-R diagram in Figure 10.21 to answer these questions.

20. This star's luminosity is roughly
    a. $0.01\ L_\odot$.   b. $1\ L_\odot$.
    c. $100\ L_\odot$.   d. $10,000\ L_\odot$.
21. This star's temperature is roughly
    a. 1000 K.   b. 10,000 K.
    c. 20,000 K.   d. 100,000 K.
22. This star's radius is roughly
    a. $0.1\ R_\odot$.   b. $1\ R_\odot$.
    c. $10\ R_\odot$.   d. $100\ R_\odot$.
23. If a star is found *directly* above the Sun on the H-R diagram, what can you conclude about its size?
    a. It is larger than the Sun.
    b. It is smaller than the Sun.
    c. It is exactly the same size as the Sun.
    d. You would need to know if it was on the main sequence to answer this question.
24. If a star is found *directly* above the Sun on the H-R diagram, what can you conclude about its luminosity?
    a. It is more luminous than the Sun.
    b. It is less luminous than the Sun.
    c. It is exactly the same luminosity as the Sun.
    d. You would need to know if it was on the main sequence to answer this question.
25. If a star has the same temperature as the Sun, what can you conclude about its mass?
    a. It is larger than the Sun.
    b. It is smaller than the Sun.
    c. It is the same size as the Sun.
    d. You would need to know if it was on the main sequence to answer this question.

Conceptual Questions

26. Make a table from the information in this chapter; show each property discussed and the methods of finding that property of an individual star. For example, one row might read:

Property	Method
Composition	Analyze the lines in the star's spectrum.

27. The light from some stars passes through dust in our galaxy before it reaches us, making stars appear dimmer than they actually are. How does this phenomenon affect parallax? How does it affect spectroscopic parallax?
28. What would happen to our ability to measure stellar parallax if we were on the planet Mars? What about Venus or Jupiter?
29. Albireo, a star in the constellation Cygnus, is a visual binary system whose two components can be seen easily with even a small amateur telescope. Viewers describe the brighter star as "golden" and the fainter one as "sapphire blue."
    a. What does this description tell you about the relative temperatures of the two stars?
    b. What does it tell you about their respective sizes?
30. Very cool stars have temperatures around 2500 K and emit blackbody spectra with peak wavelengths in the red part of the spectrum. Do these stars emit any blue light? Why or why not?
31. The stars Betelgeuse and Rigel are both in the constellation Orion. Betelgeuse appears red in color and Rigel is bluish white. To the eye, the two stars appear almost equally bright. If you can compare the temperature, luminosity, or size from just this information, do so. If not, explain why.
32. You obtain a spectrum of an object in space. The spectrum consists of a number of sharp, bright emission lines. Is this object a cloud of hot gas, a cloud of cool gas, or a star?
33. Look carefully at Figure 10.9b. In this spectrum, what tells you that there is a white light source? What tells you there is a cool cloud of gas? Suppose the cool cloud of gas were located behind the white light source. How would this spectrum be different?

34. Study Figure 10.10a. What are the differences between the ultraviolet photon and the red one? In this schematic diagram, all the photons leave the atom traveling to the right. Is that true for a real batch of hydrogen atoms in space?

35. Explain why the stellar spectral types (O, B, A, F, G, K, M) are not in alphabetical order. Also explain the sequence of temperatures defined by these spectral types.

36. In Figure 10.12, there is an absorption line at about 410 nm that is weak for O stars and weak for G stars but very strong in A stars. This particular line comes from the transition from the second excited state of hydrogen up to the sixth excited state. Why is this line weak in O stars? Why is it weak in G stars? Why is it strongest in the middle of the range of spectral types?

37. Other than the Sun, the only stars whose mass we can measure directly are those in binary systems. Explain why.

38. In Figure 10.15, two stars orbit a common center of mass.
    a. Explain why star 2 has a smaller orbit than star 1.
    b. Re-sketch this picture for the limiting case where star 1 has a very low mass, perhaps close to that of a planet.
    c. Re-sketch this picture for the limiting case where star 1 and star 2 have the same mass.

39. How do we estimate the mass of stars that are not in binary systems?

40. Compare the temperature, luminosity, and radius of stars at the lower left and upper right of the H-R diagram (for example, Figure 10.19).

Problems

41. Look at Figure 10.1b. Suppose the figure included a third star, located four times as far away as star A. How much less than star A would it appear to move each year? How much less than star B?

42. Suppose you see an object jump from side to side by half a degree as you blink back and forth between your eyes. How much farther away is an object that moves only one-third of a degree?

43. Sirius, the brightest star in the sky, has a parallax of 0.379 arcsec. What is its distance in parsecs? In light-years?

44. Betelgeuse (in Orion) has a parallax of 0.00763 ± 0.00164 arcsec, as measured by the *Hipparcos* satellite. What is the distance to Betelgeuse (in light-years) and the uncertainty in that measurement?

45. Rigel (also in Orion) has a *Hipparcos* parallax of 0.00412 arcsec. Given that Rigel and Betelgeuse appear almost equally bright in the sky, which star is actually more luminous? Betelgeuse appears reddish while Rigel appears bluish white. Which star would you say is larger, and why?

46. Sirius is actually a binary pair of two A-type stars. The brighter of the two stars is called the Dog Star, and the fainter is called the Pup Star because Sirius is in the constellation Canis Major (meaning "Big Dog"). The Dog Star appears about 6,800 times brighter than the Pup Star, even though both stars are at the same distance from us. Compare the temperatures, luminosities, and sizes of these two stars.

47. Examine Figure 10.4. This figure is plotted logarithmically on both axes. The luminosities are in units of solar luminosities.
    a. How much more luminous than the Sun is a star on the far right side of the plot?
    b. How much less luminous than the Sun is a star on the far left side of the plot?

48. A hydrogen atom makes a transition in which it loses 13.6 eV of energy. (An eV is an electron-volt, a very tiny unit of energy.) Did the atom absorb or emit a photon? How do you know? What is the energy of that photon?

49. Sirius and its companion orbit around a common center of mass with a period of 50 years. The mass of Sirius is 2.02 times the mass of the Sun.
    a. If the orbital velocity of the companion is 2.07 times greater than that of Sirius, what is the mass of the companion?
    b. What is the semimajor axis of the orbit?

50. Study Figure 10.14. If $m_1 = m_2$, where would the center of mass be located? If $m_1 = 2m_2$, where would the center of mass be located?

SMARTWORK

Norton's online homework system includes algorithmically generated versions of these questions, plus additional conceptual exercises. If your instructor assigns questions in SmartWork, log in at smartwork.wwnorton.com.

Exploration | The H-R Diagram

wwnpag.es/uou2

Visit the Student Site (wwnpag.es/uou2) and open the Hertzsprung-Russell Diagram Explorer Nebraska Simulation in Chapter 13. This simulation allows you to compare stars on the H-R diagram in two ways. You can compare an individual star (marked by a red X) to the Sun by varying its properties in the box on the left half of the window. Or you can compare groups of the nearest and brightest stars. Play around with the controls for a few minutes to familiarize yourself with the simulation.

Let's begin by exploring how changes to the properties of the individual star change its location on the H-R diagram. First, press the reset button at the top right of the window.

Decrease the temperature of the star by dragging the temperature slider to the left. Notice that the luminosity remains the same. Because the temperature has decreased, each square meter of star surface must be emitting less light. What other property of the star changes in order to keep the total luminosity of the star constant?

Predict what will happen when you slide the temperature slider all the way to the right. Now do it. Did the star behave as you expected?

1. As you move to the left across the H-R diagram, what happens to the radius?

2. What happens as you move to the right?

Press "reset," and experiment with the luminosity slider.

3. As you move up on the H-R diagram, what happens to the radius?

4. As you move down on the H-R diagram, what happens to the radius?

Press "reset" again and then predict how you would have to move the slider bars to move your star into the red giant portion of the H-R diagram (the upper right). Adjust your slider bars until the star is in that area. Were you correct?

5. How would you have to adjust the slider bars to move the star into the white dwarf area of the H-R diagram?

Press the reset button, and let's explore the right-hand side of the window. Add the nearest stars to the graph by clicking their radio button under "Plotted Stars." Using what you learned above, compare the temperatures and luminosities of these stars to the Sun (marked by the X).

6. Are the nearest stars generally hotter or cooler than the Sun?

7. Are the nearest stars generally more or less luminous than the Sun?

Press the radio button for the brightest stars. This will remove the nearest stars and add the brightest stars in the sky to the plot. Compare these stars to the Sun.

8. Are the brightest stars generally hotter or cooler than the Sun?

9. Are the brightest stars generally more or less luminous than the Sun?

10. How do the temperatures and luminosities of the brightest stars in the sky compare to the temperatures and luminosities of the nearest stars? Does this information support the claim in the chapter that there are more low-luminosity stars than high-luminosity stars? Explain.

SMARTWORK • smartwork.wwnorton.com

11 Our Star: The Sun

The Sun may be commonplace as far as stars go, but that makes it no less awesome an object on a human scale. Its mass is more than 300,000 times that of Earth, and its diameter is more than 100 times that of Earth. The Sun produces more energy in a second than all the energy generation on Earth produces in a half-million years. And because the Sun is the only star we can study at close range, much of the detailed information that we know about stars has come from studying our local star.

In the illustration on the opposite page, a student is comparing an image taken with a pinhole camera to other observations of the Sun taken at different times. In her pinhole image, she has caught the transit of Venus. Venus can be seen as a black dot near the middle-right of the Sun.

To understand the Sun, we must understand how it holds itself up under the pull of gravity, how it shines, how energy moves from the core to the surface, and the details of how it changes over time. We have a remarkable amount of information about the Sun, which we combine with physical principles to draw conclusions about what must be happening inside.

✦ LEARNING GOALS

The Sun is a main-sequence star. Its spectral class is G2. It is in the middle of its lifetime—no longer forming, but not yet running out of the fuel that keeps it shining. It is also much closer to us than any other star. A detailed understanding of the mechanisms that power the Sun can tell us much about 90 percent of the stars in the sky. The Sun's basic properties also provide a useful point of comparison. One note of caution: Although the Sun is fascinating, and its surface is one of the few astronomical objects that you can see change over the course of a day, you should never stare directly at the Sun—it is so bright that it will permanently damage your eyes. One safe way to observe the Sun is to take images through a pinhole camera and then compare them to images taken by spacecraft, as the student is doing at right. By the end of this chapter, you should be able to:

LG 1 Describe the balance between the forces that determine the structure of the Sun.

LG 2 Explain how mass is converted into energy in the Sun's core.

LG 3 List the different ways that energy moves outward from the Sun's core toward its surface.

LG 4 Sketch a physical model of the Sun's interior, and describe how observations of seismic vibrations on the surface of the Sun verify this model.

LG 5 Describe the solar activity cycles of 11 and 22 years, and explain how solar activity affects Earth.

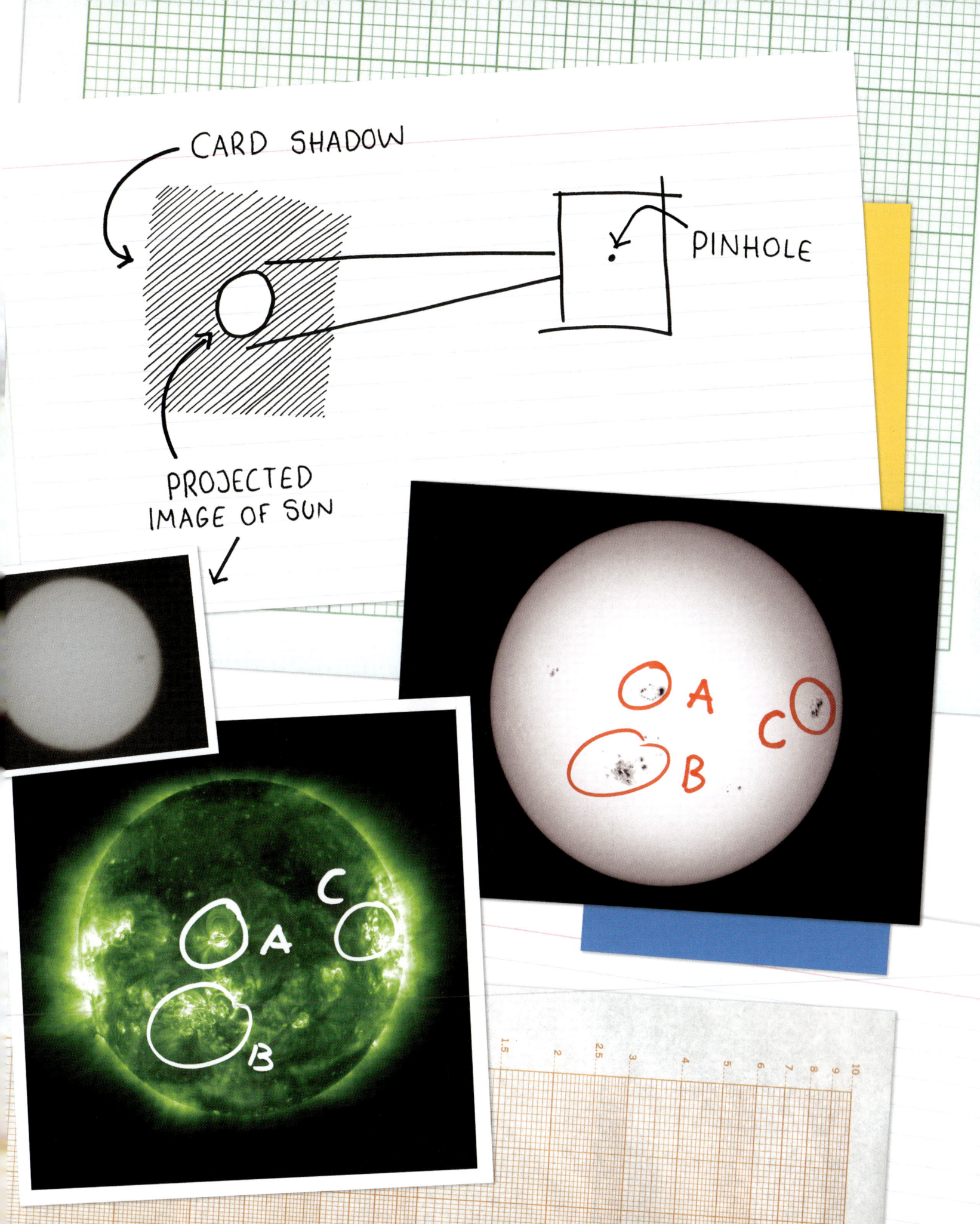

11.1 The Structure of the Sun Is a Matter of Balance

Our current model of the interior of the Sun can be summarized in a single statement: *The structure of the Sun is a matter of balance between the forces due to pressure and gravity.* The pressure is a result of energy finding its way to the surface from deep in the interior. To understand this summary statement, we need to know how these forces are produced and how they continually change to balance each other. Energy provides another sort of balance: The energy emitted by the Sun must be balanced by energy production in its core.

Hydrostatic Equilibrium

The balance between the forces due to pressure and gravity is illustrated in **Figure 11.1**. Deep in the Sun's interior, the outer layers press downward because of gravity, producing a large inward force. To maintain balance, the outward force due to pressure must be equally large. If gravity were not balanced by pressure, the Sun would collapse. If pressure were not balanced by gravity, the Sun would blow itself apart. At every point within the Sun's interior, the pressure must be just enough to hold up the weight of all the layers above that point. This balance between pressure and gravity is known as *hydrostatic equilibrium*. It is the same balance we saw in protostars in Chapter 5.

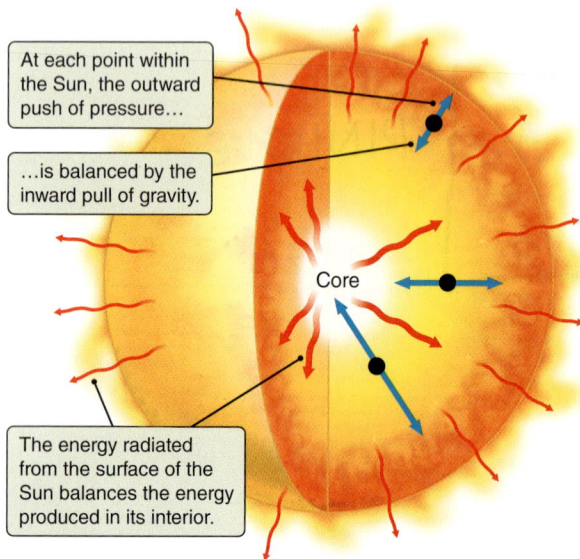

Figure 11.1 The structure of the Sun is determined by the balance between the forces of pressure and gravity and the balance between the energy generated in its core and energy radiated from its surface.

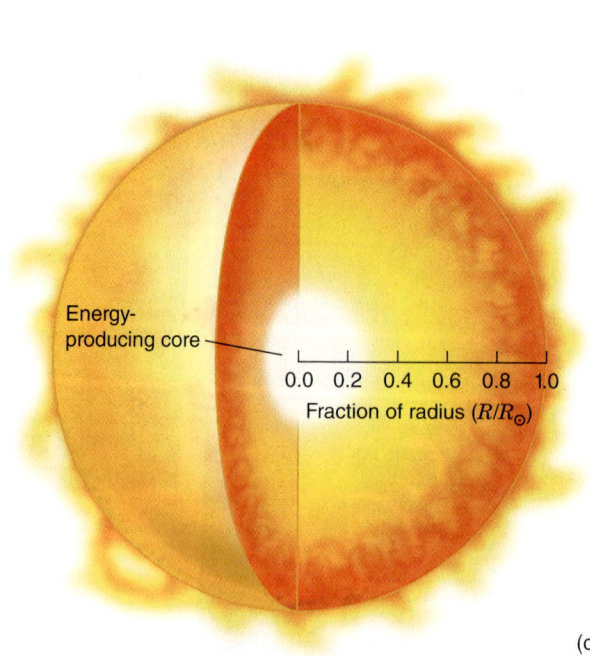

Figure 11.2 (a) This cutaway figure of the Sun shows how the fraction of radius given in the x-axis of the graphs in (b) and (c) is measured. (b) Temperature, density, and pressure increase toward the center of the Sun. (c) The energy produced by the Sun is generated in the Sun's core.

In a gas, higher pressure means higher density and/or higher temperature. **Figure 11.2** shows how conditions vary inside the Sun. Calculations show that toward the center of the Sun, the pressure climbs; and as it does, the density and temperature of the gas climb as well. As you can see in the graphs in Figure 11.2b, temperature, density, and pressure increase toward the center of the Sun.

Nuclear Fusion

Stars like the Sun are remarkably stable objects. Models of stellar evolution indicate that the luminosity of the Sun is increasing with time, but very, very slowly. The Sun's luminosity about 4.5 billion years ago was about 70 percent of its current luminosity. To remain in balance, the Sun must produce enough energy in its interior to replace the energy lost from its surface. This energy balance tells us how much energy must be produced in the interior of the Sun. The amount of energy produced by the Sun each second is truly astronomically large—3.85×10^{26} watts (W)—but this doesn't explain *how* that energy is produced. "What makes the Sun shine?" was one of the most basic questions pioneers of stellar astrophysics sought to answer.

Many possible explanations from classical physics were suggested. The answer finally came from theoretical physics and the laboratories of nuclear physicists as they began to understand *nuclear fusion*.

The nucleus of most hydrogen atoms consists of a single proton. Nuclei of all other atoms are built from a mixture of protons and neutrons. Most helium nuclei, for example, consist of two protons and two neutrons. Remember that protons have a positive electric charge, and neutrons have no charge. Because like charges repel, and the closer they are the stronger the force, all of the protons in an atomic nucleus are continually repelling each other with a tremendous force. The nuclei of atoms should fly apart due to electric repulsion—yet atoms exist. The **strong nuclear force** holds the nucleus together, overcoming this repulsion. However, the strong nuclear force acts only over very short distances, of the order 10^{-15} meter, or about a hundred-thousandth the size of an atom.

Compared to the energy required to free an electron from an atom, the amount of energy required to tear a nucleus apart is enormous. Conversely, when an atomic nucleus is formed, energy is released. **Nuclear fusion**—the process of combining two less massive atomic nuclei into a more massive atomic nucleus—occurs when atomic nuclei are brought close enough together for the strong nuclear force to overcome the force of electric repulsion, as illustrated in **Figure 11.3**. Many kinds of nuclear fusion can occur in stars. In main-sequence stars like the Sun, the primary process is the fusion of hydrogen into helium—a process called **hydrogen burning**.

The energy produced in nuclear reactions comes from the conversion of mass into energy. The exchange rate between mass and energy is given by Einstein's famous equation, $E = mc^2$, in which E is energy, m is mass, and c^2 is the speed of light squared. For any nuclear reaction, we can determine the mass that is turned into energy by calculating the mass that is lost. To find this lost mass, we subtract the mass of the outputs from the mass of the inputs. In hydrogen burning, the inputs are four hydrogen nuclei, and the output is a helium nucleus. The mass of four separate hydrogen nuclei is greater than the mass of a single helium nucleus; so when hydrogen fuses to make helium, some of the mass of the hydrogen is converted to energy (**Working It Out 11.1**).

Vocabulary Alert

burning In common language, this word is used to talk about fire, which is a chemical process of combining hydrocarbons with oxygen and releasing energy. Astronomers and nuclear physicists often talk about "hydrogen burning," which is *not* a chemical process because it alters the atoms themselves instead of just reassembling them into different molecules. However, hydrogen goes in, and energy comes out, so in that sense it is analogous to chemical burning.

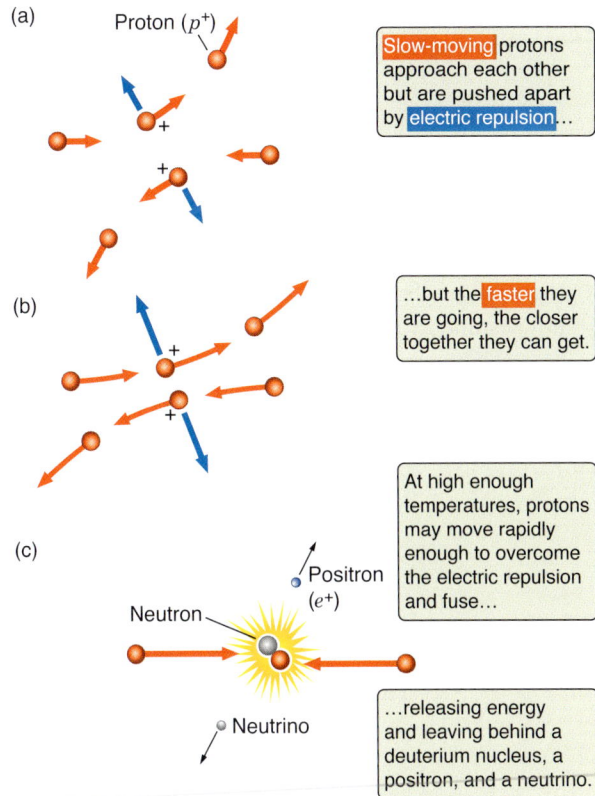

Figure 11.3 (a) Atomic nuclei are positively charged and electrically repel each other. (b) The faster that two nuclei are moving toward each other, the closer they will get before veering away. (c) At the temperatures and densities found in the centers of stars, nuclei are energetic enough to overcome this electric repulsion, so fusion takes place. (A positron is a small, positively charged particle.)

Working It Out 11.1 | How Much Longer Will the Sun "Live"?

Like all stars, the Sun's lifetime is limited by the amount of fuel available to it. We can calculate how long the Sun will live by comparing the mass involved in nuclear fusion with the amount of mass available. Converting four hydrogen nuclei (protons) into a single helium nucleus results in a loss of mass. The mass of a hydrogen nucleus is 1.6726×10^{-27} kilograms (kg). So, four hydrogen nuclei have a mass of four times that, or 6.6904×10^{-27} kg. The mass of a helium nucleus is 6.6447×10^{-27} kg, which is less than the mass of the four hydrogen nuclei. The amount of mass lost, m, is

$$m = 6.6904 \times 10^{-27} \text{ kg} - 6.6447 \times 10^{-27} \text{ kg} = 0.0457 \times 10^{-27} \text{ kg}$$

We can move the decimal point to the right and change the exponent on the 10 to rewrite the mass lost as 4.57×10^{-29} kg.

Using Einstein's equation $E = mc^2$ (and the fact that 1 kg m²/s² is 1 joule [J]), we can see that the energy released by this mass-to-energy conversion is

$$E = mc^2 = (4.57 \times 10^{-29} \text{ kg})(3.00 \times 10^8 \text{ m/s})^2 = 4.11 \times 10^{-12} \text{ J}$$

Each reaction that takes four hydrogen nuclei and turns them into a helium nucleus releases 4.11×10^{-12} J of energy, which doesn't seem like very much. But consider how small an atom is. Fusing a single gram of hydrogen into helium releases about 6×10^{11} J, which is equivalent to the energy obtained by burning 100 barrels of oil. In a sense, what this number tells you is how astonishingly big the Sun really is. For the Sun to produce as much energy as it does, it must convert roughly 600 billion kg of hydrogen into helium every second (and 4 billion kg of matter is converted to energy in the process). It has been doing this for the past 4.6 billion years. But how do we know how long the Sun will last?

Only about 10 percent of the Sun's total mass will ever be involved in fusion because the other 90 percent will never get hot enough or dense enough for the strong nuclear force to make fusion happen. Ten percent of the mass of the Sun is

$$M_{\text{fusion}} = (0.1)(M_\odot)$$
$$M_{\text{fusion}} = (0.1)(2 \times 10^{30} \text{ kg})$$
$$M_{\text{fusion}} = 2 \times 10^{29} \text{ kg}$$

That is the amount of fuel the Sun has available. How long will it last? The Sun consumes hydrogen at a rate of 600 billion kilograms per second (kg/s), and there are 3.16×10^7 seconds in a year, so each year the Sun consumes

$$M_{\text{year}} = (600 \times 10^9 \text{ kg/s})(3.16 \times 10^7 \text{ s/year})$$
$$M_{\text{year}} = 2 \times 10^{19} \text{ kg/year}$$

If we know how much fuel the Sun has (2×10^{29} kg), and we know how much the Sun burns each year (2×10^{19} kg/year), then we can divide the amount by the rate to find the lifetime of the Sun:

$$\text{Lifetime} = \frac{M_{\text{fuel}}}{M_{\text{year}}} = \frac{2 \times 10^{29} \text{ kg}}{2 \times 10^{19} \text{ kg/year}}$$

$$\text{Lifetime} = 1 \times 10^{10} \text{ years} = 10 \times 10^9 \text{ years} = 10 \text{ billion years}$$

When the Sun was formed, it had enough fuel to power it for about 10 billion years. At 4.6 billion years old, the Sun is nearly halfway through its total life span, and it will continue fusing hydrogen into helium for about 5.4 billion more years, give or take a few hundred million.

Energy is produced in the Sun's innermost region, the core, where conditions are extreme. The density is about 150 times the density of water (which is 1,000 kilograms per cubic meter [kg/m³]), and the temperature is about 15 million K. The atomic nuclei have tens of thousands of times more kinetic energy than atoms at room temperature. As illustrated in Figure 11.3c, under these conditions atomic nuclei slam into each other hard enough to overcome the electric repulsion, allowing the strong nuclear force to act.

In hotter and denser gases, collisions happen more frequently. For this reason, the rate of nuclear fusion reactions is extremely sensitive to the temperature and density of the gas, which is why these energy-producing collisions are concentrated in the Sun's core. Half of the energy produced by the Sun is generated within the inner 9 percent of the Sun's radius. This region represents less than 0.1 percent of the volume of the Sun.

Hydrogen burning is the most significant source of energy in main-sequence stars. Hydrogen is the most abundant element in the universe, so it is the most abundant source of nuclear fuel. Hydrogen burning is also the most efficient way to convert mass into energy. Most important, hydrogen is the easiest type of atom to fuse. Hydrogen nuclei—protons—have an electric charge of +1. The electric barrier that must be overcome to fuse hydrogen is the repulsion of one proton against another. To fuse carbon, for example, the repulsion of six protons in one carbon nucleus pushing against the six protons in another carbon nucleus must be overcome. The repulsion between two carbon nuclei is 36 times stronger than that between two hydrogen nuclei. Therefore, hydrogen fusion occurs at a much lower temperature than that of any other type of nuclear fusion.

▶‖ **AstroTour:** The Solar Core

The Proton-Proton Chain

In the cores of low-mass stars such as the Sun, hydrogen burns primarily through a process called the proton-proton chain. It consists of three steps, illustrated in **Figure 11.4**. Notice that the first two steps are shown twice, along the top of the figure and the bottom. These steps must occur twice, with different sets of particles, which then combine in the third step. Each step produces particles and/or energy in the form of light. Understanding each of these steps in detail allowed astronomers to test directly the proposal that the Sun shines because of nuclear fusion by searching for the products of this fusion reaction. We will begin by following the creation of the helium nucleus, and then go back to find out what happens to the other products of the reaction.

In the first step, two hydrogen nuclei fuse. During this process, one of the protons turns into a neutron. To conserve energy and charge, two particles are emitted: a positively charged particle called a positron and a neutral particle called a neutrino. Energy is also emitted in the form of light. The new nucleus

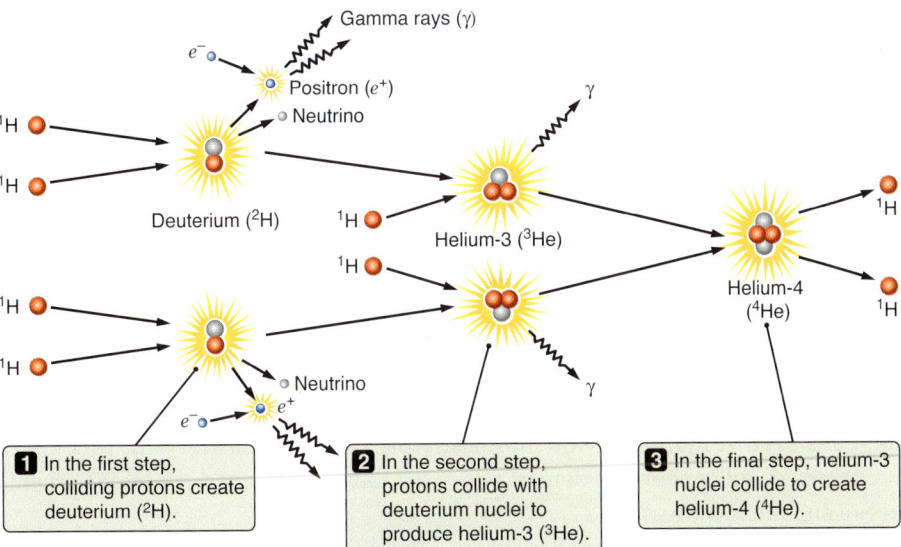

❶ In the first step, colliding protons create deuterium (^{2}H).

❷ In the second step, protons collide with deuterium nuclei to produce helium-3 (^{3}He).

❸ In the final step, helium-3 nuclei collide to create helium-4 (^{4}He).

Figure 11.4 The Sun and all other main-sequence stars get their energy by fusing the nuclei of four hydrogen atoms to make a single helium atom. In the Sun, about 85 percent of the energy produced comes from the proton-proton chain shown here.

has a proton and a neutron. It is still hydrogen because it has only one proton. Because it has more than the normal number of neutrons, we say it is an **isotope** of hydrogen; different isotopes of an element have differing numbers of neutrons. This particular isotope is called deuterium (^{2}H).

In the second step of the proton-proton chain, another proton slams into the deuterium nucleus, forming the nucleus of a helium isotope, ^{3}He, consisting of two protons and a neutron. The energy released in this step is carried away as a gamma-ray photon.

In the third and final step of the proton-proton chain, two ^{3}He nuclei collide and fuse, producing an ordinary ^{4}He nucleus and ejecting two protons in the process. The energy released in this step shows up as kinetic energy of the helium nucleus and ejected protons. Overall, four hydrogen nuclei have combined to form one helium nucleus.

Now let's go back and look at what happens to the other products of the reaction. In step 1, a positron was produced. A **positron** is a type of particle known as antimatter. Antimatter particles have the same mass as a corresponding matter particle but have opposite values of other properties, such as charge. The positron (e^+) is the antimatter counterpart of an electron (e^-). When matter (electrons) and antimatter (positrons) meet, they annihilate each other, and their total mass is converted to energy in the form of gamma-ray photons. This happens to the emitted positrons inside the Sun, and the emitted photons from the annihilation carry away part of the energy released when the two protons fused. These photons heat the surrounding gas. The gamma rays emitted in step 2 similarly heat the gas. As you will see in Section 11.2, the thermal energy produced in the core of the Sun takes 100,000 years to find its way to the Sun's surface, and so the light we see from the Sun indicates what the Sun was doing 100,000 years ago.

▶▶ **Nebraska Simulation:** Proton-Proton Animation

The neutrino emitted in step 1 has a very different fate. **Neutrinos** are particles that have no charge, very little mass, and travel at nearly the speed of light. They interact weakly with ordinary matter, so weakly that the neutrino escapes from the Sun without further interactions with any other particles. The core of the Sun lies buried beneath 700,000 kilometers (km) of dense, hot matter, yet the Sun is transparent to neutrinos—essentially all of them travel into space as if the outer layers of the Sun were not there. Because they travel at nearly the speed of light, neutrinos from the center of the Sun arrive at Earth after only 8⅓ minutes. Therefore, we can use them to probe what the Sun is doing today. If only we can catch them.

Observing the Heart of the Sun with Neutrinos

Neutrinos interact so weakly with matter that they are extremely difficult to observe. Fortunately, the extremely large number of nuclear reactions in the Sun means that the Sun produces a truly enormous number of neutrinos. As you read this sentence, about 400 trillion solar neutrinos pass through your body. This happens even at night, as neutrinos easily pass through Earth. With this many neutrinos about, a neutrino detector does not have to detect a very large percentage of them to be effective.

The first apparatus designed to detect solar neutrinos was built underground, 1,500 meters deep, within the Homestake Mine in Lead, South Dakota. Astronomers filled a tank with 100,000 gallons of dry-cleaning fluid. Over the course of 2 days, roughly 10^{22} solar neutrinos pass through the Homestake detector. Of these, on average only *one* neutrino interacts with a chlorine atom within the fluid to form an atom of argon. Even so, this instrument produces a measurable number of argon atoms over time.

The Homestake experiment detected these argon atoms—evidence of neutrinos from the Sun, confirming that nuclear fusion powers the Sun. As with many good experiments, however, these results raised new questions. After astronomers' initial joy, they noticed that there seemed to be only one-third as many solar neutrinos as predicted. The difference between the predicted and measured number of solar neutrinos was called the **solar neutrino problem**.

The solar neutrino problem in turn led particle physicists to suspect that the understanding of the neutrino itself was incomplete. The neutrino was long thought to have zero mass, like photons, and to travel at the speed of light. But if neutrinos actually do have a tiny amount of mass, then particle physics predicts that neutrinos should oscillate—alternate back and forth—among three different kinds of neutrinos: the *electron*, *muon*, and *tau neutrinos*. The solar neutrino problem supported this understanding of neutrinos. Early neutrino experiments could detect only the electron neutrino and consequently observed only about a third of the expected number of neutrinos.

Since Homestake began operating in 1965, more than two dozen additional neutrino detectors have been built, each using different reactions to detect different kinds of neutrinos. Experiments at high-energy physics labs, nuclear reactors, and neutrino telescopes around the world have shown that neutrinos *do* have a nonzero mass, and this work has uncovered evidence of neutrino oscillations.

Solving the solar neutrino problem is a good example of how science works. Experiments such as Homestake revealed a gap in our understanding. Theories of particle physics suggested possible resolutions, and more sophisticated experiments tested which one of the competing hypotheses was correct. Through this process, the solar neutrino problem led to deeper knowledge of basic physics. Perhaps this helps explain why scientists are often excited to find they are wrong. Sometimes it means that completely new knowledge is right around the corner.

11.2 Energy in the Sun's Core Moves through Radiation and Convection

In Chapter 6, we saw that although we cannot travel deep inside Earth to find out how it is structured, we are able to build a model of its interior. Similarly, we can create a model of the Sun's interior using our knowledge of the balance of forces and energy within the Sun and our understanding of how energy moves from one place to another. Waves traveling through the Sun, as well as some surface distortions, allow us to test and refine our model of its interior.

Energy Transport

Some of the energy released by hydrogen burning in the core of the Sun escapes directly into space in the form of neutrinos, but most of the energy heats the solar interior and then moves outward through the Sun to the surface. This **energy transport** is a key determinant of the Sun's structure. Energy transport can occur by conduction, convection, or radiation. Conduction is important primarily in solids. For example, when you pick up a hot object, your fingers are heated by conduction. However, because the Sun is made of gas, energy transport in the Sun moves by convection and radiation through different zones, as shown in

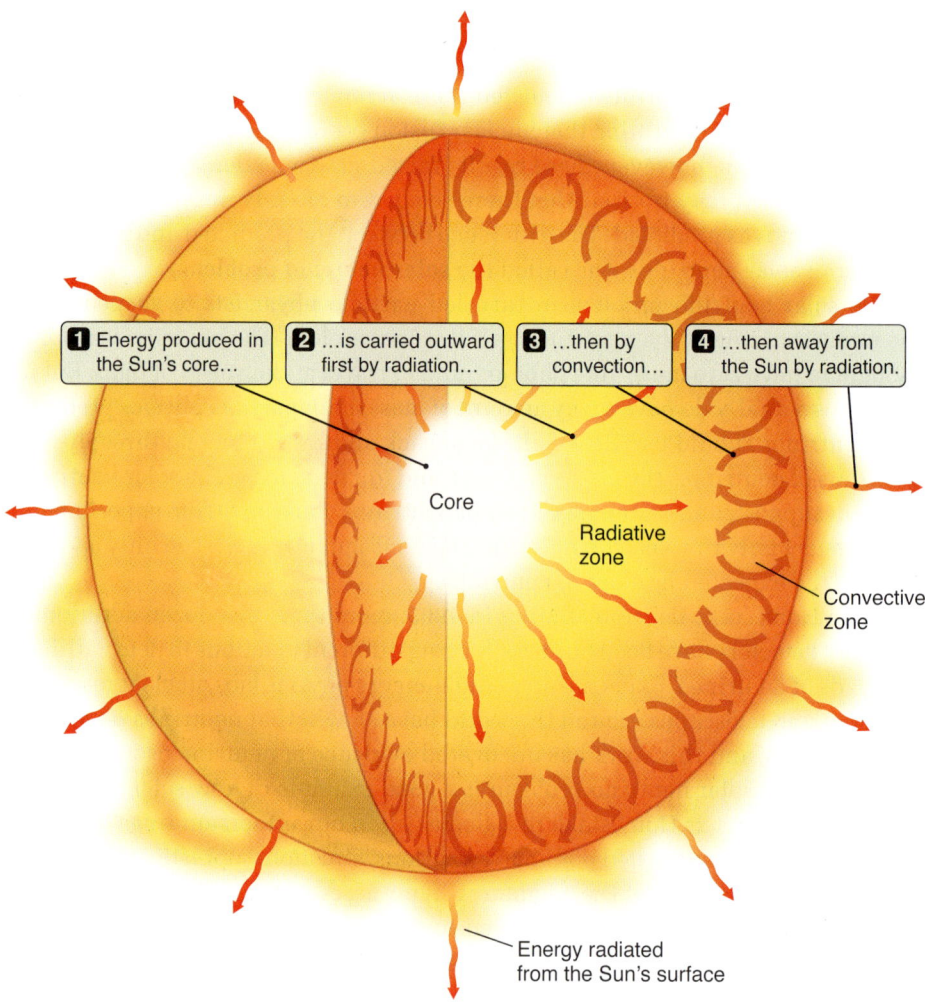

Figure 11.5 The interior structure of the Sun is divided into zones based on where energy is produced and how it is transported outward.

Back in Section 5.2 . . .

...you learned that hotter objects are bluer.

Vocabulary Alert

opacity This word is related to the word *opaque*. In common language, when something is opaque, you can't see through it *at all*. The term *opacity* allows for more subtle variations. A gas can have a low opacity, which means it is mostly transparent and allows photons to pass through, or a high opacity, which means it is mostly opaque and prevents photons from passing straight through.

Figure 11.5. First, energy moves outward through the inner layers of the Sun by radiation in the form of photons. Next, energy moves by convection in parcels of gas. Finally, energy radiates from the Sun's surface as light.

The details of energy transport from the center of the Sun outwards depend on the decreasing temperature and density as the radius increases. Near the core, radiation transfers energy from hotter to cooler regions via photons, which carry the energy with them. Consider a hotter region of the Sun located next to a cooler region, as shown in **Figure 11.6**. The hotter region contains more (and more energetic) photons than the cooler region. (Recall Wien's law from Chapter 5.) More photons will move by chance from the hotter, more crowded region to the cooler, less crowded region than in the reverse direction. A net transfer of photons and photon energy occurs from the hotter region to the cooler region, and radiation carries energy outward from the Sun's core.

The transfer of energy from one point to another by radiation also depends on how freely photons can move from one point to another within a star. The degree to which matter blocks the flow of photons through it is referred to as **opacity**. The opacity of a material depends on many things, including the density of the material, its composition, its temperature, and the wavelength of the photons moving through it.

Radiative transfer is efficient in regions with low opacity. The **radiative zone** is the region in the inner part of the Sun where the opacity is relatively low, and radiation carries the energy produced in the core outward through the star. This

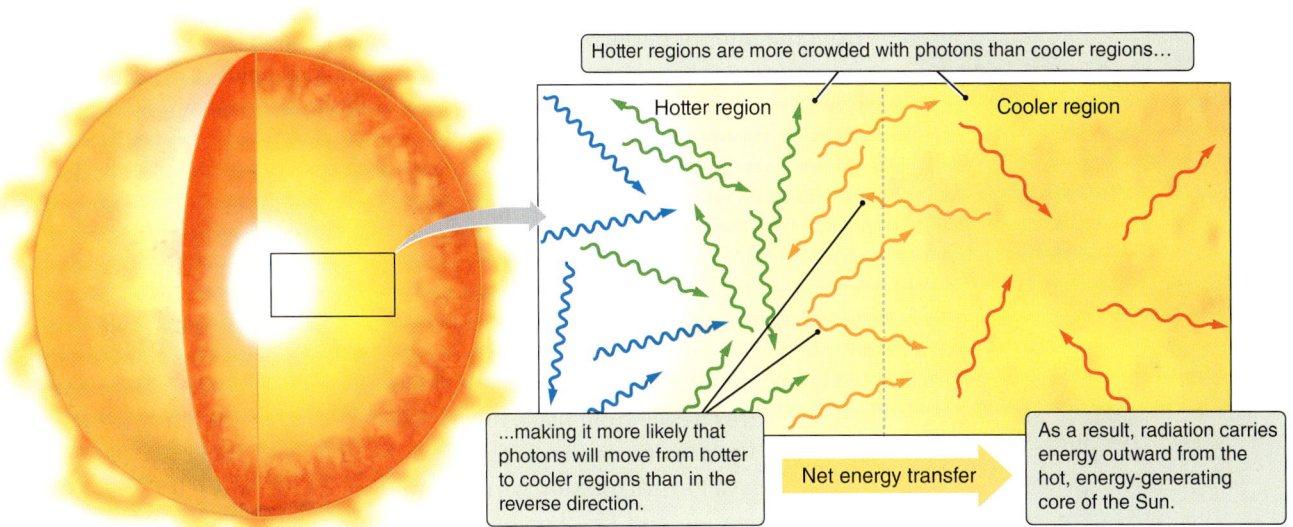

Figure 11.6 Higher-temperature regions deep within the Sun produce more radiation than do lower-temperature regions farther out. Although radiation flows in both directions, more radiation flows from the hotter regions to the cooler regions than from the cooler regions to the hotter regions. Therefore, radiation carries energy outward from the inner parts of the Sun. For clarity, we have only included in this illustration the most common photons (those at the peak of the blackbody curve). Photons of all colors are present in all regions, and there are more of all kinds in the hotter regions and fewer of all kinds in the cooler regions.

radiative zone extends about 70 percent of the way out toward the surface of the Sun. Even though this region's opacity is low enough for radiation to dominate convection as an energy transport mechanism, photons still travel only a short distance within the region before being absorbed, emitted, or deflected by matter, much like a beach ball being batted about by a crowd of people, as illustrated in **Figure 11.7**. Each interaction sends the photon in an unpredictable direction—not necessarily toward the surface of the star. The distances between interactions are so short that, on average, it takes the energy of a gamma-ray photon about 100,000 years to find its way to the outer layers of the Sun. Opacity holds energy within the interior of the Sun and lets it seep away only slowly. As it travels, the gamma-ray photon gradually becomes converted to lower-energy photons, emerging as optical and infrared radiation from the surface.

From a peak of 15 million K in the core of the Sun, the temperature falls to about 100,000 K at the outer margin of the radiative zone. At this cooler temperature, the opacity is higher, so radiation is less efficient in carrying energy from one place to another. The energy that is flowing outward through the Sun "piles up" against this edge of the radiative zone. Beyond this region, the temperature drops off very quickly.

Farther from the core of the Sun, radiative transfer becomes inefficient and the temperature changes quickly. As we have seen, convection takes over, which transports energy by moving packets of material. These packets of hot gas, like hot air balloons, become buoyant and rise up through the lower-temperature gas above them, carrying energy with them. Just as convection carries energy from the interior of planets to their surfaces, or from the Sun-heated surface of Earth upward through Earth's atmosphere, convection also plays an important role in the transport of energy outward

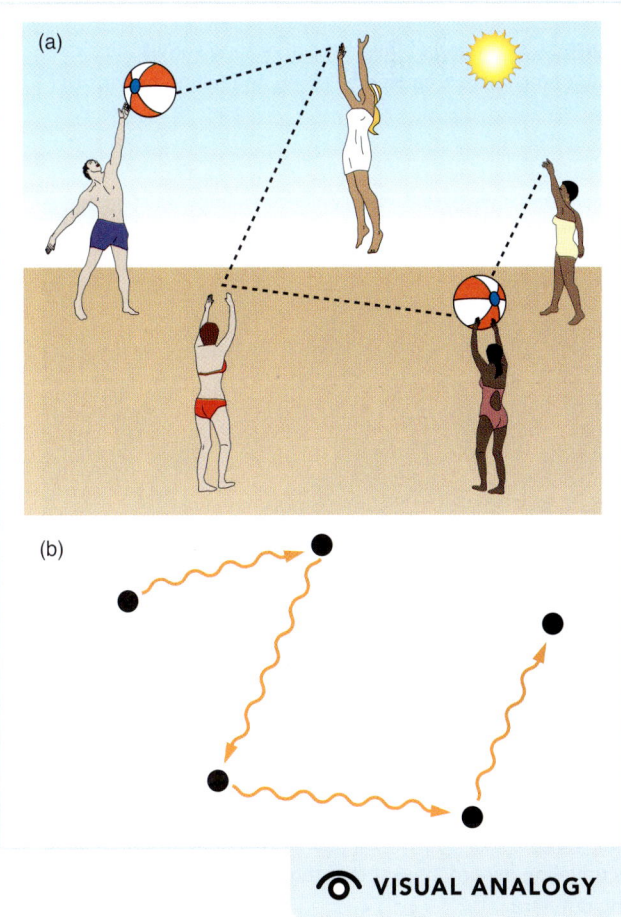

VISUAL ANALOGY

Figure 11.7 (a) When a crowd of people play with a beach ball, the ball never travels very far before someone hits it, turning it in another direction. The ball moves randomly, sometimes toward the front of the crowd, sometimes toward the back. It often takes a ball a long time to make its way from one edge of the crowd to the other. (b) When a photon travels through the Sun, it never travels very far before it interacts with an atom. The photon moves randomly, sometimes toward the center of the Sun, sometimes toward the outer edge. It takes a long time for a photon to make its way out of the Sun.

from the interior of the Sun. The solar **convective zone** (see Figure 11.6) extends from the outer boundary of the radiative zone to just below the visible surface of the Sun, where evidence of convection can be seen in the bubbling of that surface (**Figure 11.8**).

In the outermost layers of the Sun, radiation again takes over as the primary mode of energy transport, and it is radiation that transports energy from its outermost layers off into space.

Helioseismology

In Chapter 6, we found that models of Earth's interior predict how density and temperature change from place to place. These differences affect seismic waves traveling through Earth, bending the paths that they travel. To test and refine our models of Earth's interior, we compared measurements with the predictions.

The same method can be applied to the Sun. Detailed observations of the surface of the Sun show that the Sun vibrates or rings, something like a bell that has been struck. Unlike a well-tuned bell—which vibrates primarily at one frequency—the vibrations of the Sun are very complex; many different frequencies of vibrations occur simultaneously, causing some parts of the Sun to bulge outward and some to draw inward. **Figure 11.9** illustrates the motions of the different parts of the Sun, with red and blue areas moving in opposite directions. Some waves are amplified and some are suppressed, depending on how they overlap as they travel through the Sun. Just as geologists use seismic waves from earthquakes to probe Earth's interior, solar physicists use the surface oscillations of the Sun to probe the solar interior. They study these waves using the Doppler effect (see Chapter 5), which distinguishes between parts of the Sun that move toward the observer and parts that move away. The science that uses solar oscillations to study the Sun is called **helioseismology**.

To detect the disturbances of helioseismic waves on the surface of the Sun, astronomers must measure Doppler shifts of less than 0.1 meter per second (m/s) while detecting changes in brightness of only a few parts per million at any given location on the Sun. Tens of millions of different wave motions are possible within the Sun. Some waves travel around the circumference of the Sun, providing information about the density of the upper convective zone. Other waves travel through the interior of the Sun, revealing the density structure near the Sun's core. Still others travel inward toward the center of the Sun until they are bent by the changing solar density and return to the surface.

All of these wave motions are going on at the same time, so sorting out this jumble requires computer analysis of long, unbroken strings of solar observations from several sources. The Global Oscillation Network Group (GONG) is a network of six solar observation stations spread around the world, enabling astronomers to observe the surface of the Sun approximately 90 percent of the time.

Scientists compare the strength, frequency, and wavelengths of helioseismic data against predicted vibrations calculated from models of the solar interior. This technique provides a powerful test of our understanding of the solar interior, and it has led to improvements in our models. For example, some scientists had proposed that the solar neutrino problem might be solved if we had overestimated the amount of helium in the Sun. This explanation was ruled out by analysis of the waves that penetrate to the core of the Sun. Helioseismology also showed that the value for opacity used in early solar models was too low. This realization led astronomers to recalculate the location of the bottom of the convective zone. Both theory and observation now put the base of the convective

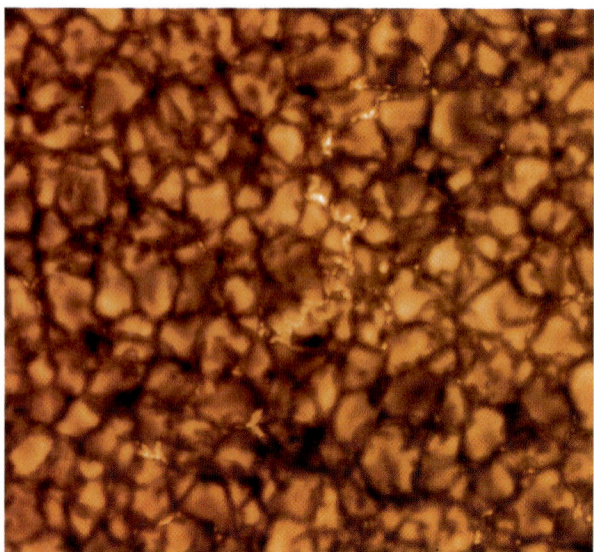

Figure 11.8 The top of the convective zone shows the bubbling of the surface caused by rising and falling packets of gas.

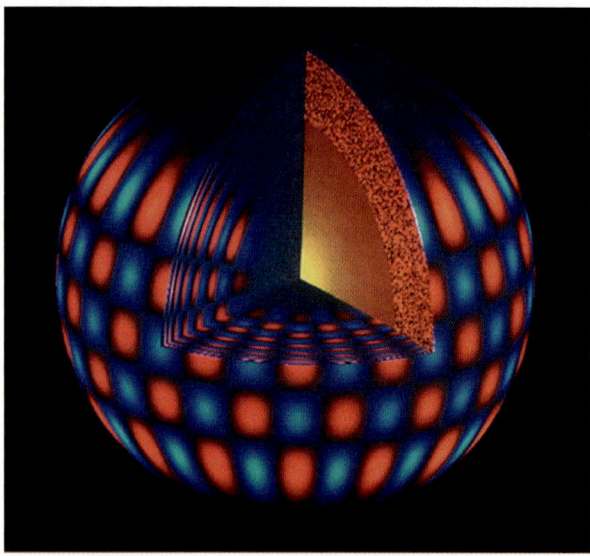

Figure 11.9 The interior of the Sun vibrates as helioseismic waves move through it. This figure shows one particular mode of the Sun's vibration. Red shows regions where gas is traveling inward; blue shows regions where gas is traveling outward.

zone at 71.3 percent of the way out from the center of the Sun, and uncertainty in this number is less than half a percent.

Working back and forth between observation and theory has enabled astronomers to probe the otherwise inaccessible interior of the Sun. We now know that the energy is produced by nuclear fusion deep in the core and that it moves outward by radiation to a point about 70 percent of the radius of the Sun. Then it travels outward by convection to the surface. We also know how the temperature, density, and pressure change with radius and how these factors change the opacity at different distances from the center. This kind of collaboration between theory and observation is essential to observational sciences like astronomy. Even though it is usually not possible to sample directly or to set up controlled experiments, the combined power of theory and observation allows us to answer many of our questions about objects we cannot sample directly.

11.3 The Atmosphere of the Sun

Beyond the convective zone lie the outer layers of the Sun, which are collectively known as the Sun's atmosphere. These onion-like layers, shown in **Figure 11.10,** include the *photosphere*, the *chromosphere*, and the *corona*. We can observe these layers of the Sun directly using telescopes, satellites, and even pinhole cameras (as in the chapter-opening illustration). Observations of the Sun's atmosphere are important because activity in the Sun's atmosphere has consequences for both human infrastructure (power grids and satellites in orbit around Earth) and for earth-bound observers.

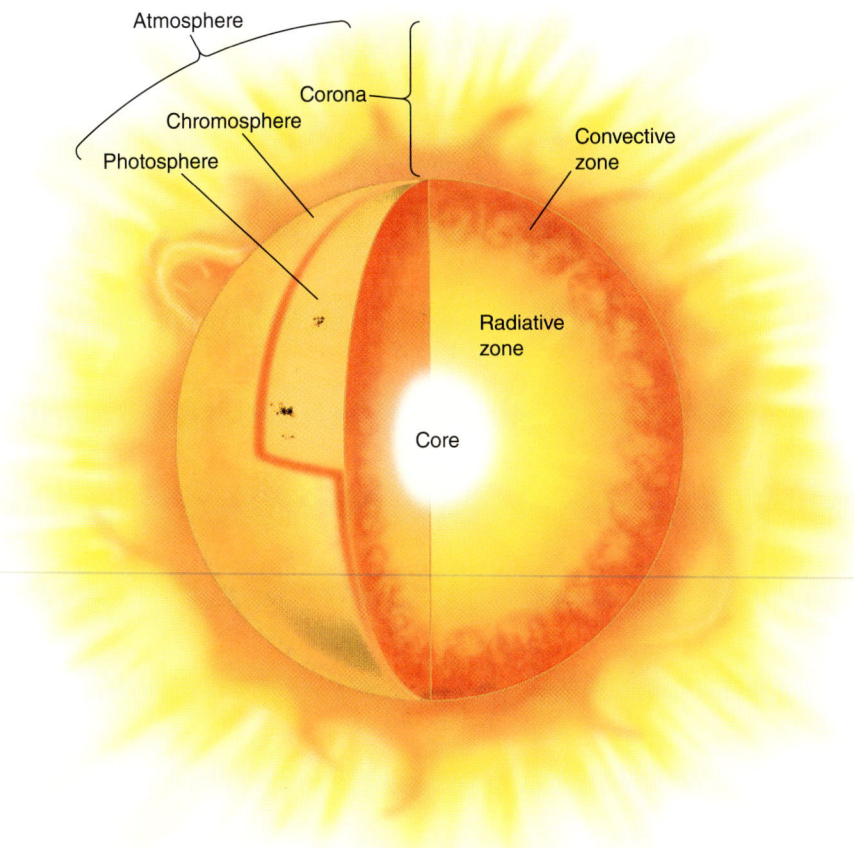

Figure 11.10 The components of the Sun's atmosphere are located above the convective zone and include the photosphere, the chromosphere, and the corona.

Observing the Sun

The Sun is a large ball of gas; unlike Earth, it has no solid surface. Its apparent surface is like a fog bank on Earth. Imagine watching a person walking into a fog bank. After the person disappears from view, you would say he is inside the fog bank, even though he never passed through a definite boundary. The apparent surface of the Sun is similar. Light from the Sun's surface can escape directly into space, so we can see it. Light from below the Sun's surface cannot escape directly into space, so we cannot see it.

The Sun's apparent surface is called the solar **photosphere**. (*Photo* means "light"; the photosphere is the sphere that light comes from.) There is no instant when you can say that you have suddenly crossed the surface of a fog bank, and by the same token there is no instant when we suddenly cross the photosphere of the Sun. The photosphere is about 500 km thick. As you can see in the graphs in **Figure 11.11**, the temperature increases sharply across the boundary between the chromosphere and the corona, while the density falls sharply across the same boundary. The Sun appears to have a well-defined surface and a sharp outline

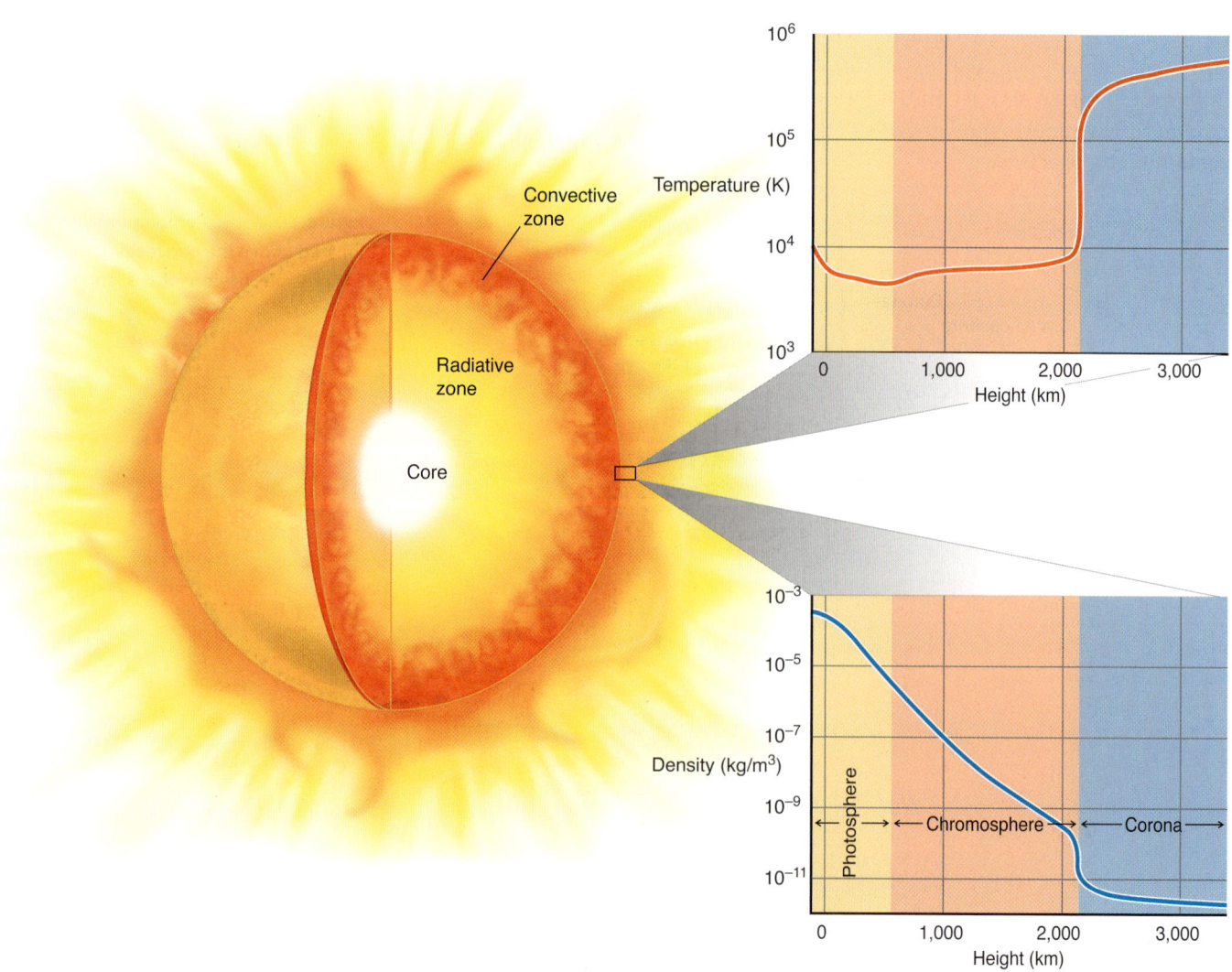

Figure 11.11 The temperature and density of the Sun's atmosphere change abruptly at the boundary between the chromosphere and the corona.

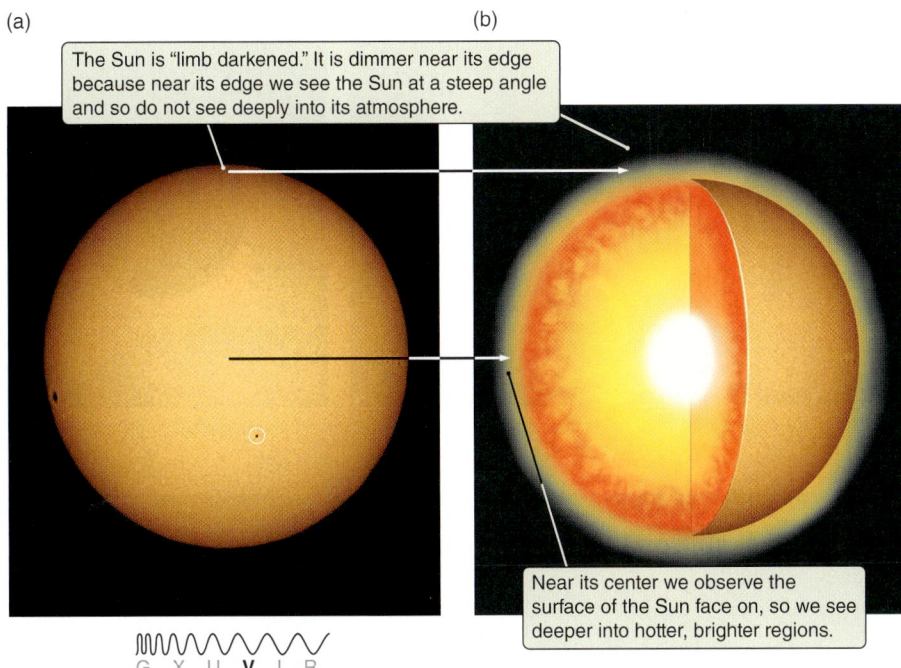

Figure 11.12 (a) When viewed in visible light, the Sun appears to have a sharp outline, even though it has no true surface. The center of the Sun appears brighter, while the limb of the Sun is darker—an effect known as limb darkening. Incidentally, the tiny black dot in the middle of the white circle is the planet Mercury, transiting the Sun. The dot near the left limb is a sunspot. (b) Looking at the middle of the Sun allows us to see deeper into the Sun's interior than we can by looking at the edge of the Sun. Because higher temperature means more luminous radiation, the middle of the Sun appears brighter than its limb.

when viewed from Earth because 500 km does not look very thick when viewed from a distance of 150 million km.

The Sun appears fainter near its edges than near its center, an effect known as **limb darkening** (**Figure 11.12a**). Limb darkening is an artifact of the structure of the Sun's photosphere. Near the edge of the Sun you are looking through the photosphere at a steep angle. As a result, you do not see as deeply into the interior of the Sun as when you are looking near the center of the Sun's disk. The light from the limb of the Sun comes from a shallower layer that is cooler and fainter (Figure 11.12b).

The Solar Spectrum

In the Sun's atmosphere, the density of the gas drops very rapidly with increasing altitude. All visible solar phenomena take place in the Sun's atmosphere. Most of the radiation from below the Sun's photosphere is absorbed by matter and reemitted at the photosphere as a blackbody spectrum.

As we examine the structure of the Sun in more detail, however, we see that this simple description is incomplete. Light from the solar photosphere must escape through the upper layers of the Sun's atmosphere, which puts its fingerprints on the spectrum that we observe. As photospheric light travels upward through the solar atmosphere, atoms in the solar atmosphere absorb the light at distinct wavelengths, forming absorption lines (**Figure 11.13**). Absorption lines from more than 70 elements have been identified. Analysis of these lines forms the basis for much of our knowledge of the solar atmosphere, including the composition of the Sun. This is also the starting point for our understanding of the atmospheres and spectra of other stars.

The Sun's Outer Atmosphere: Chromosphere and Corona

Moving upward through the Sun's photosphere, the temperature falls from 6600 K at the photosphere's bottom to 4400 K at its top. At this point, the trend reverses

Figure 11.13 This high-resolution spectrum of the Sun stretches from 400 nanometers (nm) in the lower left corner to 700 nm in the upper right corner and shows black absorption lines. This spectrum was produced by passing the Sun's light through a prism-like device, and then cutting and folding the single long spectrum (from blue to red) into rows so that it will fit in a single image from a camera.

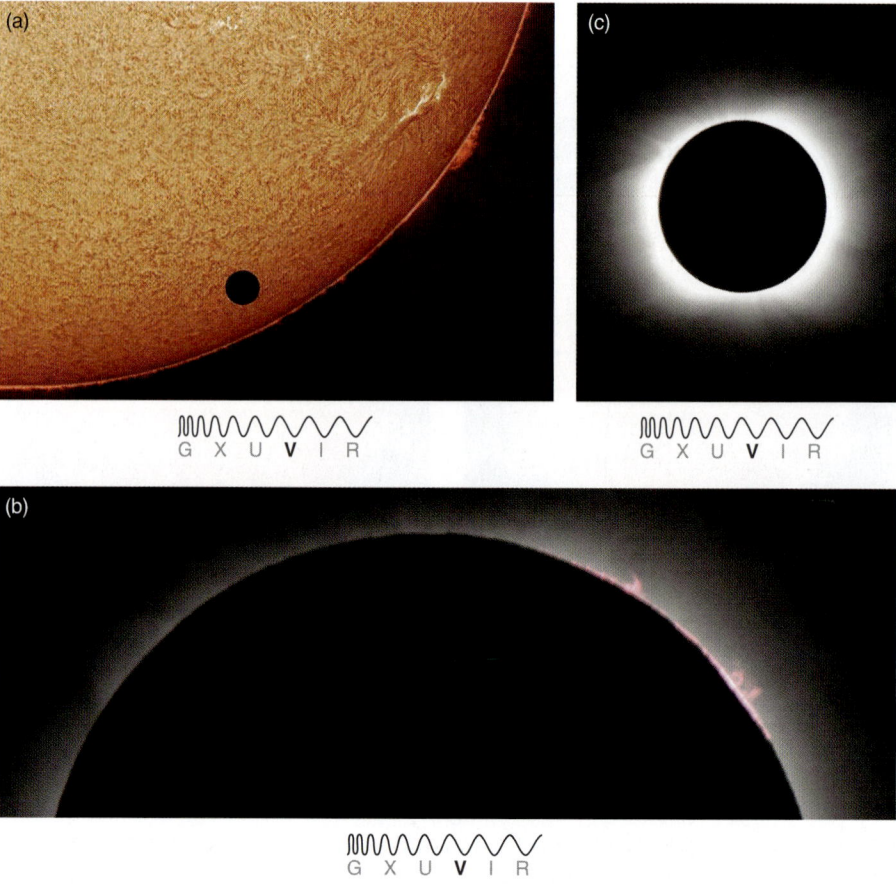

Figure 11.14 (a) This image of the Sun, taken in hydrogen alpha (Hα) light during a transit of Venus, shows structure in the Sun's chromosphere. (The planet Venus is seen in silhouette against the disk of the Sun.) (b) The chromosphere can be seen during a total eclipse. (c) This eclipse image shows the Sun's corona, consisting of million-kelvin gas that extends for millions of kilometers beyond the surface of the Sun.

and the temperature slowly begins to climb, rising to about 6000 K at a height of 1,500 km above the top of the photosphere (see Figure 11.11). This region of increasing temperature is called the **chromosphere** (**Figure 11.14a**). The reason for the chromosphere's temperature reversal with increasing height is not well understood, but it may be caused by magnetic fields propagating through the region.

The chromosphere was discovered in the 19th century during observations of total solar eclipses (Figure 11.14b). The chromosphere is seen most clearly at the solar limb as a source of emission lines, especially a particular hydrogen line that is produced when the electron falls from the third energy state to the second energy state. This line is known as the Hα line (the "hydrogen alpha line"). The deep red color of the Hα line is what gives the *chromosphere* ("the place where color comes from") its name. The element helium was discovered in 1868 from a spectrum of the chromosphere of the Sun nearly thirty years before it was recognized on Earth; helium is named after *helios*, the Greek word for "Sun."

At the top of the chromosphere, across a transition region that is only about 100 km thick, the temperature suddenly soars (see Figure 11.11), while the density abruptly drops. In the **corona**, the hot, outermost part of the Sun's atmosphere, temperatures reach 1 million to 2 million K. The corona is thought to be heated by magnetic fields in much the same way the chromosphere is, but why the temperature changes so abruptly at the transition between the chromosphere and the corona is not at all clear.

The corona is visible during total solar eclipses as an eerie glow stretching several solar radii beyond the Sun's surface (Figure 11.14c). Because it is so hot, the corona is also a strong source of X-rays, and there is so much energy in these X-ray photons that many electrons are stripped away from nuclei, leaving atoms in the corona highly ionized.

11.4 The Atmosphere of the Sun Is Very Active

The atmosphere of the Sun is a very active place. The best-known features on the surface of the Sun are relatively dark blemishes in the solar photosphere, called **sunspots**. Sunspots come and go over time, though they remain long enough for us to determine the rotation rate of the Sun. These spots are associated with active regions, loops of material and explosions that fling particles far out into the Solar System. Long-term patterns have been observed in the variations of sunspots and active regions, revealing that the magnetic field of the Sun is constantly changing.

Figure 11.15 A close-up image of the Sun shows the tangled structure of coronal loops.

Solar Activity Is Caused by Magnetic Effects

The Sun's magnetic field causes virtually all of the structure in the Sun's atmosphere. High-resolution images of the Sun reveal *coronal loops* that make up much of the Sun's lower corona (**Figure 11.15**). This texture is the result of flux tubes in the magnetic field. The corona itself is far too hot to be held in by the Sun's gravity, but over most of the surface of the Sun, coronal gas is confined by magnetic loops with both ends firmly anchored deep within the Sun. The magnetic field in the corona acts almost like a network of rubber bands that coronal gas is free to slide along but cannot cross. In contrast, about 20 percent of the surface of the Sun is covered by an ever-shifting pattern of **coronal holes**. These are apparent in extreme UV images of the Sun as dark regions (**Figure 11.16**), indicating that they are cooler and lower in density than their surroundings. In coronal holes, the magnetic field points away from the Sun, and coronal material is free to stream away into interplanetary space.

We have encountered this flow of coronal material away from the Sun before. This is the same solar wind responsible for shaping the magnetospheres of planets (see Chapters 7 and 8) and for blowing the tails of comets away from the Sun (see Chapter 9). The relatively steady part of the solar wind consists of lower-speed flows with velocities of about 350 km/s and higher-speed flows with velocities up to about 700 km/s. These higher-speed flows originate in coronal holes. Depending on their speed, particles in the solar wind take about 2–5 days to reach Earth.

Using space probes, we have been able to observe the solar wind extending out to 100 astronomical units (AU) from the Sun. But the solar wind does not go on forever. The farther it gets from the Sun, the more it spreads out. Just like radiation, the density of the solar wind follows an inverse square law. At a distance of about 100 AU from the Sun, the solar wind is probably not powerful enough to push the interstellar medium (the gas and dust that lie between stars in a galaxy and that surround the Sun) out of the way. There the solar wind stops abruptly,

Figure 11.16 X-ray images of the Sun show a very different picture of our star than do images taken in visible light. The brightest X-ray emission comes from the base of the Sun's corona, where gas is heated to temperatures of more than a million kelvins. This heating is most powerful above magnetically active regions of the Sun. The dark areas are coronal holes, regions where the Sun's corona is cooler and has lower density.

Figure 11.17 The solar wind streams away from the Sun for about 100 AU, until it finally piles up against the pressure of the interstellar medium through which the Sun is traveling. The *Voyager 1* spacecraft is now crossing this boundary.

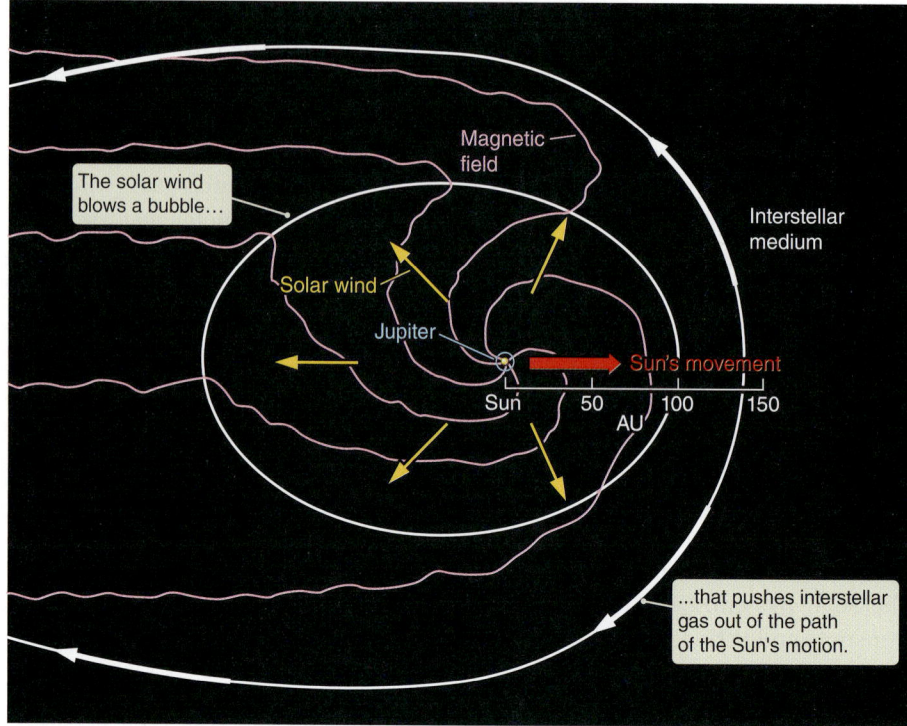

"piling up" against the pressure of the interstellar medium. **Figure 11.17** shows this region of space over which the wind from the Sun holds sway.

The boundary at which the solar wind piles up against the interstellar medium is one definition of the edge of the Solar System. Determining precisely when *Voyager 1* arrived at this boundary has been difficult, with several reports that were later retracted. In 2013, *Voyager 1* entered a region called the "magnetic highway," detected by a change in energetic particles. In this region, the Sun's magnetic field lines are connected to interstellar magnetic field lines, allowing charged particles to zoom in or out of the Solar System. This is one way to define a transition region from inside to outside the Solar System. But some specialists consider that only a change in the *direction* of the magnetic field would serve as a definitive indicator that the spacecraft has reached interstellar space. This change in direction will indicate that the spacecraft has entered a region where the magnetic field lines can no longer be traced back to the Sun.

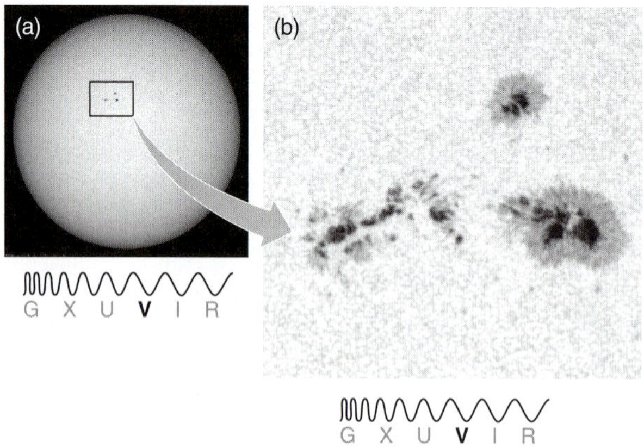

Sunspots and Changes in the Sun

Sunspots appear dark, but only in contrast to the brighter surface of the Sun (**Working It Out 11.2**). Early telescopic observations of sunspots made during the 17th century led to the discovery of the Sun's rotation, which has an average period of about 27 days as seen from Earth and 25 days relative to the stars. Because Earth orbits the Sun in the same direction that the Sun rotates, observers on Earth see a slightly longer rotation period. Observations of sunspots also show that the Sun, like Saturn, rotates more rapidly at its equator than it does at higher latitudes. This effect, referred to as **differential rotation**, is possible only because the Sun is a large ball of gas rather than a solid object.

Figure 11.18a is a photograph of a large sunspot group, and **Figure 11.18b** shows

Figure 11.18 (a) This image from the Solar Dynamics Observatory (SDO), taken in 2010, shows a large sunspot group. Sunspots are magnetically active regions that are cooler than the surrounding surface of the Sun. (b) The close-up shows a very-high-resolution view of a group of sunspots: the dark *umbra* is surrounded by the lighter *penumbra*. The solar surface around the sunspot bubbles with separate cells of hot gas, called granules. The smallest features are about 100 km across.

Working It Out 11.2 | Sunspots and Temperature

Sunspots are about 1500 K cooler than their surroundings. What does this tell us about their brightness? Think back to the Stefan-Boltzmann law in Chapter 5. The flux, $\mathcal{F}$, from a blackbody is proportional to the fourth power of the temperature, T. The constant of proportionality is the Stefan-Boltzmann constant, σ, which has a value of 5.67×10^{-8} W/(m² K⁴). We write this relationship as

$$\mathcal{F} = \sigma T^4$$

Remember that the flux is the amount of energy coming out of a square meter of surface every second. How much less energy comes out of a sunspot than out of the rest of the Sun? Let's take round numbers for the temperature of a typical sunspot and the surrounding photosphere: 4500 and 6000 K, respectively. We can set up two equations:

$$\mathcal{F}_{spot} = \sigma T^4_{spot}$$

and

$$\mathcal{F}_{surface} = \sigma T^4_{surface}$$

We could solve each of these separately and then divide the value of by $\mathcal{F}_{spot}$ by $\mathcal{F}_{surface}$ to find out how much fainter the sunspot is, but it's much easier to solve for the *ratio* of the fluxes. We divide the left side of one equation by the left side of the other equation and do the same with the right sides. Anything that is the same in both equations will divide out.

$$\frac{\mathcal{F}_{spot}}{\mathcal{F}_{surface}} = \left(\frac{\sigma T^4_{spot}}{\sigma T^4_{surface}} \right)$$

The constant σ divides out. Because both terms on the right are raised to the fourth power, we can divide them first and then take them to the fourth power:

$$\frac{\mathcal{F}_{spot}}{\mathcal{F}_{surface}} = \left(\frac{T_{spot}}{T_{surface}} \right)^4$$

Plugging in our values for T_{spot} and $T_{surface}$ gives

$$\frac{\mathcal{F}_{spot}}{\mathcal{F}_{surface}} = \left(\frac{4500 \text{ K}}{6000 \text{ K}} \right)^4$$

$$\frac{\mathcal{F}_{spot}}{\mathcal{F}_{surface}} = 0.32$$

Multiplying both sides by $\mathcal{F}_{surface}$ gives

$$\mathcal{F}_{spot} = 0.32\, \mathcal{F}_{surface}$$

So the amount of energy coming from a square meter of sunspot every second is about one-third as much as the amount of energy coming from a square meter of surrounding surface every second. In other words, the sunspot is about one-third as bright as an equal area of the surrounding photosphere. This is still extremely bright—if you could cut out the sunspot and place it elsewhere in the sky, it would be brighter than the full Moon.

---

the remarkable structure of these blemishes on the surface of the Sun. Each sunspot consists of an inner dark core called the **umbra**, which is surrounded by a less dark region called the **penumbra** that shows an intricate radial pattern, reminiscent of the petals of a flower. Sunspots are caused by magnetic fields thousands of times greater than the magnetic field at Earth's surface. They occur in pairs that are connected by loops in the magnetic field. Sunspots range in size from a few tens of kilometers across up to complex groups that may contain several dozen individual spots and span as much as 150,000 km. The largest sunspot groups are so large that they can be seen without special equipment. Look back at the chapter-opening illustration, in which a student has imaged the Sun using a pinhole camera—a tiny hole punched in a piece of cardboard and held up to the Sun. The image is projected onto a piece of paper. This is a safe way to view the Sun and image large sunspots.

Individual sunspots do not stay around for very long. Although sunspots occasionally last 100 days or longer, half of all sunspots come and go in about 2 days, and 90 percent are gone within 11 days. Sunspots have a pronounced 11-year

Figure 11.19 (a) The number of sunspots varies with time, as shown in this graph of the past few solar cycles. (b) The solar butterfly diagram shows the fraction of the Sun covered at each latitude. The data are color coded to show the percentage of the strip at that latitude that is covered in sunspots at that time: black, 0 to 0.1 percent; red, 0.1–1.0 percent; yellow, greater than 1.0 percent.

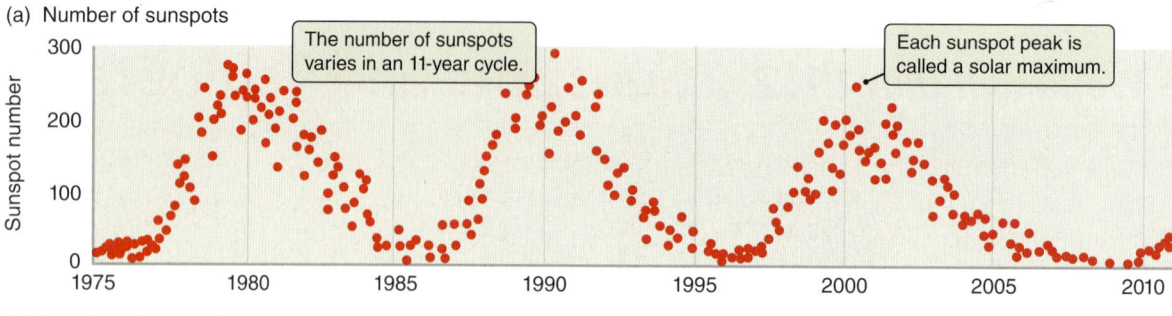

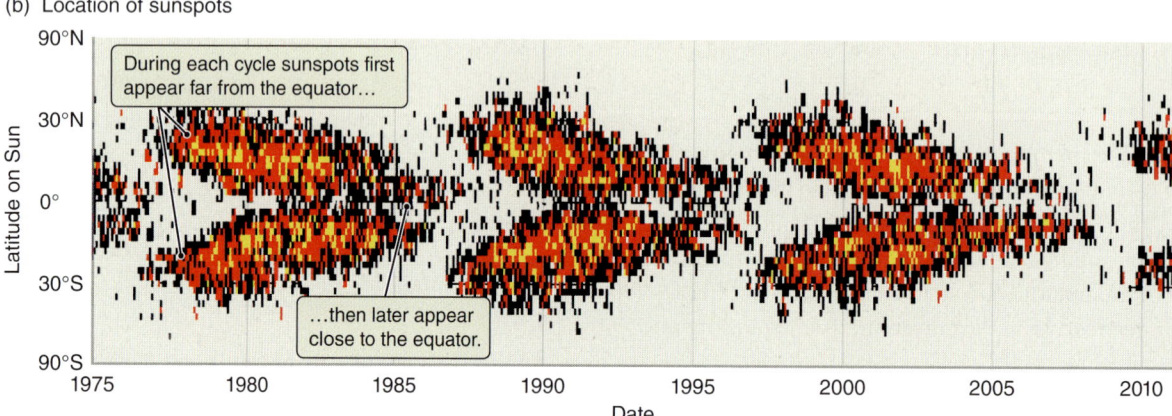

pattern called the **sunspot cycle**. **Figure 11.19a** shows data for several recent cycles. Over 11 years, both the number and the location of sunspots change. At the beginning of a cycle, sunspots appear at solar latitudes approximately 30° north and south of the solar equator. Over the following years, sunspots are found closer to the equator as their number increases to a maximum known as the **solar maximum** and then declines. As the last few sunspots approach the equator, sunspots reappear at middle latitudes and the next cycle begins. Figure 11.19b shows the number of sunspots at a given latitude plotted against time; this diagram of opposing diagonal bands is often referred to as the sunspot "butterfly diagram."

Telescopic observations of sunspots date back 400 years. We have records of sunspot observations by Chinese astronomers nearly 2,000 years before that. **Figure 11.20** shows the historical record of sunspot activity. Although the 11-year sunspot cycle is real, it is in fact neither perfectly periodic nor especially reliable. The time between peaks in the number of sunspots actually varies between about 9.7 and 11.8 years. The number of spots seen during a given cycle fluctuates as well, and there have been periods when sunspot activity has disappeared almost entirely. An extended lull in solar activity, called the **Maunder Minimum**, lasted from 1645 to 1715. Normally there would be six peaks in solar activity in 70 years, but virtually no sunspots were seen during the Maunder Minimum.

In the early 20th century, solar astronomer George Ellery Hale (1868–1938) was the first to show that the 11-year sunspot cycle is actually half of a 22-year

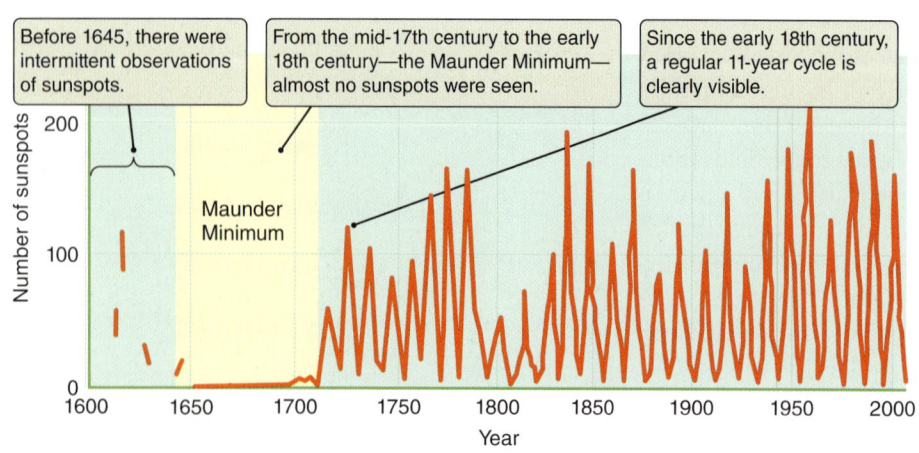

Figure 11.20 Sunspots have been observed for hundreds of years. In this plot, the 11-year cycle in the number of sunspots (half of the 22-year solar magnetic cycle) is clearly visible. Sunspot activity varies greatly over time. The period from the middle of the 17th century to the early 18th century, when almost no sunspots were seen, is called the Maunder Minimum.

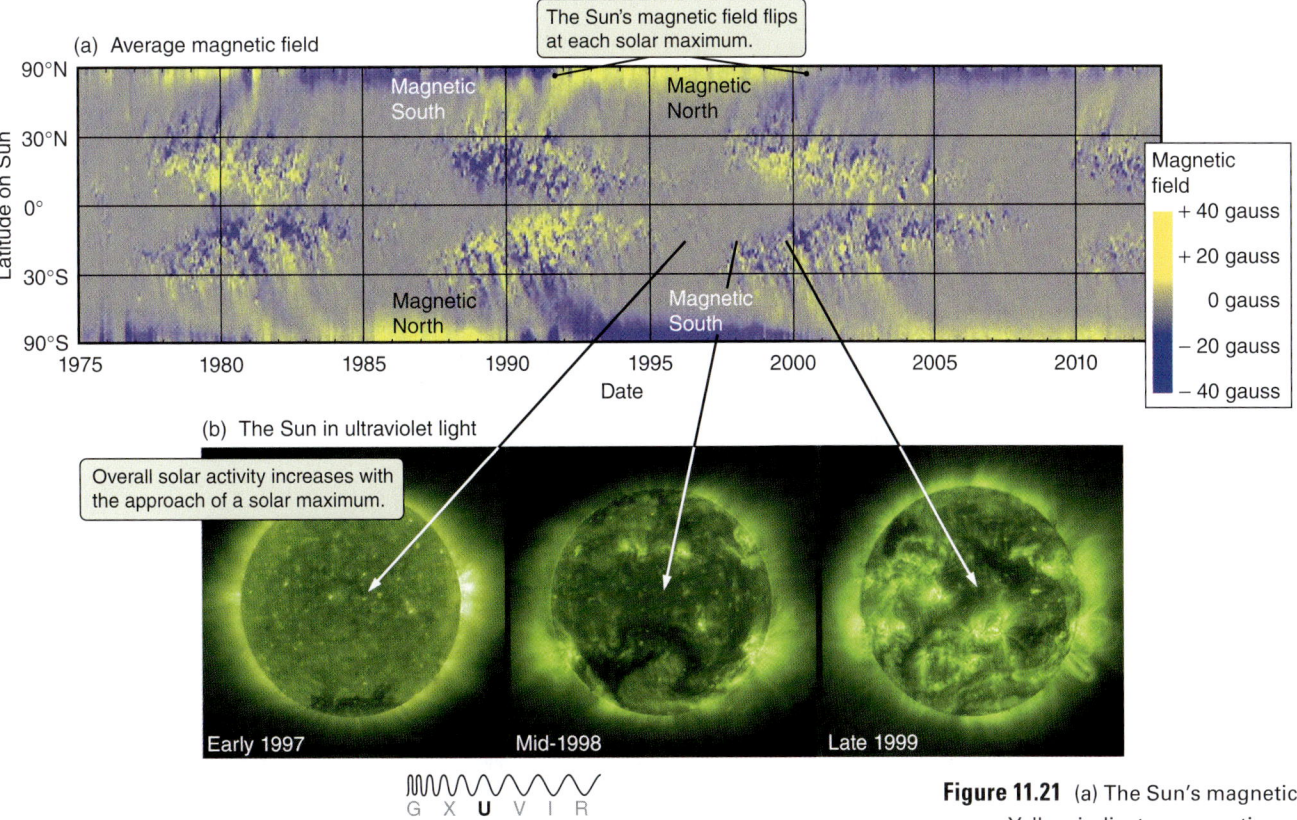

Figure 11.21 (a) The Sun's magnetic field flips every 11 years. Yellow indicates magnetic north, and blue indicates magnetic south. (b) The approach of a solar maximum is apparent in these *Solar and Heliospheric Observatory* (*SOHO*) images taken in ultraviolet light.

magnetic cycle during which the direction of the Sun's magnetic field reverses after each 11-year sunspot cycle. **Figure 11.21** shows how the average strength of the magnetic field at every latitude has changed over more than 35 years. The direction of the Sun's magnetic field flips at the maximum of each sunspot cycle. Sunspots tend to come in pairs, with one spot (the leading sunspot) in front of the other with respect to the Sun's rotation. In one sunspot cycle, the leading sunspot in each pair tends to be a north magnetic pole, whereas the trailing sunspot tends to be a south magnetic pole. In the next sunspot cycle, this polarity is reversed: the leading sunspot in each pair tends to be a south magnetic pole, whereas the trailing sunspot tends to be a north magnetic pole. The transition between these two magnetic polarities occurs near the peak of each sunspot cycle (Figure 11.21a).

Sunspots are only one of several phenomena that follow the Sun's 22-year cycle of magnetic activity. The solar maximum is a time of intense activity, as can be seen in the ultraviolet images of the Sun in Figure 11.21b. Sunspots are often accompanied by a brightening of the solar chromosphere that is seen most clearly in emission lines such as Hα. These bright regions are known as solar **active regions**. The magnificent loops arching through the solar corona, shown in **Figure 11.22**, are solar **prominences**, magnetic flux tubes of relatively cool (5000–10,000 K) gas extending through

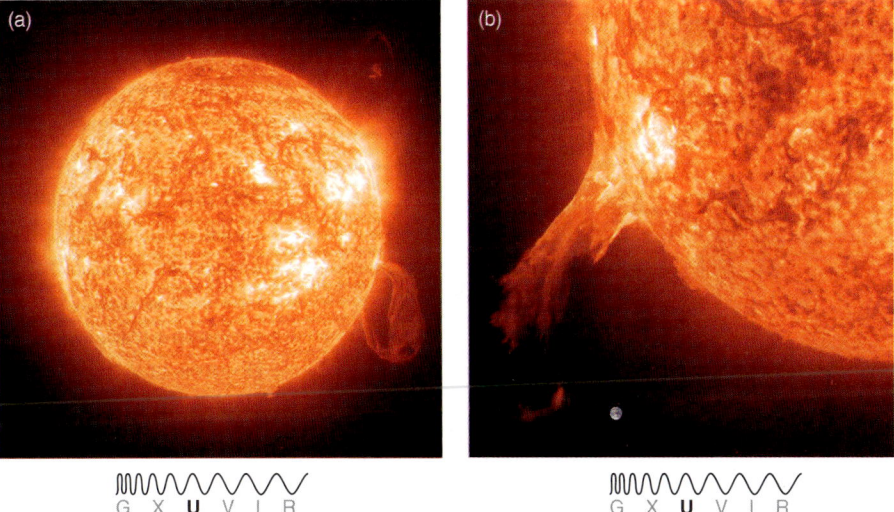

Figure 11.22 (a) Solar prominences are magnetically supported arches of hot gas that rise high above active regions on the Sun. (b) Large prominences are very large indeed, as shown in this close-up view of the base of a large prominence. Earth is shown to scale.

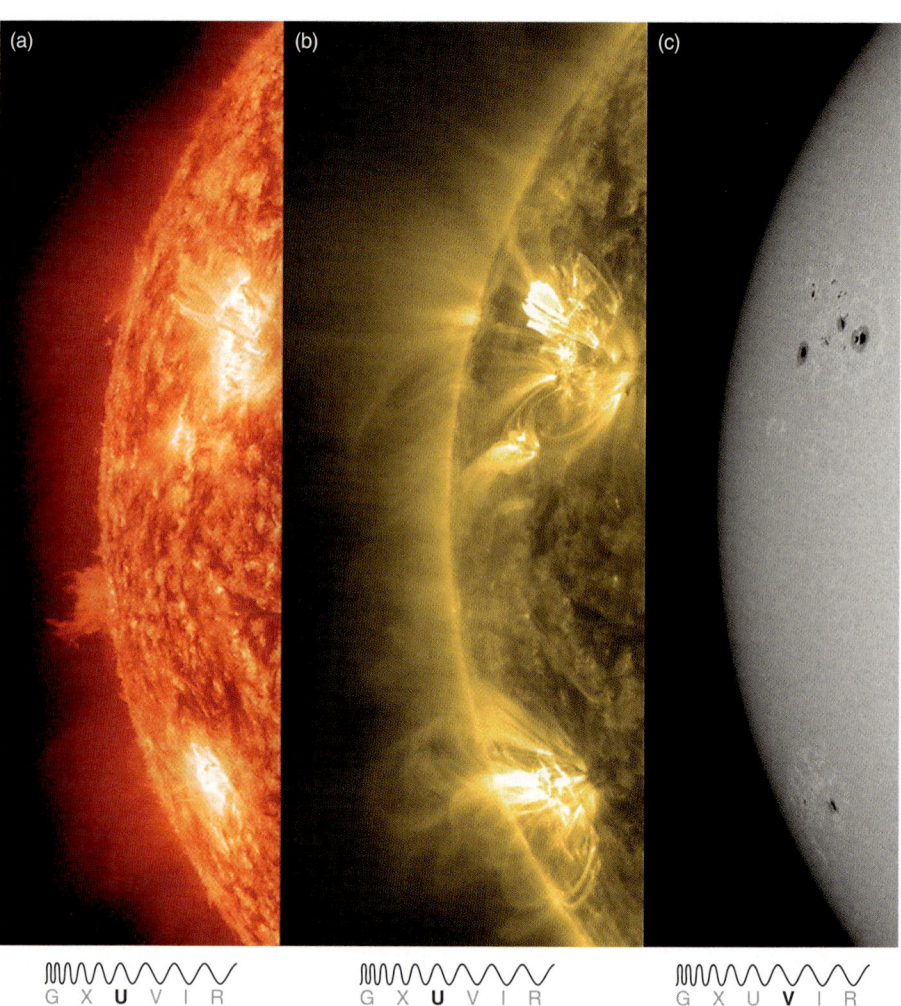

Figure 11.23 SDO observed these active regions of the Sun that produced solar flares in August 2011. (a) Activity near the surface at 60,000 K is visible in extreme ultraviolet light (along with a prominence rising up from the Sun's edge). (b) Viewed at other ultraviolet wavelengths, many looping arcs and plasma heated to about 1 million K become visible. (c) The dark spots in this image are the magnetically intense sunspots that are the sources of all the activity.

the million-kelvin gas of the corona. These prominences are anchored in the active regions. Although most prominences are relatively quiet, others can erupt out through the corona, towering a million kilometers or more over the surface of the Sun and ejecting material into the corona at speeds of 1,000 km/s.

Figure 11.23 shows **solar flares** erupting from two sunspot groups. Solar flares are the most energetic form of solar activity, violent eruptions in which enormous amounts of magnetic energy are released over the course of a few minutes to a few hours. Solar flares can heat gas to temperatures of 20 million K, and they are the source of intense X-ray and gamma-ray radiation. Hot plasma (consisting of atoms stripped of some of their electrons) moves outward from flares at speeds that can reach 1,500 km/s. Magnetic effects can then accelerate subatomic particles to almost the speed of light. Such events, called **coronal mass ejections** (**Figure 11.24**), send powerful bursts of energetic particles outward through the Solar System. Coronal mass ejections occur about once per week during the minimum of the sunspot cycle and as often as several times per day near the maximum of the cycle.

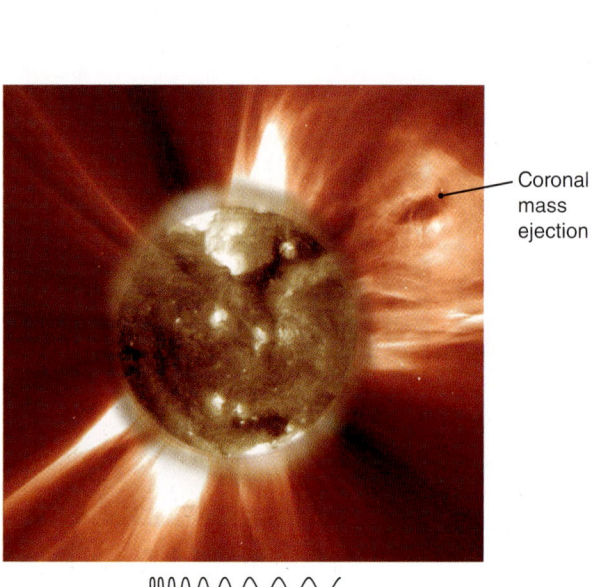

Figure 11.24 This *SOHO* image shows a coronal mass ejection (upper right), with a simultaneously recorded ultraviolet image of the solar disk superimposed.

The Effects of Solar Activity on Earth

The amount of solar radiation received at the distance of Earth from the Sun has been measured to be, on average, 1,361 watts per square meter (W/m²). As you can see in **Figure 11.25**, satellite measurements of the amount of radiation coming from the Sun show that this value can vary by as much as 0.2 percent over periods of a few weeks as dark sunspots in the photosphere and bright spots in the chromosphere move across the disk. However, the increased radiation from active regions on the Sun more than makes up for the reduction in radiation from sunspots. On average, the Sun seems to be about 0.1 percent brighter during the peak of a solar cycle than it is at its minimum.

Solar activity affects Earth in many ways. Solar active regions are the source of most of the Sun's extreme ultraviolet and X-ray emissions, energetic radiation that heats Earth's upper atmosphere and, during periods of increased solar activity, causes Earth's upper atmosphere to expand. When this happens, the swollen upper atmosphere can significantly increase the atmospheric drag on spacecraft orbiting at relatively low altitudes, such as that of the Hubble Space Telescope, causing their orbits to **decay**. Periodic boosts are necessary to keep the Hubble Space Telescope in its orbit.

Earth's magnetosphere is the result of the interaction between Earth's magnetic field and the solar wind. Recall from Chapter 7 that this interaction causes the aurora in Earth's atmosphere. Increases in the solar wind, especially coronal mass ejections, can cause magnetic storms that disrupt power grids and cause large regional blackouts. Coronal mass ejections that are emitted in the direction of Earth also hinder radio communication and navigation, and they can damage sensitive satellite electronics, including those of communication satellites. In addition, energetic particles accelerated in solar flares pose one of the greatest dangers to human exploration of space.

The *Solar and Heliospheric Observatory* (*SOHO*) spacecraft is a joint mission between NASA and the European Space Agency. *SOHO* moves in lockstep with Earth at a location approximately 1,500,000 km (0.01 AU) from Earth and almost directly in line between Earth and the Sun. *SOHO* carries 12 scientific instru-

Vocabulary Alert

decay In common language, this word often describes a process of decomposition via microbes (and usually implies a bad smell). Here we mean that the satellite loses energy, and its orbit shrinks, until the satellite crashes into the atmosphere. In other contexts, we may mean *radioactive decay*, in which an atomic nucleus spontaneously emits energy and/or particles and becomes a different element.

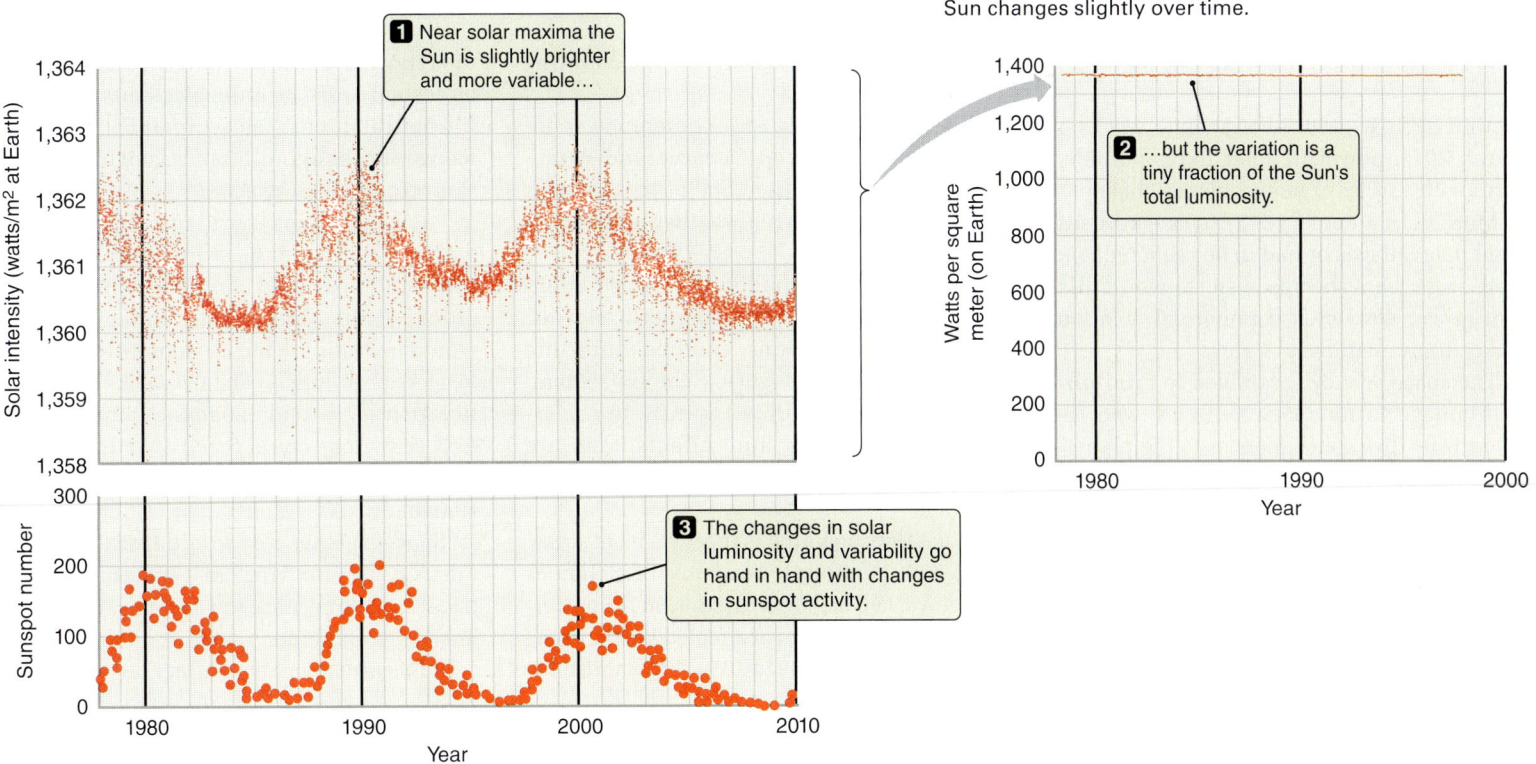

Figure 11.25 Measurements taken by satellites above Earth's atmosphere show that the amount of light from the Sun changes slightly over time.

ments that monitor the Sun and measure the solar wind upstream of Earth. Through the Solar Dynamics Observatory (SDO), NASA studies the solar magnetic field in order to predict when major solar events will occur, rather than simply responding after they happen. Detailed observations from the SDO help astronomers understand the complex nature of the solar atmosphere.

Solar activity certainly affects Earth's upper atmosphere, and we might imagine that it could affect weather patterns as well, although a mechanism to connect the two is unknown. Solar physics models suggest that observed variations in the Sun's luminosity could account for differences of only about 0.1 K in Earth's average temperature—much less than the effects due to the ongoing buildup of carbon dioxide in Earth's atmosphere. However, triggering the onset of an ice age may require a drop in global temperatures of only about 0.2–0.5 K, so the search for a definite link between solar variability and changes in Earth's climate persists.

This article describes some of the very real problems that the Sun can cause.

 **READING ASTRONOMY News**

Weather Forecast in Space: Not Sunny, with Solar Flares

By **LEDYARD KING**, Gannet Washington Bureau, ABC

SILVER SPRING, Md.—For about 2 weeks in March 2012, powerful solar flares pounded Earth with a series of geomagnetic right hooks.

More than a dozen NASA spacecraft suffered data outages or had to be rebooted, and there were fears the space storm would disrupt power networks, commercial aviation, and communications systems.

The space agency scrambled to minimize the damage to its space vehicles, and the potentially dire consequences to the planet's electronic and industrial grid never materialized. But scientists describe the episode as a wake-up call.

"This really drove the point home that we have arrived at the age of interplanetary space weather forecasting," Madhulika Guhathakurta, a heliophysics scientist at NASA headquarters, said Tuesday. "(It's) a daunting problem."

She was speaking at a conference sponsored by the National Oceanic and Atmospheric Administration, one of the agencies in charge of monitoring weather patterns on Earth and in space.

Experts say the public pays little attention to solar flares, geomagnetic storms, and other conditions lumped into the "space weather" category. Scientists themselves have begun only recently to understand its patterns and effects and are developing more reliable ways to predict activity.

What they do know is there's a lot at stake.

A 2009 National Academy of Sciences report warned of global disruption if a powerful solar-powered storm affected Earth. Using evidence from storms in 1859, 1921, and 1989, the report said massive power outages would be accompanied by radio blackouts and satellite malfunctions. Telecommunications, GPS navigation, banking, finance, and transportation would all be affected.

"Some problems would correct themselves with the fading of the storm: radio and GPS transmissions could come back online fairly quickly," the report said. "Other problems would be lasting: a burnt-out multi-ton transformer, for instance, can take weeks or months to repair. The total economic impact in the first year alone could reach $2 trillion, some 20 times greater than the costs of a Hurricane Katrina."

Space weather also would threaten NASA missions by knocking out a spacecraft's electrical or communication systems, experts at Tuesday's conference said.

It has already had an effect on aviation.

SEE WEATHER

Weather (cont.)

Officials with Delta Airlines told participants at Tuesday's conference that interference from geomagnetic and solar storms is forcing the rerouting of some long-distance flights to avoid potential communications outages. Those changes cost the company thousands of dollars per trip.

Janet Kozyra, a research professor at the University of Michigan, said there's a lot to learn about space weather.

Scientists can't predict when the Sun is going to shoot off a flare or even where it's headed once it's spotted, she said. They can't usually tell if it's going to be a "geo-effective" event that could cause havoc because they can't examine what's happening inside the flare unless there are space instruments nearby.

Even if it's not deemed geo-effective at first, it still could pose a threat, depending on how it interacts with magnetic fields and plasma it encounters.

"There are a lot of unknowns in the system that we don't understand," she said.

There has been progress, experts said.

Neal Zapp, part of the Space Radiation Group at Johnson Space Center in Texas, said it was only a decade ago scientists were lamenting there was no way to forecast space weather. Now, better instruments are giving them an opportunity to do just that.

On February 6, 2011, two NASA-launched space observatories provided researchers the first-ever complete view of the Sun.

But Guhathakurta cautions there's still much to learn about how weather develops in the 93-million-mile journey from the Sun to Earth.

"Space weather," she said, "is very much a research frontier."

Evaluating the News

1. This short article lists several sectors of the economy that were affected by solar activity in the past few years. Pick one of those, and describe how this disruption could affect you personally.
2. As with ordinary weather, predicting "space weather" does not mean controlling it. So why is it valuable to be able to predict it?
3. In the first sentence, the author refers to "geomagnetic right hooks." (The "right hook" is a particularly powerful type of punch used by boxers.) Compare this metaphorical phrase with what you have learned in this chapter about the interaction of the Sun and Earth. What does the author mean by this phrase?
4. NASA "scrambled to minimize the damage to its space vehicles." How might they do that?
5. Consider the quoted economic impact of the global disruption due to a solar storm ($2 trillion). In the story, this is compared to the cost of Hurricane Katrina, which struck the United States in 2005. How does it compare to the cost of a satellite that monitors the Sun? (You may have to do a little research to find an estimate of the costs of a NASA satellite.)

Ledyard King: "Weather Forecast in Space: Not Sunny, with Solar Flares." From *USA Today*—Gannett Washington Bureau, June 4, 2013. Used by permission of USA Today. Copyright © 2013 Gannett-USA Today. All rights reserved. Used by permission and protected by the copyright laws of the United States. The printing, redistribution, or retransmission of this Content without express permission is prohibited.

✧ SUMMARY

The forces due to pressure and gravity balance each other in hydrostatic equilibrium, maintaining the Sun's structure. Nuclear reactions converting hydrogen to helium are the source of the Sun's energy. As hydrogen fuses to helium in the core of the Sun, neutrinos are emitted. Neutrinos are elusive, almost massless particles that interact only very weakly with other matter. Observations of neutrinos confirm that nuclear fusion is the Sun's primary energy source. Energy created in the Sun's core moves outward to the surface, first by radiation and then by convection. The Sun has multiple layers, like an onion, each with a characteristic density, temperature, and pressure. The temperature of the Sun's atmosphere ranges from about 6600 K near the bottom to as much as 2 million K at the top. Material streaming away from the Sun's corona creates the solar wind. Sunspots are photospheric regions that are cooler than their surroundings, and they reveal the 11- and 22-year cycles in solar activity. Solar storms can disrupt power grids and damage satellites.

LG 1 The outward pressure of the gas inside the Sun balances the inward pull of gravity at every point. This balance is dynamically maintained. An energy balance is also maintained, with the energy produced in the core of the Sun balancing the energy lost from the surface.

LG 2 When four hydrogen atoms become one helium atom, some mass is lost. This mass is turned into energy, nearly all of which leaves the Sun either as photons (light) or as neutrinos.

LG 3 Energy moves outward through the Sun by radiation and by convection.

LG 4 The interior of the Sun is divided into zones that are defined by how energy is transported in that region. This model of the interior of the Sun has been tested by helioseismology, in much the same way that the model of Earth's interior has been tested by seismology.

LG 5 Activity on the Sun follows a cycle that peaks every 11 years but takes 22 full years for the magnetic field to reset completely. This activity, particularly ejections of mass from the corona, produces auroras and affects human technology.

SUMMARY SELF-TEST

1. The structure of the Sun is determined by both the balance between the forces due to _____ and gravity and the balance between energy generation and energy _____.
   a. pressure; production
   b. pressure; loss
   c. ions; loss
   d. solar wind; production

2. Place in order the following steps in the fusion of hydrogen into helium. If two or more steps happen simultaneously, use an equal sign (=).
   a. A positron is emitted.
   b. One gamma ray is emitted.
   c. Two hydrogen nuclei are emitted.
   d. Two ^{3}He collide and become ^{4}He.
   e. Two hydrogen nuclei collide and become ^{2}H.
   f. Two gamma rays are emitted.
   g. A neutrino is emitted.
   h. One deuterium nucleus and one hydrogen nucleus collide and become ^{3}He.

3. As energy moves out from the Sun's core toward its surface, it first travels by _____, then by _____, and then by _____.
   a. radiation; conduction; radiation
   b. conduction; radiation; convection
   c. radiation; convection; radiation
   d. radiation; convection; conduction

4. The physical model of the Sun's interior has been confirmed by observations of
   a. neutrinos and seismic vibrations.
   b. sunspots and solar flares.
   c. neutrinos and positrons.
   d. sample returns from spacecraft.
   e. sunspots and seismic vibrations.

5. Sunspots, flares, prominences, and coronal mass ejections are all caused by
   a. magnetic activity on the Sun.
   b. electric activity on the Sun.
   c. the interaction of the Sun's magnetic field and the interstellar medium.
   d. the interaction of the solar wind and Earth's magnetic field.
   e. the interaction of the solar wind and the Sun's magnetic field.

6. Ultimately, the Sun's energy comes from
   a. the mass of hydrogen nuclei.
   b. gravitational collapse.
   c. residual heat of formation.
   d. the slowing of its rotation rate.

7. Radiation transports energy by moving _____, while convection transports energy by moving _____.
   a. neutrinos; matter
   b. neutrinos; light
   c. light; matter
   d. matter; light

8. The temperature and density change abruptly at the interface between
   a. the radiative zone and the photosphere.
   b. the photosphere and the chromosphere.
   c. the chromosphere and the corona.
   d. the corona and space.

9. The solar wind
   a. makes a perfectly spherical bubble around the Solar System.
   b. does not interact with the magnetic field of the Sun.
   c. creates a teardrop-shaped bubble around the Solar System.
   d. creates a wind pressure much stronger than a wind on Earth.

10. Sunspots peak every ____ years, a consequence of the ___-year magnetic solar cycle.
    a. 11; 22
    b. 22; 11
    c. 5.5; 11
    d. 22; 44

QUESTIONS AND PROBLEMS

Multiple Choice and True/False

11. T/F: Hydrostatic equilibrium is the balance between energy production and loss.

12. T/F: Six hydrogen nuclei are involved in the proton-proton chain, although only four of them wind up in the helium nucleus.

13. T/F: Photons travel outward through the Sun because the innermost part of the Sun is more crowded with photons than the outer parts.

14. T/F: Neutrinos make it out of the Sun faster than photons. This means they travel faster than photons.

15. T/F: Sunspots can be comparable in size to Earth.

16. Hydrostatic equilibrium inside the Sun means that
    a. energy produced in the core equals energy radiated from the surface.
    b. radiation pressure balances the weight of outer layers pushing down.
    c. the Sun absorbs and emits equal amounts of energy.
    d. the Sun does not change over time.

17. The energy that is emitted from the Sun is produced
    a. at the interface between the chromosphere and the photosphere.
    b. at the top of the convection zone.
    c. in the core, by nuclear fusion.
    d. at the surface.

18. In the proton-proton chain, four hydrogen nuclei are converted to a helium nucleus. This does not happen spontaneously on Earth because the process requires
    a. vast amounts of hydrogen.
    b. very high temperatures and pressures.
    c. hydrostatic equilibrium.
    d. very strong magnetic fields.

19. The solar neutrino problem pointed to a fundamental gap in our knowledge of
    a. nuclear fusion.
    b. neutrinos.
    c. hydrostatic equilibrium.
    d. magnetic fields.

20. Sunspots appear dark because
    a. they have very low density.
    b. magnetic fields absorb most of the light that falls on them.
    c. they are cooler than their surroundings.
    d. they are regions of very high pressure.

21. Sunspots change in number and location during the solar cycle. This phenomenon is connected to
    a. the rotation rate of the Sun.
    b. the temperature of the Sun.
    c. the magnetic field of the Sun.
    d. the tilt of the axis of the Sun.

22. Suppose an abnormally large amount of hydrogen suddenly burned in the core of the Sun. Which of the following would be observed first?
    a. The Sun would become brighter.
    b. The Sun would swell and become larger.
    c. The Sun would become bluer.
    d. The Sun would emit more neutrinos.

23. The solar corona has a temperature of more than a million kelvins; the photosphere has a temperature of only about 6000 K. Why isn't the corona much, much brighter than the photosphere?
    a. The magnetic field traps the light.
    b. The corona emits only X-rays.
    c. The photosphere is closer to us.
    d. The corona has a much lower density.

24. The Sun rotates once every 25 days relative to the stars. The Sun rotates once every 27 days relative to Earth. Why are these two numbers different?
    a. The stars are farther away.
    b. Earth is smaller.
    c. Earth moves in its orbit during this time.
    d. The Sun moves relative to the stars.

25. Some engineers and physicists have been working to solve the world's energy supply problem by constructing power plants that would convert hydrogen to helium. Our Sun seems to have solved this problem. On Earth, what is the major obstacle to this solution?
    a. It is difficult to achieve the temperatures of the interior of the Sun.
    b. There is not enough hydrogen.
    c. The helium shortage is driving up prices.
    d. All the hydrogen is locked up in water already.

Conceptual Questions

26. The Sun's stability depends on hydrostatic equilibrium and energy balance. Describe how both of these work.

27. Explain how hydrostatic equilibrium acts as a safety valve to keep the Sun at its constant size, temperature, and luminosity.

28. In Figure 11.3, two protons are shown interacting at various speeds.
    a. What is different about parts (a) and (b)?
    b. Why are the blue arrows larger in (b) than in (a)? What do these blue arrows represent?

29. Radiative transfer takes photons from hot regions to cool regions, on average. Why?

30. Describe nuclear fusion and how it relates to the Sun's source of energy.

31. Two of the three atoms in a molecule of water (H_2O) are hydrogen. Why are Earth's oceans not fusing hydrogen into helium, transforming our planet into a star?

32. Sunspots are dark splotches on the Sun. Why are they dark?

33. Explain the proton-proton chain through which the Sun generates energy by converting hydrogen to helium.

34. Figure 11.4 diagrams the proton-proton chain. In this figure, a number of squiggly arrows are pointing away from the interactions. What do these squiggly arrows represent? Are they waves or particles or both?

35. The proton-proton chain is often described as "fusing four hydrogens into one helium," but actually six hydrogen nuclei are involved in the reaction. Why don't we include the other two nuclei in our description?

36. On Earth, nuclear power plants use *fission* to generate electricity. A heavy element like uranium is broken into many smaller atoms, where the total mass of the fragments is less than the original atom. Explain why fission could not be powering the Sun today.

37. In the proton-proton chain, the mass of four protons is slightly greater than the mass of a helium nucleus. Explain what happens to this "lost" mass.

38. Discuss the "solar neutrino problem" and how this problem was solved.

39. What technique do you find in common between how we probe the internal structure of the Sun and how we probe the internal structure of Earth?

40. How is the fate of the Hubble Space Telescope tied to solar activity?

Problems

41. Study Figure 11.2.
    a. For each quantity graphed (pressure, density, temperature, and energy produced), estimate the radius at which the value has dropped to half the maximum.
    b. Which of these quantities decreases fastest with radius? Which decreases slowest with radius?
    c. Explain why, even though the temperature is decreasing smoothly, the energy production drops abruptly.

42. The Sun rotates every 25 days relative to the stars.
    a. Assuming a sunspot is located at the equator of the Sun, how long does it take to go once around the Sun?
    b. How far has it traveled during this time?
    c. How fast is it traveling relative to the stars?

43. The Sun shines by converting mass into energy according to Einstein's well-known relationship $E = mc^2$. Show that if the Sun produces 3.85×10^{26} joules (J) of energy per second, it must convert 4.3 billion kg of mass per second into energy. Note that 1 J/s is a watt (W), which may be more familiar to you.

44. Assume that the Sun has been producing energy at a constant rate over its lifetime of about 4.5 billion years (1.4×10^{17} seconds).
    a. How much mass has it lost in creating energy over its lifetime?
    b. The current mass of the Sun is 2×10^{30} kg. What fraction of its current mass has been converted into energy over the lifetime of the Sun?

45. Let's examine how we know that the Sun cannot power itself by chemical reactions. Using Working It Out 11.1 and the fact that an average chemical reaction between two atoms releases 1.6×10^{-19} J of energy, estimate how long the Sun could emit energy at its current luminosity. Compare that estimate to the known age of Earth.

46. How long does it take particles in the solar wind to reach Earth from the Sun if they are traveling at an average speed of 400 km/s?

47. Examine Figure 11.11.
    a. What is the height on the x-axis measured relative to?
    b. In Figure 11.2, the density and the temperature in the Sun's interior behaved in approximately the same way (the lines had similar shapes). Here in Figure 11.11, the solar atmosphere's density and temperature do not behave the same way. What does this tell you about the relationship between density and temperature in the atmosphere of the Sun?

48. A sunspot appears only 70 percent as bright as the surrounding photosphere. The photosphere has a temperature of approximately 5780 K. Follow Working It Out 11.2 to find the temperature of the sunspot.

49. Examine Figure 11.19.
    a. Describe what is plotted in parts (a) and (b).
    b. Does the peak in sunspot number occur at the beginning or end of the sunspot cycle, or somewhere in between?
    c. In part (b) of Figure 11.19, the latitude of the Sun is plotted on the y-axis. Compare this with Figure 11.21a. Do the magnetic field strengths seem to be highest in the regions where sunspots are located?
    d. Compare the *SOHO* images in Figure 11.21b, first with the graph of the average magnetic field in Figure 11.21a, and then with the butterfly diagram in Figure 11.19b. When did this sunspot cycle begin? Why do the three images look so different?

50. Use Figure 11.25 to estimate the "tiny fraction" by which the Sun varies over its cycle.

SMARTWORK

Norton's online homework system includes algorithmically generated versions of these questions, plus additional conceptual exercises. If your instructor assigns questions in SmartWork, log in at smartwork.wwnorton.com.

Exploration | The Proton-Proton Chain

wwnpag.es/uou2

The proton-proton chain powers the Sun by fusing hydrogen into helium. As a by-product, several different particles are produced, which eventually produce energy. The process has multiple steps, and this Exploration is designed to explore these steps in detail, hopefully to help you keep them straight.

Visit the Student Site (wwnpag.es/uou2) and open the Proton-Proton Animation Nebraska Simulation in Chapter 11. Press play, and watch the animation all the way through once. Press play again, and pause the animation after the first collision. Two hydrogen nuclei (both positively charged) have collided to produce a new nucleus with only one positive charge.

1. Which particle carried away the other positive charge?

2. What is a neutrino? Did the neutrino enter the reaction or was the neutrino produced in the reaction?

Compare the interaction on the top with the interaction on the bottom.

3. Did the same reaction occur in each instance?

Press play again, and pause the animation after the second collision.

4. What two types of nuclei entered the collision? What type of nucleus resulted?

5. Was charge conserved in this reaction or was it necessary for a particle to carry charge away?

6. What is a gamma ray? Did the gamma ray enter the reaction or was it produced by the reaction?

Press play again, and allow the animation to run to the end.

7. What nuclei enter the final collision? What nuclei are produced?

8. In chemistry, a catalyst is a reaction helper. It facilitates the reaction but does not get used up in the process. Are there any nuclei that act like catalysts in the proton-proton chain?

Make a table of inputs and outputs. Which of the particles in the final frame of the animation were inputs to the reaction? Which were outputs? Fill in your table with these inputs and outputs.

9. Which of the outputs are converted into energy that leaves the Sun as light?

10. Which of the outputs could become involved in another reaction immediately?

11. Which of the outputs is likely to stay in that form for a very long time?

SMARTWORK • smartwork.wwnorton.com

12 Evolution of Low-Mass Stars

Like all main-sequence stars, the Sun gets its energy by converting hydrogen to helium. But the Sun cannot remain a main-sequence star forever. It will eventually exhaust the hydrogen fuel source in its core. New balances between gravity and energy production must constantly be found as the Sun evolves beyond the main sequence, until at last, no balance is possible. In the illustration on the opposite page, a student has taken an image of a cloud of dust and gas centered around a star like the Sun that is moving through these last evolutionary phases. He has sketched the trajectory of Sun-like stars across and down the Hertzsprung-Russell (H-R) diagram. This image fits in at the top of the diagram, as the star loses some of its mass into a nebula a thousand times the size of the Solar System. In this chapter, we will follow the fate of low-mass stars like the Sun as they run out of hydrogen and will trace out this path on the H-R diagram.

LEARNING GOALS

Within its core, the Sun loses more than 4 billion kilograms of mass each second as it fuses hydrogen to helium. Although the Sun may seem immortal by human standards, eventually it will run out of fuel, and its time on the main sequence will come to an end. By the end of this chapter, you should be able to plot the evolution of a Sun-like star along the H-R diagram, as the student has done at right. You should also be able to:

LG 1 Use the mass of a main-sequence star to estimate its lifetime.

LG 2 Explain why the Sun will grow larger and more luminous when it runs out of fuel.

LG 3 Sketch post-main-sequence evolutionary tracks on an H-R diagram.

LG 4 Describe how planetary nebulae and white dwarfs form.

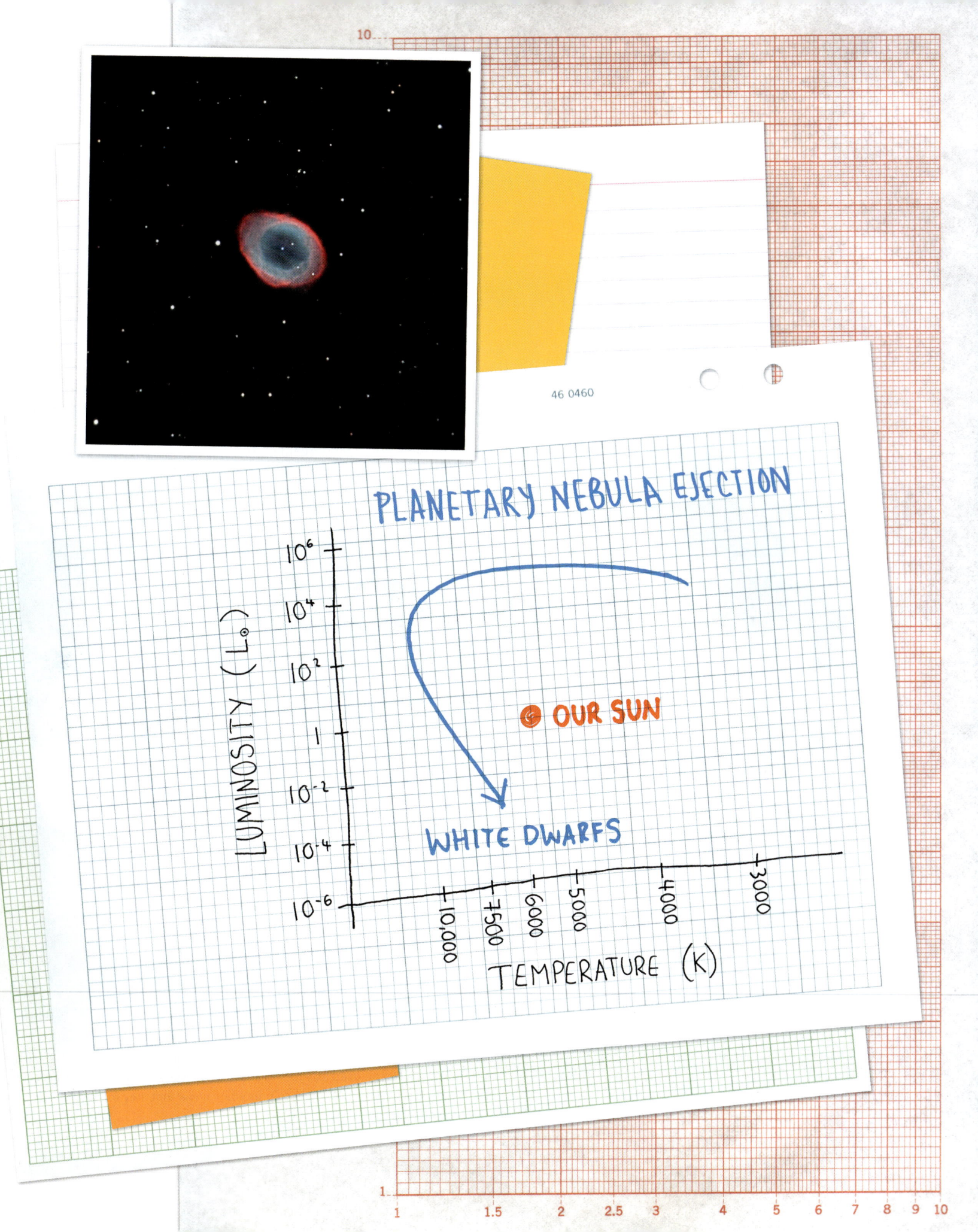

▶▶ **Nebraska Simulation:** Hertzsprung-Russell Diagram Explorer

12.1 The Life of a Main-Sequence Star Follows a Predictable Path

The evolutionary course of each star is determined when the star forms. The foremost factor in its evolution is its mass. Its chemical composition is a minor factor. Relatively small differences in the masses of two stars can sometimes result in significant differences in their fates. Nevertheless, stars can be divided roughly into two broad categories: high mass and low mass. Massive, luminous O and B stars will be discussed in the next chapter. They follow a course fundamentally different from that of the cooler, fainter, less massive stars found toward the lower right end of the main sequence. **Low-mass stars** have masses less than about 8 $M_\odot$. In this chapter, we examine the stages through which low-mass stars progress, beginning with the main-sequence evolution.

Changes in Structure

When the Sun formed, about 90 percent of its atoms were hydrogen atoms. When we discussed the collapse of a protostar in Chapter 5, we considered the idea of a changing balance between gravity and pressure. The protostar is always in bal-

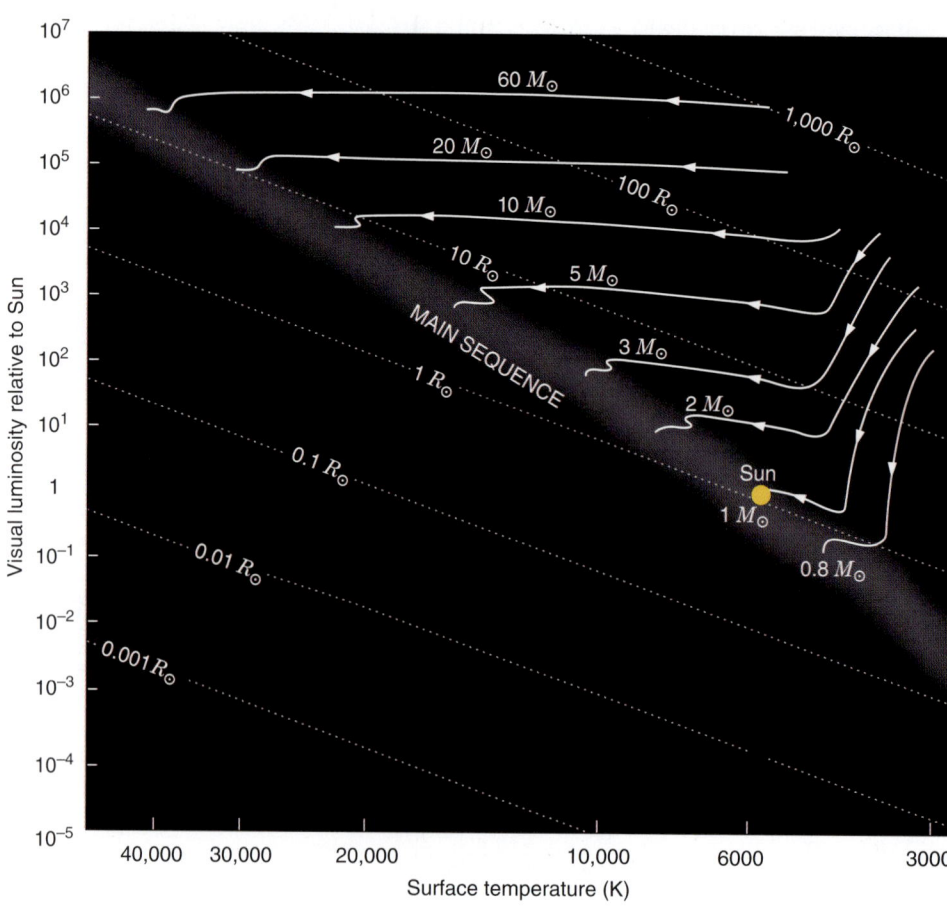

Figure 12.1 The H-R diagram can be used as a tool to show how a star evolves, or changes over time. In this case, the pre-main-sequence tracks of protostars show that they follow different paths on their way to the main sequence.

ance as it collapses, but as it radiates away thermal energy, it becomes a smaller and denser object. Since reaching the main sequence, the Sun has produced its energy by converting hydrogen into helium via the proton-proton chain. As the composition of a star changes, so must its structure.

The Hertzsprung-Russell (H-R) diagram helps us understand how a protostar changes as it becomes a main-sequence star and shows us how all stars change as they evolve throughout their lifetimes. The path a star follows across an H-R diagram as it goes through the different stages of its life is called the star's **evolutionary track**. In **Figure 12.1**, evolutionary tracks for protostars of different masses show that stars of all masses grow smaller and hotter until they reach their positions as stars on the main sequence.

Chapter 11 described how the structure of the Sun is determined by a balance between the inward-pushing force of gravity and the outward-pushing force of pressure. The pressure within the Sun is maintained by energy released by nuclear fusion in the core of the star. However, the structure of a main-sequence star must continually shift in response to the changing core composition as it uses the fuel in its core.

Compared to the events that follow, stars evolve slowly while they are on the main sequence. Between the time the Sun was born and the time it will leave the main sequence, its luminosity will roughly double. Most of this change will occur during the last billion years of its life on the main sequence.

The mass and luminosity of main-sequence stars are related. **Figure 12.2** shows that as the mass increases so does the luminosity because the mass of the star governs the rate at which nuclear reactions occur in the core. More mass means stronger gravity; stronger gravity means higher temperature and pressure in the interior; higher temperature and pressure mean faster nuclear reactions; faster nuclear reactions mean a more luminous star and a faster rate of fuel consumption. Therefore, the length of time a star spends on the main sequence is determined almost entirely by its mass with the counterintuitive result that less massive stars live longer. **Table 12.1** shows the main-sequence lifetimes for stars of different spectral types and masses.

A main-sequence star has a limited lifetime. It may seem intuitive that because a more massive star has more mass, it will live longer. Yet the star not only has more fuel but also burns it at a faster rate, and this is the factor that dominates. Stars with higher masses live shorter lives, not longer ones, because they burn their fuel faster. **Working It Out 12.1** develops this idea further.

Helium Ash in the Center of the Star

The helium "**ash**" produced in the core of a low-mass, main-sequence star does not fuse to form even heavier elements. At the temperature found at the center of a low-mass main-sequence star, atomic collisions are not energetic enough to overcome the electric repulsion between helium nuclei. This nonburning helium ash does not build up evenly throughout the interior of a star. Because the temperature and pressure are highest at the center of a main-sequence star, hydrogen burns most rapidly there. As a result, helium accumulates more rapidly at the center of the star than elsewhere.

If we could cut open a star and watch it evolve, we would see its chemi-

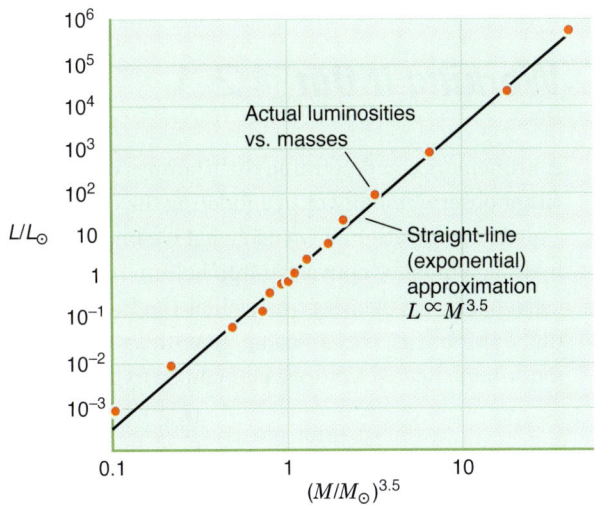

Figure 12.2 The mass and luminosity of main-sequence stars are related such that $L \propto M^{3.5}$. The exponent (3.5) is an average value over the wide range of main-sequence star masses.

Vocabulary Alert

ash In common language, this word specifically refers to the gray, dusty product of a fire, accumulating as the fire burns and collecting, for example, at the bottom of a fireplace. Astronomers use this word metaphorically to refer to the products of fusion, which collect in the core of the star.

TABLE 12.1
Main-Sequence Lifetimes

SPECTRAL TYPE	MASS ($M_\odot$)	LUMINOSITY ($L_\odot$)	MAIN-SEQUENCE LIFETIME (YEARS)
O5	60	500,000	3.6×10^5
B5	5.9	480	1.2×10^8
A5	2.0	12.3	1.8×10^9
F5	1.4	2.6	4.3×10^9
G2 (our Sun)	1.0	1.0	1.0×10^{10}
G5	0.92	0.8	1.2×10^{10}
K5	0.67	0.32	2.7×10^{10}
M5	0.21	0.008	4.9×10^{11}

Working It Out 12.1 | Estimating Main-Sequence Lifetimes

Astronomers can determine the lifetimes of main-sequence stars either observationally or by modeling the evolution of stars of a given composition. If we use what we know about how much hydrogen must be converted into helium each second to produce a given amount of energy, as well as the fraction of its hydrogen that a star burns, we can come up with a relationship that says the *main-sequence lifetime* of a star can be expressed as

$$\text{Lifetime} \propto \frac{M_{MS}}{L_{MS}}$$

where M_{MS} is the star's main-sequence mass (amount of fuel) and L_{MS} is its main-sequence luminosity (the rate at which fuel is used). We can put this relationship in quantitative terms by introducing a constant of proportionality, 1.0×10^{10}, which is the computed lifetime (in years) of a 1-$M_\odot$ star:

$$\text{Lifetime} = (1.0 \times 10^{10}) \times \frac{M_{MS}/M_\odot}{L_{MS}/L_\odot} \text{ years}$$

The relationship between the mass and the luminosity of stars is very sensitive. Relatively small differences in the masses of stars result in large differences in their main-sequence luminosities. One method for estimating luminosities of main-sequence stars is known as the **mass-luminosity relationship**, $L \propto M^{3.5}$, which is based on observed luminosities of stars of known mass (Figure 12.2). As above, we can express this relationship relative to the Sun's mass and luminosity:

$$\frac{L_{MS}}{L_\odot} = \left(\frac{M_{MS}}{M_\odot}\right)^{3.5}$$

Substituting the mass-luminosity relationship into the lifetime equation gives us

$$\text{Lifetime} = (1.0 \times 10^{10}) \times \frac{M_{MS}/M_\odot}{(M_{MS}/M_\odot)^{3.5}}$$

$$= (1.0 \times 10^{10}) \times \left(\frac{M_{MS}}{M_\odot}\right)^{-2.5} \text{ years}$$

As an example, let's look at a main-sequence K5 star. According to the masses listed in Table 12.1 (where the corresponding lifetimes are based on more detailed models than our calculations here), a K5 star has a mass that is about equal to 0.67 times that of the Sun:

$$\text{Lifetime}_{K5} = (1.0 \times 10^{10}) \times (0.67)^{-2.5} = 2.7 \times 10^{10} \text{ years}$$

A K5 star has a main-sequence lifetime 2.7 times as long as the Sun's. Even though the K5 star starts out with less fuel than the Sun, it burns that fuel much more slowly, so it lives longer.

---

cal composition changing most rapidly at its center and less rapidly as we move outward. **Figure 12.3** shows how the chemical composition inside a star like the Sun changes throughout its main-sequence lifetime. When the Sun formed, it had a uniform composition of about 70 percent hydrogen and 30 percent helium by mass. As hydrogen was fused into helium, the helium fraction in the core of the Sun climbed. Today, roughly 5 billion years later, only about 35 percent of the mass *in the core* of the Sun is hydrogen.

12.2 A Star Runs Out of Hydrogen and Leaves the Main Sequence

Eventually, a star exhausts all of the hydrogen fuel in its core. The innermost core of the star is composed entirely of helium ash. As thermal energy leaks out of the helium core into the surrounding layers of the star, no more energy is generated within the core to replace it. The balance that has maintained the structure of the star throughout its life is now broken. The star's life on the main sequence has come to an end, and its further evolution depends on temperature changes in the core, which govern fusion reactions.

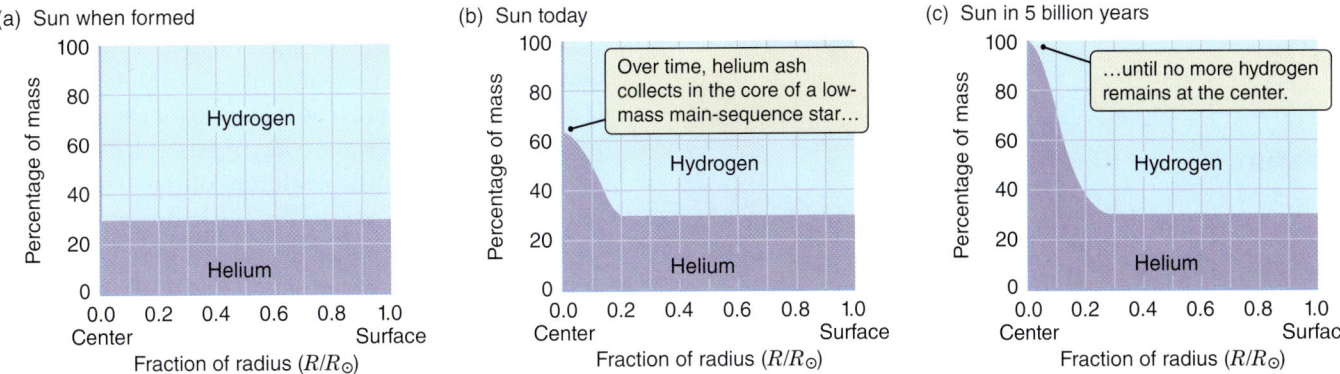

Figure 12.3 Chemical composition of the Sun is plotted here as a percentage of mass against distance from the center of the Sun. (a) When the Sun formed 5 billion years ago, about 30 percent of its mass was helium and 70 percent was hydrogen throughout. (b) Today the material at the center of the Sun is about 65 percent helium and 35 percent hydrogen. (c) The Sun's main-sequence life will end in about 5 billion years, when all of the hydrogen at the center of the Sun is gone.

Electron-Degenerate Matter in the Helium Core

All of the matter we directly experience is mostly *empty space*. An atom is mostly empty except for the tiny bit of space occupied by the nucleus and the electrons. The same is true for the matter within the Sun. At the enormous temperatures within the Sun, almost all of the electrons have been stripped away from their nuclei by energetic collisions. In other words, the gas is completely ionized—a mixture of electrons and atomic nuclei all flying about freely. Even so, the gas that makes up the Sun is still mostly empty space, and the electrons and atomic nuclei fill only a tiny fraction of the volume.

When a low-mass star like the Sun exhausts the hydrogen at its center, the situation changes. As gravity begins to win its shoving match against pressure, the helium core is crushed to an ever-smaller size and an ever-greater density, but there is a limit to how dense the core can get. More than one electron cannot occupy the same state at the same time. This limits the number of electrons that can be packed into a given volume of space at a given pressure. As the matter is compressed further and further, it finally reaches this limit. The space is now effectively "filled" with electrons that are smashed tightly together. This matter is so dense that a single cubic centimeter (about the size of a standard six-sided die) has a mass of more than 1,000 kilograms (kg). Matter that has been compressed to the point at which electrons are packed as closely as possible is called **electron-degenerate** matter. Degenerate matter acts much differently than normal matter, as we will see.

Hydrogen Shell Burning

Once a low-mass star exhausts the hydrogen at its center, nuclear burning pauses in the core. Outside the core, hydrogen continues to burn. Astronomers call this **hydrogen shell burning** because the star's hydrogen now burns only in a shell surrounding the core.

Electron-degenerate matter has a number of fascinating properties. For example, as more and more helium ash piles up on the degenerate core from the

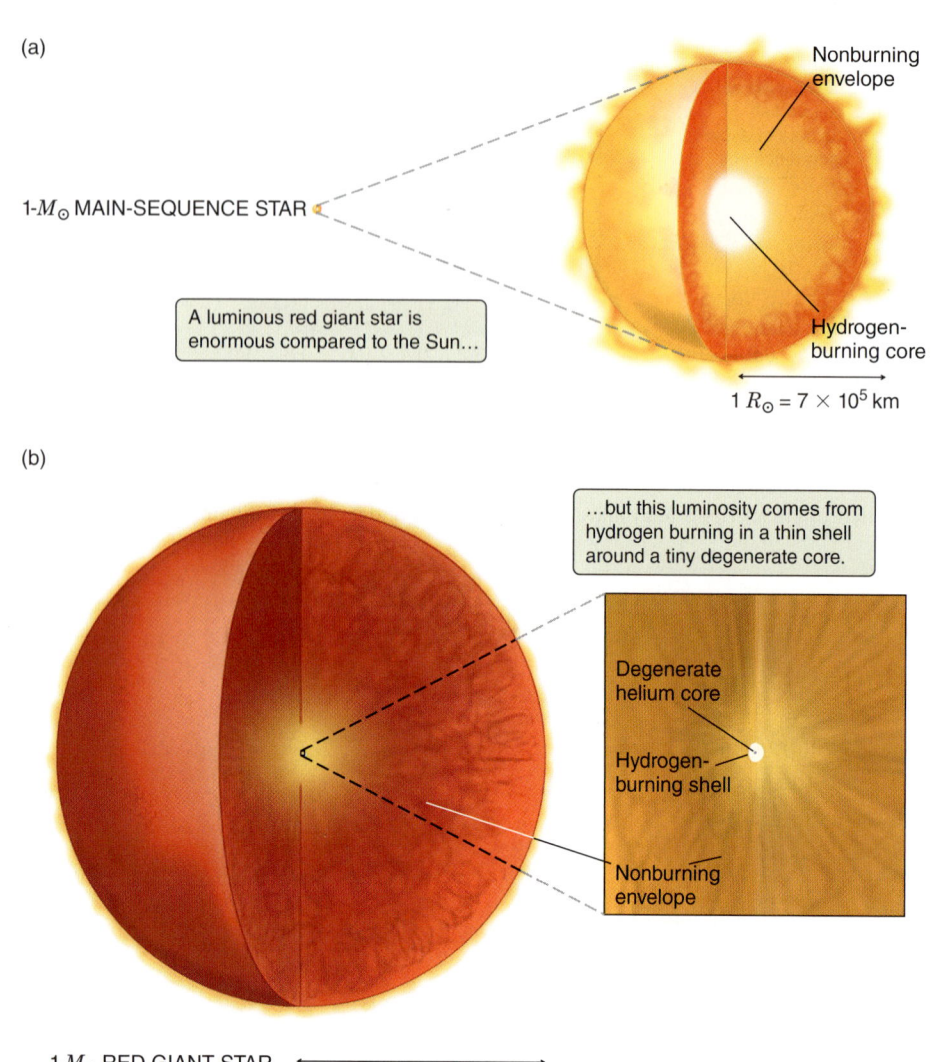

Figure 12.4 The size and structure of the Sun (a) are compared with the size and structure of a star near the top of the *red giant branch* of the H-R diagram (b). The left-hand diagrams in (a) and (b) compare the sizes of the two stars. The right-hand diagrams compare the size and structure of the Sun (a) with the size and structure of the core of the red giant (b). The diagrammatic images on the right are magnified about 50 times compared to those on the left.

hydrogen-burning shell, the core *gets smaller*. This is one of the ways that degenerate matter differs from normal matter: The more massive it is, the smaller it is. (This is noticeably not true for cows, for example!) The presence of the degenerate core triggers a chain of events that will dominate the evolution of the 1-$M_\odot$ star for the next 50 million years after the hydrogen runs out.

A degenerate core means stronger gravity, stronger gravity means higher pressure, and higher pressure means faster nuclear burning, producing greater and greater amounts of energy. This increase in energy generation heats the overlying layers of the star, causing them to expand to form a bloated, luminous giant. As illustrated in **Figure 12.4**, the internal structure of the main-sequence star (Figure 12.4a) becomes fundamentally different as the star evolves to a *red giant* (Figure 12.4b). A red giant star fuses hydrogen in a shell around a degenerate helium core and is both larger and redder than it was on the main sequence. The giant has a luminosity hundreds of times the luminosity of the Sun and a radius of more than 50 solar radii (50 $R_\odot$). Yet the core is far more compact than that of the Sun, and much of the star's mass becomes concentrated into a volume that is only a few times the size of Earth.

The star becomes larger, more luminous, and, perhaps surprisingly, cooler and redder as well. The enormous expanse of the star's surface allows it to cool very efficiently. Even though its interior grows hotter and its luminosity increases, the surface temperature of the star actually begins to drop.

The Evolution of the Star on the H-R Diagram

The H-R diagram, like the one in Figure 10.19, is a handy device for keeping track of the changing luminosity and surface temperature of the star as it evolves away from the main sequence. As soon as the star exhausts the hydrogen in its core, it leaves the main sequence and begins to move upward and to the right on the H-R diagram, growing more luminous but cooler. As the star continues to evolve, it grows larger and cooler. But after a time, its progress to the right on the H-R diagram ceases. The surface layers regulate how much radiation can escape from the star, thus preventing it from becoming any cooler.

The path that a star follows on the H-R diagram as it leaves the main sequence is like a tree "branch" growing out of the "trunk" of the main sequence, as shown in **Figure 12.5**. Astronomers refer to this track as the **red giant branch** of the H-R diagram.

As the star leaves the main sequence, the changes in its structure occur slowly at first, but then the star moves up the red giant branch faster and faster. It takes roughly 1 billion years for a star like the Sun to climb up the red giant branch. During the first half of this period, the star's luminosity increases to about 10 times the luminosity of the Sun (10 $L_\odot$). During the second half of this time, the star's luminosity skyrockets to almost 1,000 $L_\odot$. The evolution of the star, illus-

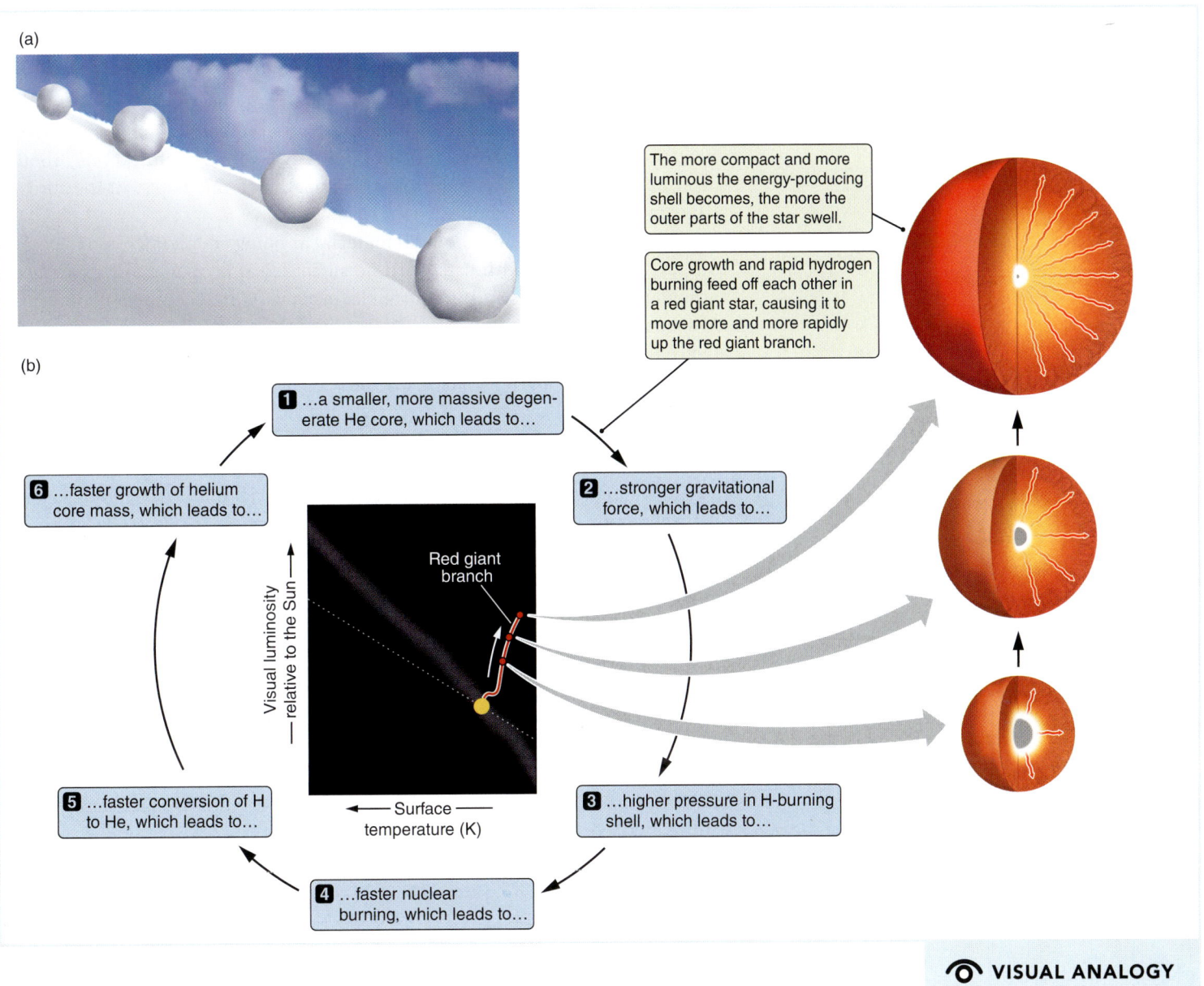

Figure 12.5 (a) As a snowball rolls down a hill, the larger it gets, the faster it grows. (b) Similarly, as a star moves up the red giant branch in the H-R diagram, the luminosity of the star grows faster and faster. The cycle burning hydrogen to helium in a shell surrounding a degenerate helium core feeds on itself.

trated in Figure 12.5, is reminiscent of the growth of a snowball rolling downhill. The larger the snowball becomes, the faster it grows; and the faster it grows, the larger it becomes.

The analogy between the evolution of a red giant star and the growth of a snowball is not perfect. The helium core of the star grows in mass—but not in radius—as hydrogen is converted to helium in the hydrogen-burning shell and the helium ash adds to the degenerate core. The increasing mass of the ever more compact helium core increases the force of gravity in the heart of the star. Stronger gravity means higher pressure, and higher pressure accelerates nuclear burning in the shell. Faster nuclear reactions in the shell convert hydrogen into helium more quickly, so the core grows more rapidly. The star has come full circle in a cycle that feeds on itself. Increasing core mass leads to ever-faster burning in the shell; and the faster hydrogen burns in the shell, the faster the core mass grows. As a result, the star's luminosity climbs at an ever-higher rate.

12.3 Helium Begins to Burn in the Degenerate Core

The growth of the red giant cannot continue forever. What will happen next? The answer lies in another unusual property of the degenerate helium core: Although the *electrons* are packed as tightly as possible, the *atomic nuclei* in the core are still able to move freely about.

We are used to thinking about all material as being equal: If a room is packed as tightly as possible with cats, we'd be surprised if people could move freely through it. But electrons and atomic nuclei may overlap to occupy the same physical space (unlike people and cats). As far as the atomic nuclei are concerned, the electron-degenerate core of the star is still mostly empty space. The nuclei behave like a normal gas, moving through the sea of degenerate electrons almost as if the electrons were not there. We can understand the fusion of helium, which comes next, by treating the nuclei as matter in a normal state. Once the pressure and temperature are high enough, these nuclei will fuse, as the hydrogen nuclei do in main-sequence stars. But the energy released will not affect the degenerate electron core in the same way as in a main-sequence star.

Helium Burning and the Triple-Alpha Process

As the star evolves up the red giant branch, its helium core grows not only smaller and more massive, but also hotter. This increase in temperature is partly due to the gravitational energy released as the core shrinks and partly due to the energy released by the ever-faster pace of hydrogen burning in the surrounding shell. The thermal motions of the atomic nuclei in the core become more and more energetic. Eventually, at a temperature of about 10^8 K, the collisions among helium nuclei in the core become energetic enough to overcome the electric repulsion. Helium nuclei are slammed together hard enough for the strong nuclear force to act, and helium burning begins.

Helium burns in a two-stage process referred to as the **triple-alpha process**, which is illustrated in **Figure 12.6**. First, two helium-4 nuclei (^{4}He) fuse to form a beryllium-8 nucleus (^{8}Be) consisting of four protons and four neutrons. The ^{8}Be nucleus is extremely unstable. Left on its own, it would break apart after only about a trillionth of a second. But if, in that short time, it collides with another ^{4}He nucleus, the two nuclei will fuse into a stable nucleus of carbon-12 (^{12}C) consisting of six protons and six neutrons. The triple-alpha process takes its name from the fact that it involves the fusion of three ^{4}He nuclei, which are referred to as alpha particles.

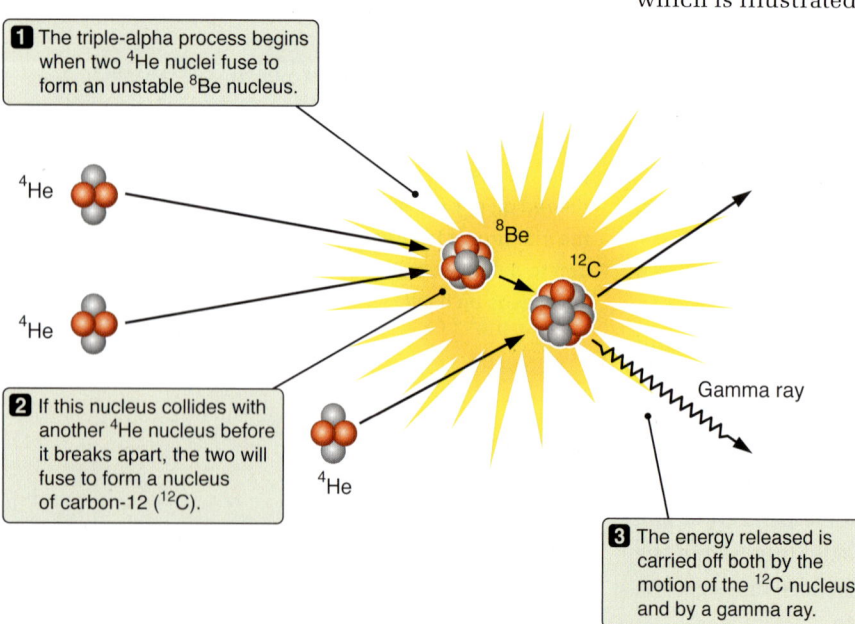

Figure 12.6 The triple-alpha process produces a stable nucleus of carbon-12.

1. The triple-alpha process begins when two ^{4}He nuclei fuse to form an unstable ^{8}Be nucleus.
2. If this nucleus collides with another ^{4}He nucleus before it breaks apart, the two will fuse to form a nucleus of carbon-12 (^{12}C).
3. The energy released is carried off both by the motion of the ^{12}C nucleus and by a gamma ray.

A Helium Flash

In the next phase of the star's evolution, the helium in the core begins burning. Degenerate material is a very good conductor of thermal energy, so any differences in temperature within the core rapidly even out. As a result, when helium burning begins at the center of the core, the energy released quickly heats the entire core. Within a few minutes, the entire core is burning helium into carbon by the triple-alpha process.

In a normal gas like the air around you, the pressure of the gas comes from the random thermal motions of the atoms. Increasing the temperature of such a gas means that the motions of the atoms become more energetic, so the pressure of the gas increases. If the helium core of a red giant star were a normal gas, the increase in temperature from helium burning would increase the pressure. The core of the star would expand; the temperature, density, and pressure would decrease; nuclear reactions would slow; and the star would settle down into a new balance between gravity and pressure. These are exactly the sorts of changes that are steadily occurring within the core of a main-sequence star like the Sun as the structure of the star shifts in response to the changing composition of the core.

However, the degenerate core of a red giant is not a normal gas. The pressure in a red giant's degenerate core comes from how densely the electrons in the core are packed together. Heating the core does not change the number of electrons that can be packed into its volume, so the core's pressure does not respond to changes in temperature. And if the pressure does not increase, the core does not expand.

The higher temperature does not change the pressure, but it does cause the helium nuclei to collide with more frequency and greater force, so the nuclear reactions become more vigorous. More vigorous reactions mean higher temperature, and higher temperature means even more vigorous reactions. Helium burning in the degenerate core runs wildly out of control as increasing temperature and increasing reaction rates feed each other. As long as the degeneracy pressure from the electrons is greater than the thermal pressure from the nuclei, this feedback loop continues.

Within seconds of helium ignition, the thermal pressure increases until it is no longer smaller than the degeneracy pressure. At this point, the helium core literally explodes in what astronomers call a **helium flash**, illustrated in **Figure 12.7**. Although the energy released in this runaway thermonuclear reaction does lift the overlying layers of the star, because the explosion is contained deep within the star, we do not see it directly. But the drama is over within a few hours because the expanded helium-burning core is no longer degenerate, and the star is on its way toward a new equilibrium.

Helium burning in the core does not cause the star to grow more luminous. The tremendous energy released during the helium flash goes into fighting gravity and "puffing up" the core. After the helium flash, the core (which is no longer degenerate) is much larger, so the force of gravity within it and the surrounding shell is much smaller. Weaker gravity means less force pushing down on the core and the shell, which means lower pressure. Lower pressure, in turn, slows the nuclear reactions. The net result is that after the helium flash, core helium burning keeps the core of the star puffed up, and the star becomes less luminous than it was as a red giant.

The star takes about 100,000 years or so to settle into stable helium burning. It then spends about 100 million years burning helium into carbon in a nondegenerate core while hydrogen burns to helium in a surrounding shell. The star is about a hundred times less luminous as it was when the helium flash occurred. The lower luminosity means that the outer layers of the

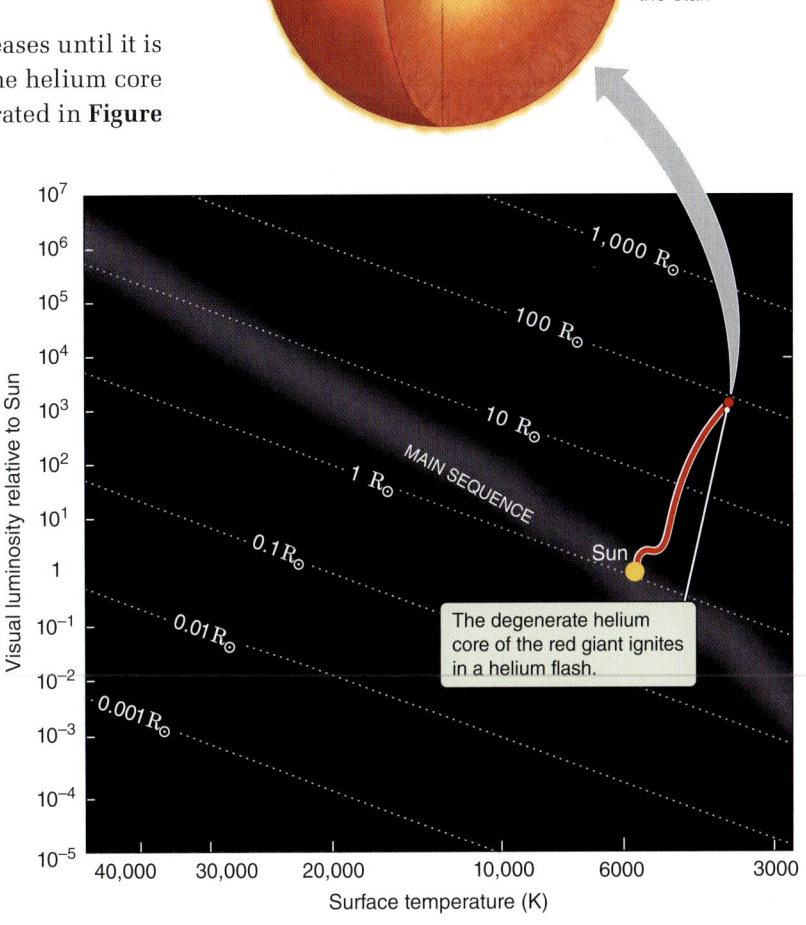

Figure 12.7 A low-mass star travels a complex path on the H-R diagram at the end of its life. The first part of that path takes it up the red giant branch to a point where helium ignites in a helium flash.

312 CHAPTER 12 Evolution of Low-Mass Stars

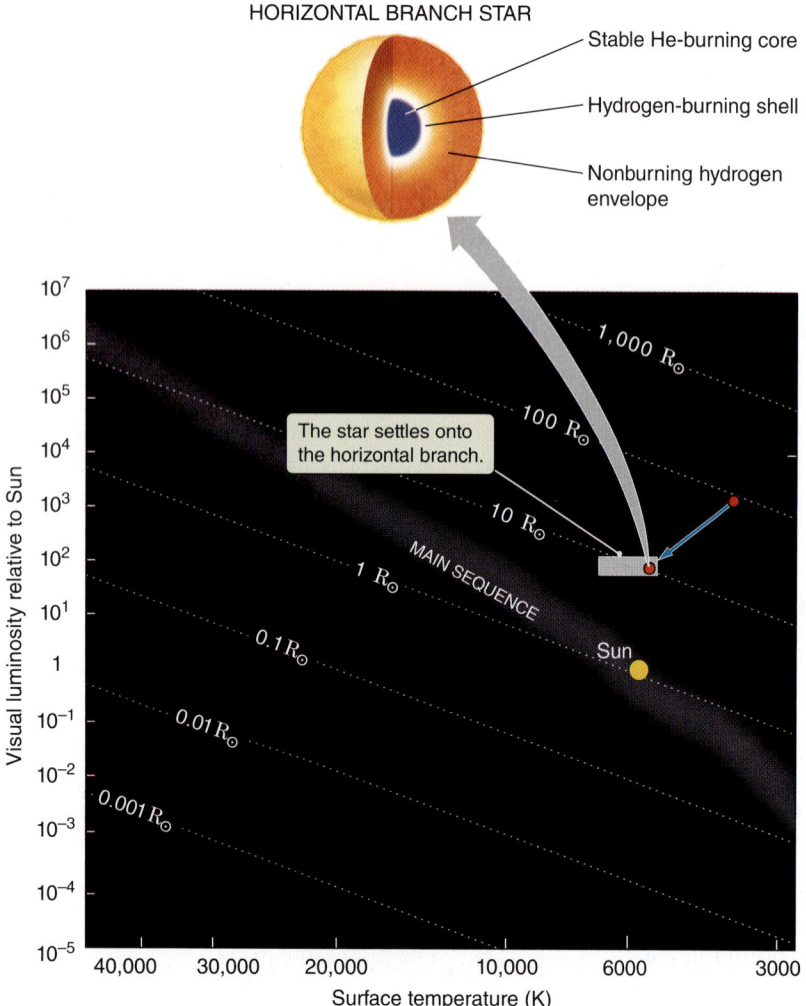

Figure 12.8 A low-mass star travels a complex path on the H-R diagram at the end of its life. The second part of that path takes the star down from the red giant branch onto the horizontal branch.

star are not puffed up as much as they were when the star was a red giant. The star shrinks, and its surface temperature climbs.

At this point in their evolution, low-mass stars with chemical compositions similar to that of the Sun lie on the H-R diagram just to the left of the red giant branch. Stars that contain much less iron than the Sun tend to distribute themselves away from the red giant branch along a nearly horizontal line on the H-R diagram. This stage of stellar evolution takes its name from this horizontal band. The star is now referred to as a **horizontal branch** star, shown in **Figure 12.8**.

12.4 The Low-Mass Star Enters the Last Stages of Its Evolution

The evolution of a star like our Sun from the main sequence through its helium flash and onto the horizontal branch is fairly well understood. Just as our understanding of the interior of the Sun comes from computer models of the physical conditions within our local star, our understanding of the evolution of a red giant comes from computer models that look at the changes in structure as the star's degenerate helium core grows. These models show that any star with a mass of about 1 $M_\odot$ will follow the march from main sequence to helium flash and then drop down onto the horizontal branch. But when we try to use computer models to understand what happens next, the road that we follow gets a bit trickier. We just noted that differences in chemical composition between stars significantly affect where they fall on the horizontal branch. From this point on, small changes in the properties of a star—mass, chemical composition, strength of the star's magnetic field, or even the rate at which the star is rotating—can lead to noticeable differences in how the star evolves.

With this caution in mind, we continue our story of the evolution of a 1-$M_\odot$ star with solar composition, following the most likely sequence of events awaiting our Sun.

Moving Up the Asymptotic Giant Branch

The behavior of a star on the horizontal branch is remarkably similar to that of a star on the main sequence. The star's life on the horizontal branch, however, is much shorter than its life on the main sequence. The star is more luminous, so it is consuming fuel more rapidly. Helium is a much less efficient nuclear fuel than hydrogen, so it has to burn fuel even faster. Even so, for 100 million years the horizontal branch star remains stable, burning helium to carbon in its core and hydrogen to helium in a shell.

The temperature at the center of a horizontal branch star is not high enough for carbon to burn, so carbon ash builds up in the heart of the star. Gravity once again begins to win as the nonburning carbon ash core is crushed by the weight of the layers of the star above it. Again, the electrons in the core are packed together as tightly as the laws of quantum mechanics allow. The carbon core is now electron-degenerate, with physical properties much like those of the degenerate helium core at the center of a red giant.

The strength of gravity in the inner parts of the star increases, which in turn drives up the pressure, which speeds up the nuclear reactions, which causes the degenerate core to grow more rapidly—we have heard this story before. The internal changes occurring within the star are similar to the changes that took place at the end of the star's main-sequence lifetime, and the path the star follows as it leaves the horizontal branch echoes that earlier phase of evolution as well. Just as the star accelerated up the red giant branch as its degenerate helium core grew, the star now leaves the horizontal branch and once again begins to grow larger, redder, and more luminous as its degenerate carbon core grows. As you can see in **Figure 12.9**, the path that the star follows, known as the **asymptotic giant branch (AGB)** of the H-R diagram, parallels the path it followed as a red giant, approaching the red giant branch as the star grows more luminous. An AGB star burns helium and hydrogen in nested concentric shells surrounding a degenerate carbon core, as the star moves once again up the H-R diagram.

Stellar Mass Loss

Building on our analogy between AGB stars and red giants, you might imagine that the next step in the evolution of an AGB star should be a "carbon flash" when carbon burning begins in the star's degenerate core. Yet a carbon flash never happens. Before the temperature in the carbon core becomes high enough for carbon to burn, the star loses its gravitational grip on itself and expels its outer layers into interstellar space.

Red giant and AGB stars are huge objects. When our Sun becomes an AGB star, its outer layers will swell to the point that they engulf the orbits of some of the innermost planets, possibly including Earth. When a star expands to such a size, the gravitational force at its surface is only 1/10,000 as strong as the gravity at the surface of the present-day Sun. It takes little extra energy to push surface material away from the star. **Stellar mass loss**—the loss of mass from the outer layers of the star as it evolves—actually begins when the star is still on the red giant branch; by the time a 1-$M_\odot$ main-sequence star reaches the horizontal branch, it may have lost 10–20 percent of its total mass. As the star ascends the asymptotic giant branch, it loses another 20 percent or even more of its total mass. And by the time it is well up this branch, the star may have lost more than half of its original mass.

Mass loss on the asymptotic giant branch can be spurred on by the star's unstable interior. The extreme sensitivity of the triple-alpha process to temperature in the core can lead to episodes of rapid energy release, which can provide the extra kick needed to expel material from the star's outer layers. Even stars that are initially quite similar can behave very differently when they reach this stage in their evolution.

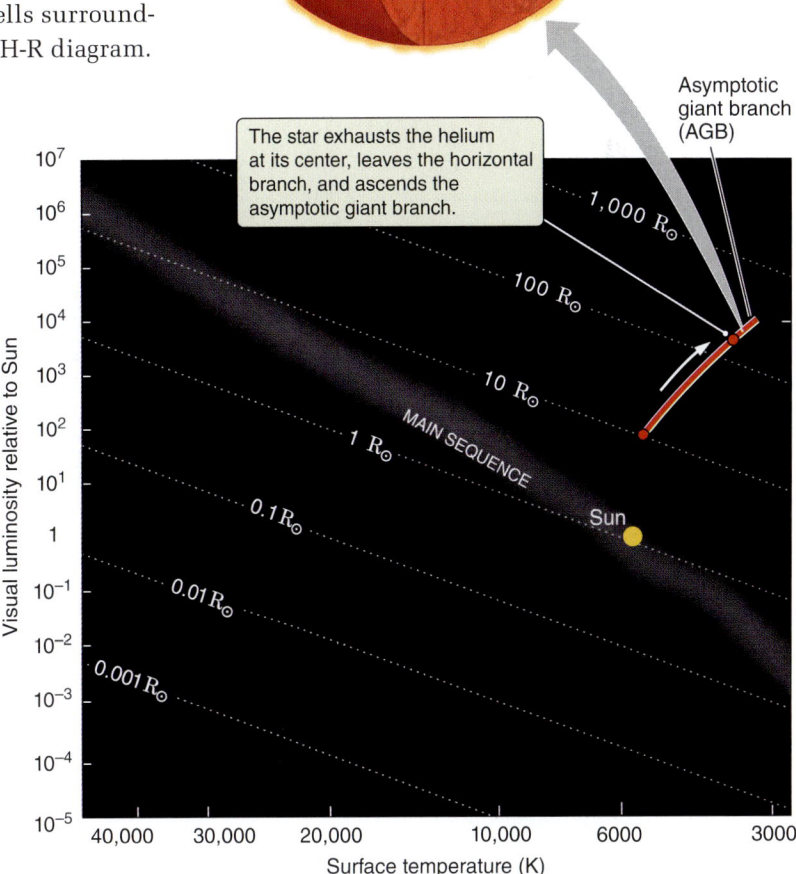

Figure 12.9 A low-mass star travels a complex path on the H-R diagram at the end of its life. The third part of that path takes the star up from the horizontal branch onto the asymptotic giant branch.

The Post-AGB Star

Toward the end of an AGB star's life, mass loss itself becomes a runaway process. When a star loses a bit of mass from its outer layers, the weight pushing down on the underlying layers of the star is reduced. Without this weight holding them down, the outer layers of the star puff up even larger than they were before. The star, which is now both less massive and larger, is even less tightly bound by grav-

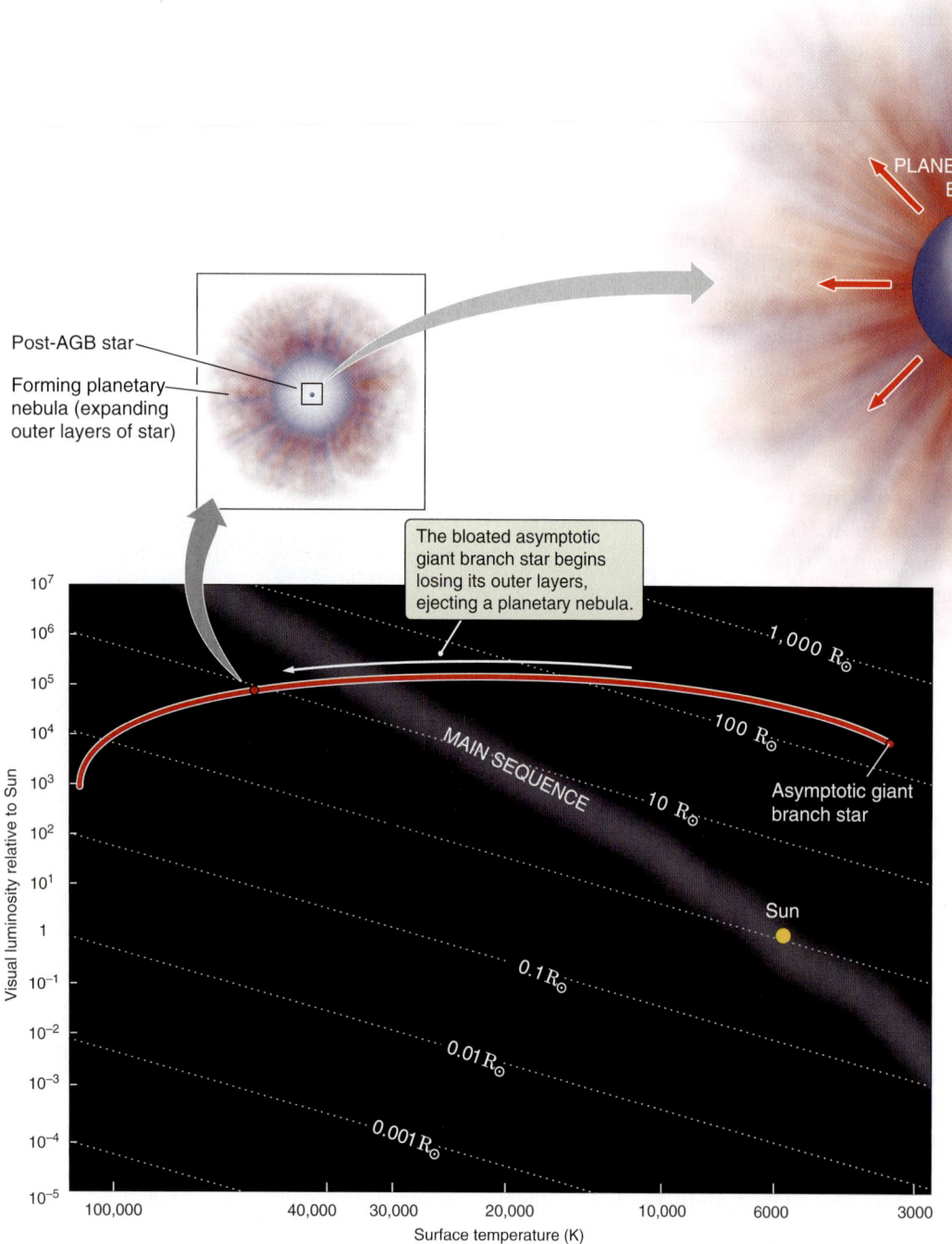

Figure 12.10 At the end of the AGB star's life, it ejects most of its mass in a planetary nebula, becoming a post-AGB star.

ity, so even less energy is needed to push its outer layers away. Much of the remaining mass of the star is ejected into space, typically at speeds of 20–30 kilometers per second (km/s).

All that is left of the low-mass star itself is a tiny, very hot, electron-degenerate carbon core surrounded by a thin envelope in which hydrogen and helium are still burning. This star is now somewhat less luminous than when it was at the top of the asymptotic giant branch, but it is still much more luminous than a horizontal branch star. The remaining hydrogen and helium in the star rapidly burn to carbon, and as more and more of the mass of the star ends up in the carbon core, the star shrinks and becomes hotter and hotter. Over the course of only 30,000 years or so after the beginning of runaway mass loss, the star moves very rapidly from right to left across the top of the H-R diagram, as shown in **Figure 12.10**.

The surface temperature of the star may eventually reach 100,000 K or more. Wien's law says that at such temperatures, most of the light from the star is in the high-energy ultraviolet (UV) part of the spectrum. The intense UV light heats and ionizes the expanding shell of gas that was recently ejected by the star, causing it to glow.

The mass ejected by the AGB star will pile up in a dense, expanding shell. If you

you were to look at such a shell through a small telescope, you would see a round or perhaps oblong patch of light, perhaps with a hole and a dot in the middle. When these glowing shells were first observed in small telescopes, they looked round, like planets; they were named **planetary nebulae** because they appeared fuzzy like nebular clouds of dust and gas, but they were approximately round, like planets. But there is nothing planetary about them (**Figure 12.11**). A planetary nebula consists of the remaining outer layers of a star, which were ejected into space as a dying gasp at the end of the star's ascent of the asymptotic giant branch. Not every star forms a planetary nebula. Stars more massive than about 8 $M_\odot$ pass through the post-AGB stage too quickly. Stars with insufficient mass take too long, so their envelope dissipates before they can illuminate it. Some astronomers think that our own Sun will not retain enough mass during its post-AGB phase to form a planetary nebula.

Planetary nebulae can be dazzling in appearance. The most distinctive ones have earned names like Owl Nebula, Clown Nebula, Cat's Eye Nebula, and Dumbbell Nebula. This extraordinary menagerie is illustrated in **Figure 12.12**. The structure of a planetary nebula tells of eras when mass loss was slower or faster and of times when mass was ejected primarily from the star's equator or its poles. The colors come from emission lines from particular atoms and ions. Compare the image of the Ring Nebula in the chapter-opening illustration with the higher-resolution image in Figure 12.11. From the colored rings, we know that we are seeing emission lines from different ions in different places. In forming this nebula, a lot of mass was lost nearly all at once. Then the mass loss ceased, resulting in a hollow shell around the central star.

Mass loss from giant stars carries the chemical elements enriching the stars' outer layers off into interstellar space. Planetary nebulae often show an overabundance of elements such as carbon, nitrogen, and oxygen compared to the outer layers of the Sun. These elements are by-products of nuclear burning. Once this chemically enriched material leaves the star, it mixes with interstellar gas, increasing the chemical diversity of the universe.

White Dwarfs

Within 50,000 years or so, a post-AGB star burns all of the fuel remaining on its surface, leaving nothing behind but a cinder—a nonburning ball of carbon. In the process, the star's position on the H-R diagram falls down the left side of the

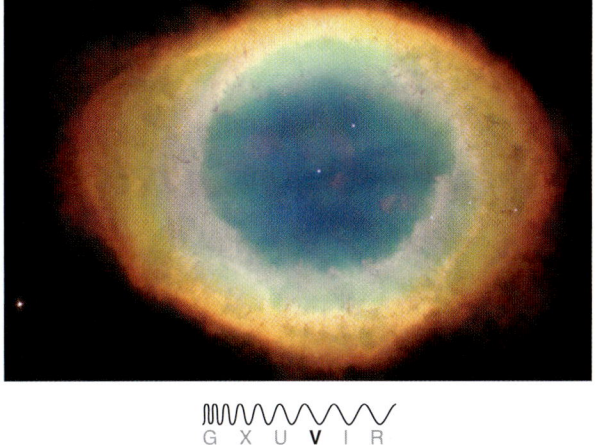

Figure 12.11 At the end of its life, a low-mass star ejects its outer layers and may form a planetary nebula consisting of an expanding shell of gas surrounding the white-hot remnant of the star. Astronomers initially thought these objects looked like planets. However, a Hubble Space Telescope image of the Ring Nebula shows the remarkable and complex structure of this expanding shell of gas.

Figure 12.12 Planetary nebulae come in a wide variety of shapes that reveal the details of the history of mass loss from the central star.

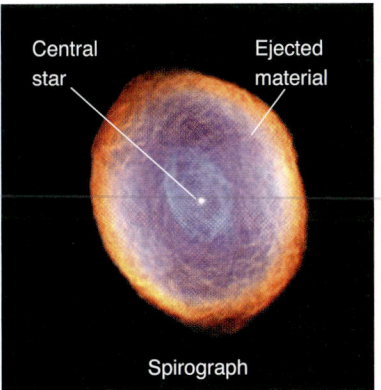

Spirograph

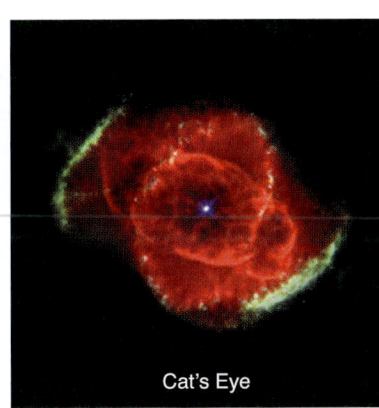

Cat's Eye

M2-9

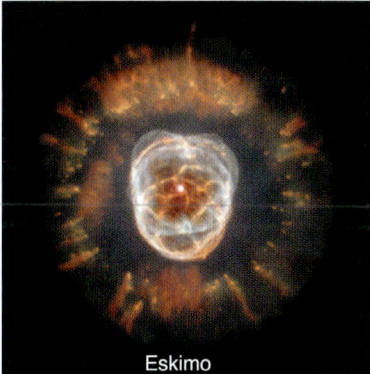

Eskimo

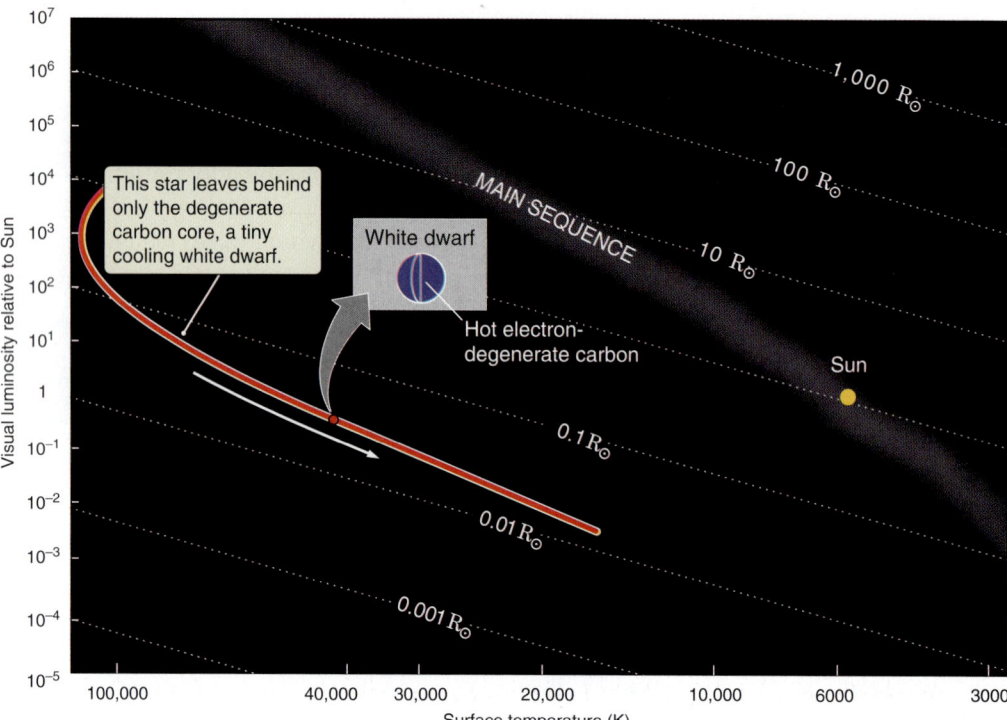

Figure 12.13 Low-mass stars finally leave behind a white dwarf—a hot, small, dense object that slowly fades from view.

H-R diagram, becoming smaller and fainter, as shown in **Figure 12.13**. Within a few thousand years the burned-out core shrinks to about the size of Earth, at which point it has become electron-degenerate and can shrink no further. This remnant of a low-mass star is called a **white dwarf**. The white dwarf, composed of nonburning electron-degenerate carbon, continues to radiate energy away into space. As it does so it cools, just like the heating coil on an electric stove once it is turned off. The cooling white dwarf moves down and to the right on the H-R diagram, following a line of constant radius. The white dwarf may remain very hot for 10 million years or so, but its tiny size means the luminosity may now be only one-thousandth that of our Sun. Many white dwarfs are known, but none can be seen without a telescope.

Figure 12.14 recaps the evolution of a solar-type 1-$M_\odot$ main-sequence star through its final existence as a 0.6-$M_\odot$ white dwarf. The star leaves the main sequence, climbs the red giant branch, falls to the horizontal branch, climbs back up the asymptotic giant branch, takes a left across the top of the diagram while ejecting a planetary nebula, and finally falls to its final resting place in the bottom left of the diagram. This process is representative of the fate of low-mass stars. Although all low-mass stars form white dwarfs at the end points of their evolution, the exact path a low-mass star follows from core hydrogen burning on the main sequence to white dwarf depends on details particular to the star.

Some 6 billion or so years from now, our Sun will become a white dwarf that will fade as it radiates its thermal energy away into space. This superdense ball—with a density of a ton per teaspoonful—actually began its life billions of years earlier as a cloud of interstellar gas billions of times more tenuous than the vacuum in the best vacuum chamber on Earth.

Once it leaves the main sequence, the Sun will travel the path from red giant to white dwarf in less than one-tenth of the time that it spent on the main sequence steadily burning hydrogen to helium in its core. Stars spend most of their lumi-

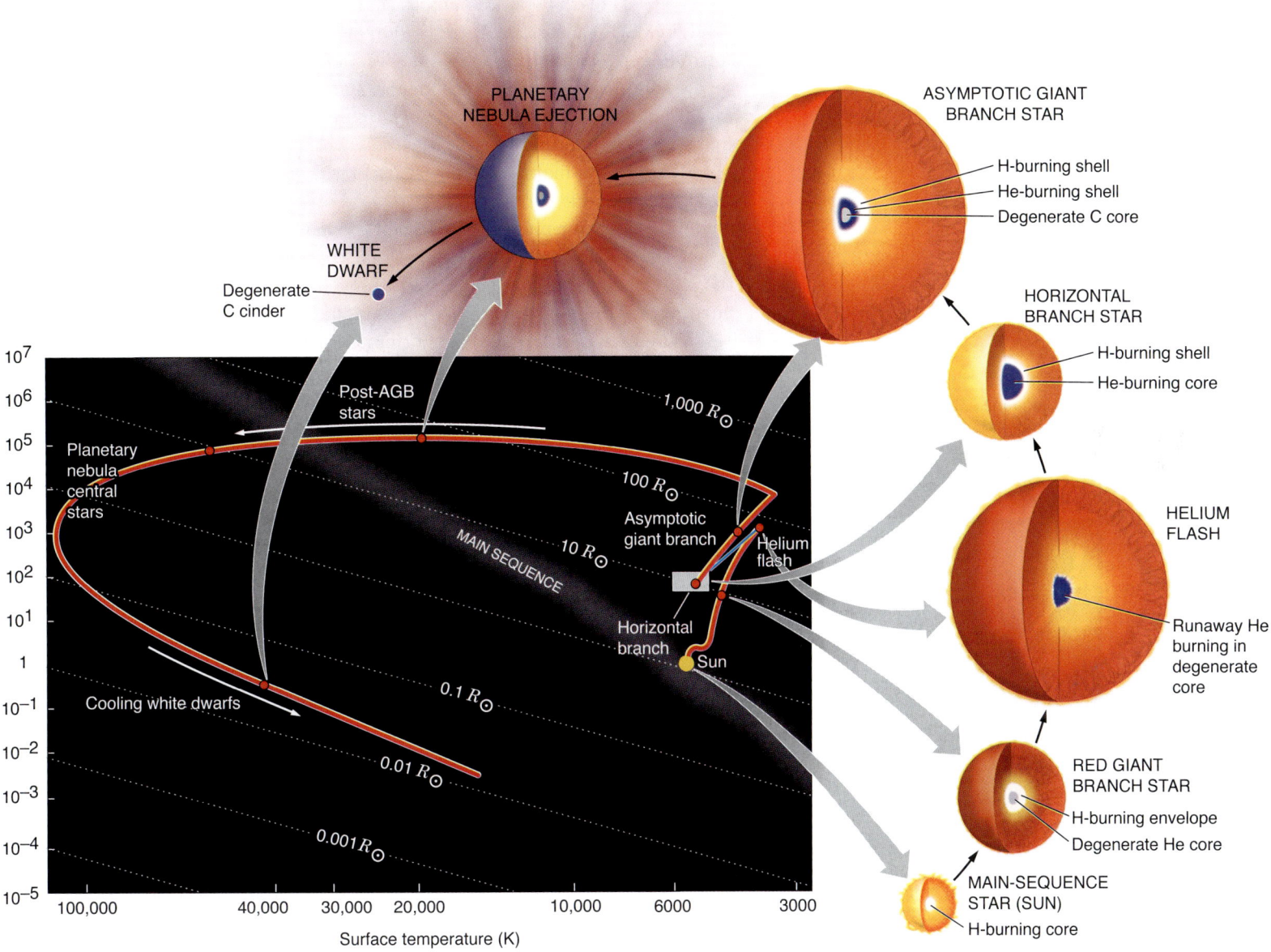

Figure 12.14 This H-R diagram summarizes the stages in the post–main sequence evolution of a 1-$M_\odot$ star.

nous lifetimes on the main sequence, which is why most of the stars that we see in the sky are main-sequence stars. In the end, though, white dwarfs will constitute the final resting place for the vast majority of stars that have been or ever will be formed.

12.5 Star Clusters Are Snapshots of Stellar Evolution

In Chapter 5, you saw that when an interstellar cloud collapses, it fragments into pieces, many stars of different masses bound together by gravity. These large groups of gravitationally bound stars are called **star clusters**. We see many such star clusters around us today, containing anywhere from a few dozen to millions of stars. Because stars in a cluster are formed together at nearly the same time, clusters are snapshots of stellar evolution. Looking at star clusters of different ages helps us learn about the process of stellar evolution.

Figure 12.15 shows the H-R diagram of a simulated cluster of 40,000 stars as it would appear at several different ages. In Figure 12.15a, stars of all masses are located on the **zero-age main sequence**, showing where they begin their lives as main-sequence stars. We would never expect to see a cluster H-R diagram that looks like the one in Figure 12.15a, however, because the stars in a cluster do not all reach the main sequence at exactly the same time. Star formation in a molecular cloud is spread out over several million years, and it takes considerable time for lower-mass stars to reach the main sequence. The H-R diagram of a very young cluster normally shows many lower-mass stars located well above the main sequence. These stars are protostars that have not yet begun nuclear burning. We did not show them here, for clarity.

The more massive a star, the shorter its life on the main sequence. After only 4 million years (Figure 12.15b), all stars with masses greater than about 20 $M_\odot$ have evolved off the main sequence and are now spread out across the top of the H-R diagram. (Details of the evolution of these high-mass stars are covered in

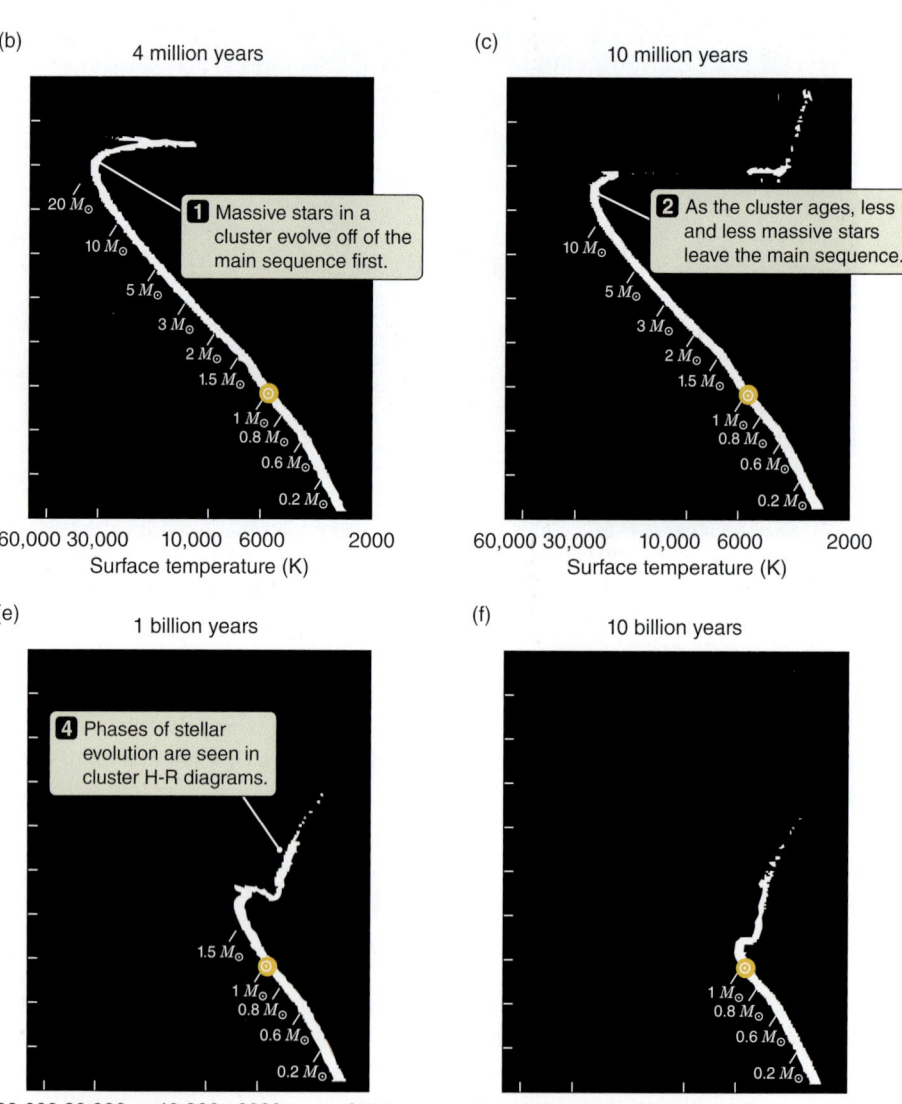

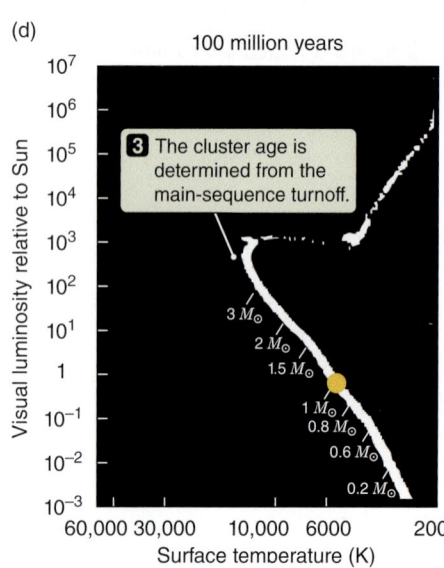

Figure 12.15 H-R diagrams of star clusters are snapshots of stellar evolution. These are H-R diagrams of a simulated cluster of 40,000 stars of solar composition seen at different times following the birth of the cluster. Note the progression of the main-sequence turnoff to lower and lower masses. Very young stars, still moving onto the main sequence, have been omitted for clarity.

Chapter 13. For now, all you need to know is that they evolve faster than low-mass stars.) As time goes on, stars of lower mass evolve off the main sequence, and the point where the stars exit the main sequence (the turnoff point) moves toward the bottom right along the main sequence. By the time the cluster is 10 million years old (Figure 12.15c), only stars with masses less than about 15 $M_\odot$ remain on the main sequence. The location of the most massive stars still on the main sequence is called the **main-sequence turnoff**. As the cluster ages, the main-sequence turnoff moves down the main sequence to stars of lower mass.

As a cluster ages further (Figures 12.15d and 12.15e), we see the details of all stages of stellar evolution. By the time the star cluster is 10 billion years old (Figure 12.15f), stars with masses of only 1 $M_\odot$ are beginning to die. Stars slightly more massive than this are seen as giants of various types. Note how few giant stars are present in any of the cluster H-R diagrams. The end phases in the evolution of a low-mass star pass so quickly in comparison with the star's main-sequence lifetime that even though the cluster started with 40,000 stars, only a handful of stars are in these phases of evolution at any given time. Similarly, even though most of the evolved stars in an old cluster are white dwarfs, all but a few of these stars will have cooled and faded into obscurity at any given time.

Figure 12.15 shows H-R diagrams as predicted by theories of stellar evolution. **Figure 12.16** shows the observed H-R diagram for the real cluster 47 Tucanae, along with a theoretical calculation of the H-R diagram for a 12-billion-year-old cluster. The fact that the predictions of models agree so well with H-R diagrams of observed star clusters is strong support for theories of stellar evolution.

Cluster evolution models are powerful tools for studying the history of star formation. When we observe a star cluster, the main-sequence turnoff immediately tells us its age. **Figure 12.17** traces the observed H-R diagrams for several real star clusters. NGC 2362 is a young cluster. Its complement of massive, young stars shows it to be only a few million years old. In contrast, NGC 752 is quite old; its main-sequence turnoff indicates a cluster age of about 7 billion years.

The lifetimes of stars should be of more than passing interest to us. We don't yet know if life exists on planets orbiting other stars, but astrobiologists think that life would be unlikely to evolve on a planet orbiting a massive star with a stable life of only a few million years (see Chapter 18). Similarly, it would be unlikely for life to develop on a planet surrounding a very-low-mass star that had yet to initiate the stable nuclear burning phase. Tracing star formation, evolution, and lifetimes through observations of stellar clusters helps pin down the types of stars to target in our search for life elsewhere.

12.6 Binary Stars Sometimes Share Mass, Resulting in Novae and Supernovae

Possibly the most significant complication in our picture of the evolution of low-mass stars arises from the fact that many stars are members of binary systems. While both members of a binary pair are on the main sequence, they usually have little effect on each other. But in some cases, if the separation between

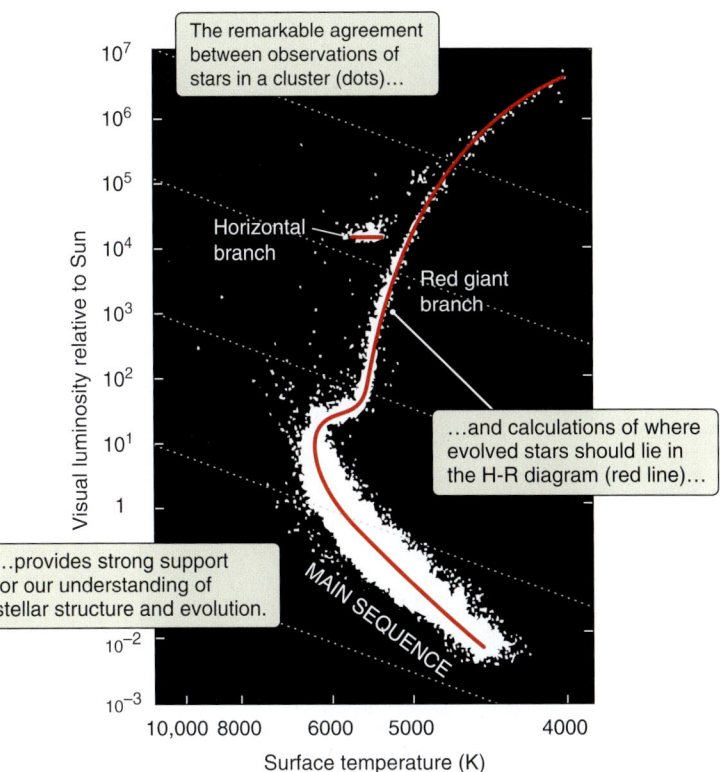

Figure 12.16 The observed H-R diagram of stars (dots) in the cluster 47 Tucanae agrees remarkably well with the theoretical calculation (red line) of the H-R diagram of a 12-billion-year-old cluster.

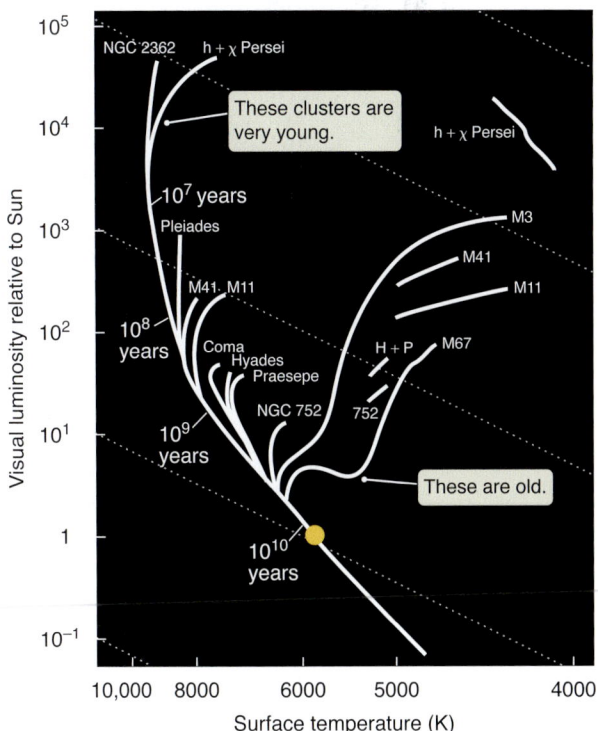

Figure 12.17 Clusters having a range of different ages are overlaid on this H-R diagram. The ages associated with the different main-sequence turnoffs are indicated.

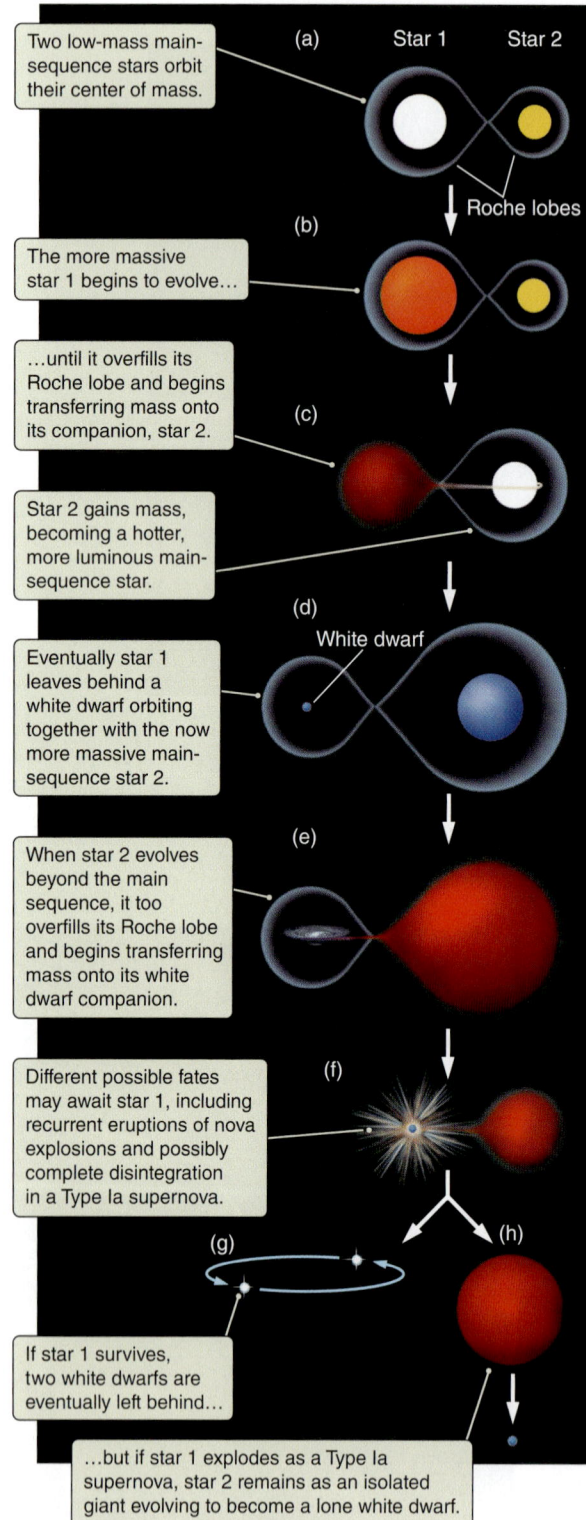

Figure 12.18 A compact binary system consisting of two low-mass stars passes through a sequence of stages as the stars evolve and mass is transferred back and forth.

the stars is small and one star is more massive than the other, their evolution may become linked. In this section, we will trace the steps from binary star through to the end point of the evolution: nova or supernova.

The Flow of Mass from an Evolving Star to Its Companion

Think for a moment about what would happen if you were to travel in a spacecraft from Earth toward the Moon. When you are still near Earth, the force of Earth's gravity is far stronger than that of the Moon. As you move away from Earth and closer to the Moon, the gravitational attraction of Earth weakens, and the gravitational attraction of the Moon becomes stronger. You eventually reach an intermediate zone where neither body has the upper hand. If you continue beyond this point, the lunar gravity begins to assert itself until you find yourself firmly in the grip of the Moon.

Exactly the same situation exists between two stars. When one star swells up, its outer layers may cross that gravitational dividing line separating the star from its companion, and any material that crosses this line no longer belongs to the first star. Some is lost to the surrounding space, but some can be pulled toward the companion. A star reaches this point if it swells to fill up its portion of an invisible dumbbell-shaped volume of space set by the force of gravity exerted by each star, as shown in **Figure 12.18a**. These regions surrounding the two stars—their gravitational domains—are called the **Roche lobes** of the system.

Evolution of a Close Binary System

The best way to understand how mass transfer affects the evolution of stars in a binary system is to apply what we have learned from the evolution of single low-mass stars. Figure 12.18a shows a close binary system consisting of two low-mass stars of somewhat different mass. The more massive of the two stars is "star 1," and the less massive of the two is "star 2." This is an ordinary binary system, and each of these stars is an ordinary main-sequence star for most of the system's lifetime.

More massive stars evolve more rapidly. Therefore, star 1 will be the first to use up the hydrogen at its center and begin to evolve off the main sequence (Figure 12.18b). If the stars are close enough to each other, star 1 will grow to overfill its Roche lobe, and material will transfer onto star 2 (Figure 12.18c). This exchange of material between the two stars is called **mass transfer**. The structure of star 2 must then change to accommodate its new status as a higher-mass star. If we plotted star 2's position on the H-R diagram during this period, we would see it move up and to the left along the main sequence, becoming larger, hotter, and more luminous.

A number of interesting things can happen at this point. For example, the transfer of mass between the two stars can result in a sort of "drag" that causes the orbits of the two stars to shrink, bringing the stars closer together and further enhancing mass loss. The two stars can even reach the point where they are effectively two cores sharing the same extended envelope of material.

Star 1, because it is losing mass to star 2, never grows large enough to move to the top of the H-R diagram as a red giant. Yet it continues to evolve, burning helium in its core on the horizontal branch, proceeding through a stage of helium shell burning, and finally losing its outer layers and leaving behind a white dwarf. Figure 12.18d shows the binary system after star 1 has completed its evolution. All that remains of star 1 is a white dwarf orbiting about its bloated main-sequence companion, star 2.

The Second Star Evolves

Figure 12.18e picks up the evolution of the binary system as star 2 begins to evolve off the main sequence. Like star 1 before it, star 2 grows to fill its Roche lobe; material from star 2 begins to pour through the "neck" connecting the Roche lobes of the two stars. Because the white dwarf is so small, the infalling material generally misses the star. Instead of landing directly on the white dwarf, the infalling mass forms an accretion disk around the white dwarf, similar in some ways to the accretion disk that forms around a protostar. As in the process of star formation, the accretion disk serves as a way station for material that is destined to find its way onto the white dwarf but starts out with too much angular momentum to hit the white dwarf directly.

A white dwarf has a mass comparable to that of the Sun but a size comparable to that of Earth. A large mass and a small radius mean strong gravity. A kilogram of material falling from space onto the surface of a white dwarf releases 100 times more energy than a kilogram of material falling from the outer Solar System onto the surface of the Sun. All of this energy is turned into thermal energy. The spot where the stream of material from star 2 hits the accretion disk is heated to millions of kelvins, where it glows in the far ultraviolet and X-ray parts of the electromagnetic spectrum.

The infalling material accumulates on the surface of the white dwarf where it is compressed by the enormous gravitational pull of the white dwarf to a density close to that of the white dwarf itself. As more and more material builds up on the surface of the white dwarf, the white dwarf shrinks (just like the core of a red giant shrinks as it grows more massive). The density increases, and at the same time the release of gravitational energy drives up the temperature of the white dwarf. The infalling material is from the outer, unburned layers of star 2, so it is composed mostly of hydrogen.

Once the temperature at the base of the white dwarf's surface layer of hydrogen reaches about 10 million K, this hydrogen begins to burn explosively. Energy released by hydrogen burning drives up the temperature, and the higher temperature drives up the rate of hydrogen burning. This runaway reaction is much like the runaway helium burning that takes place during the helium flash, except now there are no overlying layers of the star to keep things contained. An explosion called a **nova** (Figure 12.18f) occurs that blows part of the layer covering the white dwarf out into space at speeds of thousands of kilometers per second.

About 50 novae occur in our galaxy each year, but we can see only two or three of them because dust in the disk of our galaxy blocks our view. Novae reach their peak brightness in only a few hours, and for a brief time they can be several hundred thousand times more luminous than the Sun. Although the brightness of a nova sharply drops in the weeks after the outburst, it can sometimes still be seen for years. During this time, the glow from the expanding cloud of ejected material is caused by the decay of radioactive isotopes created in the explosion.

The explosion of a nova does not destroy the underlying white dwarf star. Afterward, the binary system is in much the same configuration as before; material from star 2 is still pouring onto the white dwarf (**Figure 12.19a**). This cycle can repeat itself many times (Figure 12.19b), as material builds up and ignites

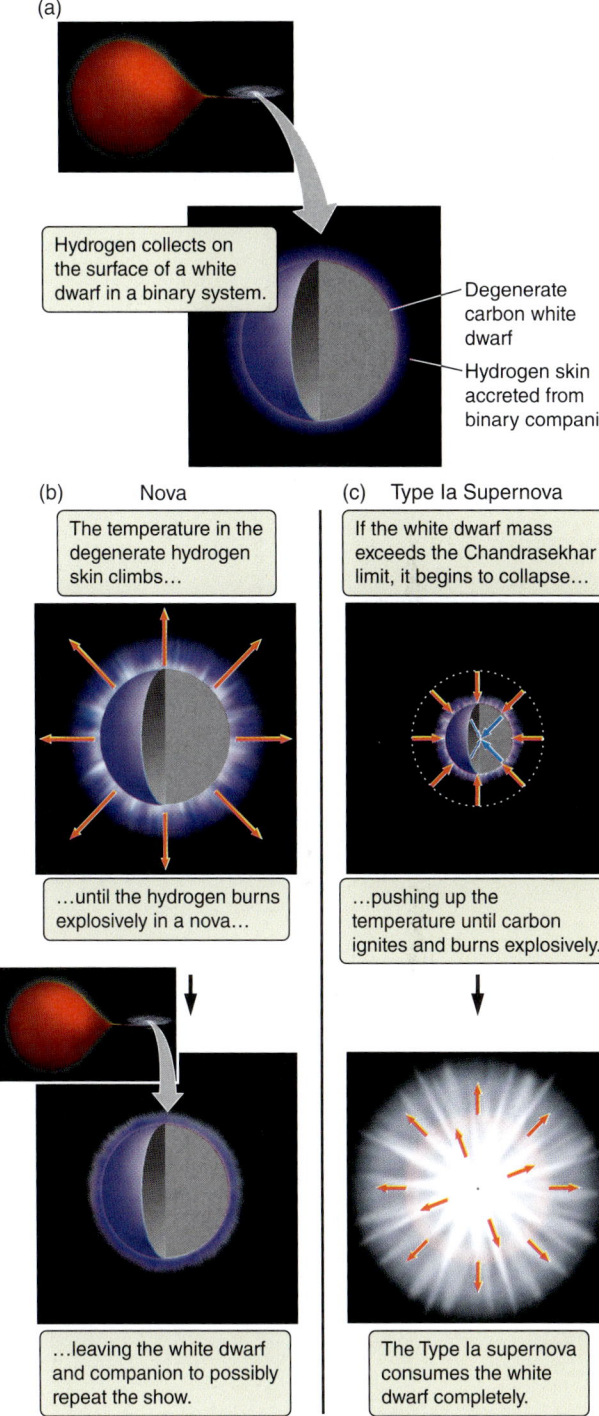

Figure 12.19 (a) In a binary system in which mass is transferred onto a white dwarf, a layer of hydrogen builds up on the surface of the degenerate white dwarf. (b) If hydrogen burning ignites on the surface of the white dwarf, the result is a nova. (c) If enough hydrogen accumulates, the white dwarf begins to collapse, carbon ignites, and the result is a Type Ia supernova.

again and again on the surface of the white dwarf. In most cases, outbursts are separated by thousands of years, so most novae have been seen only once in historical times. Some novae, however, erupt every decade or so.

A Stellar Cataclysm

It is possible that star 2 eventually will simply go on to form a white dwarf, leaving behind a stable binary system consisting of two white dwarfs, as in Figure 12.18g. There are, however, two other possibilities. Through millions of years of mass transfer from star 2 onto the white dwarf, and possibly through countless nova outbursts, the white dwarf's mass slowly increases—but it cannot have a mass greater than 1.4 $M_\odot$ and remain a white dwarf. This value is referred to as the **Chandrasekhar limit**, named for Subrahmanyan Chandrasekhar (1910–1995), who derived it. Above this mass, even the pressure supplied by degenerate electrons is no longer enough to balance gravity, and the white dwarf will begin to collapse (Figure 12.19c). But a white dwarf that is accumulating mass in this way likely does not actually reach the Chandrasekhar limit. As the star approaches the limit, core pressures and temperatures rise enough to ignite carbon and begin a simmering phase that holds off thermonuclear runaway for a while, but once the temperature reaches about 1.0×10^8 K, the star explodes. This likely happens at about 1.38 $M_\odot$, just before the white dwarf actually reaches the ultimate mass limit.

The runaway carbon burning involves the entire white dwarf. Within about a second, the whole white dwarf is consumed in the resulting explosion. In this single instant, 100 times more energy is liberated than will be given off by the Sun over its entire 10-billion-year lifetime on the main sequence. Runaway fusion reactions convert a large fraction of the mass of the star into elements such as iron and nickel, and the explosion blasts the shards of the white dwarf into space at top speeds in excess of 20,000 km/s, enriching the interstellar medium with these heavier elements. The explosion completely destroys star 1, leaving star 2 behind as a lone giant to continue its evolution toward becoming a white dwarf (Figure 12.18h). This explosion is known as a **Type Ia supernova**. Type Ia supernovae occur in a galaxy the size of the Milky Way about once a century. For a brief time they can shine with a luminosity billions of times that of our Sun, possibly outshining the galaxy itself. These objects are particularly useful to astronomers because their luminosities can be determined from a careful study of their light curves. Combining this luminosity with the apparent brightness gives the distance to the host galaxy. These supernovae are bright enough to be used to find the distance to very distant galaxies.

The other possibility is that star 1 becomes a white dwarf and then merges with star 2, producing a "super-Chandrasekhar" white dwarf. Some scientists contend that this option occurs as much as 80 percent of the time. In this case, the total mass is not constrained by the Chandrasekhar limit. If this proves to be the most common cause of Type Ia supernovae, then the use of Type Ia supernovae as accurate distance indicators is called into question, because the mass involved in the explosion varies from one supernova to the next.

 # READING ASTRONOMY News

The sheer number of stars in the sky compared to the number of astronomers on the ground means that a lot of the time, things are happening that we are simply not aware of. Even stellar explosions sometimes are missed!

Scientists May Be Missing Many Star Explosions

By **Space.com staff**

Some of the brightest stellar explosions in the galaxy may be flying under astronomers' radar, a new study suggests.

Researchers using observations from a Sun-studying satellite detected four novae—exploding stars not quite as bright or dramatic as supernovae. The scientists were able to follow the explosions in intricate detail over time, including before the novae reached maximum brightness.

While other astronomers had discovered all four novae before, two of them escaped detection until after they had reached peak luminosity, the study revealed. This fact suggests that many other stellar explosions, even some that are incredibly bright, may be occurring unnoticed, researchers said.

"So far, this research has shown that some novae become so bright that they could have been easily detected with the naked eye by anyone looking in the right direction at the right time, but are being missed, even in our age of sophisticated professional observatories," study lead author Rebekah Hounsell, a graduate student at Liverpool John Moores University (LJMU) in England, said in a statement.

The new observations are also allowing scientists to study nova explosions in unprecedented detail, according to researchers.

Hounsell and her colleagues analyzed measurements from an instrument aboard the U.S. Department of Defense's Coriolis satellite. The instrument, called the Solar Mass Ejection Imager (SMEI), was designed to detect disturbances in the solar wind. SMEI maps out the entire sky during its 102-minute orbit around Earth.

The researchers found that SMEI was also detecting star explosions, or novae. Novae occur when small, extremely dense stars called white dwarfs suck up gas from a nearby companion star, igniting a runaway thermonuclear explosion.

Unlike supernovae, novae do not result in the destruction of their stars. Stars can go nova repeatedly.

SMEI detected four novae, including one confirmed repeater called RS Ophiuchi, which is found about 5,000 light-years away in the constellation Ophiuchus. RS Ophiuchi may ultimately die in a supernova explosion—one of the brightest, most dramatic events in the universe—researchers said.

Ground-based instruments missed the peak flare-up of two of these four novae, according to researchers. That suggests that space-based instruments like SMEI might be needed to pick up many novae, after which their progress can be tracked with telescopes on the ground, researchers said.

"Two of the novae observed by SMEI have confirmed that even the brightest novae may be missed by conventional ground-based observing techniques," said co-author Mike Bode, also of LJMU.

The researchers reported their results in a recent issue of the *Astrophysical Journal*.

The new observations are giving astronomers key insights into the earlier days of novae, revealing a great deal about how they start and evolve, researchers said.

"The SMEI's very even cadences and uniformly exposed images allow us to sample the sky every 102 minutes and trace the entire evolution of these explosions as they brighten and dim," said co-author Bernard Jackson of the University of California, San Diego.

The new observations have revealed, for example, that three of the explosions faltered significantly before regaining strength and proceeding. Such a "pre-maximum halt" had been theorized before, but evidence for its existence had been inconclusive, researchers said.

Because SMEI performs a survey of the entire sky every 102 minutes, the instrument could also help astronomers understand a wide variety of transient objects and phenomena, according to the research team.

"[This] work has shown how important all-sky surveys such as SMEI are and how their sets can potentially hold the key to a better understanding of many variable objects," Bode said.

Evaluating the News

1. What kind of stellar explosion is being discussed in this article?
2. Which figure in this chapter corresponds most closely to the astronomical events being discussed in this article?
3. Can this kind of explosion occur in an isolated star? Explain your reasoning.
4. The reporter states that "Such a 'pre-maximum halt' had been theorized before, but evidence for its existence had been inconclusive." Think back to Chapter 1. Is the reporter using the word *theory* correctly here?
5. One of the main advantages of these newly available observations is that they take a picture of the same region of the sky every 102 minutes. Why is that an advantage in studying novae?
6. Why do you think ground-based observatories missed these novae?
7. Why might the astronomers quoted in the article think that RS Ophiuchi will ultimately explode as a supernova?

SUMMARY

All stars eventually exhaust their nuclear fuel as it is turned to helium ash in the cores of main-sequence stars. Less massive stars exhaust their fuel more slowly and have longer lifetimes than more massive stars. After exhausting its hydrogen, a low-mass star leaves the main sequence and swells to become a red giant, with a helium core made of electron-degenerate matter. The red giant burns helium via the triple-alpha process, until the core ignites in a helium flash, and the star then moves onto the horizontal branch. A horizontal-branch star accumulates carbon ash in its core and then moves up the asymptotic giant branch. In their dying stages, some stars eject their outer layers to form planetary nebulae. All low-mass stars eventually become white dwarfs, which are very hot but very small.

H-R diagrams of clusters give snapshots of stellar evolution. The location of the main-sequence turnoff indicates the age of the cluster. Transfer of mass within some binary systems can lead to a nuclear explosion. A nova occurs when hydrogen collects and ignites on the surface of a white dwarf in a binary system. If the mass of the white dwarf approaches 1.4 $M_\odot$, the entire star explodes in a Type Ia supernova.

LG 1 More massive stars have shorter lifetimes.

LG 2 When the Sun runs out of hydrogen in the core, it will begin to burn hydrogen in a shell around the core. This will cause it to swell onto the red giant branch.

LG 3 After leaving the main sequence, low-mass stars follow a convoluted path along the H-R diagram that includes the red giant branch, the horizontal branch, the asymptotic giant branch, and a path across the top and then down to the lower left of the diagram.

LG 4 Planetary nebulae form as the low-mass star loses mass, eventually leaving behind a white dwarf.

SUMMARY SELF-TEST

1. Which of the following stars will have the longest lifetime?
   a. a star one-tenth as massive as the Sun
   b. a star one-fifth as massive as the Sun
   c. the Sun
   d. a star five times as massive as the Sun
   e. a star 10 times as massive as the Sun

2. When the Sun runs out of hydrogen in its core, it will become larger and more luminous because
   a. it starts fusing hydrogen in a shell around a helium core.
   b. it starts fusing helium in a shell and hydrogen in the core.
   c. infalling material rebounds off the core and puffs up the star.
   d. energy balance no longer holds, and the star just drifts apart.

3. As a star leaves the main sequence, its position on the H-R diagram moves _____.
   a. up and to the left
   b. up and to the right
   c. down and to the left
   d. down and to the right

4. Planetary nebulae form when
   a. the dust and gas that makes planets disperses.
   b. the dust and gas that makes planets is collected around a protostar.
   c. a star loses mass at the end of its life, forming a cloud of dust and gas.
   d. a planet explodes at the end of its life, forming a cloud of dust and gas.

5. Degenerate matter is different from ordinary matter because
   a. degenerate matter doesn't interact with other particles.
   b. degenerate matter has no mass.
   c. degenerate matter doesn't interact with light.
   d. degenerate matter objects get smaller as they get more massive.

6. Place in order the following steps of the evolution of a low-mass star.
   a. The star moves onto the horizontal branch.
   b. The white dwarf cools.
   c. A clump forms in a giant molecular cloud.
   d. The star moves onto the red giant branch.
   e. The star moves onto the asymptotic giant branch.
   f. A protostar forms.
   g. The star sheds mass, producing a nebula.
   h. Hydrogen fusion begins.
   i. A helium flash occurs.

7. Place in order the following steps that lead some low-mass stars in binary systems to become novae or supernovae.
   a. Star 1 (the more massive star) begins to evolve off the main sequence.
   b. A white dwarf orbits a more massive main-sequence star.
   c. Two low-mass main-sequence stars orbit each other.
   d. Star 2 gains mass, becoming hotter and more luminous.
   e. Star 2 fills its Roche lobe and begins transferring mass to the white dwarf.
   f. Star 1 fills its Roche lobe and begins transferring mass to star 2.
   g. The white dwarf becomes either a nova or a supernova.

8. T/F: A planetary nebula shines because the dust and gas are hot.

9. T/F: We can find the age of the stars in a star cluster from its H-R diagram.

10. An accretion disk around a white dwarf in a binary system
    a. eventually forms planets.
    b. slowly disperses.
    c. feeds material onto the white dwarf.
    d. holds all of the mass of the star that sheds it.

QUESTIONS AND PROBLEMS

Multiple Choice and True/False

11. **T/F:** The mass-luminosity relationship says that if a star is twice as massive, it is twice as bright.

12. **T/F:** Even though their masses are the same, a red giant of 1 $M_\odot$ can have a radius 50 times as large as the Sun's.

13. **T/F:** Some types of material get smaller as the mass increases.

14. **T/F:** More than one kind of nuclear fusion can occur in low-mass stars.

15. **T/F:** A solitary low-mass star can sometimes become a supernova.

16. As a protostar moves onto the main sequence to become a low-mass star, the evolutionary track is nearly vertical until the very end of the process. During this vertical portion,
    a. the temperature and the luminosity both fall.
    b. the luminosity falls, but the temperature remains nearly constant.
    c. the temperature falls, but the luminosity remains nearly constant.
    d. the temperature and the luminosity both remain nearly constant.

17. The helium abundance in the outer 20 percent of the Sun
    a. remains nearly the same for its entire main-sequence lifetime.
    b. gradually increases over its main-sequence lifetime.
    c. gradually decreases over its main-sequence lifetime.
    d. remains the same for a long time and then increases abruptly before it leaves the main sequence.

18. When the Sun becomes a red giant, its _____ will decrease and its _____ will increase.
    a. density; luminosity
    b. density; temperature
    c. temperature; density
    d. density; rotation

19. As a low-mass star dies, it moves across the top of the H-R diagram because
    a. it is heating up.
    b. we see deeper into the star.
    c. the dust and gas around it heat up.
    d. a new fuel source is tapped.

20. Very young star clusters have main-sequence turnoffs
    a. nowhere. All the stars have already turned off in a young cluster.
    b. at the top left of the main sequence.
    c. at the bottom right of the main sequence.
    d. in the middle of the main sequence.

21. A star cluster with a main-sequence turnoff at 1 $M_\odot$ is approximately
    a. 10 million years old.
    b. 100 million years old.
    c. 1 billion years old.
    d. 10 billion years old.

22. A low-mass star might become a nova or supernova if and only if
    a. it is in a close binary system.
    b. it is really massive.
    c. it is really dense.
    d. it has a lot of heavy elements in it.

23. A white dwarf is located in the lower left of the H-R diagram. From this information alone, you can determine that
    a. it is very massive.
    b. it is very dense.
    c. it is very hot.
    d. it is very bright.

24. A red giant is located in the upper right of the H-R diagram. From this information alone, you can determine that
    a. it is very massive.
    b. it is very dense.
    c. it is very hot.
    d. it is very luminous.

25. The evolution of a star is primarily determined by its
    a. mass.
    b. composition.
    c. neighbors or lack of them.
    d. density.

Conceptual Questions

26. Why are most nearby stars low-mass, low-luminosity stars?

27. Is it possible for a star to skip the main sequence and immediately begin burning helium in its core? Explain your answer.

28. What is the primary reason that the most massive stars have the shortest lifetimes? (Note: The answer can be expressed in just a few words.)

29. Describe some possible ways in which a star might increase the temperature within its core while at the same time lowering its density.

30. Suppose a main-sequence star suddenly started burning hydrogen at a faster rate in its core. How would the star react? Discuss changes in size, temperature, and luminosity.

31. Astronomers typically say that the mass of a newly formed star determines its destiny from birth to death. However, there is a frequent environmental circumstance for which this statement is not true. Identify this circumstance, and explain why the birth mass of a star might not fully account for its destiny.

32. Is it fair to assume that stars do not change their structure while on the main sequence? Why or why not?

33. When a star runs out of nuclear fuel in its core, why does it become more luminous? Why does the surface temperature of the star cool down?

34. When a star leaves the main sequence, its luminosity increases tremendously. What does this increase in luminosity imply about the amount of time the star has left or the amount of time the star spends on any subsequent part of its evolutionary path?

35. When it is compressed, ordinary gas heats up; but degenerate gas does not. Why, then, does a degenerate core heat up as the star continues burning the shell around it?

36. Suppose you were an astronomer making a survey of the observable stars in our galaxy. What would be your chances of seeing a star undergoing the helium flash? Explain your answer.

37. Why is a horizontal branch star (which burns helium at a high temperature) less luminous than a red giant branch star (which burns hydrogen at a lower temperature)?

38. As an AGB star evolves into a white dwarf, it runs out of nuclear fuel, and one might argue that the star should cool off and move to the right on the H-R diagram. Why does the star move instead to the left?

39. The intersection of the Roche lobes in a binary system is the equilibrium point between the two stars where the gravitational attraction from both stars is equally strong and opposite in direction. Is this an example of stable or unstable equilibrium? Explain.

40. In Latin, *nova* means "new." Novae, as we now know, are not new stars. Explain how novae might have gotten their name and why they are really not new stars.

Problems

41. In the chapter, we used an approximation to relate the mass of a star to the luminosity. Examine Figure 12.2. The line is labeled "straight-line (exponential) approximation." What does this mean—how can it be a straight line and exponential at the same time?

42. For most stars on the main sequence, luminosity scales with mass at a rate proportional to $M^{3.5}$ (see Working It Out 12.1). What luminosity does this relationship predict for (a) 0.50 $M_\odot$ stars, (b) 6.0 $M_\odot$ stars, and (c) 60 $M_\odot$ stars? Compare these numbers to values given in Table 12.1.

43. Compute the main-sequence lifetimes for (a) 0.50 $M_\odot$ stars, (b) 6.0 $M_\odot$ stars, and (c) 60 $M_\odot$ stars. Compare them to the values given in Table 12.1.

44. Study Figure 12.3, which shows the percentages of hydrogen and helium throughout the Sun at three different points in its lifetime.
    a. What percentage of hydrogen did the Sun have in the center of the core originally?
    b. What percentage of hydrogen is in the center of the core today?
    c. What percentage of hydrogen will be in the center of the core 5 billion years from now?
    d. What percentage of the entire Sun was hydrogen originally?
    e. What percentage of the entire Sun is hydrogen now? (Hint: Compare the number of dark blue rectangles to the total number of rectangles.)
    f. What percentage of the entire Sun will be hydrogen 5 billion years from now?

45. The escape velocity ($v_{esc} = \sqrt{2\,GM_\odot/R_\odot}$) from the Sun's surface today is 618 km/s.
    a. What will the escape velocity be when the Sun becomes a red giant with a radius 50 times greater and a mass only 0.9 times that of today?
    b. What will the escape velocity be when the Sun becomes an AGB star with a radius 200 times greater and a mass only 0.7 times of today?
    c. How will these changes in escape velocity affect mass loss from the surface of the Sun as a red giant, and later as an AGB star?

46. An H-R diagram shows two variables on its axes, but they are not the only two that are interesting. For example, you might be interested to know how the radius of the star changes with temperature as the star moves off the main sequence to become a white dwarf. Using the information in Figures 12.7 to 12.10:
    a. Make a graph of the post-main-sequence evolution of the Sun.
    b. Label the red giant branch, the horizontal branch, the asymptotic giant branch, the ejection of a planetary nebula, and the white dwarf phase.

47. Suppose you are studying a star cluster, and you find that it has no main-sequence stars with surface temperatures hotter than 10,000 K. How old is this cluster? How do you know?

48. Roughly how large does a planetary nebula grow before it disperses? Use an expansion rate of 20 km/s and a lifetime of 50,000 years.

49. In some binary systems, a red giant can transfer mass onto a white dwarf at rates of about 10^{-9} $M_\odot$ per year. Roughly how long after mass transfer begins will the white dwarf undergo a Type Ia supernova? How does this length of time compare to the typical lifetime of a low-mass star? (Hint: Assume that a typical white dwarf starts with a mass of 0.6 $M_\odot$.)

50. A white dwarf has a density of approximately 10^9 kg/m^3. Earth has an average density of 5,500 kg/m^3 and a diameter of 12,700 km. If Earth were compressed to the density of a white dwarf, what would its radius be?

SMARTWORK

Norton's online homework system includes algorithmically generated versions of these questions, plus additional conceptual exercises. If your instructor assigns questions in SmartWork, log in at smartwork.wwnorton.com.

Exploration | Evolution of Low-Mass Stars

wwnpag.es/uou2

The evolution of a low-mass star, as discussed in this chapter, corresponds to many twists and turns on the H-R diagram. In this exploration, we return to the Hertzsprung-Russell Diagram Explorer Nebraska Simulation to investigate how these twists and turns affect the appearance of the star.

Visit the Student Site (wwnpag.es/uou2) and open the Hertzsprung-Russell Diagram Explorer Nebraska Simulation in Chapter 12. The box labeled "Size Comparison" shows an image of both the Sun and the test star. Initially these two stars have identical properties: the same temperature, the same luminosity, and the same size.

Examine the box labeled "Cursor Properties." This box shows the temperature, luminosity, and radius of a test star located at the "X" in the H-R diagram. Before you change anything, answer these questions:

1. What is the temperature of the test star?

2. What is the luminosity of the test star?

3. What is the radius of the test star?

As a star leaves the main sequence, it moves up and to the right on the H-R diagram. Grab the cursor (the X on the H-R diagram), and move it up and to the right.

4. What changes about the image of the test star next to the Sun?

5. What is the test star's temperature? What property of the image of the test star indicates that its temperature has changed?

6. What is the test star's luminosity?

7. What is the test star's radius?

8. Ordinarily, the hotter an object is, the more luminous it is. In this case, the temperature has gone down, but the luminosity has gone up. How can this be?

The star then moves around quite a lot in that part of the H-R diagram. Look at the H-R diagrams (Figures 12.7 to 12.10) in this chapter and then use the cursor to approximate the motion of the star as it moves up the red giant branch, back down and onto the horizontal branch, and then back to the right and up the asymptotic giant branch.

9. Are the changes you observe in the image of the star as dramatic as the ones you observed for question 4?

10. What is the most noticeable change in the star as it moves through this portion of its evolution?

Next, the star begins moving across the H-R diagram to the left, maintaining almost the same luminosity. Drag the cursor across the top of the H-R diagram to the left, and study what happens to the image of the star in the "Size Comparison" box.

11. What changed about the star as you dragged it across the H-R diagram?

12. How does its size now compare to that of the Sun?

Finally, the star drops to the bottom of the H-R diagram and then begins moving to the right. Move the cursor toward the bottom of the H-R diagram, where the star becomes a white dwarf.

13. What changed about the star as you dragged it down the H-R diagram?

14. How does its size now compare to that of the Sun?

To solidify your understanding of stellar evolution, press the reset button and then move the star from main sequence to white dwarf several times. This will help you remember how this part of a star's life appears on the H-R diagram.

SMARTWORK • smartwork.wwnorton.com

13 Evolution of High-Mass Stars

So far in our discussion of the lives of stars, we have concentrated on what happens to low-mass stars like our Sun. Stars with masses greater than about 8 solar masses (8 $M_\odot$) burn with luminosities thousands or even millions of times as great as the luminosity of our Sun and squander their nuclear fuel in a relatively short time—measured in millions of years, rather than billions.

The differences in the evolution of high-mass stars and low-mass stars follow from the connection among gravity, pressure, and the rate of nuclear burning: more mass means stronger gravitational force on the inner parts of the star. Greater force means higher pressure; higher pressure means faster reaction rates; and faster reaction rates mean greater luminosity. There are many differences in the evolution of low-mass and high-mass stars, but in the end they all trace back to the greater gravitational force bearing down on the interior of a high-mass star.

The most dramatic difference in evolution comes at the end of the life of the high-mass stars, when they explode with a luminosity equivalent to that of an entire galaxy of stars. In the illustration on the opposite page, a dramatic explosion has occurred in the popular astrophotography target M51. This student has obtained pictures of the host galaxy before and after the explosion, so he can easily find the supernova.

In this chapter, we will explore the evolution of high-mass stars as they leave the main sequence and become stellar corpses. To understand this process fully, you will also need to learn about Einstein's theories of relativity.

✦ LEARNING GOALS

Our Sun is quite faint compared with the brilliance of more massive stars. These stars consume their fuel very quickly and then die abruptly and spectacularly. By the end of this chapter, you should be able to explain why supernovae are so easy to find among the other stars of their host galaxies as the student has done with the images at the right. You should also be able to:

LG 1 Describe how high-mass stars differ from low-mass stars.

LG 2 List the stages that evolving high-mass stars experience, and explain the origin of chemical elements heavier than iron.

LG 3 Discuss the implications of the fact that the speed of light is a universal constant.

LG 4 Describe the key properties of supernovae, neutron stars, and black holes.

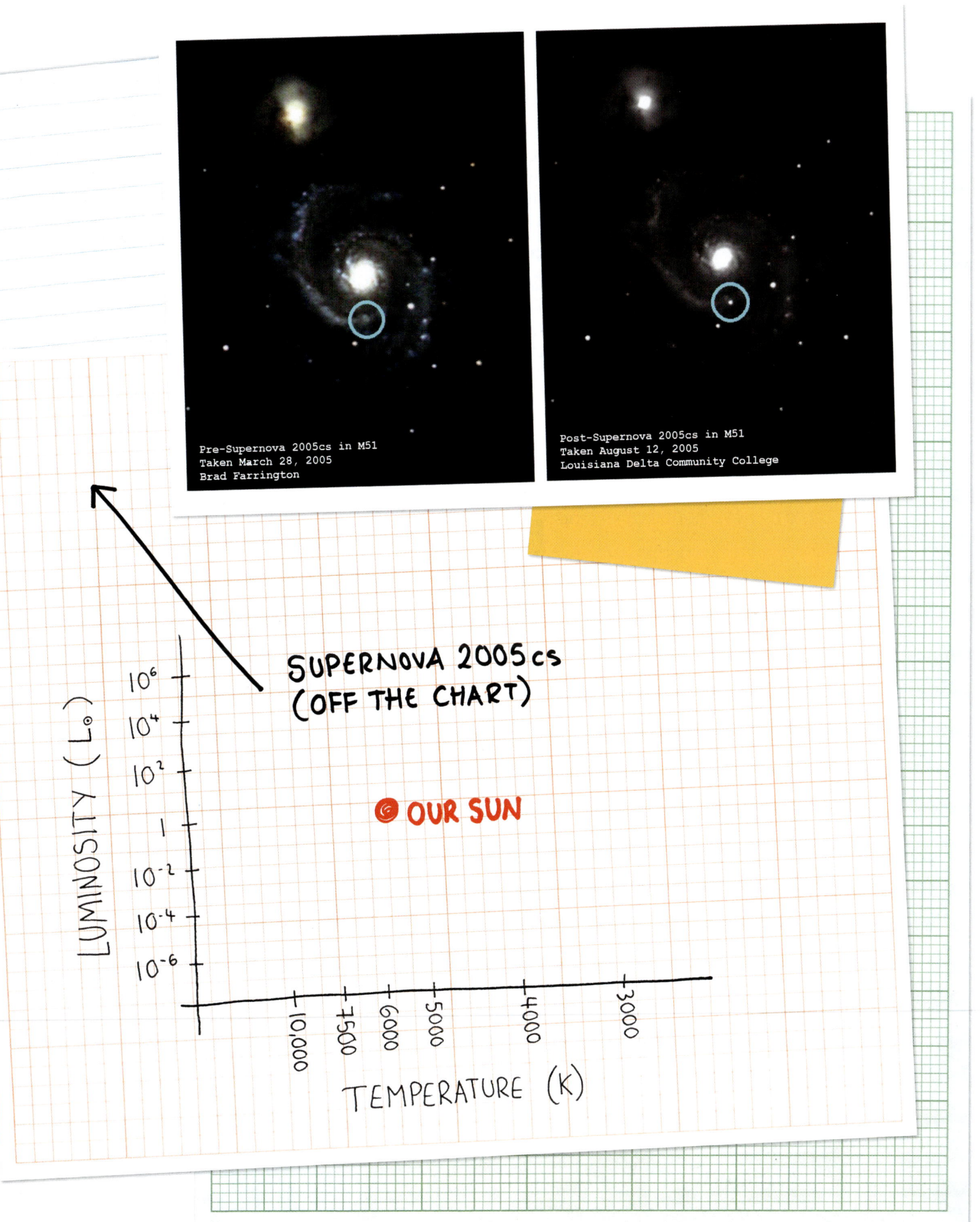

13.1 High-Mass Stars Follow Their Own Path

In the previous chapter, we followed low-mass stars along their evolutionary path. High-mass stars evolve differently as they leave the main sequence, following different paths through the Hertzsprung-Russell (H-R) diagram, because of different processes occurring in the interiors of these stars.

Leaving the Main Sequence

At the very high temperatures at the center of a high-mass star, nuclear reactions other than the proton-proton chain occur. For example, in some high-mass stars, hydrogen is fused into helium in a process called the **carbon-nitrogen-oxygen (CNO) cycle**. In this cycle, carbon is a **catalyst**: a substance that helps the reac-

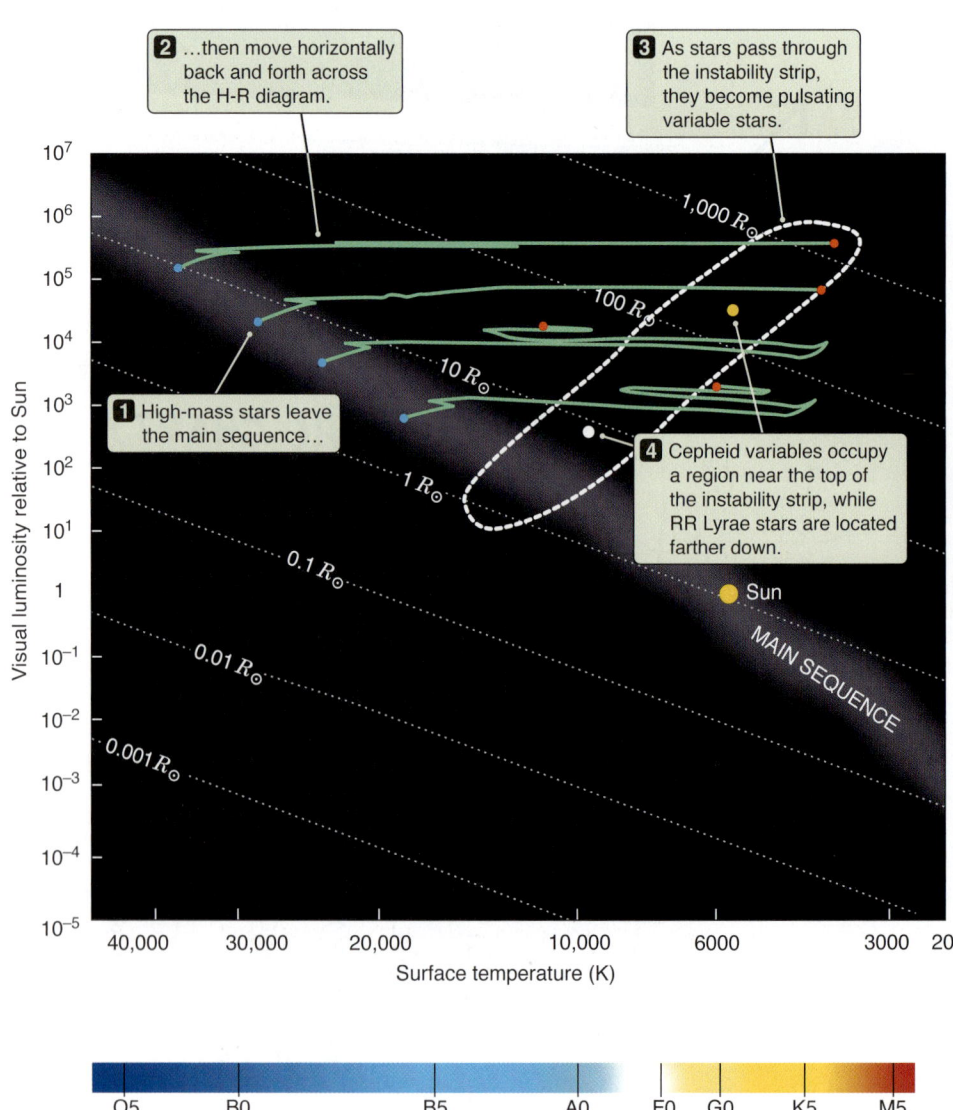

Figure 13.1 When massive stars leave the main sequence, they move horizontally across the H-R diagram.

tion proceed but is not used up in the reaction. This reaction consumes hydrogen and turns it into helium.

As the high-mass star runs out of hydrogen in its core, the weight of the overlying star compresses the core. Yet long before the core becomes electron-degenerate, the pressure and temperature become high enough for helium burning. Because the core is not made of degenerate matter, the structure of the star responds to the increase in temperature, but its luminosity changes relatively little. The star makes a fairly smooth transition from hydrogen burning to helium burning.

The high-mass star now burns helium in its core and hydrogen in a surrounding shell, like a low-mass horizontal branch star. It grows in size while its surface temperature falls. The result, illustrated in **Figure 13.1**, is that the star moves to the right on the H-R diagram, leaving the main sequence. Stars of more than 10 $M_\odot$ become red supergiants during their helium-burning phase. They have very cool surface temperatures (about 4000 K) and radii as much as 1,500 times that of the Sun.

Eventually, the high-mass star exhausts the helium in its core. As the core collapses, it reaches temperatures high enough to burn carbon. Carbon burning produces even more massive elements, including oxygen, sodium, neon, and magnesium. The star now has a carbon-burning core surrounded by a helium-burning shell surrounded by a hydrogen-burning shell. When carbon is exhausted, neon burning begins; and when neon is exhausted, oxygen begins to burn. The structure of the evolving high-mass star, shown in **Figure 13.2**, is like an onion; it has many concentric layers.

Stars on the Instability Strip

As a star evolves, it may become a **pulsating variable star**, which does not find a steady balance but alternately grows larger and smaller. Larger stars have more surface area, so they emit more light and have a higher luminosity than smaller stars of similar temperature. As a pulsating variable star's size varies, this effect causes the star's luminosity to vary as well. A pulsating variable star makes one or more passes through a region of the H-R diagram known as the **instability strip**; stars in this region pulsate with a periodic variation in luminosity (see Figure 13.1).

The most luminous pulsating variable stars are the **Cepheid variables**, named after the prototype star Delta Cephei. Cepheid variables have periods ranging from 1 to 100 days. The luminosity of a Cepheid variable is related to its period: short-period Cepheid variables are less luminous than long-period Cepheid variables. This **period-luminosity relationship** allows astronomers to use Cepheid variables to find the distances to galaxies beyond our own. By observing the period, astronomers can determine the luminosity. Combining this luminosity with the brightness in the sky gives the distance to the star. Cepheid variables are an example of a **standard candle**: an object whose luminosity can be determined independently from its distance.

Thermal energy powers the pulsations of stars like Cepheid variables. Ionization of atoms in the star alternately traps and releases thermal energy, causing the star to expand and contract. At each change, the star overshoots the equilibrium point where forces due to pressure and gravity balance each other, shrinking too far or expanding too far. These pulsations do not affect the nuclear burning in

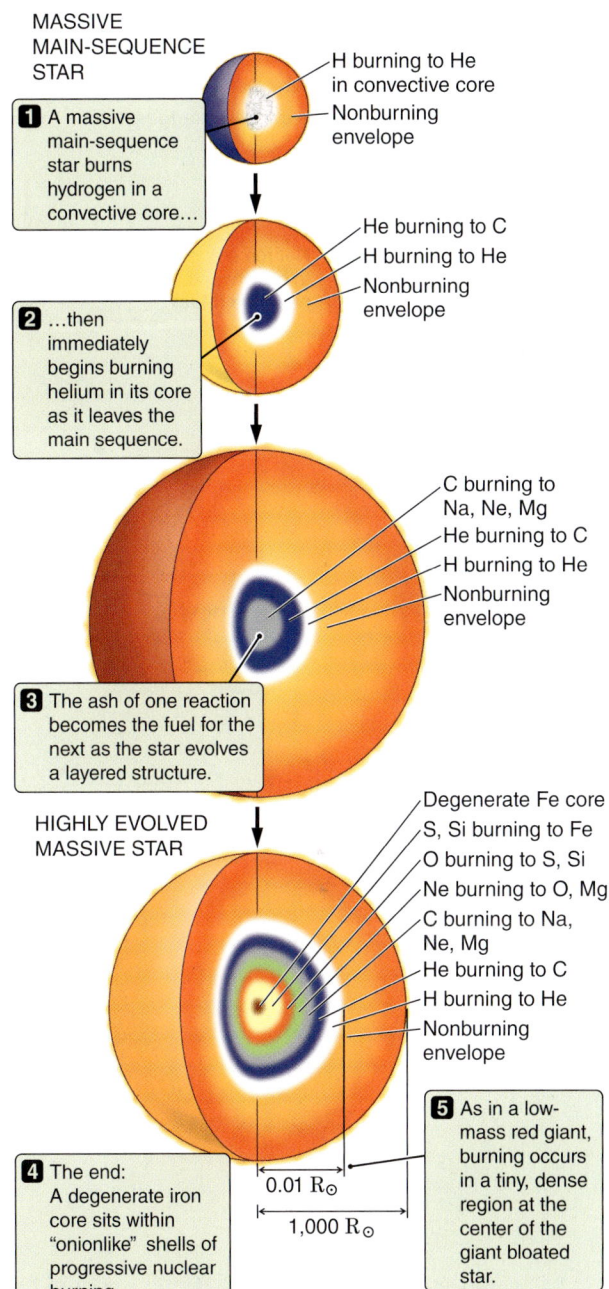

Figure 13.2 As a high-mass star evolves, it builds up a layered structure like that of an onion; progressively advanced stages of nuclear burning are found deeper and deeper within the star. Note the change in scale for the bottom image.

Figure 13.3 In a pulsating Cepheid variable, ionization of atoms in the star's interior alternately allows the surface of the star to fall inward and then pushes it back out again. (Color changes shown here are greatly exaggerated.)

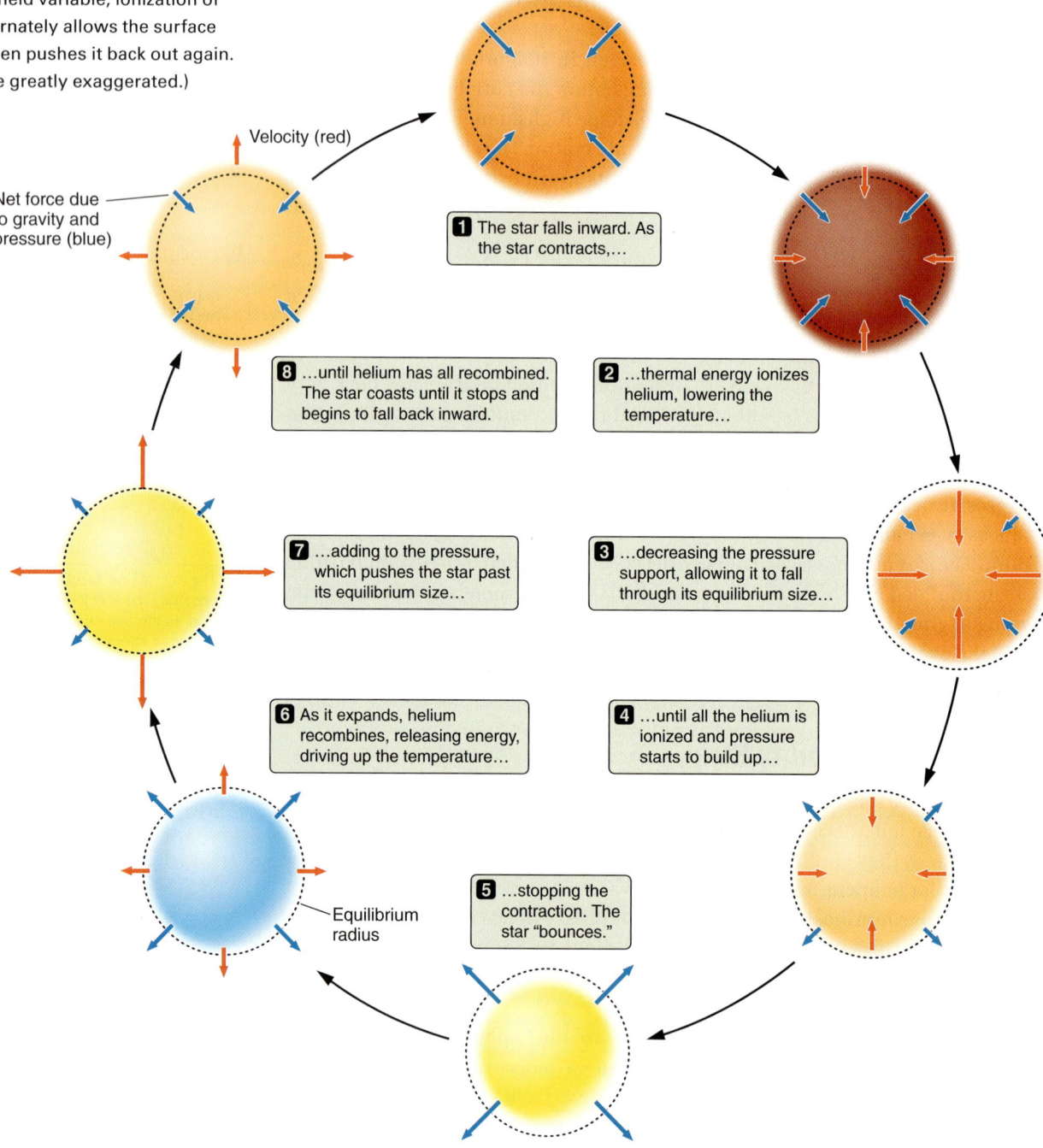

the star's interior. However, they do affect the light escaping from the star. The star is at its brightest and bluest while it expands and at its faintest and reddest as it falls back inward (see **Figure 13.3**).

The instability strip on the H-R diagram also intersects the low-mass horizontal branch. Unstable low-mass horizontal branch stars are known as **RR Lyrae variables** after their prototype star in the constellation Lyra. They are much less luminous than Cepheid variables, but they can also be used as **standard candles**. The instability strip also intersects the main sequence around spectral type A, and many type A stars do show variability. A number of other kinds of variable stars are seen elsewhere in the H-R diagram, each driven by its own type of instability.

The pressure of the intense radiation on the gas at the surface of a massive star overcomes the star's gravity and causes winds with speeds of about 3,000 kilometers per second (km/s) and mass loss ranging from about 10^{-7} to $10^{-5}\ M_\odot$ per year. These numbers may sound tiny, but over millions of years, this rate of mass loss becomes significant. Type O stars with masses of 20 $M_\odot$ lose about 20 percent of their mass while on the main sequence, and possibly more than 50 percent over their entire lifetimes. Eta Carinae (**Figure 13.4**), a star more than 100 times as massive as the Sun and as luminous as 5 million Suns, is an extreme example. Eta Carinae is currently losing 1 $M_\odot$ every 1,000 years. However, during a 19th century eruption when Eta Carinae became the second-brightest star in the sky, it shed 2 $M_\odot$ in only 20 years. Eta Carinae is expected to explode as a supernova eventually.

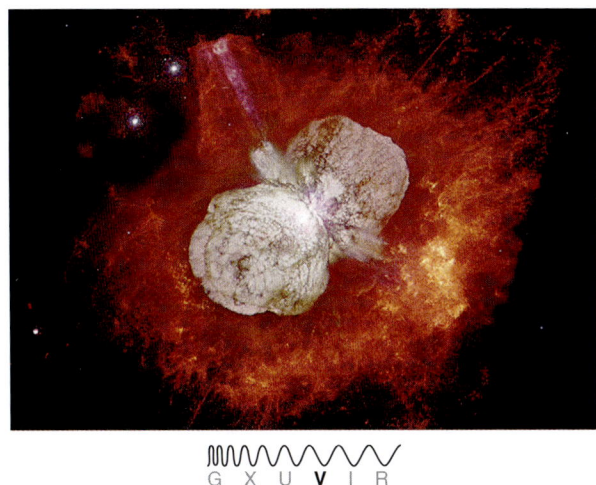

Figure 13.4 This Hubble Space Telescope image shows an expanding cloud of dusty material ejected by the luminous blue variable star Eta Carinae. The star itself, which is largely hidden by the surrounding dust, has a luminosity 5 million times that of the Sun and a mass probably in excess of 100 $M_\odot$. Dust is created when volatile material ejected from the star condenses.

13.2 High-Mass Stars Go Out with a Bang

In 1987, astronomers observed the explosion of a massive star in the Large Magellanic Cloud, a companion galaxy to the Milky Way. Even at a distance of 160,000 light-years, Supernova 1987A was so bright that it dazzled sky gazers in the Southern Hemisphere (**Figure 13.5**). Supernova 1987a was similar to the one portrayed in the chapter-opening illustration. Compare the brightness of that supernova with the brightness of the stars around it. In these events, a high-mass star came to the end of its life and exploded as an object known as a *Type II supernova*. These supernovae can be as luminous as all of the other hundreds of billions of stars in their host galaxies combined! By comparing two images of a galaxy, one before the supernova (Figure 13.5a) and one after the supernova (Figure 13.5b), you can easily identify the exploding star. In this section, you will learn how a star in balance comes to explode in such a dramatic fashion.

The Final Days in the Life of a Massive Star

Maintaining the balance in a star is somewhat like trying to keep a leaky ball inflated. The larger the leak, the more rapidly air must be pumped into the ball

Figure 13.5 Supernova 1987A (SN 1987A) was a supernova that exploded in a small companion galaxy of the Milky Way called the Large Magellanic Cloud (LMC). These images show the LMC (a) before the explosion and (b) while the supernova was near its peak.

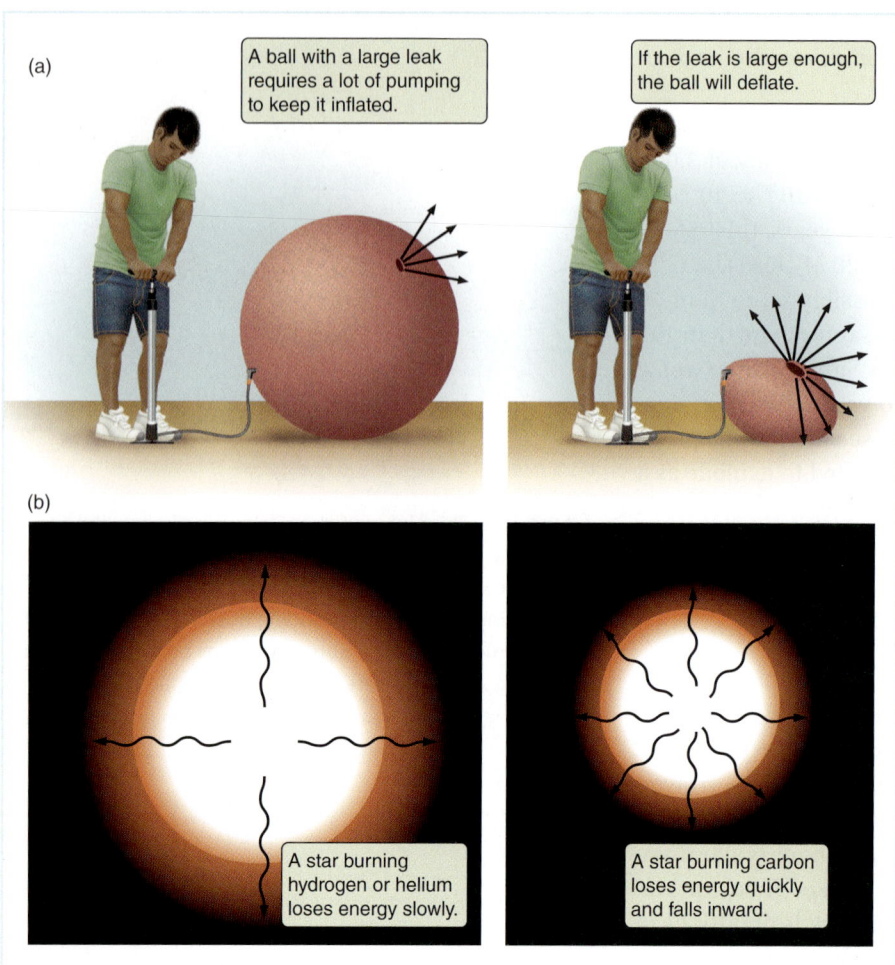

Figure 13.6 (a) If air leaves a ball faster than it can be replaced, the ball will deflate. (b) Similarly, if energy leaves a star faster than it can be replaced, the energy balance is disrupted, and the star begins to contract.

VISUAL ANALOGY

to keep it inflated. As illustrated in **Figure 13.6**, a star that is burning hydrogen or helium is like a ball with a slow leak. At the temperatures of hydrogen or helium burning, energy leaks out of the interior of a star primarily by radiation and convection. The outer layers of the star act like a thick, warm blanket. Much of the energy is kept in the star, so nuclear fuels do not need to burn very fast to support the star against gravity.

Once carbon burning begins, this balance shifts. Energy is carried primarily by neutrinos rather than radiation and convection. Recall from your study of the Sun in Chapter 11 that neutrinos escape easily, carrying energy away from the core. In a high-mass star, once cooling by neutrinos becomes significant, the outer layers of the star fall inward, drive up the star's density and temperature, and force nuclear reactions to run faster. Carbon burning supports the star for about a thousand years. Oxygen burning lasts about a year. Silicon burning lasts about a day. A silicon-burning star is not much more luminous than it was while burning helium. But because of neutrino cooling, this silicon-burning star is actually giving off about 200 million times more energy each second than it did while it was burning helium.

The Core Collapses and the Star Explodes

In Figure 13.2, we saw that an evolving high-mass star builds up its onionlike structure as hydrogen burns to helium, helium burns to carbon, carbon burns to sodium, neon, and magnesium, oxygen burns to sulfur and silicon, and silicon and sulfur burn to iron. After silicon burning, the end comes suddenly and dramatically. Many different types of nuclear reactions occur up to this point, forming almost all of the different stable isotopes of elements less massive than iron. However, because iron does not release energy when it fuses but rather absorbs energy, the chain of nuclear fusion stops with iron. When hydrogen is fused into helium, energy is released, maintaining the temperature necessary to keep the reaction going; once the reaction begins, it is self-sustaining. For iron, once the reaction starts, energy is absorbed. No longer supported by thermonuclear fusion, the iron core of the massive star begins to collapse.

Figure 13.7 shows the stages a high-mass star passes through at the end of its life. As the core collapses, the force of gravity increases and the density and temperature skyrocket. The core becomes electron-degenerate when it is about the size of Earth. The weight bearing down on the iron ash core is too great to be held up by electron degeneracy pressure (step 1 in Figure 13.7). As the core collapses, the core temperature climbs to temperatures of more than 10 billion K, while the density exceeds 10^{10} kilograms per cubic meter (kg/m^3)—10 times the density of a white dwarf.

At these temperatures, the nucleus of the star is filled with photons so energetic that they can break iron nuclei apart (step 2 in Figure 13.7). This process, called *photodisintegration*, absorbs thermal energy and begins reversing the results of nuclear fusion. At the same time, the density of the core is so great that electrons are forced into atomic nuclei, where they combine with protons to produce neutrons and neutrinos (step 3 in Figure 13.7). Both this process and photodisintegration absorb much of the energy that was holding up the dying star. Neutrinos take still more energy with them as they leave the star. The collapse of the core accelerates, reaching a speed of 70,000 km/s, or almost one-fourth the speed of

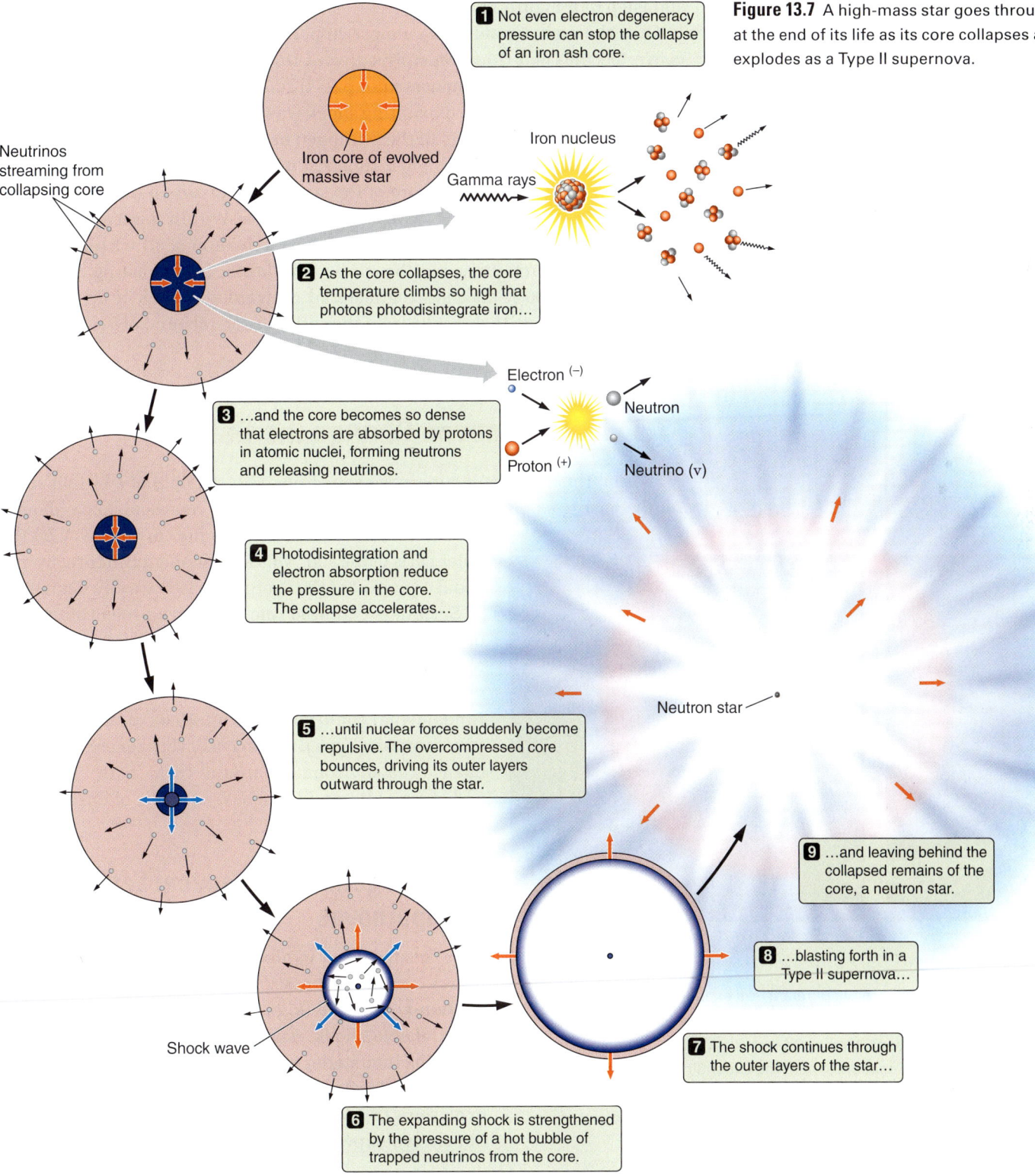

Figure 13.7 A high-mass star goes through several stages at the end of its life as its core collapses and the star explodes as a Type II supernova.

light, on its inward fall (step 4 in Figure 13.7). All of these events together take place remarkably quickly—in less than a second.

As material in the collapsing core exceeds the density of an atomic nucleus, the strong nuclear force actually becomes repulsive (step 5 in Figure 13.7). About half of the collapsing core suddenly slows its inward fall. The remaining half slams into the innermost part of the star at a significant fraction of the speed of light and "bounces," sending a tremendous shock wave back out through the star (step 6 in Figure 13.7).

Over the next second or so, almost 20 percent of the material in the core is converted into neutrinos. Most fly outward through the star; but at these phenomenal densities, not even neutrinos pass through with complete freedom. The dense material behind the expanding shock wave traps a few tenths of a percent of the neutrinos. The energy of these trapped neutrinos drives the pressure and temperature in this region higher, inflating a bubble of extremely hot gas and intense radiation around the core of the star. The pressure of this bubble adds to the shock wave moving outward through the star. Within about a minute, the shock wave has pushed its way out through the helium shell within the star. Within a few hours, it reaches the surface of the star itself, heating the stellar surface to 500,000 K and blasting material outward at speeds of up to 30,000 km/s. The explosion that results from the collapse and rebound of a massive star is called a **Type II supernova**.

13.3 Supernovae Change the Galaxy

For a brief time, a Type II supernova can shine with the light of a billion suns. Yet the energy carried away by light from a supernova represents only about 1 percent of the kinetic energy being carried away by the outer parts of the star. This kinetic energy, in turn, is only about 1 percent of the energy carried away by neutrinos.

Scientists captured one of the true scientific prizes of SN 1987A (see Figure 13.5) when neutrino detectors recorded a burst from the supernova before it was detected visually. The detection of neutrinos from SN 1987A was a fundamental confirmation of our theories about the role of neutrinos in Type II supernovae. These energetic events change the environment around them, both physically and chemically, and leave behind shells of dust and gas and dense cores.

The Energetic and Chemical Legacy of Supernovae

Type II supernova explosions leave a rich and varied legacy to the universe. Huge expanding bubbles of million-kelvin gas (**Figure 13.8a**) glow in X-rays and drive visible shock waves (Figure 13.8b) into the surrounding interstellar medium (the dust and gas between the stars). These bubbles are produced by supernova explosions that took place thousands of years previously. Supernova explosions compress nearby clouds (Figure

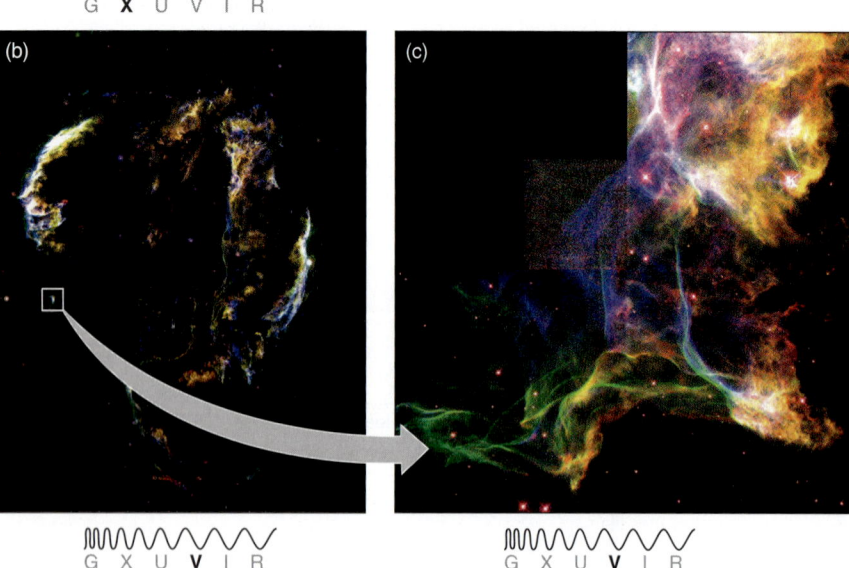

Figure 13.8 The Cygnus Loop is a supernova remnant—an expanding interstellar blast wave caused by the explosion of a massive star. (a) Gas in the interior of the cloud, with a temperature of millions of kelvins, glows in X-rays. (b) Visible light comes from the edges of the expanding bubble: locations where the expanding blast wave pushes through denser gas in the interstellar medium. (c) A Hubble Space Telescope image of a location where the blast wave is hitting an interstellar cloud. The colors indicate different types of atoms excited by the shock.

13.8c), triggering the initial collapse that begins the formation of new stars.

Only the least massive chemical elements were present at the beginning of the universe: hydrogen, helium, and trace amounts of lithium, beryllium, and boron. All of the rest of the chemical elements, including a large fraction of the atoms we are made of, were formed in the stars through nuclear reactions and then returned to the interstellar medium when the stars exploded. Low-mass stars form elements as massive as carbon and oxygen; high-mass stars produce elements as massive as iron. Yet this is not the whole story. Most naturally occurring elements are more massive than iron. If iron is the most massive element that can be formed in stars, then where do these even more massive elements come from?

Under normal circumstances, electric repulsion keeps positively charged atomic nuclei far apart. Extreme temperatures are needed to slam nuclei together hard enough to overcome this electric repulsion. Free neutrons, however, have no net electric charge, so there is no electric repulsion to prevent them from simply running into an atomic nucleus. Under normal conditions, free neutrons are rare. Under the conditions of a Type II supernova, however, free neutrons are produced in very large numbers. These are easily captured by massive atomic nuclei and later decay to become protons, forming elements more massive than iron.

Nuclear physics predicts the abundances of the elements, as shown in **Figure 13.9**. These predictions agree with abundances that have been measured in the Solar System and in the atmospheres of stars. Less massive elements are far more abundant than more massive elements. An exception to this pattern is the dip in the abundances of the light elements lithium (Li), beryllium (Be), and boron (B). Nuclear burning easily destroys these elements, and they are not produced by the common reactions involved in burning hydrogen (H) and helium (He). Conversely, carbon (C), nitrogen (N), and oxygen (O) are big winners in the triple-alpha process of helium burning and therefore are more abundant. More massive elements are progressively built up from less massive elements. The spike in the abundances of elements near iron is evidence of processes that favor these tightly bound nuclei. Even the sawtooth pattern in the abundances of even- and odd-numbered elements can be explained as a consequence of the way atomic nuclei form in stars.

Our understanding of the interiors of dying stars is confirmed by the chemical composition of the Earth under our feet. Our own bodies are formed of atoms that were made in stars. Our growing understanding of the chemical evolution of the universe and our connection to it is one of the triumphs of modern astronomy.

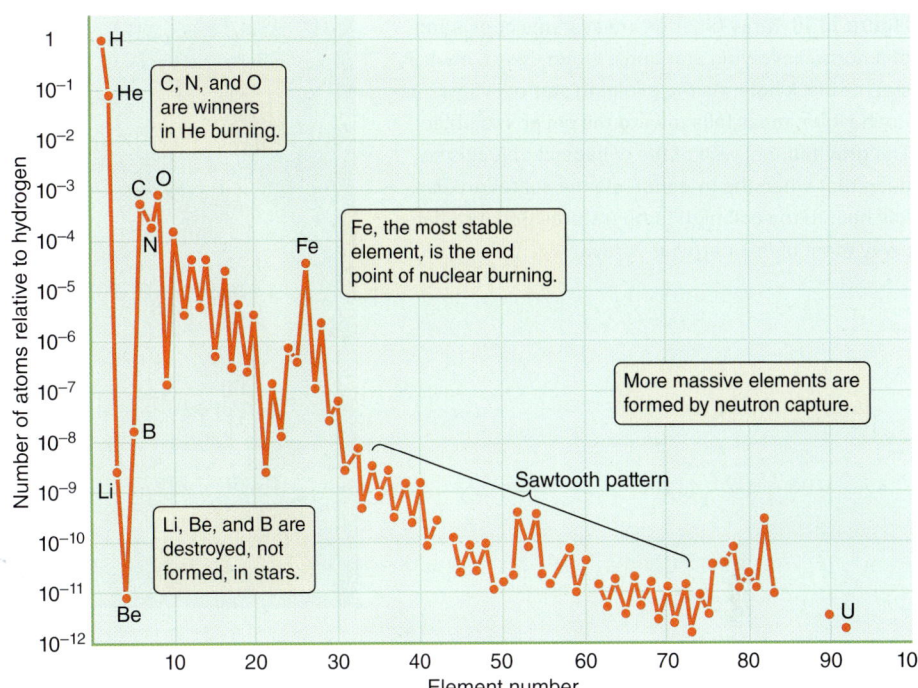

Figure 13.9 In this graph, relative abundances of different elements in the Solar System are plotted against each element's number. These abundances are compared to the number of hydrogen atoms, which is set to 1. So, for example, there are roughly one-tenth as many helium atoms as hydrogen atoms. As the atomic number increases, the number of protons (and typically also neutrons) in the nucleus increases, so the nuclei are more massive. The pattern of abundances is a direct result of the production of atomic nuclei in stars.

Neutron Stars and Pulsars

Let's now return to the remains of the star. The core has collapsed to the point where it has about the same density as the nucleus of an atom. For cores less than about 3 $M_\odot$, the collapse is halted when *neutrons* are packed as tightly together as the rules of quantum mechanics allow. This neutron-degenerate core is called

Figure 13.10 X-ray binaries are systems consisting of a normal evolving star and a white dwarf, neutron star, or black hole. As the evolving star overflows its Roche lobe, mass falls toward the collapsed object. The gravitational well of the collapsed object is so deep that when the material hits the accretion disk, it is heated to such high temperatures that it radiates away most of its energy as X-rays.

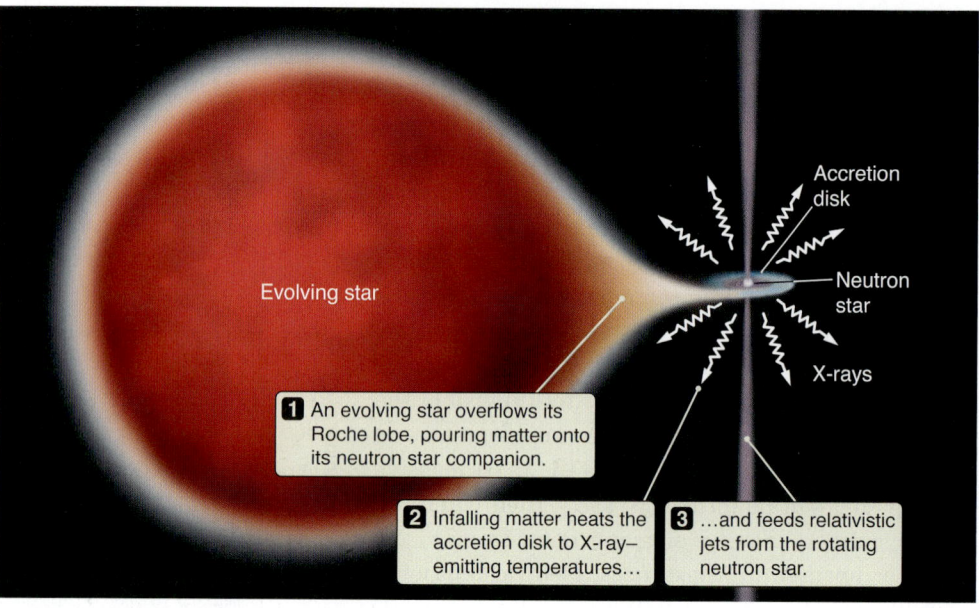

1. An evolving star overflows its Roche lobe, pouring matter onto its neutron star companion.
2. Infalling matter heats the accretion disk to X-ray–emitting temperatures…
3. …and feeds relativistic jets from the rotating neutron star.

a **neutron star**. It is roughly the size of a small city, with a diameter of about 20 km. That volume is packed with a mass between 1.4 $M_\odot$ and 3 $M_\odot$. At a density of about 10^{18} kg/m³, the neutron star is a billion times denser than a white dwarf and a thousand trillion (10^{15}) times denser than water.

If the original massive star was part of a close binary system, then the neutron star will have a companion, much like the white dwarf binary systems discussed in Chapter 12. **Figure 13.10** illustrates an **X-ray binary**, a binary system in which mass from an evolving star spills over onto a collapsed companion such as a white dwarf, neutron star, or black hole. (Black holes will be discussed later in the chapter.) As the lower-mass star evolves and overfills its Roche lobe, matter falls toward the accretion disk around the neutron star, heating it to millions of kelvins and causing it to glow brightly in X-rays. X-ray binaries sometimes develop powerful jets of material that are perpendicular to the accretion disk and carry material away at speeds close to the speed of light.

As the core of the original massive star collapses, it spins faster, because angular momentum is conserved. A main-sequence O star rotates perhaps once every few days. As a neutron star, it might rotate tens or even hundreds of times each second. The collapsing star also concentrates its magnetic field to strengths trillions of times greater than the magnetic field at Earth's surface. A neutron star has a magnetosphere just like Earth and several other planets do, except that the neutron star's magnetosphere is vastly stronger and is whipped around many times a second by the spinning star. As in planets, in stars the magnetic axis is often not aligned with the rotation axis.

Electrons and positrons move along the magnetic-field lines and are "funneled" by the field toward the magnetic poles of the system. The particles produce beams of radiation along the magnetic poles of the neutron star as shown in **Figure 13.11**. As the neutron star rotates, these beams sweep through space much like the rotating beams of a lighthouse. The neutron star appears to flash on and off with a regular period equal to the period of rotation of the star (or half the rotation period, if we see both beams). These rotating neutron stars are known as **pulsars**. As of this writing, more than 2,000 pulsars are known, and more are being discovered all the time. Most of these pulsars are radio pulsars, with a few also emitting in other regions of the electromagnetic spectrum.

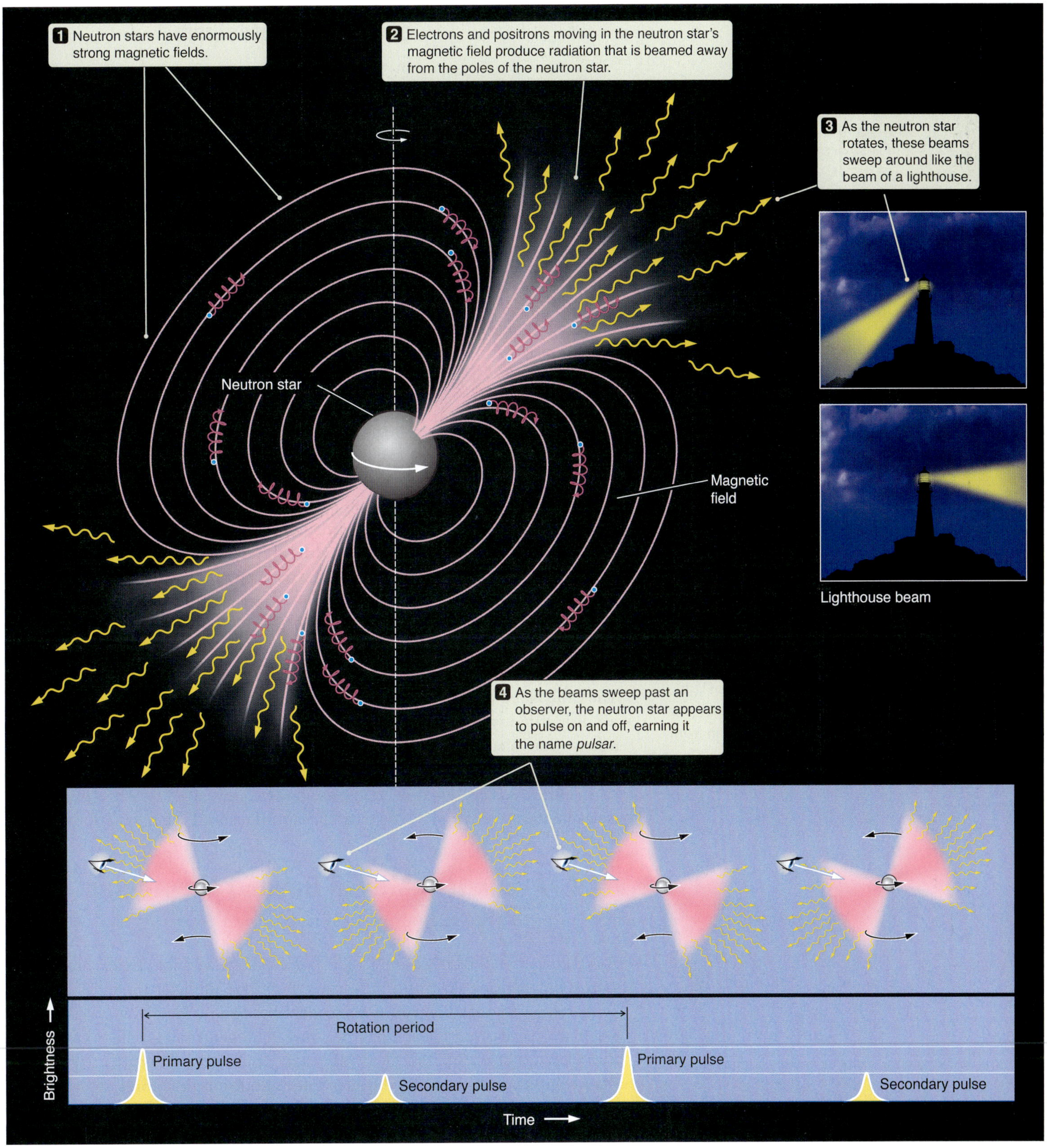

Figure 13.11 As a highly magnetized neutron star rotates rapidly, light is emitted, much like the beams from a rotating lighthouse lamp. From our perspective, as these beams sweep past us the star appears to pulse on and off, earning it the name *pulsar*.

　**VISUAL ANALOGY**

The Crab Nebula: Remains of a Stellar Cataclysm

In 1054, Chinese astronomers noticed a "guest star" in the direction of the constellation Taurus. The new star was so bright that it could be seen during the daytime for 3 weeks, and it did not fade away altogether for many months. This star was a fairly typical Type II supernova. Today, an expanding cloud of debris from this explosion occupies this place in the sky—forming an object called the Crab Nebula (**Figure 13.12a**).

The Crab Nebula has filaments of glowing gas expanding away from the central star at 1,500 km/s. These filaments contain anomalously high abundances of helium and other more massive chemical elements—the products of the nuclear reactions that took place in the supernova and its progenitor star.

The Crab pulsar at the center of the nebula flashes 60 times a second: first with a main pulse associated with one of the "lighthouse" beams, then with a fainter secondary pulse associated with the other beam. As the Crab pulsar spins 30 times a second, it whips its powerful magnetosphere around with it. At a distance from the pulsar about equal to the radius of the Moon, material in the magnetosphere must move at almost the speed of light to keep up with this rotation. Like a tremendous slingshot, the rotating pulsar magnetosphere flings particles away from the neutron star in a wind moving at nearly the speed of light. This wind fills the space between the pulsar and the expanding shell. The Crab Nebula is almost like a big balloon; but instead of being filled with hot air, it is filled with a mix of very fast particles and strong magnetic fields. The energy that accelerates these particles is exactly equal to the energy lost as the pulsar's rotation slows down. Images of the Crab Nebula such as Figure 13.12b and Figure 13.12c show this bubble as an eerie glow from synchrotron radiation (see Chapter 8) released as the particles spiral around the magnetic field.

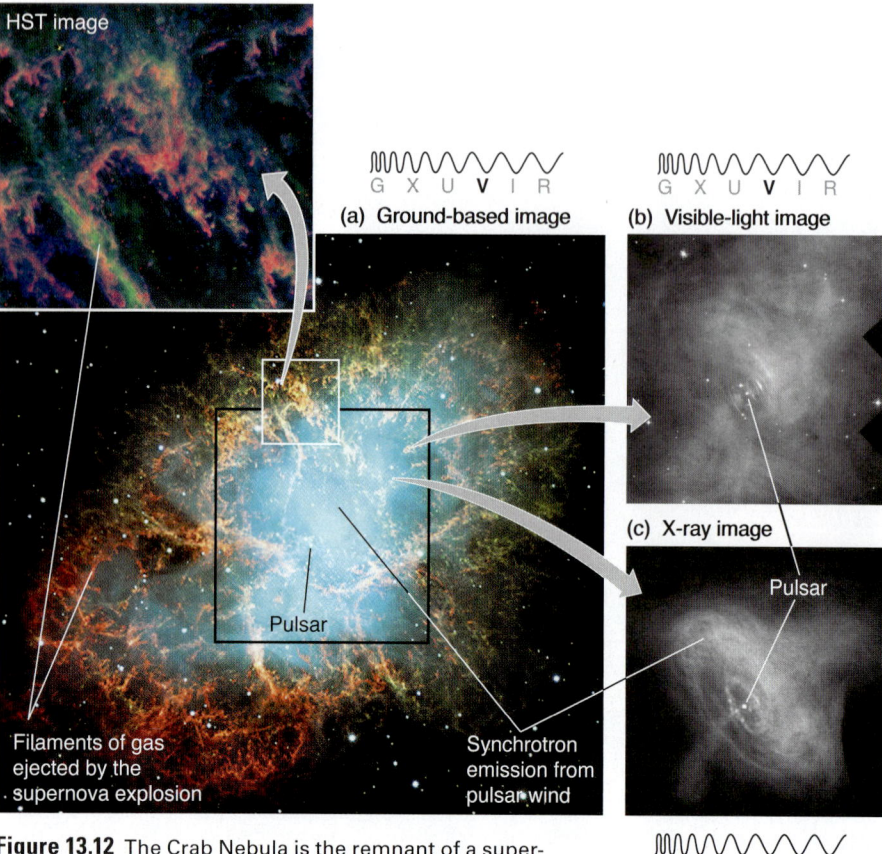

Figure 13.12 The Crab Nebula is the remnant of a supernova explosion witnessed by Chinese astronomers in 1054. (a) The object we see today is an expanding cloud of "shrapnel" from that earlier cataclysm. The spinning pulsar at the heart of the Crab Nebula sends off a "wind" of electrons and positrons moving at close to the speed of light. Inset: An expanded view of the filaments obtained by the Hubble Space Telescope. The synchrotron radiation from these particles is shown (b) in visible light and (c) in X-rays.

13.4 Einstein Moved Beyond Newtonian Physics

As you may recall from Chapter 12, white dwarfs have a maximum mass, known as the Chandrasekhar limit. Similarly, there is a limit for the mass of neutron stars. If the mass of a neutron star exceeds about 3 $M_\odot$, then gravity will be stronger than the neutron degeneracy pressure. The neutron star shrinks, and gravity increases at an ever-accelerating pace. Eventually, gravity is so strong that the escape velocity exceeds the speed of light. From this point on, nothing can escape from the collapsing object. The object is now called a **black hole**: an object with gravity so strong that even light cannot escape it. Black holes are so strange, so far from the common understanding of reality, that the laws of Newtonian physics (see Chapter 3) cannot be used to describe them. To do so, we must first look at the nature of space and time. In this section, you will learn how the constant speed of light led to the formation of the theory of special relativity, which changed our understanding of space and time.

The Speed of Light in Vacuum

Newton's laws of motion allow us to figure out how the speed of an object will be measured by different observers. Each observer inhabits a reference frame, in which they measure distances and speeds. You can follow an example in **Figure 13.13a**. Imagine that you are sitting in a car traveling at 50 miles per hour (mph) down the highway, and you throw a ball with a speed of 25 mph out the window at an oncoming car, also traveling at 50 mph. An observer standing by the side of the road watches the entire event. In your reference frame, the car is stationary; that is, the car does not move relative to you as you are moving with it. So in your reference frame, you would measure the ball traveling at 25 mph (Figure 13.13a, top). To the observer standing by the road, the ball is moving at 75 mph

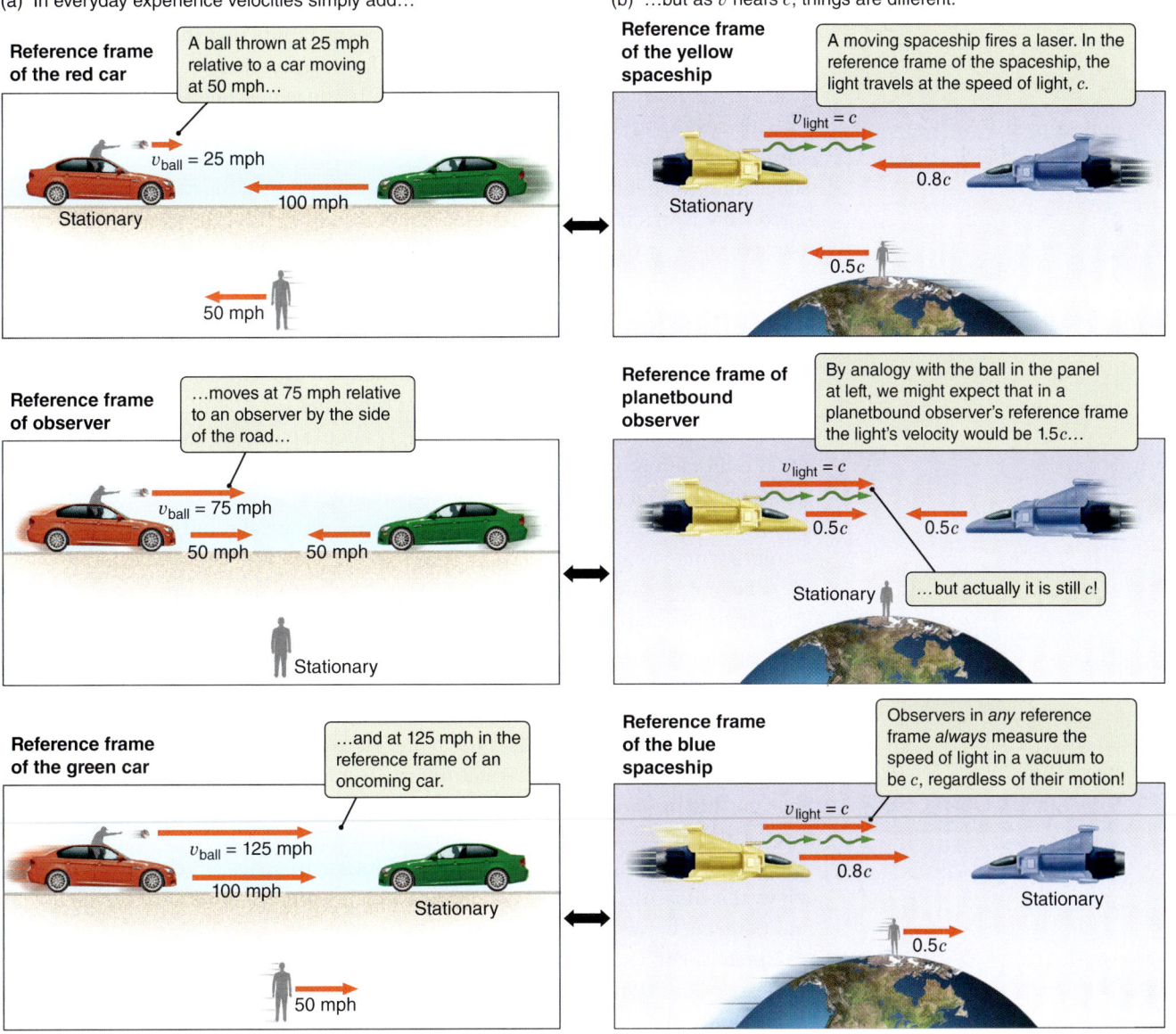

Figure 13.13 (a) The rules of motion that apply in our daily lives break down (b) when speeds approach the speed of light. The fact that light itself always travels at the same speed for any observer is the basis of special relativity. (Note that relativity also affects the relative speeds of the two spacecraft.)

(25 mph from throwing it plus 50 mph from the motion of the car; Figure 13.13a, middle). To passengers in the oncoming car moving at 50 mph, the speed of the ball is 125 mph, because the ball approaches them at 75 mph, and they approach the ball at 50 mph (Figure 13.13a, bottom). The speed of the ball depends on who is measuring it and how the observer, the car, and the ball are moving relative to one another. The velocities are added together to find the velocity of one object relative to another. This is *Galilean relativity*, and you use it in everyday life.

During the closing years of the 19th century and the early years of the 20th century, the results of laboratory experiments with light puzzled physicists. They expected that the speed of light should differ from one observer to the next as a result of the observer's motion, just like the speed of a ball in our example. Instead, they found that *all observers measure exactly the same value for the speed of a beam of light, regardless of their motion!*

Figure 13.13b demonstrates how light differs. Imagine that you are riding in a fast spaceship and you shine a beam of light forward. You measure the speed of the beam of light to be c, or 3×10^8 meters per second (m/s). That is as expected because you are holding the source of the light. But the observer on a planet you are passing *also* measures the speed of the passing beam of light to be 3×10^8 m/s. Even a passenger in an oncoming spacecraft finds that the beam from your light is traveling at exactly c in her own reference frame. Every observer always finds that light in a vacuum travels at exactly the same speed, c.

This result, so different from the example using balls and cars, had serious implications. Newton's laws of motion had withstood every experimental challenge. But this new idea challenged the understanding of the universe that developed from Newton's laws of motion. How could light have the same speed for *all* observers, regardless of their own velocity?

Time Dilation

Albert Einstein approached the problem differently from his colleagues. He started with the fact that light always travels at the same speed, and then he reasoned backward to find out what that must imply about space and time. This led to the 1905 publication of his **special theory of relativity**, sometimes called "special relativity," which describes the counterintuitive effects of traveling at constant speeds close to the speed of light. To understand special relativity, we must first explore what we mean by *event*, space, and time.

An **event** is something that happens at a particular location in space at a particular time. Snapping your fingers is an event because that action has both a time and a place. The distance between any two events depends on the reference frame of the observer. Imagine you are sitting in a moving car. You snap your fingers (event 1), and a minute later you snap your fingers again (event 2). In your reference frame, the two events happened at exactly the same place, because in your reference frame, *you* are stationary. The events were, however, separated by a minute in *time*. Now imagine an observer sitting by the road. This observer agrees that the second event happened a minute after the first, but to this observer the two events were separated from each other in space. In this "Newtonian" view, the distance between two events depends on the motion of the observer, but the time between the two events does not.

Einstein carried out a thought experiment, called the "boxcar experiment," as described in **Working It Out 13.1**. In this experiment, light travels farther in a moving boxcar than in a stationary one. The only way that all observers will

Working It Out 13.1 | The Boxcar Experiment

The special theory of relativity is a *very* counterintuitive idea, but it is so central to our modern understanding of the universe that it is worth wrestling with a bit. In the boxcar experiment, observer 1 is in a boxcar of a train moving to the right. Observer 1 has a lamp, a mirror (mounted to the roof of the boxcar), and a clock. Observer 2 is standing on the ground outside.

Figure 13.14a shows the experimental setup as seen by observer 1, who is stationary with respect to the clock. At time t_1, event 1 happens: the lamp gives off a pulse of light. The light bounces off a mirror a distance l meters away. At time t_2, event 2 happens: the light arrives at the clock. The time between events 1 and 2 is just the distance the light travels ($2l$ meters), divided by the speed of light: $t_2 - t_1 = 2l/c$.

Now let's look from the perspective of observer 2, who is standing on the ground outside the train (Figure 13.14b). In *this* observer's reference frame, he is stationary and the boxcar is moving at speed v. In observer 2's reference frame, the clock *moves* to the right between the two events, so the light has *farther to go*. (If you do not see this, use a ruler to measure the total length of the light path in Figure 13.14b and compare it with the total length of the light path in Figure 13.14a.) The time between the two events ($t_2 - t_1$) is longer for observer 2 than for observer 1 because the distance traveled is *longer* than $2l$ meters.

The two events are the *same two events*, regardless of the reference frame from which they are observed. But because the speed of light is the same for all observers, there *must* be more time between the two events when viewed by observer 2. The seconds of a moving clock are stretched. Moving clocks *must* run slow, and the passage of time *must* depend on an observer's reference frame.

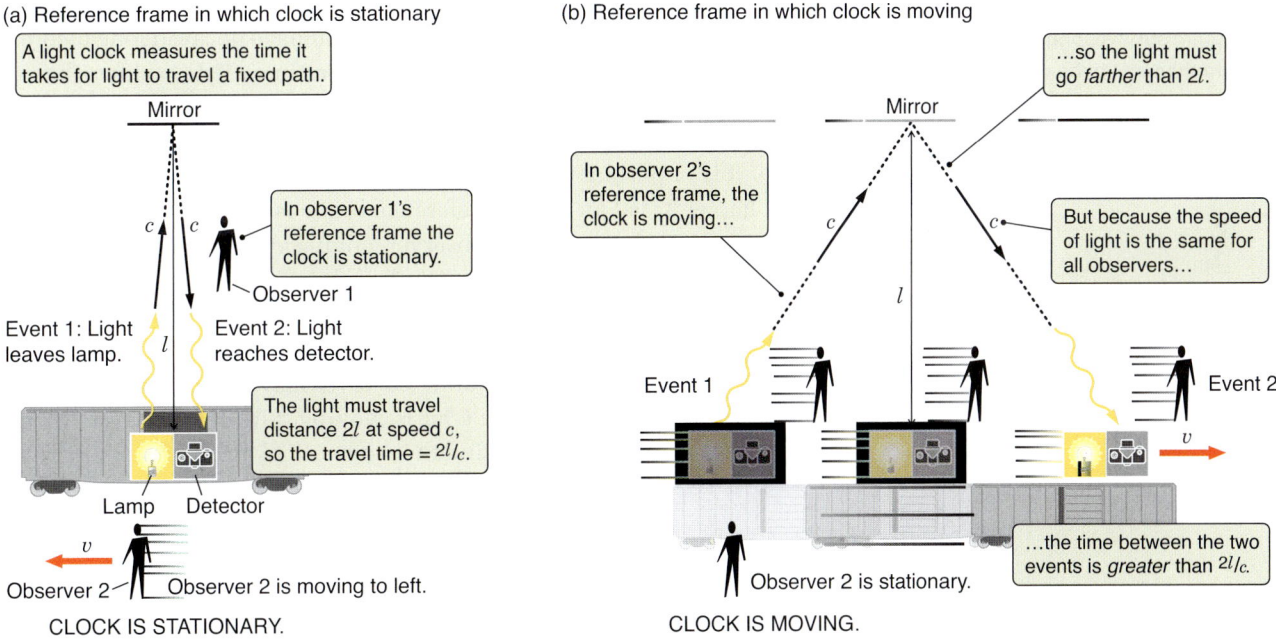

Figure 13.14 The "tick" of a light clock is different when seen in two different reference frames: (a) stationary and (b) moving. As Einstein's thought experiment demonstrates, if the speed of light is the same for every observer, then moving clocks *must* run slow.

measure the speed of light to be the same is if *the passage of time is different from one observer to the next*. For moving observers, the time is stretched out, so that each second is longer, a phenomenon known as **time dilation**. In our everyday lives, the march of time seems unchanging and absolute. But Einstein discovered that time flows differently for different observers.

By the time Einstein finished working out the implications of his insight, he had given time and space equal footing and reshaped the Newtonian three-dimensional

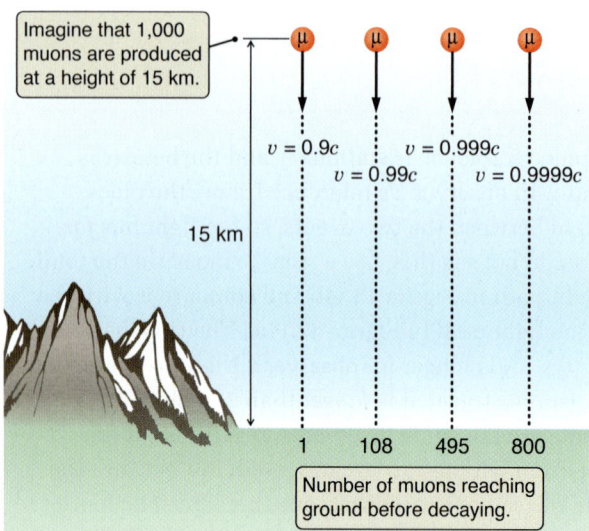

Figure 13.15 Muons created by cosmic rays high in Earth's atmosphere decay long before reaching the ground if they are not traveling at nearly the speed of light. Here we show what happens to 1,000 muons produced at an altitude of 15 km for a variety of speeds. Faster muons have slower clocks, so more of them survive long enough to reach the ground—many more than would be expected simply due to the faster speed.

universe into a four-dimensional combination of the three dimensions of space and the one dimension of time called **spacetime**. Events occur at specific locations within this four-dimensional spacetime, but how this spacetime is divided into what we perceive as "space" and what we perceive as "time" depends on our reference frame.

Einstein did not "disprove" Newtonian physics. We were not wasting our time in Chapter 3 when we studied Newton's laws of motion, because at speeds much less than the speed of light, Einstein's equations become the same equations that describe Newtonian physics. In our everyday lives, we experience a Newtonian world because we never encounter speeds approaching that of light. Even the fastest object ever made by humans (the *Helios* spacecraft) traveled at only about $0.00023c$. Only when objects approach the speed of light do our observations begin to depart noticeably from the predictions of Newtonian physics. When great velocities cause an effect different from what Newtonian physics predicts, we say a **relativistic** effect has occurred.

But can it possibly be true? Can time really pass differently for objects traveling at different speeds or is that just how it "seems"? A scientific observation demonstrates this effect. As illustrated in **Figure 13.15**, fast particles called cosmic-ray muons are produced 15 km up in Earth's atmosphere when high-energy cosmic rays strike atmospheric atoms or molecules. Muons at rest decay very rapidly into other particles. This happens so quickly that, even if they *could* move at the speed of light, virtually all muons would have decayed long before traveling the 15 km to reach Earth's surface. However, time dilation slows the muons' clocks, so the particles travel farther during their longer lifetime and reach the ground. Many other experiments have also confirmed this result.

It is important to note that no inertial reference frame is special. (Recall from Chapter 3 that an inertial reference frame is one that does not accelerate.) If you compared clocks with an observer moving at 9/10 the speed of light ($0.9c$) relative to you, you would find that the other observer's clock was running 0.44 times as fast as your clock. You might guess that to the other observer, your clock would be fast; but actually the other observer would find instead that *your* clock was running slow. To you, the other observer may be moving at $0.9c$; but to the other observer, *you* are the one who is moving. Either frame of reference is equally valid, so you would each find the other's clock to be slow compared to your own. This symmetry holds as long as neither frame accelerates.

The Implications of Relativity

Today, special relativity shapes our thinking about the motions of both the tiniest subatomic particles and the most distant galaxies. As examples, we present only four of the implications of this theory:

1. **What we think of as "mass" and what we think of as "energy" are actually two manifestations of the same thing.** Einstein's famous equation $E = mc^2$ says that even a *stationary* object has an intrinsic "rest" energy that equals the mass (m) of the object multiplied by the speed of light (c) squared. The speed of light is a very large number. This relationship between mass and energy says that a single tablespoon of water has a rest energy equal to the energy released in the explosion of more than 300,000 tons of TNT. Recall that we used this relationship between mass and energy when studying in Chapter 11 the nuclear physics that makes stars shine.

2. **The speed of light is the ultimate speed limit.** There is not enough energy in the entire universe to accelerate even one electron to the speed of light. Although getting the electron arbitrarily close to that number—$0.99999999999999\ldots \times c$—is no problem (at least in principle), there is not enough energy available to accelerate the electron beyond that to the speed of light. Nothing travels faster than light. We use the fact that light travels at a constant speed in a vacuum whenever we use the light travel time to describe astronomical distances.
3. **Time passes more slowly in a moving reference frame.** For moving objects, the seconds are stretched out by the phenomenon called time dilation. This effect is important in particle physics, such as the muons we discussed earlier. But it is also important to future space travel. This effect is represented (and misrepresented) in science fiction books and movies.
4. **An object is shorter (in the direction of motion only) than it is at rest.** Moving objects are compressed in the direction of their motion, a phenomenon referred to as **length contraction**. A meter stick moving at $0.9c$ is only 0.44 meter long. Both time dilation and length contraction are important effects in the strong gravitational fields near neutron stars and black holes.

13.5 Gravity Is a Distortion of Spacetime

Our exploration of special relativity began with the observation that the speed of light is always the same regardless of the motion of an observer or the source. This phenomenon changed the way people understood space and time. However, even this four-dimensional spacetime is itself warped and distorted by the masses it contains. In this section, you will learn about the general theory of relativity, which describes how mass affects space and time.

Free Fall and Free Float

Special relativity tells us that any inertial reference frame is as good as any other. There is no way to distinguish between sitting in an enclosed spaceship floating stationary in deep space and sitting in an enclosed spaceship traveling through our galaxy at a constant speed of 0.99999 times the speed of light. **Figure 13.16a and b** illustrates this point. These two situations do not feel any different because there is no difference between them. Each is an equally valid inertial reference frame. As long as nothing accelerates either spaceship, the laws of physics are exactly the same inside both spacecraft. But what if there is an acceleration?

Consider an astronaut inside a spaceship orbiting Earth, as shown in Figure 13.16c. This spaceship is accelerating, as the direction of its velocity is constantly changing as it orbits Earth. Still, the astronaut has no way to tell the difference between being inside the spaceship as it falls around Earth and being inside a spaceship floating through interstellar space. Even though its velocity is constantly changing as it falls, *the inside of a spaceship orbiting Earth is an inertial frame of reference just as an object drifting along a straight line through interstellar space is an inertial reference frame.* This is why an astronaut can place a wrench in the air next to him, and then pick it up again later. The reference frame is inertial, so the wrench just "floats" there and doesn't fall to the deck of the spacecraft. This statement that "free fall is the same as free float" is called the **equivalence principle**. If you close your eyes and jump off a diving board, for the brief time that you are falling freely through Earth's gravitational field,

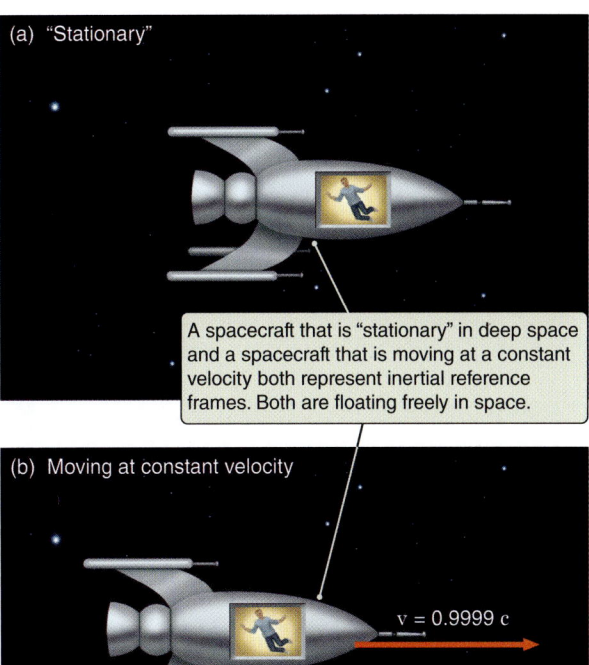

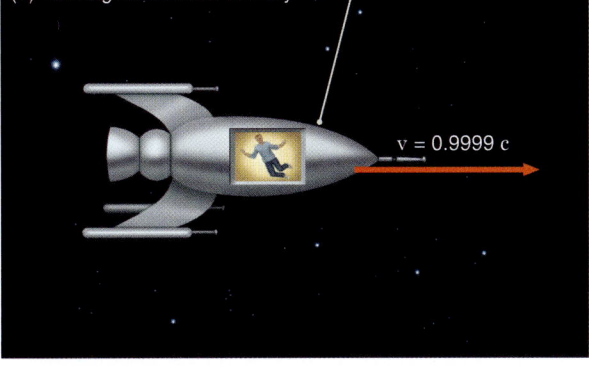

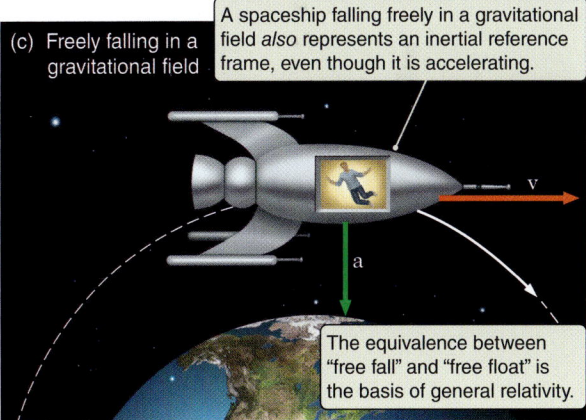

Figure 13.16 Special relativity says that there is no difference between (a) a reference frame that is floating "stationary" in space and (b) one that is moving through the galaxy at constant velocity. General relativity adds that there is no difference between these inertial reference frames and (c) an inertial reference frame that is falling freely in a gravitational field. Free fall is the same as free float, as far as the laws of physics are concerned.

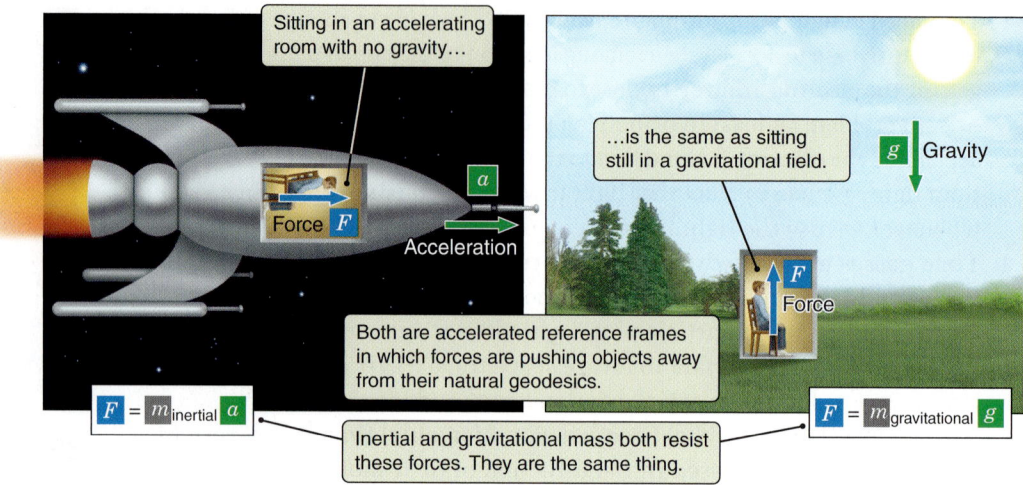

Figure 13.17 According to the equivalence principle, sitting in a spaceship accelerating at 9.8 m/s² feels the same as sitting still on Earth.

the sensation you feel is exactly the same as the sensation that you would feel floating in interstellar space.

The natural path that an object will follow through spacetime in the absence of other forces is referred to as the object's **geodesic**. In the absence of a gravitational field, an object's geodesic is a straight line, which is Newton's first law: An object will move at a constant speed in a constant direction, unless acted on by an unbalanced external force. However, the shape of spacetime becomes distorted in the presence of mass, so an object's geodesic becomes curved. Rather than thinking of gravity as a "force" that "acts on" objects, it is more accurate to say that *gravitation results from the shape of spacetime that objects move through.* The warping of spacetime leads directly to the gravity that holds you to Earth. This theory, called the **general theory of relativity**, or sometimes just "general relativity," describes how mass distorts spacetime and is another of Einstein's great contributions to science.

Imagine you are in a box inside a spaceship that is accelerating through deep space at a rate of 9.8 meters per second per second (m/s²) in the direction of the arrow shown in **Figure 13.17**. The floor of the box pushes on you to overcome your inertia and causes you to accelerate at 9.8 m/s², so you feel as though you are being pushed into the floor of the box. Now imagine instead that you are sitting in a closed box on the surface of Earth. The floor of the box pushes on you to keep you from following your curved geodesic. Again, you feel as though you are being pushed into the floor of the box.

According to the equivalence principle, the two cases are identical. There is no difference between sitting in an armchair in a spaceship with an acceleration of 9.8 m/s² and sitting in an armchair on the surface of Earth reading this book. In the first case, the force of the spaceship is pushing you away from your "floating" straight-line geodesic through spacetime. In the second case, Earth's surface is pushing you away from your curved "falling" geodesic through a spacetime that has been distorted by the mass of Earth. An acceleration is an acceleration, regardless of whether you are being accelerated off a straight-line geodesic through deep space or being accelerated off a "falling" geodesic in the gravitational field of Earth.

There is an important caveat to the equivalence principle. In an accelerated reference frame such as an accelerating spaceship, the same acceleration is experienced everywhere. In contrast, the curvature of space by a massive object changes from place to place. Tidal forces are one result of increased curvature of space closer to a gravitating body. A more careful statement of the equivalence principle is that the effects of gravity and acceleration are indistinguishable locally; that is, as long as we restrict our attention to small enough volumes of space so that changes in gravity can be ignored.

Spacetime as a Rubber Sheet

The general theory of relativity describes how mass distorts the geometry of spacetime. Imagine the surface of a tightly stretched, flat rubber sheet. A marble will roll in a straight line across the sheet. Euclidean geometry, the geometry of everyday life, applies on the surface of the sheet: If you draw a circle, the circumference is equal to 2π times its radius, r; if you draw a triangle, the angles add up to 180°; lines that are parallel anywhere are parallel everywhere.

Now place a bowling ball in the middle of the rubber sheet, creating a "well," as in **Figure 13.18**. The sheet will be stretched and distorted. If you roll a marble across the sheet, its path curves (Figure 13.18a). If you draw a circle around the bowling ball, the circumference is less than $2\pi r$ (Figure 13.18b). If you draw a triangle, the angles add up to more than 180° (Figure 13.18c). The surface of the sheet is no longer flat, and Euclidean geometry no longer applies.

Similarly, mass distorts spacetime, changing the distance between any two locations or events. You can imagine, at least in principle, stretching a rope all the way around the circumference of a circular orbit about the Sun approximately along Earth's orbit and then comparing the length of that rope with the length of a rope taken from the orbit to the center of the Sun. You might expect to find that the circumference of the orbit is equal to 2π times the radius of the orbit, just like a circle drawn on a flat piece of paper. If you carried out this experiment, however, you would find that the circumference of the orbit was 10 km shorter than 2π times the radius.

We can visualize how a rubber sheet with a bowling ball on it is stretched through a third spatial dimension, but it is impossible for most people to visualize what a curved four-dimensional spacetime would "look like." Yet experiments verify that the geometry of our four-dimensional spacetime is distorted much like the rubber sheet, whether or not we can easily picture it.

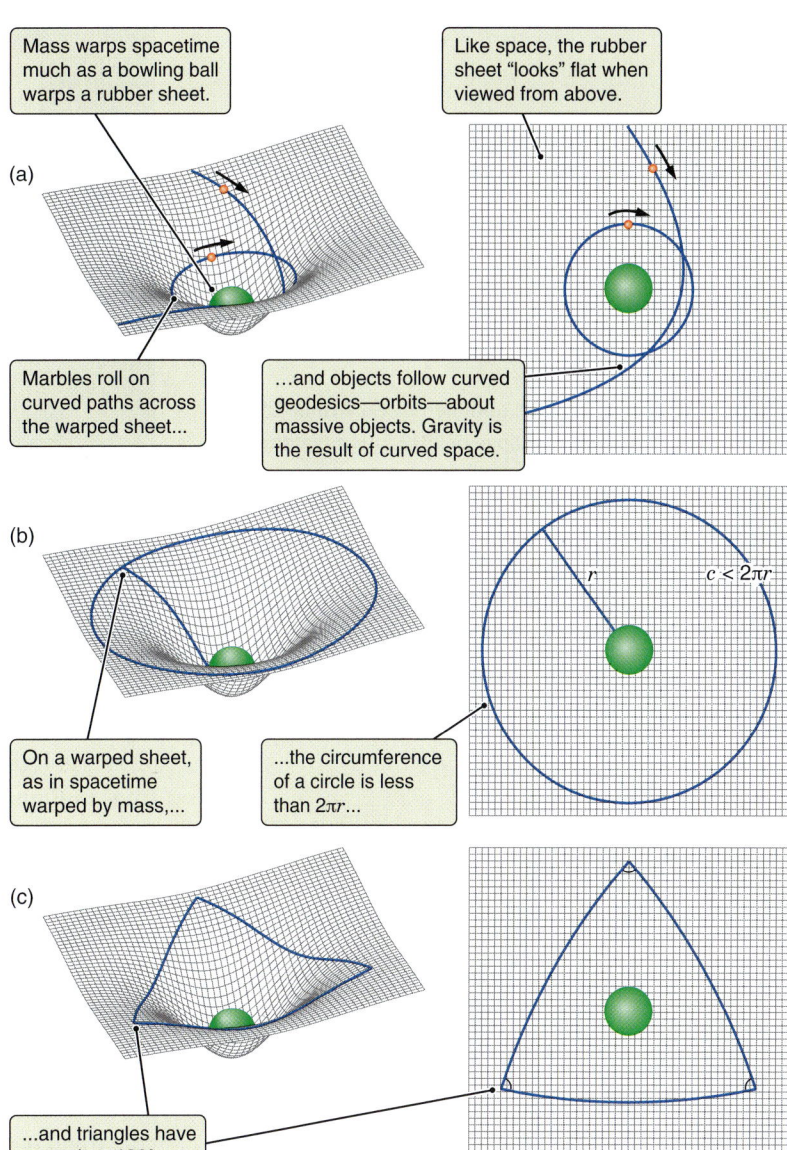

Figure 13.18 Mass warps the geometry of spacetime in much the same way that a bowling ball warps the surface of a stretched rubber sheet. This distortion of spacetime has many consequences; for example, (a) objects follow curved paths or geodesics through curved spacetime, (b) the circumference of a circle around a massive object is less than 2π times the radius of the circle, and (c) angles in triangles do not add to exactly 180°.

The Observable Consequences of General Relativity

Curved spacetime does have observable consequences. General relativity predicts that an elliptical orbit, in which the planet swings in closer and then farther from the Sun, should *precess*; that is, the long axis should slowly change its direction. For example, Newton's law of gravity predicts the major component of Mercury's precession, but even after taking this into account, there remains a very small component equal to 43 arcseconds per century that cannot be explained by Newton's laws alone. This discrepancy is explained by general relativity.

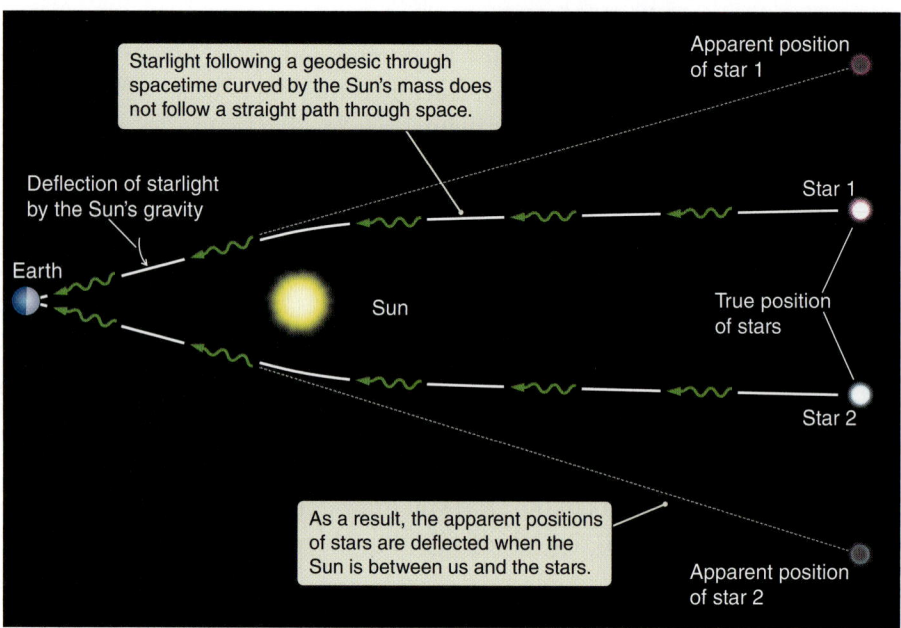

Figure 13.19 Measurements obtained by Sir Arthur Eddington during the total solar eclipse of 1919 found that the gravity of the Sun bends the light from distant stars by the amount predicted by Einstein's general theory of relativity. This is an example of gravitational lensing. Note that the "triangle" formed by the solid white lines connecting Earth and the two stars contains more than 180°, just like the triangle in Figure 13.18c.

The real-life equivalent of the triangle with more than 180° is easier to visualize. As light travels through distorted spacetime, its path bends, just like the lines in Figure 13.18c are bent by the curvature of the sheet. This bending of the light path as the light passes through bent spacetime is called **gravitational lensing** because lenses also bend light paths. But how could you ever measure this?

Several months before the total solar eclipse of 1919, Sir Arthur Stanley Eddington (1882–1944) measured the positions of stars in the direction of the sky where the eclipse would occur. He then repeated the measurements during the eclipse. **Figure 13.19** shows how the light from distant stars curved as it passed the Sun, causing the measured positions of the stars to shift outward. The stars appeared farther apart in Eddington's second measurement than in his first, as predicted by general relativity. During the eclipse, the triangle formed by Earth and any two stars contained more than 180°—just like the triangle on the surface of our rubber sheet. More recently, gravitational lensing has been used to search for unseen massive objects adrift in space: planets ejected from their star systems, and more massive objects like low-luminosity galaxies or the dark matter that surrounds them. The gravity from these objects distorts the light from background objects noticeably.

Mass distorts the geometry of time as well. Deep within the gravitational field of a massive object, clocks run more slowly from the perspective of a distant observer. This effect is called **general relativistic time dilation**. Suppose a light is attached to a clock sitting on the surface of a neutron star. The light is timed so that it flashes once a second. Because time near the surface of the star is dilated, an observer far from the neutron star perceives the light to be pulsing with a lower frequency—less than once a second. Now suppose there is an emission-line source on the surface of the neutron star. Because time is running slowly on the surface of the neutron star, the light that reaches the distant observer will have a lower frequency than when it was emitted. A lower frequency means a longer wavelength. So the light from the source will be seen at a longer, redder wavelength than the wavelength at which it was emitted.

The reddening of light as it climbs out of a gravitational well, shown in **Figure 13.20**, is called the **gravitational redshift** because the wavelengths of light from objects deep within a gravitational well are shifted to longer, redder wavelengths. The effect of gravitational redshift is similar to the Doppler redshift. In fact, there is no way to tell the difference between light that has been redshifted by gravity and light that has been Doppler redshifted.

Bringing this phenomenon a bit closer to home, a clock on the top of Mount Everest runs faster, gaining about 80 nanoseconds (80 billionths of a second) a day compared with a clock at sea level. The difference between an object on the surface of Earth and an object in orbit is even greater. Satellites in orbit travel quickly enough that special relativistic effects are measurable. Even after allowing for the slowing due to special relativity, the clocks on the satellites that make up the global positioning system (GPS) run faster than clocks on the surface of Earth. If the satellite clocks and your GPS receiver did not correct for this, then the position your GPS receiver reported would be in error by up to half a kilometer within a single hour. The fact that the GPS works is actually a strong experimental confirmation of two predictions of general relativity—gravitational redshift and general relativistic time dilation.

If you exert a force on the surface of a rubber sheet, accelerating it downward, waves will move away from where you struck it, something like ripples spreading out over the surface of a pond. Similarly, the equations of general relativity predict that if you accelerate the fabric of spacetime (for example, with the catastrophic collapse of a high-mass star), then waves in spacetime, or **gravitational waves**, will move outward at the speed of light. These gravitational waves are like electromagnetic waves in some respects. Accelerating an electrically charged particle gives rise to an electromagnetic wave. Accelerating a massive object gives rise to gravitational waves.

Gravitational waves have not yet been directly observed, but there is strong circumstantial evidence for their existence. General relativity predicts that the orbit of binary neutron stars should lose energy, which will be carried away as gravitational waves. In 1974, astronomers discovered a binary system of two neutron stars, one of which is an observable pulsar. Using the pulsar as a precise clock, astronomers accurately measured the orbits of both stars. The orbits are gradually losing energy, at the rate predicted by general relativity. The results were later duplicated with a similar binary pair with smaller obits. In 2014, astronomers announced the discovery of the signature of gravitational waves in the light coming from the very early universe—a very different avenue of investigation. These measurements from different fields of astrophysics very strongly suggest that gravitational waves exist.

Stop for a moment to recall the process of science discussed in Chapter 1. The General Theory of Relativity predicts that gravitational waves exist. This theory is scientific, because it is falsifiable. So far, the evidence of gravitational waves is circumstantial, not direct. Astrophysicists have developed a new kind of observatory, called Advanced LIGO (Laser Interferometer Gravitational-Wave

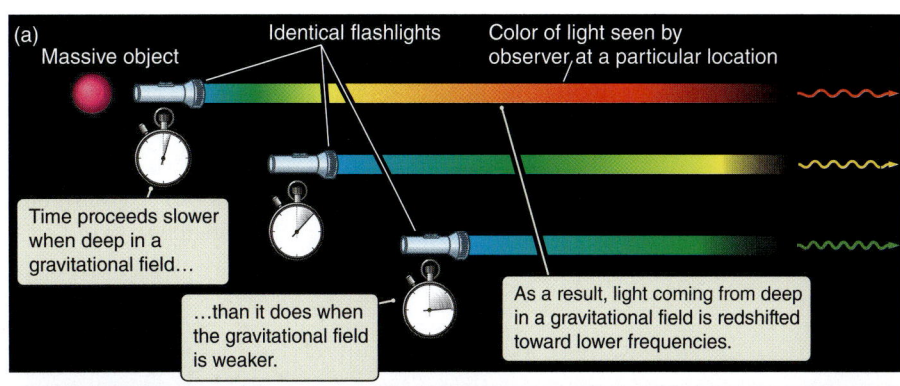
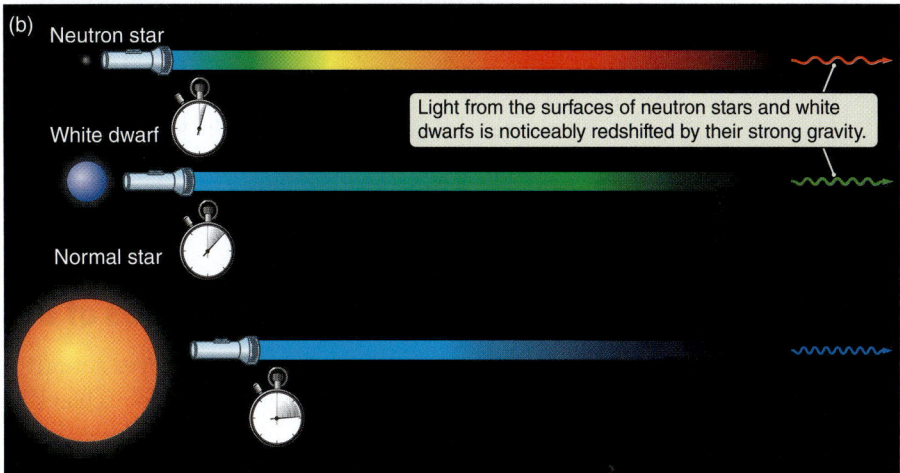

Figure 13.20 Time passes more slowly near massive objects because of the curvature of spacetime. As a result, to a distant observer light from near a massive object will have a lower frequency and longer wavelength. (a) The closer the source of radiation is to the object or (b) the more massive and compact the object is, the greater the gravitational redshift will be.

Vocabulary Alert

infinite In common language, this word sometimes just means "really, really big" or "really, really far." But astronomers mean infinite in a mathematical sense—a quantity without limit or boundary. If we say something is infinite, there is no beyond.

Observatory), to attempt to directly detect gravitational waves. If the waves are observed, the theory will be further supported. If the waves are not observed, some changes to the theory will be required to explain the result. A fundamental principle of science is that it self-corrects, by testing theories to see if they fail. As we discuss the parts of the universe where our intuition can no longer guide our understanding, it will be important to keep this particular strength of science in mind.

13.6 Black Holes Are a Natural Limit

An object on the surface of a rubber sheet causes a funnel-shaped distortion that is analogous to the distortion of spacetime by a mass. Now imagine the limit in which the funnel is **infinitely** deep—it gets narrower as we go deeper, but it has no bottom. This is the rubber-sheet analog to a black hole. The mathematics describing the shape of a black hole fail in the same way that the mathematical expression $1/x$ fails when $x = 0$. Such a mathematical anomaly is called a **singularity**. Black holes are singularities in spacetime.

Properties of Black Holes

We can never actually "see" the singularity at the center of a black hole. As the distance from a black hole decreases, the escape velocity increases, until the velocity reaches the speed of light—at this distance from the center of the black hole, even light cannot escape. The radius where the escape velocity equals the speed of light is called the **Schwarzschild radius**, named for physicist Karl Schwarzschild (1873–1916), and it is proportional to the mass of the black hole (**Working It Out 13.2**). The sphere around the black hole at this distance is called its **event horizon**. **Figure 13.21a** shows a rubber-sheet analog to a black hole, with the Schwarzschild radius and the event horizon. A black hole with a mass of 1 $M_\odot$ has a Schwarzschild radius of about 3 km. If Earth were squeezed into a black hole, it would have a Schwarzschild radius of only about a centimeter.

A black hole has only three properties: mass, electric charge, and angular momentum. The mass of a black hole determines the Schwarzschild radius. The electric charge of a black hole is the net electric charge of the matter that fell

---

Working It Out 13.2 | **Finding the Schwarzschild Radius**

The mass of a black hole determines the Schwarzschild radius:

$$R_S = \frac{2GM_{BH}}{c^2}$$

where G is the universal constant of gravitation (6.67×10^{-11} m^3/kg s^2), and c is the speed of light. Inserting values for the constants and expressing the mass in terms of the mass of the Sun gives

$$R_S = 3 \text{ km} \frac{M_{BH}}{M_\odot}$$

This relationship shows that the Schwarzschild radius is proportional to the mass, expressed in solar masses. The constant of proportionality is 3 km. A 1-$M_\odot$ black hole has a Schwarzschild radius of about 3 km. To put that in perspective, at average walking pace, it would take about 40 minutes to walk 3 km. A 3-$M_\odot$ black hole has a Schwarzschild radius of about 9 km, and a 5-$M_\odot$ black hole has a Schwarzschild radius of about 15 km.

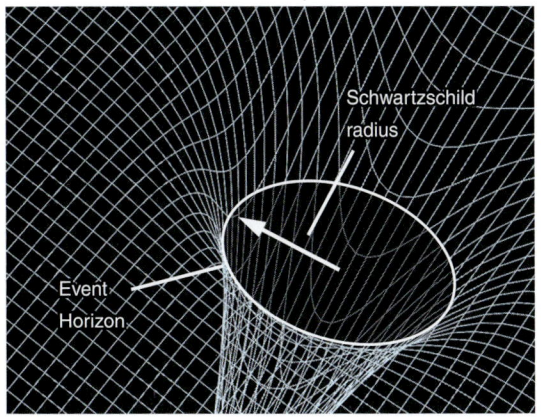
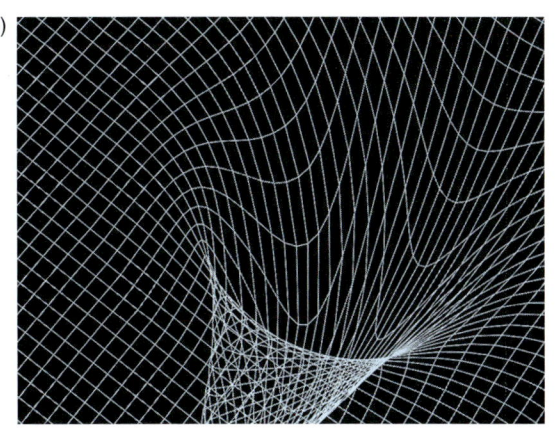

Figure 13.21 (a) A black hole's size is determined by the Schwarzschild radius and the corresponding event horizon. This image is a two-dimensional analogy for a black hole. In reality, the event horizon is a sphere. (b) If the object that formed the black hole was spinning, its angular momentum is conserved, and the black hole twists the spacetime around it.

into it. The angular momentum of a black hole twists spacetime around it (Figure 13.21b). Apart from these three properties, all information about the material that fell into the black hole is lost. Nothing of its former composition, structure, or history survives.

Imagine that an adventurer journeys into a black hole, as illustrated in **Figure 13.22**. From our perspective outside the black hole, we would see our adventurer fall toward the event horizon; but as she did, her watch would run more and more slowly and her progress toward the event horizon would slow as well. At the event horizon, the gravitational redshift becomes infinite, and clocks stop altogether. Our adventurer would approach the event horizon, but from our perspective she would never quite make it. Yet the adventurer's experience would be very different. From her perspective, there would be nothing special about the event horizon at all. She would fall past the event horizon and on, deeper into the black hole's gravitational well.

Actually, we have overlooked a rather crucial fact. Our intrepid explorer would have been torn to shreds long before she reached the black hole. Near the event horizon of a 3-$M_\odot$ black hole, the difference in gravitational acceleration between our explorer's feet and her head would be about a billion times her gravitational acceleration on the surface of Earth. In other words, her feet would be accelerating a billion times faster than her head. This is not an experiment we would ever want to perform. Although scientific theories must produce testable predictions, not all individual predictions have to be tested directly.

"Seeing" Black Holes

So far, the strongest direct evidence for black holes that result from supernovae comes from X-ray binary stars. The radio emission from Cygnus X-1 flickers rapidly, changing in as little as 0.01 second. This means that the source of the X-rays must be smaller than the distance that light travels in 0.01 second, or 3,000 km. Thus, the source of X-rays in Cygnus X-1 must be smaller than Earth. Cygnus X-1 was also identified with both a radio source and with an already cataloged optical star called HD 226868. The spectrum of HD 226868 shows that it is a normal B0 supergiant star with a mass of about 30 $M_\odot$, far too cool to produce X-ray emission. But HD 226868 is part of a binary system with a period of 5.6 days. Orbit analysis shows that the mass of the unseen compact companion of HD 226868 must be at least 6 $M_\odot$. The companion to HD 226868 is too small to be a normal star, yet it is much more massive than the Chandrasekhar limit for a white dwarf or a neutron star. Such an object can only be a black hole. The X-ray emission from

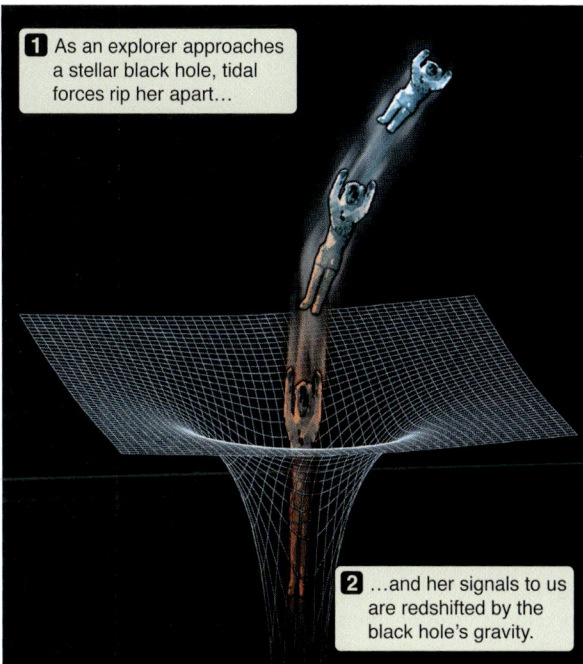

Figure 13.22 An adventurer falling into a black hole would be "spaghettified" by the extreme tidal forces.

352 CHAPTER 13 *Evolution of High-Mass Stars*

Figure 13.23 This artist's rendering of the Cygnus X-1 binary system shows material from the B0 supergiant being pulled off and falling onto an accretion disk surrounding the black hole, thereby producing X-ray emission.

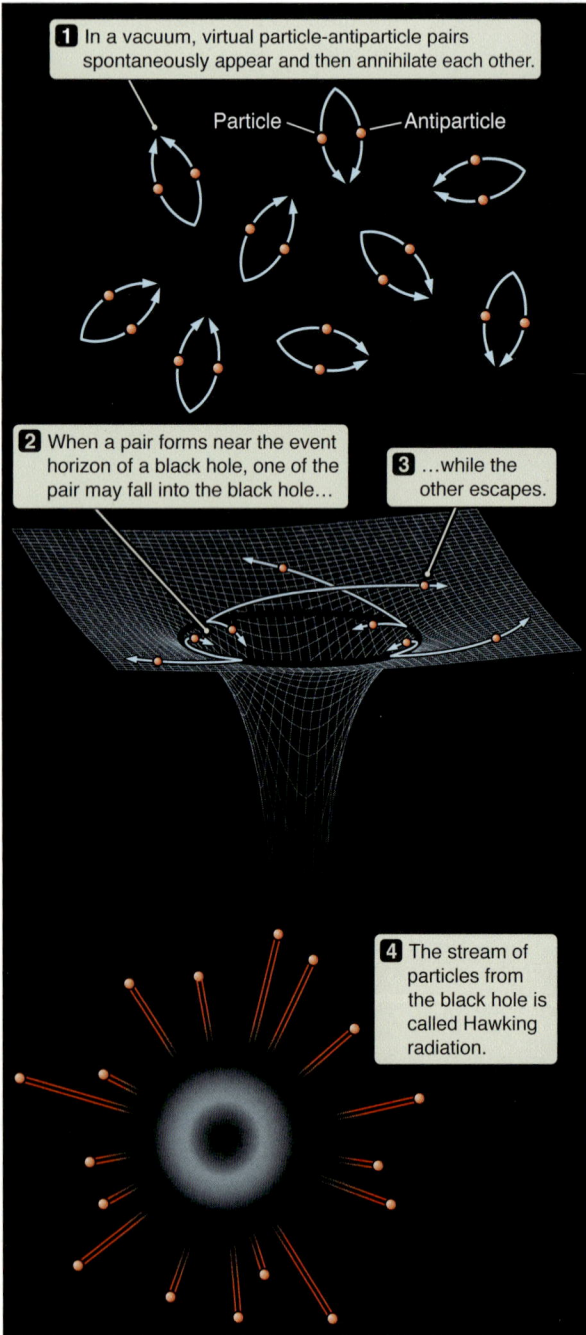

Figure 13.24 In the vacuum of empty space, particles and antiparticles are constantly being created and then annihilating each other. Near the event horizon of a black hole, however, one particle may cross the horizon before it recombines with its partner. The remaining particle leaves the black hole as light.

Cygnus X-1 arises when material from the B0 supergiant falls onto an accretion disk surrounding the black hole, as illustrated in **Figure 13.23**.

Dozens of other good candidates for stellar-mass black holes have been discovered. One such object is a rapidly varying X-ray source in our companion galaxy, the Large Magellanic Cloud. Called LMC X-3, this X-ray source orbits a B3 main-sequence star every 1.7 days, and the compact source must have a mass of at least 9 $M_\odot$. Although the evidence that these systems contain black holes is circumstantial, the arguments that lead to this conclusion seem airtight. With reports of dozens of compelling examples of such objects, the evidence is in. Black holes, once regarded as nothing more than a bizarre quirk of the mathematics describing gravitation and spacetime, exist in nature.

The most energetic explosions in the universe are likely related to black holes. Gamma-ray bursts, or GRBs, are intense bursts of gamma rays, followed by a weaker afterglow. GRBs come from all directions in the sky. Because they are not concentrated in the plane of the Milky Way, they must be so distant that they lie outside of our own galaxy. Scientists think that those of short duration—less than 2 seconds—are probably the result of two dense objects merging; either two neutron stars or a neutron star and a black hole. Longer-duration GRBs are produced after a supernova event when the rapidly spinning star collapses to form a black hole.

Can black holes ever go away? In 1974, Stephen Hawking realized that black holes should be sources of radiation. In the ordinary vacuum of empty space, particles and their antiparticle "mates" appear and then, within 10^{-21} seconds, annihilate each other. If this happens near the event horizon of a black hole, as shown in **Figure 13.24**, then one of the particles might fall into the black hole while the other particle escapes. Hawking showed that through this process, a black hole should emit a blackbody spectrum and that the effective temperature of this spectrum would increase as the black hole became smaller through this "evaporation" process. After a very, very long time (of the order 10^{61} years for a black hole with the mass of the Sun), the black hole would become small enough that it would become unstable and explode. Although the light that emerges, known as **Hawking radiation**, is of considerable interest to physicists and astronomers, in a practical sense the low intensity of Hawking radiation means it is not a useful way to "see" a black hole.

READING ASTRONOMY News

One of the great things about scientists is that when they're wrong, they get excited about it. Consider the following article about massive stars, in which a long-held belief about the largest possible size of a star is overturned by new discoveries.

What a Scorcher—Hotter, Heavier, and Millions of Times Brighter than the Sun

By **Ian Sample**, *The Guardian*

They are the most colossal stars ever seen and live short, bright, lives in faraway reaches of space before exploding in a blaze of glory.

One of the stars, now tagged R136a1, is estimated to weigh 265 times more than the Sun and to shine millions of times more brightly. Were it to replace our own star, the intensity of its rays would sterilise the Earth, leaving it lifeless.

British astronomers spotted the stars, more massive than any others on record, using the Very Large Telescope, an aptly named observatory on a mountaintop in the Atacama Desert of northern Chile.

The discovery of the stellar giants has prompted astronomers to scrap the upper limits they set on star formation, which suggested it was almost impossible for a star to grow to more than 150 times the mass of the Sun.

The team, led by Paul Crowther, an astrophysicist at Sheffield University, searched two regions of space for massive stars. The first region, known as NGC 3603, is a stellar nursery 22,000 light-years away in a region of the Milky Way called the Carina spiral arm.

The second target, RMC 136a, is a cloud of gas and dust, 165,000 light-years away in the Tarantula nebula of our neighbouring galaxy, the Large Magellanic Cloud. The astronomers were able to distinguish individual stars using exquisitely sensitive infrared instruments on the telescope and take measurements of their brightness and mass.

At least three stars examined in the first region of space weighed in at about 150 times the mass of the Sun. The record-breaking star, R136a1, was found in the second region. When born, the star could have been a staggering 320 times more massive than the Sun.

Several of the stars were found to have surface temperatures above 40,000° C, which is more than seven times hotter than the Sun.

"These stars are born heavy and lose weight as they age," said Crowther. "Being a little over a million years old, the most extreme star R136a1 is already middle-aged and has undergone an intense weight loss programme, shedding a fifth of its initial mass over that time. Owing to the rarity of these monsters I think it unlikely this new record will be broken any time soon."

If R136a1 were in our own Solar System, it would outshine the Sun as much as the Sun outshines the full Moon, the scientists said. The mass of the star is so great that it would reduce the length of an Earth year—the time it takes to circle the star—to just 3 weeks. "It would [also] bathe Earth in incredibly intense ultraviolet radiation, rendering life on our planet impossible," said Raphael Hirschi, a member of the team at Keele University.

While the latest crop of stars are the most massive and heaviest ever spotted, they are not the largest. The biggest star in the group, R136a1, is roughly 30 times as wide as the Sun. Another kind of star, known as a super red giant, can grow to many hundreds of times that size—though is considerably lighter, at only 10 times the mass of the Sun.

It is unlikely that any "alien" planets circle the massive stars that Crowther's team has studied. Radiation from the stars would obliterate any nearby cosmic material that could become compact enough to be a planet. Even if some remained, planets would take longer to form than the entire life span of a massive star.

Crowther said: "We don't really know what happens when these massive stars reach the end of their lives. When some big stars die, their cores implode and they become neutron stars or black holes, but these might be different. They might blow up in a spectacular supernova and leave no remnants behind at all." The explosions could fling the weight of 10 Suns' worth of iron into space.

The team's observations reveal what the early universe might have looked like, when many of the first stars to be born might have been cosmic monsters like R136a1.

Before the latest discovery, the most massive star known was the Peony nebula star, which, at about 175 times the mass of the Sun, could still hold the record for our own galaxy. Details of the discovery are reported in the monthly notices of the Royal Astronomical Society.

Evaluating the News

1. How much more massive is R136a1 than previous estimates of the maximum mass of a star? Is this a significant difference? What if R136a1 were only 155 times the mass of the Sun? Would the same conclusions apply?
2. In the title, the reporter states that the star is "hotter" and "millions of times brighter"

SEE SCORCHER

Scorcher (cont.)

than the Sun. Explain how this information would lead to a conclusion that the star is more massive.

3. Paul Crowther states in the article that the star is "a little over a million years old." How might he know that?

4. If Paul Crowther had stated that the star was 5 billion years old, you could immediately say, "Baloney!" Why?

5. "Stellar nurseries" are the ideal place to look for the most massive stars. Why?

6. The article states that "If R136a1 were in our own Solar System . . . The mass of the star is so great that it would reduce the length of an Earth year . . . to just 3 weeks." What physical law are the astrophysicists using to estimate this?

✦ SUMMARY

As high-mass stars evolve, their interiors form concentric shells of progressive nuclear burning. Once they leave the main sequence, they may pass through the instability strip and become pulsating variable stars. The chain of nuclear fusion reactions consists of increasingly shorter stages of burning, resulting in more massive elements up to iron. High-mass stars eventually explode as Type II supernovae, which eject newly formed massive elements into interstellar space. Some high-mass stars leave behind neutron stars, and the most massive high-mass stars leave behind black holes. In the environment surrounding black holes, relativistic effects become important. Time runs more slowly and objects deep in a black hole's gravitational well appear redshifted to an external observer. A black hole's mass determines its Schwarzschild radius: the boundary from which light cannot escape.

LG 1 The larger masses of high-mass stars allow them to fuse heavier elements than those produced in low-mass stars. This leads to a very different and more violent death that leaves massive cores behind.

LG 2 Evolving high-mass stars leave the main sequence as they burn heavier elements. Once an iron core is produced, the star becomes unstable and the core collapses, heating the material to cause photodisintegration of the iron nuclei and the merging of protons and electrons into neutrons. The outer layers bounce off of the dense core and produce a shock wave that travels outward. This shock wave causes neutrons to penetrate atomic nuclei and form more massive elements.

LG 3 The speed of light in a vacuum is the same for all observers. This is the basis of relativity, which has three other profound implications: mass and energy are equivalent; when traveling very fast or in a strong gravitational field, time runs slower; and when traveling very fast or in a strong gravitational field, lengths are shorter.

LG 4 The supernova explosion that ends the life of a massive star leaves behind a neutron star or a black hole. The mathematical singularity at the center of a black hole is still a mystery to science. However, observational evidence for black holes is very strong; scientists have identified many objects that have strong gravity but are too small to be normal matter. Black holes may, after a very, very long time, be destroyed by evaporation through Hawking radiation.

SUMMARY SELF-TEST

1. The interior of an evolved high-mass star has layers like an onion because
   a. heavier atoms sink to the bottom because stars are not solid.
   b. before the star formed, heavier atoms accumulated in the centers of clouds because of gravity.
   c. heavier atoms fuse closer to the center because the temperature and pressure are higher there.
   d. different energy transport mechanisms occur at different densities.

2. Place in order the stages of nuclear burning that evolving high-mass stars experience.
   a. helium   b. neon
   c. oxygen   d. silicon
   e. hydrogen   f. carbon

3. Iron fusion cannot support a star because
   a. iron oxidizes too quickly.
   b. iron absorbs energy when it fuses.
   c. iron emits energy when it fuses.
   d. iron is not dense enough to hold up the layers.

4. Which of the following are possible consequences of distortions of spacetime caused by mass?
   a. time dilation   b. gravity
   c. length contraction   d. tidal forces
   e. gravitational lensing f. precession

5. An astronaut who fell into a black hole would be stretched because
   a. the gravity is so strong.
   b. the gravity changes dramatically over a short distance.
   c. time is slower near the event horizon.
   d. black holes rotate rapidly, dragging spacetime with them.

6. Elements heavier than iron originate
   a. in the Big Bang.
   b. in the cores of low-mass stars.
   c. in the cores of high-mass stars.
   d. in the explosions of high-mass stars.

7. A pulsar "pulses" because
   a. its spin axis crosses our line of sight.
   b. it spins.
   c. it has a strong magnetic field.
   d. its magnetic axis crosses our line of sight.

8. X-ray binaries are similar to another type of system we have studied. This system is
   a. the Solar System.
   b. progenitors of Type Ia supernovae.
   c. progenitors of Type II supernovae.
   d. progenitors of planetary nebulae.

9. The fact that the speed of light is a universal constant forces us completely to rethink classical physics. This is an example of
   a. scientists always being completely wrong.
   b. the self-correcting nature of science.
   c. a theory becoming a hypothesis and then becoming a law.
   d. the universe changing with time.

10. If the Sun were replaced by a 1-$M_\odot$ black hole,
    a. the Solar System would collapse into it.
    b. the planets would remain in their orbits, but smaller objects would be sucked in.
    c. small objects would remain in their orbits, but larger objects would be sucked in.
    d. all objects in the Solar System would remain in their orbits.

QUESTIONS AND PROBLEMS

Multiple Choice and True/False

11. **T/F:** The end result of the CNO cycle is that four hydrogen nuclei become one helium nucleus.

12. **T/F:** When iron fuses into heavier elements, it produces energy.

13. **T/F:** Electrons and protons can combine to become neutrons.

14. **T/F:** A supernova can be as bright as its entire host galaxy.

15. **T/F:** A pulsar changes in brightness because its size pulsates.

16. In a high-mass star, hydrogen fusion occurs via the
    a. proton-proton chain.   b. CNO cycle.
    c. gravitational collapse.   d. spin-spin interaction.

17. The layers in a high-mass star occur roughly in order of
    a. atomic number.   b. decay rate.
    c. atomic abundance.   d. spin state.

18. Pulsations in a Cepheid variable star are controlled by
    a. the spin.
    b. the magnetic field.
    c. the ionization state of helium.
    d. the gravitational field.

19. Eta Carinae is an extreme example of
    a. a massive star.   b. a rotating star.
    c. a magnetized star.   d. a high-temperature star.

20. When photodisintegration starts in a star, a process begins that *always* results in a
    a. supernova.   b. neutron star.
    c. black hole.   d. pulsar.

21. Supernova remnants
    a. are viewable at all wavelengths.
    b. are viewable only at a few emission lines.
    c. are never seen in radio waves.
    d. have colors because the moving gas emits Doppler-shifted emission lines.

22. Imagine that two spaceships travel toward each other. One spaceship travels to the right at a speed of $0.9c$, and another travels to the left at $0.9c$. The pilot of the spaceship traveling right shoots a yellow laser at the spaceship traveling left. The pilot of the spaceship traveling left observes
    a. blue light traveling at c.
    b. blue light traveling at $1.9c$.
    c. yellow light traveling at c.
    d. yellow light traveling at $1.9c$.

23. Bill is standing in a small, windowless room. The force on his feet suddenly ceases. Which of the following conclusions can he draw?
    a. He is in an elevator, which is now accelerating downward.
    b. He is now in space, freely falling around Earth.
    c. The rocket he was in has stopped accelerating upward.
    d. Any of the above could be true.

24. If you could draw a very large circle in space near a black hole, its circumference
    a. would equal $2\pi r$.
    b. would be greater than $2\pi r$.
    c. would be less than $2\pi r$.
    d. can't be determined from the information given.

25. An object that passes near a black hole
    a. always falls in because of gravity.
    b. is sucked in because of vacuum pressure.
    c. is pushed in by photon pressure.
    d. is deflected by the curvature of spacetime.

Conceptual Questions

26. Look back at the chapter-opening illustration, which shows a supernova in M51. Write a brief obituary of the star that produced this supernova. Besides summarizing the important moments of its life, be sure to mention what the star leaves behind.

27. Why does the core of a high-mass star not become degenerate, as in the cases of low-mass stars?

28. For what two reasons does each post-helium-burning cycle for high-mass stars (carbon, neon, oxygen, silicon, and sulfur) become shorter than the preceding cycle?

29. Cepheids are highly luminous, variable stars in which the period of variability is directly related to luminosity. Explain why Cepheids are good indicators for determining stellar distances that lie beyond the limits of accurate parallax measurements.

30. Is it possible for a high-mass star to pass through the instability strip more than once? Use Figure 13.1 to defend your answer.

31. Identify and explain two important ways in which supernovae influence the formation and evolution of new stars.

32. Recordings show that SN 1987A was detected by neutrinos on February 23, 1987. About 3 hours later, it was detected in optical light. Why did this time delay occur?

33. Why can the accretion disk around a neutron star release so much more energy than the accretion disk around a white dwarf?

34. In Section 13.2, you learned that Type II supernovae blast material outward at 30,000 km/s, but the material in the Crab Nebula (Section 13.3) is expanding at only 1,500 km/s. What explains the difference?

35. An astronomer sees a redshift in the spectrum of an object. With no other information available, can she determine whether this is an extremely dense object (gravitational redshift) or one that is receding from us (Doppler redshift)? Explain your answer.

36. Suppose that in a speeding spaceship, the travelers are playing soccer with a perfectly round soccer ball. What is the shape of the ball according to observers outside the spacecraft?

37. Explain why more muons reach the ground traveling at $0.9999c$ than at $0.9c$.

38. Suppose astronomers discover a 3-$M_\odot$ black hole located a few light-years from Earth. Should they be concerned that its tremendous gravitational pull will lead to our planet's untimely demise?

39. If you could watch a star falling into a black hole, how would the color of the star change as it approached the event horizon?

40. Why are we not aware of the effects of special and general relativity in our everyday lives here on Earth?

Problems

41. The 100-$M_\odot$ star Eta Carinae has had episodes of mass loss during which it lost mass at the rate of 0.1 $M_\odot$ per year. Let's put that into perspective.
    a. The mass of the Sun is 2×10^{30} kg. How much mass (in kilograms) did Eta Carinae lose each minute?
    b. The mass of the Moon is 7.35×10^{22} kg. How does Eta Carinae's mass loss per minute compare with the mass of the Moon?

42. Using values given in Section 13.1, verify that an O star can lose 20 percent of its mass during its main-sequence lifetime.

43. The approximate relationship between the luminosity and the period of Cepheid variables is L_{star} ($L_\odot$ units) $= 335\, P$ (days). Delta Cephei has a cycle period of 5.4 days and a parallax of 0.0033 arcseconds (arcsec). A more distant Cepheid variable appears 1/1,000 as bright as Delta Cephei and has a period of 54 days.
    a. How far away (in parsecs) is the more distant Cepheid variable?
    b. Could the distance of the more distant Cepheid variable be measured by parallax? Explain.

44. If the Crab Nebula has been expanding at an average velocity of 3,000 km/s since it was first observed in 1054, what was its average radius in the year 2015? (Note: There are approximately 3×10^7 seconds in a year.)

45. Use Einstein's famous mass-energy equivalence formula ($E = mc^2$) to verify that 5.88×10^{13} joules (J) of energy is released from fusing 1.00 kg of helium via the triple-alpha process.

46. According to Einstein, mass and energy are equivalent. So does a cup of coffee weigh more when it is hot or when it is cold? Why? Do you think the difference is measurable?

47. Pulsars are rotating neutron stars. For a pulsar that rotates 30 times per second, at what radius in the pulsar's equatorial plane would a co-rotating satellite (revolving around the pulsar 30 times per second) have to be moving at the speed of light? Compare this to the pulsar radius of 1 km.

48. Figure 13.9 shows the relative abundance of the elements. Is this a log or a linear plot? Explain what it means that oxygen lies on the y-axis at 10^{-3}.

49. From the information given in Section 13.3, verify the claim that Earth would be roughly the size of a football stadium if it were as dense as a neutron star.

50. If a spaceship approaching us at 0.9 times the speed of light shines a laser beam at Earth, how fast will the photons in the beam be moving when they arrive at Earth?

Exploration | The CNO Cycle

wnpag.es/uou2

Nuclear reactions are quite complex, and they usually involve many steps. In a previous Exploration, you investigated the proton-proton chain. In this Exploration, you will study the CNO cycle, which is even more complex. Visit the Student Site (wnpag.es/uou2) and open the CNO Cycle Animation Nebraska Simulation in Chapter 13.

First, press play animation and watch the animation all the way through. Press reset animation to clear the screen, and then press play animation again, allowing the animation to proceed past the first collision before pressing pause.

1. Which atomic nuclei are involved in this first collision?

2. What color is used to represent the proton (hydrogen nucleus)?

3. What is represented by the blue wiggle?

4. What atomic nucleus is created in the collision?

5. The resulting nucleus is not the same type of element as either of the two that entered the collision. Why not?

Press play animation again, and then pause as soon as the yellow ball and the dashed line appear.

6. Is this a collision or a spontaneous decay?

7. What is represented by the yellow ball?

8. What is represented by the dashed line?

9. The resulting nucleus has the same number of nucleons (13), but it is a different element. What happened to the proton that was in the nitrogen nucleus but is not now in the carbon nucleus?

Proceed past the next two collisions, to ^{15}O.

10. Study the pattern that is forming. When a blue ball comes in, what happens to the number of nucleons and the type of the nucleus (that is, what happens to the "12" and the "C," or the "14" and the "N")?

11. What is emitted in these collisions?

Proceed until ^{15}N appears.

12. Is this a collision or a spontaneous decay?

13. Which previous reaction is this most like?

Now proceed to the end of the animation.

14. After this last collision, a line is drawn back to the beginning. This tells you what type of nucleus the upper red ball represents. What is this nucleus?

15. How many nucleons are not accounted for by that upper red ball? (Hint: Don't forget the ^{1}H that came into the collision.) These nucleons must be in the nucleus represented by the bottom red ball.

16. Carbon has six protons. Nitrogen has seven. How many protons are in the nucleus represented by the bottom red ball?

17. How many neutrons are in the nucleus represented by the bottom red ball?

18. What element is represented by the bottom red ball?

19. What is the net reaction of the CNO cycle? That is, what nuclei are combined and turned into the resulting nucleus?

20. Why do we not consider ^{12}C part of the net reaction?

14 Measuring Galaxies

In 1920, the astronomers Harlow Shapley (1885–1972) and Heber D. Curtis (1872–1942) engaged in a debate about the nature of the "spiral and elliptical nebulae" that we now know to be galaxies beyond our own Milky Way Galaxy. Shapley believed that our galaxy was the entire universe and therefore these nebulae must be inside the Milky Way. Curtis believed what we now know to be correct: The nebulae are separate, very distant objects that are like the Milky Way. While this "Great Debate" did not resolve the issue at the time, it set the stage for the work of Edwin P. Hubble (1889–1953), whose discoveries about the distances and velocities of galaxies fundamentally changed our understanding of the universe. In the illustration on the opposite page, a student is in the process of identifying galaxy types in the image of a cluster of galaxies. In this chapter, you will learn about the types of galaxies and their properties and composition.

LEARNING GOALS

Stars are not spread uniformly through space. Instead, they are grouped into what the philosopher Immanuel Kant called "island universes," which today we know to be galaxies. By the end of this chapter, you should be able to identify different types of galaxies, as the student has begun to do for the image of a large cluster of galaxies at right. You should be able to explain why spirals and ellipticals tend to be different colors, and you should also be able to:

LG 1 Determine a galaxy's type.

LG 2 Explain why the arms of spiral galaxies are sites of star formation.

LG 3 Describe the evidence for the existence of dark matter in galaxies.

LG 4 Explain why scientists believe that most—perhaps all—large galaxies have supermassive black holes at their centers.

LG 5 Describe how we know that we live in an expanding universe and the implications of this observation.

14.1 Galaxies Come in Many Sizes and Shapes

A **galaxy** is a gravitationally bound grouping of stars, dust, and gas. Galaxies range in size from a hundred to hundreds of thousands of parsecs across, and they are recognizable as objects distinct from their surroundings. Our galaxy, the Milky Way, contains hundreds of billions of stars. But the Milky Way is only one of hundreds of billions of galaxies that fill an observable universe vast in both space and time. These galaxies come in many shapes and a wide range of sizes. All are remarkably distant. The nearest large spiral galaxy is roughly half a million times farther from the Milky Way than the nearest star is from the Sun. Astronomers learn about distant galaxies by using all the available tools: imagery, spectroscopy, computer simulation, and reason. In this section, you will learn about the different types of galaxies in the universe.

Relating Near and Far

In Figure 1.1 in Chapter 1, we set Earth in a universal context: Earth orbits an average star within an enormous spiral galaxy, which itself is part of a group, which is part of a cluster, which is part of a supercluster of galaxies—structures that exist throughout the entire observable universe. As with most other astronomical objects, we cannot actually travel to distant galaxies to conduct experiments, thus our understanding of galaxies rests on the *cosmological principle*, which we encountered in Chapter 1: The laws of physics are the same everywhere. For example, according to the cosmological principle, gravity works the same way in distant galaxies as it does here on Earth.

As we stressed in Chapter 1, the cosmological principle is a testable scientific theory. An important prediction of the cosmological principle is that the conclusions we reach about our universe should be the same, whether we observe it from the Milky Way or from a galaxy billions of light-years away. In other words, if the cosmological principle is correct, then our universe is **homogeneous**.

It is not easy to verify the prediction of homogeneity directly. We cannot ever hope to travel to other galaxies to see whether conditions are the same. However, we can compare light arriving from closer and farther locations in the distant universe. For example, we can look at the way galaxies are distributed in distant space and ask whether that distribution is similar to the distribution nearby.

In addition to predicting that the universe is homogeneous, the cosmological principle requires that all observers (including us) have the same impression of the universe, regardless of the *direction* in which they are looking. If something is the same in all directions, then it is **isotropic**. This prediction of the cosmological principle is much easier to test directly than is homogeneity. For example, if galaxies were lined up in rows, we would get very different impressions, depending on the direction in which we looked. But the universe would still be homogeneous. In most instances, isotropy goes hand in hand with homogeneity, but the cosmological principle requires them both.

The isotropy and homogeneity of the distribution of galaxies in the universe are predictions of the cosmological principle that we can test directly. All of our observations show that the properties of the universe are the same, regardless of the direction in which we look. On very large scales, the universe appears homogeneous as well. The cosmological principle has withstood our tests and forms the basis of our study of distant galaxies and of the universe itself.

Vocabulary Alert

homogeneous This word strictly means "having at all points the same composition and properties," so clearly the universe is not truly homogeneous in an absolute sense of the word. Conditions on the surface of Earth are very different from those in deep space or in the heart of the Sun. When cosmologists speak of homogeneity of the universe, they mean that stars and galaxies in our part of the universe are on average much the same and behave in the same manner as stars and galaxies far from us. They also mean that stars and galaxies everywhere are distributed in space in much the same way as they are in our cosmic neighborhood, and that observers in those galaxies see the same properties for our universe that we do. Cosmologists apply the term *homogeneous* with a very broad brush.

The Shapes of Galaxies

Imagine taking a handful of different coins and throwing them into the air, as shown in **Figure 14.1a**. You know that all of these objects are flat and circular. When you look at the objects falling through the air, however, they do not appear to be the same. Some coins appear face-on, and they look circular. Some coins appear edge-on and look like thin lines. Most coins are seen from an angle between these two extremes. Even if this image of many coins was the only information you had, you could use it to figure out the three-dimensional shape of a coin—flat and circular.

Astronomers use a similar method to discover the true three-dimensional shapes of galaxies. Figure 14.1b shows a set of galaxies seen from various viewing angles, from face-on to edge-on. We can infer from images of the sky that, just like the coins in Figure 14.1a, some galaxies are disk-shaped and are randomly oriented on the sky. Others are more egg-like, while still others are irregularly shaped.

A major advance in understanding galaxies came from sorting these different shapes into categories. The classifications we use today date back to the 1930s,

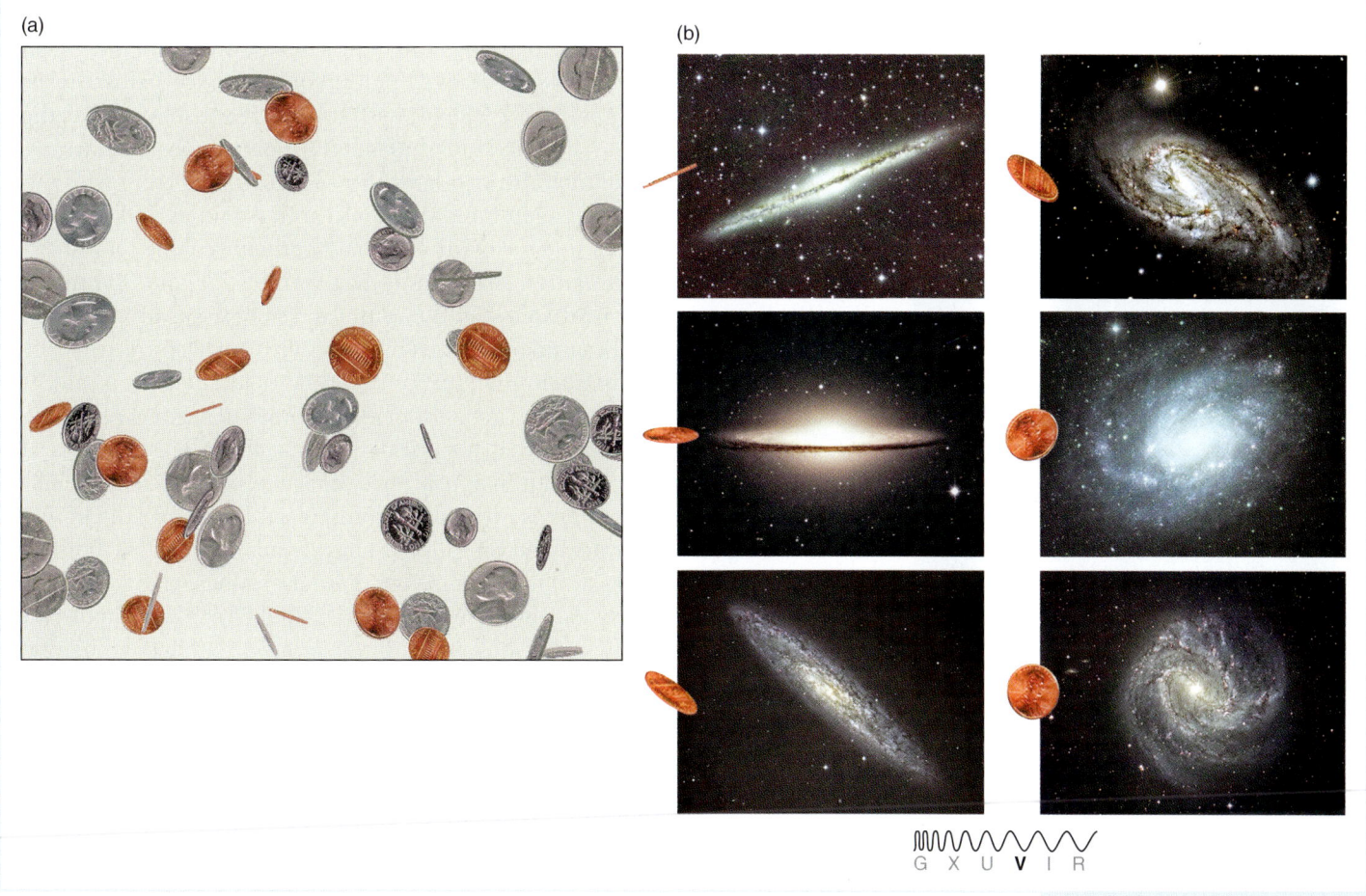

Figure 14.1 (a) A handful of coins thrown in the air provides a helpful analogy for the difficulties in identifying the shapes of certain types of galaxies. We see some face-on, some edge-on, and most somewhere in between. (b) Disk-shaped galaxies seen from various perspectives or angles. The variety of angles we see for galaxies corresponds to the range of perspectives for the coins in (a).

VISUAL ANALOGY

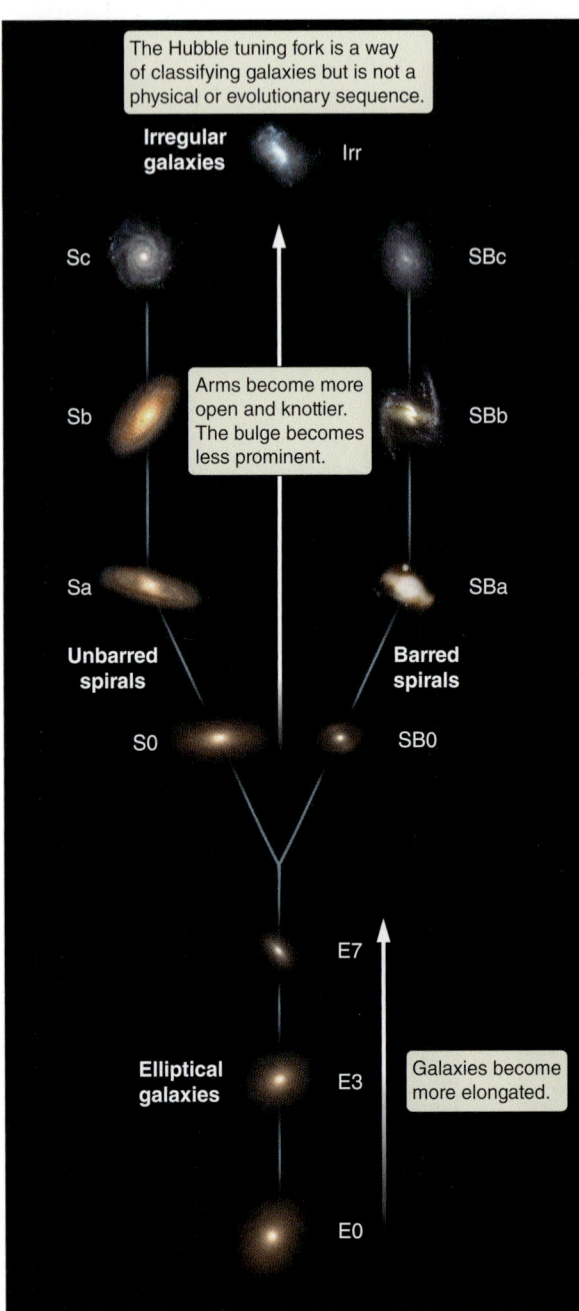

Figure 14.2 The tuning fork diagram shows Edwin Hubble's scheme for classifying galaxies according to their appearance. Elliptical galaxies form the "handle" of this tuning fork. Unbarred and barred spiral galaxies lie along the left and right tines of the fork, respectively. S0 galaxies lie along the bottom left and right tines. Irregular galaxies are not placed on the tuning fork.

when Edwin Hubble devised a scheme much like that shown in **Figure 14.2**. Hubble grouped all galaxies according to appearance and positioned them on a diagram known as a **tuning fork diagram** because it resembles the tuning forks historically used to tune musical instruments. Galaxies come in three basic types: *spirals*, *ellipticals*, and *irregulars*, the latter of which includes all other shapes. We'll look at these types one at a time. Look back to the chapter-opening illustration, in which a student has begun applying the Hubble classification scheme to the galaxies in a Hubble Space Telescope image. As you learn about the different types, you can find more galaxies of each type in that image.

On the bottom (or "handle") of the tuning fork diagram sit galaxies that are elliptical in three dimensions, something like an egg, with no outer disk. These galaxies have a circular or elliptical outline on the sky and are called **elliptical galaxies** (labeled *E* on the diagram). They have numbered subtypes ranging from nearly spherical (E0) to quite flattened (E7). Elliptical galaxies have few young stars. Many, if not most, elliptical galaxies contain small rotating disks at their centers. As with the coins tossed in the air, the appearance of an elliptical galaxy in the sky does not necessarily tell us its true shape. For example, a galaxy might actually be shaped like a jelly bean; but if we happen to see it end-on, it will look round like a gum ball.

On the two "tines" of the fork are the *spiral galaxies*. **Spiral galaxies**, designated with an initial *S*, are characterized by a flattened, rotating disk. The spiral arms that give these galaxies their name lie in this disk. In addition to disks and arms, spiral galaxies have central bulges, which extend above and below the disk.

The bulges of roughly half of all galaxies with spiral arms are bar shaped; these galaxies are known as **barred spirals** (designated *SB*). Barred spirals are arranged on the right-hand tine of the tuning fork, as shown in Figure 14.2. Spirals without bars are organized on the left-hand tine. Both spirals and barred spirals are subdivided into types a, b, and c according to the prominence of the central bulge and how tightly the spiral arms are wound. For example, Sa and SBa galaxies have the largest bulges and display tightly wound and smooth spiral arms. Sc and SBc galaxies have small central bulges and more loosely woven spiral arms, often very knotty in appearance. Our own Milky Way Galaxy is a barred spiral (SBc).

The distinction between spiral and elliptical galaxies is not always clear. Some galaxies are a combination of the two types, having stellar disks but no spiral arms, so that the disk is smooth in appearance, like an elliptical galaxy. Hubble called these intermediate types **S0 galaxies** and placed them near the junction of his tuning fork. Elliptical and S0 galaxies share other similarities: both contain small rotating disks at their centers, and neither produces many new stars.

Galaxies that fall into none of these classes are called **irregular galaxies** (designated *Irr*). As their name implies, irregular galaxies are often without symmetry in shape or structure, and they do not fit neatly on Hubble's tuning fork.

Originally, Hubble thought that his tuning fork diagram might do for galaxies what the Hertzsprung-Russell (H-R) diagram had done for stars. This has not turned out to be the case—the tuning fork diagram does not describe the evolution of galaxies—but his classification scheme organized the study of these objects and reminds us that there are three basic types.

Stellar Motions Give Galaxies Their Shapes

A galaxy is not a solid object like a coin, but rather a collection of stars, gas, and dust. In an elliptical galaxy, stars are moving in all possible directions. Unlike planets in our Solar System, which move on nearly circular orbits around the Sun, stars in an elliptical galaxy follow orbits with a wide range of shapes, as shown in **Figure 14.3**. These orbits are more complex than the orbits of planets because the gravitational field within an elliptical galaxy does not come from a single central object. Taken together, all of these stellar orbits give an elliptical galaxy its shape.

Orbital speeds are also a factor. The faster the stars are moving, the more spread out the galaxy is. If the stars in an elliptical galaxy are moving in truly random directions, the galaxy will have a spherical shape. However, if stars tend on average to move faster in one direction than in others, the galaxy will be more spread out in that direction, giving it an elongated (elliptical) shape. These differences in stellar orbits cause some elliptical galaxies to be round and others to be elongated.

The orbits of stars in the disks of spiral galaxies are quite different from those of stars in elliptical galaxies. The components of a spiral galaxy are shown in **Figure 14.4**. The defining feature of a spiral galaxy is that it has a flattened, rotating disk. Like the planets of our Solar System, most of the stars in the disk of a spiral galaxy travel in the same direction around a concentration of mass at the center of the galaxy. But the stellar orbits in a spiral galaxy's central bulge are quite different from those in the galaxy's disk. As with elliptical galaxies, the gravitational field within the bulge does not come from a single object, and the stars therefore follow a wide variety of irregular orbits. The bulges of spiral galaxies (exclusive of barred spirals) are thus roughly spherical in shape.

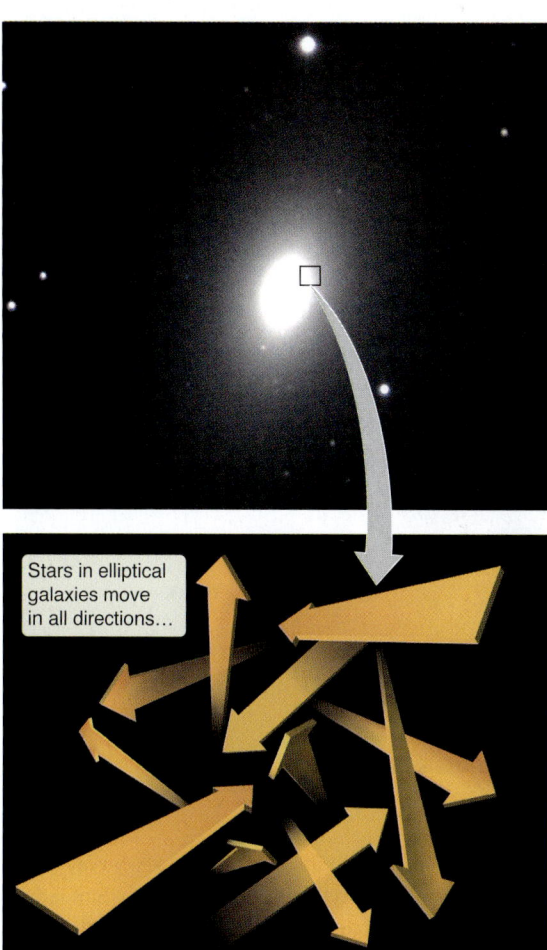
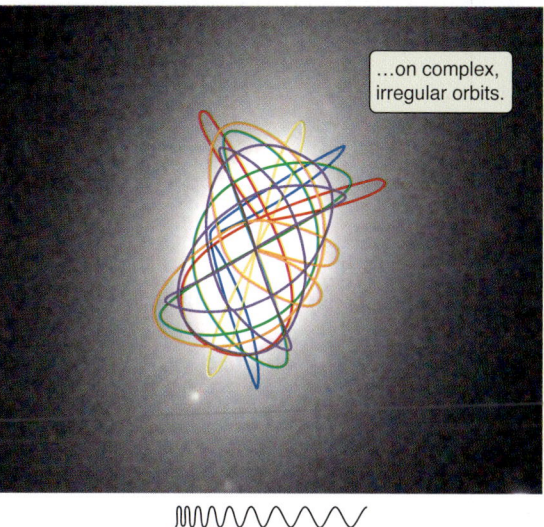

Figure 14.3 Elliptical galaxies take their shape from the orbits of the stars they contain. The colored lines superimposed on the galaxy represent the complex, irregular orbits of its stars.

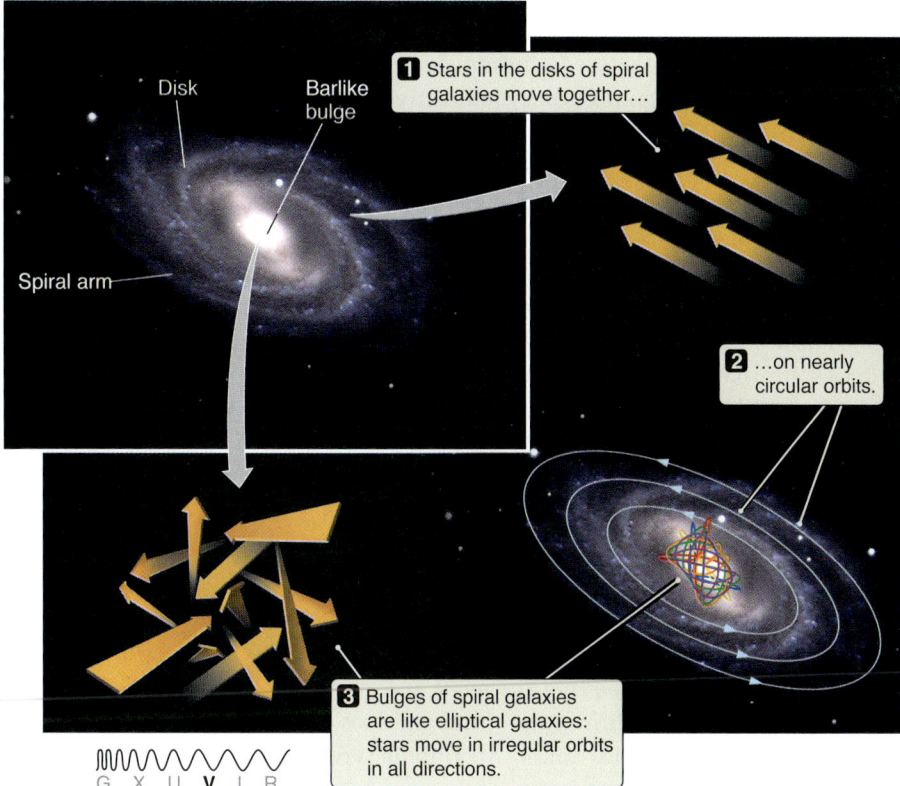

Figure 14.4 The components of a barred spiral galaxy include a disk, spiral arms, and a barlike bulge. The orbits of stars in the rotating disk and the elliptically shaped bulge are different.

Figure 14.5 (a) The dust in the plane of the Milky Way obscures our view toward the galactic center. (b) Similarly, the dust in the plane of the nearly edge-on spiral galaxy M104 is seen as a dark, obscuring band in the midplane of the galaxy.

Other Differences among Galaxies

In addition to the differences in their stellar orbits, there are other important distinctions between spiral and elliptical galaxies. These distinctions carry information not only about the way a galaxy looks, but also about the way it has evolved in the past and the way it will evolve in the future.

Gas and Dust Most spiral galaxies contain large amounts of dust and cold, dense gas concentrated in the midplanes of their disks. Just as the dust in the disk of our own galaxy is visible on a clear summer night as a dark band slicing the Milky Way in two (**Figure 14.5a**), the dust in an edge-on spiral galaxy appears as a dark, obscuring band running down the midplane of the disk (Figure 14.5b). The cold gas that accompanies the dust can also be seen in radio observations of spiral galaxies. In contrast, elliptical galaxies contain large amounts of very hot gas that we see primarily by observing the X-rays it emits.

The difference in shape between elliptical and spiral galaxies offers some insight into why the gas in ellipticals is hot, while in spirals it is cold. Just as gas settles into a disk around a forming star, cold gas settles into the disk of a spiral galaxy because of conservation of angular momentum. In contrast, elliptical galaxies do not have a net rotation, so the gas does not settle into a disk. In an elliptical galaxy, the only place that cold gas could collect is at the center. However, the density of stars in the center of elliptical galaxies is so high that evolving stars and Type Ia supernovae continually reheat the gas, preventing most of it from cooling off.

Color The colors of spiral and elliptical galaxies tell us a great deal about their star formation histories. Stars form from dense clouds of cold gas. Because the gas we see in elliptical galaxies is very hot, active star formation is not taking place in those galaxies today. The reddish colors of elliptical and S0 galaxies confirm that little or no star formation has occurred there for quite some time. The stars in these galaxies are an older population of lower-mass stars. Conversely, the bluish colors of the disks of spiral galaxies confirm that stars are forming in the cold molecular clouds contained within the disk. Even though *most* of the stars in a spiral disk are old, the massive young stars are so luminous that their blue light dominates what we see. When it comes to star formation, most irregular galaxies are like spiral galaxies. Some irregular galaxies are currently forming stars at prodigious rates, given their relatively small sizes.

Luminosity The relationship between luminosity and size among the different types of galaxies is not straightforward. Galaxies range in luminosity from about a million up to a million million solar luminosities (10^6 to $10^{12}\ L_\odot$) and in size from a hundred to hundreds of thousands of parsecs. (Recall that a parsec is 3.26 light-years.) There is no distinct size difference between elliptical and spiral galaxies; about half of both types of galaxies fall within a similar range of sizes. Although it is true that the most luminous elliptical galaxies are more luminous than the most luminous spiral galaxies, there is considerable overlap in the range of luminosities among all Hubble types.

Mass Mass is the single most important parameter in determining the properties and evolution of a star. In contrast, differences in mass and size do not lead to obvious differences among galaxies. Only subtle differences in color and concentration exist between large and small galaxies, making it difficult for us to

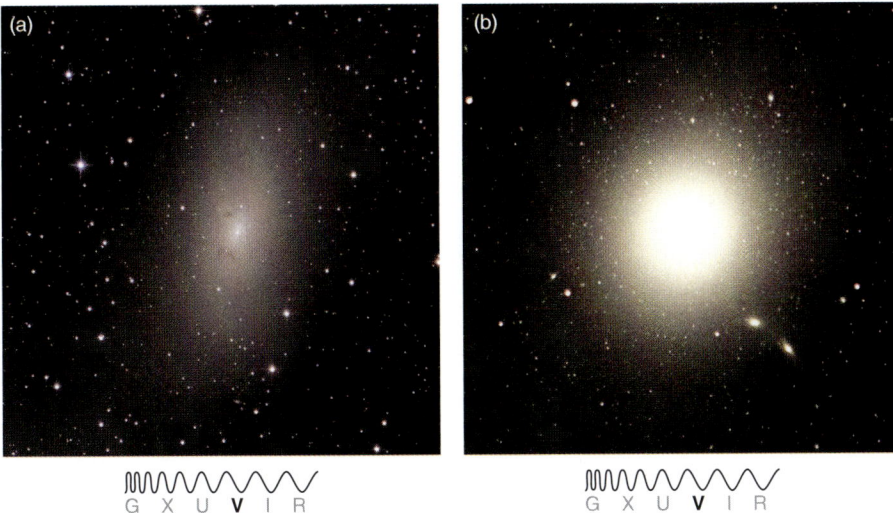

Figure 14.6 (a) Dwarf elliptical galaxies differ in appearance from (b) giant elliptical galaxies. Giant elliptical galaxies are more centrally concentrated than dwarf elliptical galaxies.

distinguish which are large and which are small. Even when a smaller, nearby spiral galaxy is seen next to a larger, distant spiral, it can be hard to tell which is which by appearance alone. Still, galaxies that have relatively low luminosity (less than 1 billion $L_\odot$) are called **dwarf galaxies**, because the brightness indicates the number of stars, and therefore the amount of stellar mass. Galaxies more than 1 billion times as luminous as the Sun are called **giant galaxies**. Only elliptical and irregular galaxies come in both types. Among spiral and S0 galaxies, we find only giants. It is relatively easy to tell the difference between a dwarf elliptical galaxy and a giant elliptical galaxy (**Figure 14.6**). Giant elliptical galaxies have a much higher density of stars, which are more centrally concentrated than stars in dwarf ellipticals.

14.2 Stars Form in the Spiral Arms of a Galaxy's Disk

From pictures of spiral galaxies outside our own spiral Milky Way Galaxy, we might have guessed that most stars in a galaxy's disk are located in the spiral arms. This turns out not to be the case. **Figure 14.7** shows images of the Andromeda Galaxy taken in ultraviolet (UV) and visible light. Notice that whereas the spiral arms are relatively prominent in the UV image (Figure 14.7a), they are less prominent when viewed in visible light (Figure 14.7b). If we carefully count the actual numbers of stars, we find that although stars are slightly concentrated in spiral arms, this concentration is not strong enough to account for the prominence of the arms. In fact, the concentration of stars in the disks of spiral galaxies varies quite smoothly as it decreases outward from the center of the disk to the edge of the galaxy. However, molecular clouds, associations of O and B stars, and other structures associated with star formation are all concentrated in spiral arms. Spiral arms look so prominent when viewed in blue or UV light because they contain significant concentrations of young, massive, luminous stars.

As you learned in Chapter 5, stars form when dense interstellar clouds become so massive and concentrated that they begin to collapse under the force of their

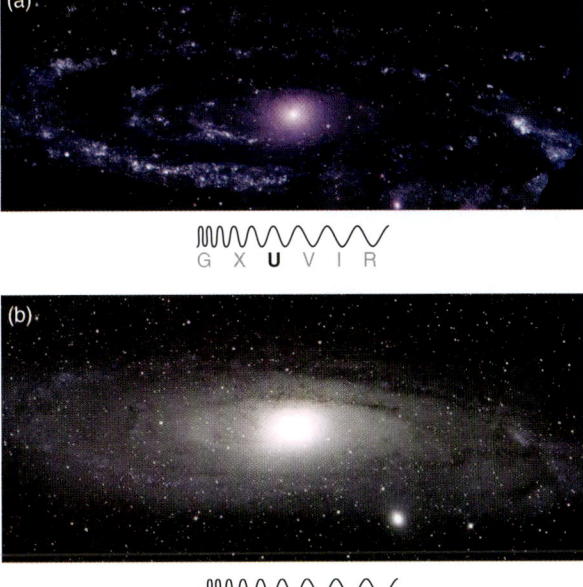

Figure 14.7 The Andromeda Galaxy is shown in (a) ultraviolet light and (b) visible light. Note that the spiral arms are most prominent in ultraviolet light, because they are dominated by young hot stars, and contain emission from interstellar clouds that are ionized by the radiation from young hot stars.

Figure 14.8 These two images of a face-on spiral galaxy show the spiral arms. (a) This visible-light image also shows dust absorption. (b) This image of 21-cm emission shows the distribution of neutral interstellar hydrogen, CO emission from cold molecular clouds, and Hα emission from ionized gas.

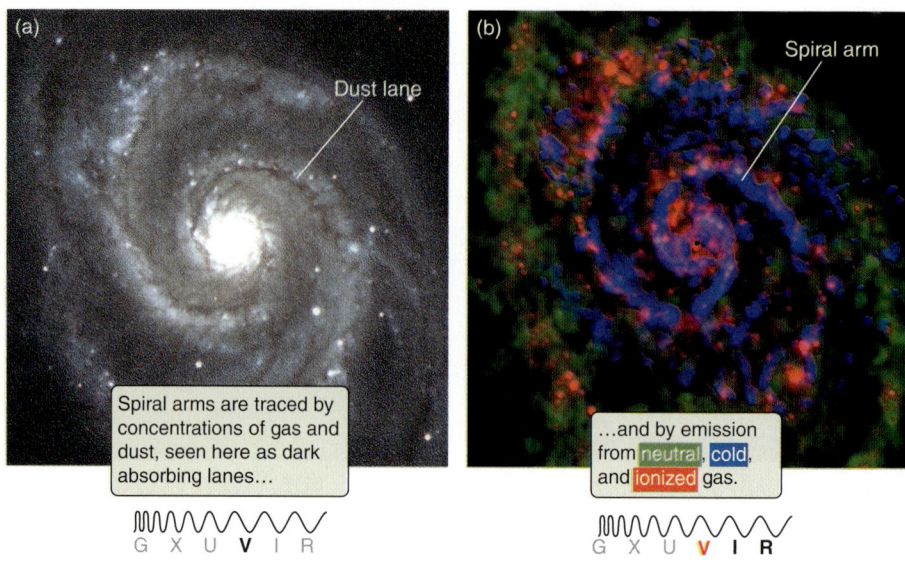

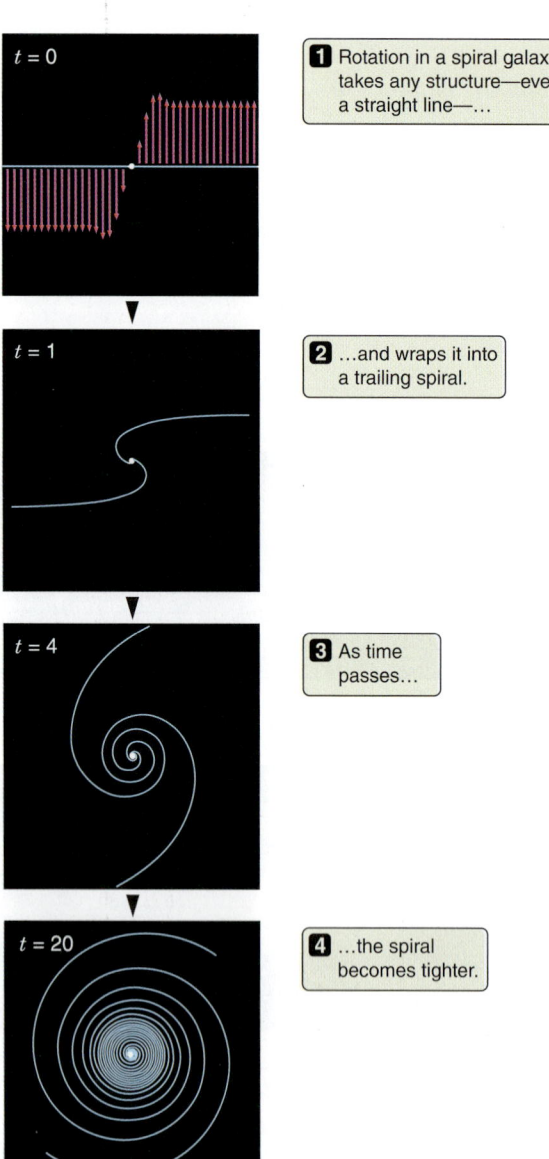

Figure 14.9 The differential rotation of a spiral galaxy will naturally take even an originally linear structure and wrap it into a progressively tighter spiral as time (*t*) passes.

own gravity. If stars form within spiral arms, then clouds of interstellar gas must pile up and compress in spiral arms. There are many ways to trace the presence of gas in the spiral arms of galaxies. Pictures of face-on spiral galaxies, such as the one featured in **Figure 14.8a**, show dark lanes where clouds of dust block starlight. These lanes provide one of the best tracers of spiral arms. Spiral arms also show up in other tracers of gas concentrations, such as radio emission from neutral hydrogen or from carbon monoxide (Figure 14.8b).

Any disturbance in the disk of a spiral galaxy will cause a spiral pattern because the disk rotates. Disks do not rotate like a solid body. Instead, material close to the center takes less time to travel around the galaxy than material farther out, and so the inner part of the disk gets ahead of the outer part. **Figure 14.9** illustrates the point: In the second frame, you can see that the outer part of the line is trailing behind the inner part. As the galaxy rotates, a straight line through the center becomes a spiral. In the time it takes for objects in the inner part of the galaxy to complete several rotations, objects in the outer parts of the galaxy may not have completed even a single rotation.

A spiral galaxy can be disturbed by, for example, gravitational interactions with other galaxies or a burst of star formation. However, a single disturbance will not produce a *stable* spiral-arm pattern. Spiral arms produced from one disturbance will wind themselves up completely in two or three rotations of the disk and then disappear. Disturbances that are repetitive sustain spiral structure indefinitely. When the bulge in the center of a spiral galaxy is elongated (as seems to be the case for most spiral galaxies), then the bulge gravitationally disturbs the disk. As the disk rotates through this disturbance, repeated episodes of star formation occur, and stable spiral arms form.

Spiral structure can also be created by star formation. Regions of star formation release energy into their surroundings through UV radiation, stellar winds, and supernova explosions. This energy compresses clouds of gas and triggers more star formation. Typically, many massive stars form in the same region at about the same time, and their combined mass outflows and supernova explosions occur over only a few million years, creating large, expanding bubbles of hot gas. These bubbles concentrate the gas into dense clouds at their edges, causing more star formation. Rotation bends the resulting strings of star-forming regions into spiral structures.

Regular disturbances in the disks of spiral galaxies are called **spiral density waves**: regions of greater mass density and increased pressure in the galaxy's interstellar medium. These waves move around a disk in the pattern of a two-armed spiral. Because they are waves, it is the disturbance that moves, not the material. As the matter in the disk orbits, material passes *through* the spiral density waves. The stars in the arm today are not the same stars that were in the arm 20 million years ago. This is roughly analogous to a traffic jam on a busy highway. The cars in the jam are changing all the time, yet the traffic jam persists as a place of higher density—where there are more cars than usual.

A spiral density wave has very little effect on stars, but it does compress the gas that flows through it. Stars form in the resulting compressed gas. Massive stars have such short lives (typically 10 million years or so) that they never drift far from the spiral arms. Less massive stars, in contrast, have plenty of time to move away from their place of birth, filling in the rest of the disk.

▶▶ **Nebraska Simulation:** Traffic Density Analogy

14.3 Galaxies Are Mostly Dark Matter

Efforts to measure the masses of galaxies during the last decades of the 20th century led to the discovery of dark matter. To understand this discovery, we first need to understand how astronomers go about measuring the mass of a galaxy.

Finding the Mass of a Galaxy

One method to find the mass of a galaxy is to add up the mass of the stars, dust, and gas that we can see. A galaxy's spectrum is primarily composed of starlight, so we can find out what types of stars are in the galaxy. Stellar evolution then tells us how to turn the luminosity of the galaxy into an estimate of the total stellar mass. The physics of radiation from interstellar gas at X-ray, infrared, and radio wavelengths enables us to estimate the mass of the gas and dust. Together, the stars, gas, and dust in a galaxy are called **luminous matter** (or simply **normal matter**) because this matter emits electromagnetic radiation.

However, this method does not allow us to determine a galaxy's total mass. Black holes, for example, would not be accounted for in this method, yet they still have mass. Fortunately, we have a method for determining mass that does not involve luminosity. Stars in disks follow orbits that are much like the Keplerian orbits of planets around their parent stars and binary stars around each other. To measure the mass of a spiral galaxy, we apply Kepler's laws, just as we do for those other systems.

Observations of Dark Matter

Astronomers would also like to know how the mass is distributed in a galaxy. They hypothesized that the mass and the light are distributed in the same way—assuming, that is, that the luminous mass in these galaxies is all the mass there is. Based on this hypothesis, they made a prediction: The light of all galaxies, including spiral galaxies, is highly concentrated toward the center (**Figure 14.10a**). It would follow, then, that nearly all the mass of a spiral galaxy is contained in its center (Figure 14.10b). This situation is much like the Solar System, where nearly all the mass is in the Sun, at the Solar System's center. This hypothesis predicts fast orbital velocities near the center of the spiral galaxy and slower orbital velocities farther out (Figure 14.10c). A graph showing how the orbital

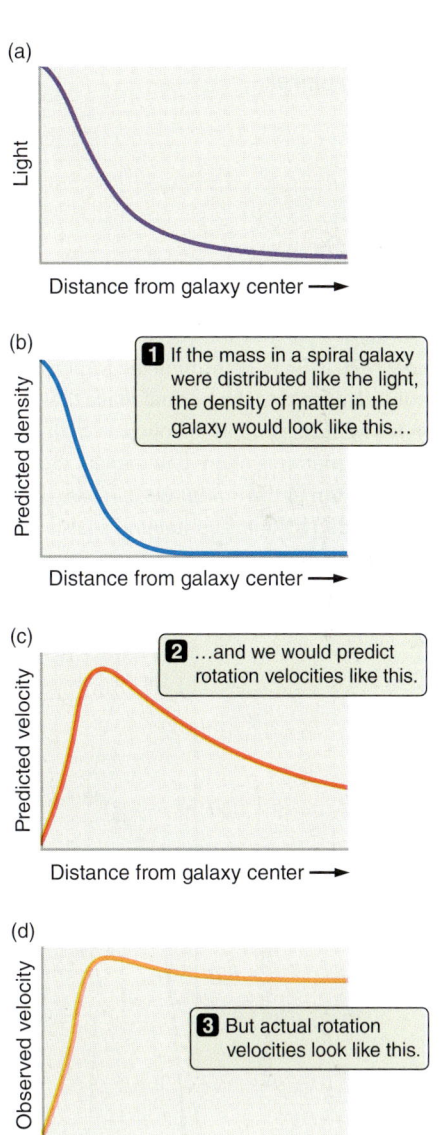

Figure 14.10 (a) The profile of visible light in a typical spiral galaxy drops off with distance from the center. (b) The mass density of stars and gas located at a given distance from the galaxy's center follows the light profile. If stars and gas accounted for the entire mass of the galaxy, then the galaxy's rotation curve would be as shown in (c). However, galaxies have observed rotation curves more like the curve shown in (d).

> **Back in Section 5.6 . . .**
>
> . . . you learned that when objects are moving away from you, their spectral lines are shifted toward the red end of the spectrum.

▶▶ **Nebraska Simulation:** Milky Way Rotational Velocity Explorer

Figure 14.11 (a) We can use the flat rotation curve of the spiral galaxy NGC 3198 to determine the total mass within a given radius. Notice that the normal mass that can be accounted for by stars and gas is only part of the needed mass. Extra dark matter is needed to explain the rotation curve. (b) In addition to the matter we can see, galaxies must be surrounded by halos containing a large amount of dark matter.

velocity of stars and gas in a galaxy changes with distance from the galaxy's center is called a **rotation curve**.

To test this prediction, astronomers used the Doppler effect to measure orbital motions of stars, gas, or dust at various distances from the galaxy's center. From these data, they made a graph of the orbital velocity versus distance as shown in Figure 14.10d.

Vera Rubin pioneered this work on galaxy rotation rates. She discovered that, contrary to earlier predictions (Figure 14.10c), the rotation velocities of spiral galaxies remain about the same out to the most distant measured parts of the galaxies (Figure 14.10d). Observations of 21-cm radiation from neutral hydrogen show that the rotation curves remain flat even well outside the extent of the visible disks. The hypothesis that mass and light are distributed in the same way is wrong.

The logical next step is to turn the question around and ask: What mass distribution would cause this unexpected rotation curve? **Figure 14.11a** shows the result of such a calculation for the spiral galaxy NGC 3198. The black line shows the speed of rotation *at* a particular radius (velocity increases vertically on the graph). The red line shows how much luminous mass is observed to be inside a particular radius (mass increases vertically on the graph). To produce a rotation curve like the one shown in black, this galaxy must have a second component consisting of matter that does not show up in our census of stars, gas, and dust. This material, which does not interact with light, and therefore reveals itself only by the influence of its gravity, is called **dark matter**. The blue line shows how much dark matter must be inside a particular radius in order to provide enough mass to make the galaxy rotate as it does.

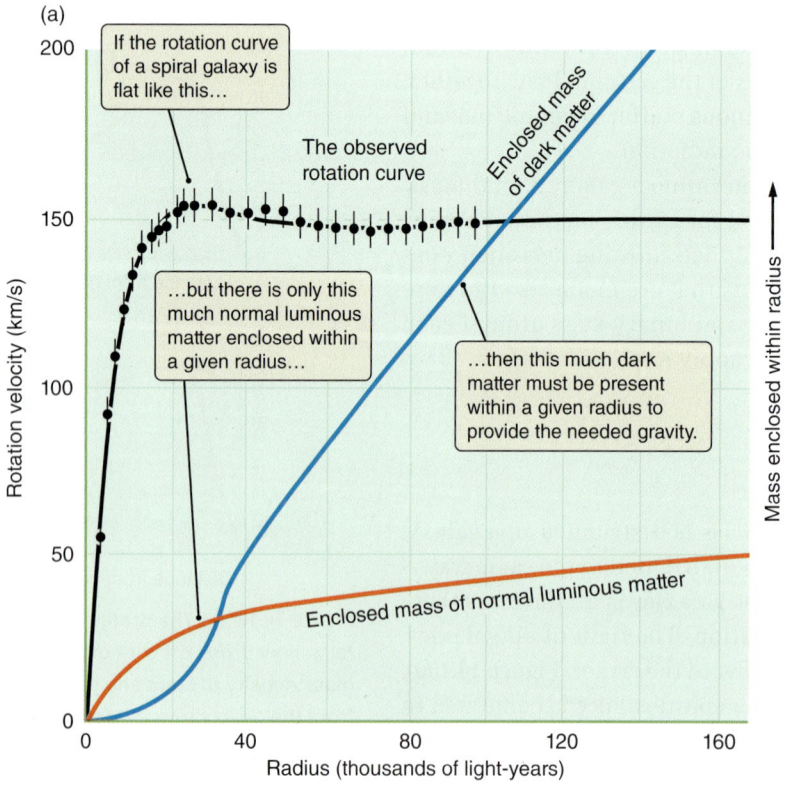

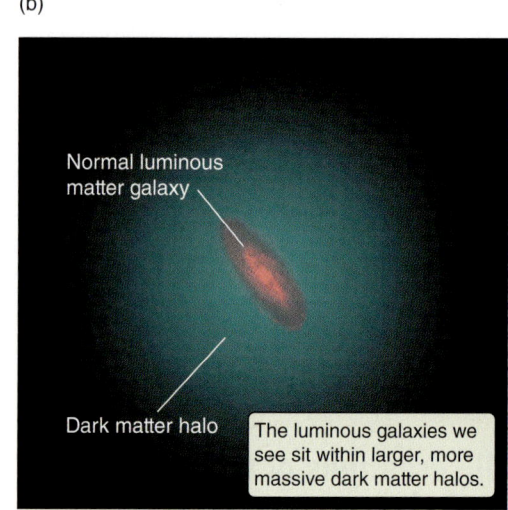

The rotation curves of the inner parts of spiral galaxies match predictions based on their luminous matter, indicating that the inner parts of spiral galaxies are mostly luminous matter. Within the entire *visual* image of a galaxy, the mix of dark and luminous matter is about half and half. However, rotation curves in the outer part of the galaxy do not match the predictions based on luminous matter, indicating that the outer parts of spiral galaxies are mostly dark matter. Astronomers currently estimate that as much as 95 percent of the total mass in some spiral galaxies consists of a greatly extended **dark matter halo** (Figure 14.11b), far larger than the visible spiral portion of the galaxy located at its center. This is a startling statement. A spiral galaxy shines only from the inner part of a much larger distribution of mass that is dominated by some type of matter we cannot see.

What about elliptical galaxies? Again, we want to compare the luminous mass measured from the light we can see, with the gravitational mass measured from the effects of gravity. Because elliptical galaxies do not have a net rotation, we cannot use Kepler's laws to measure the gravitational mass from the rotation of a disk. Instead, we measure it from the motions of individual stars, and we find that elliptical galaxies are mostly made of dark matter. Another avenue of exploration confirms this result. An elliptical galaxy's ability to hold onto its hot, X-ray-emitting gas depends on its mass: If the galaxy is not massive enough, the hot atoms and molecules will escape into intergalactic space. To find the mass of an elliptical galaxy, first we infer the total amount of gas from X-ray images, such as the blue and purple halo seen in **Figure 14.12**. Next we calculate the mass that is needed to hold onto the gas. We then compare that gravitational mass with the luminous mass. The amount of dark matter is the difference between the mass that is needed to hold onto the gas and the observed luminous mass.

Some elliptical galaxies contain up to 20 times as much mass as can be accounted for by their stars and gas alone, so they must be dominated by dark matter, just like spirals. As with spirals, the luminous matter in ellipticals is more centrally concentrated than is the dark matter. The transition from the inner parts of galaxies (where luminous matter dominates) to the outer parts (which are dominated by dark matter) is remarkably smooth. Some galaxies may contain less dark matter than others; but on average, about 90 percent of the total mass in a typical galaxy is in the form of dark matter.

What Is Dark Matter?

What is dark matter? A number of suggestions are under investigation: Jupiter-sized objects, numerous black holes, large numbers of white dwarf stars, and exotic unknown elementary particles. These candidates can be lumped into two groups: MACHOs and WIMPs.

Dark matter candidates such as small main-sequence M stars, planets, white dwarfs, neutron stars, or black holes are collectively referred to as **MACHOs**, which stands for "massive compact halo objects." If the dark matter in the Milky Way's halo consists of MACHOs, there must be a lot of these objects, and they must each exert gravitational force but not emit much light. Because they have mass, MACHOs gravitationally deflect light according to Einstein's general theory of relativity, a phenomenon called gravitational lensing (which we described in Chapter 13). If we were observing a distant star and a MACHO passed between us and the star, the star's light would be deflected and perhaps, if the geometry

Figure 14.12 In this combined visible-light and X-ray image of elliptical galaxy NGC 1132, the false-color blue and purple halo is X-ray emission from hot gas surrounding the galaxy. The hot gas extends well beyond the visible light from stars.

▶II **AstroTour:** Dark Matter

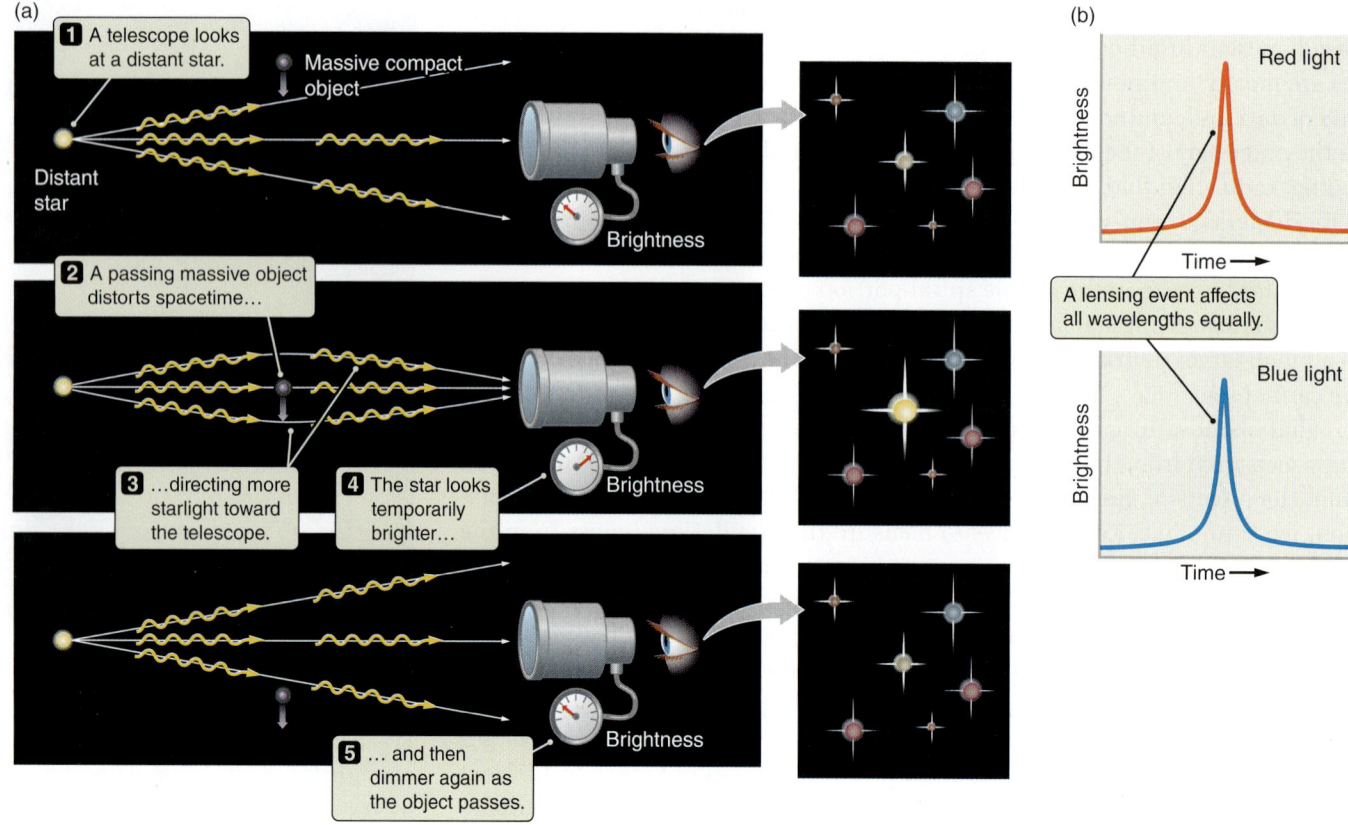

Figure 14.13 (a) The light from a distant star is affected by a compact object crossing our line of sight. (b) The observed light curves of a real star experiencing a lensing event have a distinctive shape and are the same in all colors of light.

were just right, focused by the intervening MACHO as it passed across our line of sight, as illustrated in **Figure 14.13a**. Because gravity affects all wavelengths equally, such lensing events should look the same in all colors, ruling out other causes of variability.

We would be remarkably lucky if such an event occurred just as we were observing a single distant star. However, astronomers monitored the stars in the Large and Small Magellanic Clouds (two of the small companion galaxies of the Milky Way Galaxy), observing tens of millions of stars for several years. They found a number of examples of events of the sort shown in Figure 14.13b, but not nearly enough to account for the amount of dark matter in the halo of our galaxy. Thus, it was concluded that the dark matter in the Milky Way is probably *not* composed primarily of MACHOs. The implication of this result is that dark matter in the Milky Way (and therefore other galaxies too) must be composed of something other than MACHOs.

This leaves the exotic unknown elementary particles commonly known as **WIMPs**, which stands for "weakly interacting massive particles." These particles are predicted to be something like neutrinos; they would barely interact with ordinary matter, yet would have some mass. WIMPs are currently the favored explanation because there are not enough MACHOs to account for the effects we observe. Experiments are under way at the Large Hadron Collider and on the International Space Station to prove the existence of such particles, and additional experiments are being done to detect such particles from our galactic halo as they pass through Earth.

14.4 A Supermassive Black Hole Exists at the Heart of Most Galaxies

Probing the centers of galaxies is difficult because there are so many stars and so much dust and gas in the way that we cannot get a clear picture of the center, even for nearby galaxies. Notably, our understanding of what lies at the heart of massive galaxies did not initially come from studying nearby galaxies, but instead from observing some of the most distant objects in the universe.

The Discovery of Quasars

In the late 1950s, radio surveys detected a number of bright, compact objects that at first seemed to have no optical counterparts. Improved radio positions revealed that these radio sources coincided with faint, very blue, starlike objects. Unaware of the true nature of these objects, astronomers called them "radio stars." Obtaining spectra of the first two radio stars was a laborious task, requiring 10-hour exposures with the cameras on the telescopes. Astronomers were greatly puzzled by the results. Rather than displaying the expected absorption lines characteristic of blue stars, the spectra showed only a single pair of emission lines, which were broad—indicating very rapid motions within these objects—and which did not seem to correspond to the lines of any known substances.

For several years, astronomers believed they had discovered a new type of star, until Maarten Schmidt realized that these broad spectral lines were the highly redshifted lines of ordinary hydrogen. As we will see shortly, a high redshift implies a great distance from us. These "stars" were not stars at all. They were extraordinarily luminous objects at enormous distances. These "quasi-stellar radio sources" were dubbed **quasars**. Today we know that quasars result from extreme activity in the nuclei of galaxies, which often results from interactions with other galaxies, as seen in **Figure 14.14**. Together, quasars and their less luminous but still active cousins are called **active galactic nuclei (AGNs)**.

Quasars are phenomenally powerful, pouring forth a luminosity of between a trillion and a thousand trillion (10^{12} to 10^{15}) Suns. They are also very distant—the closest really bright quasar, 3C 273, is about 750 million parsecs (750 megaparsecs [Mpc]) away. Billions of galaxies are closer to us than the nearest quasar. The distance to an object also tells us the amount of time that has passed since the light from that object left its source. Because we see quasars only at great

Figure 14.14 The environments around quasars, which are found in the centers of galaxies, often show evidence of interactions with other galaxies.

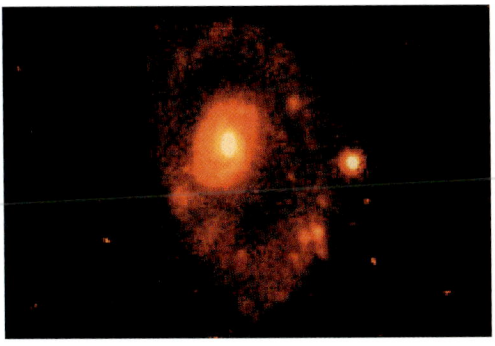
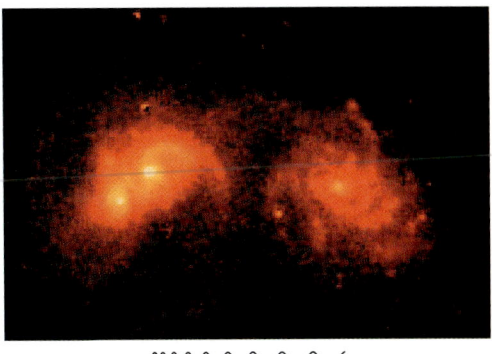
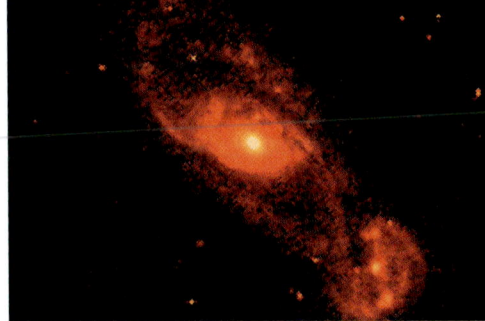

Vocabulary Alert

unresolved In common language, *unresolved* means "not solved" or "not concluded." Astronomers, however, use this term to describe a telescope's *resolution*, as discussed in Chapter 4. An *unresolved* source appears pointlike because its angular diameter is too small for the telescope to distinguish any of its parts.

distances, we know that they are quite rare in the universe at this time, but were once much more common. This discovery of the changing occurrence of quasars was one of the first pieces of evidence to demonstrate that the universe has evolved over time.

AGNs Are the Size of the Solar System

The enormous radiated power and mechanical energy of active galactic nuclei are astonishing on their own, but they are made even more spectacular because all of this power emerges from a region that can be no larger than a light-day or so across—comparable in size to our own Solar System. How can we make such a claim? For one thing, quasars and other AGNs at the centers of galaxies appear only as **unresolved** points of light, even in our most powerful telescopes. For more evidence, we turn not to the sky but to the halftime show at a local football game.

Figure 14.15 illustrates a problem faced by every director of a marching band. When a band is all together in a tight formation at the center of the field, the notes you hear in the stand are clear and crisp. But as the band spreads out across the

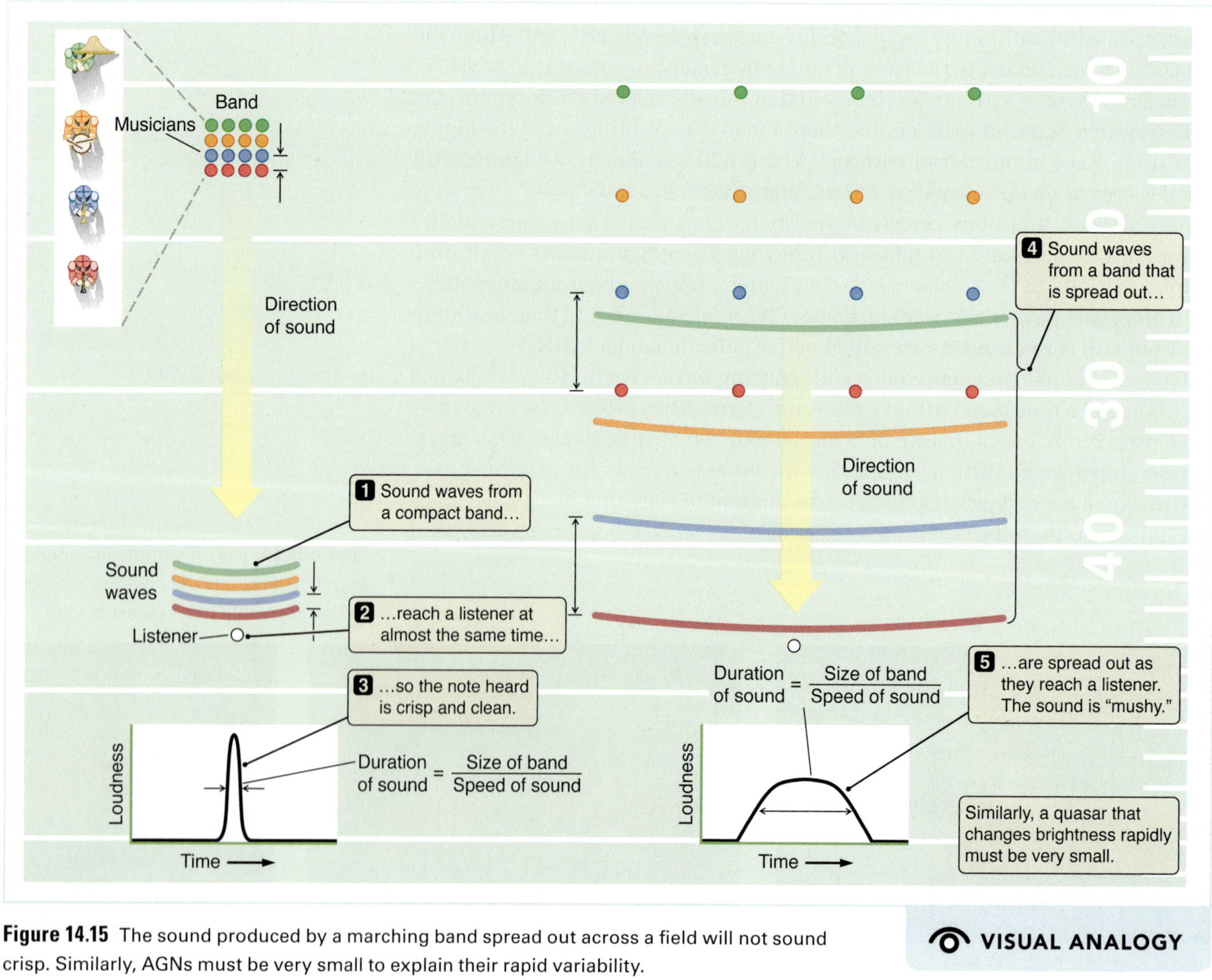

Figure 14.15 The sound produced by a marching band spread out across a field will not sound crisp. Similarly, AGNs must be very small to explain their rapid variability.

VISUAL ANALOGY

field, its sound begins to get mushy. This is not because the marchers are poor musicians. Rather, it is because sound travels at a finite speed. On a cold, dry December day, sound travels at a speed of about 330 meters per second (m/s). At this speed, it takes sound approximately one-third of a second to travel from one end of the football field to the other. Even if every musician on the field played a note at exactly the same instant in response to the director's cue, in the stands you would hear the instruments close to you first but would have to wait longer for the sound from the far end of the field to arrive.

If the band is spread from one end of the field to the other, then the beginning of a note will be smeared out over about one-third of a second, which is the time it takes for sound to travel from one end of the field to the other. If the band were spread out over two football fields, it would take about two-thirds of a second for the sound from the most distant musicians to arrive at your ear. If our marching band were spread out over a kilometer, then it would take roughly 3 seconds—the time it takes sound to travel a kilometer—for us to hear a crisply played note start and stop. Even with our eyes closed, it would be easy to tell whether the band was in a tight group or spread out across the field.

▶ll **AstroTour:** Active Galactic Nuclei

Exactly the same principle applies to the light we observe from active galactic nuclei. Quasars and other AGNs change their brightness dramatically over the course of only a day or two—and in some cases as briefly as a few hours. This rapid variability sets an upper limit on the size of an AGN, just as hearing clear music from a marching band tells us that the band musicians are close together. The AGN powerhouse must therefore be no more than a light-day or so across because if it were larger, the light we see could not possibly change in only a day or two. An AGN has the light of up to 10,000 galaxies pouring out of a region of space that would come close to fitting within the orbit of Neptune.

Supermassive Black Holes and Accretion Disks

When astronomers first discovered AGNs, they put forward a variety of ideas to explain them. Further observations revealed that AGNs had tiny sizes (compared to the entire host galaxy) and incredible energy densities. These observations imply that these galaxies contain **supermassive black holes**—black holes with masses from thousands to tens of billions of solar masses. Violent accretion disks around these black holes power AGNs. You have learned about accretion disks several times in this text. Accretion disks surround young stars, providing the raw material for planetary systems. Accretion disks around white dwarfs, fueled by material torn from their bloated evolving companions, lead to novae and Type Ia supernovae. Accretion disks around neutron stars and stellar-mass black holes a few kilometers across are seen as X-ray binary stars. Now take these examples and scale them up to a black hole with a mass of a billion solar masses and a radius comparable in size to that of the orbit of Neptune. Furthermore, imagine an accretion disk being fed by several solar masses every year rather than by small amounts of material siphoned off a star. *That is an active galactic nucleus.*

As material moves inward toward the supermassive black hole, conversion of gravitational energy heats the accretion disk to hundreds of thousands of kelvins, causing it to glow brightly in visible and ultraviolet light. Conversion of gravitational energy to thermal energy as material falls onto the accretion disk is also a source of X-rays, UV radiation, and other energetic emission. When we discussed the Sun, we marveled at the efficiency of fusion, which converts 0.7

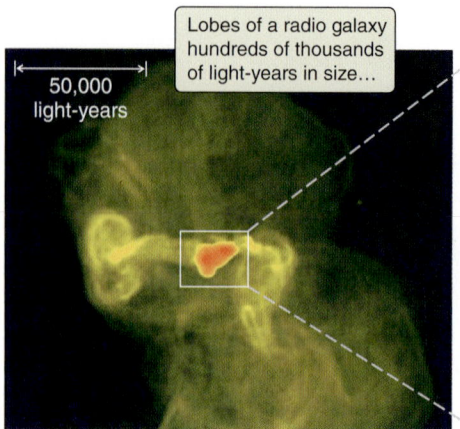
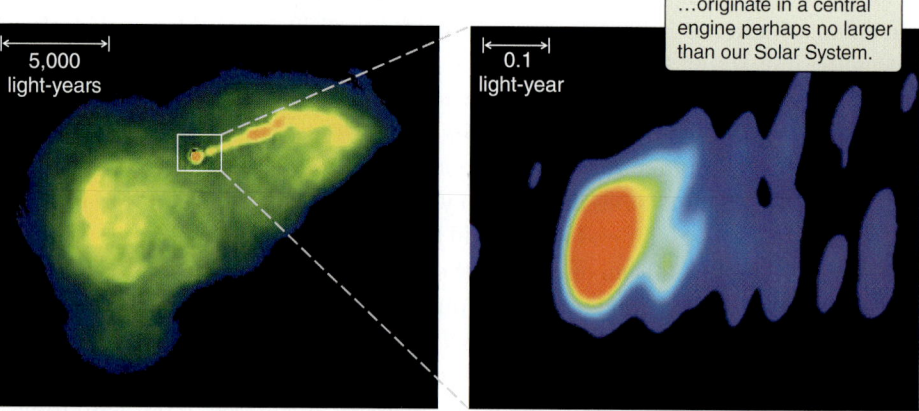

Figure 14.16 The visible jet from the galaxy M87 extends more than 30,000 parsecs, but it originates in a tiny volume at the heart of the galaxy.

percent of the mass of hydrogen into energy. In contrast, approximately 15 percent of the mass of infalling material around a supermassive black hole is converted to luminous energy. The rest of that mass falls into the black hole itself, causing it to grow even more massive.

The interaction of the accretion disk with the black hole creates powerful radio jets that emerge perpendicular to the disk (**Figure 14.16**). Throughout, twisted magnetic fields accelerate charged particles such as electrons and protons to relativistic speeds, producing a signature type of radiation called synchrotron radiation (Chapter 8). Gas in the accretion disk or in nearby clouds orbiting the central black hole at high speeds produces emission lines that are smeared out by the Doppler effect into the broad lines seen in many AGN spectra. This accretion disk surrounding a supermassive black hole is the "central engine" that produces AGNs. An outer torus, or "doughnut," of dust and gas plays a somewhat different role. Located far from the inner turmoil of the accretion disk, and far larger than the central engine, some of the outer torus is ionized by UV light from the AGN. This outer torus may obscure our view of the central engine in different ways, depending on our viewing angle. Viewing similar objects from different angles gives these objects a very different appearance, accounting for the different types of AGN.

Normal Galaxies and AGNs

The essential elements of an AGN are a central engine (an accretion disk surrounding a supermassive black hole) and a source of fuel (gas and stars flowing onto the accretion disk). Without a source of matter falling onto the black hole, an AGN would no longer be active. If we were to look at such an object, we would see a normal (not active) galaxy with a supermassive black hole sitting in its center.

Only a small percentage of present-day galaxies contain AGNs as luminous as the host galaxy. When the universe was younger, there were many more high-luminosity AGNs than exist today. If our understanding of AGNs is correct, then all the supermassive black holes that powered those dead AGNs should still be around. The number of AGNs in the past combined with ideas about how long a given galaxy remains in an active phase implies that many—perhaps even *most*—normal galaxies today contain supermassive black holes. This is a somewhat startling prediction that can be tested.

A concentration of mass at the center of a galaxy should have surrounding stars orbiting close to it. The central region of such a galaxy would be much brighter in the presence of such a mass than if stars alone were responsible for the gravitational field. Stars feeling the gravitational pull of a supermassive black hole in the center of a galaxy should also orbit at very high velocities and therefore show very large Doppler shifts. Astronomers have found evidence of this sort in every normal galaxy with a substantial bulge in which they have conducted a careful search. The masses inferred for these black holes range from 10,000 $M_\odot$ to 5 billion $M_\odot$. The mass of the supermassive black hole seems to be related to the mass of the elliptical-galaxy or spiral-galaxy bulge in which it is found. Most large galaxies probably contain supermassive black holes. The prediction is confirmed by observations. They also tell us something remarkable about the structure and history of normal galaxies.

Apparently, the only difference between a normal galaxy and an active galaxy is whether the supermassive black hole at its center is being fed at the time we see that galaxy. The rarity of present-day galaxies with very luminous AGNs does not indicate which galaxies have the potential for AGN activity. Rather, it indicates which galaxy centers are being lit up at the moment. If we were to drop a large amount of gas and dust directly into the center of any large galaxy, this material would fall inward toward the central black hole, forming an accretion disk and a surrounding torus. This process would change the nucleus of this galaxy into an AGN.

Galaxies do not exist in isolation, and they often interact. Interactions between galaxies can pull galaxies into distorted shapes in which stars and gas are drawn out into sweeping arcs and tidal tails. Galaxies that show evidence of recent interactions with other galaxies are more likely to house AGNs in their centers, because interactions can cause gas far from the center of a galaxy to fall inward to provide fuel for an AGN. During mergers, a significant fraction of a cannibalized galaxy might wind up in the accretion disk. Hubble Space Telescope images of quasars, like those in Figure 14.14, often show that quasar host galaxies are tidally distorted or are surrounded by other visible matter that is probably still falling into the galaxies. The most violent forms of AGN activity were common in the early universe because that was when galaxies were forming, and matter was drawn in by the gravity of newly formed galaxies. Interactions and mergers must have been much more prevalent when the universe was younger, which explains the larger number of AGNs that existed in the past. However, this process is still at work today.

Our understanding of AGNs is far from complete. For example, we do not know why one quasar can be a powerful radio source while another, identical in all other respects, emits no radio waves that we can detect, even with the most sensitive radio telescopes. Also, we still cannot predict how long an outburst of AGN activity will last or how often galaxies will undergo episodes of AGN activity. Any large galaxy, including our own, might be only a chance encounter away from containing an AGN.

▶‖ **AstroTour: Galaxy Interactions and Mergers**

14.5 We Live in an Expanding Universe

Studies of galaxies in the 1920s led to a revolution in the way we see the universe. The discovery that galaxies are separate from the Milky Way and located at great distances from us led astronomers to understand that the universe is much larger than previously thought. Astronomers also realized that there are far more stars than anyone had imagined. Perhaps most surprising was the knowledge that the universe changes over time—that it contains large-scale motions, that it had a beginning, and that it will have an end, of sorts. In this section, you will learn about the fundamental observation that led to the discovery that the universe is expanding.

The Discovery of Hubble's Law

In 1925, using the newly finished 100-inch telescope on Mount Wilson, high above the then small city of Los Angeles, Edwin Hubble was able to find some variable stars in the large neighboring Andromeda Galaxy. He recognized that these stars were very similar to the Cepheid variable stars in the Milky Way and the nearby Magellanic Cloud galaxies, though they appeared much fainter. Hubble used the period-luminosity relation for Cepheid variable stars (see Chapter 13) to find distances to the Andromeda Galaxy and several others. These galaxies, similar in size to our own galaxy, were located at truly immense distances.

Vesto Slipher (1875–1969) used telescopes at Lowell Observatory in Flagstaff, Arizona, to obtain spectra of these galaxies. Unsurprisingly, Slipher's galaxy spectra looked like the spectra of collections of stars with a bit of glowing interstellar gas mixed in. However, the emission and absorption lines in the spectra of these galaxies were seldom seen at the same wavelengths as in laboratory-generated spectra. Measurements of the Doppler velocities of these galaxies revealed that the spectral lines were shifted to longer, or redder, wavelengths, as shown in **Figure 14.17**. The galaxies appeared to be moving away from us, and the more distant they were, the faster they moved.

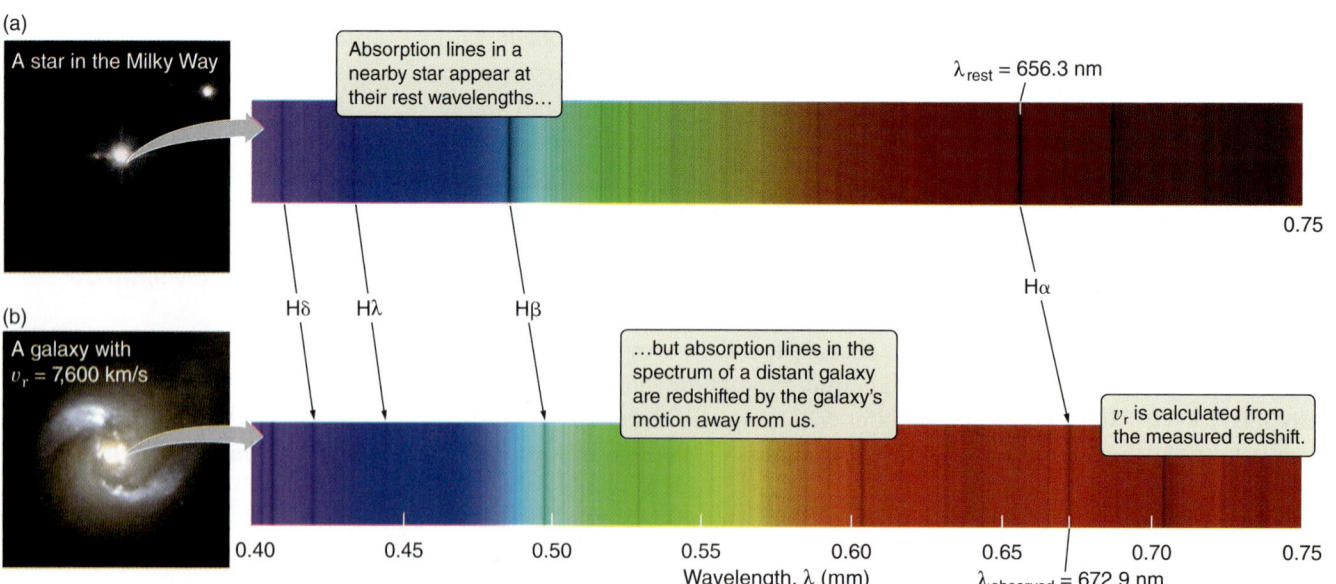

Figure 14.17 (a) The spectrum of a star in our galaxy shows absorption lines, which in this case lie at the rest wavelength. (b) A distant galaxy, shown with its spectrum at the same scale as that of the star, has lines that are redshifted to longer wavelengths. v_r is recession velocity, or radial velocity.

Working It Out 14.1 — Redshift: Calculating the Recession Velocity and Distance of Galaxies

The Doppler equation you learned for spectral lines showed that

$$v_r = \frac{\lambda_{observed} - \lambda_{rest}}{\lambda_{rest}} \times c$$

The fraction in front of the c is equal to z, the redshift. Substituting for the fraction, we get

$$v_r = z \times c$$

(Note: This correspondence works only for velocities much slower than the speed of light.)

Suppose a hydrogen line is seen in the spectrum of a distant galaxy. In the laboratory, this hydrogen line has a measured rest wavelength of 122 nanometers (nm). If the observed wavelength of the hydrogen line is 124 nm, then its redshift is

$$z = \frac{\lambda_{observed} - \lambda_{rest}}{\lambda_{rest}}$$

$$z = \frac{124 \text{ nm} - 122 \text{ nm}}{122 \text{ nm}}$$

$$z = 0.016$$

We can now calculate the recession velocity from this redshift:

$$v_r = z \times c = 0.016 \times 300{,}000 \text{ km/s} = 4{,}800 \text{ km/s}$$

How far away, though, is our distant galaxy? This is where Hubble's law and the Hubble constant ($H_0 = 70$ kilometers per second per megaparsec [km/s/Mpc]) come in. Hubble's law relates a galaxy's recession velocity to its distance and can be expressed mathematically as $v_r = H_0 \times d_G$, where d_G is the distance to a galaxy measured in millions of parsecs (that is, megaparsecs). We can divide through by H_0 to get

$$d_G = \frac{v_r}{H_0}$$

$$d_G = \frac{4{,}800 \text{ km/s}}{70 \text{ km/s/Mpc}} = 69 \text{ Mpc}$$

From a measurement of the wavelength of a hydrogen line, we have learned that the distant galaxy is approximately 69 Mpc away.

The Doppler shift causes the observed wavelengths of objects moving away from us to shift toward the red end of the spectrum. The wavelength in the laboratory is called the rest wavelength of the line, written λ_{rest}, and the redshift of a galaxy is written as z. Objects with higher values of z have greater redshifts and are therefore moving away from us more quickly. Hubble interpreted Slipher's redshifts as Doppler shifts, and he concluded that almost all of the galaxies in the universe are moving away from the Milky Way (**Working It Out 14.1**). When he combined these measurements of galaxy **recession** velocities with his own estimates of the distances to these galaxies, he made one of the greatest discoveries in the history of astronomy. Hubble found that *the apparent velocity at which a galaxy is moving away from us is proportional to the distance of that galaxy*. This simple relationship between distance and recession velocity has become known as **Hubble's law**. Hubble's law says that a galaxy at a distance of 30 Mpc from Earth moves away from us twice as fast as a galaxy at a distance of 15 Mpc from Earth.

H_0 (which astronomers pronounce as "H naught") is the constant of proportionality between the distance and the speed, so that $v = H_0 d$, and it is called the **Hubble constant**. This constant is one of the most important numbers in cosmology, and many astronomical careers have been dedicated to trying to determine its value precisely and accurately. In this text we use a value of 70 km/s/Mpc as an approximation to the best current value of 73 ± 2 km/s/Mpc.

▶▶ **Nebraska Simulation:** Galactic Redshift Simulator

▶‖ **AstroTour:** Hubble's Law

Vocabulary Alert

recession In common language, this word is associated almost entirely with economics. Astronomers use it to refer to an object that is receding, or moving away. The recession velocity is therefore the velocity at which an object moves away.

All Observers See the Same Hubble Expansion

Hubble's law is a remarkable observation about the universe that has far-reaching implications. For one thing, Hubble's law helps us test the prediction that the universe is both isotropic and homogeneous. We can confirm its isotropy by observing that galaxies in one direction in the sky obey the same Hubble law as galaxies in other directions in the sky. However, at first glance, you might think that Hubble's law suggests the universe is not homogeneous, as we seem to be sitting in a very special place: at the *center* of the universe, with everything else in the universe streaming away from us. This initial impression, however, is incorrect. As we will discuss in detail in Chapter 16, the recession velocities determined from the redshift are caused by the expansion of space between the observer and the galaxy, rather than the motion of galaxies through space. As light travels through the expanding space, it is stretched to longer, redder wavelengths. *Hubble's law actually says that we are sitting in an expanding universe and that the expansion always looks the same, regardless of our location.* To help you visualize this, we now turn to a useful model that you can build for yourself with materials you can probably find in your desk.

Figure 14.18 shows a long rubber band with paper clips attached along its length. If you stretch the rubber band, the paper clips, which represent galaxies in an expanding universe, get farther and farther apart. Imagine what this expansion would look like if you were an ant riding on paper clip A. As the rubber band is stretched, you notice that all of the paper clips are moving away from you. Clip B, the closest one, is moving away slowly. Clip C, located twice as far from you as B, is moving away twice as fast as B. Clip E, located four times as far away as B, is moving away four times as fast as B. From the perspective of an ant riding on clip A, all of the other paper clips on the rubber band are moving away with a velocity that is proportional to their distance. The paper clips located along the rubber band obey a Hubble-like law.

This demonstration of Hubble's law is a handy result, but the key insight comes from realizing there is nothing special about the perspective of paper clip A. If, instead, the ant were riding on clip E, clip D would be moving away slowly and clip A would be moving away four times as fast. Repeat this experiment for any paper clip along the rubber band, and you will arrive at the same result: The velocity at which other clips are moving away from the ant is proportional to their distance. The stretching rubber band, like the universe, is "homogeneous." The same Hubble-like law applies, regardless of where the ant is located.

The observation that nearby galaxies are carried away slowly by expanding space and distant galaxies are carried away more rapidly does not say that we are at the center of anything. *Hubble's law means that the universe is expanding uniformly.* Any observer in any galaxy will see nearby galaxies moving away slowly and more distant galaxies moving away more rapidly. The same Hubble law applies from their vantage point as applies from our vantage point on Earth.

Here is the only exception to this rule: In the case of galaxies that are gravitationally bound together, gravitational attraction dominates over the expansion of space. For example, the Andromeda Galaxy and the Milky Way are being pulled together by gravity. The Andromeda Galaxy and the Milky Way

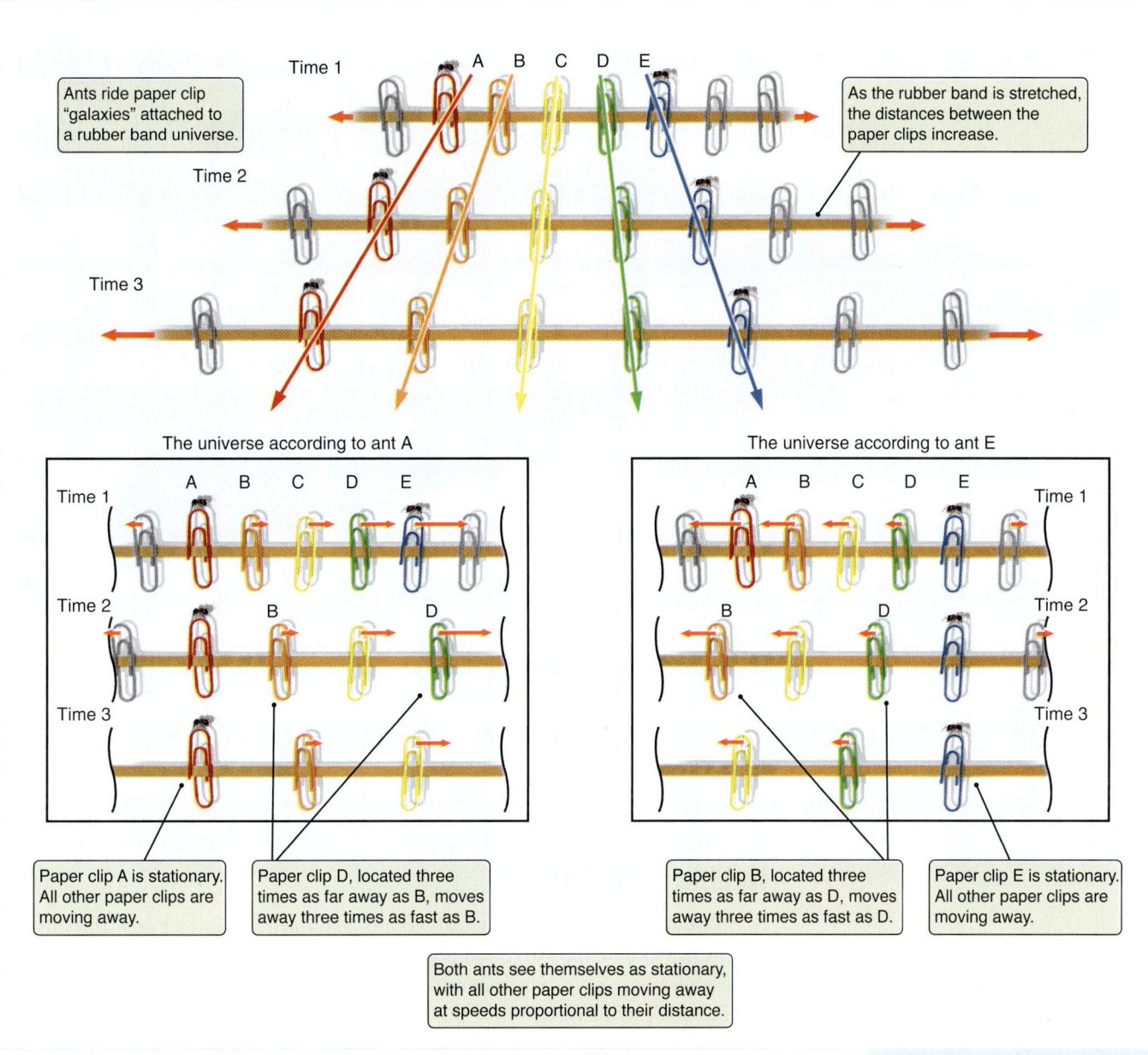

Figure 14.18 In this analogy of Hubble's law, a rubber band with paper clips evenly spaced along its length is stretched. As the rubber band stretches, an ant riding on clip A will observe clip C moving away twice as fast as clip B. Similarly, an ant riding on clip E will see clip C moving away twice as fast as clip D. Any ant will see itself as stationary, regardless of which paper clip it is riding, and it will see the other clips moving away with speed proportional to distance.

are approaching each other at about 300 kilometers per second (km/s), so light from the Andromeda Galaxy is blueshifted. This velocity is caused by the gravitational interaction between the two galaxies, rather than the expansion of space. The fact that gravitational or electromagnetic forces can overwhelm the expansion of space also explains why the Solar System is not expanding, and neither are you.

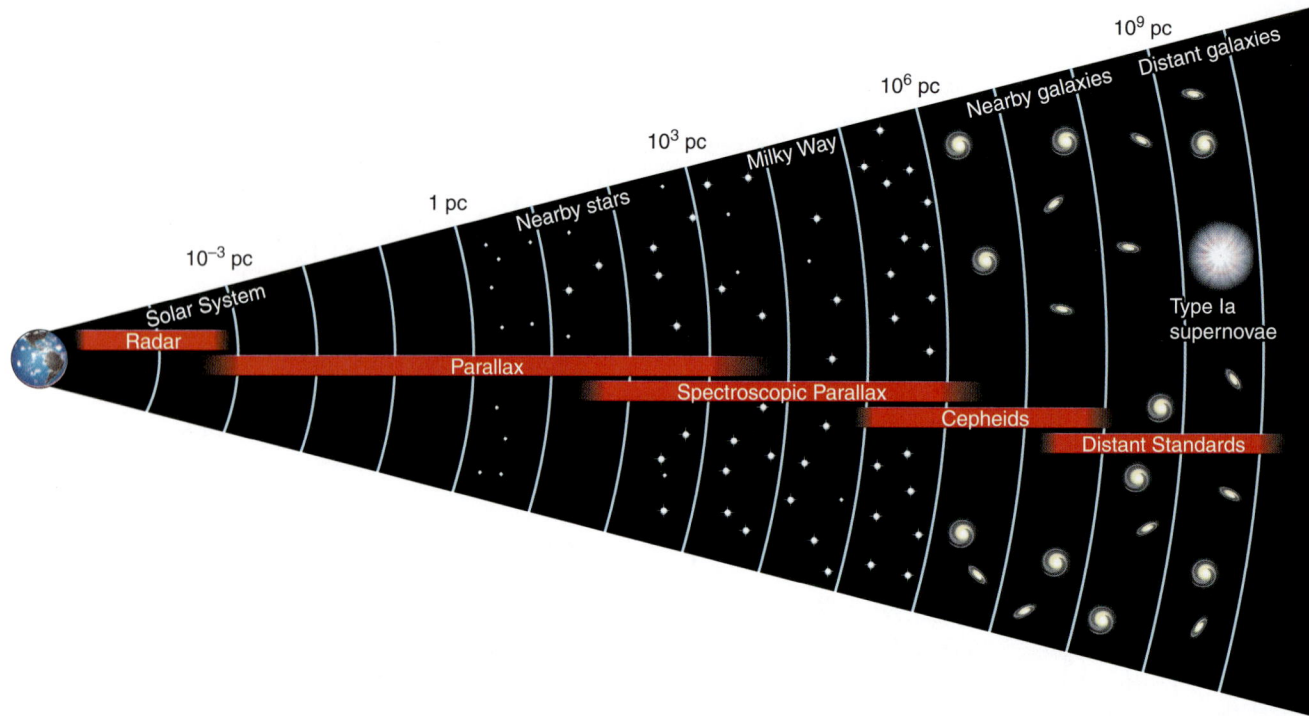

Figure 14.19 The distance ladder indicates how the distances to remote objects are achieved through a series of steps beginning with relatively nearby objects. Distances are given in parsecs (pc). Notice that the scale is logarithmic, so that each blue arc is 10 times farther than the one before it. This means that radar is only useful for a vanishingly tiny portion of the universe, and distant standards cover most of the universe.

Astronomers Build a Distance Ladder to Measure the Hubble Constant

Hubble's law tells us that our universe is expanding. But to know the current *rate* of the expansion, we need a good value for the Hubble constant, H_0. This requires knowing both the recession velocity and the distance of a large number of galaxies, including galaxies that are very far away.

Distances of remote objects are measured in a series of steps referred to as the **distance ladder**, which relates distances on a variety of scales, as illustrated in **Figure 14.19**. Within the Solar System, we can find distances using radar and signals from space probes. Once the distance to the Sun is known, we can use parallax (as discussed in Chapter 10) to measure distances to nearby stars, find their luminosities, and plot them on the H-R diagram. Astronomers use parallax to calibrate the diagram, so that we need to know only a main-sequence star's temperature (relatively easy to measure from its spectrum) to find its luminosity. This information in turn enables us to estimate the star's distance by comparing its *apparent* brightness with its luminosity. This process for finding the distance to stars is known as *spectroscopic parallax*.

Moving farther out, we can measure the distance to relatively nearby galaxies using standard candles. More distant objects of the same type, such as O stars, globular clusters, planetary nebulae, novae, and variable stars, are fainter, or smaller. Cepheid variables, an example of a standard candle, take us another step

Colliding Galaxies (cont.)

X-ray Observatory (blue), the Hubble Space Telescope (gold), and the Spitzer Space Telescope (red).

The X-ray image from Chandra shows vast clouds of hot, interstellar gas that are injected with rich deposits of elements, such as oxygen, iron, magnesium, and silicon, created in supernova explosions. The enriched gas will be incorporated into new generations of young stars and planets.

The Antennae galaxies get their name from the long, wispy antenna-like "arms" that can be detected in wide-angle views of the system. These appendage features were produced by tidal forces that were generated from the cosmic collision.

The bright, pointlike sources in the image are produced by material that is falling onto black holes and neutron stars, which are dead relics of massive stars. Some of the black holes in the Antennae galaxies may contain masses that are almost 100 times that of the Sun.

The data from the Spitzer telescope show infrared light from warm clouds of dust that have been heated by newly formed stars, with the brightest clouds lying in the overlap region between the two colliding galaxies.

The Hubble data reveal old stars and star-forming regions in gold and white, while dusty filaments appear in brown. In the optical image, many of the fainter objects denote clusters containing thousands of stars.

The Chandra image was taken in December 1999, the Spitzer image was taken in December 2003, and the Hubble image was taken in July 2004 and February 2005. Data from the three observatories were combined to make the new composite image.

Evaluating the News

1. In the article, the author states that the collision "has triggered the formation of millions of stars." How might we know that? In what part of the colliding galaxies is this star formation likely to be taking place?
2. These galaxies lie 62 million light-years (19 Mpc) away from Earth. What part of the distance ladder might have given this distance?
3. Explain why Chandra pictures show hot gas, Spitzer pictures show warm dust, and Hubble data show stars and some dust.
4. The author states that "The Antennae galaxies get their name from the long, wispy antenna-like 'arms' that can be detected in wide-angle views of the system. These appendage features were produced by tidal forces that were generated from the cosmic collision." What causes these tidal forces? Why do tidal forces produce long, wispy arms?
5. Is this image an important one for astronomers? For the public? For you? Explain the significance of this image, which was created from old data.

Figure 14.22 This beautiful composite image of two colliding galaxies was released by NASA's Great Observatories program. The collision between the Antennae galaxies, which are located about 62 million light-years (19 Mpc) from Earth, began more than 100 million years ago and is still occurring. (Credit: NASA, ESA, SAO, CXC, JPL-Caltech, and STScI)

SUMMARY

Galaxies are classified based on their shape and the types of orbits of their stars. The arms of spiral galaxies are regions of intense star formation, and the concentration of young stars makes the arms more visible. Disturbances of the disk can lead to spiral arms, but persistent spiral arms are the result of density waves. Most of the mass of a galaxy is dark matter, which interacts with light very weakly, if at all. The form of this matter is not yet known. Most galaxies have a supermassive black hole at the center, which may become an active galactic nucleus (AGN) if gas accretes onto it. The observation of galaxies led to a revolution in our thinking about the universe by showing that the universe is expanding.

LG 1 Spiral galaxies are distinguished by their flat disk and spiral arms. The stars in this disk all orbit the center of the galaxy in the same direction. Elliptical galaxies are roughly egg-shaped, and the stars orbit in all directions. Irregular galaxies are galaxies that fit neither of these classifications, usually because they are interacting with another galaxy.

LG 2 The disks of spiral galaxies contain a lot of gas and dust. The arms are locations that are disturbed by spiral density waves, which trigger star formation in the dust and gas.

LG 3 Dark matter is identified by its gravitational interaction with ordinary matter. However, it has never been observed directly, and it interacts with light either very weakly or not at all.

LG 4 Observations of distant galaxies such as quasars reveal extremely luminous, compact sources near the centers of galaxies. These active galactic nuclei are best explained as supermassive black holes, surrounded by an accretion disk and a torus of dust and gas. Stars orbiting the centers of nearby galaxies have such high velocities that there must also be a supermassive black hole at the center of each of these galaxies as well. Most (perhaps all) large galaxies contain a supermassive black hole at the center.

LG 5 Observations of the distance and the velocity of galaxies show that the two factors are related: more distant galaxies move away from us faster. This observation means that the universe is expanding, and it implies a beginning of the universe.

SUMMARY SELF-TEST

1. A particular galaxy contains only stars in disordered orbits. This type of galaxy is most likely
   a. an elliptical.
   b. a spiral.
   c. an irregular.

2. In spiral galaxies, stars form predominantly in
   a. the arms of the disk.   b. the halo.
   c. the bulge.   d. the bar.

3. The rotation curves of galaxies tell us that galaxies are mostly
   a. stars.   b. dust and gas.
   c. dark matter.   d. black holes.

4. Supermassive black holes
   a. are extremely rare. There are only a handful in the universe.
   b. are completely hypothetical.
   c. occur in most, perhaps all, large galaxies.
   d. occur only in the space between galaxies.

5. If the universe were not expanding, the relationship between the velocity of a galaxy and its distance (as in the Hubble law plot) would
   a. be horizontal.
   b. follow a downward trend.
   c. first go up, then flatten out.
   d. start out flat, then fall.

6. Which of the following are properties of dark matter?
   a. It absorbs all the light that falls on it.
   b. It reflects all the light that falls on it.
   c. It doesn't interact with light, as far as we can tell.
   d. It gravitationally attracts other matter.
   e. It gravitationally repels other matter.
   f. It doesn't interact gravitationally, as far as we can tell.

7. Supermassive black holes
   a. gradually consume their host galaxies.
   b. are "fed" by disturbed gas when galaxies interact.
   c. are a recent development in the universe's history.
   d. are an as yet untested hypothesis.

8. The light that comes to us from active galactic nuclei comes from a volume about the size of
   a. Earth.   b. the Solar System.
   c. a globular cluster.   d. the bulge of the Milky Way.

9. In every direction that astronomers look, they see the same Hubble law, with the same slope. This is an example of
   a. isotropy.
   b. homogeneity.
   c. hydrostatic equilibrium.
   d. energy balance.

10. Rank the following galaxies in order of increasing distance from the Milky Way.
    a. a galaxy with a recession velocity of +200 km/s
    b. a galaxy with a recession velocity of +400 km/s
    c. a galaxy with a recession velocity of −50 km/s
    d. a galaxy with a recession velocity of +500 km/s

QUESTIONS AND PROBLEMS

Multiple Choice and True/False

11. **T/F:** Galaxies are sometimes difficult to classify because their orientation affects their appearance.

12. **T/F:** The Hubble tuning fork diagram shows how galaxies evolve (change over time).

13. **T/F:** The orbits of stars in the bulge of a spiral galaxy are disordered.

14. **T/F:** Astronomers have evidence of dark matter in only a few galaxies to date.

15. **T/F:** Active galactic nuclei surround some supermassive black holes.

16. If a galaxy has some stars in ordered orbits, it is most likely a(n)
    a. spiral.   b. irregular.
    c. elliptical.   d. giant elliptical.

17. The orbits of stars in the bulges of spiral galaxies most closely resemble
    a. orbits of planets in the Solar System.
    b. orbits of stars in the disk of a spiral galaxy.
    c. orbits of stars in an elliptical galaxy.
    d. orbits of objects in the Kuiper Belt.

18. Most galaxies have sizes in the range of
    a. hundreds of parsecs.
    b. tens of thousands of parsecs.
    c. hundreds of thousands of parsecs.
    d. millions of parsecs.

19. The flat rotation curves of spiral galaxies imply that the distribution of mass resembles
    a. the Solar System; most mass is concentrated in the center.
    b. a wheel; the density remains the same as the radius increases.
    c. the light distribution of the galaxy; a large concentration occurs in the middle, but significant mass exists quite far out.
    d. an invisible sphere much larger than the visible galaxy.

20. No matter where you travel in space, you would observe very similar distributions of galaxies. This is an example of
    a. isotropy.
    b. homogeneity.
    c. hydrostatic equilibrium.
    d. energy balance.

21. Astronomers know active galactic nuclei are relatively small because
    a. variations in brightness occur quickly.
    b. variations in brightness occur slowly.
    c. variations in brightness are small.
    d. they have been directly imaged.

22. Emission lines from a spinning accretion disk around a black hole in an active galactic nucleus would be
    a. always redshifted.
    b. always blueshifted.
    c. both redshifted and blueshifted.
    d. it depends on the viewing angle

23. Astronomers observe two galaxies, A and B. Galaxy A has a recession velocity of 2,500 km/s, while galaxy B has a recession velocity of 5,000 km/s. This means that
    a. galaxy A is four times as far away as galaxy B.
    b. galaxy A is twice as far away as galaxy B.
    c. galaxy B is twice as far away as galaxy A.
    d. galaxy B is four times as far away as galaxy A.

24. The distances to galaxies used to establish Hubble's law are found from
    a. radar.
    b. parallax.
    c. spectroscopic parallax.
    d. standard candles, such as supernovae.

25. The Hubble constant is found from the
    a. slope of the line fit to the data in Hubble's law.
    b. y-intercept of the line fit to the data in Hubble's law.
    c. spread in the data in Hubble's law.
    d. inverse of the slope of the line fit to the data in Hubble's law.

Conceptual Questions

26. Sketch the galaxy shown in Figure 14.5b. Label the disk, the bulge, and the halo.

27. How does gas temperature differ between elliptical and spiral galaxies?

28. Describe the spiral arms in a galaxy, and explain at least one of the mechanisms that create them. Explain why star formation in spiral galaxies takes place mostly in the spiral arms.

29. What is the difference between the way dark matter interacts with electromagnetic radiation and the way normal matter does?

30. Contrast the rotation curve for a galaxy containing only normal matter and a galaxy containing a large amount of dark matter.

31. How does a spiral galaxy's dark matter halo differ from its visible spiral component?

32. The nearest bright quasar is about 750 Mpc away. Why do we not see any that are closer?

33. Contrast the size of a typical AGN with the size of our own Solar System. How do we know how big an AGN is?

34. Describe what must be happening at the centers of galaxies that contain AGNs.

35. Imagine that you are standing in the middle of a dense fog.
    a. Would you describe your environment as isotropic? Why?
    b. Would you describe it as homogeneous? Why?

36. a. Early in the 20th century, astronomers discovered that most galaxies are moving away from the Milky Way (that is, they are redshifted). What was the significance of this discovery?
    b. Edwin Hubble later made an even more important discovery: The speed with which galaxies are receding is proportional to their distance. Why was this among the more important scientific discoveries of the 20th century?

37. If you lived in a remote galaxy, would you observe distant galaxies receding from your own galaxy or would they appear to be approaching? Explain your answer.

38. Why is the Milky Way Galaxy not expanding along with the rest of the universe?

39. Suppose that the universe is not expanding but is contracting. Sketch what Figure 14.21b should look like in that case. Now, suppose that it was doing neither but is static. Sketch Figure 14.21b for the case of a static universe.

40. Examine Figure 14.19, which shows the cosmic distance ladder. Use the information in this figure to make a table that shows which distance method is used for which range of distances. In a third column, indicate which method covers the largest volume of space and which covers the smallest.

Problems

41. Suppose the number density of galaxies in the universe is, on average, 3×10^{-68} galaxies per cubic meter (m^3). If astronomers could observe all galaxies out to a distance of 10^{10} parsecs, how many galaxies would they find?

42. Figure 14.11a shows the distribution of dark and normal matter that must be present to create the observed rotation curve.
    a. Explain what is plotted in blue and in red. Is this the amount of mass present at that radius? If so, why does it continuously increase as the radius increases? If not, what is actually plotted here?
    b. At what radius is there the same enclosed mass of dark matter as normal matter?
    c. Imagine that the plot was many times wider than shown, and the radii extended billions of parsecs. What do you expect would happen to the blue line as you got very, very far from the galaxy?

43. Using the data from Figure 14.17, find the shift in wavelength of the Hα line. Use this to confirm the recession velocity of the galaxy shown.

44. Using the data from Figure 14.17, find the distance to the galaxy shown.

45. The quasar 3C 273 has a luminosity of 10^{12} $L_\odot$. Assuming that the total luminosity of a large galaxy, such as the Andromeda Galaxy, is 10 billion times that of the Sun, compare the luminosity of 3C 273 with that of the entire Andromeda Galaxy.

46. Consider a hypothetical star orbiting 3C 273 at a distance of 100,000 astronomical units (AU) with a period of 1,080 years. What is the mass of 3C 273 in solar masses?

47. A quasar has the same brightness as a foreground galaxy that happens to be 2 Mpc distant. If the quasar is 1 million times more luminous than the galaxy, what is the distance of the quasar?

48. You read in the newspaper that astronomers have discovered a "new" astronomical object that appears to be flickering with a period of 83 minutes. Having read this book, you are quickly able to estimate the maximum size of this object. How large can it be?

49. A quasar has a luminosity of 10^{41} watts (W), or joules per second (J/s), and 10^8 $M_\odot$ to feed it. Assuming constant luminosity and that 15 percent of the mass is converted to light, estimate the quasar's lifetime.

50. Examine Figure 14.21. Is this a linear plot or a logarithmic one? How do you know?

SMARTWORK

Norton's online homework system includes algorithmically generated versions of these questions, plus additional conceptual exercises. If your instructor assigns questions in SmartWork, log in at smartwork.wwnorton.com.

Exploration | Galaxy Classification

Galaxy classification sounds simple, but it can become complicated when you actually attempt it. **Figure 14.23**, taken by the Hubble Space Telescope, shows a small portion of the Coma Cluster of galaxies. The Coma Cluster contains thousands of galaxies, each containing billions of stars. Some of the objects in this image (the ones with a bright cross) are foreground stars in the Milky Way. Some of the galaxies in this image are far behind the Coma Cluster. Working with a partner, you will classify the 20 or so brightest galaxies in this cluster.

First, make a map by laying a piece of paper over the image and numbering the 20 or so brightest (or largest) galaxies in the image. Copy this map so that you and your partner each have a list of the same galaxies.

Separately, classify the galaxies (label them galaxy 1, galaxy 2, and so forth). If it is a spiral galaxy, what is its subtype: a, b, or c? If it is an elliptical galaxy, how elliptical is it? Make a table that contains the galaxy number, the type you have assigned, and any comments that will help you remember why you made that choice.

When you are done classifying, compare your list with your partner's. Now comes the fun part! Argue about the classifications until you agree—or until you agree to disagree.

If you find this activity interesting and rewarding, astronomers can use your help: Go to http://www.galaxyzoo.org to get involved in a "citizen science" project to classify galaxies, some of which have never been viewed before by human eyes.

1. Which galaxy type was easiest to classify?

2. Which galaxy type was hardest to classify?

3. What makes it hard to classify some of the galaxies?

4. Which galaxy type did you and your partner agree about most often?

5. Which galaxy type did you and your partner disagree about most often?

6. How might you improve your classification technique?

Figure 14.23 This Hubble Space Telescope image of the Coma Cluster shows a diversity of shapes.

SMARTWORK • smartwork.wwnorton.com

15 Our Galaxy: The Milky Way

We live in a universe full of galaxies of many sizes and types, visible in our most powerful telescopes all the way to the edge of the observable universe. Yet when we gaze at the night sky unaided by telescopes, we do not see this universe of galaxies. Rather, the night sky is filled with a single galaxy—our home, the galaxy we call the Milky Way. The figure on the opposite page shows a photograph of the Milky Way taken by a student. The dark lanes in the photograph imply that we live in a spiral galaxy.

Does the Milky Way have prominent spiral arms? Is the bulge barlike, and is it large or small relative to the disk? These questions are easier to ask and answer for distant galaxies than for the Milky Way, because we are inside it. Yet we have already learned far more about our galaxy than about any other. Stars, planets, and the interstellar medium—almost everything we know about them has been learned within the context of our own galactic home. It is time to merge the knowledge of other galaxies with our knowledge of our own locality to understand our Milky Way as a spiral galaxy.

✦ LEARNING GOALS

Of the hundreds of billions of galaxies in the universe, the Milky Way is the only one that we can study at close range. In this chapter, we focus on the Milky Way and how it offers clues to understanding all galaxies. By the end of this chapter, you should be able to move between visualizing the Milky Way as a band of light in the sky and visualizing it from other perspectives, such as from far above the disk. You should also be able to:

LG 1 Explain how to measure the size of the Milky Way using variable stars in globular clusters.

LG 2 Sketch the Milky Way's structure.

LG 3 Explain how differences in the age and chemical composition of groups of stars tell us about the history of star formation and the evolution of chemical composition in the Milky Way.

LG 4 Describe the evidence for dark matter and for the black hole at the center of the Milky Way.

LG 5 List the clues about galaxy formation that the Milky Way provides.

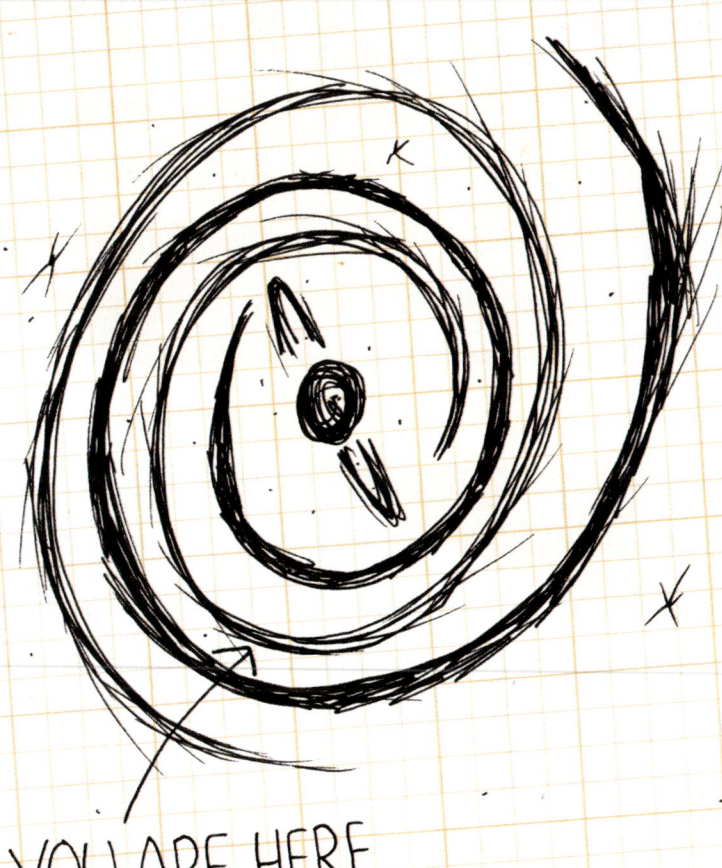

YOU ARE HERE

THE MILKY WAY

... as seen during last Saturday's star party

(120-second exposure)

15.1 Measuring the Milky Way Is a Challenge

As you learned in Chapter 14, the universe is full of galaxies of many shapes and sizes. Because Earth is embedded within the Milky Way, the details of the shape and structure of the Milky Way are actually more difficult to determine than for other galaxies. Comparing observations of the Milky Way with observations of more distant galaxies improves our understanding of the Milky Way. Astronomers use observations of much smaller objects within the Milky Way to determine its size and our location within it.

Spiral Structure in the Milky Way

Figure 15.1a shows the Milky Way in Earth's night sky. Compare this image to the image in Figure 15.1b, which shows an edge-on spiral galaxy. Simply by comparing the similarities between these two images, you might determine that the Milky Way is a spiral galaxy and that we are viewing it edge-on, from inside the disk. Confirming this hunch and finding further information about the size and shape of the Milky Way require more detailed observations in the visible, infrared, and radio regions of the electromagnetic spectrum.

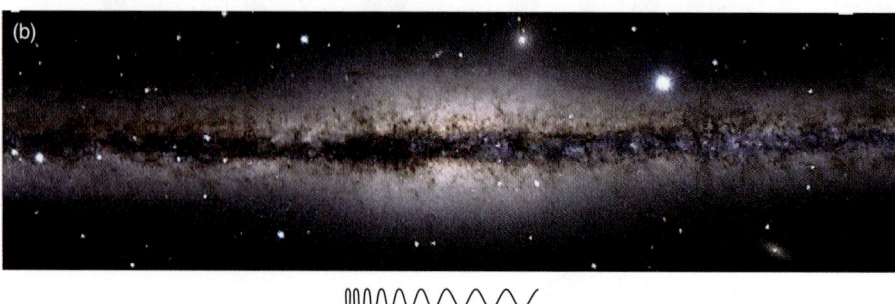

Figure 15.1 (a) The Milky Way is seen as a luminous band stretching across the night sky. Note the prominent dark lanes caused by interstellar dust that obscures the light from more distant stars. (b) The disk of the edge-on spiral galaxy NGC 891 greatly resembles the Milky Way.

Neutral hydrogen in the interstellar medium emits radiation at a wavelength of 21 cm, in the radio region of the spectrum. **Figure 15.2** shows the full sky, as mapped in 21-cm radiation from neutral hydrogen. Because of its long wavelength, 21-cm radiation freely penetrates dust in the interstellar medium, enabling us to see neutral hydrogen throughout our galaxy. This radiation was predicted in the 1940s and observed in the 1950s. By 1952, astronomers had the first maps of the neutral hydrogen in the Milky Way and other galaxies, tracing out the location of the gas in these galaxies. The maps showed spiral structure in the other galaxies and suggested spiral structure in the Milky Way. At about the same time, observations of ionized hydrogen gas in visible light showed two spiral arms with concentrations of young, hot O and B stars. These observations together confirmed that the Milky Way is a spiral galaxy.

In the 1990s, some hints emerged that the Milky Way is a barred spiral, with an elongated bulge. This was later confirmed using infrared observations from the Spitzer Space Telescope. These observations of the distribution and motion of stars toward the center of the galaxy showed a substantial bar with a modest bulge at the center (**Figure 15.3**). Two major spiral arms connect to the ends of the central bar. There are several smaller arm segments, including the Orion Spur, which contains the Sun and Solar System. The Milky Way is a giant barred spiral that is more luminous than the average spiral. If viewed from the outside, the Milky Way would look much like the barred spiral galaxy M109, shown in **Figure 15.4**.

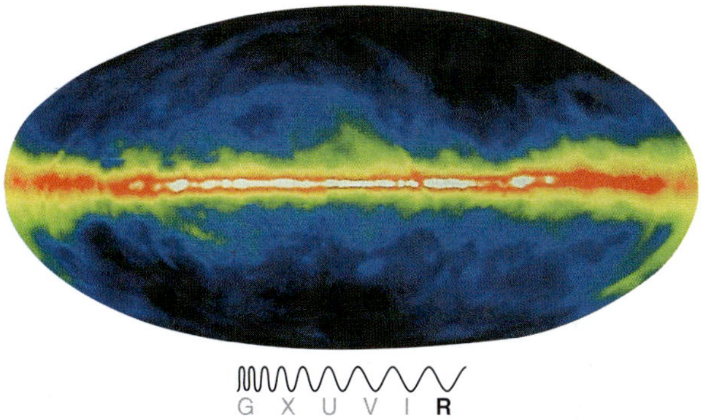

Figure 15.2 This radio image of the sky taken at a wavelength of 21 cm shows clouds of neutral hydrogen gas throughout our galaxy. Because radio waves penetrate interstellar dust, 21-cm observations are a crucial method for probing the structure of our galaxy.

The Size of the Milky Way

To find the size of the Milky Way, astronomers must observe much smaller objects and then infer the size of the Milky Way from the distribution of these objects.

Figure 15.3 Infrared and radio observations contribute to an artist's model of the Milky Way Galaxy. The galaxy's two major arms (Scutum-Centaurus and Perseus) are attached to the ends of a thick central bar. The Sun and Solar System are located within the Orion Spur, which is situated between the two major arms.

Figure 15.4 From the outside, the Milky Way would look much like this barred spiral galaxy, M109.

(a)

(b)

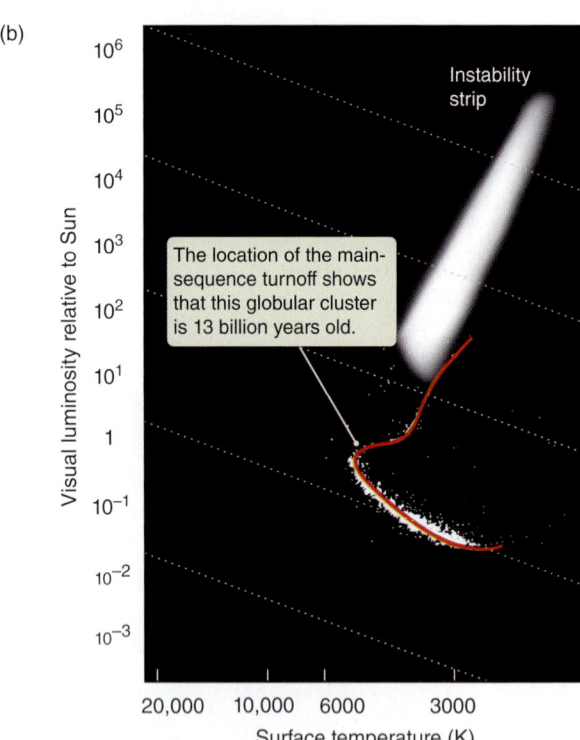

Figure 15.5 (a) The globular cluster M92 and (b) an H-R diagram of the stars it contains. The main-sequence turnoff indicates the age of the cluster.

A **globular cluster**, such as M92 shown in **Figure 15.5a**, contains between tens of thousands and a million stars in a spherical shape. Many such clusters can be seen through small telescopes. The motions of stars within a globular cluster are much like the motions of stars within an elliptical galaxy. However, globular clusters are much smaller and the stars are closer together. Because globular clusters reside within a sphere around the Milky Way, astronomers use them to map out the galaxy.

The Milky Way has more than 150 cataloged globular clusters (and there are likely many more—dust in the disk may hide them from view). The known globular clusters have luminosities ranging from a low of about 1,000 solar luminosities (1,000 $L_\odot$) to a high of about 1 million $L_\odot$. A typical globular cluster consists of 500,000 stars packed into a volume of space with a radius of only 15 light-years—much more crowded than the average density within the Milky Way.

About one-fourth of the globular clusters in the Milky Way reside in or near the disk. The rest occupy a large, spherical volume of space surrounding the disk and bulge, referred to as the halo of the Milky Way. These globular clusters lie within the dark matter halo we discussed in Chapter 14. Globular clusters are very luminous and most lie outside the dusty disk, so they can be easily seen at great distances. To find the distance to a globular cluster, we need to look at the properties of the stars they contain.

Globular clusters contain good standard candles, mainly Cepheid variables and RR Lyrae stars, as discussed in Chapter 13. Figure 15.5b shows the Hertzsprung-Russell (H-R) diagram for M92's stars. The diagram's main-sequence turnoff occurs for stars with masses of about 0.8 $M_\odot$, which corresponds to a main-sequence lifetime of close to 13 billion years. This age is typical for globular clusters, making them the oldest known objects in our galaxy. Globular clusters formed when the universe and our galaxy were very young. Compared to globular-cluster stars, our Sun, at about 5 billion years old, is a relatively young member of our galaxy.

In an H-R diagram of an old cluster, the horizontal branch crosses the instability strip. RR Lyrae stars are easy to spot in globular clusters because they have a distinctive light curve. Astronomer Henrietta Leavitt (1868–1921) determined that there is a relationship between the period and the luminosity for RR Lyrae stars. Harlow Shapley used this period-luminosity relationship to find the luminosities of RR Lyrae stars in globular clusters. He then used the inverse square law of radiation to combine these luminosities with measured brightnesses to determine the distances to globular clusters. Finally, Shapley cross-checked his results by noting that more distant clusters (measured with his standard candle) also tended to appear smaller in the sky, as expected.

Shapley made a three-dimensional map of globular clusters from these distances and the locations of the clusters in the sky. This map showed that globular clusters occupy a roughly spherical region of space with a diameter of about 300,000 light-years. These globular clusters trace out the halo of the Milky Way Galaxy, as shown in **Figure 15.6**, which reflects the modern view of the globular-cluster distribution.

The distance to the center of the galaxy cannot be measured directly, in part because it is obscured by dust, gas, and stars. Globular clusters orbit the gravitational center of the galaxy, so the center of the distribution of globular clusters is the same as the gravitational center of the galaxy. The distance from the Sun to this central point is about 27,000 light-years, placing us roughly halfway out toward the edge of the galactic disk.

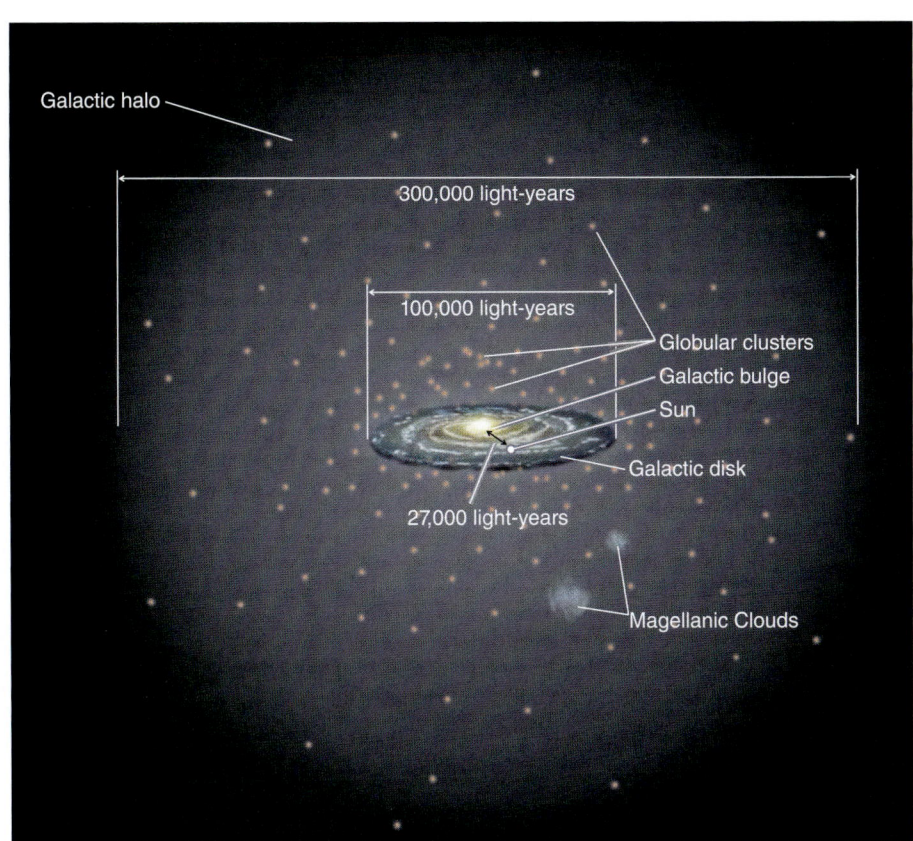

Figure 15.6 This diagram of the disk, bulge, and luminous halo of the Milky Way Galaxy also shows the distribution of globular clusters as well as two companions to the Milky Way (the Large and Small Magellanic Clouds) and the location of the Sun within the Milky Way disk.

Figure 15.7 (a) The open star cluster NGC 6530 is located 5,200 light-years away, in the disk of our galaxy. (b) The H-R diagram of stars in this cluster indicates that it has an age of a few million years or less, as the cluster does not show a main-sequence turnoff.

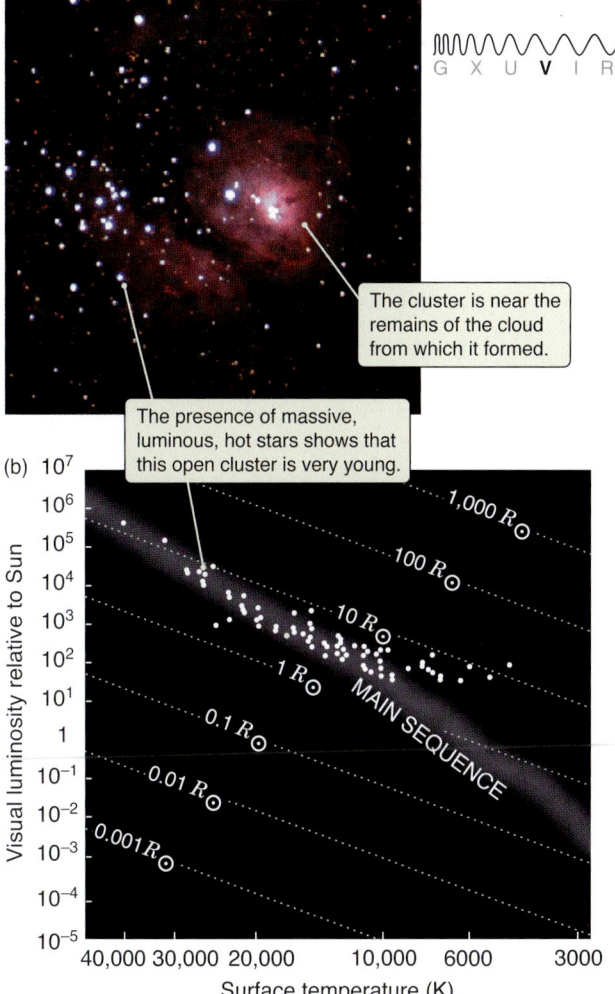

15.2 Components of the Milky Way Reveal Its Evolution

Our Sun is a middle-aged star located among other middle-aged stars that orbit around the galaxy within the disk. Yet near the Sun are other stars—most much older—that are a part of the halo of the Milky Way and whose orbits are carrying them *through* the disk. Using the ages, chemical abundances, and motions of nearby stars, we can differentiate between disk and halo stars to learn more about the galaxy's structure.

Age and Chemical Composition of Stars

Most globular clusters are in the halo. With ages of up to 13 billion years, they are among the oldest objects known; in fact, no young globular clusters have been observed. In contrast, **open clusters**, like the one shown in **Figure 15.7**, are much less tightly bound collections of a few dozen to a few thousand stars that are found in the disk of a spiral galaxy. As with globular clusters, the stars in an open cluster all formed in the same region at about the same time. Open clusters have a wide range of ages. Some contain the very youngest stars known, while others contain stars somewhat older than the Sun. Because open clusters are loosely bound together, they are easily disrupted by the tidal forces caused by the gravity from nearby objects, so they do not survive long in the disk of our galaxy. The oldest open clusters are several billion years younger than the youngest globular clusters.

The difference in ages between globular and open clusters indicates that stars in the halo formed first, but this epoch of star formation did not last long. Star formation in the disk started later but has been continuing ever since. Star formation processes in the massive, compact globular clusters also must have been much different from the processes in the less massive, more scattered open clusters.

When the universe was very young, only the least massive elements existed. All elements more massive than boron must have formed in the cores of stars. For this reason, the abundance of massive elements in the interstellar medium provides a record of all the star formation that has taken place up to the present time. Gas that is rich in massive elements has gone through a great deal of stellar processing, whereas gas that is poor in massive elements has not.

In turn, the abundance of massive elements in the atmosphere of a star provides a snapshot of the chemical composition of the interstellar medium *at the time the star formed.* (In main-sequence stars, material from the core does not mix with material in the atmosphere, so the amounts of chemical elements inferred from the spectra of a star are the same as the amounts in the interstellar gas from which the star formed.) As illustrated in **Figure 15.8**, the chemical composition of a star's atmosphere reflects the cumulative amount of star formation that has occurred up to the moment it formed.

Stars in globular clusters, among the earliest stars to form, contain only very small amounts of massive elements. Some globular-cluster stars contain only 0.5 percent as much of these massive elements as our Sun has. This relationship between age and abundances of massive elements is evident throughout much of the galaxy. Lower abundances of heavy elements characterize not just

Figure 15.8 As subsequent generations of stars form, live, and die, they enrich the interstellar medium with massive elements—the products of the formation of new elements in stars. The chemical evolution of our galaxy and other galaxies can be traced in many ways, including by the strength of interstellar emission lines and stellar absorption lines.

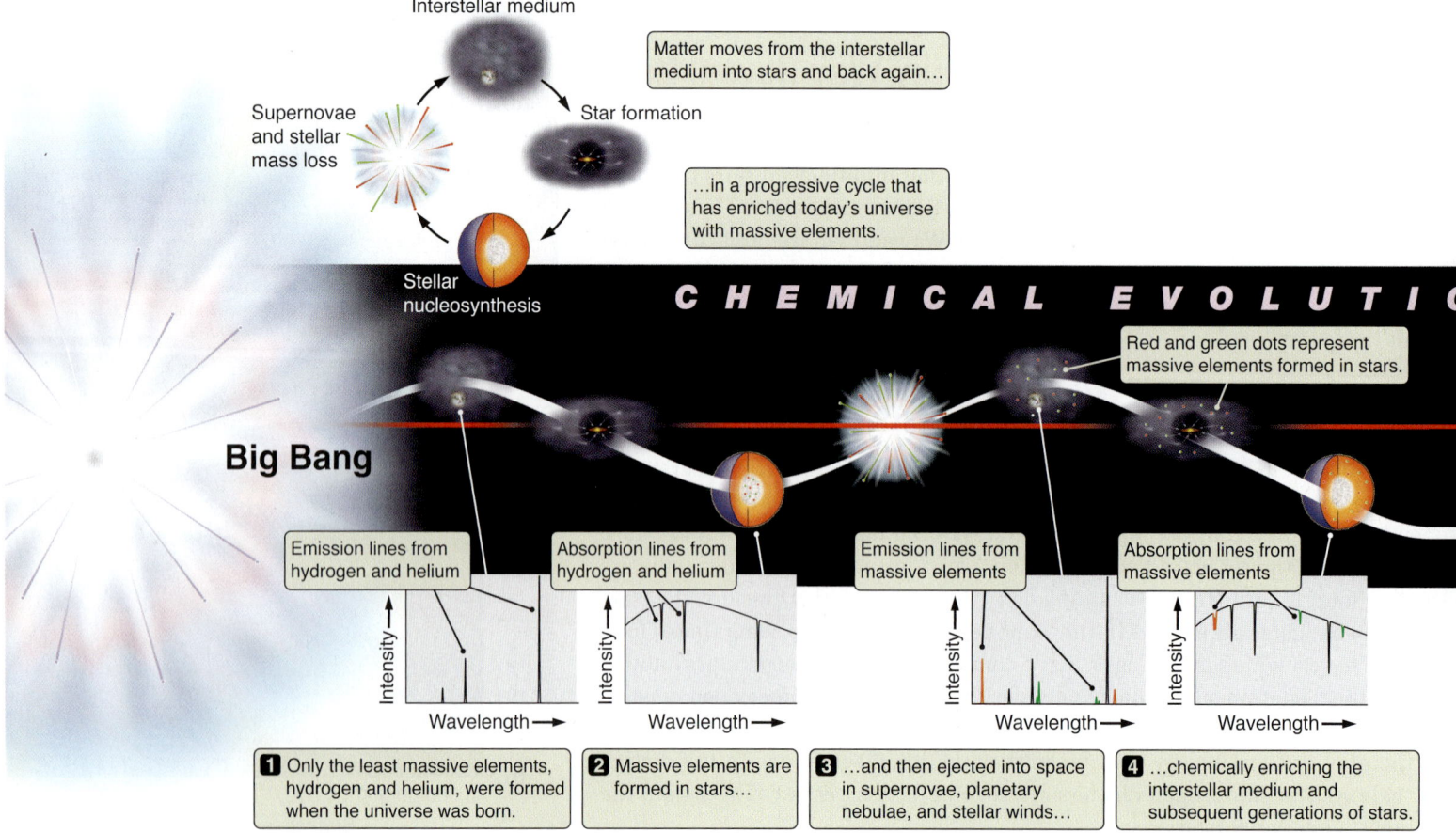

globular-cluster stars, but all of the stars in our galaxy's halo. Within the disk, the chemical evolution of the Milky Way has continued as generations of stars have further enriched the interstellar medium with the products of their nuclear fusion processes. Older disk stars are typically poorer in massive elements than younger disk stars. Similarly, older stars in the outer parts of our galaxy's bulge are poorer in massive elements than younger stars in the disk.

Star formation is generally more active in the inner part of the Milky Way than in the outer parts because interstellar gas is denser in the inner part. If such activity has continued throughout the history of our galaxy, we might predict massive elements to be more abundant in the inner parts of our galaxy than in the outer parts. Observations of chemical abundances in the interstellar medium, based both on interstellar absorption lines in the spectra of stars and on emission lines in glowing clouds of gas known as H II regions, confirm this prediction. The composition of stars also confirms this prediction. Other galaxies similarly have more abundant massive elements in the inner parts than in the outer parts.

The basic idea that higher massive-element abundances should correspond to the more prodigious star formation in the inner galaxy seems correct, but the full picture is not this simple. The chemical composition of the interstellar medium at any location depends on a wealth of factors. New material falling into the galaxy might affect the amounts of heavy elements in the interstellar medium. As illustrated in **Figure 15.9**, chemical elements produced in the inner disk might be blasted into the halo in great "fountains" powered by the energy of massive stars, only to fall back onto the disk elsewhere. Past interactions with other galaxies might have stirred the Milky Way's interstellar medium, mixing gas from those other galaxies with our own. The variations of chemical abundances within the Milky Way and other galaxies—and what these variations tell us about the history of star formation and the formation of elements—remain active topics of research.

Although the details are complex, several clear and important lessons can be

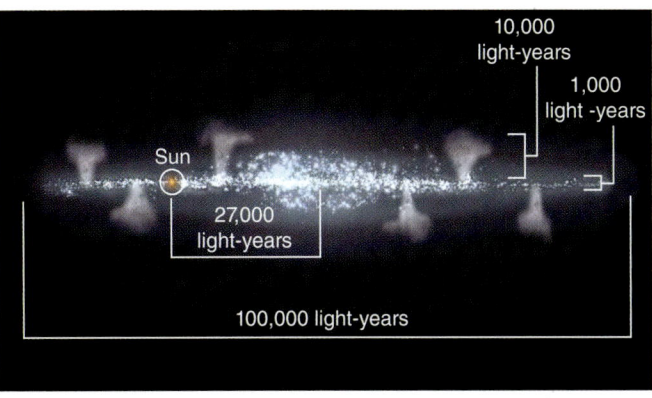

Figure 15.9 In the "galactic fountain" model of the disk of a spiral galaxy, gas is pushed away from the plane of the galaxy by energy released by young stars and supernovae and then falls back onto the disk.

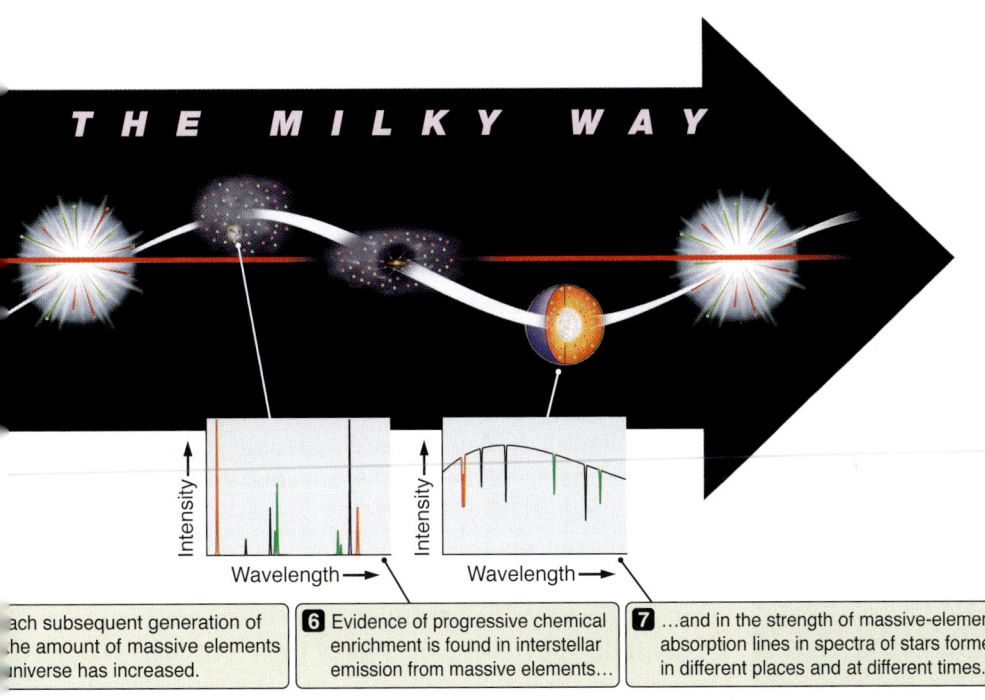

ach subsequent generation of he amount of massive elements niverse has increased.

6 Evidence of progressive chemical enrichment is found in interstellar emission from massive elements...

7 ...and in the strength of massive-element absorption lines in spectra of stars formed in different places and at different times.

learned from patterns in the amounts of massive elements in the galaxy. The first is that even the very oldest globular-cluster stars contain *some* amount of massive chemical elements. This implies that globular-cluster stars and other halo stars were not the first stars in our galaxy to form. At least one generation of massive, short-lived stars must have lived and died, ejecting newly synthesized massive elements into space, before even the oldest globular clusters formed. Further, every star less massive than about 0.8 $M_\odot$ that ever formed is still around today. Even so, we find no disk stars with exceptionally low amounts of massive elements. The gas that wound up in the plane of the Milky Way must have seen a significant amount of star formation before it settled into the disk of the galaxy and made stars.

In this discussion, we focused mainly on the variations in chemical abundances from place to place. These variations tell us a lot about the history of our galaxy and a lot about the origin of the material that we are made from. It is important to remember, however, that even a chemically "rich" star like the Sun, which is made of gas processed through approximately 9 billion years of previous generations of stars, is composed of less than 2 percent massive elements. Luminous matter in the universe is still dominated by hydrogen and helium formed long before the first stars.

Looking at a Cross Section through the Disk

The youngest stars in our galaxy are most strongly concentrated in the galactic plane, defining a disk more than 100,000 light-years across, but about 1,000 light-years thick, which is very thin. The older population of disk stars, distinguishable by lower abundances of massive elements, has a much "thicker" distribution: about 12,000 light-years thick. **Figure 15.10** illustrates how the population of stars changes with distance from the galactic plane. The youngest stars are concentrated closest to the plane of the galaxy because this is where the molecular clouds are. Older stars make up the thicker parts of the disk. There are two hypotheses for the origin of this thicker disk. One suggests that these stars formed in the midplane of the disk long ago but have since been kicked up out of the plane of the galaxy, primarily by gravitational interactions with massive molecular clouds (as shown in Figure 15.10). The other hypothesis suggests that these stars were acquired from the merging process that formed our galaxy.

A rotating cloud of gas naturally collapses into a thin disk when gas falling from one direction (above the disk) collides with gas falling from the other direction (below the disk). The same process applies to clouds of gas that are pulled by gravity toward the midplane of the disk of a spiral galaxy. Although stars are free to pass back and forth from one side of the disk to the other, cold, dense clouds of interstellar gas settle into the central plane of the disk. These clouds appear as concentrated dust lanes that lie in the middle of the disks of spiral galaxies, like tomato sauce lies between the cheese and the crust on a pizza. You can see this in the chapter-opening illustration, where the clouds of interstellar gas cut across the bright glow of the Milky Way. Compare this view to the mental picture you are building of viewing the galaxy from "outside." The thin lane slicing through the middle of the disk is the place where new stars are found and are continuing to form.

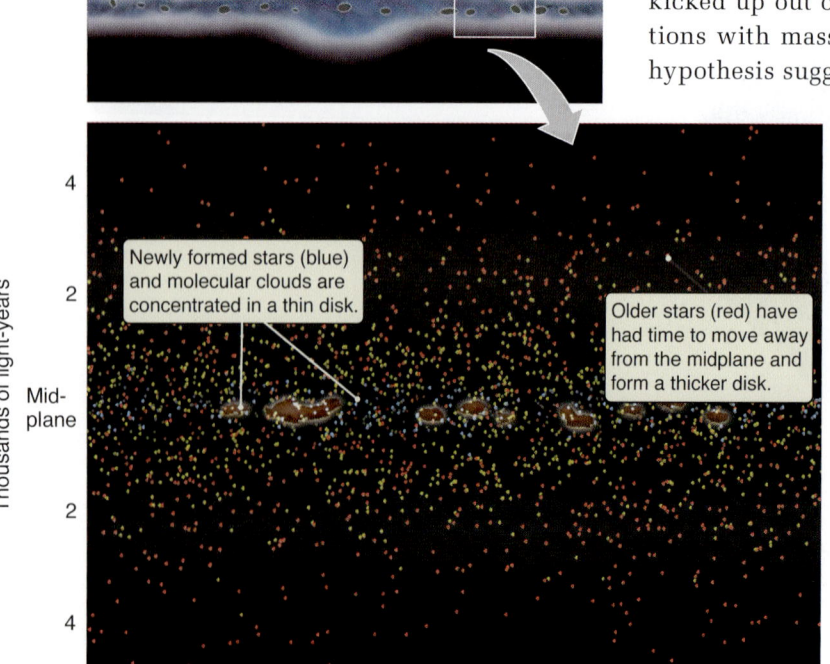

Figure 15.10 A vertical profile of the disk of the Milky Way Galaxy. Gas and young stars are concentrated in a thin layer in the center of the disk. Older populations of stars make up the thicker portions of the disk.

The interstellar medium is a dynamic place—energy from star-forming regions can shape it into impressively large structures. We mentioned earlier that energy from regions of star formation can impose interesting structure on the interstellar medium, clearing out large regions of gas in the disk of a galaxy. Many massive stars forming in the same region can blow "chimneys" out through the disk of the galaxy via a combination of supernova explosions and strong stellar winds. If enough massive stars are formed together, sufficient energy may be deposited to blast holes all the way through the plane of the galaxy. In the process, dense interstellar gas can be thrown high above this plane (see Figure 15.9). Maps of the 21-cm emission from neutral hydrogen in our galaxy and visible-light images of hydrogen emission from some edge-on external galaxies show numerous vertical structures in the interstellar medium of disk galaxies. These vertical structures are often interpreted as the "walls" of chimneys.

Other Halo Components

The globular clusters in the galactic halo tell astronomers a great deal about the history of star formation in the halo. Yet globular clusters account for only about 1 percent of the total mass of stars in the halo. As halo stars fall through the disk of the Milky Way, some (such as Arcturus, the third brightest star in the sky) pass close to the Sun, providing a sample of the halo that can be studied at closer range.

Most of the stars near the Sun are disk stars like the Sun. Astronomers can distinguish nearby halo stars in two ways. First, most halo stars are poorer in massive elements than disk stars. Second, halo stars are moving much faster, relative to the Sun, than disk stars. The disk stars and the Sun all orbit the center of the galaxy in the same direction, and those near the Sun orbit at nearly the same speed. In contrast, halo stars are analogous to comets orbiting the Sun in that they orbit the center of the galaxy in random directions. So the relative velocity between the halo stars and the Sun tends to be high. These stars are known as high-velocity stars.

By studying the orbits of high-velocity stars, astronomers have determined that the halo has two separate components: an inner halo that includes stars up to about 50,000 light-years from the center, and an outer halo that extends far beyond that. The stars in the outer halo have a lower fraction of heavier elements, implying that they formed very early, and therefore are very old. Many of them are moving in the opposite direction to the rotation of the galaxy. This suggests that the outer halo may have its origins in a merger with a small dwarf galaxy long ago.

X-ray observations suggest there is also a halo of hot (about 2 million K) gas surrounding the Milky Way. This gas halo may extend for about 300,000 light-years from the galactic center and contain as much mass as all the stars in the Milky Way. However, this is not nearly enough mass to account for the pull of gravity due to dark matter; that will be discussed in Section 15.3.

Magnetic Fields and Cosmic Rays

The interstellar medium of the Milky Way is laced with measurable magnetic fields that are wound up and strengthened by the rotation of the galaxy's disk. This interstellar magnetic field, however, is a hundred thousand times weaker than Earth's magnetic field. Charged particles spiral around magnetic field lines, moving along the field rather than across it. Conversely, magnetic fields cannot freely escape from a cloud of gas containing charged particles. The dense clouds of

Figure 15.11 The mass of interstellar clouds anchors the magnetic field of the Milky Way to the disk of the galaxy. The magnetic field in turn traps the galaxy's cosmic rays, much as a planetary magnetosphere traps charged particles.

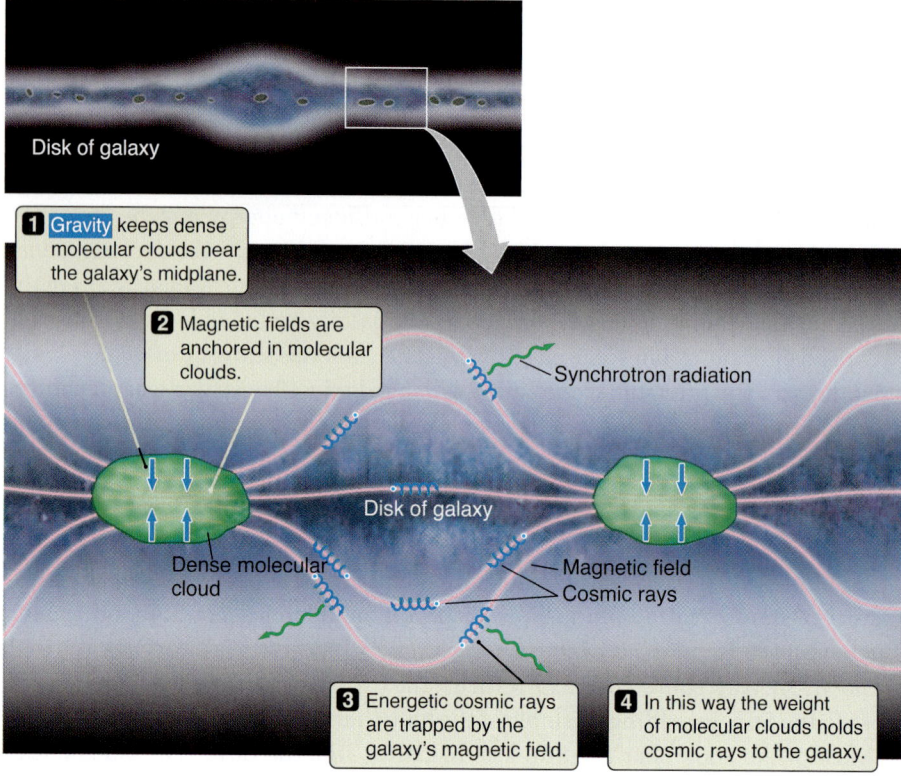

Figure 15.12 (a) In this artist's sketch, the Pierre Auger Observatory in Argentina is an array of stations designed to catch the particles that shower from collisions of cosmic rays with the upper atmosphere. (b) Each station in the array is equipped with its own particle collectors, carefully protected from the elements.

interstellar gas in the midplane of the Milky Way, shown in **Figure 15.11**, anchor the galaxy's magnetic field to the disk. These magnetic fields trap charged particles known as **cosmic rays**, which have their origin in space and travel close to the speed of light. Despite their name, cosmic rays are not a form of electromagnetic radiation, but instead are particles with mass. (They were named before their true nature was known.)

Cosmic rays are continually hitting Earth. Most cosmic-ray particles are protons, but some are nuclei of helium, carbon, and other elements. A few are high-energy electrons and other subatomic particles. Cosmic rays span an enormous range in particle energy. We can observe the lowest-energy cosmic rays, with energies as low as about 10^{-11} joules (J), by using interplanetary spacecraft. These energies correspond to the energy of a proton moving at a velocity of a few tenths of the speed of light. In contrast, the most energetic cosmic rays are 10 trillion (10^{13}) times as energetic as the lowest-energy cosmic rays. To get a better sense of just how much energy we're talking about, consider this: If you were to drop your copy of *Understanding Our Universe* (this irreplaceable textbook) from a second story window, it would hit the ground with the same energy as a *single* high-energy cosmic-ray proton! These high-energy cosmic rays are measured using the showers of elementary particles that they cause when crashing through Earth's atmosphere. These particle showers are observed by special telescopes such as the High Resolution Fly's Eye Observatory in Delta, Utah, or the Pierre Auger Observatory in Argentina (**Figure 15.12**).

Astronomers hypothesize that most cosmic rays are accelerated to high energies by the shock waves produced in supernova explosions. The very-highest-energy cosmic rays are as much as a hundred million times more energetic than any particle ever produced in a particle accelerator on Earth. These extremely high energies make them much more difficult to explain than those with lower energies.

The disk of our galaxy glows from synchrotron radiation produced by cosmic rays (mostly electrons) spiraling around the galaxy's magnetic field. Such synchrotron radiation is seen in the disks of other spiral galaxies as well, telling us that they, too, have magnetic fields and populations of energetic cosmic rays. Even so, the very-highest-energy cosmic rays are moving much too fast to be confined to our galaxy. Any such cosmic rays formed in the Milky Way soon stream away from the galaxy into intergalactic space. It is likely that some of the energetic cosmic rays reaching Earth originated in energetic events outside our galaxy.

The total energy of all of the cosmic rays in the galactic disk can be estimated from the energy of the cosmic rays reaching Earth. The strength of the interstellar magnetic field can be measured in a variety of ways, including the effect that it has on radio waves passing through the interstellar medium. These measurements indicate that in our galaxy, the magnetic-field energy and the cosmic-ray energy are about equal to each other. Both are comparable to the energy present in other energetic components of the galaxy, including the motions of interstellar gas and the total energy of electromagnetic radiation within the galaxy.

15.3 The Milky Way Is Mostly Dark Matter

As in other galaxies, the most interesting parts of the Milky Way may be the parts that can't be seen directly, but only detected by their gravitational influence on the stars around them. Dark matter accounts for the vast majority of the mass in a galaxy and extends far beyond a galaxy's visible boundary. As in all other spiral galaxies, there is compelling evidence that dark matter dominates the Milky Way.

The radio and infrared observations discussed in Section 15.2 reveal more than the Milky Way's shape. From these observations, astronomers can figure out how the disk of the Milky Way moves, and from that motion, in turn, they can determine its mass.

Measurements of the Doppler shift of the 21-cm line from neutral hydrogen indicate how fast the neutral hydrogen is moving toward us or away from us, and how quickly. The velocities of neutral interstellar hydrogen measured from 21-cm radiation are plotted versus the direction of observation in **Figure 15.13**. Looking toward the center of the galaxy, on one side hydrogen clouds are moving toward Earth, while on the other side clouds are moving away from Earth. This is the pattern of the rotation velocity of a disk. In other directions, the velocities we see are complicated by our moving vantage point within the disk and so are more difficult to interpret at a glance. Even so, observed velocities of neutral hydrogen enable us to measure our galaxy's rotation curve and even determine

> **Back in Section 5.6 . . .**
>
> . . . you learned that when objects are moving away from you, their spectral lines are shifted toward the red end of the spectrum.

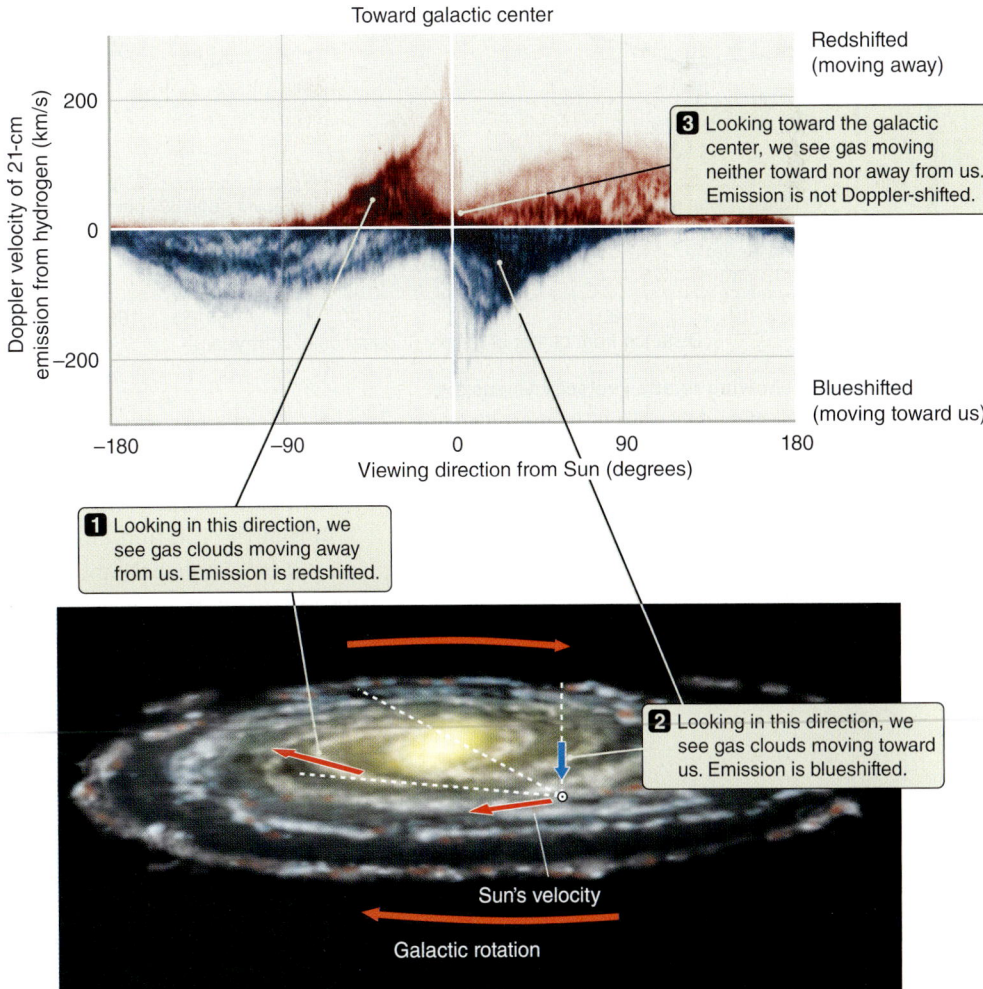

Figure 15.13 Doppler velocities measured from observations of 21-cm emission from interstellar clouds of neutral hydrogen. These velocities vary between redshift and blueshift as we look around in the plane of our galaxy. Notice that from our perspective within the Solar System (sight lines are shown as dashed white lines), we see the clear signature of a rotating disk when looking on either side of the galactic center.

Working It Out 15.1 | Finding the Mass of a Galaxy

The Sun is located about 27,000 light-years (1.7×10^9 AU, since a light-year is 63,000 astronomical units [AU]) from the center of the Milky Way, and it takes 230 million years for the Sun to orbit the center. From this information, we can find the mass of the Milky Way that lies interior to the orbit of the Sun. To find the mass of a binary system, we use Newton's version of Kepler's third law, which relates the total mass of an orbiting system ($m_1 + m_2$), in units of solar masses, to the orbital period (P) in years and the average radius of the orbit (A) in AU:

$$m_1 + m_2 = \frac{A^3_{AU}}{P^2_{years}}$$

In this case, the mass of the Sun is tiny compared to the mass of the galaxy interior to the Sun's orbit, so that $m_1 + m_2$ is essentially equal to m_1, the mass of the Milky Way Galaxy interior to the Sun's orbit. Plugging in the values for the orbit of the Sun gives

$$m_1 = \frac{A^3_{AU}}{P^2_{years}}$$

$$m_1 = \frac{(1.7 \times 10^9 \text{ AU})^3}{(230 \times 10^6 \text{ years})^2}$$

$$m_1 = 9.3 \times 10^{10} \, M_\odot$$

The mass of the Milky Way, interior to the Sun's orbit, is 93 billion solar masses. The Sun is not at the edge of the disk of the Milky Way, but rather is located partway out. Some mass is located farther from the center of the Milky Way than the Sun, so the total mass of the Milky Way is larger than the number we've just calculated. Still, this calculation indicates the vast amount of mass contained within the Milky Way.

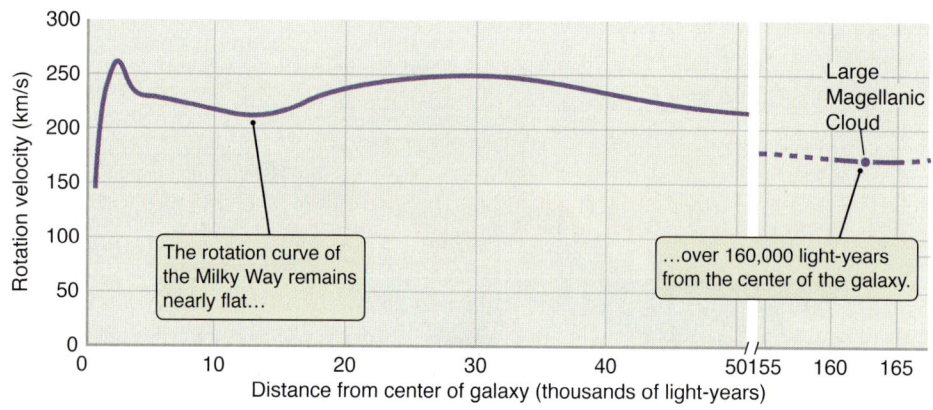

Figure 15.14 A plot showing rotation velocity versus distance from the center of the Milky Way. The most distant point comes from measurements of the orbit of the Large Magellanic Cloud. The nearly flat rotation curve indicates that dark matter dominates the outer parts of our galaxy.

the structure present throughout the disk of our galaxy.

Recall from Chapter 14 that observations of rotation curves led astronomers to conclude that the masses of spiral galaxies consist mostly of dark matter. The rotation of the Milky Way shows that dark matter dominates the Milky Way as in most spiral galaxies (**Working It Out 15.1**). **Figure 15.14** shows the rotation curve of the Milky Way as found primarily from 21-cm observations. The orbital motion of the nearby Large Magellanic Cloud is found from its motion across the sky over time. The Large Magellanic Cloud provides the outermost point in the rotation curve, at a distance of roughly 160,000 light-years from the center of the galaxy. Like other spiral galaxies, the Milky Way has a fairly flat rotation curve.

From the rotation curve of the Milky Way, we find that the galaxy's gravitational mass must be about 1.3 trillion $M_\odot$. However, the luminous mass, found from adding the masses of stars, dust, and gas, is a much lower value (roughly one-tenth of the gravitational mass). Like other spiral galaxies, the Milky Way is mostly dark matter. Like other galaxies, visible matter dominates the inner part of our galaxy, and dark matter dominates its outer parts.

15.4 The Milky Way Contains a Supermassive Black Hole

Dense clouds of dust and gas hide our visible-light view of the Milky Way's center. Yet it is here that we might expect to find a supermassive black hole, as is often

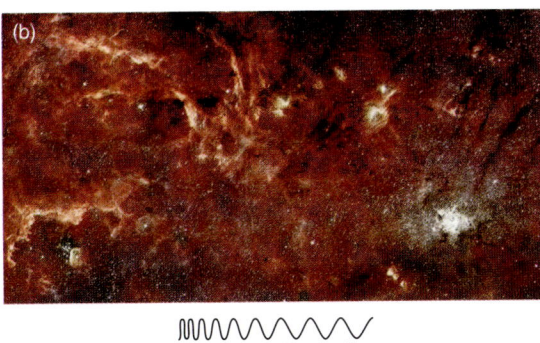
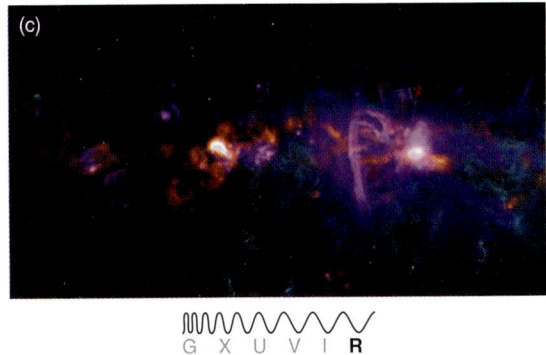

Figure 15.15 (a) An X-ray view of the Milky Way's central region shows the active source, Sgr A*, as the brightest spot at the middle of the image. Lobes of superheated gas (shown in red) are evidence of recent, violent explosions near Sgr A*. (b) This wide infrared view (890 light-years across) of the central core of the Milky Way shows hundreds of thousands of stars. The bright white spot at the lower right marks the galaxy's center, home of a supermassive black hole. (c) Radio observations of the center of the Milky Way reveal wispy molecular clouds (purple) glowing from strong synchrotron emission. Cold dust (20–30 K) associated with molecular clouds is shown in orange. Diffuse infrared emission appears in blue-green. The galactic center (Sgr A*) lies within the bright area to the right of center.

found in other galaxies with central bulges. Fortunately, infrared, radio, and some X-ray radiation passes through dust (**Figure 15.15**). The X-ray view (Figure 15.15a) shows the location of a strong radio source called Sagittarius A* (abbreviated Sgr A*), which astronomers believe lies at the exact center of the Milky Way. The infrared image (Figure 15.15b) cuts through the dust to reveal the crowded, dense core of the galaxy containing hundreds of thousands of stars. Radio observations (Figure 15.15c) reveal synchrotron emission from wisps and loops of material distributed throughout the region. This is similar to the synchrotron emission seen from active galactic nuclei (AGNs), but at far lower intensity.

The motions of stars closest to the Sgr A* source suggest a central mass very much greater than that of the few hundred stars orbiting there. Stars less than 0.1 light-year from the galaxy's center follow Kepler's laws. The closest stars studied are only about 0.01 light-year from the center, so close that their orbital periods are only about a dozen years. The positions of these stars change noticeably over time, and we can see them speed up as they whip around what can only be a supermassive black hole at the focus of their elliptical orbits (**Figure 15.16**). Using Kepler's third law, we estimate that the black hole at the center of our own galaxy is a relative lightweight, having a mass of "only" about 4 million $M_\odot$.

Clouds of interstellar gas at the galaxy's center are heated to millions of degrees by shock waves from supernova explosions and colliding stellar winds blown outward by young massive stars. Superheated gas produces X-rays, and the Chandra X-ray Observatory has detected more than 9,000 X-ray sources within the central region of the galaxy. These include frequent, short-lived X-ray flares near Sgr A*, which provide direct evidence that matter falling toward the supermassive black hole fuels the energetic activity at the galaxy's center.

The Fermi Gamma-ray Space Telescope has observed bubbles that extend about 25,000 light-years above and below the galactic plane. The bubbles may have formed after a burst of star formation a few million years ago produced massive star clusters near the center of the galaxy. If some of the gas formed stars and about 2,000 $M_\odot$ of material fell into the supermassive black hole, enough energy could have been released to power the bubbles. More recently, faint gamma-ray signals

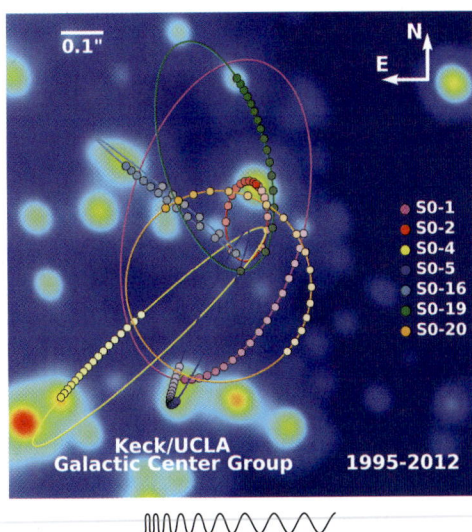

Figure 15.16 Orbits of several stars within 0.1 light-year (about 6,000 AU) of the Milky Way's center. The Keplerian motions of these stars reveal the presence of a 4-million-$M_\odot$ supermassive black hole at the galaxy's center. Colored dots show the measured positions of the stars over a 12-year interval.

Figure 15.17 The Fermi Gamma-ray Space Telescope observed gamma-ray bubbles (purple) extending about 25,000 light-years above and below the galactic plane. In this artist's conceptual view from outside of the galaxy, the gamma-ray jets (pink) are tilted with respect to the bubbles, which might imply that the accretion disk around the black hole is tilted as well.

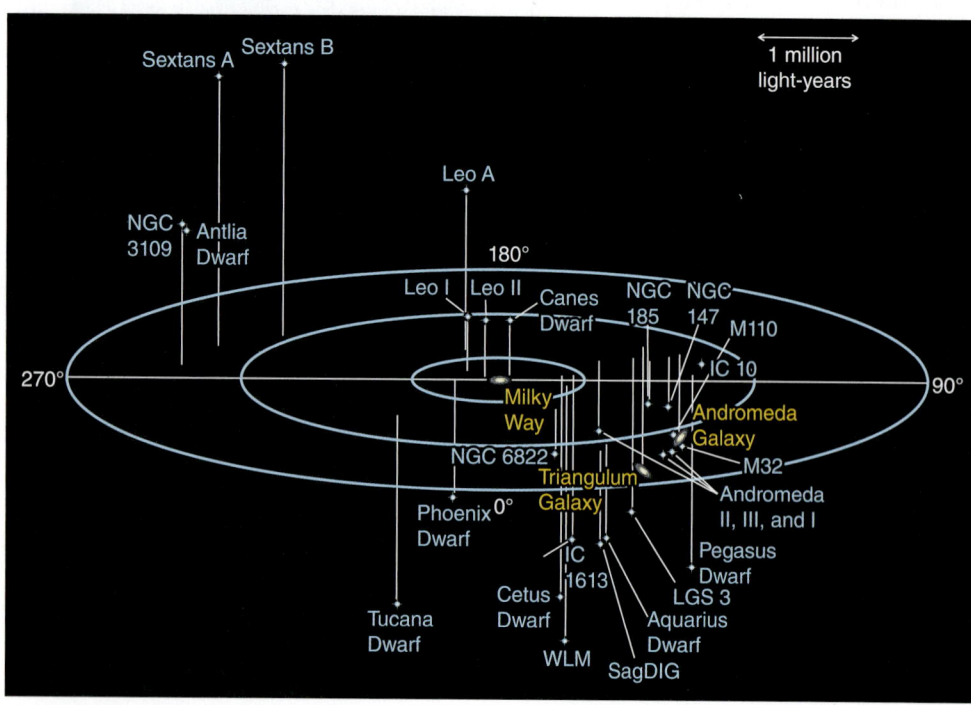

Figure 15.18 A graphical map showing the locations of the members of the Local Group of galaxies. Most are dwarf galaxies. Spiral galaxies are shown in yellow.

were observed that look like jets coming from the center, within the bubbles, shown in **Figure 15.17**. If these jets are originating from material falling into the supermassive black hole, activity might be even more recent, maybe 20,000 years ago, as observed from the distance of Earth. Some astronomers predict that gas clouds are heading toward the center and will soon be accreted by the black hole. Nevertheless, at this time, this activity is not as intense as that seen in active galaxies with central black holes. The inner Milky Way is a reminder that our galaxy was almost certainly "active" in the past and could become active once again.

15.5 The Milky Way Offers Clues about How Galaxies Form

A fundamental goal of stellar astronomy is to understand the life cycle of stars, including how stars form from clouds of interstellar gas. Galactic astronomy has the same basic goal. That is, astronomers would like very much to have a complete and well-tested theory of how our galaxy formed. The distribution of stars of different ages with different amounts of heavy elements is one clue. Additional clues come from studying other galaxies at different distances (and therefore different ages), their supermassive black holes, and their merger history.

The properties of globular clusters and high-velocity stars in the halo are important clues. These objects must have been among the first stars formed that still exist today. They are not concentrated in the disk or bulge of the galaxy, so they must have formed from clouds of gas well before those clouds had settled into the galaxy's disk. The observations that globular clusters are very old and that the youngest globular cluster is older than the oldest disk stars confirm this idea. The presence of small concentrations of massive elements in the atmospheres of halo stars indicates that at least one generation of stars must have lived and died *before* the formation of the halo stars we see today. Astronomers are looking for stars from that first generation.

Galaxies do not exist in isolation. The vast majority of galaxies are parts of gravitationally bound collections of galaxies. The smallest and most common of these are called **galaxy groups**. A galaxy group contains as many as several dozen galaxies, most of which are dwarf galaxies (see Chapter 14), in a space between 4 million and 6 million light-years across. The Milky Way is a member of the Local Group, first identified by Edwin Hubble in 1936. There are about 50 members of the Local Group, including two large barred spiral galaxies (the Milky Way and the Andromeda Galaxy, accounting for nearly 98% of the mass of the Local Group), along with a few elliptical galaxies and irregular galaxies (**Figure 15.18**), and at least 30 smaller dwarf galaxies. Many of these smaller dwarf galaxies were unknown until very recently because of their low luminosity, and there are likely more to be discovered. Most of the galaxy mass in the Local Group is in the two large barred spiral gal-

galaxies. The third largest galaxy, Triangulum, is a non-barred spiral with about one-tenth the mass of the Milky Way. Most, but not all, of the dwarf elliptical and dwarf spheroidal galaxies in the group are satellites of the Milky Way or Andromeda. The Local Group interacts with a few nearby groups.

From the properties of globular clusters and high-velocity stars, the presence of the central supermassive black hole, and the number of nearby dwarf galaxies, astronomers conclude that the Milky Way must have formed when the gas within a huge "clump" of dark matter collapsed into a large number of small protogalaxies. Some of these smaller clumps are still around today in the form of small, satellite dwarf galaxies near our own. The largest among them are the Large Magellanic Cloud and Small Magellanic Cloud (**Figure 15.19**), which are easily seen by the naked eye in the Southern Hemisphere and appear much like detached pieces of the Milky Way. Another companion, the elliptical Sagittarius Dwarf, is now plowing through the disk of the Milky Way on the other side of the bulge. Astronomers have observed streams of stars from the Sagittarius Dwarf and some other dwarf galaxies that are being tidally disrupted by the Milky Way. At some point, these dwarf galaxies will become incorporated into the Milky Way—an indication that our galaxy is still growing by cannibalizing other galaxies.

The dwarf galaxies are the lowest-mass galaxies observed, and they are dominated by an even larger fraction of dark matter than are other known galaxies. They also contain stars very low in elements more massive than helium. These ultrafaint dwarf galaxies offer clues to the formation of the Local Group. In addition, observations of the motions and speeds of the dwarf galaxies about the Milky Way will lead to improved estimates of the amount of dark matter in the Milky Way itself.

Protogalaxies merged to form the barred spiral galaxy we call the Milky Way. In this process, stars were formed in the halo, in the bulge, and in the disk. The first stars to form ended up in the halo—some in globular clusters, but many not. Gas that settled into the disk of the Milky Way quickly formed several generations of stars. A dense, concentrated mass quickly grew into the supermassive black hole at the center of the galaxy. The details of this process are still sketchy, but computer simulations indicate that so much mass was concentrated in this small region that almost any sequence of events would have led to the formation of a massive black hole.

The Milky Way offers many clues about the way galaxies form, but much of what we know of the process comes from looking beyond our local system. Images of distant galaxies (which we see as they existed billions of years ago), as well as observations of the glow left behind from the early stages of the universe itself, provide equally important pieces of the puzzle. In the next chapter, we will turn our attention to the evolution of the universe.

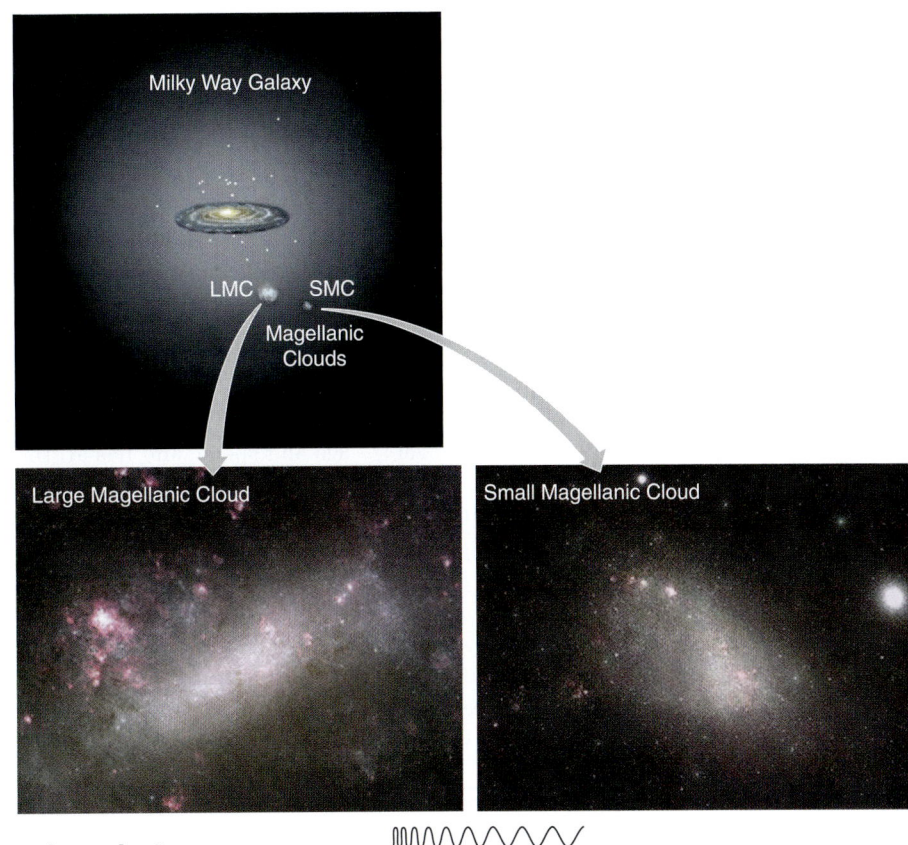

Figure 15.19 Our Milky Way is surrounded by more than 20 dwarf companion galaxies, the largest among them being the Large and Small Magellanic Clouds. The Magellanic Clouds were named for Ferdinand Magellan (c. 1480–1521), who headed an early European expedition that ventured far enough into the Southern Hemisphere to see them.

 # READING ASTRONOMY News

Once in a while, something truly odd is observed. Astronomers need to act like detectives to go back and try to figure out how it got that way.

Hyperfast Star Kicked Out of Milky Way

By **Lisa Grossman**, *Wired Science*

New Hubble observations suggest a dramatic origin story for one of the fastest stars ever detected, involving a tragic encounter with a black hole, a lost companion, and swift exile from the galaxy.

The star, HE 0437-5439, is one of just 16 so-called hypervelocity stars, all of which were thought to come from the center of the Milky Way. The Hubble observations allowed astronomers to definitively trace the star's origin to the heart of the galaxy for the first time.

Based on observations taken 3½ years apart, astronomers calculated that the star is zooming away from the Milky Way's center at a speed of 1.6 million miles per hour—three times faster than the Sun [orbits the galaxy].

"The star is traveling at an absurd velocity, twice as much as it needs to escape the galaxy's gravitational field," said hypervelocity star hunter Warren Brown of the Harvard-Smithsonian Center for Astrophysics, who found the first unbound star in 2005, in a press release. "There is no star that travels that quickly under normal circumstances—something exotic has to happen."

Earlier observations linked the star to a neighboring galaxy, the Large Magellanic Cloud. But Brown and his colleagues claim that the new Hubble observations settle the question of the star's origin squarely in favor of the Milky Way.

One reason the star's home was under debate is its bizarrely youthful appearance. Based on its speed, the star would have to be 100 million years old to have traveled from the Milky Way's center to its current location, 200,000 light-years away. But its mass—nine times that of our Sun—and blue color mean it should have burned out after only 20 million years.

The new origin story reconciles the star's age and speed, and has all the makings of a melodrama. A hundred million years ago, astronomers suggest, the runaway star was a member of a triple-star system that veered disastrously close to the galaxy's central supermassive black hole. One member of the trio was captured, and its momentum was transferred to the remaining binary pair, which was hurled from the Milky Way at breakneck speed.

As time passed, the larger star evolved into a puffy red giant and devoured its partner. The two merged into the single, massive, blue star called a blue straggler that Hubble observed.

This bizarre scenario conveniently explains why the star looks so young. By merging into a blue straggler, the two original stars managed to look like a star one-fifth its true age.

The findings were published online in a paper in *The Astrophysical Journal Letters*.

The team is hunting for the homes of four other unbound stars, all zooming around the fringes of the Milky Way.

"Studying these stars could provide more clues about the nature of some of the universe's unseen mass, and it could help astronomers better understand how galaxies form," said study coauthor Oleg Gnedin of the University of Michigan in a press release.

Figure 15.20 This Hubble Space Telescope image shows the hyperfast star HE 0437-5439.

Evaluating the News

1. What is special about the star described in the article?
2. The star is shown in **Figure 15.20**. From this image alone, would it be possible to determine that there was something special about this star?
3. In the article, Warren Brown states that the star is traveling at twice the speed needed to escape the Milky Way, and that something exotic must have happened to speed it up. Why would he make this argument? (Hint: What happens to objects moving faster than the escape velocity?)
4. What is unusual about the age of the star?
5. How might astronomers figure out the mass of the star? Why does that set a limit on how old the star should be?
6. What kind of observations could distinguish between an origin in the Large Magellanic Cloud and an origin in the Milky Way?
7. The supermassive black hole at the center of the Milky Way apparently hurled this star away from the center of the Milky Way. Is this likely to be a common occurrence? Explain your reasoning.

up the distance ladder and enable us accurately to measure distances to galaxies as far away as 30 Mpc. Even this is not far enough to determine a reliable value for the Hubble constant, but within that volume of space are many galaxies that we can use for yet more powerful distance indicators. Among the best of these are Type Ia supernovae.

Recall from Chapter 12 that Type Ia supernovae occur when gas flows from an evolved star onto its white dwarf companion, pushing the white dwarf up toward the Chandrasekhar limit for the mass of an electron-degenerate object. When this happens, the overburdened white dwarf begins to collapse and then explodes. The peak luminosities of these Type Ia supernovae can be calibrated, so that astronomers can observe their brightness and estimate their luminosity. Therefore, the distance can be determined. With a peak luminosity that outshines billions of Suns (**Figure 14.20**), Type Ia supernovae can be seen and measured with modern telescopes that can observe supernovae almost to the edge of the observable universe.

Figure 14.21 plots the measured recession velocities of galaxies against their measured distances. Because the velocity and the distance are proportional to each other, the points lie along a line on the graph with a slope equal to the proportionality constant H_0. Notice how well the data line up along the line. This tells us that the universe follows Hubble's law. Today, we have measured the Hubble constant to an accuracy of a few percent. The value is likely to be further refined in the years to come.

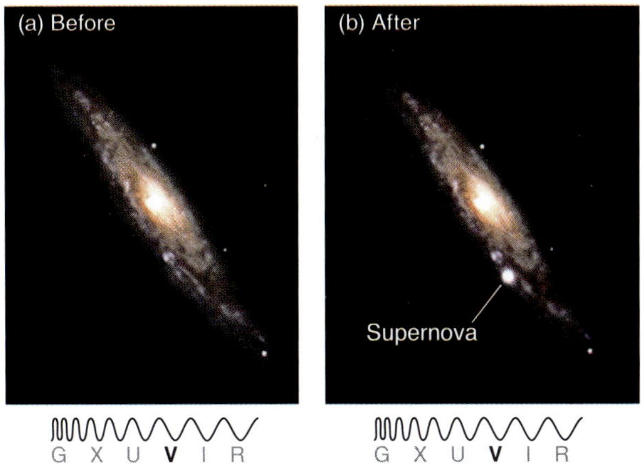

Figure 14.20 Galaxy NGC 3877 is shown (a) before and (b) after the explosion of a Type Ia supernova. Type Ia supernovae are extremely luminous standard candles.

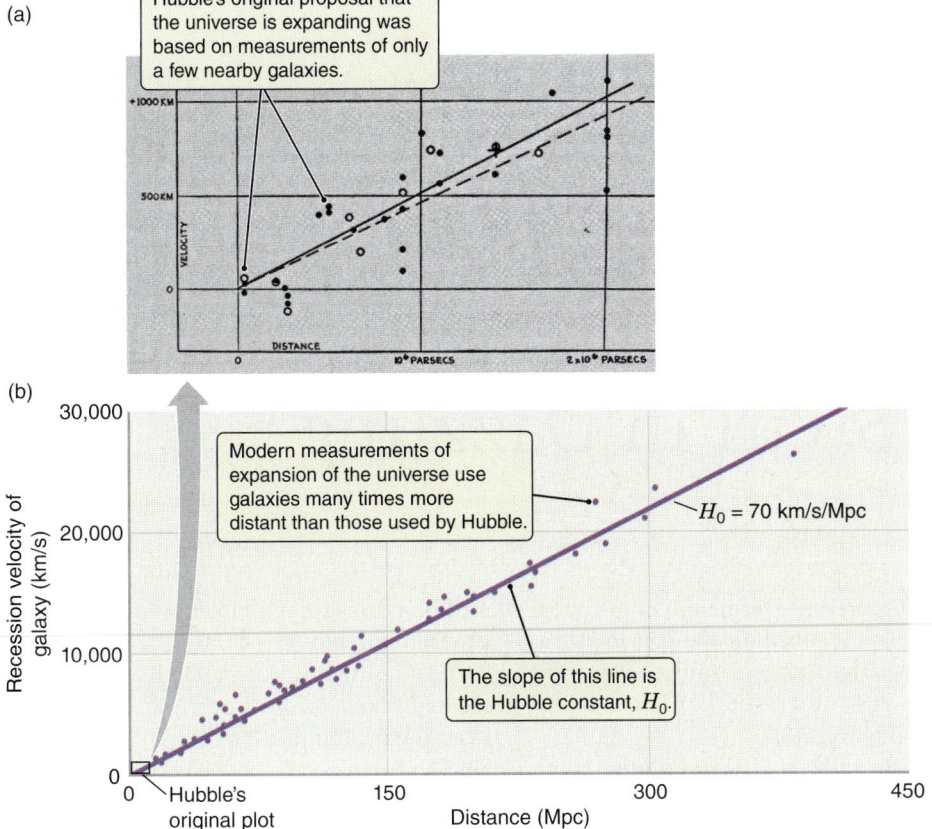

Figure 14.21 (a) Hubble's original figure illustrating that more distant galaxies are receding faster than less distant galaxies. (b) Modern data show that recession velocity is proportional to distance far into the early universe.

Hubble's Law Maps the Universe in Space and Time

Hubble's law gives us a practical tool for measuring distances to extremely remote objects. Once we know the value of H_0, we can use a straightforward measurement of the redshift of a galaxy to find its distance. In other words, once we know H_0, Hubble's law makes the once-difficult task of measuring distances in the universe relatively easy, allowing us to map the structure of the universe. This may seem like a logical impossibility, because we are using redshifts and distances to find H_0 and then using H_0 to find the distances. But we find H_0 from nearby galaxies using standard candles, and then we use that value to find the distance to a different, more distant set of galaxies.

Hubble's law does more than place galaxies in space. It also places galaxies in time. Light travels at a huge but finite speed. Remember that when we look at the Sun, we see it as it existed 8.3 minutes ago. When we look at Alpha Centauri, the nearest stellar system beyond the Sun, we see it as it existed 4.3 years ago. When we look at the center of our galaxy, the picture we see is about 27,000 years old. In general, when looking at a distant object, we speak of its **look-back time**—the time it has taken for the light from that object to reach our telescope. As we look into the distant universe, look-back times become very great indeed. The distance to a galaxy whose redshift $z = 0.1$ is about 1.4 billion light-years (assuming $H_0 = 70$ km/s/Mpc), so the look-back time to that galaxy is 1.4 billion years. The look-back time to a galaxy where $z = 0.2$ is 2.7 billion years. As we look at objects with greater and greater redshifts, we see increasingly younger stages of our universe. The most distant observed objects have a look-back time of nearly 13.8 billion years, the age of the universe.

NASA's Great Observatories program covers a wide range of wavelengths from four different instruments. Sometimes, images from these instruments are used in various combinations to produce shockingly beautiful photographs of the cosmos. Often, these are images of galaxies, as in this posting from August 2010.

READING ASTRONOMY News

Colliding Galaxies Swirl in Dazzling New Photo

By SPACE.com

A spectacular new image of two colliding galaxies shows a cosmic region teeming with stellar activity.

The cosmic smash-up, which began more than 100 million years ago but is still occurring, has triggered the formation of millions of stars in the clouds of dust and gas within the galaxies. The composite image was created using data from several different space telescopes [**Figure 14.22**].

The most massive of these young stars speed through their evolution in only a few million years, dying a violent stellar death in supernova explosions.

The colliding Antennae galaxies are located about 62 million light-years away from Earth. In addition to the photograph, NASA also released a video of the galaxy collision using the same data from the Chandra

SEE COLLIDING GALAXIES

SUMMARY

We live in the disk of the Milky Way, a barred spiral galaxy that is 100,000 light-years across. The Sun is about 27,000 light-years from the center of this galaxy. Variable stars of known luminosity yield the distances to globular clusters, which enable us to measure the size of the Milky Way's extended halo. The chemical composition of the Milky Way has evolved with time, and there must have been a generation of stars before the oldest halo and globular-cluster stars we see today formed. Star formation is still actively occurring in the disk of our galaxy, leading to complex structures within the disk. The Doppler velocities of radio lines show that the rotation curve of the Milky Way is flat, like those of other galaxies. But the inferred mass cannot be accounted for by the mass that is observed directly. This means that most of the mass of our galaxy is in the form of dark matter. At the center of the galaxy is a massive black hole, which produces rapid orbital velocities of nearby stars. The Milky Way formed from a collection of smaller protogalaxies that collapsed out of a halo of dark matter.

LG 1 The distances to globular clusters can be found from the luminosity of variable stars within them. Because these globular clusters are evenly distributed around the center of the Milky Way, the center of the distribution is located at the galaxy's center.

LG 2 The Milky Way has a disk, consisting of two parts: the thick disk and the thin disk. A sketch of the spiral structure would include an elongated bulge with a spiral arm coming from each end, and a few arm fragments (see the chapter opening figure). The halo consists of an inner halo and outer halo of stars and globular clusters, as well as a large, hot gas halo.

LG 3 The ages of stellar clusters can be determined from the main-sequence turnoff. Younger stars are abundant in heavier elements because previous generations of stars enriched the galaxy in these heavier elements as they died.

LG 4 The Milky Way's flat rotation curve is the most compelling evidence for a significant amount of dark matter in the Milky Way. Several lines of evidence point to the existence of a supermassive black hole at the center of the galaxy. Observations of stellar orbits near the black hole have allowed astronomers to determine its mass with precision.

LG 5 The Milky Way provides many clues to understand galaxy formation, The properties of stars in the halo, including those in globular clusters and high-velocity stars, indicate that these stars formed from clouds of gas well before those clouds settled into the galaxy's disk. The dwarf satellites and other neighbors in the Local Group are evidence that the Milky Way is growing through accretion. The central supermassive black hole suggests earlier periods when the nucleus was active.

SUMMARY SELF-TEST

1. The size of the Milky Way is determined from studying _____ in globular clusters.
   a. standard candles
   b. velocities
   c. standard models
   d. expansions

2. Older stars are found farther from the midplane of a galactic disk because
   a. the disk used to be thicker.
   b. the stars have lived long enough to move there.
   c. the younger stars in the thick disk were more massive and have already died.
   d. old stars come from the halo.

3. In general, older stars have lower ____ than younger stars.
   a. masses
   b. abundance of heavy elements
   c. luminosities
   d. rotation rates

4. The best evidence for dark matter in the Milky Way comes from the observation that the rotation curve
   a. is quite flat at great distances from the center.
   b. rises swiftly in the interior.
   c. falls off and then rises again.
   d. has a peak at about 2,000 light-years from the center.

5. The Milky Way formed when the gas within a clump of dark matter collapsed into a large number of _____.
   a. globular clusters.
   b. high-velocity stars.
   c. supermassive black holes.
   d. small protogalaxies.

6. Detailed observations of the structure of the Milky Way are difficult because
   a. the Solar System is embedded in the dust and gas of the disk.
   b. the Milky Way is mostly dark matter.
   c. there are too many stars in the way.
   d. the galaxy is rotating too fast (about 200 km/s).

7. Place in order the steps involved in finding the distance to the center of the Milky Way from variable stars in globular clusters.
   a. Use the luminosity and the brightness to find the distance to the globular clusters.
   b. Use the period-luminosity relation to find the luminosity of the variable stars.
   c. Find the distance to the center of the distribution of globular clusters, which is the distance to the center of the Milky Way.
   d. Find the direction to the center of the distribution of the globular clusters.
   e. Find the period of the variable stars.

8. Radio emission reveals that the Milky Way is
   a. an elliptical galaxy.
   b. an irregular galaxy resulting from a collision.
   c. a spiral with three arms.
   d. a barred spiral with two major arms.

9. In the disk of the Milky Way, stars are _____ and dust and gas are _____ than in the halo.
   a. younger; more diffuse
   b. older; more diffuse
   c. older; denser
   d. younger; denser

10. Evidence of a supermassive black hole at the center of the Milky Way comes from (choose all that apply)
    a. direct observations of stars that orbit it.
    b. X-rays from material that is falling in.
    c. strong radio emission from the region of the accretion disk.
    d. the abundance of dark matter in the galaxy.

QUESTIONS AND PROBLEMS

Multiple Choice and True/False

11. **T/F:** Interstellar dust causes stars to appear bluer than they actually are.

12. **T/F:** Ages of globular clusters put a lower limit on the age of the Milky Way.

13. **T/F:** The Milky Way contains almost no dark matter.

14. **T/F:** An old globular cluster contains no massive, luminous hot stars.

15. **T/F:** The disk of the Milky Way is threaded with magnetic field lines.

16. Looking toward the galactic center, we see no redshift or blueshift. This tells us
    a. the center of the galaxy is motionless.
    b. the Sun and the center are not moving toward or away from each other.
    c. the Milky Way is stationary with respect to other galaxies.
    d. the Sun is stationary with respect to the center of the galaxy.

17. The Milky Way Galaxy
    a. has been exactly the same since it formed.
    b. has only changed its composition, but not its dynamics, since it formed.
    c. is still evolving, consuming nearby small galaxies.
    d. is finished evolving, having consumed all nearby galaxies.

18. Why are globular clusters *uniquely* useful in determining the size of the galaxy?
    a. They are large.
    b. They are bright.
    c. They are evenly distributed around the center.
    d. They consist mostly of old stars and have been there a long time.

19. In order for a variable star to be useful as a standard candle, its luminosity must be related to its
    a. period of variation.
    b. mass.
    c. temperature.
    d. radius.

20. In general, the Milky Way
    a. has the same chemical composition as time passes.
    b. has more abundant hydrogen as time passes.
    c. has more abundant heavy elements as time passes.
    d. has less abundant heavy elements as time passes.

21. Cosmic rays are
    a. a form of electromagnetic radiation.
    b. high-energy particles.
    c. high-energy dark matter.
    d. high-energy photons.

22. In the Hubble scheme for classifying galaxies, what type of galaxy is the Milky Way?
    a. elliptical
    b. spiral
    c. barred spiral
    d. irregular

23. Where are the youngest stars in the Milky Way?
    a. in the core
    b. in the bulge
    c. in the disk
    d. in the halo

24. Halo stars are found in the vicinity of the Sun. What observational evidence distinguishes them from disk stars?
    a. the direction of their motion
    b. their speed
    c. their mass
    d. their temperature

25. Why are most of the Milky Way's satellite galaxies so difficult to detect?
    a. They are very faint.
    b. They are too far away.
    c. The halo of the Milky Way obscures the view.
    d. They are very massive, and light does not escape them.

Conceptual Questions

26. Look back at the chapter-opening illustration. Explain the logic that leads us to determine that the Milky Way is a spiral galaxy, based on images like these.

27. Why is it so difficult for astronomers to get an overall picture of the structure of our galaxy?

28. What do astronomers mean by *standard candle*? Why are stars in globular clusters so useful as standard candles in determining the size of the galaxy?

29. Describe the distribution of globular clusters within the galaxy, and explain what that implies about the size of the galaxy and our distance from its center.

30. How do we know that the stars in globular clusters are among the oldest stars in our galaxy?

31. Compare globular and open clusters.
    a. What are the main differences in the gas out of which globular and open clusters formed?
    b. Why do globular clusters have such high masses while open clusters have low masses?
    c. Are these issues related? Explain.

32. How do 21-cm radio observations reveal the rotation of our galaxy? What does the rotation curve of our galaxy say about the presence of dark matter in the galaxy?

33. Explain the observational evidence that shows we live in a spiral galaxy, not an elliptical galaxy.

34. What does the abundance of a star's massive elements tell us about the age of the star?

35. Where do we find the youngest stars in our galaxy?

36. Halo stars are found in the vicinity of the Sun. What observational evidence distinguishes them from disk stars?

37. Can a cosmic ray travel at the speed of light? Why or why not?

38. Why must we use X-ray, infrared, and 21-cm radio observations to probe the center of our galaxy?

39. Explain the evidence for a supermassive black hole at the center of the Milky Way. How does the mass of the supermassive black hole at the center of our galaxy compare with that found in most other spiral galaxies?

40. What is the origin of the Milky Way's satellite galaxies? What has been the fate of most of the Milky Way's satellite galaxies? Why are most of the Milky Way's satellite galaxies so difficult to detect?

Problems

41. The Sun completes one trip around the center of the galaxy in approximately 230 million years. How many times has our Solar System made the circuit since its formation 4.6 billion years ago?

42. Using the numbers and shapes shown in Figure 15.6, estimate the fraction of the volume of the Milky Way that is taken up by the disk. The thickness of the Milky Way's disk is, on average, about 1,000 light-years thick.

43. The Sun is located about 27,000 light-years from the center of the galaxy, and the galaxy's disk probably extends another 30,000 light-years farther out from the center. Assume that the Sun's orbit takes 230 million years to complete.
    a. With a truly flat rotation curve, how long would it take a globular cluster located near the edge of the disk to complete one trip around the center of the galaxy?
    b. How many times has that globular cluster made the circuit since its formation about 13 billion years ago?

44. Parallax measurements of the variable star RR Lyrae indicate that it is located 750 light-years from the Sun. A similar star observed in a globular cluster located far above the galactic plane appears 160,000 times fainter than RR Lyrae.
    a. How far from the Sun is this globular cluster?
    b. What does your answer to part (a) tell you about the size of the galaxy's halo compared to the size of its disk?

45. Although the flat rotation curve indicates that the total mass of our galaxy interior to about 180,000 light years is approximately 4×10^{11} $M_\odot$, electromagnetic radiation associated with normal matter suggests a mass within this radius of only 0.5×10^{11} $M_\odot$. Given this information, calculate the fraction of our galaxy's mass that is made up of dark matter.

46. Compare the H-R diagram for the young cluster NGC 6530 (see Figure 15.7) with H-R diagrams in previous chapters to determine how soon the most massive stars will explode and the mass of the protostars that are still contracting onto the main sequence.

47. Given what you have learned about the distribution of massive elements in the Milky Way and what you know about the terrestrial planets, where do you think such planets are most likely and least likely to form?

48. One of the fastest cosmic rays ever observed had a speed of $[1.0 - (1.0 \times 10^{-24})] \times c$ (very, very close to c). Assume that the cosmic ray and a photon left a source at the same instant. To a stationary observer, how far behind the photon would the cosmic ray be after traveling for 100 million years?

49. A star in a circular orbit about the black hole at the center of our galaxy ($M_{BH} = 8.2 \times 10^{36}$ kg) has an orbital radius of 0.0131 light-year (1.24×10^{14} meters). What is the average speed of this star in its orbit?

50. How large is the black hole at the center of our galaxy (that is, where is its event horizon)?

SMARTWORK

Norton's online homework system includes algorithmically generated versions of these questions, plus additional conceptual exercises. If your instructor assigns questions in SmartWork, log in at smartwork.wwnorton.com.

Exploration | The Center of the Milky Way

Adapted from Learning Astronomy by Doing Astronomy, by Ana Larson

Astronomers once thought that the Sun was at the center of the Milky Way. A more accurate picture of the size and shape of our galaxy was constructed from observations of RR Lyrae stars in globular clusters by Harlow Shapley. In this Exploration, you will repeat this experiment and find the center of the Milky Way for yourself.

Table 15.1 shows the galactic longitude and the projected distance for globular clusters in the Milky Way. To arrive at these coordinates, astronomers imagine that the plane of the Milky Way is horizontal, like a pizza, and the Sun is at the origin in the middle of it. A line is drawn straight "down" from a globular cluster to the plane. The projected distance is the distance from the Sun to the point where the line hits the plane, and the galactic longitude tells which direction that point is in. The polar graph in **Figure 15.21** goes with this type of coordinate system. Projected distance is the distance out from the center in kiloparsecs (kpc; 1 kpc = 1,000 pc). These distances are more convenient to plot in kiloparsecs than in thousands of light-years; remember that there are 3.26 light-years in a parsec. The galactic longitude is the angle measured around the outside of the graph. The galactic longitudes of significant constellations are indicated on the edge of the graph.

Plot each data pair on the graph by finding the galactic longitude indicated outside the circle, then coming in toward the center to the projected distance and making a dot. (The two globular clusters highlighted in bold in the table have already been plotted for you as examples.) After plotting all of the globular clusters, estimate the center of their distribution and mark it with an X. This is the center of the Milky Way.

1. What is the approximate distance from the Sun to the center of the Milky Way in kiloparsecs? What is this distance in light-years?

2. What is the galactic longitude of the center of the Milky Way?

3. How do we know the Sun is not at the center of the Milky Way?

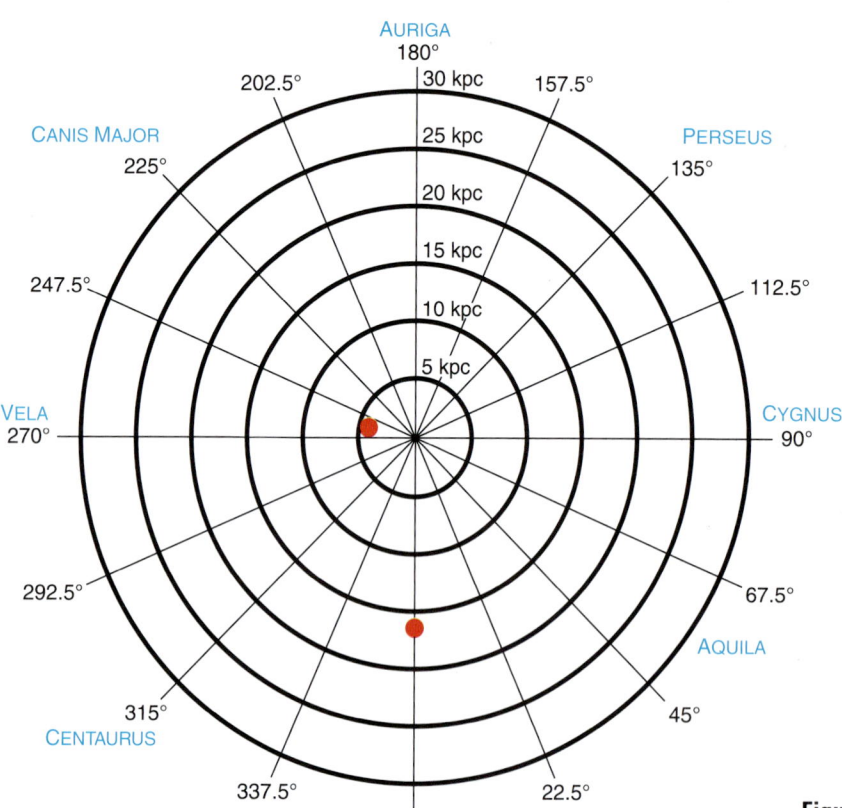

Figure 15.21 A polar graph used for plotting distance and direction.

TABLE 15.1

Globular Cluster Data

CLUSTER "NAME"	GALACTIC LONGITUDE (DEGREES)	PROJECTED DISTANCE (KPC)	CLUSTER "NAME"	GALACTIC LONGITUDE (DEGREES)	PROJECTED DISTANCE (KPC)
104	306	3.5	6273	357	7
362	302	6.6	**6287**	**0**	**16.6**
2808	283	8.9	6333	5	12.6
4147	**251**	**4.2**	6356	7	18.8
5024	333	3.4	6397	339	2.8
5139	309	5	6535	27	15.3
5634	342	17.6	6712	27	5.7
Pal 5	1	24.8	6723	0	7
5904	4	5.5	6760	36	8.4
6121	351	4.1	Pal 10	53	8.3
O 1276	22	25	Pal 11	32	27.2
6638	8	15.1	6864	20	31.5
6171	3	15.7	6981	35	17.7
6218	15	6.7	7089	54	9.9
6235	359	18.9	Pal 12	31	25.4
6266	353	11.6	288	147	0.3
6284	358	16.1	1904	228	14.4
6293	357	9.7	Pal 4	202	30.9
6341	68	6.5	4590	299	11.2
6366	18	16.7	5053	335	3.1
6402	21	14.1	5272	42	2.2
6656	9	3	5694	331	27.4
6717	13	14.4	5897	343	12.6
6752	337	4.8	6093	353	11.9
6779	62	10.4	6541	349	3.9
6809	9	5.5	6626	7	4.8
6838	56	2.6	6144	352	16.3
6934	52	17.3	6205	59	4.8
7078	65	9.4	6229	73	18.9
7099	27	9.1	6254	15	5.7

16 The Evolution of the Universe

Cosmology is the study of the universe on the very grandest of scales, including its nature, origin, evolution, and ultimate destiny. You have already learned that the universe is expanding. In this chapter, we take a closer look at the beginning of the universe itself. We will see that the universe originated in a Big Bang nearly 14 billion years ago. We will also see that it was once very hot—filled with thermal radiation that has now cooled to a temperature of 2.7 K—and that light elements in the universe were produced within the first few minutes after the Big Bang. In the illustration on the opposite page, a student is beginning to construct an ordered list of the events that occurred early in the history of the universe, combining this with images to help her remember the events.

✧ LEARNING GOALS

The universe that emerged from the Big Bang was incredibly uniform, wholly unlike today's universe of galaxies, stars, and planets. In this chapter, we will find that complex evolution is a natural, unavoidable consequence of the action of physical laws in our evolving universe. By the end of this chapter, you should be able to:

LG 1 Explain why Hubble's law and observations of the cosmic microwave background indicate that there was a hot, dense beginning to the universe.

LG 2 Describe the proposed connection between the accelerating universe and dark energy.

LG 3 Explain how the events that occurred in the earliest moments of the universe are related to the forces that operate in the modern universe.

LG 4 Describe the likely end of the universe.

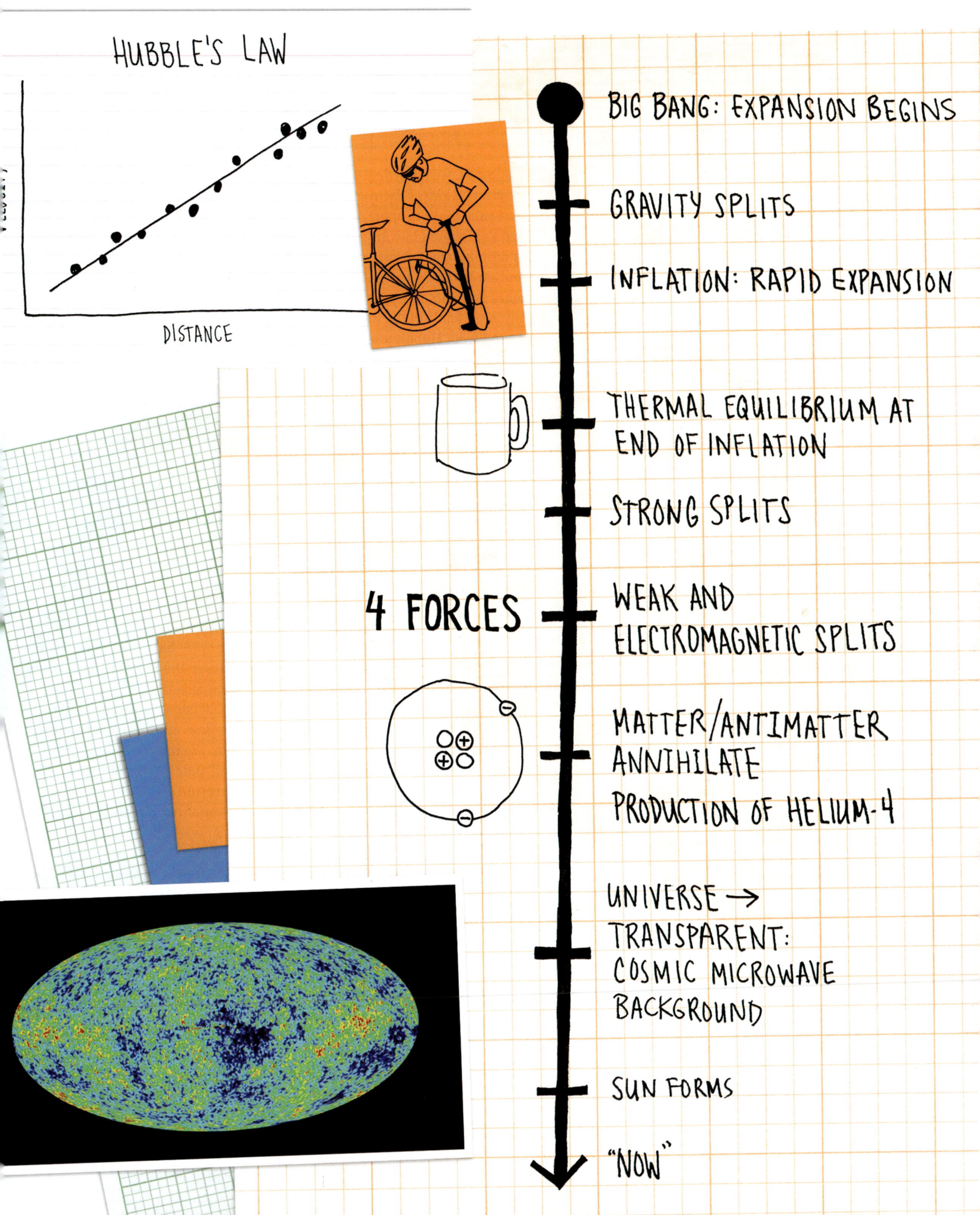

16.1 Hubble's Law Implies a Hot, Dense Beginning

In Chapter 14, we described observations of galaxy motions and distances. From these observations, Edwin Hubble created a graph that showed the galaxies in the universe moving apart, with the more distant ones moving faster—a result known as Hubble's law. Imagine watching a video of the universe, with the galaxies moving apart. Now, reverse the video, and run it backward in time. The separation between the galaxies becomes smaller and smaller as the universe becomes younger and younger. In this way, the observation of the expansion of the universe leads to the idea of a beginning to the universe at a time that can be estimated from Hubble's law. As we proceed, you will want to make a timeline of the early universe, as shown in the chapter-opening figure, where a student is illustrating key moments to make them easier to remember. In this section, we explore the implications of Hubble's law.

A Hot, Dense Beginning

Hubble's law tells us about the early universe. If we assume the speed of the expansion has always been constant, then the age of the universe can be estimated from the slope of the line in the graph of the velocity of galaxies versus their distance. This slope has units that reduce to 1/time, so its inverse has units of time. This slope is the Hubble constant, H_0, and its inverse (1 divided by H_0) is the **Hubble time**. The Hubble time is an estimate of the universe's age: 13.8 billion years (**Working It Out 16.1**).

Hubble's law shows that about 6.9 billion years ago, when the universe was half its current age, all of the galaxies in the universe were half as far apart as they are now. Twelve billion years ago, all of the galaxies in the universe were about a tenth as far apart. A little less than one Hubble time ago—13.8 billion years ago—there was almost no space between the particles that constitute today's universe. All such matter as well as energy in the universe then must have been unimaginably dense. Because expanding gases cool down, the universe then must have been much hotter than it is today in its expanded state. This hot, dense beginning, 13.8 billion years ago, is the **Big Bang** (**Figure 16.1**).

Pause for a moment to think about these astonishing conclusions. From a straightforward graph of velocity versus distance, we find that the entire universe changes over time. The inverse of the slope of this graph is an estimate of the age of the universe itself. From grade-school mathematics and a single graph, we derive a beginning to the universe, and we can determine that the beginning was hot and dense and also discover how long ago the beginning occurred.

Because it is difficult to grapple with the idea of the beginning of the universe, the idea of the Big Bang greatly troubled many astronomers in the early and middle years of the 20th century. Many astronomers tried to explain the Hubble expansion without resorting to the Big Bang. However, further observations and discoveries have only strengthened the Big Bang theory. As you will see, all the major predictions of the Big Bang theory (expansion of the universe is only one of them) have proved to be correct. The Big Bang theory

▶❙❙ **AstroTour: Hubble's Law**

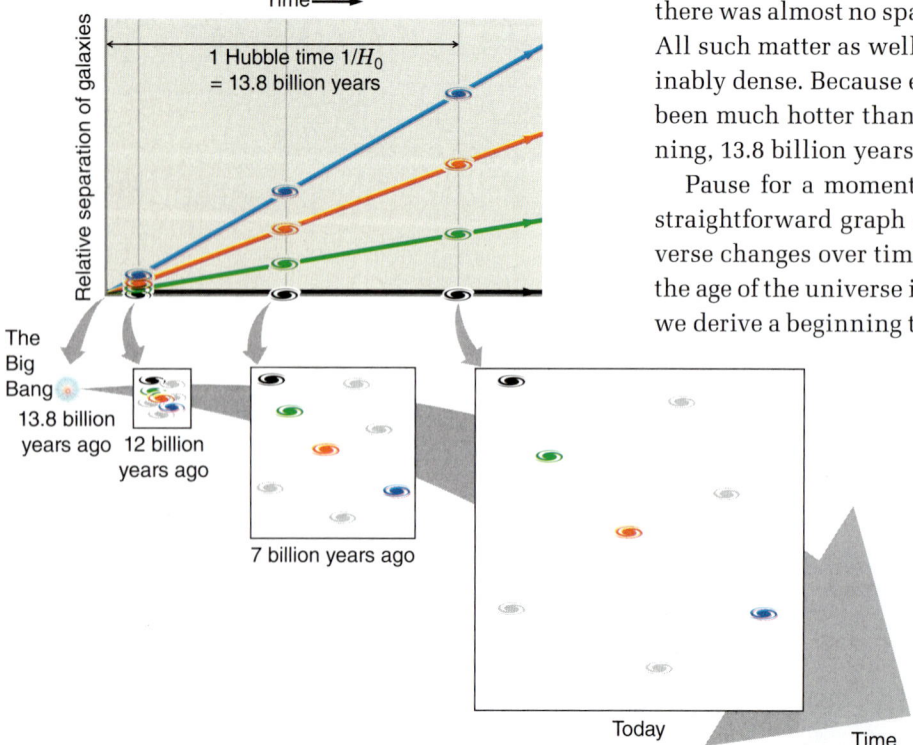

Figure 16.1 Looking backward in time, the distance between any two galaxies is smaller and smaller, until all matter in the universe is concentrated together at the same point: the Big Bang.

Working It Out 16.1 | Expansion and the Age of the Universe

We can use Hubble's law to estimate the age of the universe. Consider two galaxies located 30 Mpc ($d_G = 9.3 \times 10^{20}$ km) away from each other (**Figure 16.2**). If these two galaxies are moving apart from each other at constant speed, then at some time in the past they must have been together in the same place at the same time. According to Hubble's law, and assuming that $H_0 = 70$ km/s/Mpc, the distance between these two galaxies is increasing at the rate

$$v_r = H_0 \times d_G$$
$$v_r = 70 \text{ km/s/Mpc} \times 30 \text{ Mpc}$$
$$v_r = 2{,}100 \text{ km/s}$$

Knowing the speed at which they are traveling, we can calculate the time it took for the two galaxies to become separated by 30 Mpc (which is equal to 9.3×10^{20} km):

$$\text{Time} = \frac{\text{Distance}}{\text{Speed}} = \frac{9.3 \times 10^{20} \text{ km}}{2{,}100 \text{ km/s}} = 4.4 \times 10^{17} \text{ s}$$

Dividing by the number of seconds in a year (about 3.2×10^7 s/yr) gives

$$\text{Time} = 1.4 \times 10^{10} \text{ yr}$$

In other words, *if* expansion of the universe has been constant, two galaxies that today are 30 Mpc apart started out at the same place about 14 billion years ago.

Now let's do the same calculation with two galaxies that are 60 Mpc apart (see Figure 16.1). These two galaxies are twice as far apart, but the distance between them is increasing twice as rapidly:

$$v_r = H_0 \times d_G = 70 \text{ km/s/Mpc} \times 60 \text{ Mpc} = 4{,}200 \text{ km/s}$$

Therefore,

$$\text{Time} = \frac{\text{Distance}}{\text{Speed}} = \frac{19 \times 10^{20} \text{ km}}{4{,}200 \text{ km/s}} = 4.5 \times 10^{17} \text{ s}$$

Dividing by the number of seconds in a year gives

$$\text{Time} = 1.4 \times 10^{10} \text{ yr}$$

We find that these galaxies also took about 14 billion years to reach their current locations. We could do this calculation again and again for any pair of galaxies in the universe today. (Small differences in the intermediate steps of these calculations are due to rounding and are not significant to the argument.) The most precise measurements give a result of 13.8 billion years.

If we work out the example using words instead of numbers, we can see why the answer is always the same. Because the velocity we are calculating comes from Hubble's law, velocity equals Hubble's constant multiplied by distance. Writing this out as an equation, we get

$$\text{Time} = \frac{\text{Distance}}{\text{Velocity}}$$

$$\text{Time} = \frac{\text{Distance}}{H_0 \times \text{Distance}}$$

Distance divides out to give

$$\text{Time} = \frac{1}{H_0}$$

We define $1/H_0$ as the *Hubble time*. This way of estimating the age of the universe assumes it has always expanded at the same rate.

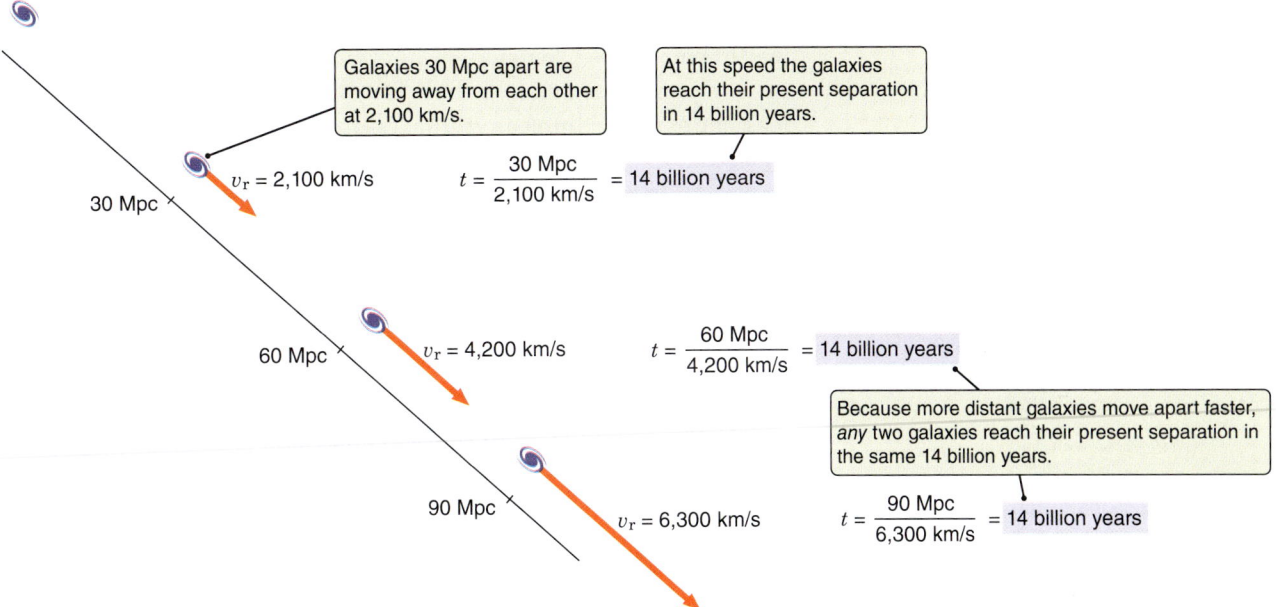

Figure 16.2 These three galaxies have three different apparent speeds. They are also located at three different distances from the Milky Way. Dividing each distance by each speed gives the same elapsed time: about 14 billion years. This is the amount of time that they have been traveling: the age of the universe.

for the origin of our universe is now so well corroborated that it has truly earned the title of scientific theory.

The implications of Hubble's law are striking and have changed our concept of the origin and history of the universe in which we live. At the same time, Hubble's law has pointed to many new questions about the universe. To address them, we need to consider precisely what the term *expanding universe* means.

Carried Along by the Expansion

At this point in our discussion, you may be picturing the expanding universe as a cloud of debris from an explosion flying outward through surrounding space. This is a common depiction of the Big Bang in movies and television shows where they show a tiny bright spot that explodes to fill the screen. However, this depiction is completely incorrect. The Big Bang is not an explosion in the usual sense of the word, and there is no surrounding space into which the universe expands.

To understand the Big Bang, we need to think about one concept at a time. To begin with, picture the universe as infinite. Infinity is a difficult concept for most people—it's not something you can see. But if you think about all the whole numbers (0, 1, 2, 3, and so forth) marked on a line, extending off to your right, you begin to grasp infinity. You could walk and walk and walk, and never come to the end of these numbers. No matter how large a number you walked to, you could add 1 to it, and then 2, and so on, and the line would continue on as far in front of you as you can imagine. This is infinity. To make the line infinite in both directions, place the negative whole numbers (−1, −2, −3, and so forth) to the left of the zero. **Figure 16.3a** should help you visualize this line extending in both directions to infinity.

Now, draw a second line (the *y*-axis) perpendicular to the first (the *x*-axis) that also extends to infinity in both directions. Then draw a third line, forward and back (the *z*-axis), extending to infinity. This is the size of an infinite universe: It is infinite in size in all directions, and it always was—it was infinite the moment it formed.

So now, consider this question: What is on the outside of the universe? This very question is problematic. Before you can explain *what* is outside, you would need to know *where* "outside" is. But if space extends infinitely in any direction, it never comes to an end, so there is no "outside." The universe expands, but it does not expand "into" anything, because there are no edges. This reasoning leads to an important point. If there are no edges to the universe, there is also no center of the universe. Think about how you would find the center of the room you are sitting in. You would first measure from wall to wall, right to left. Then you would measure from front to back, and top to bottom. You would find the center by taking half of all these distances. But if there are no walls, there is no center. An infinite universe has no edges and, consequently, it has no center.

"But wait!" you are thinking, "if the universe is infinite, how can it expand? How can it get bigger?" Consider again the line of numbers that extends infinitely to your left and right. There are an infinite number of whole numbers on this line. Now imagine the number of fractions between 0 and 1— to make it more manageable, consider the fractions that are formed from whole numbers: ½, ⅓, ¼, ⅕, and so forth. There are an infinite number of these fractions. Figure 16.3b should help you visualize the fractions between the whole numbers. As you "zoom in" on those fractions,

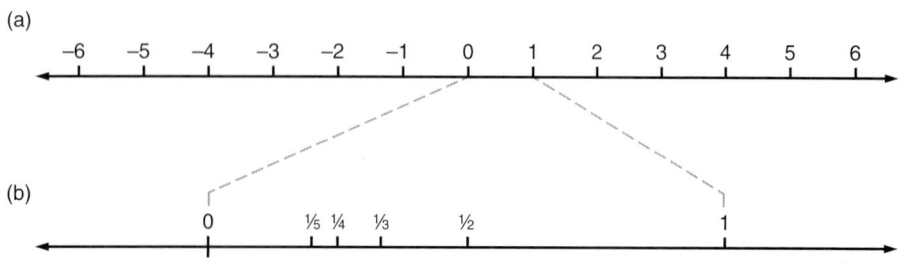

Figure 16.3 (a) The number of whole numbers is infinite, so arranging them along a line extending left and right results in an infinite line. (b) Fitting in an infinite number of fractions between 0 and 1 requires visually stretching the space between 0 and 1. This is analogous to the expansion of space. The whole numbers are like galaxies, carried along by the stretching of the space between them.

the space between 0 and 1 gets bigger to allow the number of fractions to fit. This is a more correct picture of the expansion of space: just as the space between 0 and 1 gets bigger, the spaces between the galaxies get bigger. As the 0 and the 1 are carried along by the expansion of the line, the galaxies are carried along by the expansion of space, not moving *through* space (recall Figure 14.18, in which the ant is carried along on the rubber band).

This picture of the Big Bang and the expansion of space helps answer another common question about the Big Bang: "Where did it take place?" Take a moment to see if you can figure out the answer for yourself.

The Big Bang took place *everywhere*. Wherever you are in the universe today, you are sitting at the site of the Big Bang. The entire universe formed at once, and the space between places in the universe has been stretching ever since, carrying the galaxies along in the expansion. Hubble's law is an observable consequence of the expanding space between galaxies.

When thinking about the expanding universe, remember that the laws of physics are themselves unchanged by the expansion. For example, when we stretched out the line of numbers, we did not change the properties of the numbers 1 or 2 or 4. Similarly, as space expands, the sizes and other physical properties of atoms, stars, and galaxies within it also remain unchanged. In addition, on relatively small scales, the expansion is extremely weak compared to the pull of gravity. Locally, the Solar System is not expanding, because gravity overwhelms the expansion and holds the Solar System together. Neither is the Milky Way expanding, because gravity overwhelms the expansion. Recall that the Andromeda Galaxy is actually approaching the Milky Way. Gravity is winning there as well. But over larger distances, where the gravitational attraction between galaxies is small, the expansion of space carries galaxies apart.

We must draw a distinction between the universe and the *observable universe*. The **observable universe** is the part of the universe that we can see. The observable universe extends 13.8 billion light years in every direction. This limit exists because that is the length of time the universe has been around. The light from more distant regions has not yet had time to travel to us, and so we cannot see it yet.

The Shape and Size of the Universe

We have already seen an example of curved space, near a massive object such as a black hole. That type of curvature is called *positive curvature* and follows geometric rules similar to those on the surface of Earth. For example, parallel lines converge in positively curved space. Lines of longitude are parallel to each other at the equator, but eventually converge to a point at the North Pole. *Negative curvature* is like the shape of the center of a Pringle potato chip and follows geometric rules opposite to those on the surface of Earth. Parallel lines diverge away from each other in negatively curved space. *Flat space* follows the type of geometry you learned in school. Parallel lines remain parallel forever.

As you will see in Section 16.5, current observations indicate that our universe is flat on the largest scales, to a quite high degree of precision. (It is so flat, in fact, that its very flatness is a problem!) From the cosmological principle, we may reason that if the universe is flat in the part of the universe we can see, it is flat everywhere. If that is true, then two parallel lines would remain parallel forever and never converge or diverge. This is another way to say that the universe is infinite.

We do not know for certain that the universe is exactly flat, because measurements are not infinitely precise. Therefore, we also do not know if it is truly infinite or just very much larger than the part of it that is observable. Even if the universe

416 CHAPTER 16 *The Evolution of the Universe*

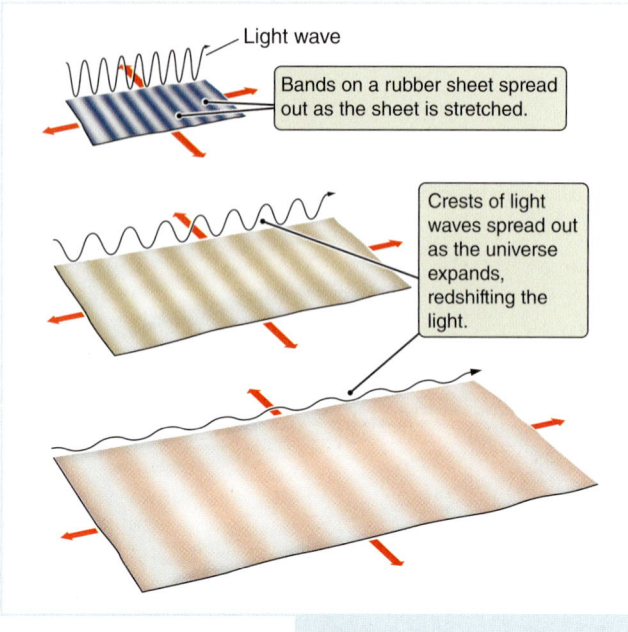

Figure 16.4 Bands drawn on a rubber sheet represent the positions of the crests of a light wave in space. As the rubber sheet is stretched—that is, as the universe expands—the wave crests get farther apart. The light is redshifted.

> **Back in Section 5.6 . . .**
> . . . you learned that when objects are moving away from you, their spectral lines are shifted toward the red end of the spectrum.

is finite, it is still not expanding "into" anything. One way to understand this is to consider that the universe is, by definition, everything. So even if there were something outside for it to expand into, that would be part of the universe too.

There is another way to visualize a finite, expanding universe with no "outside." Consider a universe that is finite, yet unbounded. This means that it is like the two-dimensional surface of Earth. As a tiny creature, you are more or less confined to the surface of Earth. This surface is finite, yet you could walk around it and around it and around it, and never find an edge. The universe as a whole object might be similarly finite, yet without boundaries. If you traveled far enough through the universe in one direction, you might find yourself back where you started. There is nothing for such a universe to expand into; there is no "outside." We may never know if the universe is flat everywhere or if it eventually curves back upon itself. But even a finite universe encompasses all there is. There is no outside.

In an infinite universe, there is no center, because there are no edges. What about the finite, unbounded universe? This universe has no center, because, again, there are no edges. Just as there is no central point on the surface of Earth, there is no center of a finite, unbounded universe.

Redshift and the Expansion of the Universe

Recall that Hubble's law is the relationship between the velocity and distance of galaxies. These velocities were determined from the redshift of the galaxies using the Doppler effect. However, we have seen that the galaxies are not moving through space; rather, the space between them is growing. If the galaxies are stationary, why is their light redshifted? Although it is true that the distance between galaxies is increasing as a result of the expansion of the universe and that we can use the *equation* for Doppler shifts to express galaxy redshifts in terms of velocity, these redshifts are not due to Doppler shifts.

Recall the rubber sheet analogy that we used when discussing black holes in Chapter 13. In that case, we imagined space as a two-dimensional rubber sheet. **Figure 16.4** uses this rubber-sheet analogy to explain why the increasing distance between galaxies causes the light to be redshifted. If we draw a series of bands on the rubber sheet to represent the crests of a light wave, we can watch what happens to the wave as the sheet is stretched. As light comes toward us from distant galaxies, the space through which the light travels is stretching, and the light is also "stretched out" as the space through which it travels expands. The farther light travels through expanding space, the redder light becomes. Distant galaxies are redder than nearby ones, as the light has traveled farther through expanding space to reach us. The redshift of light from distant galaxies is therefore a direct measure of how much the universe has expanded since the time the light left its source.

16.2 The Cosmic Microwave Background Confirms the Big Bang

It is quite remarkable that we live in a time when we are finding real, testable answers to cosmological questions using the empirical methods of science. What evidence, apart from the observed expansion of the universe itself, causes us to accept that the Big Bang actually took place? One piece of evidence comes from observations of the early universe across the entire sky. In this section, you will learn about the cosmic microwave background, one of the major confirming observations of the theory that the universe had a beginning.

Radiation from the Big Bang

In the late 1940s, cosmologists Ralph Alpher (1921–2007), Robert Herman (1914–1997), and George Gamow (1904–1968) were thinking about the implications of Hubble expansion. They reasoned that because a compressed gas cools as it expands, the expanding universe should also be cooling. When the universe was very young and small, it must have consisted of an extraordinarily hot, dense gas. This hot, dense early universe would have been awash in blackbody radiation.

Gamow and Alpher took this idea a step further. They reasoned that as the universe expanded, this radiation would have been redshifted to longer and longer wavelengths. Recall Wien's law from Chapter 5, which states that the temperature associated with blackbody radiation is inversely proportional to the peak wavelength. Shifting the wavelength of blackbody radiation to longer wavelengths is like shifting the temperature to lower values. As illustrated in **Figure 16.5**, doubling the wavelength of the photons in a blackbody spectrum by stretching space in the universe is equivalent to cutting the temperature of the blackbody spectrum in half. The conclusion was that the radiation from the early universe should still be visible today and should have a blackbody spectrum with a temperature of about 5–10 K.

Early attempts to detect this radiation were unsuccessful. However, in the early 1960s, two physicists at Bell Laboratories, Arno Penzias (b. 1933) and Robert Wilson (b. 1936), were trying to bounce radio signals off Echo 1, a newly launched satellite. This hardly seems much of a feat today, when we routinely use cell phones and GPS systems that communicate directly with satellites. But at the time, the effort pushed technology to its limits. Penzias and Wilson needed a very sensitive microwave telescope for their work. Any spurious signals coming from the telescope itself might wash out the faint signals bounced off a satellite. Wilson

Back in Section 5.1 . . .

. . . you learned that hotter objects are bluer.

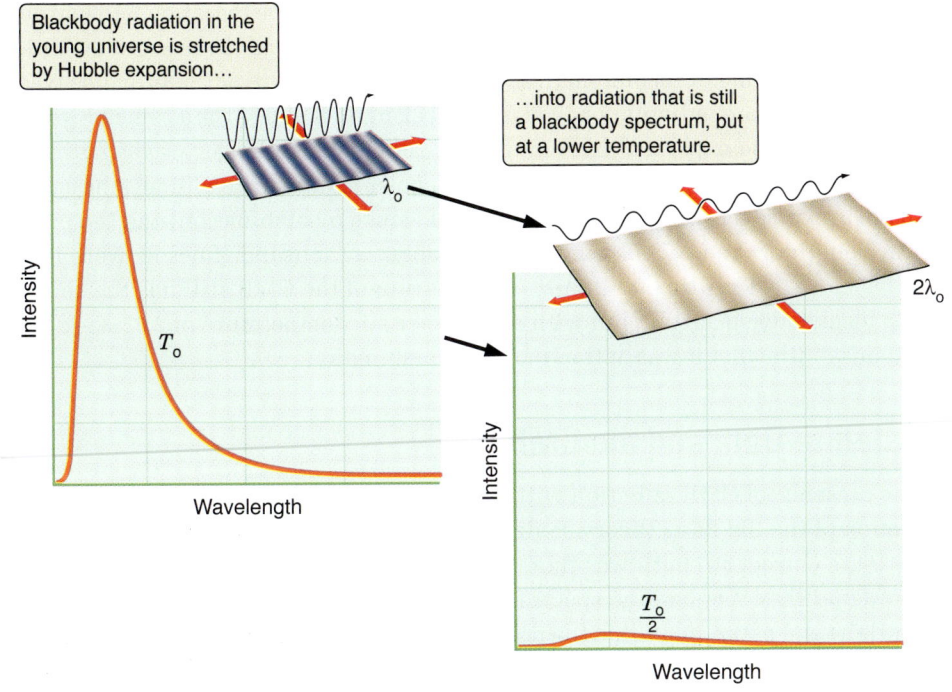

Figure 16.5 As the universe expanded, blackbody radiation left over from the hot young universe was redshifted to longer wavelengths. Redshifting a blackbody spectrum is equivalent to lowering its temperature.

Figure 16.6 Robert Wilson (left) and Arno Penzias (right) are shown next to the Bell Labs radio telescope antenna with which they discovered the cosmic microwave background. This antenna in Holmdel, New Jersey, is now a U.S. National Historic Landmark.

and Penzias, pictured in **Figure 16.6** along with their radio telescope, worked tirelessly to eliminate all possible sources of interference originating from within their instrument. This work included such endless and menial tasks as keeping the telescope free of bird droppings and other extraneous material. Even so, Penzias and Wilson found that no matter how hard they tried to eliminate sources of extraneous noise, they could still detect a faint microwave signal when they pointed the telescope at the sky. Eventually, they came to accept that the signal they were detecting was real. The sky faintly glows in microwaves.

In the meantime, physicist Robert Dicke (1916–1997) and his colleagues at Princeton University also predicted a hot early universe, arriving independently at the same basic conclusions that Alpher and Gamow had reached two decades earlier. When Dicke and colleagues heard of the signal that Penzias and Wilson had found, they interpreted it as the radiation left behind by the hot early universe. The strength of the detected signal was consistent with the glow from a blackbody with a temperature of about 3 K, very close to the predicted value. Penzias and Wilson's results, published in 1965, reported the discovery of the glow left behind by the Big Bang. Penzias and Wilson shared the 1979 Nobel Prize in Physics for their remarkable discovery.

The radiation left over from the early universe is called **cosmic microwave background (CMB) radiation**. When the universe was young, it was hot enough that all of the atoms in the universe were ions. Free electrons in such conditions interact strongly with radiation, blocking its progress. As illustrated in **Figure 16.7a**, the universe was an opaque blackbody. At that time in the early universe, conditions within the universe were much like the conditions within a star: hot, dense, and opaque.

As the universe expanded, the gas filling it cooled. By the time the universe was several hundred thousand years old, the universe was about a thousandth of its current size, and the temperature had dropped to a few thousand kelvins. At this time, **recombination** occurred: Hydrogen and helium nuclei combined with electrons to form neutral atoms.

Hydrogen atoms block radiation much less effectively than free electrons, so when recombination occurred, the universe suddenly became transparent to radiation, as illustrated in Figure 16.7b. Since that time, the radiation left behind from the Big Bang has been able to travel largely unimpeded throughout the universe. At the time of recombination, when the temperature of the universe was about 3000 K, the radiation peaked at a wavelength of about 1 micrometer (μm). As the universe expands, this radiation is redshifted to longer wavelengths—and therefore, as in Figure 16.5, represents cooler temperatures. Today, the scale of the universe has increased a thousandfold since recombination, and the peak wavelength of the CMB has increased by about a thousand times as well, to a value close to 1 millimeter (mm). The spectrum of the CMB still has the shape of a blackbody spectrum, but with a characteristic temperature of 2.73 K—only a thousandth what it was at the time of recombination.

Satellite Data and the CMB

The presence of CMB radiation with a blackbody spectrum is a very strong prediction of the Big Bang theory. Penzias and Wilson had confirmed that a signal with the correct strength was there, but they could not say for certain whether the signal they saw had the spectral shape of a blackbody spectrum. It was not until the end of the 1980s that the predictions of Big Bang cosmology for the CMB were put to the ultimate test. A satellite called the Cosmic Background Explorer,

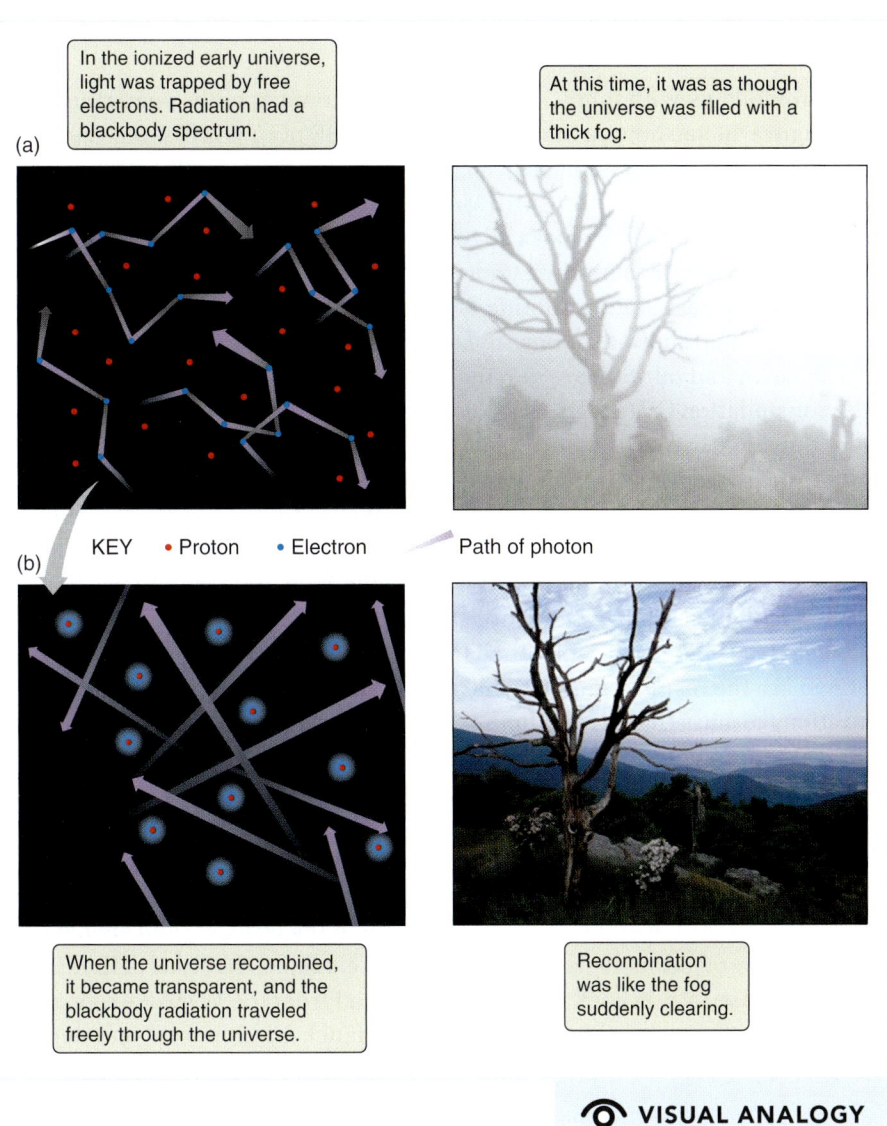

Figure 16.7 The cosmic microwave background radiation originated at the moment the universe became transparent. (a) Before recombination, the universe was like a foggy day, except that the "fog" was a sea of hydrogen atoms. Radiation interacted strongly with free electrons and so could not travel far. The trapped radiation had a blackbody spectrum. (b) When the universe recombined, the fog cleared and this radiation was free to travel unimpeded.

or COBE, was launched in 1989. COBE made extremely precise measurements of the CMB at many wavelengths, from a few micrometers out to 1 centimeter (cm). In January 1990, hundreds of astronomers gathered in a large conference room in Washington, D.C., at the winter meeting of the American Astronomical Society to hear the COBE team present its first results. Security surrounding the new findings had been tight, so the atmosphere in the room was electric. The tension did not last long; presentation of a single graph brought the room's occupants to their feet in a spontaneous ovation.

The data shown on that graph are reproduced in **Figure 16.8**. The small dots in the figure are the COBE measurements of the CMB at different frequencies. The uncertainty in each measurement is far less than the size of each dot. The line in the figure, which runs perfectly through the data points, is a blackbody spectrum with a temperature of 2.73 K. The agreement between theoretical prediction and observation is truly remarkable. The observed spectrum so perfectly matches the one predicted by Big Bang cosmology that there can be no real doubt we are seeing the residual radiation left behind from the hot, dense beginning of the early universe. John Mather (b. 1946) and George Smoot (b. 1945) were awarded a Nobel Prize for this work in 2006.

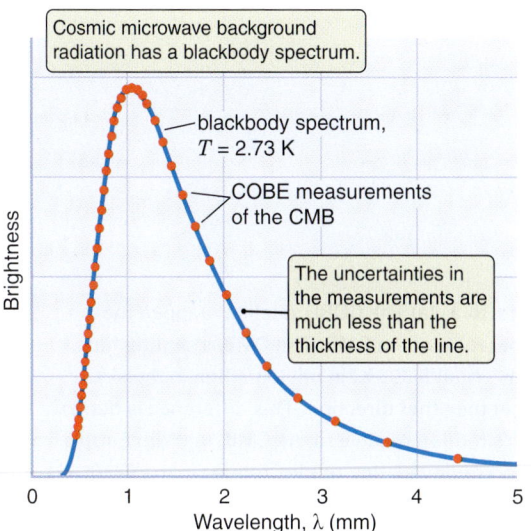

Figure 16.8 The spectrum of the CMB as measured by the Cosmic Background Explorer (COBE) satellite is shown by the red dots. The uncertainty in the measurement at each wavelength is much less than the size of a dot. The line running through the data is a blackbody spectrum with a temperature of 2.73 K.

Variations in the CMB

COBE provided us with much more than a measurement of the spectrum of the CMB. **Figure 16.9a** shows a map obtained by COBE of the CMB from the entire sky. The different colors in the map correspond to variations of less than 0.1 percent in the temperature of the CMB. Most of this range of temperature is present because one side of the sky looks slightly warmer than the opposite side of the sky. This difference has nothing to do with the structure of the universe itself, but rather is the result of the motion of Earth relative to the CMB.

The COBE map shows that one side of the sky is slightly hotter than the other because Earth and our Sun are moving at a velocity of 368 km/s in the direction of the constellation Crater. Radiation coming from the direction in which we are moving is slightly blueshifted by our motion, whereas radiation coming from the opposite direction is Doppler-shifted toward the red. Our motion is due to a combination of factors, including the motion of our Sun around the center of the Milky Way and the motion of our galaxy relative to the CMB.

If we subtract this asymmetry from the COBE map, only slight variations in the CMB remain, as shown in Figure 16.9b. The brighter parts of this image are only about 0.001 percent brighter than the fainter parts. These tiny fluctuations in the CMB are the result of gravitational redshifts (see Chapter 13) caused by concentrations of mass that existed in the early universe. These concentrations later gave rise to galaxies and the rest of the structure that we see in the universe today.

Subsequent observations support the COBE findings. Beginning in 2001, the Wilkinson Microwave Anisotropy Probe (WMAP) satellite made more precise measurements of the variations of the CMB. Figure 16.9c shows the ripples measured by WMAP. These much-higher-resolution maps have profound implications for our understanding of the origin of structure in the universe and enable us to determine several cosmological parameters. For example, the value of the Hubble constant we use in this book is the value inferred from the WMAP experiment.

In 2013, the European Space Agency (ESA) Planck mission went one step further, producing maps in even higher resolution. Overall, these highly detailed

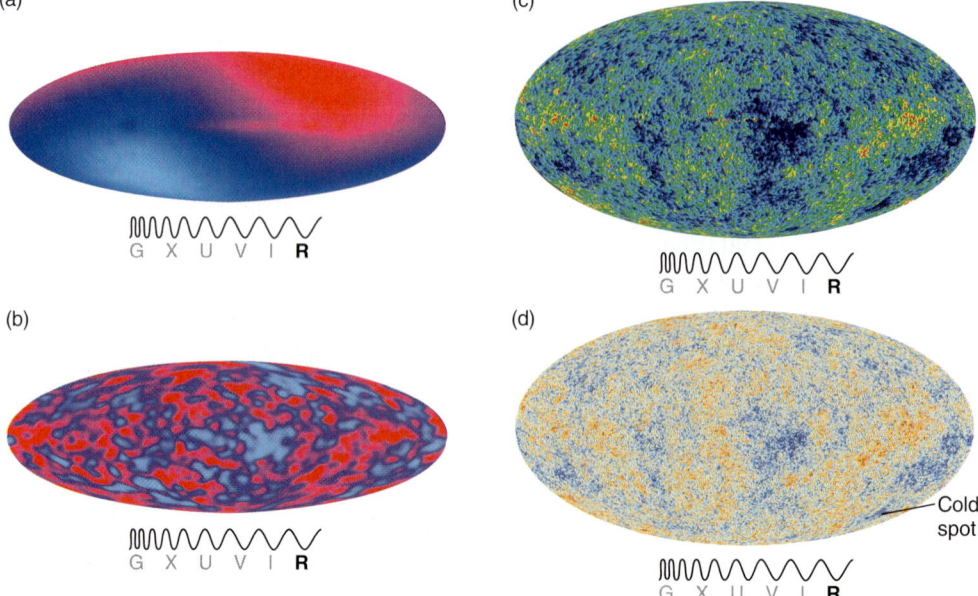

Figure 16.9 (a) The COBE satellite mapped the temperature of the CMB. The CMB is slightly hotter (by about 0.003 K) in one direction in the sky than in the other direction. This difference is due to Earth's motion relative to the CMB. (b) With Earth's motion removed, tiny ripples remain in the CMB. (c) WMAP confirmed the fundamentals of cosmological theory at small and intermediate scales. (d) The Planck mission has provided the highest resolution yet of the CMB and has detected some surprises, such as the "cold spot." The radiation seen in this image was emitted less than 400,000 years after the Big Bang.

observations support the existing theories of Big Bang cosmology. Intriguingly, however, early results show that variations at large scales are not quite as strong as expected from the theoretical predictions. Also, as shown in Figure 16.9d, a "cold spot" first detected in the WMAP data was confirmed to be real. This is exciting news: There is something new to figure out, and details may need to be added to existing theories to account for these improved observations.

Because the Big Bang theory is so fundamental to our contemporary understanding of the universe, it is one of the most challenged theories in all of science. Astronomers, particle physicists, and other scientists are continually making predictions from the theory and then testing those predictions. So far, the theory has resisted all efforts to falsify it. As we will see in Chapter 17, there are other confirmed predictions of the Big Bang beyond the CMB.

16.3 The Expansion of the Universe Is Speeding Up

The simplest approach to Big Bang cosmology is to assume that the expansion occurs at a constant rate, neither speeding up nor slowing down. We made this assumption in Section 16.1. But what if it's not true? For example, the universe contains mass and therefore gravity that pulls galaxies together. Does this mass slow the expansion of the universe? How could we know? Astronomers asked those very questions, and the answers were surprising.

Gravity and the Expansion of the Universe

To see how gravity affects the expansion of the universe, it will help to recall gravity's effects on the motion of projectiles. The fate of a projectile fired straight up from the surface of a planet depends on the planet's mass and its radius. If the planet is massive enough, it will slow the projectile, stop it, and pull it back down. However, if the planet is not massive enough, the projectile will slow down but still escape to space. The size of the planet is also important. If the planet is smaller, so the mass is more densely packed, the projectile begins closer to the center of the planet, so the gravitational pull is stronger, and the planet can pull the projectile down. However, if the planet has lower density, so the projectile at the surface starts farther from the center of the planet, the gravitational pull is weaker, and the projectile will escape. Whether the projectile escapes depends on both the planet's mass and the distance between the projectile and the planet's center.

Just as the mass of the planet slows the climb of a projectile, the mass in the universe slows the expansion of the universe. The fate of the universe is determined by the universe's mass and the separation between masses. If the mass is packed closely together, so that the average density is high, the expansion will slow, stop, and reverse. If the mass is very spread out, so that the average density is low, the universe will expand forever.

Astronomers define a **critical density** for which the mass in the universe would cause it to just barely stop expanding after a very long time. This critical density determines the dividing line between two possible fates of the universe: expanding forever or collapsing. The critical density of a universe depends on the speed of the expansion—a faster expansion requires a higher density to stop it. Current measures of the expansion of our universe give a critical density to about

Vocabulary Alert

critical In common language, this word often means "indispensable." Here, astronomers use *critical* in the sense of a turning point or boundary between two cases.

Figure 16.10 There are three possible fates of the universe, based on the critical density of the universe if the expansion has occurred at constant speed in the past.

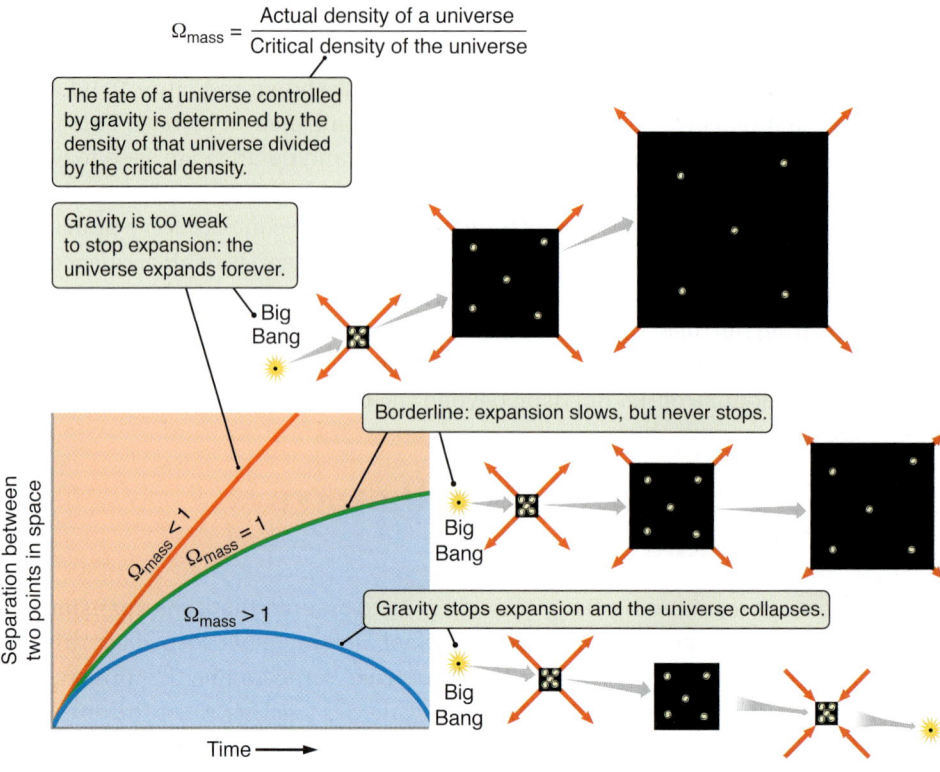

8×10^{-27} kilograms per cubic meter (kg/m³). Rather than trying to keep track of such awkward numbers, astronomers talk about the *ratio* of the actual density of the universe to this critical density. This ratio is called Ω_{mass} (pronounced "omega mass"). If Ω_{mass} is larger than 1, the universe has a density higher than the critical density, and the universe will collapse. If Ω_{mass} is less than 1, the universe has a density lower than the critical density, and the universe will expand forever. The dividing line, where Ω_{mass} equals 1, corresponds to a universe in which the expansion slows down but never quite stops. The expansion of different possible mass-dominated universes is shown in **Figure 16.10**, which plots the expansion of the universe versus time for different values of Ω_{mass}.

Until the closing years of the 20th century, most astronomers thought that gravity was all there was to the question of expansion and collapse. Researchers carefully measured the mass of galaxies and collections of galaxies in the hope that this would reveal the density and therefore the fate of the universe. The luminous matter gives a value for Ω_{mass} of about 0.02. Galaxies contain about 10 times as much dark matter as normal matter, so adding in the dark matter in galaxies pushes the value of Ω_{mass} up to about 0.2. When we include the mass of dark matter *between* galaxies, Ω_{mass} could increase to 0.3 or a bit higher. By this accounting there is only about one-third as much mass in the universe as is needed to stop the universe's expansion. Many astronomers were convinced that the universe was "open"—that the expansion would never be zero, but it would slow down due to gravity. They were wrong.

The Accelerating Universe

If the expansion of the universe is slowing down, then the expansion must have been faster in the past. If this is the case, objects that are very far away (so that we see them as they were long ago) should therefore have larger redshifts than the local Hubble law would lead us to expect.

During the 1990s, astronomers tested this prediction. They measured the brightness of Type Ia supernovae in very distant galaxies and compared each of those brightnesses with the expected brightness based on the redshifts of those galaxies. The findings of these studies were startling. Rather than showing that the expansion of the universe is slowing, the data indicated that the expansion is *speeding up*. To describe this increasing expansion rate, people often say "the universe is accelerating." This does not mean that the universe is zooming through space faster and faster like a car along a road, but instead means that the expansion is happening faster and faster. In order for the expansion rate to increase, a force that is stronger than gravity—a previously unknown force—must be acting. Naturally, astronomers repeated the experiment many times, in different ways, and continued to improve their certainty in the result. Results from the WMAP experiment early in the 21st century confirmed the result independently, and in 2011, the Nobel Prize was awarded to the original discoverers: Saul Perlmutter, Brian P. Schmidt, and Adam G. Reiss. The rate of expansion of the universe is speeding up. But the origins of the repulsive force remained unknown.

The idea of a repulsive force that opposes gravity was not new. Nearly a century earlier, more than a decade before Hubble discovered the expansion of the universe, Einstein developed the theory of general relativity. At that time, Einstein believed that the universe was stationary—that it neither expands nor collapses. When Einstein first used general relativity to determine the structure of the universe, he was greatly troubled. The theory clearly indicated that any universe containing mass could not sit still, any more than a ball can hang motionless in the air. Einstein inserted a "fudge factor," called the **cosmological constant**, into his equations. The cosmological constant acts as a repulsive force that opposes gravity. If it has just the right value, the cosmological constant can lead to a static universe in which galaxies remain stationary despite their mutual gravitational attraction.

When Hubble announced his discovery that the universe is expanding, Einstein realized his mistake. Einstein could have *predicted* that the universe must either be expanding or contracting with time, but instead forced his equations to comply with conventional wisdom. He called the introduction of the cosmological constant the "biggest blunder" of his scientific career. However, the discovery of an accelerating universe restores the cosmological constant to general relativity. A repulsive force like the force associated with the cosmological constant is just what is needed to describe a universe that is accelerating. Today, we write the cosmological constant as Λ. The fraction of the critical density provided by the cosmological constant is written as Ω_Λ (pronounced "omega lambda"). Because the universe's expansion is accelerating, we know Ω_Λ is not zero.

While a universe is young and densely packed, gravity dominates the effect of Ω_Λ. As a universe expands, gravity gets weaker because the mass spreads out. The cosmological constant (because it *is* a constant) becomes increasingly important as Ω_{mass} declines. Unless gravity is able to turn the expansion around, the cosmological constant wins in the end, causing the expansion to continue accelerating forever. Even if the density of the universe is greater than the critical density, so that Ω_{mass} is greater than 1, a large enough Ω_Λ could overwhelm gravity and make the universe expand forever. **Figure 16.11** shows plots of the expansion versus time that are similar to those shown in Figure 16.10, but now we have included the effects of Ω_Λ. If Ω_{mass} is sufficiently large, a universe will collapse back on itself, regardless of the value of Ω_Λ. However, if a universe expands forever, its future will depend on Ω_Λ.

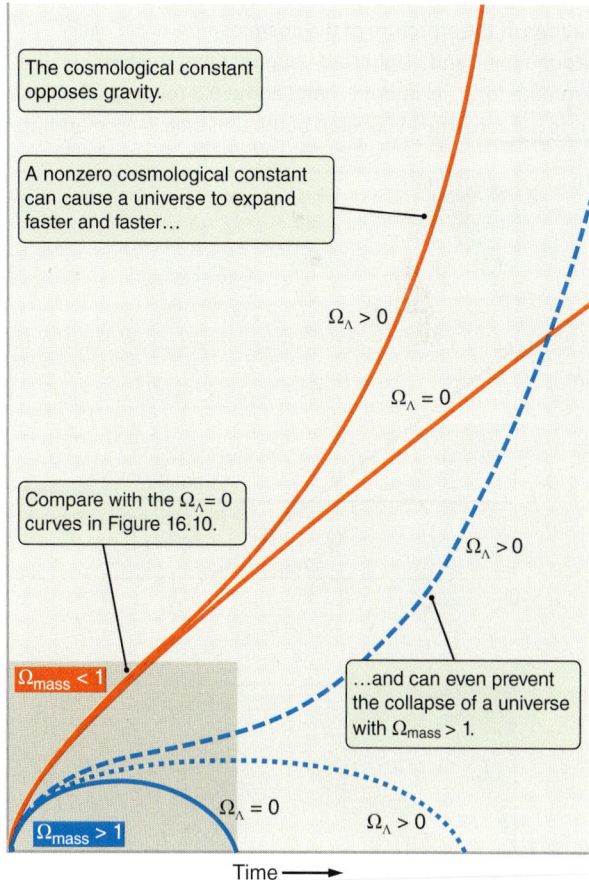

Figure 16.11 The distance between two points in the universe changes with time. This is shown here for cosmologies with and without a cosmological constant. If there is enough mass in a universe, gravity could still overcome the cosmological constant and cause that universe to collapse. Any universe without enough mass eventually to collapse will instead end up expanding at an ever-increasing rate because of a nonzero cosmological constant

When Einstein added the cosmological constant to his equations of general relativity, he considered it a new fundamental constant, similar to Newton's universal gravitational constant, G. Today we realize that empty space can have distinct physical properties of its own. For example, space can have a nonzero energy even in the total absence of matter. This energy of empty space produces exactly the kind of repulsive force that could accelerate the expansion of the universe, and we call it **dark energy**. The cosmological constant is an example of dark energy that accelerates the expansion of the universe. But while the cosmological constant does not change, other versions of dark energy can evolve over time. Dark energy is a very active area of study, as scientists work to figure out where this energy comes from, its form and properties, and whether the amount changes over time.

Testing the Hypothesis

One fundamental rule of science is that hypotheses must be tested, over and over. And the findings from different experiments must agree with each other and with the hypothesis, otherwise the hypothesis is flawed and must be modified or thrown out altogether. Observations of the expansion of the universe provide a nice example of scientific method in practice, in which multiple lines of evidence from multiple experiments can be compressed into one graph. The graph in **Figure 16.12a**, which plots Ω_Λ versus Ω_{mass}, shows all possible universes. Universes dominated by dark energy (upper left of the diagram) do not need a Big Bang; universes dominated by mass (lower portion of the diagram) will recollapse eventually, and so on.

Measurements allow only certain regions of the graph, as shown in Figure 16.12b. The supernova data are shown in the yellow region, and so our universe must have values of Ω_Λ and Ω_{mass} that lie in this yellow region. These observations of supernovae have constrained possible values of Ω_{mass} to be less than about 1.2 and possible values of Ω_Λ to be between about 0 and about 2. All other values outside of the yellow region of the graph are excluded. Our universe did have a Big Bang (the "No Big Bang" region is outside the yellow region), and it will not recollapse eventually (the yellow region lies above the "Recollapses eventually" line). Additional data sets further constrain the data, as shown in Figure 16.12c, where four separate lines of inquiry, following four separate types of experiments, observing very different types of objects, overlap. We will discuss some of these in more detail in this chapter and the next.

Figure 16.12 (a) A plot of Ω_Λ versus Ω_{mass} shows all possible types of universes. (b) Observations constrain our universe to lie in a particular region. Here, supernova data show that Ω_{mass} must be less than about 1.2, and Ω_Λ must be somewhere between about 0.2 and 2. (c) Combining current observations from different sources narrow down those ranges—Type Ia supernovae (yellow), measurements of mass in galaxies and clusters (orange), and detailed observations of the structure of the cosmic microwave background (pink and bright red)—suggest that the best current estimate for Ω_Λ is about 0.7 and about 0.3 for Ω_{mass}. Because Ω_Λ is not zero, the expansion of the universe is accelerating.

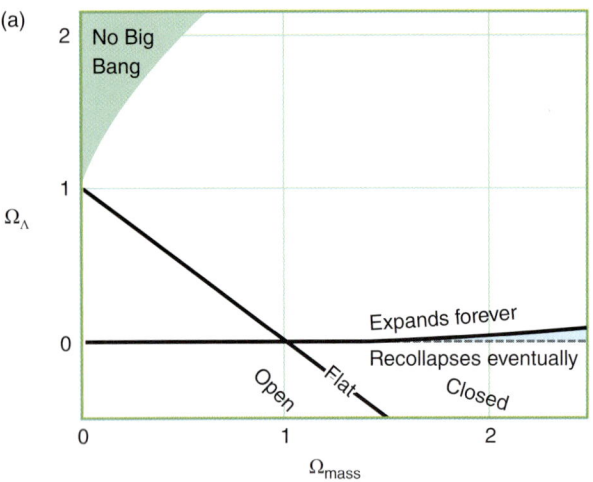

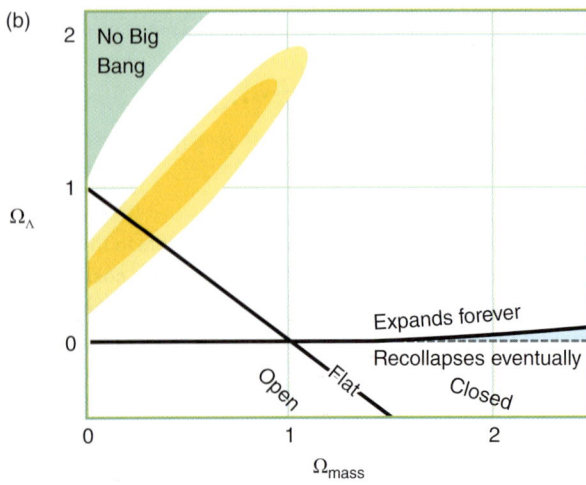

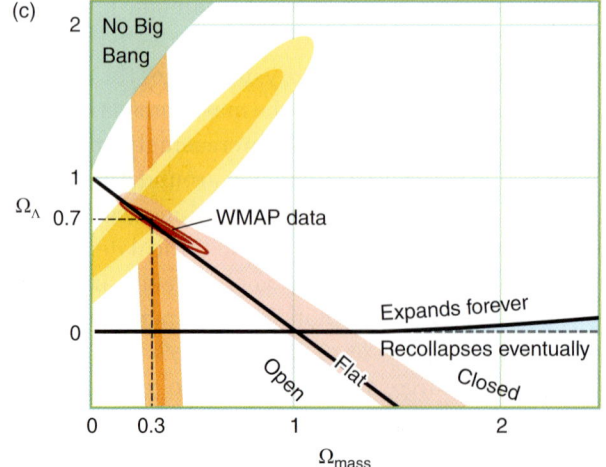

Figure 16.12c shows the range of values for Ω_{mass} and Ω_Λ that are allowed by observations. Each colored region represents the data from a different experiment. Values of Ω_{mass} and Ω_Λ outside of these regions are ruled out by these experiments, so the allowed values of Ω_{mass} and Ω_Λ must lie within the area on the graph where *all* of these regions overlap. The allowed values for Ω_{mass} and Ω_Λ are about 0.3 and 0.7, respectively. These values are most tightly constrained by the data from WMAP, in bright red. The expansion of our universe is dominated by the effect of dark energy.

The dark diagonal line in Figure 16.12 shows where $\Omega_{mass} + \Omega_\Lambda = 1$. The experiments tightly constrain the value of $\Omega_{mass} + \Omega_\Lambda$ to lie along this line. This means that on very large scales, the universe we live in is "flat"—it is only locally warped by mass, as in the rubber sheet analogy in Chapter 13. Space is not warped on the largest scales. Even more recent data, for example from the Planck mission, are consistent with this conclusion.

Multiple lines of evidence, from multiple types of objects, from many different scientists all lead to the same answer, which increases in precision as more data are collected. This is the strongest possible type of evidence that a scientific theory can have. The conclusion that we live in an accelerating universe dominated by dark energy and having a hot, dense beginning about 13.8 billion years ago is as strong a conclusion as can be made. For any alternate conclusion to be valid, it must provide a better explanation not only of each data set individually, but also of their agreement with each other.

16.4 The Earliest Moments of the Universe Connect the Very Largest Size Scales to the Very Smallest

A century ago, astronomers were struggling just to get a handle on the size of the universe. Today we have a comprehensive theory that ties together many diverse facts about nature: the constancy of the speed of light, the properties of gravity, the motions of galaxies, and even the origins of the very atoms we are made of. Notably, when it comes to understanding the earliest moments of the universe, we must study the universe on the smallest possible scales. Particle physics is the study of subatomic particles, which are smaller than atoms. In this section, you will learn about the earliest moments of the universe, when atoms had not yet formed, and how the interaction among these subatomic particles governed the conditions and events that took place.

The Standard Model

There are four fundamental forces in nature, and everything in the universe is a result of their action. The **electromagnetic force,** which includes both electric and magnetic interactions, acts on charged particles like protons and electrons. This force governs not only chemistry, but also light. The *strong nuclear force* that binds together the protons and neutrons in the nuclei of atoms governs reactions like the fusion reactions in the heart of the Sun, discussed in Chapter 11. The **weak nuclear force** governs the decay of a neutron into a proton, an electron, and an antineutrino. Finally, *gravity*, which plays a major role in astronomy, governs how matter affects the geometry of spacetime. In models of particle physics, the first three forces combine in a single force, leaving only gravity to stand alone.

To understand how these four forces formed and came to govern the universe today, we must explore backward in time, toward the Big Bang itself.

The underlying concept of the standard model is that forces between particles are caused by the exchange of carrier particles. For example, quantum electrodynamics (QED) treats charged particles almost as if they were baseball players engaged in an endless game of catch. As baseball players throw and catch baseballs, they experience forces. Similarly, in QED, charged particles "throw" and "catch" an endless stream of "virtual photons." The force that results from the average of all these exchanges acts, over large scales, like the electromagnetic force that we observe in chemistry. QED is one of the most accurate, well tested, and precise branches of physics. As of this writing, not even the tiniest measurable difference between the predictions of QED and the outcome of an actual experiment has been found.

In **electroweak theory**, the electromagnetic force and the weak nuclear force have been combined into a single force that predicts the existence of three carrier particles that mediate the weak nuclear force. The combined force is known as the electroweak force. In the 1980s, physicists identified these carrier particles in laboratory experiments, thus confirming the essential predictions of this theory.

The strong nuclear force is described by a third theory in which particles such as protons and neutrons are composed of more fundamental building blocks, called **quarks**. These quarks are bound together by the exchange of another type of carrier particle.

At high temperatures, the different forces are indistinguishable, because the different carrier particles have such high energy. Therefore, our universe started out with all the forces unified, as described by one (as yet unknown) theory of everything. As the universe expanded and cooled, symmetry between the particles was broken, and carrier particles for different forces became distinguishable. You may have heard of the *Higgs field*, which is responsible for breaking the symmetry between different kinds of carrier particles. This field is mediated by the *Higgs boson*, an elementary particle. The existence of this particle was predicted in 1964 and finally detected at the Large Hadron Collider in 2012. In 2013, Peter Higgs and François Englert shared the Nobel Prize for their work predicting this particle. As the universe cooled, the carrier particles (and also other particles like electrons) gained mass due to the Higgs field, and so the forces also began to be distinguishable.

Together, these theories of particle physics are referred to as the **standard model** of particle physics. Excluding gravity, the standard model explains all the observed interactions of matter and has made many predictions that have been confirmed by laboratory experiments.

A Universe of Particles and Antiparticles

In the standard model, every particle in nature has an **antiparticle** that is identical in mass, but opposite in charge, to the particle. For example, the positron emitted during the proton-proton chain, discussed in Chapter 11, is the antiparticle of an electron. A positron is identical to an electron except that it has a positive charge instead of a negative charge. If a particle-antiparticle pair meet, for example a positron and an electron, the two particles will annihilate each other, and their energy will be carried away by photons. The reverse process is also possible: Two photons may collide to produce a particle and its antiparticle by a process called **pair production**.

As the universe cooled, energetic photons constantly collided and underwent pair production, creating matter-antimatter pairs. Early on, swarms of protons and antiprotons formed. When the universe was less than about 100 seconds old, it had a temperature greater than a billion kelvins, and electron-positron pairs formed at this lower temperature. These electron-positron pairs constantly annihilated each other, creating pairs of gamma-ray photons. The two processes reached equilibrium, determined strictly by temperature, in which pair creation and pair annihilation exactly balanced each other. Rather than being filled only with a swarm of photons, at this time the universe was filled with a swarm of photons, electrons, and positrons, as illustrated in **Figure 16.13a**.

As the universe cooled, there was no longer enough energy to create particle pairs, so the swarm of particles and antiparticles that filled the early universe annihilated each other and were not replaced. When this cooling happened, every proton should have been annihilated by an antiproton, and a positron should have annihilated every electron. This was almost the case, but not quite. For every electron in the universe today, there were 10 billion and one electrons in the early universe, but only 10 billion positrons. This one-part-in-10-billion excess of electrons over positrons meant that when electron-positron pairs finished annihilating each other, some electrons were left over—enough to account for all the electrons in all the atoms in the universe today (Figure 16.13b). Similarly, there was an excess of protons over antiprotons in the early universe, and the protons we see today are all that is left from the annihilation of proton-antiproton pairs.

If the standard model of particle physics were a complete description of nature, then the imbalance of one part in 10 billion between matter and antimatter would not have been present in the early universe. The symmetry between matter and antimatter would have been complete. No matter would have survived into today's universe, and we would not exist. The fact that you are reading this page demonstrates that something more needs to be added to the model.

Grand Unified Theories

Several competing ideas seek to explain why the amounts of matter and antimatter were not equal in the early universe. One set of ideas is called **grand unified theories (GUTs)** because they join the electromagnetic force, weak nuclear force, and strong nuclear force together into a single force. When the universe was *very* young (younger than about 10^{-35} second) and *very* hot (hotter than about 10^{27} K), enough energy was available for particles associated with a GUT to be freely created. During this time, the distinction among the electromagnetic, weak nuclear, and strong nuclear forces had not yet taken place. There was only the one unified force. Several GUTs are among the leading candidates to unify these three forces, and only the very simplest of GUTs have been ruled out.

How does gravity fit into this scheme? General relativity provides a beautifully successful description of gravity that correctly predicts the orbits of planets, describes the ultimate collapse of stars, and even enables us to calculate the structure of the universe. Yet general relativity's description of gravity is very different from our theories of the other three forces. Unlike electromagnetism, there is no theory of gravity alone involving the exchange of particles. By itself, gravity describes only the bending of spacetime.

We might be tempted to say, "Oh well. Gravity works one way and the other forces work another way, and that is how the universe happens to be." In practice, astronomers consider the forces separately when considering stars or galaxies or black holes. Even the era of GUTs is described perfectly if we treat gravity as a separate force. As we push back even closer to the moment of the Big Bang, however, it is not known how to combine gravity with the other forces.

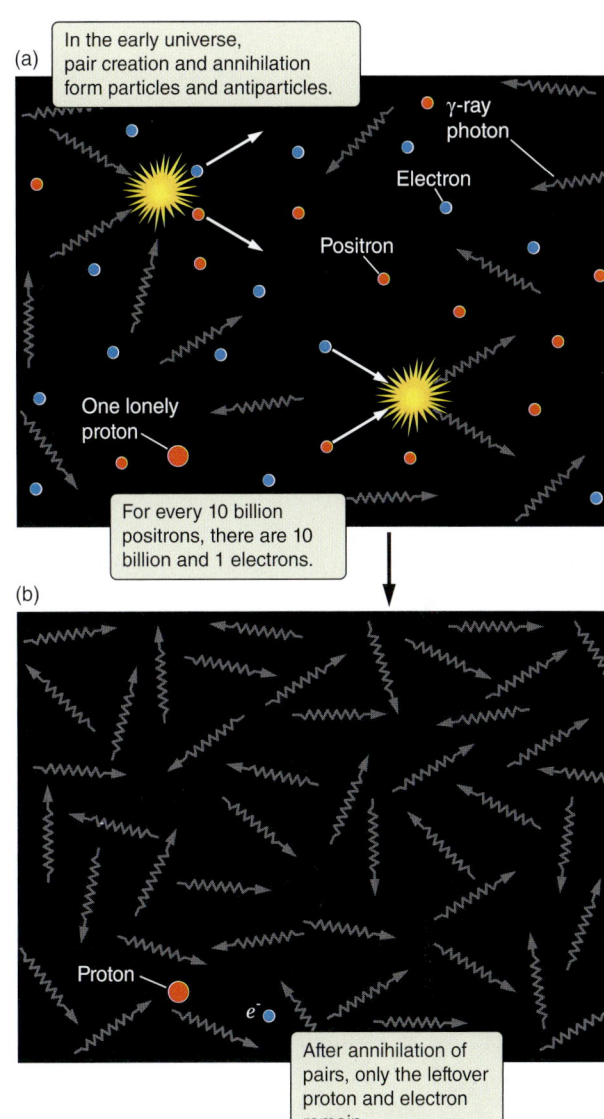

Figure 16.13 (a) A swarm of electrons, positrons, and photons filled the very early universe. For every one billion positrons, there are one billion and one electrons. Before this time, an era of proton-antiproton annihilation left this small piece of universe with only one proton and no antiprotons. (b) After the electrons and positrons annihilate, only the one electron is left, along with the remaining proton and many photons.

Toward a Theory of Everything

Even earlier in the universe than the time when the three forces were unified, when the universe was younger than about 10^{-43} second, its density was incomprehensibly high. The *observable* universe was so small that 10^{60} of them would have fit into the volume of a single proton. Under these extreme conditions, general relativity can no longer describe spacetime. This era in the history of the universe—the early time, just after the Big Bang, when the universe as a whole must be described with quantum mechanics—is referred to as the **Planck era**.

Known physics can take us back to a time when the universe was a 10-millionth of a trillionth of a trillionth of a trillionth of a second old; but to push back any further, we need something new. We need a theory that combines general relativity and our understanding of particles into a single theoretical framework unifying all four of the fundamental forces—a **theory of everything** (**TOE**). To understand the earliest moments of the universe, we need a TOE, but we don't have one.

A successful TOE would tell us which of the possible GUTs is correct and would tell us the nature of dark matter and dark energy. A successful TOE would also explain the how, when, and why of an early, rapid expansion, known as inflation. Physicists are currently grappling with what a TOE might look like. One contender for the title is **string theory**, in which elementary particles are viewed not as points but as tiny loops called "strings." According to string theory, different types of elementary particles are like different "notes" played by vibrating loops of string. This is analogous to the way in which a guitar string vibrates in one way to play an F, another way to play a G, and a third way to play an A.

To make string theory work, we have to imagine that these tiny loops of string are vibrating in a universe with nine spatial dimensions. (Adding time to the list would make our universe 10-dimensional.) How can that be, when we clearly experience only three spatial dimensions? Whereas the three spatial dimensions that we know spread out across the vastness of our universe, the other six spatial dimensions predicted by string theory wrap tightly around themselves as illustrated in **Figure 16.14**. These six spatial dimensions extend no further today than they did a brief instant after the Big Bang.

To visualize the bizarre notion of six dimensions, imagine what it would be like to live in a three-dimensional universe, like the one we experience, in which one of those dimensions extended for only a tiny distance. Living in such a universe would be like living within a thin sheet of paper that extends billions of light-years in two directions but was far smaller than an atom in the third. In such a universe, we would easily be aware of length and width—we could move in those directions at will. In contrast, we would have no freedom to move in the third dimension at all, and we might not even recognize that the third dimension existed. Perhaps our only inkling of the true nature of space would come from the fact that in order to explain the results of particle physics experiments, we would have to assume that particles extended into a third, unseen dimension. If string theory is correct, we see three spatial dimensions extending possibly forever, but we are unaware of the fact that each point in our three-dimensional space also has a tiny extent in six other dimensions.

String theory is only a pale shadow of the sort of well-tested theories that we have made use of throughout this book. In some respects, string theory is no more than a promising idea providing direction to theorists searching for a TOE. This is an inconsistent use of the term "theory" that we have tried to be so careful about. "String hypothesis" or "string idea" would be more consistent

Figure 16.14 It is virtually impossible to visualize six spatial dimensions wrapped up into structures far smaller than the nucleus of an atom. Here such geometries are projected onto the two-dimensional plane of the paper, with additional dimensions "rolled up" into little balls.

with the definitions in Chapter 1. We will probably never be able to build particle accelerators that enable us to search directly for the most fundamental particles predicted by a TOE. The energies required are simply too high. Fortunately, some progress may be made by studying the ultimate particle accelerator: the Big Bang itself.

The Cooling Universe

To understand the very earliest moments in the history of the universe, we have been looking backward at earlier and earlier times, and consequently to higher and higher energies. Now let's organize the events the other way, beginning at the beginning.

Figure 16.15 illustrates when the four fundamental forces emerged in the evolving universe. In the first 10^{-43} second after the Big Bang, the physics of elementary particles and the physics of spacetime were the same. As the universe expanded and cooled, gravity separated from the forces described by the GUT, and spacetime took on the properties described by general relativity.

As the universe continued to expand and its temperature fell further, less and less energy was available for the creation of particle-antiparticle pairs. When the particles responsible for GUT interactions could no longer form, the strong force split off from the others. Somewhere along the line, as the unity of the original TOE was lost, the symmetry between matter and antimatter was broken. As a result, the universe ended up with more matter than antimatter.

The next big change took place when the particles responsible for unifying the electromagnetic and weak nuclear forces split out, leaving these two forces independent of each other. All four fundamental forces of nature that govern today's universe were now separate. At one 10-trillionth of a second, the temperature of the universe had fallen to 10^{16} K. It was a full minute or two before the universe cooled to the billion-kelvin mark, below which not even pairs of electrons and positrons could form.

Although the universe was now too cool for pair production, it was still hot enough for nuclear reactions to take place. These reactions formed the least massive elements, including helium, lithium, beryllium, and boron, but could not create more massive elements.

The nuclear reactions had come to an end by the time the universe was 5 minutes old, and the temperature of the universe had dropped below about 800 million K. The density of the universe at this point in time had fallen to about only a tenth that of water. Normal matter in the universe now consisted of atomic nuclei, electrons, and photons. The universe remained this way for the next several hundred thousand years, until finally the temperature dropped so far that electrons were able to combine with atomic nuclei to form neutral atoms. This was the era of recombination, which we see directly when we look at the cosmic microwave background. It was the moment at which the universe became transparent, and light could move freely through it for the first time.

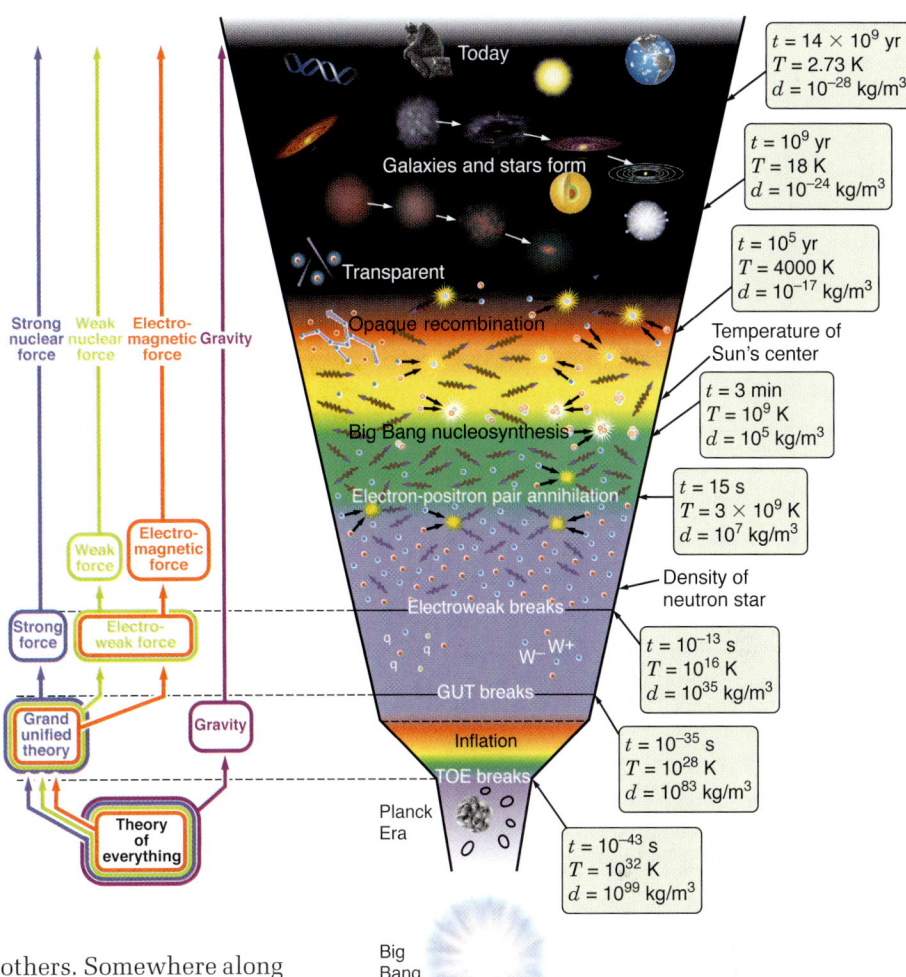

Figure 16.15 As the universe expanded and cooled after the Big Bang, it went through a series of phases determined by what types of particles (for example, exotic W or q particles or electron-positron pairs) could be created freely at that temperature. Later, the structure of the universe was set by the gravitational collapse of material to form galaxies and stars and by the chemistry made possible by elements formed in stars.

16.5 Inflation Solves Several Problems in Cosmology

The case for the Big Bang is compelling. The Big Bang explains observations at all size scales, from the behavior of particles to the expansion of the universe. Even so, as our knowledge of the expansion of the universe has grown and our observations of the cosmic microwave background have improved, a number of puzzles have arisen. These puzzles have forced us to consider some remarkable ideas about how our universe expanded when it was very young. The concept of inflation solves a number of these problems, two of which we discuss here.

The Flatness Problem

The first puzzle that we run into when observing our universe is that it is too flat. In fact, the universe is much too close to being exactly flat for this to have happened by chance. Any deviation from flatness would grow over time, so if the universe originally had a value of $\Omega_{mass} + \Omega_\Lambda$ even slightly different from 1, the value would by now be drastically different and easily detectable. For the present-day value of $\Omega_{mass} + \Omega_\Lambda$ to be as close to 1 as it is, when the universe was 1 second old, $\Omega_{mass} + \Omega_\Lambda$ must have been equal to 1 all the way out to at least the tenth decimal place. At even earlier times, it had to be much flatter still. This is too special a situation to be the result of chance—a fact referred to in cosmology as the **flatness problem**: The universe is much too flat. It is so flat that something about the early universe must have *forced* $\Omega_{mass} + \Omega_\Lambda$ to have a value incredibly close to 1, but what?

The Horizon Problem

The second problem faced by our cosmological models is that the CMB is surprisingly smooth. After the discovery of the CMB in the 1960s, many observers turned their attention to mapping this background glow. At first they were reassured as result after result showed that the temperature of the CMB is remarkably constant (variations of less than one part in 3,000), regardless of where one looks in the sky. Yet over time, this strong confirmation of Big Bang cosmology came to challenge our view of the early universe. Once we remove our motion relative to the CMB from the picture, the CMB is not just smooth—it is *too* smooth.

To understand this, we might think about carefully adding a tablespoon of cold water to hot water, as illustrated in **Figure 16.16**. At that instant, there is a large temperature difference between the cold water and the hot water. Then the hot water heats up the cold water, and the entire cup full of liquid eventually reaches thermal equilibrium; that is, it comes to the same temperature. But this takes time. If the cup of water somehow grows much larger before the temperature has evened out, there will be cold spots and hot spots that grow as the cup of water grows. Similarly, in the early universe, conditions would have fluctuated in unpredictable ways. Some regions should have been hotter than others, and there would not have been time for the whole universe to come to the same temperature if the expansion was uniform. The fact that the CMB is so smooth and shows the whole universe at the same temperature is referred to as the **horizon problem** in cosmology: Different parts of the universe are too much like other parts of the universe—distant parts that should have been too far away ("over the horizon") for the temperature to become the same by the time the CMB was emitted.

> **Back in Section 7.2 . . .**
> . . . you learned that objects reach thermal equilibrium when they exchange energy and come to the same temperature.

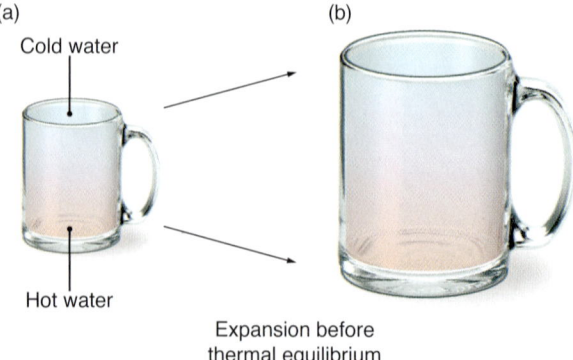

Figure 16.16 (a) When cold water is carefully added to hot water, initially the cold and hot water are at very different temperatures. (b) As the cup grows, the cold spot persists.

Inflation: Early, Rapid Expansion

In the early 1980s, Alan Guth (b. 1947) offered a solution to cosmology's flatness and horizon problems. Guth suggested that the universe has not expanded uniformly, but that it started out much more compact than uniform expansion would predict. Then, for a brief time, the young universe expanded at a rate *far* in excess of the speed of light. This rapid expansion of the universe is called **inflation**. Between about 10^{-35} and 10^{-33} seconds after the Big Bang, the space between every two points in the universe increased by a factor of at least 10^{30} and perhaps very much more. This is a very short period of time—about a billionth of the time it takes light to cross the nucleus of an atom. And it is a very large increase in the size of space; for example, a fine grain of sand is about a factor of 10^{30} smaller than the current observable universe. During inflation, space expanded so rapidly that the distances *between* points in space increased faster than the speed of light. Note that inflation does not violate the rule that nothing can travel *through* space at greater than the speed of light, because space itself was expanding.

To understand how inflation solves the flatness problem, imagine that you are an ant living in the two-dimensional universe defined by the surface of a golf ball, as shown in **Figure 16.17**. This universe is positively curved, like the surface of Earth.

Vocabulary Alert

inflation In common language, this word has two meanings. The increase in the cost of goods due to the decrease in the value of money is known as inflation. Also, the blowing up of a balloon is inflation. Here astronomers are using *inflation* in the second sense, as an analogy to the behavior of the universe. Keep in mind, however, that the universe is not a balloon. So, for example, the universe does not inflate *into* anything.

Figure 16.17 If a round, lumpy golf ball were suddenly inflated to the size of Earth, it would seem extraordinarily flat and smooth to an ant on its surface. Similarly, after undergoing inflation, any universe would seem both extremely flat and extremely smooth, regardless of the exact geometry and irregularities it started with. Notice that the ant does not expand with the golf ball, just as galaxies do not expand with the universe—gravity keeps galaxies from expanding.

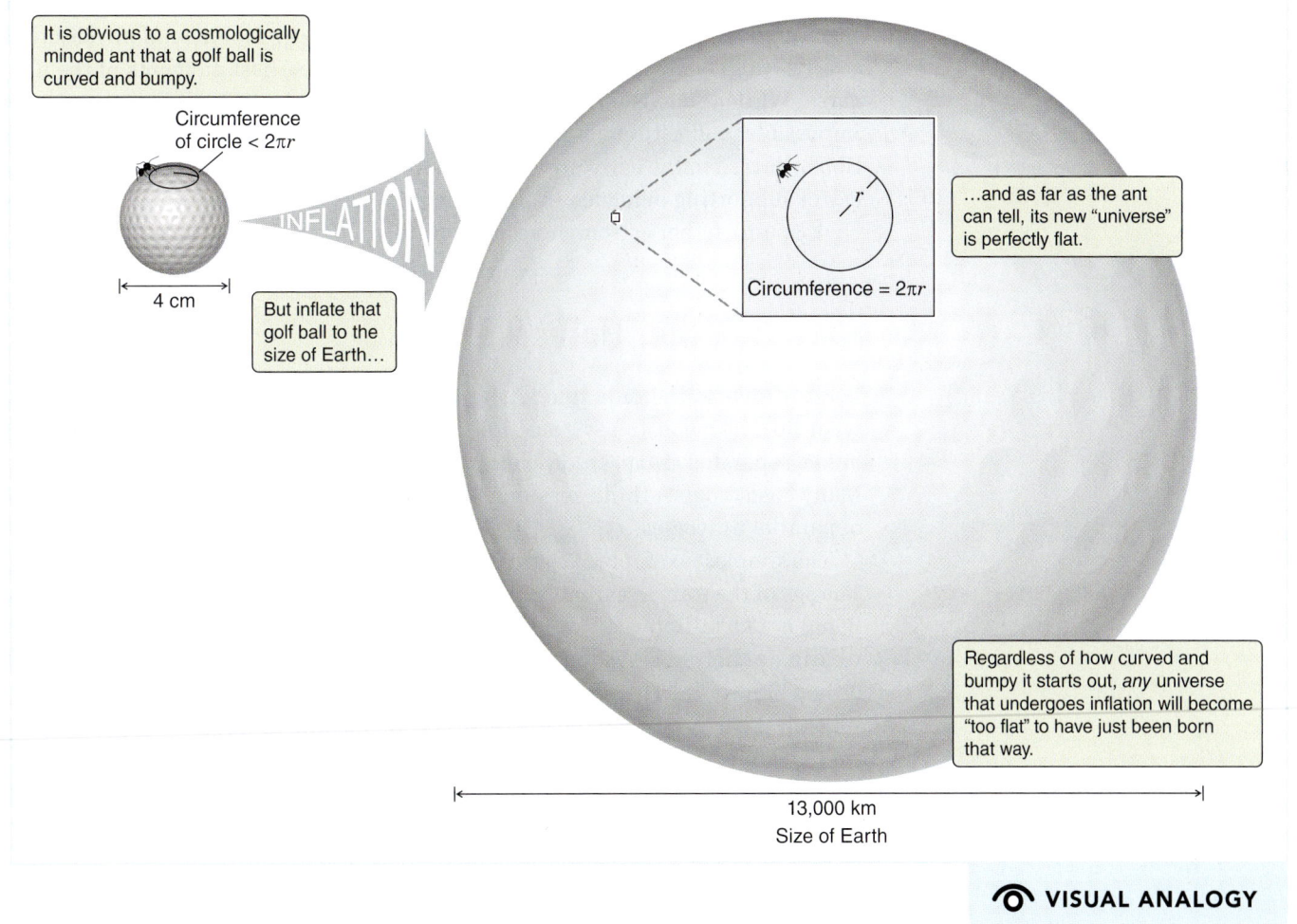

VISUAL ANALOGY

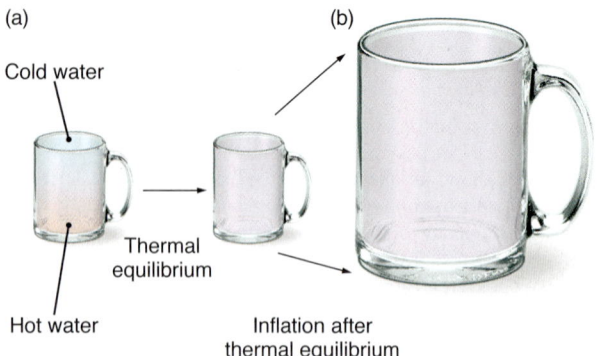

Figure 16.18 (a) If cold water is added to a small mug of hot water, the system comes to thermal equilibrium quickly. (b) If the cup (and the liquid) then proceeds to grow very fast, the system stays in thermal equilibrium because it had time to reach equilibrium before the inflation started.

Now imagine that your golf-ball universe suddenly grew to the size of Earth. The curvature of your universe would no longer be apparent. An ant (or person) walking around on the surface of Earth would be hard-pressed to tell that Earth is not flat. In the case of inflationary cosmology, the universe after inflation would be extraordinarily flat (that is, with $\Omega_{mass} + \Omega_\Lambda$ extraordinarily close to 1) *regardless* of what the geometry of the universe was before inflation. Because the universe was inflated by a factor of at least 10^{30}, $\Omega_{mass} + \Omega_\Lambda$ immediately after inflation must have been equal to 1 to within one part in 10^{60}, which is flat enough for $\Omega_{mass} + \Omega_\Lambda$ to remain close to 1 today. If inflation occurred, then today's universe is not flat by chance. It is flat because *any* universe that underwent inflation would become flat.

What about the horizon problem? Consider again adding cold water to hot water, illustrated in **Figure 16.18**. If the cup was small when the cold water was added, the temperature of all of the fluid in the cup would rapidly equalize. Inflating the cup (and the fluid inside) would cause the fluid in the larger cup to become the same temperature. The universe prior to inflation was so much more compact that there was time for the variations in temperature to smooth out before the CMB was emitted. There was time for the universe to come to thermal equilibrium before it began its rapid expansion. The flattening of the dimples in the golf ball in Figure 16.17 is a second way to picture this process.

An early era of inflation in the history of the universe offers a handy way of solving the horizon and flatness problems, but it seems quite remarkable that the universe should have undergone a period during which it expanded at such an "astronomical" rate. The cause of inflation lies in the fundamental physics that governed the behavior of matter and energy at the earliest moments of the universe. While the existence of an inflationary epoch is difficult to test, it is not impossible, and astronomers are currently devising ways to test whether inflation occurred in the early universe. In 2014, astronomers announced the discovery of supporting evidence in the orientation of light in the cosmic microwave background. Other astronomers are working to independently test this result.

16.6 Other Universes?

Is our universe the only one? As we've defined the universe as "everything," what does it mean to say "multiple universes"? Are there parallel universes, either separated in space or even occupying exactly the same space as ours? Many cosmologists think seriously about the idea of **multiverses**, or collections of parallel universes.

The simplest example of such parallel universes is illustrated in **Figure 16.19**. The age of the universe (that is, the amount of time that has passed since the Big Bang) is 13.8 billion years. Therefore, our observable universe—everything that we can possibly observe today—must be within a sphere having a radius of 13.8 billion light-years. The observational evidence suggests that the geometry of space is flat. If this is true, then the universe is truly infinite in size and must therefore contain an infinite number of similar spheres. As dark energy causes the universe to expand faster and faster, the separate observable universes move farther apart faster than light can travel. These parallel universes are simply too far away for us ever to be able to observe them, and they are moving farther away all the time.

What are these other parallel universes like? Because of the cosmological principle, on large scales each of these observable universes should look similar to our own, although details may be different. Still, in a truly infinite universe there must be an infinite number of observable universes exactly like ours, with an exact copy of you reading an identical version of *Understanding Our Universe*. We know this because there are no more than 10^{118} particles in the observable universe, and there are only so many ways that this finite number of particles can be arranged. If you then ask how far you must go before you will find an observable universe just like—and we mean *identical to*—our own, the answer is about $10^{10^{118}}$ Mpc. Yes, that's 10 raised to the power 10^{118}. So, in an infinite universe—as enormous as it might be—we still know how far away an identical copy of you must be. There must also be a universe in which the only difference from ours is that your copy of *Understanding Our Universe* has an extra chapter. In general, nearly all other universes would be different from our own. Some would be only slightly different, and some dramatically so.

Other types of multiverses include those in which a universe undergoes eternal inflation, with no beginning or end to the inflation. If such a universe exists, then small fluctuations in the universe may cause some regions to expand more slowly than the rest of the universe. As a result, such a region may form a bubble whose inflating phase will soon end. In this scenario, we would be living inside such a region, and our Big Bang would just be the condensation of our bubble within the eternally inflating universe. This type of multiverse neatly answers the question of what there was before the Big Bang. Because the universe has been inflating and will continue to inflate forever, there is no beginning or end. Our own bubble, or parallel universe, separated from the rest of the universe at a time we call the Big Bang, but other bubbles are constantly separating and becoming their own parallel-universe big bangs.

In one version of multiverses derived from the physics of the smallest scales, each event in the universe spawns multiple universes in which each possible outcome of the event exists. While cosmologists consider this at the level of particle interactions, it is more simply explained at the human scale: You made a decision about breakfast this morning. That event caused two universes: one in which you did have breakfast and one in which you didn't. Which one are you in while you read this book? It is fun to wonder what is happening for the "other you." Multiverses are common in science fiction and in popular science books. Multiverses form a type of common mythology about alternative realities that is explored with enthusiasm in popular culture.

Is the idea of parallel universes, or multiverses, really science? Throughout this book we have emphasized that any legitimate scientific theory must be testable and ultimately falsifiable. There is considerable debate within the scientific community as to whether models such as eternal inflation can be truly falsified. Are there tests capable of proving these multiverse ideas to be wrong? Possibly. The eternal inflation model is difficult to test, because we will never directly observe its parallel universes (although some scientists interpret the latest Planck data as hinting at this type of universe). But if we obtain a theory of everything that predicts eternal inflation, and if that theory of everything is itself falsifiable, then there will be a connection between eternal inflation and observation.

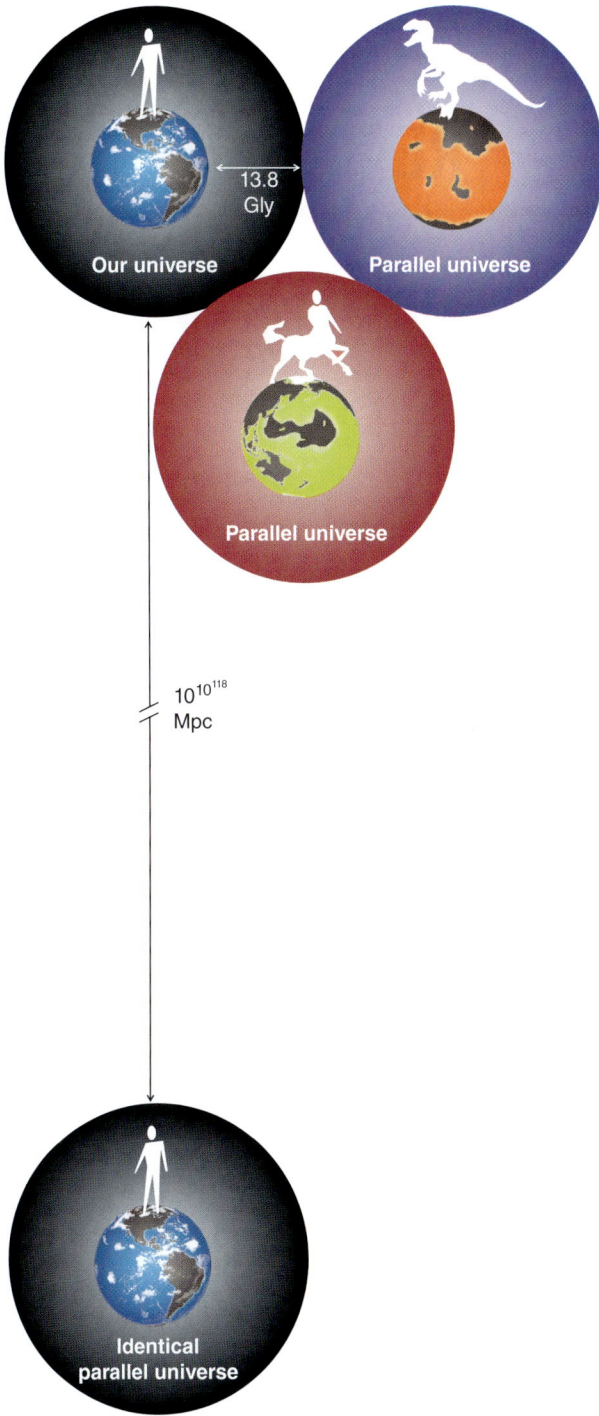

Figure 16.19 The observable universe is a sphere with a radius equal to the distance light has traveled since the Big Bang (13.8 billion light-years, or 13.8 giga-light-years [Gly]). If the universe is infinite, there must be an infinite number of similar spheres. The rules of probability dictate that some of these are *exactly* like our own.

# READING ASTRONOMY News

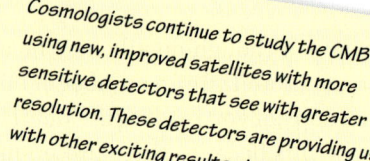

Cosmologists continue to study the CMB using new, improved satellites with more sensitive detectors that see with greater resolution. These detectors are providing us with other exciting results along the way.

Planck: Big Bang's Afterglow Reveals Older Universe, More Matter

By **Amina Khan**, *Los Angeles Times*

The universe has hidden its age well. The European Space Agency's Planck space telescope has scanned the skies for the Big Bang's fingerprint and discovered that the universe is about 100 million years older than thought, and that there's more normal matter and dark matter filling the cosmos.

The findings announced Thursday by ESA and NASA peg the universe's age at 13.8 billion years and produced a multicolored map showing the tiny temperature fluctuations that reveal the seeds of the universe's future structure [see Figure 16.9d].

"It might look a little bit like a dirty rugby ball or a piece of modern art," ESA's George Efstathiou, a Cambridge University astrophysicist, said in a press conference early Thursday morning. "But I can assure you there are cosmologists who would have hacked our computers or maybe even given up their children to get hold of a copy of this map."

The Planck mission is a successor to NASA's Wilkinson Microwave Anisotropy Probe (WMAP) and the Cosmic Background Explorer (COBE), and it takes our snapshot of the early universe's afterglow to unprecedented clarity.

"It's as if we've gone from a standard television to a high-definition television," NASA astrophysics director Paul Hertz said in a later press conference. "New and important details have become crystal clear."

The map represents the first 15.5 months of observation by the Planck space telescope, which looked at the universe's cosmic microwave background—that extremely cold, barely noticeable glow left after the Big Bang when the universe was just a cosmic baby, about 380,000 years old.

Although it's cold now, this afterglow is an imprint from when the young universe's soup of particles heated to about 4,900 degrees Fahrenheit, stretched across the skies during a period of expansion known as inflation, and cooled over the intervening billions of years to just a few degrees above absolute zero.

Because the light's path is affected by all the mass around it, the new radiation map also allows scientists to create a map of all the mass in the universe, scientists said.

"We can see the subtle effects of the gravitational pulls from literally everything in the universe," said U.S. Planck project scientist Charles Lawrence, based at Jet Propulsion Laboratory. "That tells us about . . . everything in the universe."

The researchers found that they could raise the estimates of normal matter in the universe to 4.9 percent, up from 4.6 percent. The dark matter share rose to 26.8 percent, up from 24 percent. The overwhelming majority, dark energy—that strange force that's causing the universe to expand faster and faster—shrunk accordingly, from 71.4 percent to 68.3 percent.

And with less dark energy to push things apart, the universe isn't expanding quite as quickly as thought, the scientists found.

The map is colored in shades from blue (for colder spots) to yellow to red (for warmer spots). But the differences between cold and warm are infinitesimally small, of the order 100 millionths of a degree.

"The contrasts are turned way up," Lawrence said.

The Planck spacecraft's data also highlight two other mysteries: why the heat from the background radiation isn't evenly distributed, and what could have caused a strange cold spot in the map, now revealed to be even larger than previous estimates.

Evaluating the News

1. The article states that the Planck data show that "the universe is about 100 million years older than thought." What percentage of the age of the universe is this? Is this a large percentage?
2. George Efstathiou explains that cosmologists were very excited to see the map—they might even give up their children to get a copy of it! Why were cosmologists so excited?
3. The article compares the temperature of the universe when the light was emitted (4,900 degrees Fahrenheit) to the temperature of the afterglow now (just a few degrees above absolute zero). Comment on the fact that the reporter changed temperature scales within a single sentence. Why did the author do that? Is it likely to cause confusion for readers?
4. The new radiation map gives information about "everything in the universe." Explain how the mass in the universe would affect light as it traveled toward Earth.
5. The estimate of the proportion of dark energy has fallen by a few percent because of the Planck observations. The article states that this means that "the universe isn't expanding quite as quickly" as previously estimated. Explain why this statement is not quite a correct expression of how dark energy affects the expansion.

✦ SUMMARY

The universe has been expanding since the Big Bang, which occurred nearly 14 billion years ago. The Big Bang happened everywhere; it is not an explosion spreading out from a single point. The universe has no center and no edge, although the observable universe has limits. Observed redshifts of distant galaxies result from the increasing space between them. Observations of the cosmic microwave background independently confirm that the universe had a hot, dense beginning. Both gravity and the cosmological constant (or dark energy) determine the fate of the universe. Observations indicate that rather than slowing down, the expansion of the universe is accelerating. The best explanation for this phenomenon is dark energy. During the very earliest moments in the universe, the four fundamental forces of nature were unified. The very early universe may have gone through a brief but dramatic period of exceptionally rapid expansion, known as inflation. If this is true, inflation would explain both the flatness and the uniformity of the universe we see today. Ideas about multiple universes have been proposed, but none are yet directly testable.

LG 1 Hubble's law indicates that the universe was once very hot and very dense. This hot, dense beginning is known as the Big Bang. Big Bang theory predicts that we should be able to observe the radiation from a few hundred thousand years after the Big Bang. This radiation should have the same spectrum in every direction. Later observations of the cosmic microwave background confirm precisely this prediction.

LG 2 Recent observations suggest that the expansion of the universe is speeding up, implying that there must be a force acting to increase the expansion rate. This force may come from dark energy: the energy of empty space. Because the rate of expansion of the universe is increasing, the universe will likely expand forever.

LG 3 During the earliest moments of the universe, the four forces split off, each becoming separate at a different time. Modern physicists search for the combined theory of all four forces: the theory of everything.

LG 4 Inflation was an early epoch of rapid expansion. Proposed to solve several problems in cosmology, observational confirmation of inflation is accumulating.

SUMMARY SELF-TEST

1. Hubble's law and Big Bang theory are related because
    a. Hubble's law is an observation that led to the development of the theory.
    b. Big Bang theory predicted Hubble's law, which was then observed.
    c. Hubble's law is a hypothesis that eventually became the Big Bang theory.
    d. Hubble's law is a fact, and Big Bang theory is an untested hypothesis.

2. The CMB and Big Bang theory are related because
    a. the CMB is an observation that led to the development of the theory.
    b. Big Bang theory predicted the existence of the CMB.
    c. The Big Bang is a hypothesis that eventually became known as the CMB.
    d. The CMB is a fact, and Big Bang theory is an untested hypothesis.

3. Dark energy has been hypothesized to solve which problem?
    a. The universe is expanding.
    b. The CMB is too smooth.
    c. The expansion of the universe is accelerating.
    d. Stars orbit the centers of galaxies too fast.

4. The flatness problem relates to (choose all that apply)
    a. the density of the universe.
    b. the shape of the universe.
    c. the size of the universe.
    d. the expansion of the universe.
    e. the acceleration of the universe.

5. The standard model describes (select all that apply)
    a. the unification of all of the forces in the early universe.
    b. the behavior of particles smaller than atoms.
    c. the carrier particles that are exchanged when particles interact.
    d. how gravity fits in with the other three forces.
    e. how particles and forces behave under all conditions.

6. Parallel universes are
    a. an observation.   b. a law.
    c. a hypothesis.   d. a theory.

7. Which of the following are properties of the young universe revealed by the cosmic microwave background? Select all that apply.
    a. It was hot.   b. It was cold.
    c. It was dense.   d. It was diffuse.
    e. It was uniform   f. It was "clumpy" on
       on large scales.   large scales.

8. Place the following forces in order of their "freeze-out" in the first moments after the Big Bang.
    a. gravity   b. strong nuclear force
    c. weak nuclear force   d. electromagnetic force

9. Which of the following adjectives describes the Big Bang? (Choose all that apply.)
    a. hot   b. dense
    c. loud   d. tiny
    e. vast   f. slow
    g. fast

10. Which of the following adjectives describes the early universe, as observed through the CMB? (Choose all that apply.)
    a. hot   b. cold
    c. dense   d. diffuse
    e. uniform on large scales   f. uniform on small scales
    g. rapidly expanding   h. slowly expanding

QUESTIONS AND PROBLEMS

Multiple Choice and True/False

11. **T/F:** The cosmological constant makes the expansion of the universe accelerate.

12. **T/F:** In our universe, $\Omega_{mass} + \Omega_\Lambda$ is as close to zero as we can measure.

13. **T/F:** Inflation is the theory that the universe is expanding today.

14. **T/F:** The early universe was filled with almost equal numbers of matter and antimatter particles.

15. **T/F:** Cosmological redshift causes a change in the spectrum of a galaxy similar to the change that Doppler shift causes in the spectrum of a star.

16. If Ω_{mass} is 0.7 and Ω_Λ is 0.5 today, then the universe will
    a. expand forever.
    b. expand and then contract.
    c. expand, but gradually slow down.
    d. remain static.

17. If a universe is dominated by dark energy, it will
    a. expand forever.
    b. expand and then contract.
    c. expand, but gradually slow down.
    d. remain static.

18. If a universe is dominated by matter, it will
    a. expand forever.
    b. expand and then contract.
    c. expand, but gradually slow down.
    d. remain static.

19. Our universe will
    a. expand forever.
    b. expand for a long time and then collapse.
    c. expand, but gradually slow down.
    d. neither expand nor contract.

20. A positron is related to an electron in that
    a. it has all the same properties except opposite mass.
    b. it has all the same properties except opposite charge.
    c. it has all the same properties except opposite spin.
    d. it has all the same properties except how it interacts with light.

21. The flatness problem states that
    a. the ratio of matter to antimatter is too large.
    b. the cosmological constant is too close to 1.
    c. the cosmic microwave background is too bumpy.
    d. the cosmic microwave background is too uniform.

22. The cosmic microwave background looks like the spectrum of a blackbody at the low temperature of 2.73 K because
    a. this was the temperature of the universe at the Big Bang.
    b. the light has been redshifted since the Big Bang.
    c. the light has been reddened by dust since the Big Bang.
    d. Earth is receding slowly from the Big Bang.

23. If there is enough mass in the universe that the density is higher than the critical density, the universe will
    a. expand forever.
    b. expand, but gradually slow down.
    c. eventually collapse.
    d. neither expand nor contract.

24. When astronomers discovered that the universe was _____, they had to revive Einstein's cosmological constant.
    a. accelerating
    b. expanding
    c. contracting
    d. decelerating

25. The cosmological redshift in Hubble's law is distinct from the redshift arising from the Doppler effect because
    a. the redshift happens as the photon moves through expanding space.
    b. the redshift happens as the galaxy moves through space.
    c. it is used for the light, not for sound.
    d. the two are not distinct; they are the same.

Conceptual Questions

26. As applied to the universe, what is the meaning of *critical density*?

27. What set of circumstances would cause an expanding universe to reverse its expansion and collapse?

28. Describe what astronomers mean by dark energy and the observational evidence that suggests dark energy exists. If the universe is being forced apart by dark energy, why isn't our galaxy, Solar System, or planet being torn apart?

29. What is the flatness problem, and why has it been a problem for cosmologists?

30. During the period of inflation, the universe may have briefly expanded at 10^{30} (a million trillion trillion) or more times the speed of light. Why did this ultrarapid expansion not violate Einstein's special theory of relativity, which says that neither matter nor communication can travel faster than the speed of light?

31. Why is the physics of tiny particles important to our understanding of the early universe?

32. Name the four fundamental forces in nature. Of these, which one depends on electric charge?

33. Describe what happens when you bring a particle and an antiparticle together. Describe the process of pair creation.

34. What are the basic differences between a grand unified theory (GUT) and a theory of everything (TOE)?

35. Consider the term *string theory* in light of the discussion in Chapter 1. Many scientists object to using the word *theory* to describe string theory. Why?

36. As the sensitivity of our instrumentation increases, we are able to look ever farther into space and, therefore, ever further back in time. However, we can see no further back in time than the era of recombination. Explain why.

37. What is meant by *Hubble time*?

38. Examine Figure 16.2. The black galaxy is at 0 distance at time $t = 0$ and then remains at 0 distance. Why?

39. As the universe expands after the Big Bang, we know that galaxies are not actually flying apart from one another. What is really happening? Why is this distinction important?

40. What is the origin of the CMB? Why is it significant that the CMB displays a blackbody spectrum? What is the significance of the tiny brightness variations that are observed in the CMB?

Problems

41. Consider Figure 16.11.
    a. What do the orange lines represent?
    b. What do the blue lines represent?
    c. The uppermost blue line shoots off toward the top of the graph. What does this say about the universe described by the line?
    d. The bottom blue line rises and then drops back to zero. What does this say about the universe this line describes?
    e. Which of the five lines on the graph most likely describes our actual universe today?

42. Figure 16.12c is an important graph in cosmology. It places the data from three independent experiments on the same graph.
    a. What conclusions can we draw from the observation that all of the data overlap near the line labeled "Flat?"
    b. Suppose that the latest data, represented by bright red ellipses, were located at $\Omega_\Lambda = 2$. What would this mean about our prior models of the universe?
    c. The upper left of the diagram is the region in which a universe would not have required a Big Bang. Do the data allow us to entertain this possibility for our own universe?
    d. According to the data represented on this graph, will our universe eventually recollapse?

43. How many hydrogen atoms need to be in 1 cubic meter (m³) of space to equal the critical density of the universe?

44. The proton and antiproton each have the same mass, $m_p = 1.67 \times 10^{-27}$ kilograms (kg). What is the energy (in joules) of each of the two gamma rays created in a proton-antiproton annihilation?

45. Suppose you brought together a gram of ordinary-matter hydrogen atoms (each composed of a proton and an electron) and a gram of antimatter hydrogen atoms (each composed of an antiproton and a positron). Keep in mind that 2 grams is less than the mass of a dime.
    a. Calculate how much energy (in joules) would be released as the ordinary matter and antimatter hydrogen atoms annihilated one another.
    b. Compare this amount of energy with the energy released by a 1-megaton hydrogen bomb (1.6×10^{14} J).

46. One GUT predicts that on average a proton will decay in about 10^{31} years, which means if you have 10^{31} protons, you should see one decay per year. The Super-Kamiokande observatory in Japan holds about 20 million kg of water in its main detector, and it did not see any decays in 5 years of continual operation. How many protons are in the main detector at Super-Kamiokande? How many decays were expected in 5 years of continuous observations? Are the experimental results consistent with GUT predictions?

47. The spectrum of the CMB is shown as the red dots in Figure 16.8, along with a blackbody spectrum for a blackbody at temperature of 2.73 K. From the graph, determine the peak wavelength of the CMB spectrum. Use Wien's law to find the temperature of the CMB. How does this rough measurement that you just made compare to the accepted temperature of the CMB?

48. Observations of the CMB show that our Solar System is moving in the direction of the constellation Crater at a speed of about 370 km/s relative to the cosmic reference frame. What is the blueshift (negative value of z) associated with this motion?

49. The average density of normal matter in the universe is 4×10^{-28} kg/m³. The mass of a hydrogen atom is 1.66×10^{-27} kg. On average, how many hydrogen atoms are there in each cubic meter in the universe?

50. To get a feeling for the emptiness of the universe, compare its density (about 9.9×10^{-27} kg/m³) with that of Earth's atmosphere at sea level (1.2 kg/m³). How much denser is our atmosphere? Write this ratio using standard notation.

SMARTWORK

Norton's online homework system includes algorithmically generated versions of these questions, plus additional conceptual exercises. If your instructor assigns questions in SmartWork, log in at smartwork.wwnorton.com.

Exploration | Hubble's Law for Balloons

The expansion of the universe is extremely difficult to visualize, even for professional astronomers. In this Exploration, you will use the surface of a balloon to get a feel for how an "expansion" changes distances between objects. Throughout this Exploration, you must remember to think of the surface of the balloon as a two-dimensional object, much like the surface of Earth is a two-dimensional object for most people. The average person can move east or west, or north or south, but into Earth and out to space are not options. For this Exploration, you will need a balloon, 11 small stickers, a piece of string, and a ruler. A partner is helpful as well. **Figure 16.20** shows some of the steps in this process.

Blow up the balloon partially, and do not tie it shut. Stick the 11 stickers on the balloon (these represent galaxies) and number them. Galaxy 1 is the reference galaxy.

Measure the distance between the reference galaxy and each of the numbered galaxies. The easiest way to do this is to use your piece of string. Lay it along the balloon between the two galaxies and then measure the string. Record these data in the "Distance 1" column of a table like the one on the following page.

Simulate the expansion of your balloon universe by *slowly* blowing up the balloon the rest of the way. Have your partner count the number of seconds it takes you to do this, and record this number in the "Time Elapsed" column of the table (each row has the same time elapsed, because the expansion occurred for the same amount of time for each galaxy). Tie the balloon shut. Measure the distance between the reference galaxy and each numbered galaxy again. Record these data in the "Distance 2" column of the table.

Subtract the first measurement from the second. Record the difference in the table.

Divide this difference, which represents the distance traveled by the galaxy, by the time it took to blow up the balloon. A distance divided by a time gives you an average speed.

Make a graph of the velocity (on the *y*-axis) versus Distance 2 (on the *x*-axis) to get "Hubble's law for balloons." You may wish to roughly fit a line to these data to clarify the trend.

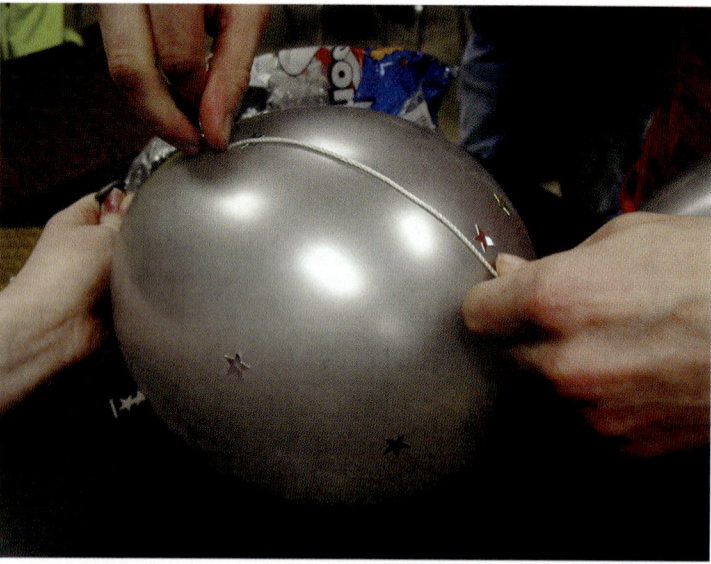 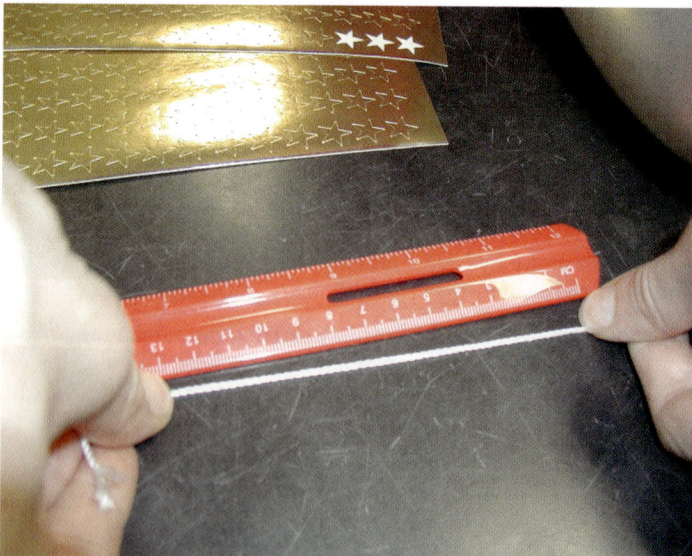

Figure 16.20 The easiest way to measure the distance around a curved balloon is to use a string.

GALAXY NUMBER	DISTANCE 1	DISTANCE 2	DIFFERENCE	TIME ELAPSED	VELOCITY
1 (reference)	0	0	0		0
2					
3					
4					
5					
6					
7					
8					
9					
10					
11					

1. Describe your data. If you fit a line to them, would it be horizontal, trend upward, or trend downward?

2. Was there anything special about your reference galaxy? Was it different in any way from the others?

3. If you had picked a different reference galaxy, would the trend of your line be different? If you are not sure about the answer to this question, get another balloon and try it!

4. The expansion of the universe behaves similarly to the movement of the galaxies on the balloon. We don't want to carry the analogy too far, but there is one more thing to think about. In your balloon, you probably have some areas that expanded less than others because the material was thicker—there was more "balloon stuff" holding it together. How is this similar to some places in the actual universe?

SMARTWORK • smartwork.wwnorton.com

17 Formation of Structure

The universe that emerged from the Big Bang was incredibly uniform. But the modern universe has structure with planets, stars, and galaxies grouped into larger and larger structures. Cosmologists study how the universe evolved from the uniformity of the earliest times to the non-uniform structure of today. In the illustration on the opposite page, a student is studying the structure of the universe on the largest scales—shown in the top image—and working on an Earth-bound analogy. In this chapter, we take a closer look at the nature of the universe, how its structure has evolved over time, and contemplate its ultimate fate.

✦ LEARNING GOALS

In this chapter, we will find that complex structure is a natural consequence of the physical laws in our evolving universe. By the end of the chapter, you should be able to visualize the large-scale structure of the universe and develop Earth-bound analogies, as the student is doing with the illustration on the opposite page. You should also be able to:

LG 1 Describe the distribution of galaxies in the universe.

LG 2 Explain how galaxies formed in the early universe.

LG 3 Distinguish between the formation of the first stars and the formation of stars today.

LG 4 Place in order the steps that led to large-scale structure.

PROF KEEPS SAYING IT'S LIKE SOAP BUBBLES...

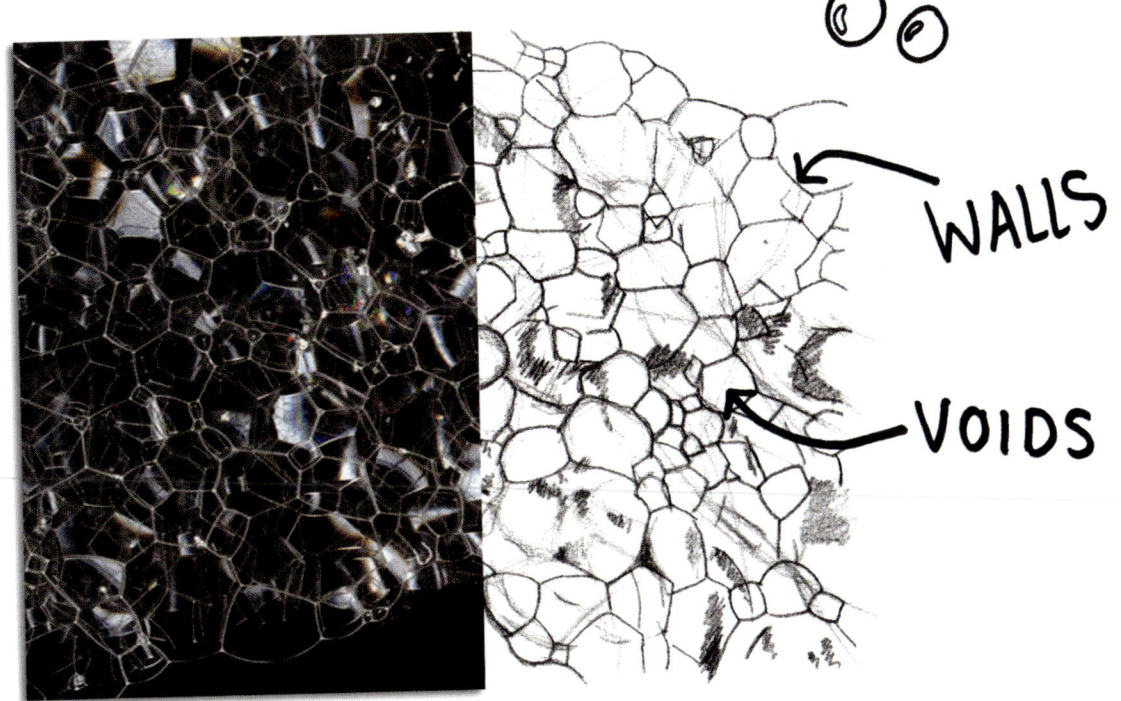

WALLS

VOIDS

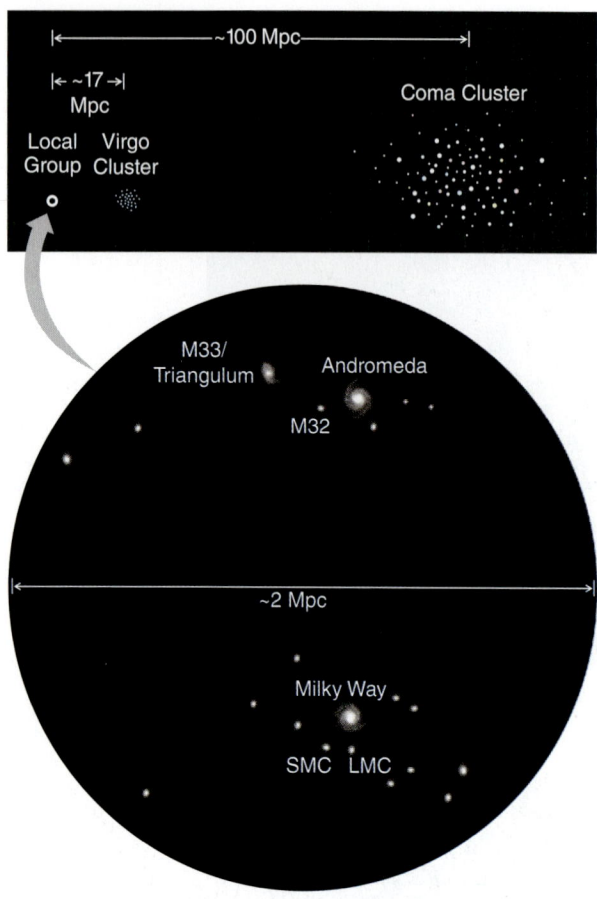

Figure 17.1 This portion of the Local Group (bottom) includes three spiral galaxies and many smaller dwarf galaxies. The local group is 17 Mpc from the Virgo Cluster and 100 Mpc from the Coma Cluster. LMC, Large Magellanic Cloud; SMC, Small Magellanic Cloud.

17.1 Galaxies Form Groups, Clusters, and Larger Structures

Galaxies are not distributed evenly throughout the universe, but instead are clumped together into structures of varying size scales. To understand the formation of these larger structures, we need consider only the way that galaxy-sized clumps of matter fall together under the force of gravity.

Types of Galaxy Structures

The vast majority of galaxies are parts of gravitationally bound collections of galaxies. As we saw in Chapter 15, galaxy groups are the smallest and most common galaxy collections and contain up to several dozen galaxies, most of them dwarf galaxies. We also saw that the Milky Way is a member of the Local Group, which consists of two giant spiral galaxies—the Milky Way Galaxy and the Andromeda Galaxy—along with more than 30 smaller dwarf galaxies in a volume of space roughly 2 megaparsecs (Mpc) in diameter (**Figure 17.1**). Most of the galaxy mass (both luminous and dark matter) in the Local Group resides in the two giant galaxies.

Galaxy clusters are gravitationally bound collections of hundreds to thousands of galaxies, typically 3–5 Mpc (about 10 million to 15 million light-years) across. Galaxy clusters often have a more regular structure than is found in galaxy groups. Galaxy clusters are larger than groups. Figure 17.1 shows the Local Group's position relative to two well-known clusters, the Virgo Cluster and the Coma Cluster. In a cluster, as in groups, the number of dwarf galaxies is larger than the number of giant galaxies. However, most of the galaxy *mass* in galaxy clusters resides in the giant galaxies. Although spiral galaxies are common in most systems, giant elliptical galaxies are prevalent in only about one-fourth of galaxy clusters. The Virgo Cluster, located 17 Mpc from the Local Group, is an example of a cluster containing mostly spiral galaxies. The more distant Coma Cluster is dominated by giant elliptical and S0 galaxies.

Galaxy clusters and groups of galaxies themselves bunch together to form enormous **superclusters**, which contain tens of thousands or even hundreds of thousands of galaxies and span regions of space typically more than 30 Mpc in size. Our Local Group is part of the Virgo Supercluster, which also includes the Virgo Cluster.

Observing Components of Galaxy Structures

Understanding how clusters form and evolve requires observations beyond the visible. Clusters are bright in the X-ray region of the spectrum (**Figure 17.2a**), indicating that they are rich in hot gas. The amount of visible, luminous mass is not enough to keep this hot gas from escaping. Like stars in the disks of galaxies, cluster galaxies orbit the center of mass of the cluster much faster than can be accounted for by the luminous mass we observe. These observations indicate large amounts of dark matter, which has other effects that can be observed directly. The total luminous and dark mass in clusters acts as a gravitational lens for light coming from galaxies in the background. These gravitational lenses warp the images of background galaxies into distorted arcs. In Figure 17.2b, the blue-white glow overlying the image taken in the visible part of the spectrum shows the distribution of the cluster's dark matter inferred from the arcs.

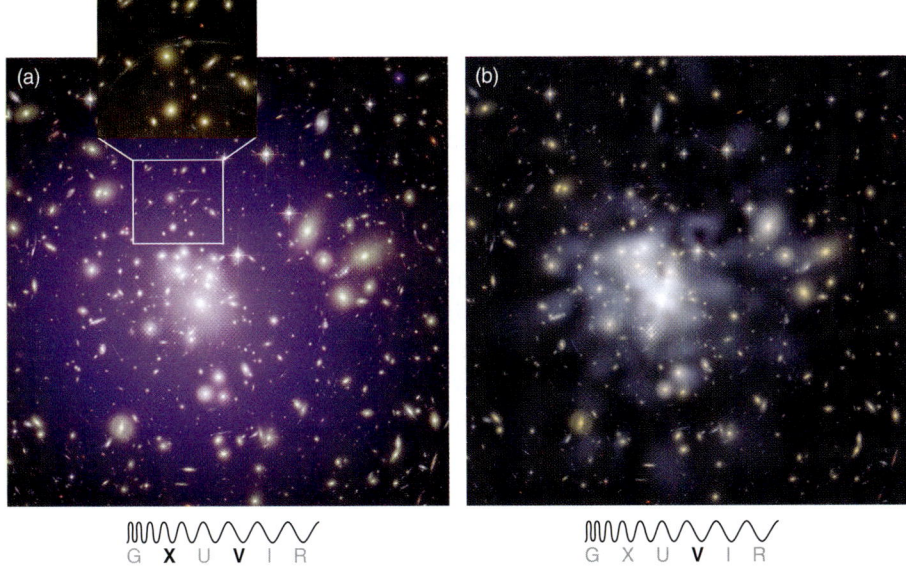

G X U V I R   G X U V I R

Figure 17.2 (a) Galaxy clusters like Abell 1689 are rich in hot, X-ray-emitting gas (shown here in purple overlying the image taken in the visible part of the spectrum). Gravitational lensing warps images of background galaxies into arcs. Astronomers use these arcs to trace the distribution of dark matter in the cluster. The inset zooms in on an arc, shown here in visible light only. (b) This is also Abell 1689, with the inferred dark matter distribution shown in blue-white.

Mapping Galaxy Structures

Hubble's law of the expansion of the universe is a powerful tool for mapping the distribution of galaxies, groups, clusters, and superclusters in space. The distance to a galaxy can be obtained by measuring the redshift from a single observation of the galaxy's spectrum. We now know the redshifts to more than 1 million galaxies, and therefore we know the approximate distances to these galaxies. From this information, we can develop a map of the structure of the universe on the largest scales.

The Harvard-Smithsonian Center for Astrophysics conducted the first large redshift survey in 1986 and presented the astronomical community with a "slice of the universe," as displayed in **Figure 17.3a**. The observations show that clusters and superclusters are not scattered randomly through space but are linked in an intricate network of relatively thin structures known as filaments and walls. These concentrations of galaxies surround large regions of space with very few galaxies, known as **voids**. Voids are some of the largest structures seen in the universe. Though the voids may seem empty, we do not know that they are empty of matter—only that they have very few observable galaxies.

Clusters and superclusters are located within walls and filaments. This structure is not limited to the "nearby" universe. Subsequent surveys have looked at much larger volumes of space. Figure 17.3b shows the results of one such survey conducted with the Anglo-Australian Telescope at the Siding Spring Observatory in Australia. Results from a more recent survey, the Sloan Digital Sky Survey, are shown in Figure 17.3c. For as far out as our observations can currently measure, the universe has a porous structure reminiscent of a sponge or a pile of soap bubbles. Take a moment to visualize and sketch this type of structure, as shown in the chapter-opening figure. Together, galaxies and the larger groupings in which they are found are referred to as **large-scale structure**.

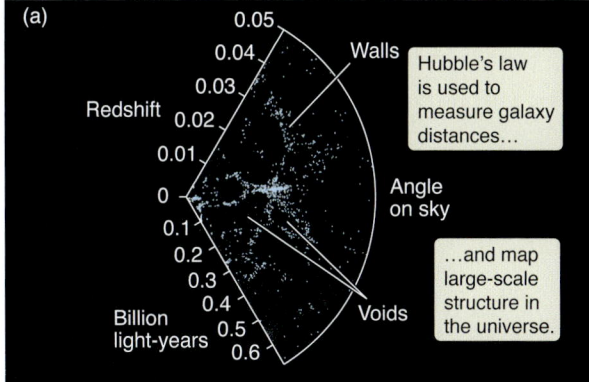
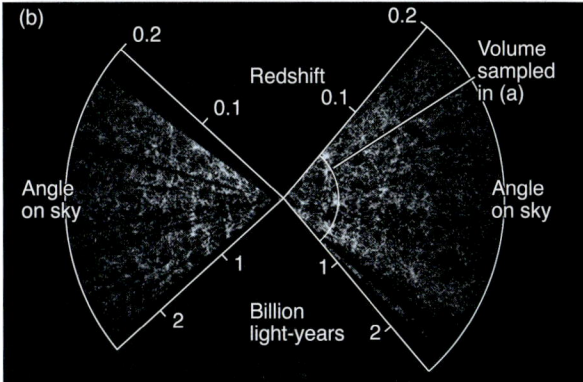
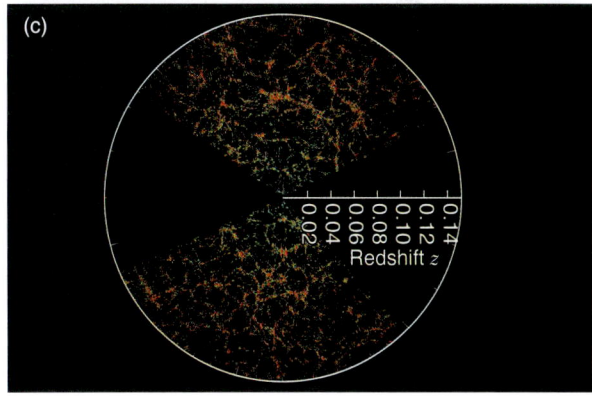

Figure 17.3 Redshift surveys use Hubble's law to map the universe. (a) In 1986, the Harvard-Smithsonian Center for Astrophysics redshift survey, called "A Slice of the Universe," was the first to show that clusters and superclusters of galaxies are part of even larger-scale structures. (b) The Two-degree-Field (2dF) Galaxy Redshift Survey, completed in 2003, shows similar structures at even larger distances. (c) The 2008 Sloan Digital Sky Survey map of the universe extends outward to a distance of about 600 Mpc. Shown here is a sample of 67,000 galaxies colored according to the ages of their stars; the redder, more strongly clustered points show galaxies that are made of older stars. The redshift scale is shown on all three images, so that you might easily compare the relative sizes of the maps.

Recall from Chapters 1 and 14 that the cosmological principle states that the universe is homogeneous (the same everywhere) and isotropic (the same in every direction). These observations of the large-scale structure of the universe provide direct observational evidence for the cosmological principle, because they show that on this very largest scale, the structure of the universe is the same everywhere, in every direction. If this had not been true, the cosmological principle would have been proved false. If, for example, the universe in the top half of the multicolored image of Figure 17.3c showed a uniform distribution of galaxies while the universe in the bottom half of the image showed walls and voids, then the universe would not be isotropic. All conclusions that are based on the cosmological principle would have been called into doubt. As it is, however, observations support this underlying principle of cosmology.

17.2 Gravity Forms Large-Scale Structure

What causes the large-scale structure of walls and voids? Cosmologists have proposed a number of ideas. Early on, it was suggested that voids were the result of huge expanding blast waves from tremendous explosions that might have occurred in the early universe. The correct answer has turned out to be less fanciful, but far more satisfying: large-scale structure is caused by gravity.

Gravitational Instabilities

In Chapter 5, you learned about gravitational instabilities involved in star formation. Star formation can begin in a molecular cloud with clumps inside it. Gravity causes those clumps to collapse faster than their surroundings: gravity can turn density variations of clouds into stars. The same gravitational instability can turn density variations of the universe into galaxies.

As discussed in Chapter 16, space missions such as COBE, WMAP, and Planck have revealed variations in the background radiation of about one part in 100,000. Theoretical models show that these variations are far too small to explain the structure we see in today's universe. These models indicate that for ripples in the density of the universe to have formed today's galaxies, the variations in the cosmic microwave background (CMB) today should be at least 30 times larger than what is observed. At first glance this might seem to be irreconcilable with our understanding of the origin of structure in the universe, but it is instead a crucial result that leads us to reconsider the role of dark matter in the universe, as we will discuss shortly.

The slight ripples in the CMB result from tiny (on the scale of subatomic particles) variations that imprinted structure on the early universe at the time of inflation. These variations provided the "clumps" or "seeds" from which galaxies and collections of galaxies grew. The physics that governs atomic nuclei, atoms, and molecules is responsible for seeding the very largest structures that we can see in our universe.

It is one thing to say that galaxies and larger structures formed from gravitational instabilities that began with slight irregularities in the early universe. It is quite another to turn this statement into a real scientific theory with testable predictions. To accomplish that, we combine previous observations with the ideas we want to test and the laws of physics. We then construct a model or a computer simulation and then compare the predictions of that model with other observations of the universe.

Building a Testable Model of Large-Scale Structure Formation

To build a model of the formation of large-scale structure, we need three key pieces of information. First, we have to decide what universe we are going to model—what values of Ω_{mass} and Ω_Λ (that is, what densities of matter and energy) we are going to assume. These values determine how rapidly the universe expands. The more rapidly a universe expands, or the less mass it contains, the more difficult it will be for gravity to pull material together into galaxies and larger-scale structures. As discussed in Chapter 16, these values have been constrained by observations, such that Ω_{mass} is about 0.3 and Ω_Λ is about 0.7. These observational conclusions are inputs to the theoretical models and simulations of the formation of large-scale structure.

Second, we need to know how large and how concentrated the early clumps were. Structure in the CMB shows what the early clumps in the universe must have looked like. Additionally, models of inflation predict the structure that will emerge after the rapid expansion. These predictions are especially important to test because they tie together the large-scale structure of today's universe with our most basic ideas about what the universe was like in the briefest instant after the Big Bang. Currently, we know enough to say that the early universe was "clumpier" on smaller (galaxy-sized) scales than it was on the scales of the clusters, superclusters, filaments, and voids. Therefore, smaller structures formed first and larger structures formed later. This idea of small structures forming first and larger structures forming later is referred to as **hierarchical clustering**.

Third, we need a complete list of the types and amounts of ingredients that existed in the early universe. We need to know the balance between radiation, normal matter, and dark matter. We also need to make some choices about the nature of the dark matter that we use in our model. To understand dark matter at this level, we need some more information about matter in the universe.

The Composition of Dark Matter

When the universe was only a few minutes old, its temperature and density were high enough for nuclear reactions to take place. Collisions between protons in the early universe built up low-mass nuclei, including deuterium (heavy hydrogen) and isotopes of helium, lithium, beryllium, and boron in a process called **Big Bang nucleosynthesis**. These nuclear reactions determined the final chemical composition of the matter that emerged from the hot phase of the Big Bang. Even in this short time after the Big Bang, the density of the universe had fallen too low for reactions such as the triple-alpha process—which forms carbon in the interiors of stars—to occur (see Chapter 12). Therefore, all elements more massive than boron, including most of the atoms that make up our planet and us, formed in subsequent generations of stars.

Figure 17.4 shows the observed and calculated predictions of deuterium (^{2}H) and the very stable helium isotope (^{4}He), plotted as a function of the present-day density of normal matter in the universe. Observations of current abundances are shown as horizontal bands. Theoretical predictions, which depend on the density

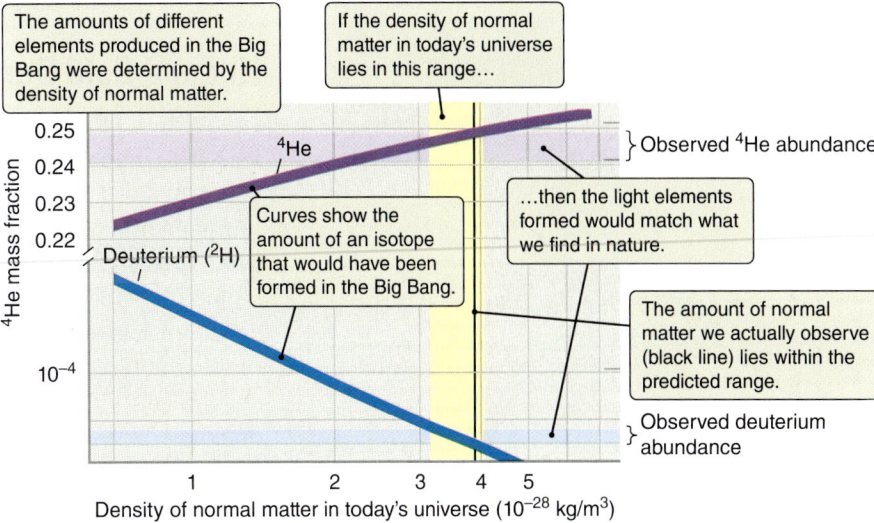

Figure 17.4 Observed and calculated abundances of ^{2}H and ^{4}He, plotted against the density of normal matter in today's universe. Big Bang nucleosynthesis correctly predicts the amounts of these isotopes found in the universe today.

of the universe, are shown as darker, thick lines. Big Bang nucleosynthesis predicts that about 24 percent of the mass of the normal matter formed in the early universe should have ended up in the form of the very stable isotope ^{4}He. Indeed, when we look about us in the universe today, we find that about 24 percent of the mass of normal matter in the universe is in the form of ^{4}He, in complete agreement with the prediction of Big Bang nucleosynthesis. In fact, this agreement between theoretical predictions and observation provides powerful evidence that the universe began in a Big Bang.

Unlike helium, most isotope abundances depend on the density of normal matter in the universe, so comparing current abundances with models of isotope formation in the Big Bang pins down the density of the early universe. From these comparisons, astronomers can find the density of normal matter in the universe today. The location on the graph where the observations and the theoretical predictions intersect gives the average density of normal matter in the universe today: about 3.9×10^{-28} kilograms per cubic meter (kg/m^3). This value lies well within the range predicted by the observations for many different light isotopes.

Cosmologists can also compare the density of normal matter inferred from Big Bang nucleosynthesis with the density of normal matter we actually observe in the universe. They find that the density of normal matter inferred from nucleosynthesis is about the same as the density of normal matter in today's universe and much less than the density of dark matter. This agreement provides a powerful constraint on the nature of dark matter, which dominates the mass in the universe. Dark matter cannot consist of normal matter made up of neutrons and protons. If it did, the density of neutrons and protons in the early universe would have been much higher, and the resulting abundances of light elements in the universe would have been very different from what we actually observe.

Dark matter must instead be something that has no electric charge, and therefore does not interact with electromagnetic radiation. We cannot see dark matter in the CMB, and dark matter does not interact with radiation. Because clumps of such dark matter in the early universe did not interact with radiation or normal matter, they were not smoothed out by radiation pressure. **Figure 17.5** shows how the distribution of normal matter and the distribution of dark matter differ due to the difference in how they interact with radiation. Unlike normal matter, which smoothes out over time, dark matter remains clumpy. Dark matter solves the problems of the formation of galaxies and clusters of galaxies by providing a stronger gravitational attraction without being smoothed out by radiation in the early universe.

Hot and Cold Dark Matter

Here is the story of galaxy formation in a nutshell. Dark matter in the early universe was much more strongly clumped than normal matter. Within a few million years after recombination, these dark matter clumps pulled in the surrounding normal matter. Later, gravitational instabilities caused these clumps to collapse. The normal matter in the clumps went on to form visible galaxies. This story seems plausible enough, but the details of how this happened depend greatly on the properties of dark matter itself. Recall that in Chapter 14, we discussed the search for MACHOs and WIMPs as possible explanations of dark matter and concluded that MACHOs could not explain the missing mass. Even though we do not yet know exactly what dark matter is made of, based on how dark matter behaves, we can categorize two broad classes: cold dark matter and hot dark matter.

▶️ **AstroTour:** Big Bang Nucleosynthesis

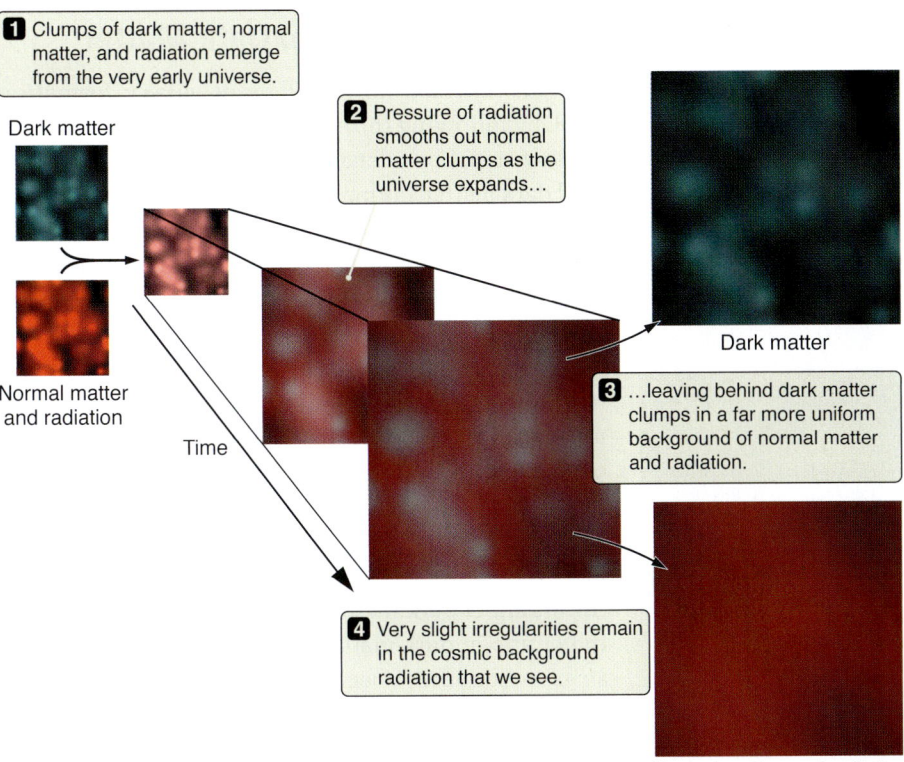

Figure 17.5 Radiation pressure in the early universe smoothed out variations in normal matter, but irregularities in the dark matter survived to become the seeds of galaxy formation.

Cold dark matter consists of feebly interacting particles that are moving about relatively slowly, like the atoms and molecules in a cold gas. There are several candidates for cold dark matter. Most likely, cold dark matter consists of an unknown elementary particle. One candidate is the **axion**, an as yet undetected particle first proposed to explain some observed properties of neutrons. Axions should have very low mass, and they would have been produced in great abundance in the Big Bang. Another candidate is the **photino**, an elementary particle related to the photon. Some theories of particle physics predict that the photino exists and has a mass about 10,000 times that of the proton. Our state of knowledge about the particles that make up cold dark matter could soon change: Photinos might be detected in current particle accelerators, and experiments are under way to search for axions and photinos that are trapped in the dark matter halo of our galaxy.

Hot dark matter consists of particles that are moving so rapidly that gravity cannot confine them to the same region as the luminous matter in the galaxy. Neutrinos are one example of hot dark matter. We have seen that neutrinos interact with matter so feebly that they are able to flow freely outward from the center of the Sun. There is no question that the universe is filled with neutrinos, which might account for as much as 5 percent of its mass. Although this percentage is not high enough to account for all of the dark matter in the universe, it may still have had a noticeable effect on the formation of structure.

Slow-moving particles are more easily held by gravity than fast-moving particles, so particles of cold dark matter clump together more easily into galaxy-sized structures than do particles of hot dark matter. On the largest scales of massive superclusters, both hot dark matter and cold dark matter can form the kinds of structures we see; but on much smaller scales, only cold dark matter can clump enough to produce structures like the galaxies we see filling the universe. To account for the formation of today's galaxies, we need cold dark matter.

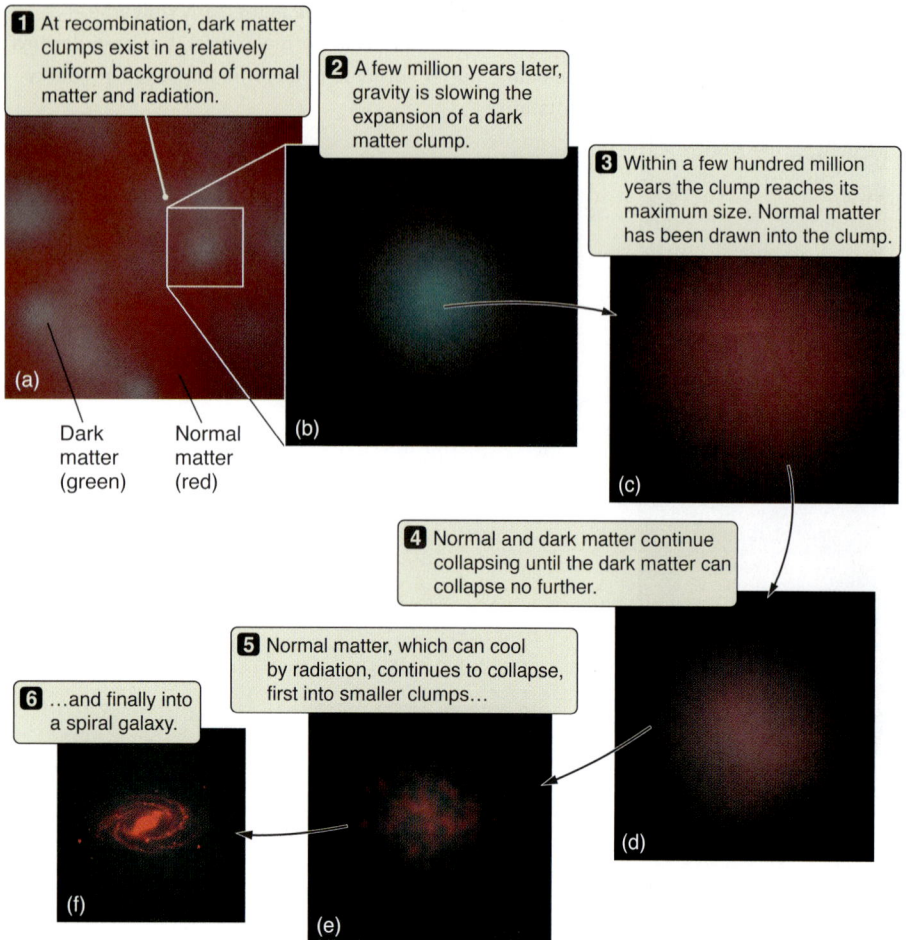

Figure 17.6 A spiral galaxy passes through roughly six stages as it forms from the collapse of a clump of cold dark matter.

Collapsing Clumps of Dark Matter

We can best see how models of galaxy formation work by following the events predicted by the models step-by-step. Consider a universe made up primarily of cold dark matter, clumped together with normal matter in a manner consistent with observations of the cosmic microwave background. On the scale of an individual galaxy, the effect of the cosmological constant is so small that it can be ignored.

Figure 17.6a shows a model simulation of one clump of dark matter at the time of recombination. The dark matter is less uniformly distributed than normal matter, but overall the distribution of matter is still remarkably uniform. By a few million years after recombination, shown in Figure 17.6b, the universe in the simulation has expanded severalfold. Spacetime is expanding, so the clump of dark matter is also expanding. However, the clump of dark matter is not expanding as rapidly as its surroundings because its self-gravity has slowed down its expansion. The clump now stands out more with respect to its surroundings. The gravity of the dark matter clump has begun to pull in normal matter as well. By the stage shown in Figure 17.6c, normal matter is clumped in much the same way as dark matter.

A ball thrown in the air will slow, stop, and then fall back to Earth. Similarly, the clump of dark matter will stop expanding when its own self-gravity slows and then stops its initial expansion. Eventually, the clump of dark matter has reached its maximum size and is beginning to collapse (see Figure 17.6c). The collapse of the dark matter clump stops when the clump is about half its maximum size, however, because the particles making up the cold dark matter are moving too rapidly to be pulled in any closer (Figure 17.6d). The clump of cold dark matter is now given its shape by the orbits of its particles, in the same way that an elliptical galaxy is given its shape by the orbits of the stars it contains.

Unlike dark matter (which cannot emit radiation), the normal matter in the clump is able to radiate away energy, cool, and collapse. Small concentrations of normal matter within the dark matter collapse under their own gravity to form clumps of normal matter that range from the size of globular clusters to the size of dwarf galaxies. These clumps of normal matter then fall inward toward the center of the dark matter clump, as shown in Figure 17.6e. According to models, gas in our universe can cool quickly enough to fall in toward the center of the dark matter clump only if the clump has a mass of 10^8 to 10^{12} solar masses ($M_\odot$). This is just the range of masses of observed galaxies. This agreement between theory and observation is an important success of the theory that galaxies form from cold dark matter.

Gravitational interaction between irregular clumps tugs on each protogalactic clump, so that it has a little bit of rotation when it begins its collapse. As normal matter falls inward toward the center of the dark matter clump, this rotation forces much of the gas to settle into a rotating disk (Figure 17.6f), just as the collapsing cloud around a protostar settles into an accretion disk. The disk formed by the collapse of each protogalactic clump becomes the disk of a spiral galaxy.

17.3 The First Stars and Galaxies Form

Recall from Chapter 16 that the universe first became transparent about 400,000 years after the Big Bang, when the temperature dropped low enough for atoms to form. The CMB was emitted at this moment, and we can observe it today. Between about 200 million and 600 million years after the Big Bang, the first stars began forming from the elements created in the Big Bang. As these stars formed, they heated up until they emitted ultraviolet (UV) photons with enough energy to reionize neutral hydrogen in interstellar space. During this **reionization** stage, when the electrons were stripped from the hydrogen atoms, the hydrogen began to glow at visible wavelengths. The reionization started between 200 million and 600 million years after the Big Bang and continued with star formation in the first low-luminosity galaxies and with radiation from the first supermassive black holes. Reionization was completed about 900 million years after the Big Bang.

Only within the past few years have astronomers begun detecting objects from the first billion years of the universe. Many astronomers were surprised to find galaxies, quasars, and gamma-ray bursts at these early times because it was thought that they did not form until later. For example, gamma-ray bursts result from the explosive deaths of massive stars, and so there must have been massive stars that had already died by 650 million years after the Big Bang. Similarly, supermassive black holes must have formed in less than 750 million years after the Big Bang. The study of these very early objects is one of the most dynamic topics in astronomy today. New telescopes are regularly detecting objects from earlier and earlier times.

The First Stars

The first stars must have formed from the elements created in the Big Bang: hydrogen, helium, and a tiny amount of lithium. Stars with only this limited set of elements have not yet been detected, but observers are searching for them. Old stars in the halo of the Milky Way contain very low abundances of heavy elements, but not quite as low as zero. Astronomers use computer simulations that combine data from these old stars with the conditions in the early universe to figure out what happened before those old stars formed.

The lack of heavy elements during the formation of the first stars affected their formation in multiple ways. Because there were no heavy elements, there was no dust, and also no molecular clouds full of cold, dense gas. Instead, the first stars formed inside dark matter mini-halos that were about 1 million $M_\odot$ and 100 parsecs (pc) across (**Figure 17.7**). These mini-halos formed a few hundred million years after the Big Bang. Neutral hydrogen trapped in these mini-halos combined to form molecular hydrogen, cooling the gas. As the gas cooled, it collapsed to the center of the mini-halo, forming a tiny protostar that accreted more gas to become a star. Theoretical models predict that these first stars were likely to be hot (so their luminosity was high and peaked in the UV) and massive (10–100 $M_\odot$ for single stars and 10–40 $M_\odot$ for double stars). Large clusters likely didn't form, and these stars were singles, doubles, or small multiples.

The first stars were much more massive than today's average star. Therefore, their main-sequence lifetimes were very short: 10 million years or less. These stars ended their short lives as supernovae or black holes. If the core of such a star rotated rapidly at the time of the explosion, it might have emitted a gamma-ray burst.

Some of these stars were massive enough to become black holes. Black holes with companions can become energetic X-ray binary systems, as mass falls onto

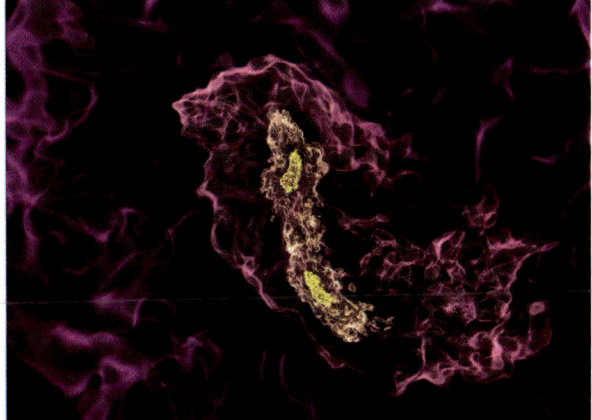

Figure 17.7 In this image from a supercomputer simulation of early star formation, two massive stars form a few hundred astronomical units apart. The brighter yellow regions are higher density than the purple regions.

the accretion disk of the black hole as the companion evolves. Because these stars were all so massive, it is likely that both stars in a binary system (or all of them in a multiple system) would become black holes, and these black holes could then merge. Gravitational waves emitted during a black hole merger might be detectable in future experiments. Some think that these merged black holes might have become the seeds for the supermassive black holes found in galaxies, but other models suggest it would take too long for these stellar black holes to build up to a mass of 1 million to 1 billion $M_\odot$.

The explosions of these massive first stars scattered heavy elements, such as carbon, nitrogen, and oxygen, into nearby gas clouds. Some of these elements condensed into dust grains, which further cooled the clouds, so that the next generation of stars formed in a manner similar to the way stars form today. These "second-generation stars" had very low amounts of heavy elements, but measurably more than the first stars. Because they formed in a cooler environment than the first stars, lower mass stars could form. All stars less massive than about 0.8 $M_\odot$ have such long lifetimes that they are still burning hydrogen on the main sequence today. These stars are not very luminous, but a few have been found in the halo of the Milky Way. Their spectra show small but measurable amounts of many of the heavier elements on the periodic table. Astronomers are very interested in studying these second-generation stars because they offer clues about the nature of the first stars and the conditions of the young Milky Way.

The First Galaxies

How did the first galaxies form? Almost by definition of the word *galaxy*, it seems likely that the first galaxy was made up of the first system of stars that were gravitationally bound in a dark matter halo. These stars may have been first- or second-generation stars. The properties of the first galaxies were shaped by the first stars: their radiation, their production of heavier elements, and the black holes resulting from them. The masses of these galaxies are thought to have been about a hundred million $M_\odot$, and they were built up hierarchically from the merging of mini-halos.

One piece of evidence for this model comes from infrared observations, shown in **Figure 17.8**. Figure 17.8a includes the usual nearby stars and galaxies, but when these are all subtracted, a glow remains. This remaining structure, seen in Figure 17.8b, likely arose from the first stars and galaxies, roughly 500 million years after the Big Bang.

Another piece of evidence comes from the discovery of the most distant—therefore we see them when they were very young—galaxies and quasars. These observations constrain the timeline by indicating how soon the first galaxies—and the first supermassive black holes—formed after the Big Bang. The peak of the spectrum of these galaxies has been cosmologically redshifted into the infrared (**Figure 17.9**). The images of these distant objects look like small, faint dots, with none of the detail seen in closer galaxies. Astronomers are excited by such images because just the detection of these objects contributes to an understanding of when and how the first galaxies formed.

The first galaxies are thought to have formed by about 400 million years after the Big Bang. The heavier elements created from the first stars cooled the gas in larger dark matter halos, which then collapsed, probably to a disk, and stars formed. These youngest galaxies appear to be small, 20 times smaller than the Milky Way, which adds support for the hierarchical model of galaxy formation.

Recall that the Local Group has small, faint dwarf galaxies orbiting the Milky

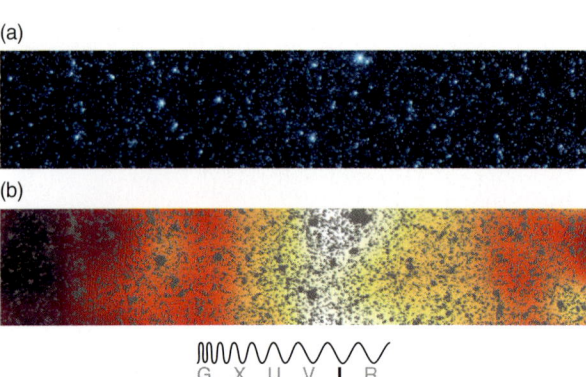

Figure 17.8 (a) A standard infrared image from the Spitzer Space Telescope shows stars and some galaxies in this strip of sky. (b) In this image, the nearby stars and galaxies have been subtracted out (gray) and the remaining glow enhanced, showing some structure from the time when the earliest stars and galaxies were forming.

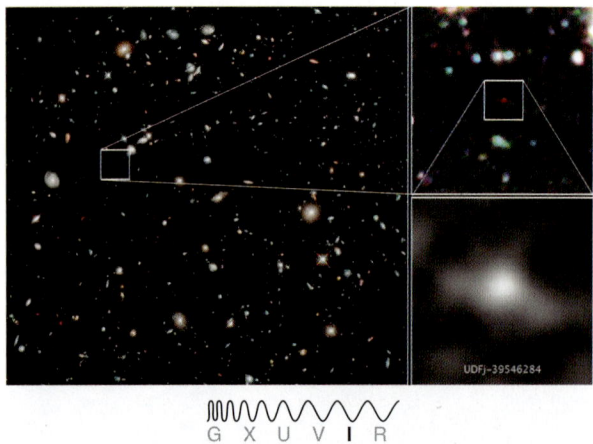

Figure 17.9 The earliest, highest-redshifted galaxies are observable in infrared light. This Hubble Space Telescope image shows a compact faint galaxy as it existed about 480 million years after the Big Bang. At least 100 of these small galaxies would be needed to build up a galaxy like the Milky Way.

Way. Streams of material from the dwarfs are falling onto the Milky Way, indicating that the Milky Way is still gaining mass. Recently, the known number of these galaxies has doubled. About a dozen of these small galaxies are called **ultrafaint dwarf galaxies**, because they are less than 100,000 times more luminous than the Sun. These galaxies contain mostly old, faint stars with low abundances of heavy elements, hinting that they may have contributed to building the Milky Way's halo. The ultrafaint dwarf galaxies may not have had any further star formation after the first stars, and they may be the oldest galaxies around. They may even be the fossil remains of the first galaxies or of the first mini-halos. Because they are more massive than their luminosity suggests, we can determine that even these very old galaxies are dominated by dark matter.

17.4 Galaxies Evolve

Galaxies continued to evolve hierarchically, with smaller fragments merging to form larger objects. The early universe was denser, so early fragments were closer together and mergers were more common. Computer simulations indicate that small concentrations of normal matter within the dark matter would have collapsed under their own gravity as they radiated and cooled, forming clumps that ranged from the size of globular clusters to the size of dwarf galaxies.

In a large spiral galaxy like the Milky Way, faint dwarf spheroidal galaxies (with dark matter) and the oldest globular clusters (without much dark matter) may be leftover protogalactic fragments. The gas collapsed to form a rotating disk as it cooled. A recent simulation that included dark matter, gravity, star formation, and supernova explosions was able to reproduce a Milky Way–like galaxy with a large disk and a small bulge (**Figure 17.10**). Observationally, astronomers have used photos of 400 spiral galaxies at various stages of evolution to understand how galaxies like the Milky Way form. They combine this information with simulations and observations of the oldest parts of the Milky Way to find that the oldest globular clusters are about 13.5 billion years old, but the halo itself may be only about 11.5 billion years old. Most of the stars in the Milky Way formed between 11 billion and 7 billion years ago. The disk and bulge formed at about that same time.

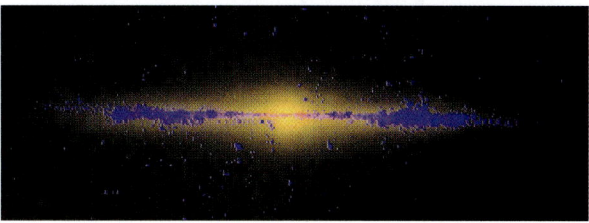

Figure 17.10 This simulation of the formation of a Milky Way–like galaxy has been able to reproduce the small bulge and big disk. Blue colors indicate recent star formation, while older stars are redder.

The Most Distant Galaxies

Figure 17.11 shows images of galaxies throughout the history of the universe. The galaxies observed in the very early universe, before about 11 billion years ago, are

Figure 17.11 A comparison of the Hubble classification of galaxies today with galaxies throughout the history of the universe. There were more peculiar galaxies in the past, indicating that it took some time for spirals to form.

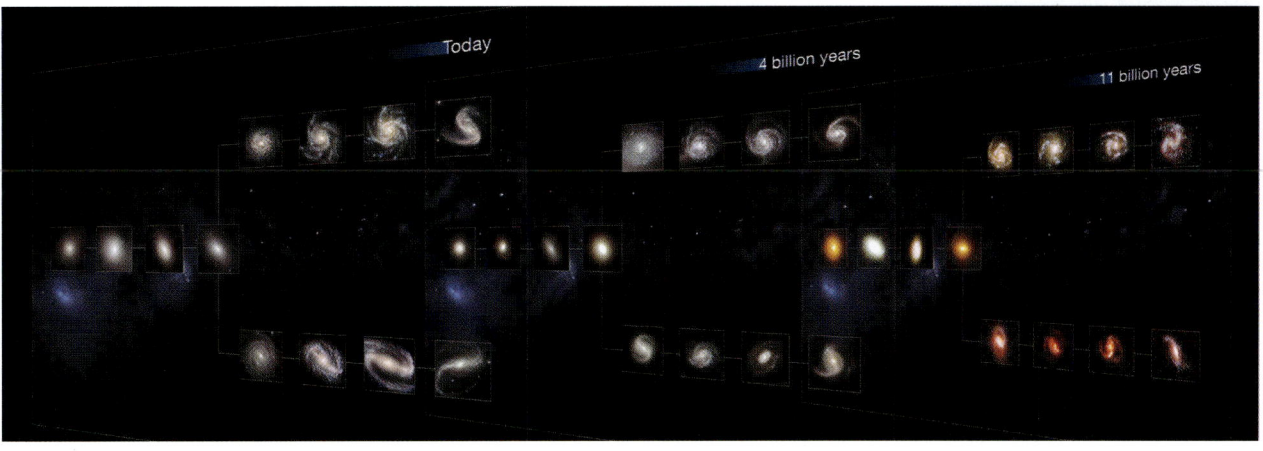

Figure 17.12 Chandra X-ray observations (red, orange, and yellow), combined with Hubble Space Telescope observations (blue and white) of NGC 6240 show that it has two black holes less than 1,000 pc apart. The black holes at the center of the white features will likely merge in about 100 million years.

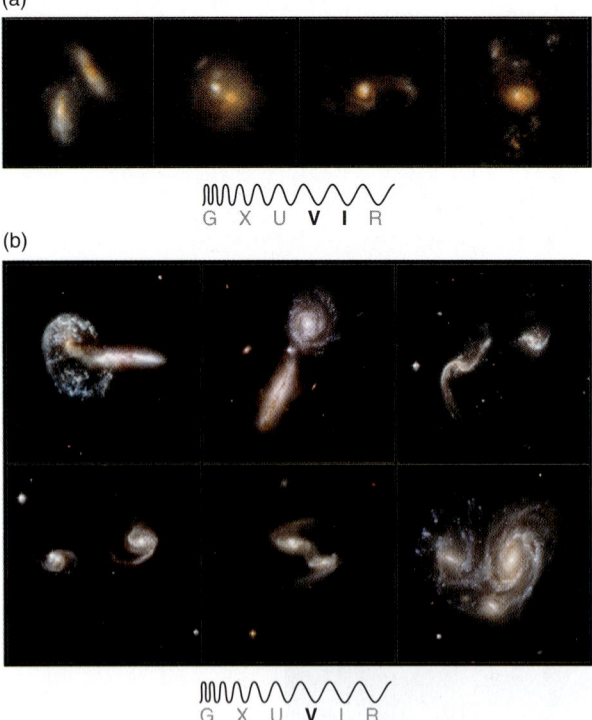

Figure 17.13 The merging galaxies in (a) are between 730 and 2,000 Mpc from Earth, and observations show them at a younger age than the merging galaxies shown in (b), which are located closer to Earth.

so faint that no structure can be seen. Observations of galaxies at about 11 billion years ago have shown visible structure that is much less regular than that of galaxies today. Even at 4 billion years ago, galaxies were much more irregular. Early peculiar galaxies are merging galaxies, which fits with our ideas that in the early universe, when galaxies were closer together, there were more mergers. Most of today's galaxies conform to the Hubble classification, and only about 10 percent are peculiar. However, 4 billion years ago, more than half of the galaxies were peculiar, the number of ellipticals and S0 galaxies was about the same, and there were many fewer spirals. This difference in galaxy types at different times suggests that it took time for spirals to form. These mergers likely produced spiral galaxies over time.

Hierarchical merging also may have triggered the formation of supermassive black holes at the center of galaxies. The first supermassive black holes, which power distant quasars, could have grown from the merging of mini-halos with stellar black holes left after the first stars. Or they could have formed through the accretion of gas from the material between the galaxies during mergers of the first galaxies or through rapid collapse from hot, dense gas at the center of the first galaxies. In nearby galaxies, the mass of the supermassive black hole and the bulge properties are related, suggesting that the growth of the black hole and the bulge might have been linked when they were younger. Supermassive black holes could have grown even more massive from the mergers of large galaxies too. **Figure 17.12** shows a nearby galaxy with two supermassive black holes about 900 pc (roughly 3,000 light-years) apart, which are in the process of merging.

The hierarchical merging and growth of galaxies also affects star formation in the evolving galaxies. The tidal interactions between the galaxies and the collisions between gas clouds trigger many regions of star formation throughout the combined system. In the past, star formation increased over time, peaking about 3 billion years after the Big Bang, and then decreasing again to the current star formation rate.

Merging galaxies at many different distances have been observed by the Hubble Space Telescope (**Figure 17.13a**), showing how mergers differ at various times in the history of the universe. Ellipticals are now thought to result from the merger of two or more spiral galaxies. The dark matter halos of the galaxies merge, and the stars eventually take on the blob-like shape of an elliptical galaxy. Elliptical galaxies are more common in dense clusters, where mergers are more frequent. Compare the young mergers in Figure 17.13a with those of closer, older galaxies (Figure 17.13b).

Just as galaxies merge, clusters of galaxies also merge. **Figure 17.14** shows the high-speed collision and merging of two galaxy clusters in the Bullet Cluster. Optical images show the individual galaxies; hot gas is seen in X-rays (shown in red), and the distribution of the total mass can be found by studying gravitational lenses. The ordinary matter slowed down in the collision, but the dark matter (shown in blue) did not, providing evidence for dark matter in clusters. Clusters of galaxies also evolve hierarchically, growing from smaller structures to larger ones over time. As with galaxies themselves, younger, distant clusters are messier than older, nearby ones. This is additional evidence that the formation of structure in the universe was hierarchical.

Simulating Structure

Supercomputer simulations of the universe start with billions of particles of dark matter and use the most recent observations of the CMB. The simulations model the formation and evolution of dark matter clumps and halos, filaments and voids, small and large galaxies, and galaxy groups and clusters. These simulations also simulate the flow of ordinary gas within these structures as stars form, and they are used to create images of what the universe should look like at different times. These images are then compared to images of the actual universe. This comparison sets limits on the parameters of the universe: the amount of mass, for example, or the type of dark matter. If the inputs to the simulation are correct, the two sets of images should look very similar. If not, the two sets of images look very different.

In 2009, a simulation was run on NASA supercomputers that shows that slight variations in density after inflation led to higher-density regions that became the seeds for the growth of structure. These density variations are observed in the variations of the CMB. During the first few billion years, dark matter fell together into structures comparable in size to today's clusters of galaxies. The sponge-like filaments, walls, and voids became well defined later. The similarities between images produced by the simulation and images of the observed universe are quite remarkable. **Figure 17.15** compares the simulated view with the observed slice of the universe from the Sloan Digital Sky Survey. These results constrain the combination of mass, CMB variations, types of dark matter, dark matter halos, and values for the cosmological constant in the universe. This is a very important result. This model contains assumptions consistent with observational and theoretical knowledge of the early universe and predicts the formation of large-scale structure similar to what is actually observed in today's universe.

17.5 Astronomers Think about the Deep Future

How will our universe evolve into the future? Using well-established physics, we can calculate how the existing structures in the universe will evolve over

Figure 17.14 The Bullet Cluster of galaxies is an example of the later stages of the merging of two giant galaxy clusters. The smaller cluster of galaxies (top) appears to have passed through the larger cluster (bottom) like a bullet.

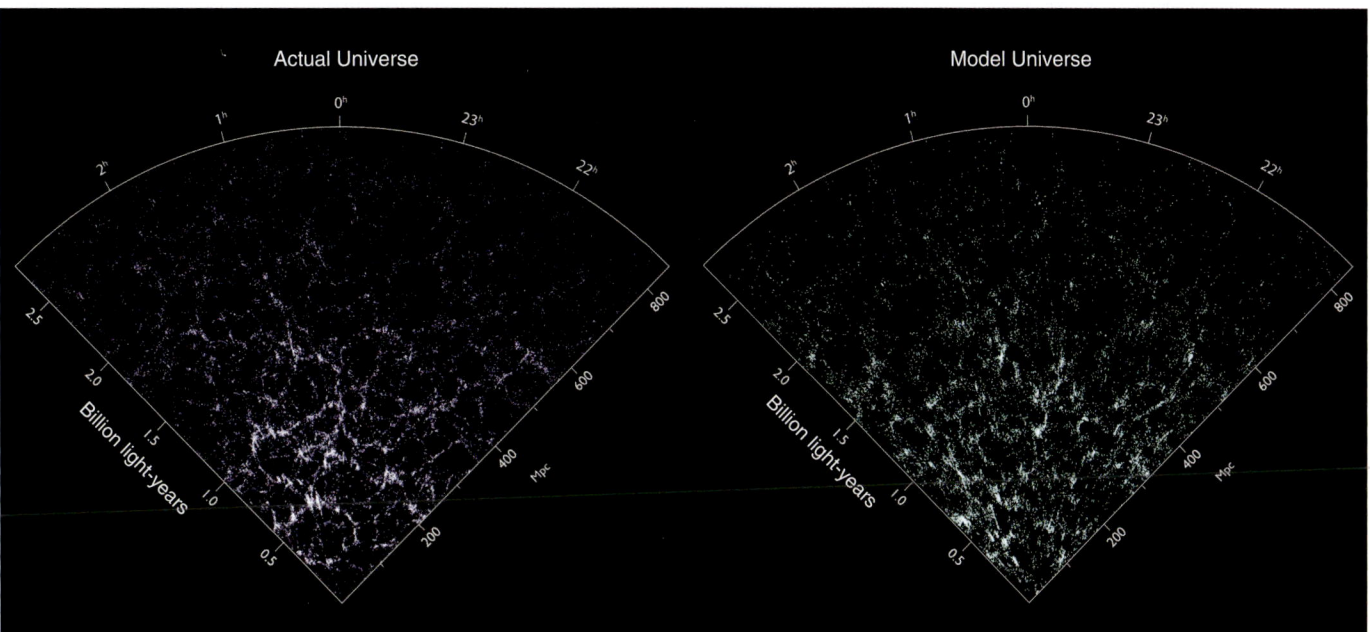

Figure 17.15 The large-scale structure of dark matter halos produced by this simulation (right) is remarkably similar to the distribution of galaxies observed in the Sloan Digital Sky Survey (left).

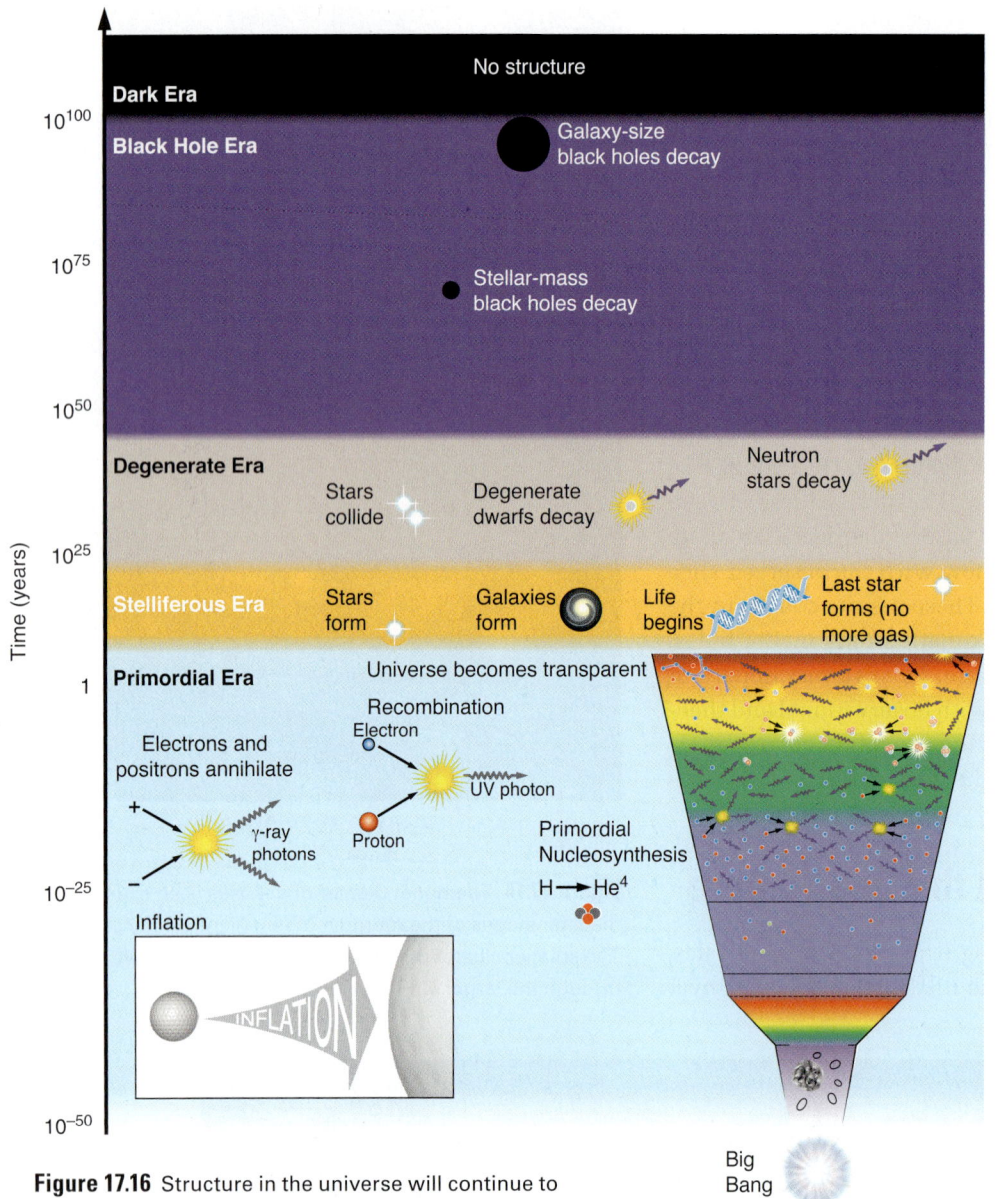

Figure 17.16 Structure in the universe will continue to evolve as it passes from the Primordial Era to the present and then to the future Dark Era.

a very long time. The result of one such calculation is illustrated in **Figure 17.16**. During the first era of the universe, the Primordial Era—the first several hundred thousand years after the Big Bang and before recombination—the universe was a swarm of radiation and elementary particles. Today, we live during the second era, the Stelliferous Era ("Era of Stars"), but this era, too, will end. Some 100 trillion (10^{14}) years from now (about 10,000 times as long as the current age of the universe), the last molecular cloud will collapse to form stars, and a mere 10 trillion years later, the least massive of these stars will evolve to form white dwarfs.

After the Stelliferous Era, most of the normal matter in the universe will be locked up in brown dwarfs and degenerate stellar objects: white dwarfs and neutron stars. During this Degenerate Era, the occasional star will still flare up as ancient substellar brown dwarfs collide, merging to form low-mass stars that burn out in a trillion years or so. However, the main source of energy during this era will come from the decay of particles like protons and neutrons and the annihilation of particles of dark matter. Even these processes will eventually run out of fuel. In 10^{39} years, white dwarfs will have been destroyed by proton decay, and neutron stars will have been destroyed by the beta decay of neutrons.

As the Degenerate Era comes to an end, the only significant concentrations of mass left will be black holes. These will range from small ones with the masses of single stars to greedy monsters that grew during the Degenerate Era to have masses as large as those of galaxy clusters. During the period that follows—the Black Hole Era—these black holes will slowly evaporate into elementary particles through the emission of Hawking radiation. A black hole with a mass of a few solar masses will evaporate into elementary particles in 10^{65} years, and galaxy-sized black holes will evaporate in about 10^{98} years. By the time the universe reaches an age of 10^{100} years, even the largest of the black holes will be gone. A universe vastly larger than ours will contain little but photons with colossal wavelengths, neutrinos, electrons, positrons, and other waste products of black hole evaporation. The Dark Era will have arrived as the universe continues to expand forever—into the long, cold, dark night of eternity.

READING ASTRONOMY News

Star formation, supermassive black holes, and the mergers of galaxies are all connected together. A complete understanding of these processes requires the study of thousands of galaxies across a wide range of luminosities to find the connections.

Massive Black Holes Sidle Up to Other Galaxies

By **STUART GARY**, *ABC Science*

Galaxies containing supermassive black holes tend to cluster closer to other galaxies, according to a new study.

The discovery, reported in the *Astrophysical Journal*, also reveals how gravitational changes caused by these nearby galaxies could affect star formation.

Supermassive giant black holes are millions to billions of times the mass of the Sun and lurk deep in the hearts of active galaxies.

Scientists believe galaxies interacting with, or colliding into, active galaxies could increase the rate at which gas and other matter falls into the supermassive black hole, effectively shutting down the formation of new stars.

"Astronomers have been scratching their heads to think of what mechanism can shut down star formation in some galaxies," says Dr. Heath Jones from Melbourne's Monash University, who is one of the study's authors.

"In recent years there's been ideas around whether or not these active galactic nuclei, the supermassive black holes, actually control star formation in a galaxy . . . like a switch."

Fuelling Monster Hunger

Jones and colleagues used the Australian Six-degree-Field [6dF] Galaxy Survey conducted by the Siding Spring Observatory to find a sample of 2,178 active galaxies with strong radio wavelength signatures.

"We wanted to find out if mergers between galaxies are responsible for producing these rare active galaxies or whether it's something more random," says Jones.

The researchers compared the active galaxies to a sample of less active galaxies of the same size, brightness, and stellar composition.

They found the most active galaxies are located in clusters and within 522,000 light-years of a neighbouring galaxy.

The authors say the findings suggest that mergers play a role in forming the most powerful active galaxies. However, the findings only apply to active galaxies that are at least 200 times more powerful than their quieter counterparts.

"With the very strongest ones, there is a difference, but over the full range of radio power the differences we measure get smaller and smaller the less powerful you go," says Jones.

"What that tells us is that this is by no means the end of the story."

"Our understanding of exactly what mechanism lies at the heart of these galaxies, to give them this active nuclei, is not as straightforward as simply merging galaxies."

Evaluating the News

1. Recall the discussion of active galaxies in Chapter 14. Has the author of this article given a good description of an active galaxy? Why or why not?
2. Consider the number of galaxies in this study. Is this a significant number? What does this mean for the significance of the result?
3. The Australian 6dF Galaxy Survey covered 41 percent of the Southern Hemisphere. What percentage of the entire sky does this represent? Is this a significant fraction of the sky?
4. What conclusions have researchers drawn from the study described in the article?
5. How will this study help astronomers understand mergers and their effects on active galaxies?

SUMMARY

Galaxies are not distributed uniformly, but rather are clumped into groups, clusters, superclusters, and walls. Galaxies develop due to gravitational instabilities in the presence of cold dark matter. The first stars formed in mini-halos of dark matter, while the first galaxies formed later, in larger dark matter halos. Over time, smaller galaxy fragments merged to form larger galaxies. Mergers still happen today. On the longest timescales, the universe will eventually become cold and dark and expand forever.

LG 1 Galaxies are hierarchically gathered into groups, clusters, superclusters, and walls. The walls surround voids in which very few galaxies are present.

LG 2 Cold dark matter halos provided the gravitational assist necessary to form galaxy fragments in the early universe.

LG 3 The formation of the first stars occurred in dark matter mini-halos, rather than in clouds of dust and gas. Because of the absence of heavy elements in these dark matter halos, star formation and evolution were different than in the nearby universe.

LG 4 Large-scale structure began with the formation of dark matter halos in the early universe. These halos caused the collapse of normal matter into stars and clusters of stars. Gravitationally bound groupings of stars formed, which were the earliest galaxies. These earliest galaxies merged to become larger galaxies, which in turn accumulated into clusters. The clusters grew hierarchically, through mergers, to become larger clusters, superclusters, and walls.

SUMMARY SELF-TEST

1. Place the following types of galaxy collections in order of increasing size.
   a. wall
   b. cluster
   c. group
   d. supercluster

2. Voids cannot be filled with dark matter because
   a. the light from the far side of the void would not reach us.
   b. galaxies would be drawn toward the center of the voids.
   c. all the dark matter is in the walls.
   d. the walls do not rotate fast enough to require a lot of dark matter.

3. Galaxy fragments in the early universe formed
   a. before the first stars.
   b. in dark matter mini-halos.
   c. due to gravity from large dark matter halos.
   d. in places where the universe was slightly less dense than average.

4. The dominant force in the formation of galaxies is
   a. gravity
   b. angular momentum
   c. the strong nuclear force
   d. the electromagnetic force

5. Hierarchical models of structure formation are those in which
   a. the smallest structures form first.
   b. the largest structures form first.
   c. formation follows an arc from smallest to largest to smallest.
   d. formation follows an arc from largest to smallest to largest.

6. Our universe will
   a. expand forever.
   b. expand for a long time and then collapse.
   c. expand, but gradually slow down.
   d. neither expand nor contract.

7. Which of the following are candidates for the composition of cold dark matter? Select all that apply.
   a. neutrinos
   b. electrons
   c. axions
   d. neutrons
   e. photinos
   f. tiny black holes
   g. protons

8. Following are the three major observations related to the theory of the Big Bang that have been discussed in the past few chapters. Which led to the development of the theory, and which were predicted by the theory and subsequently observed?
   a. Hubble's law
   b. cosmic microwave background radiation
   c. the abundance of helium

9. Place the following in the order in which they occurred or may occur.
   a. Degenerate Era
   b. Primordial Era
   c. Black Hole Era
   d. Stelliferous Era
   e. the Big Bang

10. What differentiates galaxy groups from galaxy clusters? (Choose all that apply.)
    a. the volume they occupy
    b. the number of galaxies
    c. the total mass of the galaxies

QUESTIONS AND PROBLEMS

Multiple Choice and True/False

11. **T/F:** Most galaxies are isolated, and do not interact gravitationally with other galaxies.

12. **T/F:** Gravity caused ripples in the CMB in the early universe.

13. **T/F:** The first stars were much more massive than stars forming today.

14. **T/F:** Collections of galaxies form "bottom-up"—with groups merging to form larger structures.

15. **T/F:** The universe became transparent when it cooled below the surface temperature of the Sun.

16. On the largest scales, galaxies in the universe are distributed
    a. uniformly.
    b. along filaments and walls.
    c. in disconnected clumps.
    d. along lines extending radially outward from the center.

17. Galaxies in the young universe were _____ galaxies in the universe today.
    a. just like
    b. smaller and more irregular than
    c. far more numerous than
    d. larger and more prototypical than

18. The helium abundance in the current universe
    a. is due partly to the Big Bang and partly to stellar evolution.
    b. is due only to the Big Bang.
    c. is due only to stellar evolution.
    d. is unknown at large scales.

19. Galaxy formation is similar to star formation because both
    a. are the result of gravitational instabilities.
    b. are dominated by the influence of dark matter.
    c. end with the release of energy through fusion.
    d. result in the formation of a disk.

20. Dark matter cannot consist of protons, neutrons, and electrons, because if it did
    a. the abundances of isotopes would not be the same as those observed.
    b. it would have interacted with light in the early universe.
    c. stars and galaxies would be much more massive.
    d. both a and b.

21. Neutrinos are an example of
    a. hot dark matter.   b. cold dark matter.
    c. charged particles.   d. both a and c.

22. Dark matter clumps stop collapsing because
    a. angular momentum must be conserved.
    b. they are not affected by normal gravity.
    c. fusion begins, and radiation pressure stops the collapse.
    d. the clumps cannot radiate away any energy.

23. According to the definitions in Chapter 1, the concept of dark matter is classified as
    a. an idea.   b. a law.
    c. a theory.   d. a principle.

24. Astronomers have never observed a star that has no elements heavier than boron. What does this imply about the first stars?
    a. They must have died before galaxies were fully formed.
    b. The first stars did not form until after galaxies formed.
    c. The first stars must have had very low masses.
    d. The first stars must have been enriched in heavy elements.

25. Giant elliptical galaxies come from
    a. the gravitational collapse of clouds of normal and dark matter.
    b. the collision of smaller elliptical galaxies.
    c. the fragmentation of large clouds of normal and dark matter.
    d. the merging of two or more spiral galaxies.

Conceptual Questions

26. What is the difference between a galaxy cluster and a supercluster? Is our galaxy part of either? How do we know this?

27. Can a galaxy be located inside a void? Explain.

28. Suppose you could view the early universe when galaxies were first forming. How would it be different from the universe we see today?

29. Imagine that there are galaxies in the universe composed mostly of dark matter with relatively few stars or other luminous normal matter. If this were true, how might we learn of the existence of such galaxies?

30. How are the processes of star formation and galaxy formation similar? How do they differ?

31. What is the origin of large-scale structure?

32. Why is dark matter so essential to the galaxy formation process?

33. Why does the current model of large-scale structure require that we include the effects of dark matter?

34. Why do we think that some hot dark matter exists?

35. How does a roughly spherical cloud of gas collapse to form a disk-like, rotating spiral galaxy?

36. How do we know that the dark matter in the universe must be composed mostly of cold—rather than hot—dark matter?

37. Describe the process of structure formation in the universe, starting at recombination (half a billion years after the Big Bang) and ending today.

38. How can we be certain that gravity, and not the other fundamental forces, is responsible for large-scale structure?

39. Previous chapters painted a fairly comprehensive picture of how and why stars form. Why, then, is it difficult to model the star formation history of a young galaxy? Is this difficulty a failure of our theories?

40. What important characteristics of the early universe are revealed by today's observed abundances of various isotopes, such as ^{2}H and ^{3}He?

Problems

41. Figure 17.1 shows a few members of the Local Group.
    a. What is the approximate distance from the Milky Way to Andromeda in light years? (Hint: 1 pc = 3.26 light-years.)
    b. What is the approximate distance from the Milky Way to the LMC?
    c. Andromeda and the Milky Way are approaching each other at about 120 km/s. How long will it be before these two galaxies reduce the distance between them to zero?
    d. Will the two galaxies meet before or after the Sun dies (about 5 billion years from now)?

42. If new observations suggested that the mass fraction of helium remains at 24 percent but the mass fraction of deuterium (^{2}H) is really 10^{-6}, how would our estimates of the density of normal matter in the universe be affected?

43. If 300 million neutrinos fill each cubic meter of space, and if neutrinos account for only 5 percent of the mass density (including dark energy) of the universe, estimate the mass of a neutrino.

44. What is the approximate mass of
    a. an average group of galaxies?
    b. an average cluster?
    c. an average supercluster?

45. The lifetime of a black hole varies in direct proportion to the cube of the black hole's mass. How much longer does it take a supermassive black hole of 3 million $M_\odot$ to decay compared to a stellar black hole of 3 $M_\odot$?

46. Figure 17.3 shows three slices of the universe at three different scales. The scale represented in part (a) has been marked in part (b). Resketch the pie shapes from part (b), and add a line to represent the volume sampled in part (c). How many billion light-years is it to the outer edge of the circle in part (c)?

47. Figure 17.15 shows real data in the left panel and simulated data in the right panel. These two panels are not in exact agreement. Do these differences indicate a significant problem in the simulation's ability to represent reality? Why or why not?

48. Figure 17.9 is a piece of a famous picture called the Hubble UltraDeep Field IR.
    a. Estimate the number of galaxies in the large panel of Figure 17.9.
    b. This image in the right-hand panel of Figure 17.9 has an area of roughly 8 square arcminutes (it has been cropped from its original size). The entire sky has an area of about 148,510,660 square arcminutes. From this information, estimate the number of galaxies in the entire sky.
    c. Explain how you used the cosmological principle in the above calculation.

49. The Bullet Cluster image in Figure 17.14 shows the collision of two galaxy clusters. Estimate the number of galaxies you can see in each cluster.

50. The timescales shown on Figure 17.16 are truly mind-boggling.
    a. Is the vertical (time) axis linear or logarithmic?
    b. Expressed as a power of 10, what is our approximate current time?
    c. In what era does the current time lie on the figure?
    d. Expressed as a power of 10, what is the approximate time of the end of the Primordial Era? (Hint: Think about the CMB.)
    e. Using what you know about logarithms, determine which of the first four eras is the longest.

SMARTWORK

Norton's online homework system includes algorithmically generated versions of these questions, plus additional conceptual exercises. If your instructor assigns questions in SmartWork, log in at smartwork.wwnorton.com.

Exploration | The Story of a Proton

Now that you have surveyed the current astronomical understanding of the universe, you are prepared to put the pieces together to make a story of how you came to be sitting in your chair, holding this book, and reading these pages. It is valuable to take a moment to work your way backward through the book, from the Big Bang through all the intervening steps that had to occur, to the beginning of the book, which started with looking at the sky.

In the Big Bang, how is a proton formed?

How might that proton become part of one of the first stars?

Suppose that proton later becomes part of a carbon atom in a 4-$M_\odot$ star. Through what type of nebula does it pass before returning to the interstellar medium?

Suppose that carbon atom then becomes part of the molecular-cloud core forming the Sun and the Solar System. What two physical processes dominate the core's collapse as the Solar System forms and that carbon atom becomes part of a planet?

5. Beginning with the Big Bang, create a timeline that traces the full history of a proton that becomes a part of the nucleus of a carbon atom on Earth.

This is in essence the astronomical story of how Earth formed. We will address what comes next—how you have come to be on Earth at this time, studying the sky—in Chapter 18.

SMARTWORK • smartwork.wwnorton.com

18 Life in the Universe

We have followed the origin of structure in the universe from the earliest moments after the Big Bang to the formation of galaxies and other large-scale structure visible today. We have seen how stars, including our Sun, formed from clouds of gas and dust within these galaxies and how planets, including our Earth, formed around those stars. We looked at the geological processes that shaped early Earth into the planet we know. In short, we have traced the origin of structures in the universe from the instant the universe came into existence to today. But no discussion of how structure evolved in the universe would be complete without some consideration of the origin and evolution of *life*. In the illustration on the opposite page, a student who had traveled to Australia snapped a picture of stromatolites, one of the earliest forms of life to appear on Earth. Studying life on Earth gives context to our search for life elsewhere in the universe.

LEARNING GOALS

Throughout this book, we have developed an understanding of the universe and all it contains. Still, one very important and very personal component remains: you, the reader of this book. In this chapter, we will investigate the scientific understanding of the origin and evolution of life on Earth and explore its possible existence elsewhere. The illustration on the opposite page shows colonies of microbes known as stromatolites. By the end of the chapter, you will understand the significance of these and other microbial colonies to the story of life on Earth, and you will be able to identify their location on the tree of life. After reading this chapter, you should also be able to:

LG 1 Explain our current understanding of how life began on Earth and how life evolved to reach today's complexity.

LG 2 Describe the limits that the chemistry of known life places on possible habitats in our Solar System.

LG 3 Define a habitable zone.

LG 4 Explain the significance of the discoveries of extrasolar planets to searches for extraterrestrial life.

LG 5 Explain why all life on Earth will eventually end.

STROMATOLITES
Shark Bay, Australia
EARTH IS FORMED
5 BILLION YEARS AGO
FIRST STROMATOLITES
3.5 BILLION YEARS AGO
CAMBRIAN EXPLOSION
540 MILLION YEARS AGO
EXTINCTION OF DINOSAURS
65 MILLION YEARS AGO
FIRST HUMAN CIVILIZATIONS
10,000 YEARS AGO
TODAY

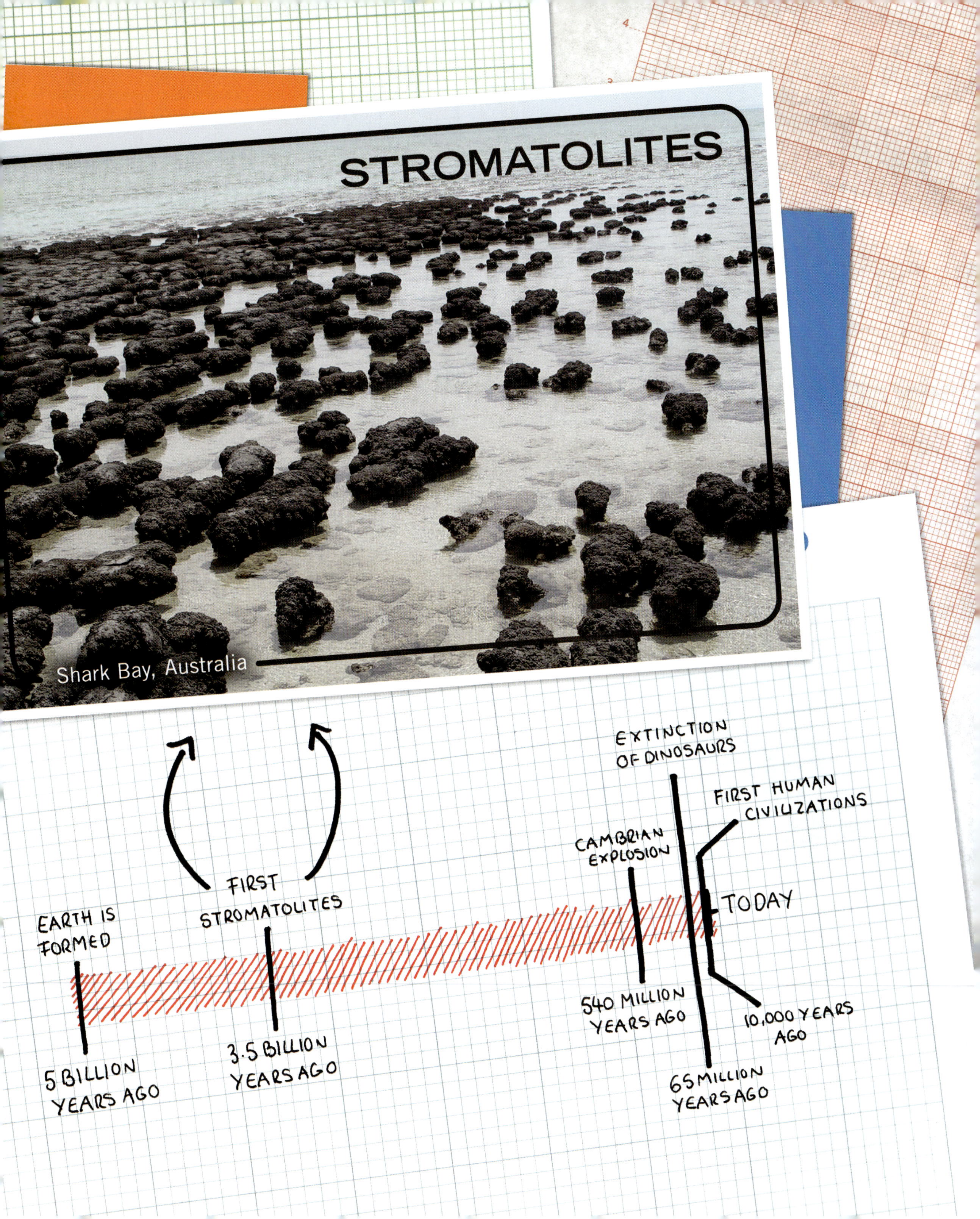

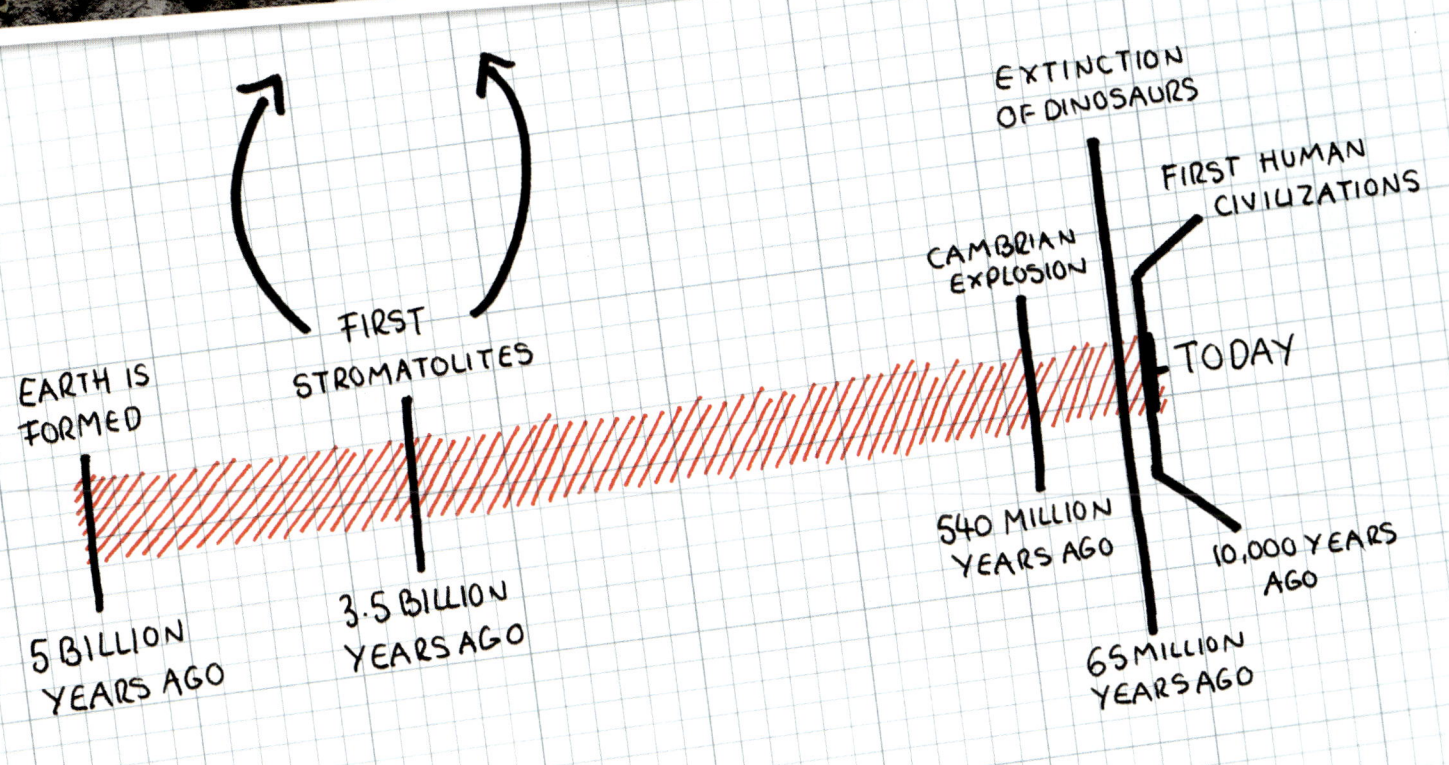

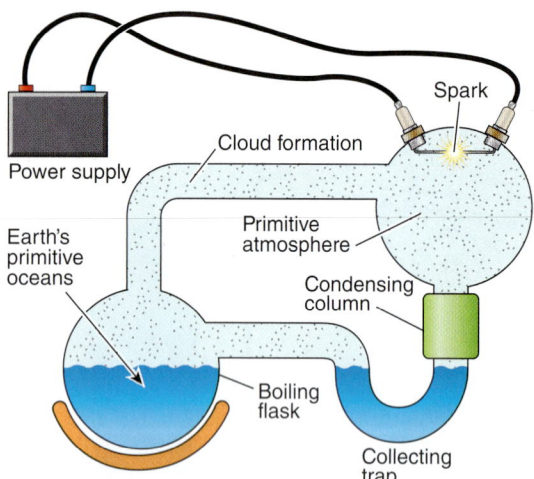

Figure 18.1 The Urey-Miller experiment was designed to simulate conditions in an early-Earth atmosphere.

Figure 18.2 (a) Life on Earth may have arisen near oceanic geothermal vents like this one. Similar environments might exist elsewhere in the Solar System. (b) Living organisms around such vents, such as the giant tube worms shown here, rely on geothermal rather than solar energy for their survival.

18.1 Life on Earth Began Early and Evolved Over Time

How do we define *life*? Many scientists suggest that there is no single definition of life that would encompass all the life we may find in the universe. A complete definition would have to take into account life-forms that scientists know nothing about. At present, we have one example—terrestrial life—so it makes sense to begin a discussion of life by reviewing what we know about the origin and evolution of life on Earth.

Origins of Life on Earth

From studying terrestrial life, we conclude that **life** is a complex biochemical process that draws energy from the environment to survive and reproduce. With the assistance of special molecules such as ribonucleic acid (RNA) and deoxyribonucleic acid (DNA), organisms are able to evolve. All terrestrial life involves carbon-based chemistry and uses liquid water as its biochemical "solvent."

With at least a basic idea of what we mean by *life*, we turn to the question of its origins. We have seen that Earth's secondary atmosphere was formed in part by carbon dioxide and water vapor that poured forth as a product of volcanism; a heavy bombardment of comets probably added large quantities of water, methane, and ammonia to the mix. Although these are all simple molecules, early Earth had abundant sources of energy (such as lightning and ultraviolet solar radiation) that could tear these molecules apart into fragments that could reassemble into molecules of greater mass and complexity. As rain carried the heavier organic molecules out of the atmosphere, they ended up in Earth's oceans, forming a "primordial soup."

In 1952, American chemists Harold Urey (1893–1981) and Stanley Miller (1930–2007) attempted to create conditions similar to what they thought existed on early Earth. They placed water in a laboratory apparatus to represent the ocean and then added methane, ammonia, and hydrogen as a primitive atmosphere; electric sparks simulated lightning as a source of energy (**Figure 18.1**). Within a week, the Urey-Miller experiment yielded 11 of the 20 basic amino acids that link together to form proteins, the structural molecules of life. Other organic molecules that are components of nucleic acids, the precursors of RNA and DNA, also appeared in the mix. More recent experiments that included carbon dioxide, nitrogen, and hydrogen sulfide as components of the atmosphere have produced results similar to those of Urey and Miller. From laboratory experiments such as these, scientists have developed various models of how life might have begun in environments like the ocean that were rich in the organic molecules that may be necessary for life. However, the details of where and how these molecules evolved into the molecules of life are not yet fully understood.

Some biologists think life began in the ocean depths, where volcanic vents provided the hydrothermal energy needed to create the highly organized molecules responsible for biochemistry (**Figure 18.2**). Others think that life originated in tide pools, where lightning and ultraviolet radiation supplied the energy (**Figure 18.3**). In either case, short strands of molecules that could replicate (copy) themselves may have formed first, later evolving into RNA and finally into DNA, the huge molecule that serves as the biological "blueprint" for reproducing organisms.

A few scientists think that perhaps life on Earth was "seeded" from space in the form of microbes brought by meteoroids or comets. Although this hypothesis might tell us how life came to Earth, it does not explain its origin elsewhere in the Solar System or beyond. There is no scientific evidence at this time to support the seeding hypothesis.

The First Life

If life did indeed get its start in Earth's oceans, when did it happen? In Chapter 5, we described how young Earth suffered severe bombardment by Solar System debris for several hundred million years after its formation roughly 4.6 billion years ago. These were hardly the conditions under which life could form and gain a foothold. However, once the bombardment abated and Earth's oceans appeared, there were opportunities for living organisms to evolve. The earliest *indirect* evidence for terrestrial life is carbonized material found in Greenland in rocks dating back to 3.85 billion years ago. Stronger and more direct evidence for early life appears in the form of stromatolites, which are masses of simple microbes such as cyanobacteria (single-celled organisms commonly known as blue-green algae). Fossilized stromatolites that date back to about 3.5 billion years ago have been found in western Australia and southern Africa, and living examples still exist today (see the chapter-opening illustration; as we proceed through the chapter, illustrate a similar timeline of the history of life on Earth, to help you keep the events in order). We may never know the precise date when life first appeared on Earth, but current evidence suggests that the earliest life-forms appeared within a billion years after the formation of the Solar System, and shortly after the end of young Earth's catastrophic bombardment by leftover planetesimals.

The earliest organisms were extremophiles, life-forms that not only survive but thrive under extreme environmental conditions. **Extremophiles** include organisms adapted to live in subfreezing environments or in water temperatures as high as 120°C, which occur in the vicinity of deep-ocean hydrothermal vents. Other extremophiles are found under the severe conditions of extraordinary salinity, pressure, dryness, acidity, or alkalinity. Among these early life-forms was an ancestral form of cyanobacteria. Cyanobacteria photosynthesize carbon dioxide and release oxygen as a waste product. Oxygen, however, is a highly reactive gas, and at the time, the newly released oxygen was quickly removed from Earth's atmosphere by the oxidation of surface minerals. Once the exposed minerals could no longer absorb more oxygen, atmospheric levels of oxygen began to rise. Oxygenation of Earth's atmosphere and oceans began about 2 billion years ago, and the current level was reached only about 250 million years ago (see Chapter 7). Without cyanobacteria and other photosynthesizing organisms, Earth's atmosphere would be as oxygen-free as the atmospheres of Venus and Mars.

Biologists comparing DNA sequences find that terrestrial life is divided into two types: prokaryotes and eukaryotes. Prokaryotes, which include Bacteria and Archaea, are simple organisms that consist of free-floating DNA inside a cell wall; as shown in **Figure 18.4a**, they lack both cell structure and a nucleus. Eukaryotes, like the cells in animals, plants, and fungi, have a more complex form of DNA contained within the cell's membrane-enclosed nucleus, illustrated in Figure 18.4b. The first eukaryote fossils date from about 2 billion years ago, coincident with the rise of free oxygen in the oceans and atmosphere, although the first *multicellular* eukaryotes did not appear until a billion years later.

Figure 18.3 Life may have begun in tide pools.

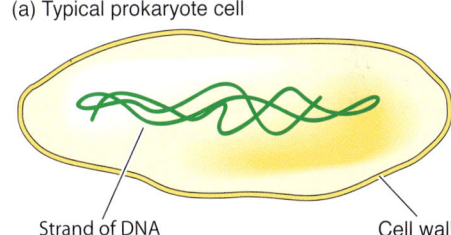

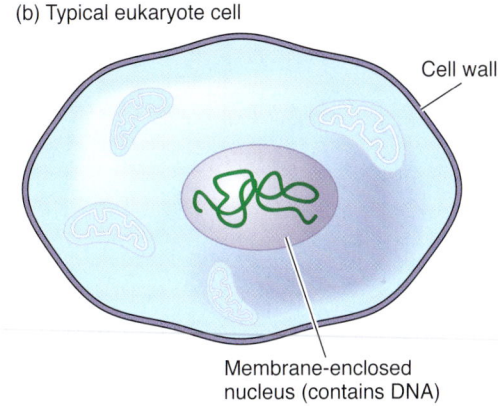

Figure 18.4 (a) A simple prokaryotic cell contains little more than the cell's genetic material. (b) A eukaryotic cell contains several membrane-enclosed structures, including a nucleus, which houses the cell's genetic material.

Life Becomes More Complex

All life on Earth, whether prokaryotic or eukaryotic, shares a similar genetic code that originated from a common ancestor. DNA sequencing enables biologists to trace backward to the time when different types of life first appeared on Earth and to identify the species from which these life-forms evolved. Scientists have used DNA sequencing to establish what is known as the "phylogenetic tree of life", shown in **Figure 18.5**. This complex tree describes the evolutionary interconnectivity of all species of Bacteria, Archaea, and Eukarya, and has revealed some interesting relationships. For example, Archaea were initially thought to be the same as Bacteria, but genetic studies show they diverged long ago, and they have genes and metabolic pathways that are more similar to those of Eukarya than to those of Bacteria. On the macroscopic scale, the phylogenetic tree places animals closest to fungi, which branched off the evolutionary tree after slime molds and plants. Along the "animals" branch, the earliest primates branched off from other mammals about 70 million years ago, and the great apes (gorillas, chimpanzees, bonobos, and orangutans) split off from the lesser apes about 20 million years ago. DNA tests show that humans and chimpanzees share about 98 percent of their DNA; the two groups are believed to have evolved from a common ancestor about 6 million years ago. For comparison, all humans share 99.9 percent of their DNA.

Living creatures in Earth's oceans remained much the same—a mixture of single-celled and relatively primitive multicellular organisms—for more than 3 billion years after the first appearance of terrestrial life. Then, between 540 million and 500 million years ago, the number and diversity of biological species increased spectacularly. Biologists call this event the Cambrian explosion. The trigger of this sudden surge in biodiversity remains unknown, but possibilities include rising oxygen levels, an increase in genetic complexity, major climate change, or a combination of these. The "Snowball Earth" hypothesis suggests that before the Cambrian explosion, Earth was in a period of extreme cold between about 750 million and 550 million years ago and was covered almost entirely by ice. During this period, many animals may have died out, thus making it easier

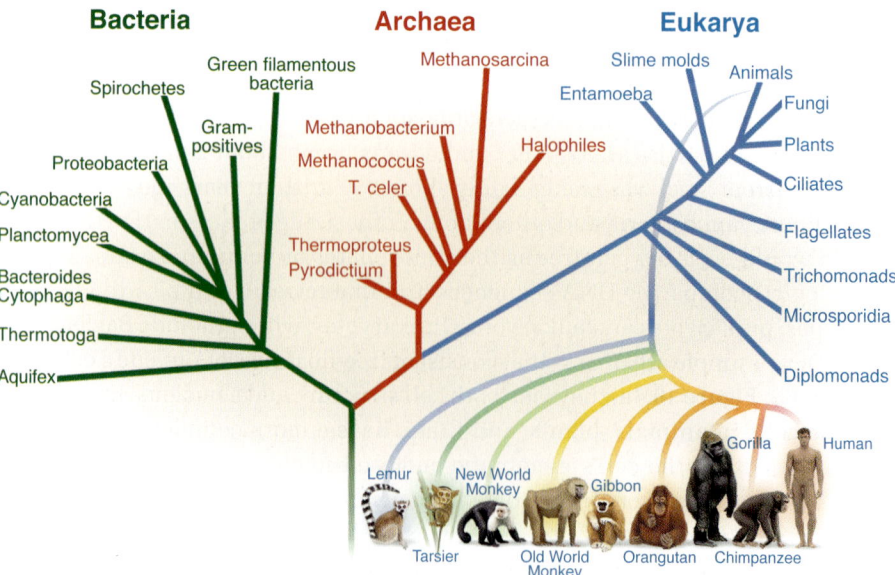

Figure 18.5 This simplified version of the phylogenetic tree has been constructed from analysis of the DNA strands of different life-forms. Humans are included in the "Animals" twig on the Eukarya branch. The primate branch, which includes humans, is shown in the inset.

for new species to adapt and thrive. Another possibility is that a marked increase in atmospheric oxygen (see Figure 7.5) would have been accompanied by a corresponding increase in stratospheric ozone that, as we learned in Chapter 7, shields us from deadly solar ultraviolet radiation. With a protective ozone layer in place, life was free to leave the oceans and move to land (**Figure 18.6**).

The first plants appeared on land about 475 million years ago. Large forests and insects go back 360 million years. The age of dinosaurs began 230 million years ago and ended abruptly 65 million years ago, when a small asteroid or comet collided with Earth. The collision threw so much dust into the atmosphere that the sunlight was dimmed for months, causing the extinction of more than 70 percent of all existing plant and animal species. Mammals were the big winners in the aftermath. Our earliest human ancestors appeared a few million years ago, and the first civilizations occurred a mere 10,000 years ago. Our industrial society is barely more than two centuries old.

Figure 18.6 *Tiktaalik roseae*, a fish with both ribs and limb-like fins, was an animal in a midevolutionary step of leaving the water for dry land.

Humans are here today because of a series of events that have occurred throughout the history of the universe. Some of these events are common in the universe, such as the creation of heavy elements by earlier generations of stars and the formation of planets, including Earth. Other events in Earth's history may have been less likely to happen elsewhere, such as the formation of a planet with life-supporting conditions like Earth or the development of self-replicating molecules that led to Earth's earliest life. A few events stand out as random, such as the impact of a piece of space debris 65 million years ago that led to the extinction of many species. This event made possible the evolution of advanced forms of mammalian life and, ultimately, human beings.

Evolution: A Means of Change

Imagine that just once during the first few hundred million years after the formation of Earth, a single molecule formed somewhere in Earth's oceans. That molecule had a very special property: Chemical reactions between that molecule and other molecules in the surrounding water caused the molecule to reproduce itself. The molecule became "self-replicating." Chemical reactions would produce copies of each of these two molecules, making four. Four became eight, eight became 16, 16 became 32, and so on. By the time the original molecule had copied itself just 100 times, more than a *million trillion trillion* (10^{30}) copies would exist. That is 100 million times more of these molecules than there are stars in the observable universe. As it happens, such unconstrained replication is highly unlikely because of the limited availability of raw materials needed for reproduction and the competition from similar molecules. However, there were ample resources present to allow replication on a smaller scale—enough to create life as we know it.

The molecules of DNA (**Figure 18.7**) that make up the chromosomes in the nuclei of the cells throughout your body are direct descendants of those early self-replicating molecules that flourished in the oceans of a young Earth. Although the DNA in your body is far more elaborate than those early molecules in Earth's oceans, the fundamental rules of biochemistry remain the same. *Mutation* and *heredity* lead to *natural selection* and to organisms that *evolve* over time.

Chemical reactions are not always error-free. Sometimes when a molecule replicates, the new molecule is not quite a duplicate of the old one. The likelihood that a copying error will occur while a molecule is replicating increases significantly with the number of copies being made. For DNA, which contains

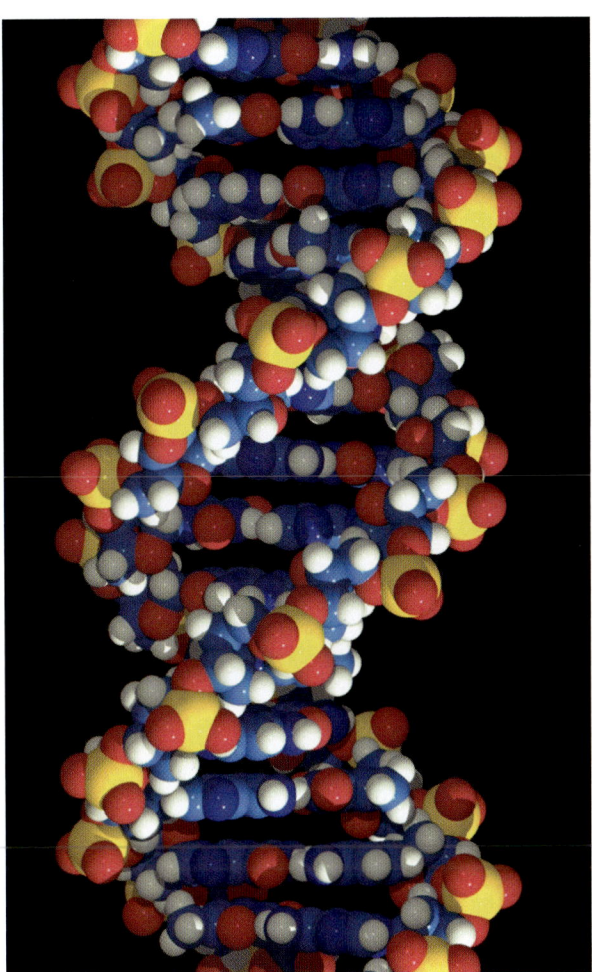

Figure 18.7 DNA is the molecular basis for heredity. This model of a DNA molecule shows the two strands that are wound together into a shape known as a double helix.

the genetic code for entire organisms, the change in the code is called a **mutation**. Sometimes such an error leads to an organism that is no longer suited to its environment. Organisms with these changes will not reproduce and flourish. Other times a mutation is helpful, leading to a molecule or an organism that is *better* suited to the environment than the original. Organisms with these mutations will reproduce successfully. Even if changes in the copying process crop up only once every 100,000 times that an organism reproduces its DNA, and even if only one out of 100,000 of these errors turns out to be beneficial, after only 100 generations there will still be a hundred million trillion (10^{20}) mutations that, by chance, might improve on the original. Copies of each improved molecule will inherit the change. Mutations affect the genetic **heredity** of the organism—the ability of one generation to pass on its genetic code to future generations.

Cells have a membrane that encloses all the molecular machinery and protects the structures inside from the environment. This membrane may have formed from fatty molecules that enclosed bubbles of water, along with self-replicating molecules to make the first organic cells. These early cells were likely fragile, dissolving and reforming in response to environmental factors. Over time, cells with sturdier membranes reproduced more successfully than those with weaker membranes. Biologists today are learning how these membranes contribute to the survival of cells in varied environments. For example, **thermophiles** are cells that thrive at high temperatures. These cells have a larger fraction of saturated fatty acids in their membranes, which makes the membrane stable at higher temperatures. Two familiar fats, butter and olive oil, are different for the same reason. Butter is high in saturated fat and is a solid at room temperature, but olive oil is low in saturated fat and is a liquid at room temperature.

As the organisms of the early Earth continued to interact with their surroundings and make copies of themselves, mutations caused them to diversify into many different species. In some cases, the resources they needed to reproduce became scarce. In the face of this scarcity of resources, varieties that were more successful reproducers became more numerous. Competition for resources, predation of one variety on another, and cooperation between organisms became important to the survival of different varieties. Some varieties were more successful and reproduced to become more numerous, while less successful varieties became less and less common. This process, in which better-adapted organisms reproduce and thrive while less-well-adapted ones become extinct is called **natural selection**.

Four billion years is a long time—enough time for the combined effects of heredity and natural selection to shape the descendants of that early self-copying molecule into a huge variety of complex, competitive, successful structures (**Working it Out 18.1**). Geological processes on Earth have preserved a fossil record of the history of these structures (**Figure 18.8**). Among these descendants are "structures" capable of thinking about their own existence and unraveling the mysteries of the stars.

18.2 Life beyond Earth Is Possible

The story of the formation and evolution of life cannot be separated from the narrative of astronomy. We know that we live in a universe full of stars, and that systems of planets orbit many—probably most—of those stars. The evolution of life on Earth is but one of many examples that we have encountered of

Figure 18.8 Fossils, such as this Parasaurolophus ("near crested lizard"), record the history of the evolution of life on Earth. This plant-eating dinosaur lived in North America about 75 million years ago.

Working It Out 18.1 | Exponential Growth

Exponential growth means that a population grows by a percentage (say, 7 percent) each time period, rather than by a fixed amount (say, increasing by 7) each time period. The difference is important because in exponential growth, the percentage represents a larger number of individuals in each time step: The population adds more individuals with time. Biological systems, because they are self-replicating, grow by exponential growth until a resource limit is reached.

Assume a single hypothetical self-replicating molecule that makes one copy of itself each minute. The doubling time, then, is 1 minute. How many molecules will exist after an hour? We will use P_0 to represent the population at time $t = 0$. In this case, $P_0 = 1$ (a single molecule). P_F is the final population after an hour has elapsed. Population increases by 100 percent—a factor of 2—each minute for 60 minutes. So to find the final population, we must double the initial population (multiply it by 2) 60 times. That would be tedious, but there is a shortcut. Doubling the population 60 times is the same as multiplying by 2 to the 60th power:

$$P_F = 2^{60} \times P_0$$
$$P_F = 2^{60} \times 1$$
$$P_F = 2^{60}$$

Plugging 2^{60} into the calculator gives 1.2×10^{18}. Therefore, there are a billion billion of these molecules after 1 hour.

In general, we use n to represent the number of doubling times, so the more general expression is

$$P_F = 2^n \times P_0$$

Now, suppose that a change in the copying process occurs only once every 50,000 times that a molecule reproduces itself, and one out of 200,000 of these changes is beneficial. After 100 generations, how many new molecules with these beneficial changes will form? First, we must find the final number of molecules. Again, $P_0 = 1$, but now n is 100:

$$P_F = 2^{100} \times 1$$
$$P_F = 1.3 \times 10^{30}$$

To find the number of these molecules that have changes, P_C, we must multiply by the fraction of times a change occurs. Because a change occurs one time in 50,000, we must multiply by 1/50,000:

$$P_C = 1.3 \times 10^{30} \times \frac{1}{50,000}$$
$$P_C = 2.6 \times 10^{25}$$

To find the number of molecules with beneficial changes, P_B, we must account for the fact that only 1/200,000 of these changes is beneficial. So we multiply by 1/200,000:

$$P_B = 2.6 \times 10^{25} \times \frac{1}{200,000}$$
$$P_B = 1.3 \times 10^{20}$$

There will be roughly 10^{20} (100 million trillion) molecules in this generation that have a beneficial mutation. Because this number does not count earlier beneficial changes that themselves replicated, the total number of molecules with beneficial changes will be much larger!

---

the emergence of structure in an evolving universe. This point leads naturally to one of the more profound questions that we can ask about the universe: Has life arisen elsewhere? To explore this question, we need to take a closer look at the chemistry of life on Earth.

The Chemistry of Life

When we speak about the chemistry of life, what we really mean is life itself. All known organisms are composed of a more or less common suite of complex chemicals, and very complex ones at that. Consider the human body. Approximately two-thirds of the atoms in our bodies are hydrogen (H), about one-fourth are oxygen (O), a tenth are carbon (C), and a few hundredths are nitrogen (N). The remaining elements, and there are several dozen of them, make up only 0.2 percent of the total inventory of the atoms in our bodies. All known living creatures are an assemblage of molecules composed almost entirely of these four elements, sometimes called CHON, along with small amounts of phosphorus and sulfur. Some of these molecules are enormous. Consider DNA, which is responsible for our genetic code. DNA is made entirely from only *five* elements: CHON and phosphorus. But the DNA in each cell of our bodies is composed of combinations of *tens of billions* of atoms of these same five elements. Then there are proteins, the huge molecules responsible for the structure and function of living organisms. Proteins are long chains of smaller molecules called amino acids. Terrestrial life uses 20 specific amino acids, which also contain no more than five elements—in this case CHON plus sulfur instead of phosphorus.

The chemistry of life requires more than a mere half-dozen elements. Many others are present in smaller amounts but are essential to the complicated chemical processes of life. These include sodium, chlorine, potassium, calcium, magnesium, iron, manganese, and iodine. Finally, there are the so-called trace elements, such as copper, zinc, selenium, and cobalt. Trace elements also play a crucial role in biochemistry but are needed in only tiny amounts.

You have learned that the infant universe was composed basically of hydrogen and helium and very little else. After 9 billion years of stellar nucleosynthesis, all the heavier chemical elements essential to life were present and available in the molecular cloud that gave birth to our Solar System. As you saw in Chapters 12 and 13, those heavy elements—up to and including iron—were created during earlier generations of low-mass and high-mass stars and were then dispersed into space. At times, this dispersal was passive. For example, low-mass stars, such as dying red giants, lose their gravitational grip on their overly extended atmospheres. Along with hydrogen and helium, the newly created heavy elements are blown off into space, eventually finding their way into molecular clouds. Other dispersals were more violent. Most of the trace elements essential to biology are more massive than iron, so they are not produced in the interiors of main-sequence stars. They are instead created within a matter of minutes during the violent supernova explosions that mark the death of high-mass stars. These elements, too, are part of the chemical mix found in molecular clouds.

Terrestrial life is based on carbon. Carbon is the lightest among the tetravalent atoms, which can bond with as many as four other atoms or molecules (**Figure 18.9**). (*Tetra* means "four," and *valence* refers to an atom's ability to attach to other atoms or molecules.) If the attached molecules also contain carbon, the result can be an enormous variety of long-chain molecules. This great versatility enables carbon to form the complex molecules that provide the basis for terrestrial life's chemistry.

There could be carbon-based forms of extraterrestrial life that have chemistries quite different from our own. For example, there are countless varieties of amino acids beyond the 20 used in terrestrial life. Furthermore, molecules other than RNA and DNA may be capable of self-replication.

Science-fiction writers often speculate about silicon-based life-forms because silicon, like carbon, is tetravalent—each silicon atom can bond with as many as four other atoms, so that a large number of combinations are possible. As a potential life-enabling atom, silicon has both advantages and disadvantages compared to carbon. An important advantage is that silicon-based molecules remain stable at much higher temperatures than carbon-based molecules, perhaps enabling possible silicon-based life to thrive in high-temperature environments, such as on planets that orbit close to their parent star. But silicon has a serious disadvantage that makes silicon-based life less likely; it is a larger and more massive atom than carbon, and it cannot form long chains of atoms as complex as those based

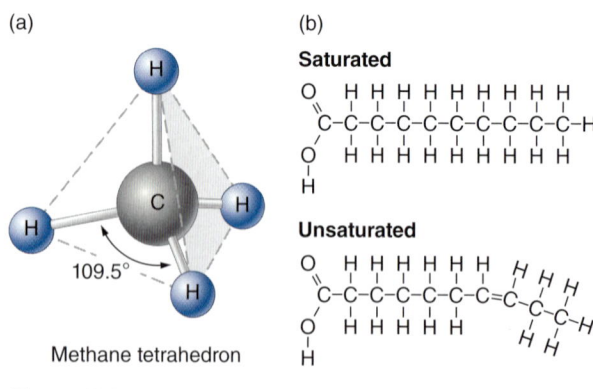

Figure 18.9 (a) Carbon is tetravalent, meaning it can bond to as many as four other atoms, as in methane. (b) This property allows carbon to form an enormous variety of long-chain molecules, as in saturated and unsaturated fatty acids.

on carbon. Any silicon-based life likely would be simpler than life-forms here on Earth, but it might exist in high-temperature niches somewhere in the universe.

Although carbon's unique properties make it readily adaptable to the chemistry of life on Earth, we don't know what other chemistries life might adopt. Life on Earth is highly adaptable and tenacious; when it comes to the form that extraterrestrial life might take, nothing should surprise us.

Life within Our Solar System

The logical place to start looking for extraterrestrial life is right here in our own Solar System. Early conjectures about life in our Solar System seem naïve, considering what we now know about the Solar System. Two centuries ago, the eminent astronomer Sir William Herschel, discoverer of Uranus, proclaimed, "We need not hesitate to admit that the Sun is richly stored with inhabitants." In 1877, Italian astronomer Giovanni Schiaparelli (1835–1910) observed what appeared to be linear features on Mars and dubbed them *canali*, meaning "channels" in Italian. In one of astronomy's great ironies, the famous American observer of Mars, Percival Lowell (1855–1916), misinterpreted Schiaparelli's *canali* as "canals," suggesting that they were constructed by intelligent beings.

During the mid-20th century, ground-based telescopes discovered that Mars possesses an atmosphere and water, both considered essential for any terrestrial-type life to get its start and evolve. During the 1960s, the United States and the Soviet Union sent reconnaissance spacecraft to the Moon, Venus, and Mars, but the instrumentation carried aboard these spacecraft was more suited to learning about the physical and geological properties of these bodies than to searching for life. Serious efforts to look for signs of life—past or present—require more advanced spacecraft with specialized bio-instrumentation.

In the meantime, astronomers and biologists alike were discussing where to look and what to look for as part of a new science called **astrobiology**: the study of the origin, evolution, distribution, and future of life in the universe. Because Mercury and the Moon lacked atmospheres, astrobiologists determined they were not conducive to life. The giant planets and their moons were thought to be too remote and too cold to sustain life. Venus was far too hot, but Mars seemed just right. In the mid-1970s, two American *Viking* spacecraft were sent to Mars with detachable landers containing a suite of instruments designed to find evidence of a terrestrial type of life. When the *Viking* landers failed to find evidence of life on Mars, hopes faded for finding life on any other body orbiting our Sun.

Since that time, however, there is renewed optimism. A better understanding of the history of Mars indicates that at one time the planet was wetter and warmer, leading many scientists to believe that fossil life or even living microbes might yet be buried under the planet's surface. The first decade of the 21st century saw a return to Mars and a continuation in the search for evidence of current or preexisting life. In 2008, NASA's *Phoenix* spacecraft landed at a far northern latitude, inside the planet's arctic circle, where specialized instruments dug into and analyzed the martian water-ice permafrost. *Phoenix* found that the martian arctic soil has a chemistry similar to the Antarctic dry valleys on Earth, where life exists deep below the surface at the ice-soil boundary. Minerals that form in water, such as calcium carbonate, reveal that ancient oceans were once present on the planet. However, *Phoenix* did not find direct evidence of life.

The most recent Mars mission, the Mars Science Laboratory rover *Curiosity* landed in Gale Crater on Mars in 2012. This large rover—about the size of a small car—studies the rocks and soil of Mars to provide data for a better understanding of the history of the planet's climate and geology. Shortly after landing, *Curiosity* found evidence that a stream of liquid water had once flowed in the crater. The rover observed rounded, gravelly pebbles stuck together, which have been interpreted as coming from a stream roughly 30 centimeters deep and flowing at 1 meter per second. The presence of liquid, flowing water on the surface of Mars could be seen as a positive indication that life might have arisen there. Conversely, *Curiosity* has failed to find evidence of methane on Mars. Previous data had indicated that there might be methane, which is a possible indicator of life although it is also made by geologic processes. This result implies that methane-producing life is not present on Mars now.

The *Mars Atmosphere and Volatile Evolution Mission* (*MAVEN*) arrived at Mars in September 2014. This mission's purpose is to study the upper atmosphere to learn more about the escape of carbon dioxide, hydrogen, and nitrogen from the planet's atmosphere. The loss of these gases affects surface pressure and the existence of liquid water.

NASA's instrumented robots reached the outer Solar System in the 1980s, and what they found surprised many astrobiologists. Although the outer planets themselves did not appear to be habitats for life, some of their moons became objects of special interest. Jupiter's moon Europa is covered with a layer of water ice that appears to overlie a great ocean of liquid water. Impacts by comet nuclei may have added a mix of organic material, another essential ingredient for life. Once thought to be a frozen, inhospitable world, Europa is now a candidate for biological exploration. Recently, scientists using the Hubble Space Telescope observed water geysers taller than Mount Everest erupting from the icy surface. Ejected material from these geysers may make it possible to search for life on Europa without drilling down through the ice.

Saturn's moon Titan appeared to be rich in organic chemicals, many of which are thought to be precursor molecules of a type that existed on Earth before life appeared here. The *Cassini* mission currently orbiting Saturn found additional evidence for a variety of molecules that might be necessary for life in Titan's atmosphere, as well as a liquid lake of methane on the surface and probably a liquid-water ocean under the surface. In addition, it identified another potential site for life: Saturn's tiny moon Enceladus. The spacecraft detected water-ice crystals spouting from ice volcanoes near the south pole of Enceladus. Liquid water must lie beneath its icy surface, and Enceladus therefore joins Europa as a possible habitat of extremophile life, perhaps life similar to that found near geothermal vents deep within Earth's oceans.

Efforts to find evidence of life on other bodies within our own Solar System have so far been unsuccessful, but the quest continues. The discovery of life on even one Solar System body beyond Earth would be exciting: If life arose independently *twice* in the same planetary system, then life might be common and exist throughout the universe.

Habitable Zones

Searching for life within our own Milky Way Galaxy is a daunting task. Nevertheless, astronomers are narrowing the possibilities by searching for planets with environments conducive to the formation and evolution of life as we understand it, while eliminating planets that are clearly unsuitable.

One criterion astrobiologists look for is planetary systems that are stable.

Planets in stable systems remain in nearly circular orbits that preserve relatively uniform climatological and oceanic environments. Planets in very elliptical orbits can experience wild temperature swings that could be detrimental to the survival of life. A stable temperature that maintains the existence of water in a liquid state might be important. We know that liquid water was essential for the formation and evolution of life on Earth. Of course, we don't know if liquid water is an absolute requirement for life elsewhere, but it's a good starting point.

The region around a star that provides a range of temperatures in which liquid water can exist is called the star's **habitable zone**. On planets that are too close to their parent star, water would exist only as a vapor—if at all. On planets that are too far from their star, water would be permanently frozen as ice. Even if a planet is located in the habitable zone, we cannot yet determine if it is actually inhabited. We only know that liquid water could exist on the surface. Still another consideration is planet size. Large planets such as Jupiter retain most of their light gases—hydrogen and helium—during formation and so become gas giants without a surface. Planets that are very small may have insufficient surface gravity to retain their atmospheric gases and so end up like our Moon. Calculating whether any particular planet is habitable in this sense is quite complicated. Many astrobiologists are working on models of the climates of known extrasolar planets to figure out whether they are in their star's habitable zone.

In our own Solar System, Venus, which orbits at 0.7 times Earth's distance from the Sun, has become an inferno because of its runaway greenhouse effect (see Chapter 7). Any liquid water that might once have existed on Venus has long since evaporated and been lost to space. Mars orbits about 1.5 times farther from the Sun than Earth, and nearly all of the water that we see on Mars today is frozen. But the orbit of Mars is more elliptical and variable than Earth's, giving the planet a greater variety of climate, including long-term cycles that might occasionally permit liquid water to exist. Most astrobiologists put the habitable zone of our Solar System at about 0.9–1.4 astronomical units (AU), which includes Earth but just misses Venus and Mars. Yet this range may be too narrow because of liquid water under ice as seen in the moons of the outer solar system. For example, extremophiles could be thriving in liquid water beneath the surfaces of some icy moons of Jupiter and Saturn.

Astronomers must also think about the type of star they are observing in their search for life-supporting planets. **Figure 18.10** shows that stars that are less massive than the Sun and thus cooler will have narrower habitable zones, lessening the chance that life will form on an orbiting planet. Stars that are more massive than the Sun are hotter and will have a larger habitable zone. However, recall that a star's main-sequence lifetime depends on its mass. For example, a star of 2 $M_\odot$ would enjoy relative stability on the main sequence for only about a billion years before the helium flash incinerated everything around it. On Earth, a billion years was long enough for bacterial life to form and cover the planet, but insufficient for anything more advanced to evolve. Of course, we don't know if evolution might happen at a different pace elsewhere; we have only our one terrestrial case as an example. Still,

▶▶ **Nebraska Simulation:** Circumstellar Habitable Zone Simulator

Figure 18.10 The habitable zone changes with the mass and temperature of a star. Habitable zones around hot, high-mass stars are more distant and wider than habitable zones around cooler, lower-mass stars. The Sun and Solar System are shown for comparison.

main-sequence lifetime is a sufficiently strong consideration that most astronomers prefer to focus their efforts on stars with longer lifetimes—specifically, spectral types F, G, K, and M.

Finally, some astronomers think about a "galactic habitable zone," referring to a star's location within the Milky Way Galaxy. Stars that are situated too far from the galactic center may not have enough heavy elements —such as oxygen and silicon (silicates), iron, and nickel—in their protoplanetary disks to form rocky planets like Earth. Conversely, regions too close to the galactic center experience less star formation and therefore fewer opportunities to gather heavy elements into planetary environments. Perhaps more serious is the high-energy radiation environment near the galactic center (X-ray and gamma ray), which is damaging to RNA and DNA. Even so, for many stars the galactic habitable zone may not remain as a permanent home if they tend to have an orbit within the galaxy that changes their distance from the galactic center over time. In short, astronomers try to narrow down the vast numbers of stars as they conduct their search, but they acknowledge that these types of arguments—based on what worked well for planet Earth—might not be applicable when we are looking at other planetary systems.

▶▶ **Nebraska Simulation:** Milky Way Habitability Explorer

▶▶ **Nebraska Simulation:** Exoplanet Radial Velocity Simulator

▶▶ **Nebraska Simulation:** Exoplanet Transit Simulator

Searching for Earth-like Planets

The search for Earth-like planets is already well under way. Currently, there are more than 100 projects from the ground and from space focused on searching for extrasolar planets using the methods discussed in Chapter 5. Each week, it seems, a new system makes the news, either because it's more like Earth than any other yet discovered or because it's unusual in some way. The field is still young and is changing rapidly: Scientists have moved from having data on one system to having data on thousands of systems in fewer than 15 years.

In 2006, the European Space Agency (ESA) launched COROT, the first space telescope dedicated to extrasolar planet detection. COROT monitored nearby stars for transiting extrasolar planets. The ESA spacecraft made several discoveries, including one planet that is only 1.7 times the size of Earth. In 2013, COROT was decommissioned due to an onboard computer failure.

In 2009, NASA launched the Kepler space telescope into a solar orbit that trailed Earth in its own orbit around the Sun. This location enabled uninterrupted monitoring of Kepler's target stars, as Earth never got in the way of Kepler's cameras. Over the next several years, Kepler's photometer simultaneously and continually monitored the brightness of more than 100,000 stars, looking for transiting extrasolar planets. Kepler could detect planets as small as 0.8 times the size of Earth. By mid-2013, Kepler had discovered nearly 4,000 suspected extrasolar planetary systems, roughly 1,000 of which have subsequently been confirmed. Many more candidates will likely be found upon closer examination of the data, and many more will likely be confirmed as follow-up observations by other ground-based or orbiting instruments proceed. Many of these systems include multiple planets. The star Gliese 667C, for example, has seven planets in orbit around it, three of which are super-Earths in the habitable zone. This is the first system found with a fully packed habitable zone. In 2013, an equipment failure on the Kepler telescope caused it to be shut down. Astronomers have developed a new method of guiding the telescope's observations and are working to validate the method in order to resume the hunt for new extrasolar planets.

The data from all these projects gives a sense of how fast the field is moving. As of this writing, there are about 4,000 planets and planetary candidates. If these discoveries had been spread over the 20 years in which astronomers have been finding extrasolar planets, they would have been finding one every other day. Instead, most have been found in just the past few years with Kepler. Of these objects, roughly 1,000 have been confirmed to be planets, around more than 500 stars. Determining how many are habitable is difficult and is a current focus of astrobiologists, who model the climate for each planet as information about its orbit becomes known. From observations so far, astronomers estimate that habitable planets are very common in the Milky Way. Estimates range as high as one in five stars like the Sun have an Earth-sized planet in the habitable zone. There are about 40 billion Sun-like stars in the Milky Way, so this estimate implies that there could be as many as 8 billion Earth-sized planets in the habitable zones of their stars. Astronomers are working on the rest of those planetary candidates, following up to determine whether they are, in fact, planets.

18.3 Scientists Search for Signs of Intelligent Life

Scientists would find any evidence of extraterrestrial life exciting, even if it were the fossil remnants of blue-green algae. But many people are only interested in extraterrestrial life if it is sufficiently advanced that we could communicate with it. Searching for signs of intelligent life in the universe requires exploring a large number of questions that are not astronomical in nature: Given the right conditions, does life always arise? Among life-forms, how common is intelligence? Or language? Or complex technology such as radio transmitters or lasers that can send messages to space? Framing the discussion of the search for intelligent life requires addressing all of these questions and more. But while scientists explore these questions, they have already sent out messages in case someone, somewhere, is listening.

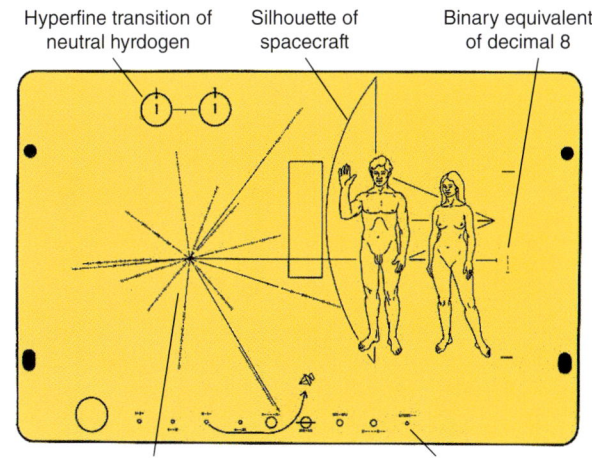

Figure 18.11 This plaque was attached to the *Pioneer 10* and *Pioneer 11* probes, which were launched in the early 1970s and will eventually leave the Solar System to travel through the millennia in interstellar space.

Saying "Hello"

During the 1970s, humans made preliminary efforts to send a message to alien life. The *Pioneer 10* and *Pioneer 11* spacecraft, which will probably spend eternity drifting through interstellar space, each carry the plaque shown in **Figure 18.11**. It describes ourselves and our location to any traveler who might happen to find it. Also, the two *Voyager* spacecraft each carry a different message prepared on an identical type of phonograph record: The message contains greetings from planet Earth in 60 languages, samples of music, animal sounds, and a message from then-President Jimmy Carter. Notably, these messages created concern among some politicians and nonscientists who felt that scientists were dangerously advertising our location in the galaxy, even though radio signals had been broadcast into space for nearly 80 years at the time. The messages were also criticized by some philosophers, who claimed that we were making ridiculous anthropomorphic assumptions about the aliens being sufficiently like us to decode these messages.

Sending messages on spacecraft is not the most efficient way to make contact with aliens. A somewhat more practical effort was made in 1974, when astronomers used the 300-meter-wide dish of the Arecibo radio telescope to beam a message (**Figure 18.12**) toward the globular cluster M13. If someone from M13 answers, we will know in about 50,000 years. In 2008, a radio telescope in Ukraine sent a message to the extrasolar planet Gliese 581c. The message was composed of 501

Figure 18.12 The binary-encoded message we beamed toward the globular cluster M13 in 1974 contains the binary symbols for the numbers 1–10, diagrams of hydrogen and carbon atoms, some interesting molecules, DNA, a human with description, basics of our Solar System, and basics of the Arecibo telescope. A reply may be forthcoming in about 50,000 years.

digitized images and text messages selected by users on a social networking site and will arrive at Gliese 581c in 2029.

The Drake Equation

It took less than a billion years for life to form in Earth's oceans. The earliest life was primitive, and it took another 3.5 billion years for modern humans to develop. Humans have reached that stage at a time when our star is only halfway through its lifetime. As noted in Section 18.2, many stars do not live long enough for life to evolve. If we are to search for signs of intelligent life, we can start by looking at planets surrounding stars of about a solar mass or smaller—stars of type F, G, K, or M.

The first serious effort to search for intelligent extraterrestrial life was made by astronomer Frank Drake in 1960. Drake used what was then astronomy's most powerful radio telescope to listen for signals from two nearby stars. Although his search revealed nothing unusual, it prompted him to develop an equation that bears his name. This equation is different from other equations in this book because the values for each variable are quite uncertain. Still, it is a very useful way to organize our thoughts about whether intelligent life might exist elsewhere, as it includes the things we need to know to make an informed estimate about how many intelligent civilizations there might be. The **Drake equation** estimates the number (N) of intelligent, communicating civilizations that may exist within the Milky Way Galaxy:

$$N = R^* \times f_p \times n_e \times f_l \times f_i \times f_c \times L$$

The seven factors on the right side of the equation are the conditions that Drake thought must be met for a civilization to exist:

1. R^* is the number of stars that form in our galaxy each year that are suitable for the development of intelligent life. Astronomers consider these to be F, G, K, or M spectral-type stars because their lifetimes are sufficiently long. This is roughly seven per year and is the best understood of all these terms.
2. f_p is the fraction of those stars that form planetary systems. The abundance of extrasolar planets indicates that planets form as a natural by-product of star formation and that many—perhaps most—stars have planets. We will assume that f_p is between 0.5 and 1.
3. n_e is the number of planets and moons, per stellar system, with an environment suitable for life. If we look at the Solar System, we might decide this number is about 2. Including Europa and Enceladus increases the number to 4. Earth definitely did develop life, and Mars might have. Of course, the Solar System is just one example, and we don't really know what this term should be, but a number between 1 and 10 seems reasonable.
4. f_l is the fraction of suitable planets and moons on which life actually arises. Remember that just a single self-replicating molecule may be enough to get the ball rolling. Some biochemists now believe that if the right chemical and environmental conditions are present, then life *will* develop, but others disagree. Estimates range from 100 percent (life always develops) to 1 percent (life is rare). Astronomers use a range of 0.01 to 1.
5. f_i is the fraction of those planets harboring life that eventually develop intelligent life. Intelligence is certainly the kind of survival trait that might often be strongly favored by natural selection. However, on Earth it took about

4 billion years—roughly half the expected lifetime of our star—to evolve tool-building intelligence. The correct value for f_i might be close to 0.01 or it might be closer to 1. The truth is, we just don't know.

6. f_c is the fraction of intelligent life-forms that develop technologically advanced civilizations; that is, civilizations that send communications into space. With only one example of a technological civilization to work with, f_c is hard to estimate. We will consider estimates for f_c between 0.1 and 1.

7. L is the number of years that such civilizations exist. This factor is certainly the most difficult of all to estimate because it depends on the long-term stability of advanced civilizations. We have had a technological civilization on Earth for less than 100 years, and during that time we have developed, deployed, and used weapons with the potential to eradicate our civilization and render Earth hostile to life for many years to come. We have also so degraded our planet's ecosystem that many respectable biologists and climatologists wonder whether Earth is in danger of reducing or losing its habitability. Do all technological civilizations destroy themselves within a thousand years? Conversely, if most technological civilizations learn to use their technology for survival rather than self-destruction, might they instead survive for a million years? For L in our calculation, we will use a range between 1,000 years and 1 million years.

As illustrated in **Figure 18.13**, the conclusions we draw using the Drake equation depend a great deal on the assumptions we make. For the most pessimistic of our estimates, the Drake equation sets the number of technological civilizations in our galaxy at about 1. If this is correct, then we are the *only* technological civilization in the Milky Way. (Because we are here, we know that N must be at least 1.) Nevertheless, such a universe would still be full of intelligent life: With a hundred billion galaxies in the observable universe, even these pessimistic assumptions would mean that a hundred billion technological civilizations exist out there somewhere. Yet, we would have to go a *very* long way (10 million parsecs or so) to find our nearest neighbors.

At the other extreme, what if we take the most optimistic view, assuming that intelligent life arises and survives everywhere it gets the chance? The Drake equation then says that there should be roughly 15 million technological civilizations in our galaxy alone. In this case, our nearest neighbors may be "only" 40 or 50 light-years away. If scientists in that civilization are listening to the universe with their own radio telescopes, hoping to answer the question of other life in the universe for themselves, then as you read this page they may be puzzling over an episode of the original *Star Trek* television series that began in 1966.

If we did run across another technologically advanced civilization, what would it be like? Looking back at the Drake equation, we can see it is highly unlikely that we have neighbors nearby, unless civilizations typically live for many thousands or even millions of years. In this case, any civilization that we encountered would almost certainly have been around for much longer than we have.

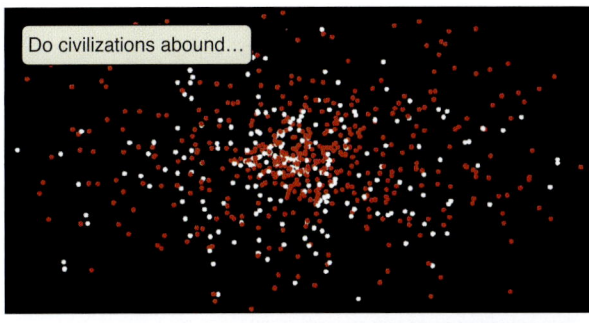
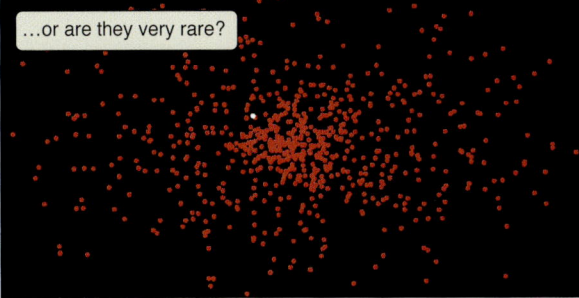

Figure 18.13 Optimistic and pessimistic estimates of the existence of intelligent civilizations in our galaxy based on the Drake equation are very different because uncertainty about the seven factors (see the text) affects the estimated prevalence of intelligent extraterrestrial life. White dots indicate stars with possible civilizations.

Technologically Advanced Civilizations

During a lunch with colleagues, the famous physicist Enrico Fermi (1901–1954), a firm believer in extraterrestrial beings, is reported to have asked, "If the universe is teeming with aliens . . . where is everybody?" Fermi's question—first posed in 1950 and sometimes called the Fermi paradox—remains unanswered. Consider also the following closely related question: If intelligent life-forms are common but interstellar travel is difficult or impossible, why don't we detect their signals? We have failed so far to detect any—but it is not for lack of trying.

Drake's original project of listening for radio signals from intelligent life around two nearby stars has grown over the years into a much more elaborate program that is referred to as the Search for Extraterrestrial Intelligence, or **SETI**. Scientists from around the world have thought carefully about what strategies might be useful for finding life in the universe. Most of these have focused on the idea of using radio telescopes to listen for signals from space that bear an unambiguous signature of an intelligent source. Some have listened intently at certain key frequencies, such as the frequency of the interstellar 21-centimeter line from hydrogen gas. The assumption behind this approach is that if a civilization wanted to be heard, its denizens would tune their broadcasts to a channel that astronomers throughout the galaxy should be listening to. More recent searches have made use of advances in technology to record as broad a wavelength range of radio signals from space as possible. Analysts then use computers to search these databases for patterns that might suggest the signals are intelligent in origin.

Unlike much astronomical research, SETI receives its funding from private rather than government sources, and SETI researchers have found ingenious ways to continue the search for extraterrestrial civilizations with limited resources. One project, known as SETI@home, involves the use of hundreds of thousands of personal computers around the world to analyze the institute's data. SETI programs installed on personal computers worldwide download radio observations from the SETI Institute over the Internet, analyze these data while computer owners are not using their equipment, and then report the results of their searches back to the institute. Perhaps the first sign of intelligent life in the universe might be found by a computer sitting on your desk.

The SETI Institute's Allen Telescope Array (ATA) is named for Microsoft cofounder Paul Allen, who provided much of the initial financing for the project. A joint venture between the SETI Institute and the University of California, the ATA consists of a "farm" of small, inexpensive radio dishes like those used to capture signals from orbiting communication satellites (**Figure 18.14**). One of the key projects of the ATA is to observe planets discovered by the Kepler mission in search of signs of intelligent life. Each dish has a diameter of 6.1 meters. All the telescopes working together have a total signal-receiving area greater than that of a 100-meter radio telescope. Just as your brain can sort out sounds coming from different directions, this array of radio telescopes is able to determine the direction a signal is coming from, allowing it to listen to many stars at the same time. Over several years' time, astronomers using the ATA are expected to survey as many as a million stars, hoping to find a civilization that has sent a signal in our direction. If reality is anything like the more optimistic of the assumptions we used in evaluating the Drake equation, this project will stand a good chance of success.

Finding even one other nearby civilization in our galaxy would make scientists more optimistic that the universe holds a great deal more intelligent life. SETI may not be in the mainstream of astronomy, and the likelihood of its success is difficult to predict, but its potential payoff is enormous. Few discoveries would do more to change our understanding of ourselves than certain knowledge that we are not alone.

Science fiction is filled with tales of humans who leave Earth to "seek out new

Figure 18.14 When complete, the Allen Telescope Array will listen for evidence of intelligent life from as many as a million stellar systems.

life and new civilizations." Unfortunately, these scenarios are not scientifically realistic. The distances to the stars and their planets are enormous; to explore a significant sample of stars would require extending the physical search over tens or hundreds of light-years. As we discussed in Chapter 13, special relativity limits how fast we can travel. The speed of light is the limit, and even at that rate it would take more than 4 years to get to the *nearest* star. Time dilation would favor the astronauts themselves, and they would return to Earth younger than if they had stayed at home. But suppose they visited a star 15 light-years distant. Even if they traveled at speeds close to that of light, by the time they returned to Earth, 30 years would have passed. Some science-fiction enthusiasts get around this problem by invoking "warp speed" or "hyperdrive" to travel faster than the speed of light or by using "wormholes" as shortcuts across the galaxy—but there is absolutely no evidence that any one of these shortcuts is possible.

Life as We Do Not Know It

So far, we have confined our discussion to the search for life as we know it, as it may exist now. We do this for sound scientific reasons, primarily having to do with falsifiability. Recall Chapter 1, where we defined scientific ideas as being first and foremost falsifiable. There are no tests we can apply to prove that very different forms of life did not (or will not) exist; therefore, the idea behind them is not a scientific hypothesis. Very different forms of life, or life in the distant past or future, are purely speculative and nonscientific. This does not mean these ideas are uninteresting. They are just not falsifiable, and therefore not science.

Now consider more common examples. Some people claim that aliens have already visited us: tabloid newspapers, books, and websites are filled with tales of UFO sightings, government conspiracies and cover-ups of alien crash-landings, alleged alien abductions, and UFO religious cults. However, none of these reports meet the basic standards of science. They are not falsifiable—they lack verifiable evidence and repeatability—and we must conclude that there is no scientific evidence for any alien visitations.

When you encounter discussions of past, future, or very different alien life-forms, it is useful to ask whether the ideas are falsifiable and whether the evidence is verifiable and repeatable. If not, the discussions are speculation, not science.

18.4 The Fate of Life on Earth

In this book, we have used our understanding of physics and cosmology to look back through time and watch as structure formed throughout the universe. We have peered into the future and seen the ultimate fate of the universe, from its enormous clusters of galaxies down to its tiniest components. Now we examine our own destiny and contemplate the fate that awaits Earth, humanity, and the star on which all terrestrial life depends.

Eventually, the Sun Must Die

About 5 billion years from now, the Sun will end its long period of relative stability. Shedding its identity as the passive star that has nurtured life on Earth, the Sun will expand to become a red giant and later an asymptotic giant branch

(AGB) star, swelling to hundreds of times its current size. The giant planets, orbiting outside the extended red giant atmosphere, may survive in some form. Even so, they will suffer the blistering radiation from a Sun grown thousands of times more luminous than it is today.

The terrestrial planets will not fare as well. Some—perhaps all—of the worlds of the inner Solar System will be engulfed by the expanding Sun. Just as an artificial satellite is slowed by drag in Earth's tenuous outer atmosphere and eventually falls to the ground in a dazzling streak of white-hot light, so, too, will a terrestrial planet caught in the Sun's atmosphere be consumed by the burgeoning star. If this is Earth's fate, our home world will leave no trace other than a slight increase in the amount of massive elements in the Sun's atmosphere. As the Sun loses more and more of its atmosphere in an AGB wind, our atoms may be expelled back into the reaches of interstellar space from which they came, perhaps to become incorporated into new generations of stars, planets, and even life itself.

Another planetary fate is possible, however. In this scenario, as the red giant Sun loses mass in a powerful wind, its gravitational grasp on the planets will weaken, and the orbits of both the inner and outer planets will spiral outward. If Earth moves out far enough, it may survive as a seared cinder, orbiting the white dwarf that the Sun will become. Barely larger than Earth and with its nuclear fuel exhausted, the white dwarf Sun will slowly cool, eventually becoming a cold, inert sphere of degenerate carbon, orbited by what remains of its collection of planets. Thus the ultimate outcome for our Earth—consumed in the heart of the Sun or left behind as a frigid, burned rock orbiting a long-dead white dwarf—is not yet known.

The Future of Life on Earth

Life on our planet has even less of a future, for it will not survive long enough to witness the Sun's departure from the main sequence. Well before that cataclysmic event takes place, the Sun's luminosity will begin to rise. As solar luminosity increases, so will temperatures on all the planets, including our own. The inner edge of the Sun's habitable zone will move out past the orbit of Earth. Eventually, Earth's temperatures will climb so high that all animal and plant life will perish. Even the extremophiles that inhabit the oceanic depths will die as the oceans boil away. Models of the Sun's evolution are still too imprecise to predict with certainty when that fatal event will occur, but the end of all terrestrial life may be only 1 billion or 2 billion years away. That is, of course, a comfortably distant time from now, but it is well short of the Sun's departure from the main sequence.

It is far from certain, however, that the descendants of today's humanity will even be around a billion years from now. Some of the threats that await us come from beyond Earth. For the remainder of the Sun's life, the terrestrial planets, including Earth, will continue to be bombarded by asteroids and comets. Perhaps a hundred or more of these impacts will involve kilometer-sized objects, capable of causing the kind of devastation that eradicated the dinosaurs and most other species 65 million years ago. Although these events may create new surface scars, they will have little effect on the integrity of Earth itself. Earth's geological record is filled with such events, and each time they happen, life manages to recover and reorganize.

It seems likely, then, that some form of life will survive to see the Sun begin its march toward instability. However, individual species do not necessarily fare so well when faced with cosmic cataclysm. We are rapidly developing technology that could enable us to detect most threatening asteroids and modify their orbits well before they could strike Earth. However, comets are more difficult to guard against because long-period comets appear from the outer Solar System with little warning. To offer protection, various means of defending ourselves would have to be in place, ready to be used on very short notice. Although impacts from kilometer-sized objects are infrequent, smaller objects only a few dozen meters in size, carrying the punch of a several-megaton bomb, strike Earth more than once every 100 years. There have been three such events in the past 105 years: Tunguska, Sikhote-Alin, and Chelyabinsk.

We might protect ourselves from the fate of the dinosaurs, but in the long run the descendants of humanity will either leave this world or die out. Planetary systems surround other stars, and all that we know tells us that many other Earth-like planets may well exist throughout our galaxy. Colonizing other planets is currently the stuff of science fiction, but if our descendants are ultimately to survive the death of our home planet, off-Earth colonization must become science fact at some point in the future.

Although humankind may soon be capable of protecting Earth from life-threatening comet and asteroid impacts, in other ways we are our own worst enemy. Human activities are dramatically affecting the balances of atmospheric gases. The climate and ecosystem of Earth constitute a finely balanced, complex system. But the record of past climatic shifts shows that the system is capable of large changes in response to even small disturbances. Drastic changes in climate and resource availability would certainly have consequences for our own survival. For the first time in human history, we possess the means to unleash nuclear, chemical, or biological disasters that could threaten the very survival of our species. In the end, the fate of humanity may depend more than anything on how well we accept responsibility for our actions and the effect we have on each other and our planet.

Figure 18.15 shows an image of Earth as taken by the *Voyager 1* spacecraft at a distance of more than 40 AU (beyond Neptune's orbit). The beams in the picture are sunlight scattered off the spacecraft. The arrow points to a tiny dot, which is Earth, the only place in the entire universe where life is confirmed to exist. Compare the size of that dot to the size of the universe. Compare the history of life on Earth to the history of the universe. Compare Earth's future with the fate of the universe. Astronomy is humbling. We occupy a tiny part of space and time. And yet we are unique, as far as we know. Think for a moment about what that means to you. This may be the most important lesson the universe has to teach us.

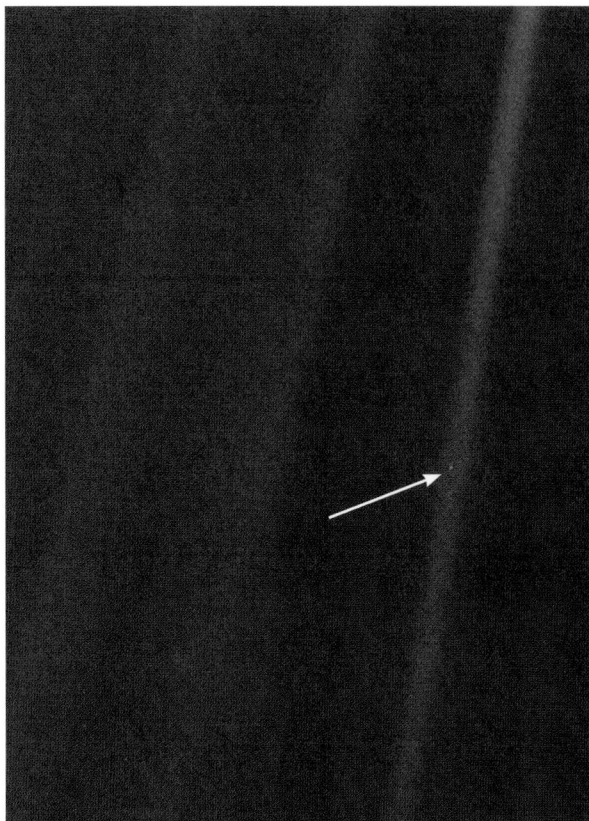

Figure 18.15 This image from the *Voyager 1* spacecraft shows the Earth from a distance of 3.8 billion miles, well past the orbit of Neptune. The streaks in this image are scattered sunlight. The "pale blue dot" in the rightmost streak is Earth.

READING ASTRONOMY News

> Astronomers often use the data from large surveys in ways that were not part of the original mission. In this case, Geoff Marcy proposes a "fishing expedition" to go looking for something he may never find.

Astronomer Uses Kepler Telescope's Data in Hunt for Spacecraft from Other Worlds

By **PETER BRANNEN**, *Washington Post*

In the field of planet hunting, Geoff Marcy is a star. After all, the astronomer at the University of California at Berkeley found nearly three-quarters of the first 100 planets discovered outside our Solar System. But with the hobbled planet-hunting Kepler telescope having just about reached the end of its useful life and reams of data from the mission still left uninvestigated, Marcy began looking in June for more than just new planets. He's sifting through the data to find alien spacecraft passing in front of distant stars.

He's not kidding—and now he has the funding to do it.

Last fall, the Templeton Foundation, a philanthropic organization dedicated to investigating what it calls the "big questions"—which, unsurprisingly, include "Are we alone?"—awarded Marcy $200,000 to pursue his search for alien civilizations.

As far as Marcy, an official NASA researcher for the Kepler mission, is concerned, that question has a clear answer: "The universe is simply too large for there not to be another intelligent civilization out there. Really, the proper question is: 'How far away is our nearest intelligent neighbor?' They could be 10 light-years, 100 light-years, a million light-years or more. We have no idea."

To answer that question, Marcy has begun to sift through the Kepler data and to search the heavens for a galactic laser Internet that might be in use somewhere out there. (More on that in a bit.)

Launched in 2009, Kepler was designed as a 4-year mission to detect planets—habitable or otherwise—around distant stars by measuring the dimming of those stars as orbiting bodies pass in front of them. In May, a component of the spacecraft designed to keep it pointing precisely failed, dealing a crushing blow to Marcy and his colleagues who last year convinced NASA to extend funding for the mission into 2016, which Marcy says would have allowed researchers to further refine the number of known Earth-like planets in our galactic neighborhood.

"It's a heartbreaker," he says. "People are reacting a little bit as if a close family member died. There's this combination of severe depression and confusion, coupled with denial."

100 Billion Exoplanets?

Kepler has been wildly successful in its 4 years. To date, it has found 132 exoplanets—that is, planets outside our solar system—and possibly 3,216 more that await confirmation. Researchers have extrapolated from Kepler data that our Milky Way Galaxy alone contains at least 100 billion exoplanets, as many planets as there are stars. Still, with the telescope—which is 40 million miles from Earth—having collected data on 150,000 star systems, researchers are only beginning to pick through all the information.

Marcy hopes that hiding within it will be hints about intelligent life abroad. What if, say, the dimming of a star that Kepler observes is caused by something even more fanciful than the passage of extrasolar planets? Something synthetic, perhaps? Marcy admits that even he's not certain what he's looking for.

"I do know that if I saw a star that winked out, then at some point it winked back on again, then winked out for a long, long time and then blinked on again, that that would be so weird," he says. "Obviously that wouldn't constitute the detection of an advanced civilization yet, but it would at least alert us that follow-up observations are warranted."

Such an irregular pattern might signal the leisurely and unpredictable passage of massive spacecraft in front of the star. But, perhaps more likely, it might indicate the presence of a Dyson sphere, a mainstay of science fiction first proposed by physicist Freeman Dyson in 1960.

The concept is simple. The energy needs of a civilization a thousand or a million years more advanced than our own would probably be vastly greater than those of even the most profligate Earthlings. The greatest source of energy in a solar system is its star, and that energy could be captured by building a massive structure tiled with solar panels enveloping the star—the ultimate green jobs initiative.

Under the second law of thermodynamics, the structure would produce incredible amounts of waste heat in the form of infrared radiation. In September, a Penn State team led by astrophysics professor Jason Wright began searching the sky for just that by combing through data from NASA's Wide-field Infrared Survey Explorer, or WISE. The Penn State work is also being funded by Templeton.

If Dyson spheres pop up in the data, Marcy thinks they would more likely appear as a patchwork of solar panels rather than a solid sphere. Perhaps the dimming of a star would be erratic or quasi-periodic, unlike the regular transit of planets.

To detect such aberrant dimming patterns,

SEE HUNT FOR SPACECRAFT

Hunt for spacecraft (cont.)

Marcy's Templeton grant is funding the salary of a Berkeley student to write software that will chew through the Kepler data. "Writing the computer code is not easy," Marcy says. "There's no prescription in any computer science book about how to search for aliens."

Beyond Radio Waves

The rest of the $200,000 grant is buying Marcy time on the Keck Observatory in Hawaii, the largest telescope in the world, to search for—what else?—a galactic laser Internet.

While the movie *Contact*, based on Carl Sagan's book of the same name, popularized the idea of aliens dozens of light-years away picking up an old telecast of the 1936 Berlin Olympics that was unintentionally transmitted into space, our civilization has become quieter to any outside observers in recent decades. As our civilization makes the jump from analog to digital, communication is increasingly carried by fiber-optic cables and relatively weak cellphone repeaters rather than powerful broadcast transmitters. Rather than spilling out messy radio transmissions, Marcy posits that alien civilizations would use something much more precise and efficient than radio waves to stay connected, and lasers fit the bill. At the Keck Observatory, he hopes to spy an errant beam flashing from a distant star system, an observation that would be strikingly obvious on a spectrum.

This shift to new ways for finding E.T. is in part due to the failure of traditional SETI (Search for Extraterrestrial Intelligence) to pick up radio signals from deep space. Federal funding for SETI projects ended in 1995, but private benefactors have stepped up to support the search for alien radio transmissions, including Microsoft cofounder Paul Allen, who has sunk more than $30 million into a giant radio telescope array now under construction northeast of San Francisco.

Nevertheless, the silence underscores the question once posed by Nobel Prize–winning physicist Enrico Fermi: If intelligent life is common in the galaxy, "where is everybody?"

Marcy admits that this so-called Fermi paradox poses a powerful counterargument to the prospect of success for any search for extraterrestrial intelligence. But what if, even if the chances are vanishingly remote, he is successful? More disturbingly, what if (as some respected physicists fear) he finds a Death Star?

"The first thing we do is transmit a message to them that says, 'We taste bad.'"

Evaluating the News

1. What does Geoff Marcy want to do and why?
2. Marcy states that "Really, the proper question is: 'How far away is our nearest intelligent neighbor?' They could be 10 light-years, 100 light-years, a million light-years or more. We have no idea." The Drake equation is one way in which astronomers try to get a handle on the number of intelligent civilizations. Given that, why do you think that Geoff Marcy says "We have no idea" about the distance of our nearest intelligent neighbor?
3. Kepler has collected information on 150,000 star systems. Is that a significant fraction of the stars in the Milky Way?
4. What is a Dyson sphere? What is its purpose? Could you envision humans constructing a Dyson sphere? Why or why not?
5. Consider the definitions related to the scientific method that you learned about in Chapter 1. Where does Geoff Marcy's project fall in the continuum of scientific ideas and theories?
6. Compare and contrast the topic of this article with the Reading Astronomy News article in Chapter 5. How well does the reporter cover the topic? How accessible is it to a general audience?
7. This project, along with traditional SETI, is now privately funded. Why do you think federal funding has not been granted for these projects?

SUMMARY

Life on Earth is a form of complex carbon-based chemistry, made possible by self-replicating molecules. Life likely formed in Earth's oceans, then evolved chemically from simple molecules into self-replicating organisms. Life-forms that are very different from ours, including those based on silicon chemistry, cannot be ruled out. Space-based instruments have begun the search for Earth-like planets, and astronomers focus their search for extraterrestrial life on Earth-like planets in habitable zones. Although our galaxy may be teeming with intelligent life, none has yet been detected. The Drake equation organizes our thoughts about the likelihood of intelligent life in the Milky Way Galaxy. Long before the Sun ends its period of stability on the main sequence, all terrestrial life will have perished after an increasingly luminous Sun makes Earth uninhabitable.

LG 1 Life on Earth likely began in the oceans because they were rich in the building blocks of RNA and DNA. Life then evolved through a combination of mutation and heredity to form the complex ecosystems that exist today.

LG 2 The chemistry of life as we know it requires liquid water. Europa and Enceladus have liquid water now, and Mars had liquid water in the past.

LG 3 A stellar habitable zone is defined as the region around a star in which a planet might have liquid water on the surface: neither too hot nor too cold. The galactic habitable zone is not too close to the high-energy radiation that dominates the center of the Milky Way, yet it is still close enough that there is a sufficient amount of the heavy elements that make up life.

LG 4 Scientists have found extrasolar planets to be common. This indicates that the term in the Drake equation representing the fraction of stars with planets is high. Extraterrestrial life has not yet been discovered, but this null result is not definitive, given the limitations on the search so far.

LG 5 Eventually, the Sun will move off of the main sequence on the H-R diagram. Life on Earth will not even last until that time, as the Sun becomes more luminous, moving the inner edge of the habitable zone out past Earth's orbit.

SUMMARY SELF-TEST

1. Scientists think that terrestrial life probably originated in Earth's oceans because (select all that apply)
   a. all the chemical pieces were in the ocean.
   b. energy was available in the ocean.
   c. the earliest evidence for life on Earth is from fossils of ocean-dwelling organisms.
   d. the deepest parts of the ocean have hydrothermal vents.

2. The term *primordial soup* refers to
   a. conditions just after the Big Bang.
   b. conditions during the collapse of the nebula that formed the Sun.
   c. conditions just after the formation of Earth.
   d. conditions in the early Earth's oceans.

3. The Urey-Miller experiment produced _____ in a laboratory jar.
   a. life
   b. RNA and DNA
   c. amino acids
   d. proteins

4. Europa and Enceladus are both outside of the Sun's habitable zone. However, they contain liquid water because they receive _____ from their parent planets.
   a. tidal energy
   b. reflected sunlight
   c. magnetic heating
   d. electrical energy through flux tubes

5. A star in the halo of the Milky Way is likely outside of the galactic habitable zone because
   a. there is too much dark matter.
   b. there is too much high-energy radiation.
   c. there are not enough supernovae nearby to start star formation.
   d. there are not enough heavy elements.

6. *Natural selection* means that
   a. poorly adapted forms of life die, while well-adapted forms reproduce.
   b. well-adapted forms of life are chosen and encouraged.
   c. forms of life are designed for their environment.
   d. environments are designed for the life-forms that inhabit them.

7. Any system with heredity, mutation, and natural selection
   a. will change over time. (Choose all that apply.)
   b. will become larger over time.
   c. will become more complex over time.
   d. will develop intelligence.

8. The fact that we have not detected alien civilizations yet tells us that
   a. they are not there.
   b. they are rare.
   c. we are in a "blackout," and they are not talking to us.
   d. we don't know enough yet to draw any conclusions.

9. All life on Earth *must* eventually come to an end because
   a. humans will make the planet uninhabitable.
   b. asteroids will make the planet uninhabitable.
   c. new species will arise that we won't recognize as life.
   d. the Sun will make the planet uninhabitable.

10. The study of life and the study of astronomy are connected because (select all that apply)
    a. life may be quite commonplace in the universe.
    b. studying other planets may help explain why there is life on Earth.
    c. explorations of extreme environments on Earth suggest where to look for life elsewhere.
    d. life is a structure that evolved through physical processes, and life on Earth might not be unique.
    e. life elsewhere is most likely to be found by astronomers.

QUESTIONS AND PROBLEMS

Multiple Choice and True/False

11. T/F: All that is required for life to begin is a solitary, self-replicating molecule.

12. T/F: All life on Earth evolved from a common ancestor.

13. T/F: Astronomers do not consider the possibility of non-carbon-based life.

14. T/F: There are several places other than Earth where life might exist in our Solar System.

15. T/F: It is likely that we will discover many civilizations that are not advanced.

16. The Cambrian explosion
    a. killed the dinosaurs.
    b. produced the carbon that is now here on Earth.
    c. was a sudden increase in biodiversity.
    d. released a lot of carbon dioxide into the atmosphere.

17. The difference between prokaryotes and eukaryotes is that
    a. prokaryotes have no DNA.
    b. prokaryotes have no cell wall.
    c. prokaryotes have no nucleus.
    d. prokaryotes do not exist today.

18. In the phrase "theory of evolution," the word *theory* means that evolution
    a. is an idea that can't be tested scientifically.
    b. is an educated guess to explain natural phenomena.
    c. probably doesn't happen anymore.
    d. is a well-tested, well-corroborated scientific explanation of natural phenomena.

19. Mutations are
    a. changes to an organism's DNA.
    b. changes in the appearance of the organism.
    c. deadly to the organism's children.
    d. changes in the whole species.

20. The habitable zone is the place around a star where
    a. life has been found.
    b. atmospheres can contain oxygen.
    c. liquid water exists.
    d. liquid water can exist on the surface of a planet.

21. The length of time an intelligent, communicating civilization lasts affects _____ in the Drake equation.
    a. the value of R^*
    b. the value of f_i
    c. the value of f_c
    d. the value of L

22. Carbon is a favorable base for life because
    a. it can bond to many other atoms in long chains.
    b. it is nonreactive.
    c. it forms weak bonds that can be readily reorganized as needed.
    d. it is organic.

23. A thermophile is an organism that lives in extremely _____ conditions.
    a. salty   b. hot
    c. cold   d. dry

24. Life first appeared on Earth _____ of years ago.
    a. thousands
    b. hundreds of thousands
    c. millions
    d. billions

25. Astronomers think that intelligent life is more likely to be found around stars of types F, G, K, and M because
    a. those stars are hot enough to have planets and moons with liquid water.
    b. those stars are cool enough to have planets and moons with liquid water.
    c. those stars live long enough for life to begin and evolve.
    d. those stars produce no UV radiation or X-rays.

Conceptual Questions

26. Why do we generally talk about molecules such as DNA or RNA when discussing life on Earth?

27. How do we suspect that the building blocks of DNA first formed on Earth?

28. Today, most known life enjoys moderate climates and temperatures. Compare this environment to some of the conditions in which early life developed.

29. What are the similarities and differences between prokaryotes and eukaryotes?

30. Tracing back on the evolutionary tree, what kinds of life do we find that animals are most similar to?

31. Why do you suppose plants and forests appeared in large numbers before large animals did?

32. Why was evolution inevitable on Earth?

33. What were the general conditions needed on our planet for life to arise?

34. Is biological evolution under way on Earth today? If so, how might humans continue to evolve?

35. Where did all the atoms in your body come from?

36. The *Viking* spacecraft did not find evidence of life on Mars when it visited that planet in the late 1970s, nor did the *Phoenix* lander when it examined the martian soil in 2009. Does this imply that life never existed on the planet? Why or why not?

37. What is a habitable zone? What defines its boundaries?

38. In searching for intelligent life elsewhere, why is listening with radio telescopes currently our favored method?

39. Why is it likely that life on Earth as we know it will end long before the Sun runs out of nuclear fuel?

40. Why do you think we sent a coded radio signal to the globular cluster M13 in 1974—rather than, say, to a nearby star?

Problems

41. Suppose that an organism replicates itself each second. What will the final population be after 10 seconds?

42. Suppose that an organism replicates itself each second. After how many seconds will the population increase by a factor of 1,024?

43. As a rule of thumb, you can find the doubling time for exponential growth by dividing 70 by the rate of increase. So, if the population increases by 7 percent per year, the doubling time is 10 years (70/7 = 10). Earth's human population grows by 1 percent annually. What is the doubling time? How much time will pass before there are four times as many humans on Earth?

44. Use the Drake equation to calculate your own "most likely" number of intelligent civilizations. To do this, choose numbers for each parameter that you consider most likely.

45. Use the Drake equation to calculate your own "least likely" number of intelligent civilizations. To do this, choose numbers for each parameter that you consider least likely. If you calculate a number less than 1, you know that you have been too pessimistic. Why?

46. Which factors in the Drake equation are affected by the Kepler mission's search for planets in the habitable zones of stars? How will the final number *N* be affected if astronomers find that most stars have planets in their habitable zones?

47. Study Figure 18.12. The four rows of white blocks at the top represent the numbers 1–10: The bottom row is a placeholder, and the top three rows are the actual counting. Explain the "rule" for the kind of counting shown here. (For example, how do three white blocks represent the number 7?)

48. The time for a sample of *Escherichia coli* to double is 20 minutes. You become infected when just 10 bacteria enter your system. How many *E. coli* bacteria are in your body 12 hours after infection begins?

49. Suppose astronomers announce the discovery of a new planet around a star with a mass equal to that of the Sun. This planet has an orbital period of 87 days. Is this planet in the habitable zone for a Sun-like star?

50. If a self-replicating molecule has begun replicating, and seven doubling times have passed, how many molecules are there?

SMARTWORK

Norton's online homework system includes algorithmically generated versions of these questions, plus additional conceptual exercises. If your instructor assigns questions in SmartWork, log in at smartwork.wwnorton.com.

Exploration | Fermi Problems and the Drake Equation

The Drake equation is a way of organizing our thoughts about whether there might be other intelligent, communicating civilizations in the galaxy. This type of thinking is very useful for getting estimates of a value, particularly when analyzing systems for which counting is not actually possible. The types of problems that can be solved in this way are often called Fermi Problems after Enrico Fermi, who was mentioned in this chapter. For example, we might ask, "What is the circumference of Earth?"

You could Google this, or you could already "know" it, or you might look it up in this textbook. Alternatively, you could very carefully measure the shadow of a stick in two locations at the same time on the same day. Or you could drive around the planet. Or you could reason this way:

How many time zones are between New York and Los Angeles?
  3 time zones. You know this from traveling or from television.
How many miles is it from New York to Los Angeles?
  3,000 miles. You know this from traveling or from living in the world.
So, how many miles per time zone?
  3,000/3 = 1,000
How many time zones in the world?
  24, because there are 24 hours in a day, and each time zone marks an hour.
So what is the circumference of Earth?
  24,000 miles, because there are 24 time zones, each 1,000 miles wide.

The measured circumference is 24,900 miles, which agrees to within 4 percent.

Following are several Fermi Problems. Time yourself for an hour, and work as many of them as possible. (You don't have to do them in order!)

1. How much has the mass of the human population on Earth increased in the past year?

2. How much energy does a horse consume in its lifetime?

3. How many pounds of potatoes are consumed in the United States annually?

4. How many cells are there in the human body?

5. If your life earnings were given to you by the hour, how much is your time worth per hour?

6. What is the weight of solid garbage thrown out by American families each year?

7. How fast does human hair grow (in feet per hour)?

8. If all the people on Earth were crowded together, how much area would we cover?

9. How many people could fit on Earth if every person occupied 1 square meter of land?

10. How much carbon dioxide (CO_2) does an automobile emit each year?

11. What is the mass of Earth?

12. What is the average annual cost of an automobile including overhead (maintenance, looking for parking, cleaning, and so forth)?

13. How much ink was used printing all the newspapers in the United States today?

SMARTWORK • smartwork.wwnorton.com

APPENDIX 1

PERIODIC TABLE OF THE ELEMENTS

Group	1 (1A)	2 (2A)	3 (3B)	4 (4B)	5 (5B)	6 (6B)	7 (7B)	8 (8B)	9 (8B)	10 (8B)	11 (1B)	12 (2B)	13 (3A)	14 (4A)	15 (5A)	16 (6A)	17 (7A)	18 (8A)
1	1 H Hydrogen 1.00794																	2 He Helium 4.002602
2	3 Li Lithium 6.941	4 Be Beryllium 9.012182											5 B Boron 10.811	6 C Carbon 12.0107	7 N Nitrogen 14.0067	8 O Oxygen 15.9994	9 F Fluorine 18.9984032	10 Ne Neon 20.1797
3	11 Na Sodium 22.98976928	12 Mg Magnesium 24.3050											13 Al Aluminum 26.9815386	14 Si Silicon 28.0855	15 P Phosphorus 30.973762	16 S Sulfur 32.065	17 Cl Chlorine 35.453	18 Ar Argon 39.948
4	19 K Potassium 39.0983	20 Ca Calcium 40.078	21 Sc Scandium 44.955912	22 Ti Titanium 47.867	23 V Vanadium 50.9415	24 Cr Chromium 51.9961	25 Mn Manganese 54.938045	26 Fe Iron 55.845	27 Co Cobalt 58.933195	28 Ni Nickel 58.6934	29 Cu Copper 63.546	30 Zn Zinc 65.38	31 Ga Gallium 69.723	32 Ge Germanium 72.64	33 As Arsenic 74.92160	34 Se Selenium 78.96	35 Br Bromine 79.904	36 Kr Krypton 83.798
5	37 Rb Rubidium 85.4678	38 Sr Strontium 87.62	39 Y Yttrium 88.90585	40 Zr Zirconium 91.224	41 Nb Niobium 92.90638	42 Mo Molybdenum 95.96	43 Tc Technetium [98]	44 Ru Ruthenium 101.07	45 Rh Rhodium 102.90550	46 Pd Palladium 106.42	47 Ag Silver 107.8682	48 Cd Cadmium 112.411	49 In Indium 114.818	50 Sn Tin 118.710	51 Sb Antimony 121.760	52 Te Tellurium 127.60	53 I Iodine 126.90447	54 Xe Xenon 131.293
6	55 Cs Cesium 132.9054519	56 Ba Barium 137.327	57 La Lanthanum 138.90547	72 Hf Hafnium 178.49	73 Ta Tantalum 180.94788	74 W Tungsten 183.84	75 Re Rhenium 186.207	76 Os Osmium 190.23	77 Ir Iridium 192.217	78 Pt Platinum 195.084	79 Au Gold 196.966569	80 Hg Mercury 200.59	81 Tl Thallium 204.3833	82 Pb Lead 207.2	83 Bi Bismuth 208.98040	84 Po Polonium [209]	85 At Astatine [210]	86 Rn Radon [222]
7	87 Fr Francium [223]	88 Ra Radium [226]	89 Ac Actinium [227]	104 Rf Rutherfordium [261]	105 Db Dubnium [262]	106 Sg Seaborgium [266]	107 Bh Bohrium [264]	108 Hs Hassium [277]	109 Mt Meitnerium [268]	110 Ds Darmstadtium [271]	111 Rg Roentgenium [272]	112 Cn Copernicium [285]	113 Uut Ununtrium [284]	114 Fl Flerovium [289]	115 Uup Ununpentium [288]	116 Lv Livermorium [292]	117 Uus Ununseptium [294]	118 Uuo Ununoctium [294]

Key:
- 1 — Atomic number
- H — Symbol
- Hydrogen — Name
- 1.00794 — Average atomic mass
- Metals, Metalloids, Nonmetals

6 Lanthanides

58 Ce Cerium 140.116	59 Pr Praseodymium 140.90765	60 Nd Neodymium 144.242	61 Pm Promethium [145]	62 Sm Samarium 150.36	63 Eu Europium 151.964	64 Gd Gadolinium 157.25	65 Tb Terbium 158.92535	66 Dy Dysprosium 162.500	67 Ho Holmium 164.93032	68 Er Erbium 167.259	69 Tm Thulium 168.93421	70 Yb Ytterbium 173.05	71 Lu Lutetium 174.967

7 Actinides

90 Th Thorium 232.03806	91 Pa Protactinium 231.03588	92 U Uranium 238.02891	93 Np Neptunium [237]	94 Pu Plutonium [244]	95 Am Americium [243]	96 Cm Curium [247]	97 Bk Berkelium [247]	98 Cf Californium [251]	99 Es Einsteinium [252]	100 Fm Fermium [257]	101 Md Mendelevium [258]	102 No Nobelium [259]	103 Lr Lawrencium [262]

We have used the U.S. system as well as the system recommended by the International Union of Pure and Applied Chemistry (IUPAC) to label the groups in this periodic table. The system used in the United States includes a letter and a number (1A, 2A, 3B, 4B, etc.), which is close to the system developed by Mendeleev. The IUPAC system uses numbers 1–18 and has been recommended by the American Chemical Society (ACS). While we show both numbering systems here, we use the IUPAC system exclusively in the book. Elements with atomic numbers higher than 112 have been reported but not yet fully authenticated.

APPENDIX 2

PROPERTIES OF PLANETS, DWARF PLANETS, AND MOONS

Physical Data for Planets and Dwarf Planets

PLANET	EQUATORIAL RADIUS (km)	(R/R⊕)	MASS (kg)	(M/M⊕)	AVERAGE DENSITY (RELATIVE TO WATER*)	ROTATION PERIOD (DAYS)	TILT OF ROTATION AXIS (DEGREES, RELATIVE TO ORBIT)	EQUATORIAL SURFACE GRAVITY (RELATIVE TO EARTH†)	ESCAPE VELOCITY (km/s)	AVERAGE SURFACE TEMPERATURE (K)
Mercury	2,440	0.383	3.30×10^{23}	0.055	5.427	58.65	0.01	0.378	4.3	340 (100, 725)§
Venus	6,052	0.949	4.87×10^{24}	0.815	5.243	243.02‡	177.36	0.907	10.36	737
Earth	6,378	1.000	5.97×10^{24}	1.000	5.513	1.000	23.44	1.000	11.19	288 (183, 331)§
Mars	3,396	0.533	6.42×10^{23}	0.107	3.934	1.0260	25.19	0.377	5.03	210 (133, 293)§
Ceres	487.3	0.075	9.47×10^{20}	0.0002	2.100	0.378	3.0	0.029	0.51	200
Jupiter	71,492	11.209	1.90×10^{27}	317.83	1.326	0.4135	3.13	2.528	59.5	165
Saturn	60,268	9.449	5.68×10^{26}	95.16	0.687	0.4440	26.73	1.065	35.5	134
Uranus	25,559	4.007	8.68×10^{25}	14.536	1.270	0.7183‡	97.77	0.889	21.3	76
Neptune	24,764	3.883	1.02×10^{26}	17.148	1.638	0.6713	28.32	1.14	23.5	58
Pluto	1,184	0.182	1.30×10^{22}	0.0021	2.030	6.387‡	122.53	0.083	1.23	40
Haumea	~650	0.11	4.0×10^{21}	0.0007	~3	0.163	?	0.045	0.84	<50
Makemake	750	0.12	4.18×10^{21}	0.0007	~2	0.32	?	0.048	0.8	~30
Eris	1,200	0.188	1.5×10^{22} (est.)	0.0025 (est.)	~2	>0.3?	?	0.082	~1.3	30

*The density of water is 1,000 kg/m³.
†The surface gravity of Earth is 9.81 m/s².
‡Venus, Uranus, and Pluto rotate opposite to the directions of their orbits. Their north poles are south of their orbital planes.
§Where given, values in parentheses give extremes of recorded temperatures.

Orbital Data for Planets and Dwarf Planets

PLANET	MEAN DISTANCE FROM SUN (A*)		ORBITAL PERIOD (P) (SIDEREAL YEARS)	ECCENTRICITY	INCLINATION (DEGREES, RELATIVE TO ECLIPTIC)	AVERAGE SPEED (km/s)
	(10^6 km)	(AU)				
Mercury	57.9	0.387	0.2408	0.2056	7.005	47.36
Venus	108.2	0.723	0.6152	0.0068	3.395	35.02
Earth	149.6	1.000	1.0000	0.0167	0.000	29.78
Mars	227.9	1.524	1.8808	0.0934	1.850	24.08
Ceres	413.7	2.765	4.6027	0.079	10.587	17.88
Jupiter	778.3	5.203	11.8626	0.0484	1.304	13.06
Saturn	1,426.7	9.537	29.4475	0.0539	2.485	9.64
Uranus	2,870.7	19.189	84.0168	0.0473	0.772	6.80
Neptune	4,495.1	30.070	164.7913	0.011	1.769	5.43
Pluto	5,906.4	39.48	247.9207	0.2488	17.14	4.72
Haumea	6,428.1	43.0	281.9	0.198	28.22	4.50
Makemake	6,789.7	45.3	305.3	0.164	29.00	4.39
Eris	10,183	68.05	561.6	0.4339	43.82	3.43

*A is the semimajor axis of the planet's elliptical orbit.

Properties of Selected Moons*

PLANET	MOON	ORBITAL PROPERTIES		PHYSICAL PROPERTIES		
		P (DAYS)	A (10^3 km)	R (km)	M (10^{20} kg)	RELATIVE DENSITY† (g/cm³) (WATER = 1.00)
Earth (1 moon)	Moon	27.32	384.4	1,737.5	735	3.34
Mars (2 moons)	Phobos	0.32	9.38	13.4 × 11.2 × 9.2	0.0001	1.9
	Deimos	1.26	23.46	7.5 × 6.1 × 5.2	0.00002	1.5
Jupiter (64 known moons)	Metis	0.30	127.97	21.5	0.00012	3
	Amalthea	0.50	181.40	131 × 73 × 67	0.0207	0.8
	Io	1.77	421.80	1,822	893	3.53
	Europa	3.55	671.10	1,561	480	3.01
	Ganymede	7.15	1,070	2,631	1,482	1.94
	Callisto	16.69	1,883	2,410	1,080	1.83
	Himalia	250.56	11,461	85	0.067	2.6
	Pasiphae	744‡	23,624	30	0.0030	2.6
	Callirrhoe	759‡	24,102	4.3	0.00001	2.6

(*continued*)

Properties of Selected Moons* (continued)

PLANET	MOON	ORBITAL PROPERTIES		PHYSICAL PROPERTIES		
		P (DAYS)	A (10^3 km)	R (km)	M (10^{20} kg)	RELATIVE DENSITY[†] (g/cm³) (WATER = 1.00)
Saturn (62 known moons)	Pan	0.58	133.58	14	0.00005	0.42
	Prometheus	0.61	139.38	74.0 × 50.0 × 34.0	0.0016	0.5
	Pandora	0.63	141.70	55 × 44 × 31	0.0014	0.5
	Mimas	0.94	185.54	198	0.38	1.15
	Enceladus	1.37	238.04	252	1.08	1.6
	Tethys	1.89	294.67	533	6.17	0.97
	Dione	2.74	377.42	562	11.0	1.48
	Rhea	4.52	527.07	764	23.1	1.23
	Titan	15.95	1,222	2,575	1,346	1.88
	Hyperion	21.28	1,501	205 × 130 × 110	0.0559	0.54
	Iapetus	79.33	3,561	736	18.1	1.08
	Phoebe	550.3[‡]	12,948	107	0.08	1.6
	Paaliaq	687.5	15,024	11	0.0001	2.3
Uranus (27 known moons)	Cordelia	0.34	49.80	20	0.0004	1.3
	Miranda	1.41	129.90	236	0.66	1.21
	Ariel	2.52	190.90	579	12.9	1.59
	Umbriel	4.14	264.96	585	12.2	1.46
	Titania	8.71	436.30	789	34.2	1.66
	Oberon	13.46	583.50	761	28.8	1.56
	Setebos	2,225[‡]	17,420	24	0.0009	1.5
Neptune (13 known moons)	Naiad	0.29	48.0	48 × 30 × 26	0.002	1.3
	Larissa	0.55	73.5	108 × 102 × 84	0.05	1.3
	Proteus	1.12	117.6	218 × 208 × 201	0.5	1.3
	Triton	5.88[‡]	354.8	1,353	214	2.06
	Nereid	360.13	5,513.82	170	0.3	1.5
Pluto (5 known moons)	Charon	6.39	17.54	604	15.2	1.65
Haumea (2 moons)	Namaka	18	25.66	85	0.018	~1
	Hi'iaka	49	49.88	170	0.179	~1
Eris	Dysnomia	15.8	37.4	50–125?	?	?

*Innermost, outermost, largest, and/or a few other moons for each planet.
[†]The density of water is 1,000 kg/m³.
[‡]Irregular moon (has retrograde orbit).

APPENDIX 3

NEAREST AND BRIGHTEST STARS

Stars within 12 Light-Years of Earth

NAME*	DISTANCE (ly)	SPECTRAL TYPE†	RELATIVE VISUAL LUMINOSITY‡ (SUN = 1.000)	APPARENT MAGNITUDE	ABSOLUTE MAGNITUDE
Sun	1.58×10^{-5}	G2V	1.000	−26.74	4.83
Alpha Centauri C (Proxima Centauri)	4.24	M5.5V	0.000052	11.05	15.48
Alpha Centauri A	4.36	G2V	1.5	0.01	4.38
Alpha Centauri B	4.36	K0V	0.44	1.34	5.71
Barnard's star	5.96	M4Ve	0.00044	9.57	13.25
Luhman 16	6.59	L8	?	10.7**	?
CN Leonis	7.78	M5.5	0.000020	13.53	16.64
BD +36-2147	8.29	M2.0V	0.0056	7.47	10.44
Sirius A	8.58	A1V	22	−1.43	1.47
Sirius B	8.58	DA2	0.0025	8.44	11.34
BL Ceti	8.73	M5.5V	0.000059	12.61	15.40
UV Ceti	8.73	M6.0	0.000039	12.99	15.85
V1216 Sagittarii	9.68	M3.5V	0.00050	10.43	13.07
HH Andromedae	10.32	M5.5V	0.00010	12.29	14.79
Epsilon Eridani	10.52	K2V	0.28	3.73	6.19
Lacaille 9352	10.74	M1.5V	0.011	7.34	9.75
FI Virginis	10.92	M4.0V	0.00033	11.13	13.51
Wise 1506+7027	11.1	T6	?	14.3**	?
EZ Aquarii A	11.26	M5.0V	0.000063	13.03	15.33
EZ Aquarii B	11.26	M5e	0.000050	13.27	15.58
EZ Aquarii C	11.26	—	0.00001	15.07	17.37
Procyon A	11.40	F5IV-V	7.38	0.38	2.66
Procyon B	11.40	DA	0.00054	10.70	12.98

(continued)

Stars within 12 Light-Years of Earth (continued)

NAME*	DISTANCE (ly)	SPECTRAL TYPE†	RELATIVE VISUAL LUMINOSITY‡ (SUN = 1.000)	APPARENT MAGNITUDE	ABSOLUTE MAGNITUDE
61 Cygni A	11.40	K5.0V	0.086	5.21	7.49
61 Cygni B	11.40	K7.0V	0.040	6.03	8.31
Gliese 725 A	11.52	M3.0V	0.0029	8.90	11.16
Gliese 725 B	11.52	M3.5V	0.0014	9.69	11.95
Groombridge 34 A	11.62	M1.5V	0.0063	8.08	10.32
Groombridge 34 B	11.62	M3.5V	0.00041	11.06	13.30
Epsilon Indi A	11.82	K5Ve	0.15	4.69	6.89
Epsilon Indi B (brown dwarf)	11.82	T1.0	—	—	—
Epsilon Indi C (brown dwarf)	11.82	T6.0	—	—	—
DX Cancri	11.82	M6.0V	0.000012	14.9	17.10
Tau Ceti	11.88	G8.5V	0.46	3.49	5.68
Gliese 1061	11.99	M5.5V	0.000067	13.09	15.26

*Stars may carry many names, including common names (such as Sirius), names based on their prominence within a constellation (such as Alpha Canis Majoris, another name for Sirius), or names based on their inclusion in a catalog (such as BD +36-2147). Addition of letters A, B, and so on, or superscripts indicates membership in a multiple-star system.

†Spectral types such as M3 are discussed in Chapter 10. Other letters or numbers provide additional information. For example, V after the spectral type indicates a main-sequence star, and III indicates a giant star. Stars of spectral type T are brown dwarfs.

‡*Luminosity* in this table refers only to radiation in "visual" light.

**Brown dwarf, detected in IR J-band only.

The 25 Brightest Stars in the Sky

NAME	COMMON NAME	DISTANCE (ly)	SPECTRAL TYPE	RELATIVE VISUAL LUMINOSITY* (SUN = 1.000)	APPARENT VISUAL MAGNITUDE	ABSOLUTE VISUAL MAGNITUDE
Sun	Sun	1.58×10^{-5}	G2V	1.000	−26.8	4.82
Alpha Canis Majoris	Sirius	8.60	A1V	22.9	−1.47	1.42
Alpha Carinae	Canopus	313	F0II	13,800	−0.72	−5.53
Alpha Bootis	Arcturus	36.7	K1.5IIIFe-0.5	111	−0.04	−0.29
Alpha^1 Centauri	Rigel Kentaurus	4.39	G2V	1.50	−0.01	4.38
Alpha Lyrae	Vega	25.3	A0Va	49.7	0.03	0.58
Alpha Aurigae	Capella	42.2	G5IIIe+G0III	132	0.08	−0.48
Beta Orionis	Rigel	770	B8Ia	40,200	0.12	−6.69
Alpha Canis Minoris	Procyon	11.4	F5IV-V	7.38	0.34	2.65
Alpha Eridani	Achernar	144	B3Vpe	1,090	0.50	−2.77
Alpha Orionis	Betelgeuse	427	M1-2Ia-Iab	9,600	0.58	−5.14
Beta Centauri	Hadar	350	B1III	8,700	0.60	−5.03
Alpha Aquilae	Altair	16.8	A7V	11.1	0.77	2.21
Alpha Tauri	Aldebaran	65.1	K5+III	151	0.85	0.63
Alpha Virginis	Spica	262	B1III-IV+B2V	2,230	1.04	−3.55
Alpha Scorpii	Antares	604	M1.5Iab-Ib+B4Ve	11,000	1.09	−5.28
Beta Geminorum	Pollux	33.7	K0IIIb	31.0	1.15	1.09
Alpha Piscis	Fomalhaut	25.1	A3V	17.2	1.16	1.73
Alpha Cygni	Deneb	3,200	A2Ia	260,000	1.25	−8.73
Beta Crucis	Becrux	353	B0.5III	3,130	1.30	−3.92
Alpha^2 Centauri	Alpha Centauri B	4.39	K1V	0.44	1.33	5.71
Alpha Leonis	Regulus	77.5	B7V	137	1.35	−0.52
Alpha Crucis	Acrux	321	B0.5IV	4,000	1.40	−4.19
Epsilon Canis Majoris	Adara	431	B2II	3,700	1.51	−4.11
Gamma Crucis	Gacrux	88.0	M3.5III	142	1.59	−0.56

Sources: Data from *The Hipparcos and Tycho Catalogues*, 1997, European Space Agency SP-1200; SIMBAD Astronomical Database (http://simbad.u-strasbg.fr/simbad); and Research Consortium on Nearby Stars (www.chara.gsu.edu/RECONS).

Luminosity in this table refers only to radiation in "visual" light.

APPENDIX 4

STAR MAPS

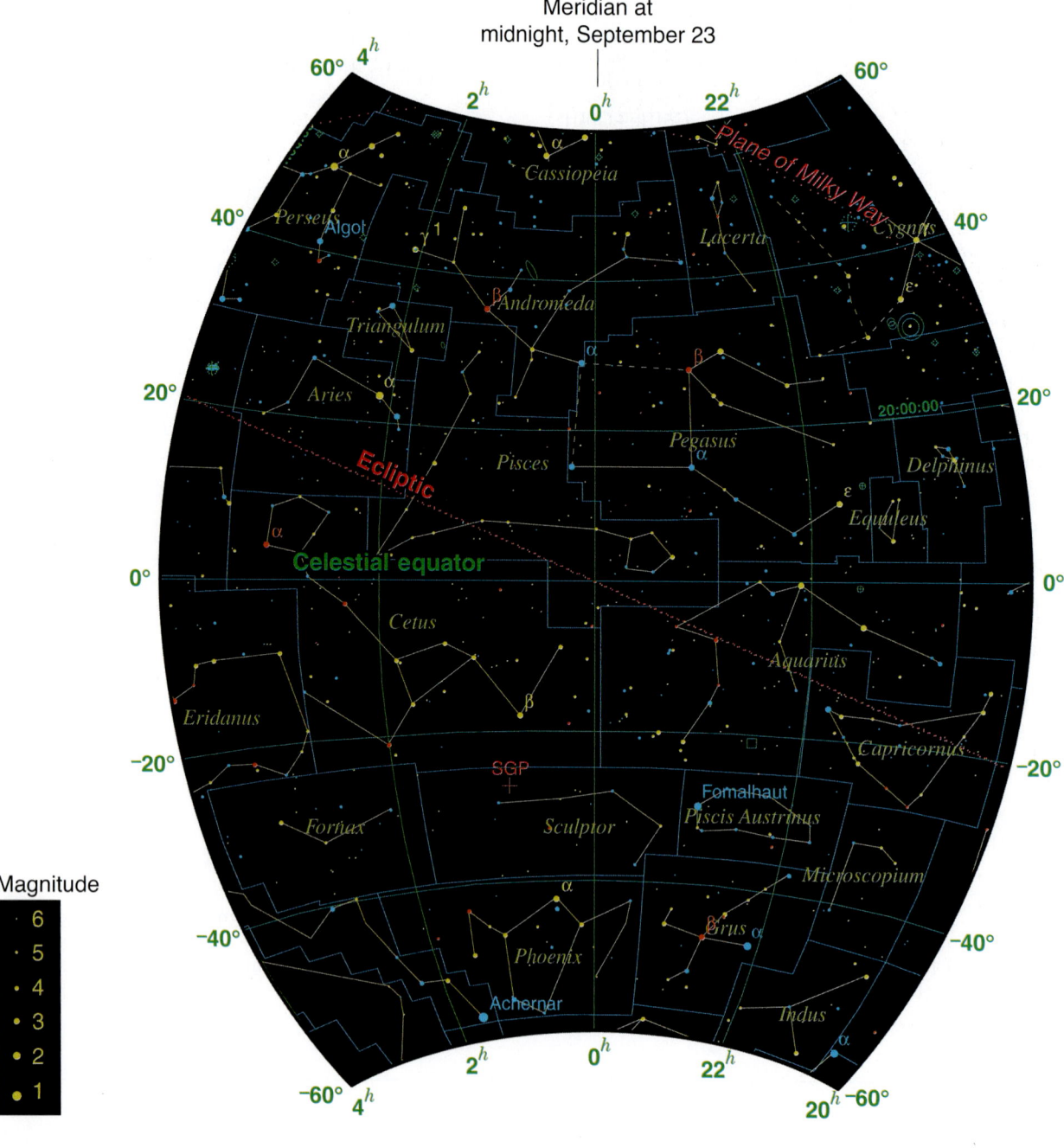

Figure A4.1 The sky from right ascension 20ʰ to 04ʰ and declination −60° to +60°.

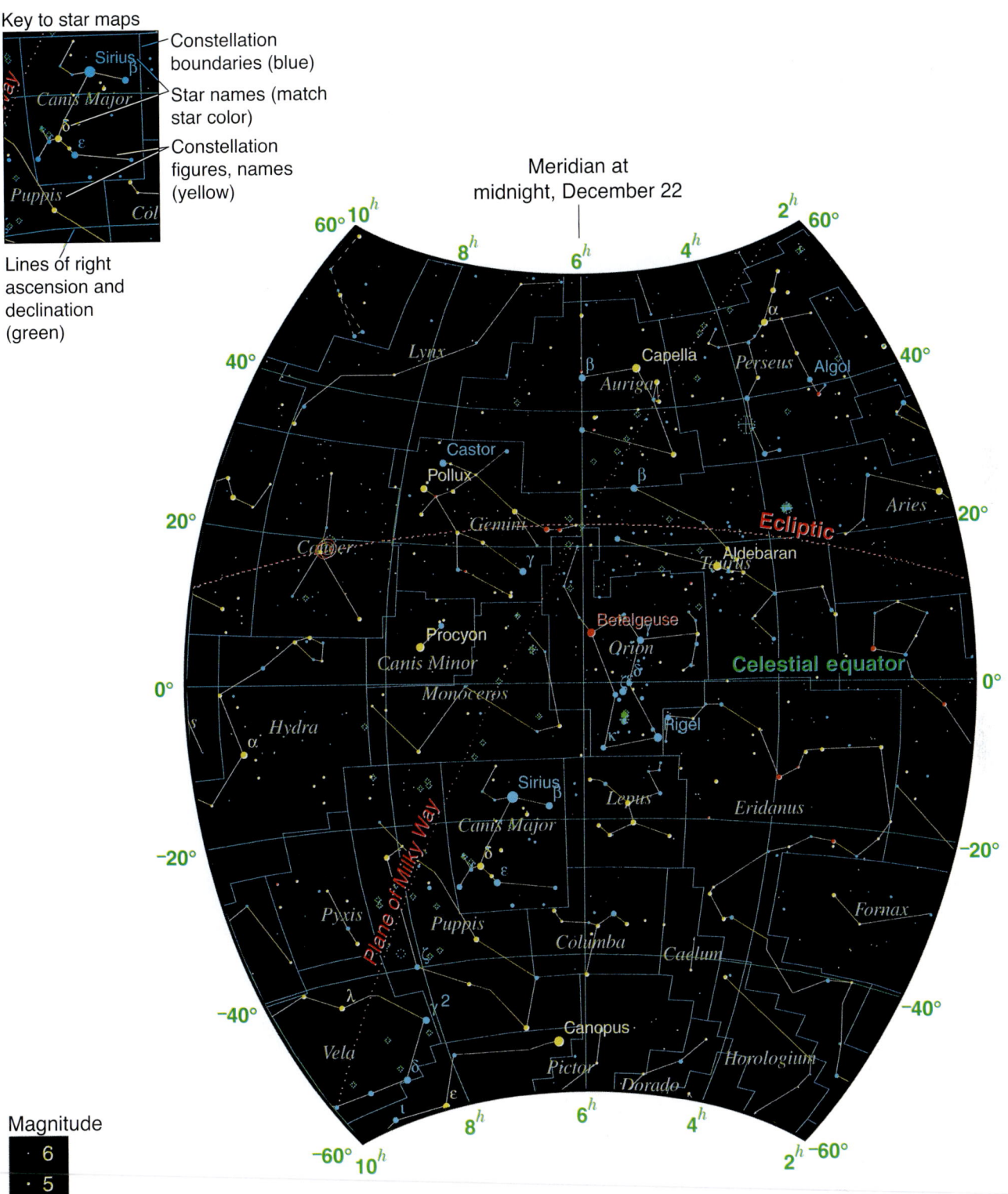

Figure A4.2 The sky from right ascension 02^h to 10^h and declination −60° to +60°.

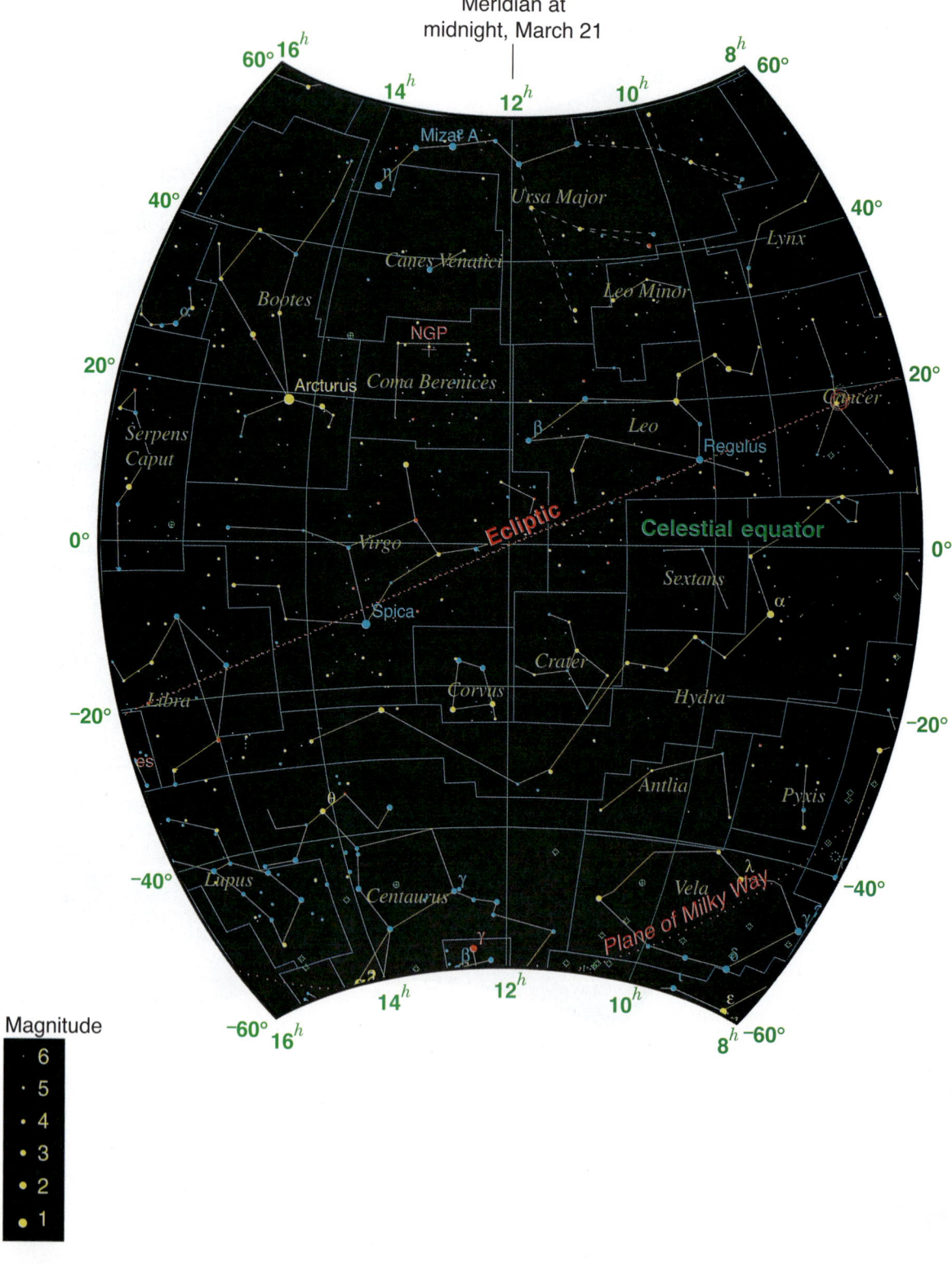

Figure A4.3 The sky from right ascension 08h to 16h and declination −60° to +60°.

APPENDIX 4   A-11

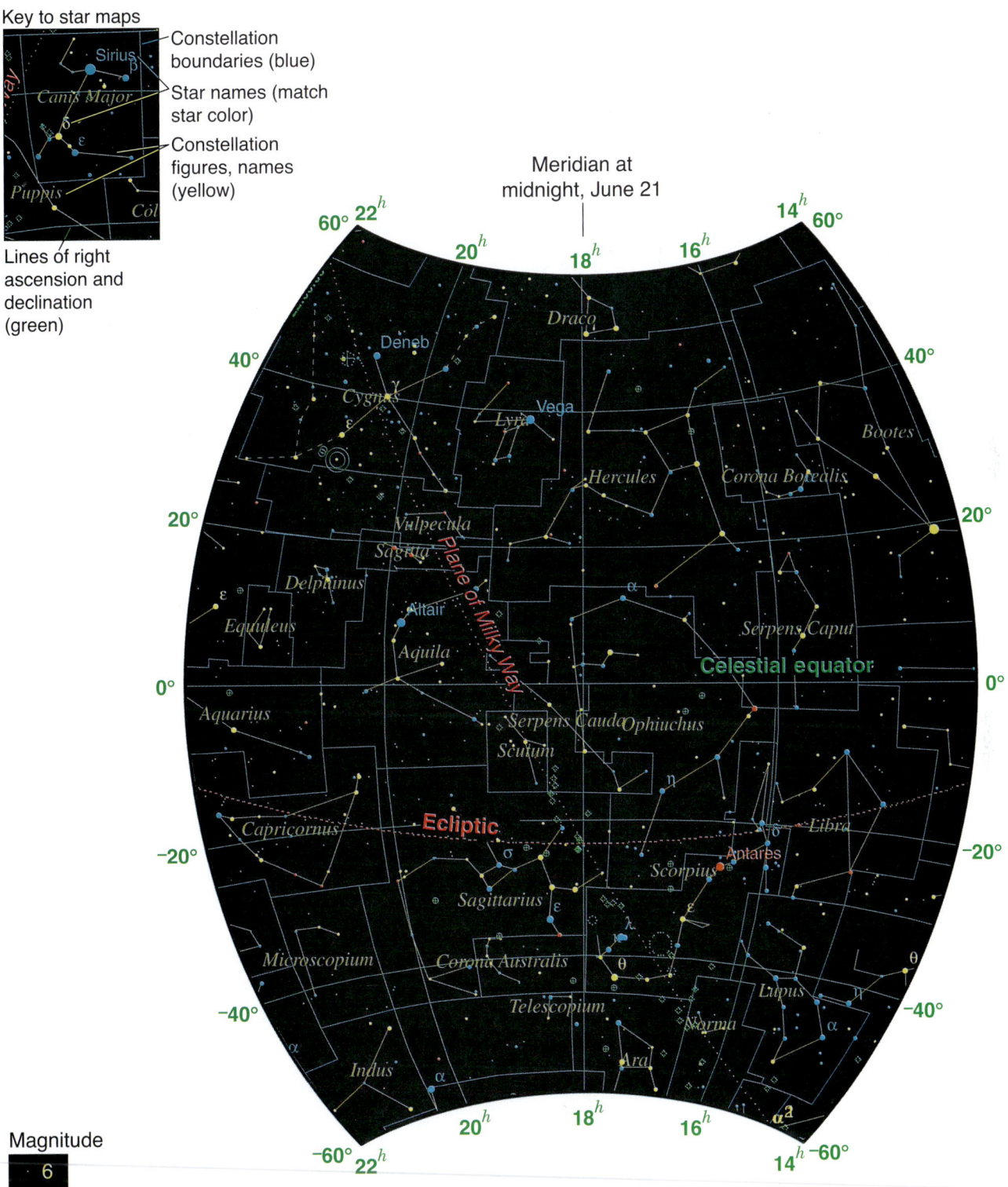

Figure A4.4 The sky from right ascension 14^h to 22^h and declination −60° to +60°.

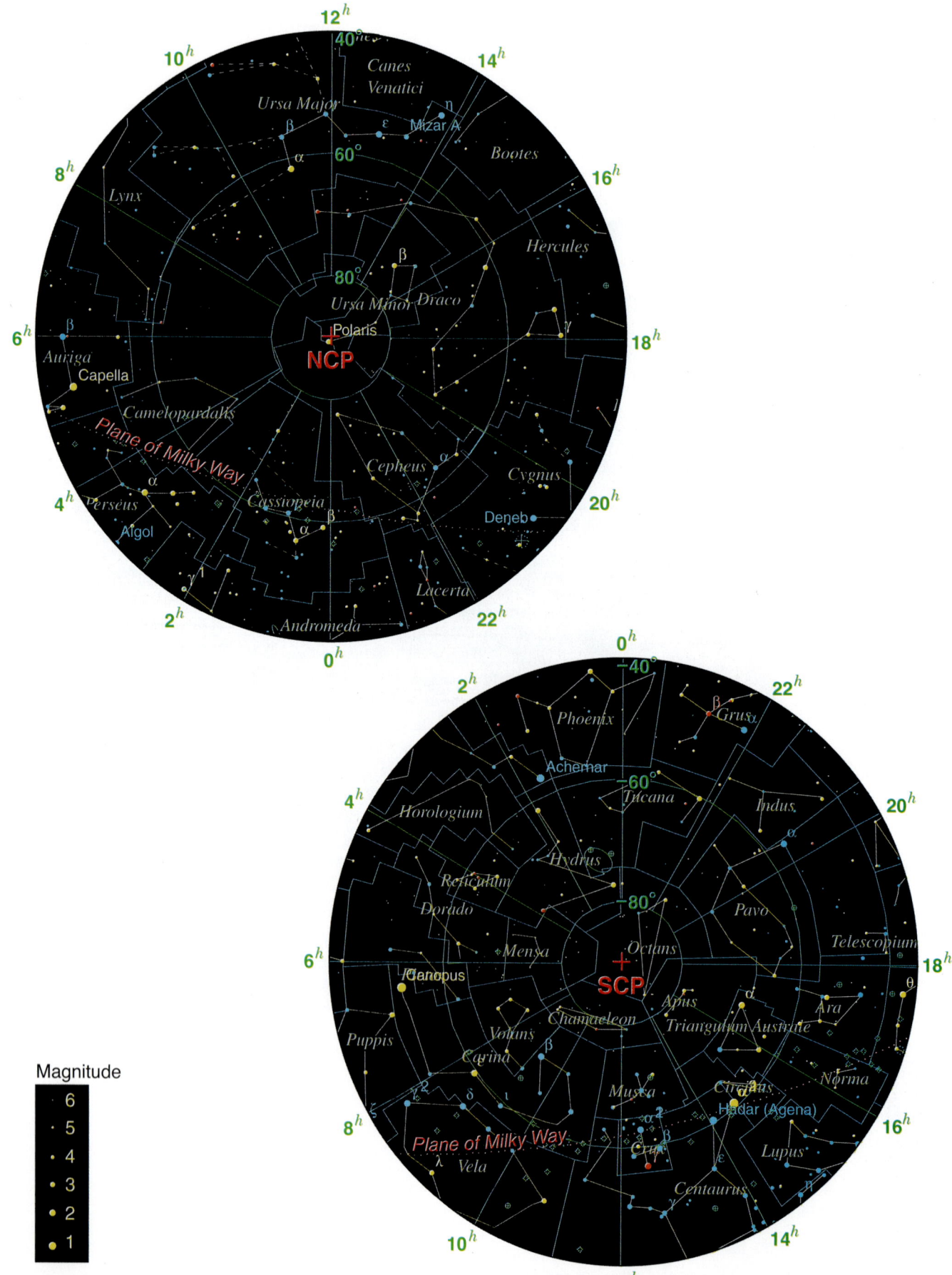

Figure A4.5 The regions of the sky north of declination +40° and south of declination −40°. NCP, north celestial pole; SCP, south celestial pole.

GLOSSARY

A

aberration of starlight The apparent displacement in the position of a star that is due to the finite speed of light and Earth's orbital motion around the Sun.

absolute magnitude A measure of the intrinsic brightness of a celestial object, generally a star. Specifically, the apparent brightness of an object, such as a star, if it were located at a standard distance of 10 parsecs (pc). Compare *apparent magnitude*.

absolute zero The temperature at which thermal motions cease. The lowest possible temperature. Zero on the Kelvin temperature scale.

absorption The capture of electromagnetic radiation by matter. Compare *emission*.

absorption line An intensity minimum in a spectrum that is due to the absorption of electromagnetic radiation at a specific wavelength determined by the energy levels of an atom or molecule. Compare *emission line*.

absorption spectrum A spectrum showing absorption lines.

abundance A measure (by mass, number, or volume) of how much of an element is present, typically measured relative to hydrogen.

acceleration The rate at which the speed and/or direction of an object's motion is changing.

accretion disk A flat, rotating disk of gas and dust surrounding an object, such as a young stellar object, a forming planet, a collapsed star in a binary system, or a black hole.

achondrite A stony meteorite that does not contain chondrules. Compare *chondrite*.

active comet A comet nucleus that approaches close enough to the Sun to show signs of activity, such as the production of a coma and tail.

active galactic nucleus (AGN) A highly luminous, compact galactic nucleus whose luminosity may exceed that of the rest of the galaxy.

active region An area of the Sun's chromosphere anchoring bursts of intense magnetic activity.

adaptive optics Electro-optical systems that largely compensate for image distortion caused by Earth's atmosphere.

AGB See *asymptotic giant branch*.

age (1) The length of time that an object has existed. (2) The length of time since a planetary surface (for example) was last extensively modified.

AGN See *active galactic nucleus*.

albedo The fraction of electromagnetic radiation incident on a surface that is reflected by the surface.

algebra A branch of mathematics in which numeric variables are represented by letters.

alpha particle A ^{4}He nucleus, consisting of two protons and two neutrons. Alpha particles get their name from the fact that they are given off in the type of radioactive decay referred to as "alpha decay."

altitude The location of an object above the horizon, measured by the angle formed between an imaginary line from an observer to the object and a second line from the observer to the point on the horizon directly below the object.

Amors A group of asteroids whose orbits cross the orbit of Mars but not the orbit of Earth. Compare *Apollos* and *Atens*.

amplitude In a wave, the maximum deviation from its undisturbed or relaxed position. For example, in a water wave, the amplitude is the vertical distance from the crest to the undisturbed water level.

angular momentum A conserved property of a rotating or revolving system whose value depends on the velocity and distribution of the system's mass.

angular resolution The minimum angular distance between distinguishable objects in the focal plane of an imaging device, such as a telescope.

annular solar eclipse The type of solar eclipse that occurs when the apparent diameter of the Moon is less than that of the Sun, leaving a visible ring of light ("annulus") surrounding the dark disk of the Moon. Compare *partial solar eclipse* and *total solar eclipse*.

Antarctic Circle The circle on Earth with latitude 66.5° south, marking the northern limit where at least 1 day per year is in 24-hour daylight. Compare *Arctic Circle*.

anticyclonic motion The rotation of a weather system resulting from the Coriolis effect as air moves outward from a region of high atmospheric pressure. Compare *cyclonic motion*.

antimatter Matter made from antiparticles.

antiparticle An elementary particle of antimatter identical in mass but opposite in charge and all other properties to its corresponding ordinary matter particle.

aperture The clear diameter of a telescope's objective lens or primary mirror.

aphelion (pl. **aphelia**) The point in a solar orbit that is farthest from the Sun. Compare *perihelion*.

Apollos A group of asteroids whose orbits cross the orbits of both Earth and Mars. Compare *Amors* and *Atens*.

apparent magnitude A measure of the apparent brightness of a celestial object, generally a star. Compare *absolute magnitude*.

apparent retrograde motion A motion of the planets with respect to the "fixed stars," in which the planets appear to move westward for a period of time before resuming their normal eastward motion. The heliocentric model explains this effect much more simply than the geocentric model.

arcminute (arcmin) A minute of arc ('), a unit used for measuring angles. An arcminute is 1/60 of a degree of arc.

arcsecond (arcsec) A second of arc ("), a unit used for measuring very small angles. An arcsecond is 1/60 of an arcminute, or 1/3,600 of a degree of arc.

Arctic Circle The circle on Earth with latitude 66.5° north, marking the southern limit where at least 1 day per year is in 24-hour daylight. Compare *Antarctic Circle*.

ash The products of fusion that collect in the core of a star.

asteroid Also called *minor planet*. A primitive rocky or metallic body (planetesimal) that has survived planetary accretion. Asteroids are parent bodies of meteoroids.

asteroid belt Also called *main asteroid belt*. The region between the orbits of Mars and Jupiter that contains most of the asteroids in our Solar System.

astrobiology An interdisciplinary science combining astronomy, biology, chemistry, geology, and physics to study life in the cosmos.

astrology The belief that the positions and aspects of stars and planets influence human affairs and characteristics, as well as terrestrial events.

astrometry Precision measurement of the position and motion of astronomical objects.

astronomical seeing A measurement of the degree to which Earth's atmosphere degrades the resolution of a telescope's view of astronomical objects.

astronomical unit (AU) The average distance from the Sun to Earth: approximately 150 million kilometers (km).

astronomy The scientific study of planets, stars, galaxies, and the universe as a whole.

astrophysics The application of physical laws to the understanding of planets, stars, galaxies, and the universe as a whole.

asymptotic giant branch (AGB) The path on the H-R diagram that goes from the horizontal branch toward higher luminosities and lower temperatures, asymptotically approaching and then rising above the red giant branch.

Atens A group of asteroids whose orbits cross the orbit of Earth but not the orbit of Mars. Compare *Amors* and *Apollos*.

atmosphere The gravitationally bound, outer gaseous envelope surrounding a planet, moon, or star.

atmospheric greenhouse effect A warming of planetary surfaces produced by atmospheric gases that transmit optical solar radiation but partially trap infrared radiation.

atmospheric probe An instrumented package designed to provide in situ measurements of the chemical and/or physical properties of a planetary atmosphere.

atmospheric window A region of the electromagnetic spectrum in which radiation is able to penetrate a planet's atmosphere.

atom The smallest piece of a chemical element that retains the properties of that element. Each atom is composed of a nucleus (neutrons and protons) surrounded by a cloud of electrons.

AU See *astronomical unit*.

aurora Emission in the upper atmosphere of a planet from atoms that have been excited by collisions with energetic particles from the planet's magnetosphere.

autumnal equinox 1. One of two points where the Sun crosses the celestial equator. 2. The day on which the Sun appears at this location, marking the first day of autumn (about September 22 in the Northern Hemisphere and March 20 in the Southern Hemisphere). Compare *vernal equinox*.

axion A hypothetical elementary particle first proposed to explain certain properties of the neutron and now considered a candidate for cold dark matter.

B

backlighting Illumination from behind a subject as seen by an observer. Fine material such as human hair and dust in planetary rings stands out best when viewed under backlighting conditions.

bar A unit of pressure. One bar is equivalent to 10^5 newtons per square meter—approximately equal to Earth's atmospheric pressure at sea level.

barred spiral A spiral galaxy with a bulge having an elongated, barlike shape.

basalt Gray to black volcanic rock, rich in iron and magnesium.

beta decay 1. The decay of a neutron into a proton by emission of an electron (beta ray) and an antineutrino. 2. The decay of a proton into a neutron by emission of a positron and a neutrino.

Big Bang The event that occurred 13.8 billion years ago that marks the beginning of time and the universe.

Big Bang nucleosynthesis The formation of low-mass nuclei (H, He, Li, Be, B) during the first few minutes after the Big Bang.

Big Rip A hypothetical cosmic event in which all matter in the universe, from stars to subatomic particles, is progressively torn apart by expansion of the universe.

binary star A system in which two stars are in gravitationally bound orbits around their common center of mass.

binding energy The minimum energy required to separate an atomic nucleus into its component protons and neutrons.

biosphere The global sum of all living organisms on Earth (or any planet or moon). Compare *hydrosphere* and *lithosphere*.

bipolar outflow Material streaming away in opposite directions from either side of the accretion disk of a young star.

black hole An object so dense that its escape velocity exceeds the speed of light; a singularity in spacetime.

blackbody An object that absorbs and can reemit all electromagnetic energy it receives.

blackbody spectrum Also called *Planck spectrum*. The spectrum of electromagnetic energy emitted by a blackbody per unit area per second, which is determined only by the temperature of the object.

blueshift The Doppler shift toward shorter wavelengths of light from an approaching object. Compare *redshift*.

Bohr model A model of the atom, proposed by Niels Bohr in 1913, in which a small positively charged nucleus is surrounded by orbiting electrons, similar to a miniature solar system.

bolide A very bright, exploding meteor.

bound orbit A closed orbit in which the velocity is less than the escape velocity. Compare *unbound orbit*.

bow shock 1. The boundary at which the speed of the solar wind abruptly drops from supersonic to subsonic in its approach to a planet's magnetosphere; the boundary between the region dominated by the solar wind and the region dominated by a planet's magnetosphere. 2. The interface between strong collimated gas and dust outflow from a star and the interstellar medium.

brightness The apparent intensity of light from a luminous object. Brightness depends on both the *luminosity* of a source and its distance. Units at the detector: watts per square meter (W/m^2).

brown dwarf A "failed" star that is not massive enough to cause hydrogen fusion in its core. An object whose mass is intermediate between that of the least massive stars and that of supermassive planets.

bulge The central region of a spiral galaxy that is similar in appearance to a small elliptical galaxy.

burning In stellar nuclear physics, the fusion of light elements into heavier ones. For example, the fusion of hydrogen into helium is often called "hydrogen burning".

C

C See *Celsius*.

C-type asteroid An asteroid made of material that has largely been unmodified since the formation of the Solar System; the most primitive type of asteroid. Compare *M-type asteroid* and *S-type asteroid*.

caldera The summit crater of a volcano.

carbon-nitrogen-oxygen (CNO) cycle One of the ways in which hydrogen is converted to helium (hydrogen burning) in the interiors of main-sequence stars. See also *proton-proton chain* and *triple-alpha process*.

carbon star A cool red giant or asymptotic giant branch star that has an excess of carbon in its atmosphere.

carbonaceous chondrite A primitive stony meteorite that contains chondrules and is rich in carbon and volatile materials.

Cassini Division The largest gap in Saturn's rings, discovered by Jean-Dominique Cassini in 1675.

catalyst An atomic and molecular structure that permits or encourages chemical and nuclear reactions but does not change its own chemical or nuclear properties.

CCD See *charge-coupled device*.

celestial equator The imaginary great circle that is the projection of Earth's equator onto the celestial sphere.

celestial sphere An imaginary sphere with celestial objects on its inner surface and Earth at its center. The celestial sphere has no physical existence but is a convenient tool for picturing the directions in which celestial objects are seen from the surface of Earth.

Celsius (C) Also called *centigrade scale*. The arbitrary temperature scale—defined by Anders Celsius (1701–1744)—that defines 0°C as the freezing point of water and 100°C as the boiling point of water at sea level. Unit: °C. Compare *Fahrenheit* and *Kelvin scale*.

center of mass The location associated with an object system at which we may regard the entire mass of the system as being concentrated. The point in any isolated system that moves according to Newton's first law of motion.

centigrade scale See *Celsius*.

centripetal force A force directed toward the center of curvature of an object's curved path.

Cepheid variable An evolved high-mass star with an atmosphere that is pulsating, leading to variability in the star's luminosity and color.

Chandrasekhar limit The upper limit on the mass of an object supported by electron degeneracy pressure; approximately 1.4 solar mass ($M_\odot$).

chaos Behavior in complex, interrelated systems in which tiny differences in the initial configuration of a system result in dramatic differences in the system's later evolution.

charge-coupled device (CCD) A common type of solid-state detector of electromagnetic radiation that transforms the intensity of light directly into electric signals.

chemistry The study of the composition, structure, and properties of substances.

chondrite A stony meteorite containing chondrules. Compare *achondrite*.

chondrule A small, crystallized, spherical inclusion of rapidly cooled molten droplets found inside some meteorites.

chromatic aberration A detrimental property of a lens in which rays of different wavelengths are brought to different focal distances from the lens.

chromosphere The region in the Sun's atmosphere located between the photosphere and the corona.

circular velocity (v_{circ}) The orbital velocity needed to keep an object moving in a circular orbit.

circumpolar Referring to the part of the sky, near either celestial pole, that can always be seen above the horizon from a specific location on Earth.

circumstellar disk See *protoplanetary disk*.

classical mechanics The science of applying Newton's laws to the motion of objects.

classical planets The eight major planets of the Solar System: Mercury, Venus, Earth, Mars, Jupiter, Saturn, Uranus, and Neptune.

climate The state of an atmosphere averaged over an extended time. Compare *weather*.

closed universe A finite universe with a curved spatial structure such that the sum of the angles of a triangle always exceeds 180°. Compare *flat universe* and *open universe*.

cloud 1. In planetary astronomy, a concentration of condensed volatiles in a planetary atmosphere. 2. In the interstellar medium, a volume of dust and gas that is denser than the surroundings.

CMB See *cosmic microwave background radiation*.

CNO cycle See *carbon-nitrogen-oxygen cycle*.

cold dark matter Particles of dark matter that move slowly enough to be gravitationally bound even in the smallest galaxies. Compare *hot dark matter*.

color index The color of a celestial object, generally a star, based on the ratio of its brightness in blue light (b_B) to its brightness in "visual" (or yellow-green) light (b_V). The difference between an object's blue (B) magnitude and visual (V) magnitude, $B-V$.

coma (pl. **comae**) The nearly spherical cloud of gas and dust surrounding the nucleus of an active comet.

comet A complex object consisting of a small, solid, icy nucleus; an atmospheric halo; and a tail of gas and dust.

comet nucleus A primitive planetesimal composed of ices and refractory materials that has survived planetary accretion. The "heart" of a comet, containing nearly the entire mass of the comet. A "dirty snowball."

comparative planetology The study of planets through comparison of their chemical and physical properties.

complex system An interrelated system capable of exhibiting chaotic behavior. See also *chaos*.

composite volcano A large, cone-shaped volcano formed by viscous, pasty lava flows alternating with pyroclastic (explosively generated) rock deposits. Compare *shield volcano*.

compound lens A lens made up of two or more elements of differing refractive index, the purpose of which is to minimize chromatic aberration.

conservation law A physical law stating that the amount of a particular physical quantity (such as energy or angular momentum) of an isolated system does not change over time.

conservation of angular momentum The physical law stating that the amount of angular momentum of an isolated system does not change over time.

conservation of energy The physical law stating that the amount of energy of an isolated, closed system does not change over time.

constant of proportionality The multiplicative factor by which one quantity is related to another.

constellation An imaginary image formed by patterns of stars; any of 88 defined areas on the celestial sphere used by astronomers to locate celestial objects.

constructive interference A state in which the amplitudes of two intersecting waves reinforce one another. Compare *destructive interference*.

continental drift The slow motion (centimeters per year) of Earth's continents relative to each other and to Earth's mantle. See also *plate tectonics*.

continuous radiation Electromagnetic radiation with intensity that varies smoothly over a wide range of wavelengths.

convection The transport of thermal energy from the lower (hotter) to the higher (cooler) layers of a fluid by motions within the fluid driven by variations in buoyancy.

convective zone A region within a star in which energy is transported outward by convection.

conventional greenhouse effect The solar heating of air in an enclosed space, such as a closed building or car, resulting primarily from the inability of the hot air to escape.

core 1. The innermost region of a planetary interior. 2. The innermost part of a star.

core accretion A process for forming giant planets, whereby large quantities of surrounding hydrogen and helium are gravitationally captured onto a massive rocky core.

Coriolis effect The apparent displacement of objects in a direction perpendicular to their true motion as viewed from a rotating frame of reference. On a rotating planet, different latitudes rotating at different speeds cause this effect.

corona The hot, outermost part of the Sun's atmosphere.

coronal hole A low-density region in the solar corona containing "open" magnetic field lines along which coronal material is free to stream into interplanetary space.

coronal mass ejection An eruption on the Sun that ejects hot gas and energetic particles at much higher speeds than are typical in the solar wind.

cosmic microwave background (CMB) radiation Isotropic microwave radiation from every direction in the sky having a 2.73-kelvin (K) blackbody spectrum. The CMB is residual radiation from the Big Bang.

cosmic ray A very-fast-moving particle (usually an atomic nucleus); cosmic rays fill the disk of our galaxy.

cosmological constant A constant, introduced into general relativity by Einstein, that characterizes an extra, repulsive force in the universe due to the vacuum of space itself.

cosmological principle The (testable) assumption that the same physical laws that apply here and now also apply everywhere and at all times, and that there are no special locations or directions in the universe.

cosmological redshift (z) The redshift that results from the expansion of the universe rather than from the motions of galaxies or gravity (see *gravitational redshift*).

cosmology The study of the large-scale structure and evolution of the universe as a whole.

Crab Nebula The remnant of the Type II supernova explosion witnessed by Chinese astronomers in A.D. 1054.

crescent Any phase of the Moon, Mercury, or Venus in which the object appears less than half illuminated by the Sun. Compare *gibbous*.

Cretaceous-Tertiary (K-T) boundary The boundary between the Cretaceous and Tertiary periods in Earth's history. This boundary corresponds to the time of the impact of an asteroid or comet and the extinction of the dinosaurs.

critical density The value of the mass density of the universe that, ignoring any cosmological constant, is just barely capable of halting expansion of the universe.

crust The relatively thin, outermost, hard layer of a planet, which is chemically distinct from the interior.

cryovolcanism Low-temperature volcanism in which the magmas are composed of molten ices rather than rocky material.

cyclone See *hurricane*.

cyclonic motion The rotation of a weather system resulting from the Coriolis effect as air moves toward a region of low atmospheric pressure. Compare *anticyclonic motion*.

Cygnus X-1 A binary X-ray source and probable black hole.

D

dark energy A form of energy that permeates all of space (including the vacuum) producing a repulsive force that accelerates the expansion of the universe.

dark matter Matter in galaxies that does not emit or absorb electromagnetic radiation. Dark matter is thought to constitute most of the mass in the universe. Compare *luminous matter*.

dark matter halo The centrally condensed, greatly extended dark matter component of a galaxy that contains up to 95 percent of the galaxy's mass.

daughter product An element resulting from radioactive decay of a more massive *parent element*.

day The time for Earth to rotate around its axis. A sidereal day is the time to rotate relative to the fixed stars (23h 56m). A solar day is the time to rotate relative to the Sun (24h).

decay 1. The process of a radioactive nucleus changing into its daughter product. 2. The process of an atom or molecule dropping from a higher-energy state to a lower-energy state. 3. The process of a satellite's orbit losing energy.

deep In astronomy, far from Earth, distant.

density The measure of an object's mass per unit of volume. Units: kilograms per cubic meter (kg/m³).

destructive interference A state in which the amplitudes of two intersecting waves cancel one another. Compare *constructive interference*.

differential rotation Rotation of different parts of a system at different rates.

differentiation The process by which materials of higher density sink toward the center of a molten or fluid planetary interior.

diffraction The spreading of a wave after it passes through an opening or past the edge of an object.

diffraction limit The limit of a telescope's angular resolution caused by diffraction.

diffuse ring A sparsely populated planetary ring spread out both horizontally and vertically.

direct imaging A technique for detecting extrasolar planets by observing them directly with telescopes.

dispersion The separation of rays of light into their component wavelengths.

distance ladder A sequence of techniques for measuring cosmic distances; each method is calibrated using the results from other methods that have been applied to closer objects.

Doppler effect The change in wavelength of sound or light that is due to the relative motion of the source toward or away from the observer.

Doppler redshift See *redshift*.

Doppler shift The amount by which the wavelength of light is shifted by the Doppler effect.

Drake equation A prescription for estimating the number of intelligent civilizations existing elsewhere.

dust devil A small tornado-like column of air containing dust or sand.

dust tail A type of comet tail consisting of dust particles that are pushed away from the comet's head by radiation pressure from the Sun. Compare *ion tail*.

dwarf galaxy A small galaxy with a luminosity ranging from 1 million to 1 billion solar luminosities ($L_\odot$). Compare *giant galaxy*.

dwarf planet A body with characteristics similar to those of a classical planet except that it has not cleared smaller bodies from the neighboring regions around its orbit. Compare *planet* (definition 2).

dynamic equilibrium A state in which a system is constantly changing but its configuration remains the same because one source of change is exactly balanced by another source of change. Compare *static equilibrium*.

dynamo A device that converts mechanical energy into electric energy in the form of electric currents and magnetic fields. The "dynamo effect" is thought to create magnetic fields in planets and stars by electrically charged currents of material flowing within their cores.

E

eccentricity (e) A measure of the departure of an ellipse from circularity; the ratio of the distance between the two foci of an ellipse to its major axis.

eclipse 1. The total or partial obscuration of one celestial body by another. 2. The total or partial obscuration of light from one celestial body as it passes through the shadow of another celestial body.

eclipse season Any time during the year when the Moon's line of nodes is sufficiently close to the Sun for eclipses to occur.

eclipsing binary A binary system in which the orbital plane is oriented such that the two stars appear to pass in front of one another as seen from Earth. Compare *spectroscopic binary* and *visual binary*.

ecliptic 1. The apparent annual path of the Sun against the background of stars. 2. The projection of Earth's orbital plane onto the celestial sphere.

ecliptic plane The plane of Earth's orbit around the Sun. The ecliptic is the projection of this plane on the celestial sphere.

effective temperature The temperature at which a black body, such as a star, appears to radiate.

ejecta 1. Material thrown outward by the impact of an asteroid or comet on a planetary surface, leaving a crater behind. 2. Material thrown outward by a stellar explosion.

electric field A field that is able to exert a force on a charged object, whether at rest or moving. Compare *magnetic field*.

electric force The force exerted on a charged particle by an electric field. Compare *magnetic force*.

electromagnetic force The force, including both electric and magnetic forces, that acts on electrically charged particles. One of four fundamental forces of nature. The force mediated by photons.

electromagnetic radiation A traveling disturbance in the electric and magnetic fields caused by accelerating electric charges. In quantum mechanics, a stream of photons. Light.

electromagnetic spectrum The spectrum made up of all possible frequencies or wavelengths of electromagnetic radiation, ranging from gamma rays through radio waves and including the portion our eyes can use.

electromagnetic wave A wave consisting of oscillations in the electric-field strength and the magnetic-field strength.

electron (e^-) A subatomic particle having a negative charge of 1.6×10^{-19} coulomb (C), a rest mass of 9.1×10^{-31} kilograms (kg), and mass-equivalent energy of 8×10^{-14} joules (J). The antiparticle of the *positron*. Compare *proton* and *neutron*.

electron-degenerate Describing the state of material compressed to the point at which electron density reaches the limit imposed by the rules of quantum mechanics.

electroweak theory The quantum theory that combines descriptions of both the electromagnetic force and the weak nuclear force.

element One of 92 naturally occurring substances (such as hydrogen, oxygen, and uranium) and more than 20 human-made ones (such as plutonium). Each element is chemically defined by the specific number of protons in the nuclei of its atoms.

elementary particle One of the basic building blocks of nature that is not known to have substructure, such as the *electron* and the *quark*.

ellipse A conic section produced by the intersection of a plane with a cone when the plane is passed through the cone at an angle to the axis other than 0° or 90°.

elliptical galaxy A galaxy of Hubble type "E" class, with a circular to elliptical outline on the sky containing almost no disk and a population of old stars. Compare *irregular galaxy*, *S0 galaxy*, and *spiral galaxy*.

emission The release of electromagnetic energy when an atom, molecule, or particle drops from a higher-energy state to a lower-energy state. Compare *absorption*.

emission line An intensity peak in a spectrum that is due to sharply defined emission of electromagnetic radiation in a narrow range of wavelengths. Compare *absorption line*.

emission spectrum The spectrum of the light emitted from an object. May contain emission lines.

empirical Derived directly from observations or evidence, rather than upon logic or theoretical inference.

empirical science Scientific investigation that is based primarily on observations and experimental data. It is descriptive rather than based on theoretical inference.

energy The conserved quantity that gives objects and systems the ability to do work. Units: joules (J).

energy transport The transfer of energy from one location to another. In stars, energy transport is carried out mainly by radiation or convection.

entropy A measure of the disorder of a system related to the number of ways a system can be rearranged without its appearance being affected.

equator The imaginary great circle on the surface of a body midway between its poles that divides the body into northern and southern hemispheres. The equatorial plane passes through the center of the body and is perpendicular to its rotation axis. Compare *meridian*.

equilibrium The state of an object in which physical processes balance each other so that its properties or conditions remain constant.

equinox Literally, "equal night." 1. One of two positions on the ecliptic where it intersects the celestial equator. 2. Either of the two times of year (the *autumnal equinox* and *vernal equinox*) when the Sun is at one of these two positions. At this time, night and day are of the same length everywhere on Earth. Compare *solstice*.

equivalence principle The principle stating that there is no difference between a frame of reference that is freely floating through space and one that is freely falling within a gravitational field.

erosion The degradation of a planet's surface topography by the mechanical action of wind and/or water.

escape velocity The minimum velocity needed for an object to achieve a parabolic trajectory and thus permanently leave the gravitational grasp of another mass.

eternal inflation The idea that a universe might inflate forever. In such a universe, quantum effects could randomly cause regions to slow their expansion, eventually stop inflating, and experience an explosion resembling our Big Bang.

event A particular location in spacetime.

event horizon The effective "surface" of a black hole. Nothing inside this surface—not even light—can escape from a black hole.

evolutionary track The path that a star follows across the H-R diagram as it evolves through its lifetime.

excited state An energy level of a particular atom, molecule, or particle that is higher than its ground state. Compare *ground state*.

exoplanet See *extrasolar planet*.

extrasolar planet Also called *exoplanet*. A planet orbiting a star other than the Sun.

extremophiles Microbes that are adapted to tolerate extreme conditions, such as high or low temperatures.

F

F See *Fahrenheit*.

Fahrenheit (F) The arbitrary temperature scale—defined by Daniel Gabriel Fahrenheit (1686–1736)—that defines 32°F as the melting point of water and 212°F as the boiling point of water at sea level. Unit: °F. Compare *Celsius* and *Kelvin scale*.

falsified Proven wrong.

fault A fracture in the crust of a planet or moon along which blocks of material can slide.

filter An instrument element that transmits a limited wavelength range of electromagnetic radiation. For the optical range, such elements are typically made of different kinds of glass and take on the hue of the light they transmit.

first quarter Moon The phase of the Moon in which only the western half of the Moon, as viewed from Earth, is illuminated by the Sun. It occurs about a week after a new Moon. Compare *third quarter Moon*.

fissure A fracture in the planetary lithosphere from which magma emerges.

flat rotation curve A rotation curve of a spiral galaxy in which rotation rates do not decline in the outer part of the galaxy, but remain relatively constant to the outermost points.

flat universe An infinite universe whose spatial structure obeys Euclidean geometry, such that the sum of the angles of a triangle always equals 180°. Compare *closed universe* and *open universe*.

flatness problem The surprising result that the sum of $\Omega_{mass} + \Omega_\Lambda$ is extremely close to unity in the present-day universe; equivalent to saying that it is surprising the universe is so close to being exactly flat.

fluid A liquid or a gas; a substance that flows to take on the shape of its container.

flux The total amount of energy passing through each square meter of a surface each second. Units: watts per square meter (W/m²).

flux tube A strong magnetic field contained within a tubelike structure. Flux tubes are found in the solar atmosphere and connecting the space between Jupiter and its moon Io.

flyby A spacecraft that first approaches and then continues flying past a planet or moon. Flybys can visit multiple objects, but they remain in the vicinity of their targets only briefly. Compare *orbiter*.

focal length The optical distance between a telescope's objective lens or primary mirror and the plane (called the focal plane) on which the light from a distant object is focused.

focal plane The plane, perpendicular to the optical axis of a lens or mirror, in which an image is formed.

focus (pl. **foci**) 1. One of two points that define an ellipse. 2. A point in the focal plane of a telescope.

force A push or a pull on an object.

frame of reference A coordinate system within which an observer measures positions and motions.

free fall The motion of an object when the only force acting on it is gravity.

frequency The number of times per second that a periodic process occurs. Unit: hertz (Hz), 1/s.

full Moon The phase of the Moon in which the near side of the Moon, as viewed from Earth, is fully illuminated by the Sun. It occurs about 2 weeks after a *new Moon*.

G

galaxy A gravitationally bound system that consists of stars and star clusters, gas, dust, and dark matter; typically greater than 1,000 light-years across and recognizable as a discrete, single object.

galaxy cluster A large, gravitationally bound collection of galaxies containing hundreds to thousands of members; typically 3–5 Mpc (about 10 million to 15 million light-years) across. Compare *galaxy group* and *supercluster*.

galaxy group A small, gravitationally bound collection of galaxies containing from several to a hundred members; typically 4 million to 6 million light-years across. Compare *galaxy cluster* and *supercluster*.

gamma ray Electromagnetic radiation with higher frequency, higher photon energy, and shorter wavelength than all other types of electromagnetic radiation.

gas giant A giant planet formed mostly of hydrogen and helium. In our Solar System, Jupiter and Saturn are the gas giants. Compare *ice giant*.

gauss A basic unit of magnetic flux density.

general relativistic time dilation The verified prediction that time passes more slowly in a gravitational field than in the absence of a gravitational field. Compare *time dilation*.

general relativity See *general theory of relativity*.

general theory of relativity Sometimes referred to as simply *general relativity*. Einstein's theory explaining gravity as the distortion of spacetime by massive objects, such that particles travel on the shortest path between two events in spacetime. This theory deals with all types of motion. Compare *special theory of relativity*.

geocentric A coordinate system having the center of Earth as its center. Compare *heliocentric*.

geodesic The path an object will follow through spacetime in the absence of external forces.

geometry A branch of mathematics that deals with points, lines, angles, and shapes.

giant galaxy A galaxy with luminosity greater than about 1 billion solar luminosities ($L_\odot$). Compare *dwarf galaxy*.

giant molecular cloud An interstellar cloud composed primarily of molecular gas and dust, having hundreds of thousands of solar masses.

giant planet One of the largest planets in the Solar System (Saturn, Jupiter, Uranus, or Neptune), typically 10 times the size and many times the mass of any *terrestrial planet* and lacking a solid surface.

gibbous Any phase of the Moon, Mercury, or Venus in which the object appears more than half illuminated by the Sun. Compare *crescent*.

global circulation The overall, planet-wide circulation pattern of a planet's atmosphere.

globular cluster A spherically symmetric, highly condensed group of stars, containing tens of thousands to a million members. Compare *open cluster*.

gluon The particle that carries (or, equivalently, mediates) interactions due to the strong nuclear force.

gossamer ring An extremely tenuous planetary ring found beyond Jupiter's main ring.

grand unified theory (GUT) A unified quantum theory that combines the strong nuclear, weak nuclear, and electromagnetic forces but does not include gravity.

granite Rock that is cooled from magma and is relatively rich in silicon and oxygen.

grating An optical surface containing many narrow, closely and equally spaced parallel grooves or slits that spectrally disperse reflected or transmitted light.

gravitational force Force due to the gravitational interaction between two or more objects.

gravitational lens A massive object that gravitationally focuses the light of a more distant object to produce multiple brighter, magnified, possibly distorted images.

gravitational lensing The bending of light by gravity. Can be used to detect extrasolar planets.

gravitational potential energy The stored energy in an object that is due solely to its position within a gravitational field.

gravitational redshift The shifting to longer wavelengths of radiation from an object deep within a gravitational well.

gravitational wave A wave in the fabric of spacetime emitted by accelerating masses.

gravity 1. The mutually attractive force between massive objects. 2. An effect arising from the bending of spacetime by massive objects. 3. One of four fundamental forces of nature.

great circle Any circle on a sphere that has as its center the center of the sphere. The celestial equator, the meridian, and the ecliptic are all great circles on the sphere of the sky, as is any circle drawn through the zenith.

Great Red Spot The giant, oval, brick-red anticyclone seen in Jupiter's southern hemisphere.

greenhouse effect See *atmospheric greenhouse effect* and *conventional greenhouse effect*.

greenhouse gas One of a group of atmospheric gases such as carbon dioxide that are transparent to visible radiation but absorb infrared radiation.

Gregorian calendar The modern calendar. A modification of the Julian calendar decreed by Pope Gregory XIII in 1582. By this time, the less accurate Julian calendar had developed an error of 10 days over the 13 centuries since its inception.

ground state The lowest possible energy state for a system or part of a system, such as an atom, molecule, or particle. Compare *excited state*.

GUT See *grand unified theory*.

H

H II region A region of interstellar gas that has been ionized by UV radiation from nearby hot massive stars.

H-R diagram The Hertzsprung-Russell diagram, which is a plot of the luminosities versus the surface temperatures of stars. The evolving properties of stars are plotted as tracks across the H-R diagram.

habitable zone The distance from its star at which a planet must be located in order to have a temperature suitable for life; often assumed to be temperatures at which water exists in a liquid state.

Hadley circulation A simplified, and therefore uncommon, atmospheric global circulation that carries thermal energy directly from the equator to the polar regions of a planet.

half-life The time it takes half a sample of a particular radioactive parent element to decay to a daughter product.

halo The spherically symmetric, low-density distribution of stars and dark matter that defines the outermost regions of a galaxy.

harmonic law See *Kepler's third law*.

Hawking radiation Radiation from a black hole.

Hayashi track The path that a protostar follows on the H-R diagram as it contracts toward the main sequence.

head The part of a comet that includes both the nucleus and the inner part of the coma.

heat death The possible eventual fate of an open universe, in which entropy has triumphed and all energy– and structure-producing processes have come to an end.

heavy element Also called *massive element*. 1. In astronomy, any element more massive than helium. 2. In other sciences (and sometimes also in astronomy), any of the most massive elements in the periodic table, such as uranium and plutonium.

Heisenberg uncertainty principle The physical limitation that the product of the position and the momentum of a particle cannot be smaller than a well-defined value, Planck's constant (h).

heliocentric A coordinate system having the center of the Sun as its center. Compare *geocentric*.

helioseismology The use of solar oscillations to study the interior of the Sun.

helium flash The runaway explosive burning of helium in the degenerate helium core of a red giant star.

Herbig-Haro (HH) object A glowing, rapidly moving knot of gas and dust that is excited by bipolar outflows in very young stars.

heredity The process by which one generation passes on its characteristics to future generations.

hertz (Hz) A unit of frequency equivalent to cycles per second.

Hertzsprung-Russell diagram See *H-R diagram*.

HH object See *Herbig-Haro object*.

hierarchical clustering The "bottom-up" process of forming large-scale structure. Small-scale structure first produces groups of galaxies, which in turn form clusters, which then form superclusters.

high-mass star A star with a main-sequence mass greater than about 8 solar masses ($M_\odot$). Compare *low-mass star*.

high-velocity star A star belonging to the halo found near the Sun, distinguished from disk stars by moving far faster and often in the direction opposite to the rotation of the disk and its stars.

homogeneous In cosmology, describing a universe in which observers in any location would observe the same properties.

horizon The boundary that separates the sky from the ground.

horizon problem The puzzling observation that the cosmic background radiation is so uniform in all directions, despite the fact that widely separated regions should have been "over the horizon" from each other in the early universe.

horizontal branch A region on the H-R diagram defined by stars burning helium to carbon in a stable core.

hot dark matter Particles of dark matter that move so fast that gravity cannot confine them to the volume occupied by a galaxy's normal luminous matter. Compare *cold dark matter*.

hot Jupiter A large, Jovian-type extrasolar planet located very close to its parent star.

hot spot A place where hot plumes of mantle material rise near the surface of a planet.

Hubble constant (H_0) The constant of proportionality relating the recession velocities of galaxies to their distances. See also *Hubble time*.

Hubble time An estimate of the age of the universe from the inverse of the Hubble constant, $1/H_0$.

Hubble's law The law stating that the speed at which a galaxy is moving away from us is proportional to the distance of that galaxy.

hurricane Also called *cyclone* or *typhoon*. A large tropical cyclonic system circulating counterclockwise in the Northern Hemisphere and clockwise in the Southern Hemisphere. Hurricanes can extend outward from their center to more than 600 kilometers (km) and generate winds in excess of 300 kilometers per hour (km/h).

hydrogen burning The release of energy from the nuclear fusion of four hydrogen atoms into a single helium atom.

hydrogen shell burning The fusion of hydrogen in a shell surrounding a stellar core that may be either degenerate or fusing more massive elements.

hydrosphere The portion of Earth that is largely liquid water. Compare *biosphere* and *lithosphere*.

hydrostatic equilibrium The condition in which the weight bearing down at a particular point within an object is balanced by the pressure within the object.

hypothesis A well-thought-out idea, based on scientific principles and knowledge, that leads to testable predictions. Compare *theory*.

Hz See *hertz*.

I

ice The solid form of a volatile material; sometimes the *volatile material* itself, regardless of its physical form.

ice giant A giant planet formed mostly of the liquid form of volatile substances (ices). In our Solar System, Uranus and Neptune are the ice giants. Compare *gas giant*.

ideal gas law The relationship of pressure (P) to number density of particles (n) and temperature (T) expressed as $P = nkT$, where k is Boltzmann's constant.

igneous activity The formation and action of molten rock (magma).

impact crater The scar of the impact left on a solid planetary or moon surface by collision with another object. Compare *secondary crater*.

impact cratering A process involving collisions between solid planetary objects.

index of refraction (n) The ratio of the speed of light in a vacuum (c) to the speed of light in an optical medium (v).

inert Non-reactive.

inert gas A gaseous element that combines with other elements only under conditions of extreme temperature and pressure. Examples include helium, neon, and argon.

inertia The tendency for objects to retain their state of motion.

inertial frame of reference 1. A frame of reference that is not accelerating. 2. In general relativity, a frame of reference that is falling freely in a gravitational field.

infinite Limitless; extending without end.

inflation An extremely brief phase of ultrarapid expansion of the very early universe. After inflation, the standard Big Bang models of expansion apply.

infrared (IR) radiation Electromagnetic radiation with frequencies and photon energies occurring in the spectral region between those of visible light and microwaves. Compare *ultraviolet radiation*.

instability strip A region of the H-R diagram containing stars that pulsate with a periodic variation in luminosity.

integration time The time interval during which photons are collected and added up in a detecting device.

intensity (of light) The amount of radiant energy emitted per second per unit area. Units for electromagnetic radiation: watts per square meter (W/m^2).

intercloud gas A low-density region of the interstellar medium that fills the space between interstellar clouds.

interference The interaction of two sets of waves producing high and low intensity, depending on whether their amplitudes reinforce (*constructive interference*) or cancel (*destructive interference*).

interferometer Also called *interferometric array*. A group or array of separate but linked optical or radio telescopes whose overall separation determines the angular resolution of the system.

interferometric array See *interferometer*.

interstellar cloud A discrete, high-density region of the interstellar medium made up mostly of atomic or molecular hydrogen and dust.

interstellar dust Small particles or grains (0.01–10 micrometers [μm]) of matter, primarily carbon and silicates, distributed throughout interstellar space.

interstellar extinction The dimming of visible and ultraviolet light by interstellar dust.

interstellar gas The tenuous gas, far less dense than air, comprising 99 percent of the matter in the interstellar medium.

interstellar medium The gas and dust that fill the space between the stars within a galaxy.

inverse square law The rule that a quantity or effect diminishes with the square of the distance from the source.

ion An atom or molecule that has lost or gained one or more electrons.

ion tail A type of comet tail consisting of ionized gas. Particles in the ion tail are pushed directly away from the comet's head in the antisunward direction at high speeds by the solar wind. Compare *dust tail*.

ionize The process by which electrons are stripped free from an atom or molecule, resulting in free electrons and a positively charged atom or molecule.

ionosphere A layer high in Earth's atmosphere in which most of the atoms are ionized by solar radiation.

IR Infrared.

iron meteorite A metallic meteorite composed mostly of iron-nickel alloys. Compare *stony-iron meteorite* and *stony meteorite*.

irregular galaxy A galaxy without regular or symmetric appearance. Compare *elliptical galaxy*, *S0 galaxy*, and *spiral galaxy*.

irregular moon A moon that has been captured by a planet. Some irregular moons revolve in the opposite direction from the rotation of the planet, and many are in distant, unstable orbits. Compare *regular moon*.

isotopes Forms of the same element with differing numbers of neutrons.

isotropic In cosmology, describing a universe whose properties observers find to be the same in all directions.

J

J See *joule*.

jansky (Jy) The basic unit of flux density. Units: watts per square meter per hertz (W/m²/Hz).

jet 1. A stream of gas and dust ejected from a comet nucleus by solar heating. 2. A collimated linear feature of bright emission extending from a protostar or active galactic nucleus.

joule (J) A unit of energy or work. 1 J = 1 newton meter.

Jy See *jansky*.

K

K See *kelvin*.

K-T boundary See *Cretaceous-Tertiary boundary*.

KBO See *Kuiper Belt object*.

kelvin (K) The basic unit of the Kelvin scale of temperature.

Kelvin temperature scale The temperature scale—defined by William Thomson, better known as Lord Kelvin (1824–1907)—that uses Celsius-sized degrees, but defines its zero point (that is, 0 K) as absolute zero instead of as the melting point of water. Compare *Celsius* and *Fahrenheit*.

Kepler's first law A rule of planetary motion, inferred by Johannes Kepler, stating that planets move in orbits of elliptical shapes with the Sun at one focus.

Kepler's laws The three rules of planetary motion inferred by Johannes Kepler from the data acquired by Tycho Brahe.

Kepler's second law Also called *law of equal areas*. A rule of planetary motion, inferred by Johannes Kepler, stating that a line drawn from the Sun to a planet sweeps out equal areas in equal times as the planet orbits the Sun.

Kepler's third law Also called *harmonic law*. A rule of planetary motion inferred by Johannes Kepler that describes the relationship between the period of a planet's orbit and its distance from the Sun. The law states that the square of the period of a planet's orbit, measured in years, is equal to the cube of the semimajor axis of the planet's orbit, measured in astronomical units: $(P_{years})^2 = (A_{AU})^3$.

kinetic energy (E_K) The energy of an object due to its motions. $E_K = ½ mv^2$. Units: joules (J).

Kirkwood gap A gap in the main asteroid belt related to orbital resonances with Jupiter.

Kuiper Belt A disk-shaped population of comet nuclei extending from Neptune's orbit to perhaps several thousand astronomical units (AU) from the Sun. The highly populated innermost part of the Kuiper Belt has an outer edge approximately 50 AU from the Sun.

Kuiper Belt object (KBO) Also called *trans-Neptunian object*. An icy planetesimal (comet nucleus) that orbits within the Kuiper Belt beyond the orbit of Neptune.

L

Lagrangian equilibrium point One of five points of equilibrium in a system consisting of two massive objects in nearly circular orbit around a common center of mass. Only two Lagrangian points (L_4 and L_5) represent stable equilibrium. A third smaller body located at one of the five points will move in lockstep with the center of mass of the larger bodies.

lander An instrumented spacecraft designed to land on a planet or moon. Compare *rover*.

large-scale structure Observable aggregates on the largest scales in the universe, including galaxy groups, clusters, and superclusters.

latitude The angular distance north (+) or south (−) from the equatorial plane of a nearly spherical body.

law of equal areas See *Kepler's second law*.

law of gravitation See *universal law of gravitation*.

leap year A year that contains 366 days. Leap years occur every 4 years when the year is divisible by 4, correcting for the accumulated excess time in a normal year, which is approximately 365¼ days long.

length contraction The relativistic compression of moving objects in the direction of their motion.

Leonids A November meteor shower associated with the dust debris left by comet Tempel-Tuttle. Compare *Perseids*.

libration The apparent wobble of an orbiting body that is tidally locked to its companion (such as Earth's Moon) resulting from the fact that its orbit is elliptical rather than circular.

life A biochemical process in which living organisms can reproduce, evolve, and sustain themselves by drawing energy from their environment. All terrestrial life involves carbon-based chemistry, assisted by the self-replicating molecules ribonucleic acid (RNA) and deoxyribonucleic acid (DNA).

light All electromagnetic radiation, which comprises the entire electromagnetic spectrum.

light-year (ly) The distance that light travels in 1 year—about 9 trillion kilometers (km).

limb The outer edge of the visible disk of a planet, moon, or the Sun.

limb darkening The darker appearance caused by increased atmospheric absorption near the limb of a planet or star.

line of nodes 1. A line defined by the intersection of two orbital planes. 2. The line defined by the intersection of Earth's equatorial plane and the plane of the ecliptic.

lithosphere The solid, brittle part of Earth (or any planet or moon), including the crust and the upper part of the mantle. Compare *biosphere* and *hydrosphere*.

lithospheric plate A separate piece of Earth's lithosphere capable of moving independently. See also *continental drift* and *plate tectonics*.

Local Group The small group of galaxies of which the Milky Way and the Andromeda galaxies are members.

long-period comet A comet with an orbital period of greater than 200 years. Compare *short-period comet*.

longitudinal wave A wave that oscillates parallel to the direction of the wave's propagation. Compare *transverse wave*.

look-back time The time that it has taken the light from an astronomical object to reach Earth.

low-mass star A star with a main-sequence mass of less than about 8 solar masses ($M_\odot$). Compare *high-mass star*.

luminosity The total flux emitted by an object. Unit: watts (W). See also *brightness*.

luminosity class A spectral classification based on stellar size, from the largest supergiants to the smallest white dwarfs.

luminosity-temperature-radius relationship A relationship among these three properties of stars indicating that if any two are known, the third can be calculated.

luminous Shining.

luminous matter Also called *normal matter*. Matter in galaxies—including stars, gas, and dust—that emits electromagnetic radiation. Compare *dark matter*.

lunar eclipse An eclipse that occurs when the Moon is partially or entirely in Earth's shadow. Compare *solar eclipse*.

lunar tide A tide on Earth that is due to the differential gravitational pull of the Moon. Compare *solar tide*. See also *tide* (definition 2).

ly See *light-year*.

M

M-type asteroid An asteroid that was once part of the metallic core of a larger, differentiated body that has since been broken into pieces; made mostly of iron and nickel. Compare *C-type asteroid* and *S-type asteroid*.

MACHO Literally, "massive compact halo object." MACHOs include brown dwarfs, white dwarfs, and black holes, which are candidates for being considered dark matter. Compare *WIMP*.

magma Molten rock, often containing dissolved gases and solid minerals.

magnetic field A field that is able to exert a force on a moving electric charge. Compare *electric field*.

magnetic force A force associated with, or caused by, the relative motion of charges. Compare *electric force*. See also *electromagnetic force*.

magnetosphere The region surrounding a planet that is filled with relatively intense magnetic fields and plasmas.

magnitude A system used by astronomers to describe the brightness or luminosity of stars. The brighter the star, the smaller its magnitude.

main asteroid belt See *asteroid belt*.

main sequence The strip on the H-R diagram where most stars are found. Main-sequence stars are fusing hydrogen to helium in their cores.

main-sequence lifetime The amount of time a star spends on the main sequence, fusing hydrogen into helium in its core.

main-sequence turnoff The location on the H-R diagram of a single-aged stellar population (such as a star cluster) where stars have just evolved off the main sequence. The position of the main-sequence turnoff is determined by the age of the stellar population.

mantle The solid portion of a rocky planet that lies between the crust and the core.

mare (pl. **maria**) A dark region on the Moon composed of basaltic lava flows.

mass 1. Inertial mass: the property of matter that resists changes in motion. 2. Gravitational mass: the property of matter defined by its attractive force on other objects. According to general relativity, the two are equivalent.

mass-luminosity relationship An empirical relationship between the luminosity (L) and mass (M) of main-sequence stars expressed as a power law—for example, $L \propto M^{3.5}$.

mass transfer The transfer of mass from one member of a binary star system to its companion. Mass transfer occurs when one of the stars evolves to the point that it overfills its Roche lobe, so that its outer layers are pulled toward its binary companion.

massive Containing mass.

massive element See *heavy element*.

matter 1. Objects made of particles that have mass, such as protons, neutrons, and electrons. 2. Anything that occupies space and has mass.

Maunder Minimum The period from 1645 to 1715, when very few sunspots were observed.

medium The material through which a wave travels.

megabar A unit of pressure equal to 1 million bars.

meridian The imaginary arc in the sky running from the horizon at due north through the zenith to the horizon at due south. The meridian divides the observer's sky into eastern and western halves. Compare *equator*.

mesosphere The layer of Earth's atmosphere immediately above the stratosphere, extending from an altitude of 50 kilometers (km) to about 90 km.

meteor The incandescent trail produced by a small piece of interplanetary debris as it travels through the atmosphere at very high speeds. Compare *meteorite* and *meteoroid*.

meteor shower A larger-than-normal display of meteors, occurring when Earth passes through the orbit of a disintegrating comet, sweeping up its debris.

meteorite A *meteoroid* that survives to reach a planet's surface. Compare *meteor* and *meteoroid*.

meteoroid A small cometary or asteroid fragment ranging in size from 100 micrometers (μm) to 100 meters. When entering a planetary atmosphere, the meteoroid creates a *meteor*, which is an atmospheric phenomenon. Compare *meteor* and *meteorite*; also *planetesimal* and *zodiacal dust*.

microlensing Gravitational lensing by relatively small objects like planets (rather than galaxies).

micrometer (μm) Also called *micron*. 10^{-6} meter; a unit of length used for the wavelength of electromagnetic radiation. Compare *nanometer*.

micron See *micrometer*.

microwave radiation Electromagnetic radiation with frequencies and photon energies occurring in the spectral region between those of infrared radiation and radio waves.

Milky Way Galaxy The galaxy in which our Sun and Solar System reside.

minor planet See *asteroid*.

minute of arc See *arcminute*.

μm See *micrometer*.

model 1. A representation (often mathematical) of objects and the interaction between them. 2. In computing, a simulation to reproduce the behavior of a system, in one, two or three dimensions.

modern physics Usually, the physical principles, including relativity and quantum mechanics, that have been developed since James Maxwell's equations were published.

molecular cloud An interstellar cloud composed primarily of molecular hydrogen.

molecular-cloud core A dense clump within a molecular cloud that forms as the cloud collapses and fragments. Protostars form from molecular-cloud cores.

molecule Generally, the smallest particle of a substance that retains its chemical properties and is composed of two or more atoms. A very few types of molecules, such as helium, are composed of single atoms.

momentum The product of the mass and velocity of a particle. Units: kilograms times meters per second (kg m/s).

moon A less massive satellite orbiting a more massive object. Moons are found around planets, dwarf planets, asteroids, and Kuiper Belt objects.

multiverse A collection of parallel universes that together comprise all that is.

mutation In biology, an imperfect reproduction of self-replicating material.

N

N See *newton*.

nadir The point on the celestial sphere located directly below an observer, opposite the *zenith*.

nanometer (nm) One billionth (10^{-9}) of a meter; a unit of length used for the wavelength of light. Compare *micrometer*.

natural selection The process by which forms of structure—ranging from molecules to whole organisms—that are best adapted to their environment become more common than less-well-adapted forms.

NCP See *north celestial pole*.

neap tide An especially weak tide that occurs around the time of the first- or third-quarter Moon when the gravitational forces of the Moon and the Sun on Earth are at right angles to each other, thus producing the least pronounced tides. Compare *spring tide*. See also *tide* (definition 2).

near-Earth asteroid An asteroid whose orbit brings it close to the orbit of Earth. See also *near-Earth object*.

near-Earth object (NEO) An asteroid, comet, or large meteoroid whose orbit intersects Earth's orbit.

nebula (pl. nebulae) A cloud of interstellar gas and dust, either illuminated by stars (bright nebula) or seen in silhouette against a brighter background (dark nebula).

nebular hypothesis The first plausible theory of the formation of the Solar System, proposed by Immanuel Kant in 1734. Kant hypothesized that the Solar System formed from the collapse of an interstellar cloud of rotating gas.

NEO See *near-Earth object*.

neutrino A very-low-mass, electrically neutral particle emitted during beta decay. Neutrinos interact with matter only very feebly and so can penetrate through great quantities of matter.

neutrino cooling The process in which thermal energy is carried out of the center of a star by neutrinos rather than by electromagnetic radiation or convection.

neutron A subatomic particle having no net electric charge and a rest mass and rest energy nearly equal to that of the proton. Compare *electron* and *proton*.

neutron star The neutron-degenerate remnant left behind by a Type II supernova.

new Moon The phase of the Moon in which the Moon is between Earth and the Sun, and from Earth we see only the side of the Moon not being illuminated by the Sun. Compare *full Moon*.

newton (N) The force required to accelerate a 1-kilogram (kg) mass at a rate of 1 meter per second per second (m/s^2). Units: kilograms times meters per second squared (kg m/s^2).

Newton's first law of motion The law, formulated by Isaac Newton, stating that an object will remain at rest or will continue moving along a straight line at a constant speed until an unbalanced force acts on it.

Newton's laws See *Newton's first law of motion*, *Newton's second law of motion*, and *Newton's third law of motion*.

Newton's second law of motion The law, formulated by Isaac Newton, stating that if an unbalanced force acts on a body, the body will have an acceleration proportional to the unbalanced force and inversely proportional to the object's mass: $a = F/m$. The acceleration will be in the direction of the unbalanced force.

Newton's third law of motion The law, formulated by Isaac Newton, stating that for every force there is an equal and opposite force.

nm See *nanometer*.

normal matter See *luminous matter*.

north celestial pole (NCP) The northward projection of Earth's rotation axis onto the celestial sphere. Compare *south celestial pole*.

North Pole The location in the Northern Hemisphere where Earth's rotation axis intersects the surface of Earth. Compare *South Pole*.

nova (pl. novae) A stellar explosion that results from runaway nuclear fusion in a layer of material on the surface of a white dwarf in a binary system.

nuclear burning Release of energy by fusion of low-mass elements.

nuclear fusion The combination of two less massive atomic nuclei into a single more massive atomic nucleus.

nucleosynthesis The formation of more massive atomic nuclei from less massive nuclei, either in the Big Bang (Big Bang nucleosynthesis) or in the interiors of stars (stellar nucleosynthesis).

nucleus (pl. nuclei) 1. The dense, central part of an atom. 2. The central core of a galaxy, comet, or other diffuse object.

O

objective lens The primary optical element in a telescope or camera that produces an image of an object.

oblate The flattening of an otherwise spherical planet or star caused by its rapid rotation.

obliquity The inclination of a celestial body's equator to its orbital plane.

observable universe The part of the universe that can be observed, because light has had time to travel to Earth.

observational uncertainty The fact that real measurements are never perfect; all observations are uncertain by some amount.

Occam's razor The principle that the simplest hypothesis is the most likely; named after William of Occam (circa 1285–1349), the medieval English cleric to whom the idea is attributed.

ocean A vast expanse of liquid, not necessarily water.

Oort Cloud A spherical distribution of comet nuclei stretching from beyond the Kuiper Belt to more than 50,000 astronomical units (AU) from the Sun.

opacity A measure of how effectively a material blocks the radiation going through it.

open cluster A loosely bound group of a few dozen to a few thousand stars that formed together in the disk of a spiral galaxy. Compare *globular cluster*.

open universe An infinite universe with a negatively curved spatial structure (much like the surface of a saddle) such that the sum of the angles of a triangle is always less than 180°. Compare *closed universe* and *flat universe*.

orbit The path taken by one object moving around another object under the influence of their mutual gravitational or electric attraction.

orbital resonance A situation in which the orbital periods of two objects are related by a ratio of small integers.

orbiter A spacecraft that is placed in orbit around a planet or moon. Compare *flyby*.

organic Describing a substance, not necessarily of biological origin, that contains the element carbon.

P

P wave See *primary wave*.

pair production The creation of a particle-antiparticle pair from a source of electromagnetic energy.

paleomagnetism The record of Earth's magnetic field as preserved in rocks.

parallax 1. The apparent shift in the position of one object relative to another object, caused by the changing perspective of the observer. 2. In astronomy, the displacement in the apparent position of a nearby star caused by the changing location of Earth in its orbit.

parent element A radioactive element that decays to form more stable *daughter products*.

parsec (pc) The distance to a star with a parallax of 1 arcsecond using a base of 1 astronomical unit (AU). One parsec is approximately 3.26 light-years.

partial lunar eclipse A lunar eclipse in which the Moon passes through the penumbra of Earth's shadow. Compare *total lunar eclipse*.

partial solar eclipse The type of eclipse that occurs when Earth passes through the penumbra of the Moon's shadow, so that the Moon blocks only a portion of the Sun's disk. Compare *annular solar eclipse* and *total solar eclipse*.

pc See *parsec*.

peculiar velocity The motion of a galaxy relative to the overall expansion of the universe.

penumbra (pl. penumbrae) 1. The outer part of a shadow, where the source of light is only partially blocked. 2. The region surrounding the *umbra* of a sunspot. The penumbra is cooler and darker than the surrounding surface of the Sun but not as cool or dark as the umbra of the sunspot.

perihelion (pl. perihelia) The point in a solar orbit that is closest to the Sun. Compare *aphelion*.

period The time it takes for a regularly repetitive process to complete one cycle.

period-luminosity relationship The relationship between the period of variability of a pulsating variable star, such as a Cepheid or RR Lyrae variable, and the luminosity of the star. Longer-period Cepheid or RR Lyrae variables are more luminous than their shorter-period cousins.

Perseids A prominent August meteor shower associated with the dust debris left by comet Swift-Tuttle. Compare *Leonids*.

perturb Move an astronomical object from its undisturbed path or location.

phase One of the various appearances of the sunlit surface of the Moon or a planet caused by the change in viewing location of Earth relative to both the Sun and the object. Examples include crescent phase and gibbous phase.

photino An elementary particle related to the photon. One of the leading candidates for cold dark matter.

photochemical reaction A chemical reaction driven by the absorption of electromagnetic radiation.

photodissociation The breaking apart of molecules into smaller fragments or individual atoms by the action of photons.

photoelectric effect An effect whereby electrons are emitted from a substance illuminated by photons above a certain critical frequency.

photometry The process of measuring the brightness of a source of light, generally over a specific range of wavelength.

photon Also called *quantum of light*. A discrete unit or particle of electromagnetic radiation. The energy of a photon is equal to Planck's constant (h) multiplied by the frequency (f) of its electromagnetic radiation: $E_{photon} = h \times f$. The photon is the particle that mediates the electromagnetic force.

photosphere The apparent surface of the Sun as seen in visible light.

physical law A broad statement that predicts a particular aspect of how the physical universe behaves and that is supported by many empirical tests. See also *theory*.

pixel The smallest picture element in a digital image array.

Planck era The early time, just after the Big Bang, when the universe as a whole must be described with quantum mechanics.

Planck spectrum See *blackbody spectrum*.

Planck's constant (h) The constant of proportionality between the energy of a photon and the frequency of the photon. This constant defines how much energy a single photon of a given frequency or wavelength has. Value: $h = 6.63 \times 10^{-34}$ joule-second.

planet 1. A large body that orbits the Sun or other star that shines only by light reflected from the Sun or star. 2. In the Solar System, a body that orbits the Sun, has sufficient mass for self-gravity to overcome rigid body forces so that it assumes a spherical shape, and has cleared smaller bodies from the neighborhood around its orbit. Compare *dwarf planet*.

planet migration The theory that a planet can move from its formation distance around its parent star to a different distance though gravitational interactions with other bodies or loss of orbital energy from interaction with gas in the protoplanetary disk.

planetary nebula The expanding shell of material ejected by a dying asymptotic giant branch star. A planetary nebula glows from fluorescence caused by intense ultraviolet light coming from the hot, stellar remnant at its center.

planetary system A system of planets and other smaller objects in orbit around a star.

planetesimal A primitive body of rock and ice, 100 meters or more in diameter, that combines with others to form a planet. Compare *meteoroid* and *zodiacal dust*.

plasma A gas that is composed largely of charged particles but also may include some neutral atoms.

plate tectonics The geological theory concerning the motions of lithospheric plates, which in turn provides the theoretical basis for continental drift.

positron A positively charged subatomic particle; the antiparticle of the *electron*.

power The rate at which work is done or at which energy is delivered. Unit: watts (W) or joules per second (J/s).

precession of the equinoxes The slow change in orientation between the ecliptic plane and the celestial equator caused by the wobbling of Earth's axis.

pressure Force per unit area. Units: newtons per square meter (N/m^2) or bars.

primary atmosphere An atmosphere, composed mostly of hydrogen and helium, that forms at the same time as its host planet. Compare *secondary atmosphere*.

primary mirror The principal optical mirror in a reflecting telescope. The primary mirror determines the telescope's light-gathering power and resolution. Compare *secondary mirror*.

primary wave Also called *P wave*. A longitudinal seismic wave, in which the oscillations involve compression and decompression parallel to the direction of travel (that is, a pressure wave). Compare *secondary wave*.

principle A general idea or sense about how the universe is that guides us in constructing new scientific theories. Principles can be testable theories.

prograde motion 1. Rotational or orbital motion of a moon that is in the same sense as the planet it orbits. 2. The counterclockwise orbital motion of Solar System objects as seen from above Earth's orbital plane. Compare *retrograde motion*.

prominence An archlike projection above the solar photosphere often associated with a sunspot.

proportional Describing two things whose ratio is a constant.

proton (p or p^+) A subatomic particle having a positive electric charge of 1.6×10^{-19} coulomb (C), a mass of 1.67×10^{-27} kilograms (kg), and a rest energy of 1.5×10^{-10} joules (J). Compare *electron* and *neutron*.

proton-proton chain One of the ways in which hydrogen burning can take place. This is the most important path for hydrogen burning in low-mass stars such as the Sun. See also *carbon-nitrogen-oxygen cycle* and *triple-alpha process*.

protoplanetary disk Also called *circumstellar disk*. The remains of the accretion disk around a young star from which a planetary system may form.

protostar A young stellar object that derives its luminosity from the conversion of gravitational energy to thermal energy, rather than from nuclear reactions in its core.

pulsar A rapidly rotating neutron star that beams radiation into space in two searchlight-like beams. To a distant observer, the star appears to flash on and off, earning its name.

pulsating variable star A variable star that undergoes periodic radial pulsations.

Q

QCD See *quantum chromodynamics*.

QED See *quantum electrodynamics*.

quantized Describing a quantity that exists as discrete, irreducible units.

quantum chromodynamics (QCD) The quantum mechanical theory describing the strong nuclear force and its mediation by gluons. Compare *quantum electrodynamics*.

quantum efficiency The fraction of photons falling on a detector that actually produces a response in the detector.

quantum electrodynamics (QED) The quantum theory describing the electromagnetic force and its mediation by photons. Compare *quantum chromodynamics*.

quantum mechanics The branch of physics that deals with the quantized and probabilistic behavior of atoms and subatomic particles.

quantum of light See *photon*.

quark The building block of protons and neutrons.

quasar Short for *quasi-stellar radio source*. The most luminous of the active galactic nuclei, seen only at great distances from our galaxy.

R

radial velocity The component of velocity that is directed toward or away from the observer.

radian The angle at the center of a circle subtended by an arc equal to the length of the circle's radius. Therefore, 2π radians equals $360°$, and 1 radian equals approximately $57.3°$.

radiant The direction in the sky from which the meteors in a meteor shower seem to come.

radiation 1. Energy which has been emitted as particles or waves. 2. Light of all wavelengths.

radiation belt A toroidal ring of high-energy particles surrounding a planet.

radiative transfer The transport of energy from one location to another by electromagnetic radiation.

radiative zone A region in the interior of a star through which energy is transported outward by radiation.

radio galaxy A type of elliptical galaxy with an active galactic nucleus at its center and having very strong emission (10^{35}–10^{38} watts [W]) in the radio part of the electromagnetic spectrum. Compare *Seyfert galaxy*.

radioisotope A radioactive isotope.

radio telescope An instrument for detecting and measuring radio-frequency emissions from celestial sources.

radio wave Electromagnetic radiation in the extreme long-wavelength region of the spectrum, beyond the region of microwaves.

radioisotope A radioactive element.

radiometric dating Use of radioactive decay to measure the ages of materials such as minerals.

ratio The relationship in quantity or size between two or more things.

ray 1. A beam of electromagnetic radiation. 2. A bright streak emanating from a young impact crater.

recession Moving away, receding.

recombination 1. The combining of ions and electrons to form neutral atoms. 2. An event early in the evolution of the universe in which hydrogen and helium nuclei combined with electrons to form neutral atoms. The removal of electrons caused the universe to become transparent to electromagnetic radiation.

red giant A low-mass star that has evolved beyond the main sequence and is now fusing hydrogen in a shell surrounding a degenerate helium core.

red giant branch A region on the H-R diagram defined by low-mass stars evolving from the main sequence toward the horizontal branch.

reddening The effect by which stars and other objects, when viewed through interstellar dust, appear redder than they actually are. Reddening is caused by the fact that blue light is more strongly absorbed and scattered than red light.

redshift Also called *Doppler redshift*. The shift toward longer wavelengths of light by any of several effects, including Doppler shifts, gravitational redshift, or cosmological redshift. Compare *blueshift*.

reflecting telescope A telescope that uses mirrors for collecting and focusing incoming electromagnetic radiation to form an image in their focal planes. The size of a reflecting telescope is defined by the diameter of the primary mirror. Compare *refracting telescope*.

reflection The redirection of a beam of light that is incident on, but does not cross, the surface between two media having different refractive indices. If the surface is flat and smooth, the angle of incidence equals the angle of reflection. Compare *refraction*.

refracting telescope A telescope that uses objective lenses to collect and focus light. Compare *reflecting telescope*.

refraction The redirection or bending of a beam of light when it crosses the boundary between two media having different refractive indices. Compare *reflection*.

refractory material Material that remains solid at high temperatures. Compare *volatile material*.

regular moon A moon that formed together with the planet it orbits. Compare *irregular moon*.

reionization The process that reionized matter in the early universe after the Big Bang.

relative humidity The amount of water vapor held by a volume of air at a given temperature compared (stated as a percentage) to the total amount of water that could be held by the same volume of air at the same temperature.

relative motion The difference in motion between two individual frames of reference.

relativistic Describing physical processes that take place in systems traveling at nearly the speed of light or located in the vicinity of very strong gravitational fields.

relativistic beaming The effect created when material moving at nearly the speed of light beams the radiation it emits in the direction of its motion.

remote sensing The use of images, spectra, radar, or other techniques to measure the properties of an object from a distance.

resolution The ability of a telescope to separate two point sources of light. Resolution is determined by the telescope's aperture and the wavelength of light it receives.

rest wavelength The wavelength of light we see coming from an object at rest with respect to the observer.

retrograde motion 1. Rotation or orbital motion of a moon that is in the opposite sense to the rotation of the planet it orbits. 2. The clockwise orbital motion of Solar System objects as seen from above Earth's orbital plane. Compare *prograde motion*.

revolve Orbit.

ring An aggregation of small particles orbiting a planet or star. The rings of the four giant planets of the Solar System are composed variously of silicates, organic materials, and ices.

ring arc A discontinuous, higher-density region within an otherwise continuous, narrow ring.

ringlet A narrowly confined concentration of ring particles.

Roche limit The distance at which a planet's tidal forces exceed the self-gravity of a smaller object, such as a moon, asteroid, or comet, causing the object to break apart.

Roche lobe The hourglass-shaped or figure eight–shaped volume of space surrounding two stars, which constrains material that is gravitationally bound by one or the other.

rotation curve A plot showing how the orbital velocity of stars and gas in a galaxy changes with radial distance from the galaxy's center.

rover A remotely controlled instrumented vehicle designed to traverse and explore the surface of a terrestrial planet or moon. Compare *lander*.

RR Lyrae variable A variable giant star whose regularly timed pulsations are good predictors of its luminosity. RR Lyrae stars are used for distance measurements to globular clusters.

S

S-type asteroid An asteroid made of material that has been modified from its original state, likely as the outer part of a larger, differentiated body that has since broken into pieces. Compare *C-type asteroid* and *M-type asteroid*.

S wave See *secondary wave*.

S0 galaxy A galaxy with a bulge and a disk-like spiral, but smooth in appearance like ellipticals. Compare *elliptical galaxy*, *irregular galaxy*, and *spiral galaxy*.

satellite 1. An object in orbit about a more massive body. 2. A moon.

scale factor (R_U) A dimensionless number proportional to the distance between two points in space. The scale factor increases as the universe expands.

scattering The random change in the direction of travel of photons, caused by their interactions with molecules or dust particles.

Schwarzschild radius The distance from the center of a nonrotating, spherical black hole at which the escape velocity equals the speed of light.

scientific method The formal procedure—including hypothesis, prediction, and experiment or observation—used to test (attempt to falsify) the validity of scientific hypotheses and theories.

scientific notation The standard expression of numbers with one digit (which can be zero) to the left of the decimal point and multiplied by 10 to the exponent required to give the number its correct value. Example: $2.99 \times 10^8 = 299,000,000$.

SCP See *south celestial pole*.

second law of thermodynamics The law stating that the entropy or disorder of an isolated system always increases as the system evolves.

second of arc See *arcsecond*.

secondary atmosphere An atmosphere that formed—as a result of volcanism, comet impacts, or another process—sometime after its host planet formed. Compare *primary atmosphere*.

secondary crater A crater formed from ejected material thrown from an *impact crater*.

secondary mirror A small mirror placed on the optical axis of a reflecting telescope that returns the beam back through a small hole in the *primary mirror*, thereby shortening the mechanical length of the telescope.

secondary wave Also called *S wave*. A transverse seismic wave, which involves the sideways motion of material. Compare *primary wave*.

seeing The quality of observed images resulting from atmospheric distortions.

seismic wave A vibration due to an earthquake, a large explosion, or an impact on the surface that travels through a planet's interior.

seismometer An instrument that measures the amplitude and frequency of seismic waves.

self-gravity The gravitational attraction among all the parts of the same object.

semimajor axis Half of the longer axis of an ellipse.

SETI The Search for Extraterrestrial Intelligence project, which uses advanced technology combined with radio telescopes to search for evidence of intelligent life elsewhere in the universe.

Seyfert galaxy A type of spiral galaxy with an active galactic nucleus at its center; first discovered in 1943 by Carl Seyfert. Compare *radio galaxy*.

shepherd moon A moon that orbits close to rings and gravitationally confines the orbits of the ring particles.

shield volcano A volcano formed by very fluid lava flowing from a single source and spreading out from that source. Compare *composite volcano*.

short-period comet A comet with an orbital period of less than 200 years. Compare *long-period comet*.

sidereal day The length of time the Earth takes to rotate to the same position relative to the stars: 23^h56^m. Compare *solar day*.

sidereal period An object's orbital or rotational period measured with respect to the stars. Compare *synodic period*.

silicate One of the family of minerals composed of silicon and oxygen in combination with other elements.

singularity The point where a mathematical expression or equation becomes meaningless, such as the denominator of a fraction approaching zero. See also *black hole*.

solar abundance The relative amount of an element detected in the atmosphere of the Sun, expressed as the ratio of the number of atoms of that element to the number of hydrogen atoms.

solar day The 24-hour period of Earth's axial rotation that brings the Sun back to the same local meridian where the rotation started. Compare *sidereal day*.

solar eclipse An eclipse that occurs when the Sun is partially or entirely blocked by the Moon. Compare *lunar eclipse*.

solar flare Explosions on the Sun's surface associated with complex sunspot groups and strong magnetic fields.

solar maximum (pl. maxima) The time, occurring about every 11 years, when the Sun is at its peak activity, meaning that sunspot activity and related phenomena (such as prominences, flares, and coronal mass ejections) are at their peak.

solar neutrino problem The historical observation that only about a third as many neutrinos as predicted by theory seemed to be coming from the Sun.

Solar System The gravitationally bound system made up of the Sun, planets, dwarf planets, moons, asteroids, comets, and Kuiper Belt objects, along with their associated gas and dust.

solar tide A tide on Earth that is due to the differential gravitational pull of the Sun. Compare *lunar tide*. See also *tide* (definition 2).

solar wind The stream of charged particles emitted by the Sun that flows at high speeds through interplanetary space.

solstice Literally, "sun standing still." 1. One of the two most northerly and southerly points on the ecliptic. 2. Either of the two times of year (the *summer solstice* and *winter solstice*) when the Sun is at one of these two positions. Compare *equinox*.

south celestial pole (SCP) The southward projection of Earth's rotation axis onto the celestial sphere. Compare *north celestial pole*.

South Pole The location in the Southern Hemisphere where Earth's rotation axis intersects the surface of Earth. Compare *North Pole*.

spacetime The four-dimensional continuum in which we live, and which we experience as three spatial dimensions plus time.

special relativity See *special theory of relativity*.

special theory of relativity Sometimes referred to as simply *special relativity*. Einstein's theory explaining how the fact that the speed of light is a constant affects nonaccelerating frames of reference. Compare *general theory of relativity*.

spectral type A classification system for stars based on the presence and relative strength of absorption lines in their spectra. Spectral type is related to the surface temperature of a star.

spectrograph A device that spreads out the light from an object into its component wavelengths. See also *spectrometer*.

spectrometer A *spectrograph* in which the spectrum is generally recorded digitally by electronic means.

spectroscopic binary A binary star pair whose existence and properties are revealed only by the Doppler shift of its spectral lines. Most spectroscopic binaries are close pairs. Compare *eclipsing binary* and *visual binary*.

spectroscopic parallax Use of the spectroscopically determined luminosity and the observed brightness of a star to determine the star's distance.

spectroscopic radial velocity method A technique for detecting extrasolar planets by examining Doppler shifts in the light from stars.

spectroscopy The study of electromagnetic radiation from an object in terms of its component wavelengths.

spectrum (pl. **spectra**) 1. The intensity of electromagnetic radiation as a function of wavelength. 2. Waves sorted by wavelength.

speed The rate of change of an object's position with time, without regard to the direction of movement. Units: meters per second (m/s) or kilometers per hour (km/h). Compare *velocity*.

spherically symmetric Describing an object whose properties depend only on distance from the object's center, so that the object has the same form viewed from any direction.

spin-orbit resonance A relationship between the orbital and rotation periods of an object such that the ratio of their periods can be expressed by simple integers.

spiral density wave A stable, spiral-shaped change in the local gravity of a galactic disk that can be produced by periodic gravitational kicks from neighboring galaxies or from nonspherical bulges and bars in spiral galaxies.

spiral galaxy A galaxy of Hubble type "S" class, with a discernible disk in which large spiral patterns exist. Compare *elliptical galaxy*, *irregular galaxy*, and *S0 galaxy*.

spoke One of several narrow radial features seen occasionally in Saturn's B ring. Spokes appear dark in backscattered light and bright in forward, scattering light, indicating that they are composed of tiny particles. Their origin is not well understood.

sporadic meteor A meteor that is not associated with a specific meteor shower.

spreading center A zone from which two tectonic plates diverge.

spring tide An especially strong tide that occurs near the time of a new or full Moon, when lunar tides and solar tides reinforce each other. Compare *neap tide*. See also *tide* (definition 2).

stable equilibrium An equilibrium state in which the system returns to its former condition after a small disturbance. Compare *unstable equilibrium*.

standard candle An object whose luminosity either is known or can be predicted in a distance-independent way, so its brightness can be used to determine its distance via the inverse square law of radiation.

standard model The theory of particle physics that combines electroweak theory with quantum chromodynamics to describe the structure of known forms of matter.

star A luminous ball of gas that is held together by gravity. A normal star is powered by nuclear reactions in its interior.

star cluster A group of stars that all formed at the same time and in the same general location.

static equilibrium A state in which the forces within a system are all in balance so that the system does not change. Compare *dynamic equilibrium*.

Stefan-Boltzmann constant (σ) The proportionality constant that relates the flux emitted by an object to the fourth power of its absolute temperature. Value: 5.67×10^{-8} W/(m² K⁴) (W = watts, m = meters, K = kelvin).

Stefan-Boltzmann law The law stating that the amount of electromagnetic energy emitted from the surface of a body, summed over the energies of all photons of all wavelengths emitted, is proportional to the fourth power of the temperature of the body.

stellar mass loss The loss of mass from the outermost parts of a star's atmosphere during the course of its evolution.

stellar occultation An event in which a planet or other Solar System body moves between the observer and a star, eclipsing the light emitted by that star.

stellar population A group of stars with similar ages, chemical compositions, and dynamic properties.

stereoscopic vision The way an animal's brain combines the different information from its two eyes to perceive the distances to objects around it.

stony-iron meteorite A meteorite consisting of a mixture of silicate minerals and iron-nickel alloys. Compare *iron meteorite* and *stony meteorite*.

stony meteorite A meteorite composed primarily of silicate minerals, similar to those found on Earth. Compare *iron meteorite* and *stony-iron meteorite*.

stratosphere The atmospheric layer immediately above the *troposphere*. On Earth it extends upward to an altitude of 50 km.

strength 1. Magnitude of a force. 2. In spectroscopy: brightness of an emission line or depth of an absorption line.

string theory The theory that conceives of particles as strings in 10 dimensions of space and time; the current contender for a theory of everything.

strong nuclear force The attractive short-range force between protons and neutrons that holds atomic nuclei together; one of the four fundamental forces of nature, mediated by the exchange of gluons. Compare *weak nuclear force*.

subduction zone A region where two tectonic plates converge, with one plate sliding under the other and being drawn downward into the interior.

subgiant A giant star smaller and lower in luminosity than normal giant stars of the same spectral type. Subgiants evolve to become giants.

subgiant branch A region of the H-R diagram defined by stars that have left the main sequence but have not yet reached the red giant branch.

sublimation The process in which a solid becomes a gas without first becoming a liquid.

subsonic Moving within a medium at a speed slower than the speed of sound in that medium. Compare *supersonic*.

summer solstice 1. One of two points where the Sun is at its greatest distance from the celestial equator. 2. The day on which the Sun appears at this location, marking the first day of summer (about June 21 in the Northern Hemisphere and December 21 in the Southern Hemisphere). Compare *winter solstice*.

sungrazer A comet whose perihelion is within a few solar diameters of the surface of the Sun.

sunspot A cooler, transitory region on the solar surface produced when loops of magnetic flux break through the surface of the Sun.

sunspot cycle The approximate 11-year cycle during which sunspot activity increases and then decreases. This is one-half of a full 22-year cycle, in which the magnetic polarity of the Sun first reverses and then returns to its original configuration.

supercluster A large conglomeration of galaxy clusters and galaxy groups; typically, more than 100 million light-years (more than 30 Mpc) in size and containing tens of thousands to hundreds of thousands of galaxies. Compare *galaxy cluster* and *galaxy group*.

superluminal motion The appearance (though not the reality) that a jet is moving faster than the speed of light.

supermassive black hole A black hole of 1,000 solar masses ($M_\odot$) or more that resides in the center of a galaxy, and whose gravity powers active galactic nuclei.

supernova (pl. supernovae) A stellar explosion resulting in the release of tremendous amounts of energy, including the high-speed ejection of matter into the interstellar medium. See also *Type Ia supernova* and *Type II supernova*.

supersonic Moving within a medium at a speed faster than the speed of sound in that medium. Compare *subsonic*.

superstring theory See *string theory*.

surface The outermost layer of something. In astronomy, this is often defined to be the *visible surface*.

surface brightness The amount of electromagnetic radiation emitted or reflected per unit area.

surface wave A seismic wave that travels on the surface of a planet or moon.

symmetry 1. The property that an object has if the object is unchanged by rotation or reflection about a particular point, line, or plane. 2. In theoretical physics, the correspondence of different aspects of physical laws or systems, such as the symmetry between matter and antimatter.

synchronous rotation The case in which the period of rotation of a body on its axis equals the period of revolution in its orbit around another body. A special type of spin-orbit resonance.

synchrotron radiation Radiation from electrons moving at close to the speed of light as they spiral in a strong magnetic field; named because this kind of radiation was first identified on Earth in particle accelerators called synchrotrons.

synodic period An object's orbital or rotational period measured with respect to the Sun. Compare *sidereal period*.

T

T Tauri star A young stellar object that has dispersed enough of the material surrounding it to be seen in visible light.

tail A stream of gas and dust swept away from the coma of a comet by the solar wind and by radiation pressure from the Sun.

tectonism Deformation of the lithosphere of a planet.

telescope The basic tool of astronomers. Working over the entire range from gamma rays to radio, astronomical telescopes collect and concentrate electromagnetic radiation from celestial objects.

temperature A measure of the average kinetic energy of the atoms or molecules in a gas, solid, or liquid.

terrestrial planet An Earth-like planet, made of rock and metal and having a solid surface. In our Solar System, the terrestrial planets are Mercury, Venus, Earth, and Mars. Compare *giant planet*.

theoretical model A detailed description of the properties of a particular object or system in terms of known physical laws or theories. Often, a computer calculation of predicted properties based on such a description.

theory A well-developed idea or group of ideas that are tied solidly to known physical laws and make testable predictions about the world. A very well-tested theory may be called a *physical law*, or simply a fact. Compare *hypothesis*.

theory of everything (TOE) A theory that unifies all four fundamental forces of nature: strong nuclear, weak nuclear, electromagnetic, and gravitational forces.

thermal conduction The transfer of energy in which the thermal energy of particles is transferred to adjacent particles by collisions or other interactions. Conduction is the most important way that thermal energy is transported in solid matter.

thermal energy The energy that resides in the random motion of atoms, molecules, and particles, by which we measure their temperature.

thermal equilibrium The state in which the rate of thermal-energy emission by an object is equal to the rate of thermal-energy absorption.

thermal motion The random motion of atoms, molecules, and particles that gives rise to thermal radiation.

thermal radiation Electromagnetic radiation resulting from the random motion of the charged particles in every substance.

thermophiles A type of extremophile that tolerates high temperatures (between 45 and 122 °C).

thermosphere The layer of Earth's atmosphere at altitudes greater than 90 km, above the mesosphere. Near its top, at an altitude of 600 km, the temperature can reach 1000 K.

third quarter Moon The phase of the Moon in which only the eastern half of the Moon, as viewed from Earth, is illuminated by the Sun. It occurs about 1 week after the full Moon. Compare *first quarter Moon*.

tidal bulge A distortion of a body resulting from tidal stresses.

tidal locking Synchronous rotation of an object caused by internal friction as the object rotates through its tidal bulge.

tidal stress Stress due to differences in the gravitational force of one mass on different parts of another mass.

tide 1. The deformation of a mass due to differential gravitational effects of one mass on another because of the extended size of the masses. 2. On Earth, the rise and fall of the oceans as Earth rotates through a tidal bulge caused by the Moon and Sun.

time dilation The relativistic "stretching" of time. Compare *general relativistic time dilation*.

TNO See *trans-Neptunian object*.

TOE See *theory of everything*.

topographic relief The differences in elevation from point to point on a planetary surface.

tornado A violent rotating column of air, typically 75 meters across with 200– kilometer-per-hour (km/h) winds. Some tornadoes can be more than 3 km across, and winds up to 500 km/h have been observed.

torus A three-dimensional, doughnut-shaped ring.

total lunar eclipse A lunar eclipse in which the Moon passes through the umbra of Earth's shadow. Compare *partial lunar eclipse*.

total solar eclipse The type of eclipse that occurs when Earth passes through the umbra of the Moon's shadow, so that the Moon completely blocks the disk of the Sun. Compare *annular solar eclipse* and *partial solar eclipse*.

transform fault The actively slipping segment of a fracture zone between lithospheric plates.

transit method A technique for detecting extrasolar planets by observing a star's decrease in brightness when a planet passes in front of it.

trans-Neptunian object (TNO) See *Kuiper Belt object*.

transverse wave A wave that oscillates perpendicular to the direction of the wave's propagation. Compare *longitudinal wave*.

triple-alpha process The nuclear fusion reaction that combines three helium nuclei (alpha particles) together into a single nucleus of carbon. See also *carbon-nitrogen-oxygen cycle* and *proton-proton chain*.

Trojan asteroid One of a group of asteroids orbiting in the L_4 and L_5 Lagrangian points of Jupiter's orbit.

tropical year The time between one crossing of the vernal equinox and the next. Because of the precession of the equinoxes, a tropical year is slightly shorter than the time that it takes for Earth to orbit once about the Sun.

Tropics The region on Earth between latitudes 23.5° south and 23.5° north, and in which the Sun appears directly overhead twice during the year.

tropopause The top of a planet's troposphere.

troposphere The convection-dominated layer of a planet's atmosphere. On Earth, the atmospheric region closest to the ground within which most weather phenomena take place. Compare *stratosphere*.

tuning fork diagram The two-pronged diagram showing Hubble's classification of galaxies into ellipticals, S0s, spirals, and barred spirals.

turbulence The random motion of blobs of gas within a larger cloud of gas.

Type Ia supernova A supernova explosion in which no trace of hydrogen is seen in the ejected material. Most Type Ia supernovae are thought to be the result of runaway carbon burning in a white dwarf star onto which material is being deposited by a binary companion.

Type II supernova A supernova explosion in which the degenerate core of an evolved massive star suddenly collapses and rebounds.

typhoon See *hurricane*.

U

ultrafaint dwarf galaxies A dwarf galaxy of very low luminosity.

ultraviolet (UV) radiation Electromagnetic radiation with frequencies and photon energies greater than those of visible light but less than those of X-rays and with wavelengths shorter than those of visible light but longer than those of X-rays. Compare *infrared radiation*.

umbra (pl. **umbrae**) 1. The darkest part of a shadow, where the source of light is completely blocked. 2. The darkest, innermost part of a sunspot. Compare *penumbra*.

unbalanced force The nonzero net force acting on a body.

unbound orbit An orbit in which the velocity is greater than the escape velocity. Compare *bound orbit*.

uncertainty A description of the accuracy of a measurement, sometimes expressed as a percentage, more often as an interval. The uncertainty gives the range over which one might expect to obtain measurements if the experiment were repeated multiple times.

uncertainty principle See *Heisenberg uncertainty principle*.

unified model of AGNs A model in which many different types of activity in the nuclei of galaxies are all explained by accretion of matter around a supermassive black hole.

uniform circular motion Motion in a circular path at a constant speed.

unit A fundamental quantity of measurement—for example, metric units or English units.

universal gravitational constant (G) The constant of proportionality in the universal law of gravity. Value: $G = 6.673 \times 10^{-11}$ newtons times meters squared per kilogram squared (N m²/kg²), equivalently 6.673×10^{-11} meters cubed per kilogram seconds squared [m³/(kg s²)].

universal law of gravitation The law stating that the gravitational force between any two objects is proportional to the product of their masses and inversely proportional to the square of the distance between them: $F \propto (m_1 m_2 / r^2)$.

universe All of space and everything contained therein.

unresolved Images overlapping to the extent that they cannot be distinguished from one another.

unstable equilibrium An equilibrium state in which a small disturbance will cause a system to move away from equilibrium. Compare *stable equilibrium*.

UV Ultraviolet.

V

vacuum A region of space that contains very little matter. In quantum mechanics and general relativity, however, even a perfect vacuum has physical properties.

vaporize Turn to gas.

variable star A star with varying luminosity. Many periodic variables are found within the instability strip on the H-R diagram.

velocity The rate and direction of change of an object's position with time. Units: meters per second (m/s) or kilometers per hour (km/h). Compare *speed*.

vernal equinox 1. One of two points where the Sun crosses the celestial equator. 2. The day on which the Sun appears at this location, marking the first day of spring (about March 20 in the Northern Hemisphere and September 22 in the Southern Hemisphere). Compare *autumnal equinox*.

virtual particle A particle that, according to quantum mechanics, comes into existence only momentarily. According to theory, fundamental forces are mediated by the exchange of virtual particles.

visual binary A binary system in which the two stars can be seen individually from Earth. Compare *eclipsing binary* and *spectroscopic binary*.

void A region in space containing little or no matter. Examples include regions in cosmological space that are largely empty of galaxies.

volatile material Sometimes called *ice*. Material that remains gaseous at moderate temperature. Compare *refractory material*.

volcanism The occurrence of volcanic activity on a planet or moon.

vortex (pl. **vortices**) Any circulating fluid system. Specifically, 1. an atmospheric anticyclone or cyclone; 2. a whirlpool or eddy.

W

W See *watt*.

waning The changing phases of the Moon as it becomes less fully illuminated between full Moon and new Moon as seen from Earth. Compare *waxing*.

watt (W) A measure of *power*. Units: joules per second (J/s).

wave A disturbance moving along a surface or passing through a space or a medium.

wavefront The imaginary surface of an electromagnetic wave, either plane or spherical, oriented perpendicular to the direction of travel.

wavelength The distance on a wave between two adjacent points having identical characteristics. The distance a wave travels in one period. Unit: meter.

waxing The changing phases of the Moon as it becomes more fully illuminated between new Moon and full Moon as seen from Earth. Compare *waning*.

weak nuclear force The force underlying some forms of radioactivity and certain interactions between subatomic particles. It is responsible for radioactive beta decay and for the initial proton-proton interactions that lead to nuclear fusion in the Sun and other stars. One of the four fundamental forces of nature, mediated by the exchange of *W* and *Z* particles. Compare *strong nuclear force*.

weather The state of an atmosphere at any given time and place. Compare *climate*.

weight 1. The force equal to the mass of an object multiplied by the local acceleration due to gravity. 2. In general relativity, the force equal to the mass of an object multiplied by the acceleration of the frame of reference in which the object is observed.

white dwarf The stellar remnant left at the end of the evolution of a low-mass star. A typical white dwarf has a mass of 0.6 solar mass ($M_\odot$) and a size about equal to that of Earth; it is made of nonburning, electron-degenerate carbon.

Wien's law A relationship describing how the peak wavelength, and therefore the color, of electromagnetic radiation from a glowing blackbody changes with temperature.

WIMP Literally, "weakly interacting massive particle." A hypothetical massive particle that interacts through gravity but not with electromagnetic radiation and is a candidate for dark matter. Compare *MACHO*.

winter solstice 1. One of two points where the Sun is at its greatest distance from the celestial equator. 2. The day on which the Sun appears at this location, marking the first day of winter (about December 21 in the Northern Hemisphere and June 21 in the Southern Hemisphere). Compare *summer solstice*.

X

X-ray Electromagnetic radiation having frequencies and photon energies greater than those of ultraviolet light but less than those of gamma rays and having wavelengths shorter than those of ultraviolet light but longer than those of gamma rays.

X-ray binary A binary system in which mass from an evolving star spills over onto a collapsed companion, such as a neutron star or black hole. The material falling in is heated to such high temperatures that it glows brightly in X-rays.

Y

year The time it takes Earth to make one revolution around the Sun. A solar year is measured from equinox to equinox. A sidereal year, Earth's true orbital period, is measured relative to the stars.

Z

zenith The point on the celestial sphere located directly overhead from an observer. Compare *nadir*.

zero-age main sequence The strip on the H-R diagram plotting where stars of all masses in a cluster begin their lives.

zodiac The constellations lying along the plane of the ecliptic.

zodiacal dust Particles of cometary and asteroidal debris less than 100 micrometers (μm) in size that orbit the inner Solar System close to the plane of the ecliptic. Compare *meteoroid* and *planetesimal*.

zodiacal light A band of light in the night sky caused by sunlight reflected by zodiacal dust.

zonal wind The planet-wide circulation of air that moves in directions parallel to the planet's equator.

✦ SELECTED ANSWERS

Chapter 1
Summary Self-Test
1. Earth–Sun–Solar System–Milky Way Galaxy–Local Group–Virgo Supercluster–universe
2. c
3. b
4. falsify
5. a
6. radius of Earth–light-minute–distance from Earth to Sun–light-hour–radius of Solar System–light-year
7. its age
8. b-d-a-c-e
9. $1.60934 \times 10^3 \rightarrow 1{,}609.34$ and $9.154 \times 10^{-3} \rightarrow 0.009154$
10. $86{,}400 \rightarrow 8.64 \times 10^4$ and $0.0123 \rightarrow 1.23 \times 10^{-2}$

Conceptual Questions
28. For example, the distance from the Sun to Neptune is about the time needed to fly from New York City to London.
30. 2.5 million years. This is the amount of time it takes for light that leaves Andromeda to reach us.
33. A *theory* is generally understood to mean an idea a person has, whether or not there is any proof, evidence, or way to test it. A *scientific theory* is an explanation for an occurrence in nature, must be based on observations and data, and must make testable predictions.

Problems
46. 5.9×10^{17} miles
49. 5,333 km/h, or 6.7 times faster
50. about 600 billion to 1 trillion stars

Chapter 2
Summary Self-Test
1. a
2. a, b, c
3. a
4. a
5. c
6. the tilt of the Earth's axis
7. b
8. a
9. b
10. a, b

Conceptual Questions
28. Polaris was never above the horizon.
36. Eclipses occur only when the Earth-Sun and Earth-Moon planes line up.
39. Moonlight is reflected sunlight.

Problems
42. 50 minutes later
46. (a) 23.5° above the horizon at noon. (b) 23.5° above the horizon at midnight
49. The tropics would be at latitudes ±10° and the circles at ±80°.

Chapter 3
Summary Self-Test
1. (a) closest; farthest. (b) star; nothing. (c) larger.
2. b
3. c
4. b = e < c < d < a
5. c-b-d-a
6. How "oval" or elongated the orbit is.
7. b-a-c
8. d
9. a, b, c, d
10. a

Conceptual Questions
36. The Moon's gravity is one-seventh that of Earth's.
37. A *bound* orbit will repeat itself; an *unbound* orbit never returns.
39. The object came from outside the Solar System.

Problems
44. (a) 6.94 m/s². (b) 8,328 N. (c) The road pushes against the tires.
46. (a) 2m²/s. (b) 5s.
50. (a) 0.71 year. (b) 41.7 km/s.

Chapter 4
Summary Self-Test
1. a
2. d
3. a, b, c
4. a, b, c
5. c
6. a
7. d
8. a, e
9. e-c-b-d-a
10. (a)-3, (b)-4, (c)-2, (d)-6, (e)-7, (f)-1, (g)-5

Conceptual Questions
27. distance
29. chromatic aberration
33. Refraction happens when a wave travels at an angle through media with different speeds of propagation.

Problems
42. 380 meters; 3.05 meters
44. 2 million times greater
46. 480 meters. Our pupil's tiny size means we would have no angular resolution at all.

Chapter 5
Summary Self-Test
1. b, e, f
2. c
3. c
4. b
5. a
6. b
7. a
8. a
9. c
10. d

Conceptual Questions
28. There would be no stars in the universe.
29. A system must be acted upon by an outside force to change a conserved quantity.
40. Most of the stars would be orbited by planets.

Problems
47. 1 percent
48. (a) 0.70 AU. (b) This planet would receive more sunlight than Earth.
49. (a) 4.9 times the volume. (b) 4.9 times the mass.

Chapter 6
Summary Self-Test
1. b
2. b
3. c, d, e
4. c
5. c

SA-1

6. d
7. a, d
8. c
9. b, c, e
10. c, d, e

Conceptual Questions
30. The rock with the least of the parent and the most of the daughter nuclei is the oldest, as decay was occurring for the longest amount of time. Rock B is oldest.
31. radioactive decay and friction
40. Mars shows rivers, canyons, and teardrop-shaped erosions around craters, indicating that water was flowing and eroding the surface. Mars is too cold to have significant amounts of liquid water on its surface today.

Problems
42. (a) Volume = 1.08×10^{21} m^3. (b) Area = 5.11×10^{14} m^2. (c) Volume would become eight times larger and the surface area four times larger.
46. 110 W/m^2
50. 120 million years ago

Chapter 7

Summary Self-Test
1. d-b-c-f-a-e
2. d-e-b-c-a
3. c
4. c
5. All are correct.
6. b
7. c
8. a, b, c
9. b, c, d
10. d

Conceptual Questions
28. Impacts from icy comets
31. Venus's atmosphere is extremely dense, while that of Mars is very tenuous. There are many more greenhouse molecules in Venus to trap heat, and very few in the atmosphere of Mars.
38. Extreme heat

Problems
43. (a) 64 pennies. (b) 9.2×10^{15}.
46. 1,334 pounds per square inch.
47. (a) 271.7 K. (b) 156.9 K.

Chapter 8

Summary Self-Test
1. gas, ice
2. d
3. b
4. temperature; pressure
5. a
6. b
7. d
8. b
9. c
10. a, b, d

Conceptual Questions
28. Rapid rotation creates strong centripetal forces.
38. The rings are inside the Roche limit and will never make one ring.
40. Rings do not last forever because gravitational forces on the particles cause them to move in or out. The E Ring is fed by Enceladus.

Problems
42. 632 AU. This is 21 times farther than Neptune's orbit.
44. (a) 1,320 Earths.
(b) $\dfrac{r_J}{r_\oplus} = \dfrac{M_J/V_J}{M_\oplus/V_\oplus} = \dfrac{M_J}{M_\oplus}\dfrac{V_\oplus}{V_J} = \dfrac{318 M_\oplus}{M_\oplus}\dfrac{V_\oplus}{1{,}320 V_\oplus} = 0.24$.
46. Pressure increases more rapidly on Neptune than Uranus. Saturn has the slowest rate of change, and Neptune has the fastest.

Chapter 9

Summary Self-Test
1. Dwarf planets; moon; asteroids; comets
2. b, d
3. massive
4. b
5. d
6. All are correct.
7. a, b, d, e
8. b, e
9. b
10. nucleus, coma, tail

Conceptual Questions
30. Europa's surface is covered with water ice. Its density and magnetic readings are consistent with a liquid water ocean. The surface shows evidence of liquid seepage through cracks.
33. Gravitational "stirring" from nearby Jupiter keeps the objects in the asteroid belt well mixed.
35. Kuiper Belt objects are too far from the Sun for their volatile gases to sublimate into gas.

Problems
41. (a) 2.5 km/s. (b) 2.5 times faster than the speed of material in Io's vents.
42. (a) Surface area of 4.1×10^{13} m^2 and a volume of 2.5×10^{19} m^3. (b) 1.2×10^{11} m^3. (c) 2×10^8 years. (d) At least 20 times.
45. (a) 2.2 AU (Encke), 17.9 AU (Halley), 186 AU (Hale-Bopp). (b) 35.8 AU (Halley), 371 AU (Hale-Bopp). (c) Hale-Bopp is most pristine because it made the fewest passes around the Sun; Encke is the least pristine because it made the most passes.

Chapter 10

Summary Self-Test
1. b
2. a
3. b
4. a, b
5. velocity; period; distance
6. b
7. a
8. a, b, c, d
9. b
10. d

Conceptual Questions
28. From Mars and Jupiter, we could measure the distance to stars farther away. From Venus, we could measure only the distance to closer ones.
34. Red photons are shown with a long wavelength, while ultraviolet photons have short wavelengths. In a real cloud of gas, photons will travel randomly in all directions and not just leave in one direction.
37. We can directly measure stellar masses only by using their gravitational influence on an orbiting body; for example, by using Kepler's third law.

Problems
45. Rigel is twice as distant as Betelgeuse, so to appear the same brightness, Rigel must be about four times more luminous. Betelgeuse is red and so much cooler than Rigel, so it must be much larger.
49. (a) 1 $M_\oplus$. (b) 20 AU.
50. Halfway between the two masses; two-thirds of the way toward the heavier mass

Chapter 11

Summary Self-Test
1. b
2. e = a = g; f; h = b; d = c
3. c
4. a
5. a
6. a
7. c
8. c
9. c
10. a

Conceptual Questions
31. The oceans are not hot enough.
36. There are very few heavy elements within the Sun or any other star.
37. The extra mass (or mass difference) between four hydrogens and one helium is converted to energy (remember Einstein's formula $E = mc^2$, which tells us mass and energy are equivalent).

Problems

42. (a) 25 days. (b) 4.4×10^9 meters. (c) 2.04 km/s.
45. A few hundred thousand years. This is so much shorter than the age of Earth that it is not a viable mechanism.
47. (a) Height above the base of the photosphere. (b) In the upper atmosphere of the Sun, temperature and density do not track each other, meaning that the emission of energy is not a blackbody process.

Chapter 12

Summary Self-Test

1. a
2. a
3. b
4. c
5. d
6. c-f-h-d-i-a-e-g-b
7. c-a-f-d-b-e-g
8. true
9. true
10. c

Conceptual Questions

26. Low-mass, low-luminosity stars are more common because they are longer lived.
30. Faster hydrogen burning means the star will grow larger and more luminous to allow the extra energy created to escape; but because the temperature of the core remains constant, the star's surface temperature will drop.
34. Higher luminosity implies faster rates of nuclear burning, which means the star has much less time to live.

Problems

43. (a) 56 billion years. (b) 110 million years. (c) 360,000 years. These ages are of the same order as the actual stars in Table 12.1.
45. (a) 83 km/s. (b) 37 km/s. (c) Mass loss increases with increasing size, assince the surface gravity drops.
50. 112 km

Chapter 13

Summary Self-Test

1. c
2. e-a-f-b-c-d
3. b
4. a, b, d, e, f
5. b
6. d
7. d
8. b
9. b
10. d

Conceptual Questions

28. In each successive burning cycle, there are fewer nuclei to be fused, and the increasing core temperature burns through the fuel more rapidly.
37. Time dilation at high speeds. The faster the muon travels, the longer it travels before its internal clock says it is time to decay.
40. The effects of relativity show themselves only in extreme circumstances, like traveling close to the speed of light or being right next to a black hole or neutron star.

Problems

41. (a) 3.8×10^{23} kg/min. (b) 5.25 times the mass of the Moon per minute.
43. (a) 30,300 pc. (b) We cannot use parallax to measure the distance to the farther star because it is more than 1,000 pc away.
47. 1,592 km. This is significantly larger than the pulsar's size.

Chapter 14

Summary Self-Test

1. a
2. a
3. c
4. c
5. a
6. c, d
7. b
8. b
9. a
10. c-a-b-d

Conceptual Questions

29. Dark matter is believed to interact only gravitationally with other matter, but does not emit or absorb light.
37. The observer would see all galaxies expanding away, just as we do.
38. The gravitational forces that hold our galaxy together are far stronger than the expansion of the universe, thus the Milky Way does not expand with the Hubble flow.

Problems

41. 3.6×10^{12} galaxies
43. 16.6 nm, 7,588 km/s
48. 10 AU across

Chapter 15

Summary Self-Test

1. a
2. b
3. c
4. a
5. d
6. a
7. e-b-a-d-c
8. d
9. d
10. a, b, c

Conceptual Questions

32. 21-cm emission reveals that our rotation curve is flat to very large distances from the galactic center, implying that up to 90 percent of the Milky Way is dark matter.
35. In regions of active star formation, which are generally in the spiral arms.
40. The Milky Way's satellite galaxies are remnants of the small protogalaxies that first formed in our dark-matter halo. Most coalesced to form the Milky Way. They are hard to find because they have very low surface brightness.

Problems

42. About 0.02 percent.
44. (a) 300,000 light-years. (b) The halo is about three times larger than the disk.
49. 2,094 km/s

Chapter 16

Summary Self-Test

1. a
2. b
3. c
4. a, b
5. b, c
6. c
7. a, c, e
8. a-b-c-d
9. a, b, d, e, g
10. a, c, e

Conceptual Questions

28. Just as our planet, galaxy, or Solar System are not being torn apart by the expansion of space, so too are they stable against the repulsive force of dark energy: gravity holds them together. Also, a large collection of matter is not a vacuum, so the amount of dark energy present in say, the Sun, is very low.
32. Gravity, strong nuclear, weak nuclear, electromagnetic. Only the latter depends on charge.
37. The Hubble time is roughly the age of the universe.

Problems

43. Five atoms per cubic meter.
45. (a) 1.8×10^{14} J. (b) Approximately the same as a 1-megaton hydrogen bomb.
48. −0.00123.

Chapter 17

Summary Self-Test

1. c-b-d-a

2. b
3. c
4. a
5. a
6. a
7. c, e, f
8. (a) led to the theory; (b) and (c) were predicted and observed.
9. e-b-d-a-c
10. b

Conceptual Questions
27. A single galaxy could be found in a void, but a cluster could not. Voids do not mean "100 percent empty" but rather that most of the mass (that is, galaxies) is clustered outside that space.
33. Inhomogeneities in the density of normal matter were never strong enough to cause the structure seen today.
38. The nuclear forces are too short ranged to work over cosmic distances, and electromagnetic forces work only for charged particles; most matter in the universe is neutral.

Problems
42. We would predict a normal matter density for the universe that is about 10 times greater than we actually observe.
43. About 2×10^{-36} kg
44. (a) 10^{13} solar masses. (b) 10^{14} to 10^{15} solar masses. (c) 10^{16} to 10^{18} solar masses.

Chapter 18
Summary Self-Test
1. a, b, c
2. d
3. c
4. a
5. d
6. a
7. a
8. d
9. d
10. All are correct.

Conceptual Questions
30. We are similar to chimpanzees (6 million years ago), which were preceded by primates (70 million years ago), which were preceded by the arrival of animals and fungi (1 billion years ago).
33. A reasonably stable environment, abundant building blocks, a "solvent" in which the chemistry could work, and a lot of time.
35. H and He are mostly from the Big Bang. Everything else is from stellar evolution.

Problems
44. Using optimistic values from the text ($R^* = 7$, $f_p = 1$, $n_e = 2$, $f_l = 1$, $f_i = 1$, $f_c = 1$, $L = 10^6$), there should be about 14 million advanced civilizations in our galaxy.
45. Using pessimistic values from the text ($R^* = 7$, $f_p = 0.5$, $n_e = 1$, $f_l = 0.1$, $f_i = 0.1$, $f_c = 0.1$, $L = 10^3$), there should be about three advanced civilizations in our galaxy.
48. 6.87×10^{11}

CREDITS

Text Credits

ABC 4 News: "ATK Building World's Largest Space Telescope in Magna." From 4Utah.com (ABC 4 News), March 19, 2013. Used by permission of Nexstar Broadcasting, Inc.

Peter Brannen: "Astronomer uses Keplar Telescope's Data in Hunt for Spacecraft from Other Worlds." By Peter Brannen, first published in *The Washington Post*, July 22, 2013. Copyright © Peter Brannen. Used by permission of the author.

Denise Chow: "Giant Propellers Discovered in Saturn's Rings." From Space.com, July 8, 2010. Used by permission of Tech Media Network.

Rachel Courtland: "Keplar's Continuing Mission." From IEEE Spectrum, October 2013. Used by permission of the Institute of Electrical and Electronics Engineers via the Copyright Clearance Center.

Stuart Gary: "Massive Black Holes Sidle Up to Other Galaxies." From ABC.net.au, Australian Broadcasting Corporation, July 22, 2013. Used by permission of the Australian Broadcasting Corporation.

Lisa Grossman: "Hyperfast Star Kicked Out of Milky Way." From Wired.com, July 22, 2010. Copyright © 2010 Condé Nast Publications. All rights reserved. Originally published in Wired.com. Reprinted by permission.

Amina Khan: "'Moon is Wetter, Chemically More Complex than Thought' NASA Says." From *Los Angeles Times*, October 22, 2010, "Curiosity Rover Sees Signs of Vanishing Martian Atmosphere." From *Los Angeles Times*, April 9, 2013, and "Planck: Big Ban's Afterglow Reveals Older Universe, More Matter." From *Los Angeles Times*, March 21, 2013. Used by permission of the Los Angeles Times.

Ledyard King: "Weather Forecast in Space: Not Sunny, with Solar Flares." From *USA Today* – Gannett Washington Bureau, June 4, 2013. Used by permission of USA Today. Copyright © 2013 Gannett-USA Today. All rights reserved. Used by permission and protected by the copyright laws of the United States. The printing, redistribution, or retransmission of this Content without express permission is prohibited.

John P. Millis: "Mystery of SS Cygni Star System Finally Resolved." From RedOrbit.com, May 23, 2013. Used by permission of Dr. John P. Millis for redOrbit.com - Your Universe Online"

Ian O'Neill: "Curiosity, Interrupted: Sun Makes Mars Go Dark." From Discovery.com, April 15, 2013. Used by permission of Discovery Communications, LLC.

Dennis Overbye: "Vote Makes It Official: Pluto Isn't What It Used to Be." From *The New York Times*, August 25, 2006. Copyright © 2006 The New York Times. All rights reserved. Used by permission and protected by the copyright laws of the United States. The printing, redistribution, or retransmission of this Content without express permission is prohibited.

Phil Plait: "Two Eclipses, Two Stories." From Slate.com, March 27, 2013. Used by permission of Slate.

Ian Sample: "What a Scorcher—Hotter, Heavier and Millions of Times Brighter than the Sun." From *The Guardian*, July 22, 2010. Copyright Guardian News & Media Ltd. 2010.

IPCC: Climate Change 2013: The Physical Science Basis. Working Group I Contribution to the Fifth Assessment Report of the Intergovernmental Panel on Climate Change, Figure SPM.1 (a). Cambridge University Press. Used by permission of the Intergovernmental Panel on Climate Change.

Space.com Staff: "Scientists May be Missing Many Star Explosions." From Space.com, Space.com, December 27, 2010, and "Colliding Galaxies Swirl in Dazzling New Photo." From August 5, 2010. Used by permission of Tech Media Network.

Photo Credits

Chapter 1
Page 3 (photos): Ron Proctor; (Topographic Map): USGS / Utah Geological Survey; p. 5 (from top down): 1.2a and 1.2b: Neil Ryder Hoos (2); 1.2c: Owen Franken/Corbis; Monkey Business Images/Dreamstime.com; PhotoDisc/Getty Images; American Museum of Natural History (2); NASA/JPL/Caltech; p. 6 (top): NASA/ESA and J. Hester (Arizona State University); (bottom): Michael J. Tuttle (NASM/NASA); p. 7 (top): Courtesy Douglas Finkbeiner; http://halpha.skymaps.info and Swinburne Astronomy Productions; (bottom): Courtesy CERN; p. 9: Courtesy of the Archives, California Institute of Technology; p. 10: Matheisl/Getty Images; p. 12: Craig Lovell/Corbis.

Chapter 2
Page 21 (clockwise from top): John W. Bova /Science Source; Eckhard Slawik / Science Source; John W. Bova/Science Source (5); p. 25 (left): Pekka Parviainen / Science Source; (right): David Nunuk/Science Source; p. 38: Nick Quinn; p. 39 (a): Johannes Schedler; (b): Anthony Ayiomamitis (TWAN); p. 40 (both): NASA/SDO.

Chapter 3
Page 47 (all): NASA; p. 48: Tunc Tezel; p. 53: SSPL/Jamie Cooper /The Image Works; p. 66: NASA/JPL-Caltech.

Chapter 4
Page 73: Dreamstime; p. 78: Stockbyte/Getty Images; p. 81 (left): Science Photo Library; (right): Hulton-Deutsch Collection/Corbis; p. 82: Jean-Charles Cuillandre (CFHT); 84 (center): ASU Physics Instructional Resource Team; (bottom): © Russell Kightley; p. 86 (top): National Radio Astronomy Observatory; (bottom): NAIC-Arecibo Observatory, a facility of the NSF; 87 (top): Dave Finley/National Radio Astronomy Observatory; (bottom): ESO; p. 88 (top): HST/STSci/NASA; (bottom): ASU Physics Instructional Resource Team; p. 90: Courtesy Heidi B. Hammel, Imke de Pater and the Keck Observatory.

Chapter 5
Page 97 (clockwise from left): NASA, ESA, N. Smith (University of California, Berkeley), and the Hubble Heritage Team (STScI/AURA); NASA, ESA, M. Robberto (STScI), and the HST Orion Nebula Team; NASA, ESA, M. Robberto (STScI), and the HST Orion Nebula Team (2); p. 102: Jeff Hester and Paul Scowen (Arizona State University), Bradford Smith (University of Hawaii), Roger Thompson (University of Arizona), and NASA; p. 105 (top left): STScI photo: Karl Stapelfeldt (JPL); (top right): D. Padgett (IPAC/Caltech), W. Brandner (IPAC), K. Stapelfeldt (JPL) and NASA; (bottom): Photograph by Richard Pelisson, SaharaMet; p. 107: Reuters/Corbis; p. 109 (bottom left): Karl Stapelfeldt/JPL/NASA; (bottom right): NASA/JPL-Caltech/UIUC; p. 112: NASA/Johns Hopkins University/Applied Physics; p. 115: Courtesy Dawn Ellner; p. 118 (top): C. Marois, National Research Council, Canada; (bottom): NASA, ESA and P. Kalas (University of California, Berkeley and Seti Institute.)

Chapter 6
Page 127: ESO; p. 128 (top): NASA; (bottom, left to right): NASA/Johns Hopkins University Applied Physics Laboratory/Carnegie Institution of Washington; NASA-JPL; M-Sat Ltd / Science Source; NASA/USGS; Science Source; p. 129: NASA/JSC; p. 130 (top): Photograph by D. J. Roddy and K. A. Zeller, USGS, Flagstaff, AZ; (bottom): NASA/JPL; p. 131 (all): NASA/JPL/Caltech; p. 137: Art Directors & TRIP / Alamy; p. 139: Donald Duckson/Visuals Unlimited/Corbis; p. 143 (top left): ESA/DLR/FU Berlin (G. Neukum); (bottom): NASA/Magellan Image/JPL; (top right): NSSDC/NASA; p. 145 (top): Courtesy Tomas Janececk; (bottom): Shutterstock; p. 146 (top): NASA/JSC; (bottom): K. C. Pau; p. 147: NASA/JPL/Malin Space Science Systems; p. 148: NASA/JPL/University of Arizona; p. 149 (left): NASA/JPL-Caltech/MSSS and PSI; (right): NASA/JPL-Caltech/University of Arizona/Texas A&M University; p. 150: NASA/JPL-Caltech/University of Arizona/Texas A&M University.

Chapter 7
Page 157: USAF Senior Airman Joshua Strang/Wikimedia Commons; p. 163 (all): NASA/NSSDC/GSFC; p. 165 (from left to right): Bob Blaylock; http://creativecommons.org/licenses/by-sa/3.0/deed.en; Zorkun 2008, Wikidoc; http://creativecommons.org/licenses/by-sa/3.0/; (2005) The Evolution of Self-Fertile Hermaphroditism: The Fog Is Clearing. PLoS Biol 3(1): e30. doi:10.1371/journal.pbio.0030030; http://creativecommons.org/licenses/by/2.5/deed.en; Peter Halasz; http://creativecommons.org/licenses/by-sa/2.5/deed.en; De Agostini Picture Library/De Agostini/Getty Images; Photo by Shizhao; http://creativecommons.org/licenses/by-sa/2.5/deed.en; Photo by Figaro, 2005, released in pd. Wikimedia; p. 167: Courtesy Alan Lakritz; p. 169 (left): NASA; (right): LOOK Die Bildagentur der Fotografen GmbH/Alamy; p. 172: NASA/NSSDC; p. 173: NASA/JPL; 174 (left): NASA; (right): NASA/STScI; p. 176 : IPCC, 2013: Summary for Policymakers. In: Climate Change 2013: The Physical Science Basis. Contribution of Working Group I to the Fifth Assessment Report of the Intergovernmental Panel on Climate Change [Stocker, T.F., D. Qin, G.-K. Plattner, M. Tignor, S.K. Allen, J. Boschung, A. Nauels, Y. Xia, V. Bex and P.M. Midgley (eds.)]. Cambridge University Press, Cambridge, United Kingdom and New York, NY, USA, in press.

Chapter 8
Page 183 (both): NASA/JPL (Cassini); p. 184 (all): NASA/JPL; p. 188: NASA, ESA, and the Hubble Heritage Team (STScI/AURA); p. 189 (left): NASA/JPL/University of Ari-

zona; (right): NASA/JPL/Caltech; p. 190 (top): CICLOPS/NASA/JPL/University of Arizona; (bottom): NASA/JPL/Space Science Institute; p. 191 (top left):NASA/JPL/Space Science Institute; (inset, top right): Courtesy of Cassini Imaging Central Laboratory for Operations; (bottom, both): NASA, L. Sromovsky, and P. Fry (University of Wisconsin-Madison); NASA, L. Sromovsky, and P. Fry (University of Wisconsin-Madison); p. 196 (top): NASA/JPL/Caltech; (bottom): NASA, ESA, L. Sromovsky and P. Fry (University of Wisconsin), H. Hammel (Space Science Institute), and K. Rages (SETI Institute); p. 201: http://www.freenaturepictures.com; p. 202: Nick Schneider, University of Colorado, Boulder; (left): J. Clarke (University of Michigan), NASA; (right): Courtesy of John Clark, Boston University and NASA/STScI; p. 206 NASA/JPL/Caltech; p. 207: NASA/JPL/Space Science Institute; p. 208: NASA/JPL/Caltech; p. 209: NASA/JPL/Space Science Institute; p. 213: NASA/JPL.

Chapter 9

Page 215: Original artwork by Ron Proctor; p. 216 (top): Dr. R. Albrecht, ESA/ESO Space Telescope European Coordinating Facility/NASA; (bottom): M. Brown (Caltech), C. Trujillo (Gemini), D. Rabinowitz (Yale), NSF, NASA; p. 217 (top): NASA/ESA/J. PARKER; (bottom, all): NASA/JPL/Caltech; p. 218 NASA/JPL/Caltech 219 (both): NASA/JPL/Space Science Institute; p. 220 (left, both): NASA/JPL/Caltech; (right): NASA/JPL; p. 221 (all): NASA/JPL/Space Science Institute; p. 222 (top, left): NASA/JPL; (top, right): NASA/JPL/ESA/University of Arizona; (bottom): ESA/NASA/JPL/University of Arizona; p. 223: NASA; p. 224: NASA/JPL/Caltech; p. 225 (top): NEAR Project, NLR, JHUAPL, Goddard SVS, NASA; (bottom, left): NASA/JPL/University of Arizona; (bottom, right): ESA/DLR/FU Berlin (G. Neukum); p. 228: Courtesy of Terry Acomb; p. 229 (left): Dr. Robert McNaught; (right): © Don Goldman; p. 230 (top): NASA/NSSDC/GSFC; (bottom): NASA/JPL/Caltech; p. 231 (all): NASA/JPL-Caltech/UMD; p. 232: Science Source; p. 233 (left): Camera Press/Ria Novosti/Redux; (right): Xinhua/RIA/Xinhua/Landov; p. 235 (left): Tony Hallas/Science Faction/Corbis; (right): Noel Power/Shutterstock; p. 236 (a): Courtesy Arizona Skies Meteorites; (b): Collection and photo courtesy Thomas Witzke; (c): Suzanne Morrison/Aerolite Meteorites; (d): Dr. Svend Buhl; p. 237: NASA/JPL/Cornell.

Chapter 10

Page 245: Roberto Mura/Wikimedia Commons; p. 257: Chris Stoughton.

Chapter 11

Page 275 (clockwise from top left): Courtesy Dr. Gerald Gwinner, Dept. of Physics & Astronomy, University of Manitoba; SOHO/NASA/ESA (2); p. 284 (top): Hinode JAXA/NASA/PPARC; (bottom): NOAO/AURA/NSF; p. 287 (top): Photo by Brocken Inaglory, 2006; http://creativecommons.org/licenses/by-sa/3.0/deed.en; (bottom): Nigel Sharp, NOAO/NSO/Kitt Peak FTS/AURA/NSF; p. 288 (top, left): Stefan Seip, www.astromeeting.de; (top, right): © Laura Kay; (bottom): © Laura Kay; p. 289 (top): Science Source; p. (bottom): SOHO/ESA/NASA; p. 290 (both): NASA /SDO / Solar Dynamics Observatory; p. 293 (top): NASA; (center, all): SOHO-EIT Consortium, ESA, NASA; (bottom): SOHO/ESA/NASA; p. 293: SOHO/ESA/NASA; p. 294 (top, a-c): NASA SDO; (bottom): SOHO/ESA and NASA.

Chapter 12

Page 303: Anne, Karen and David Rajala/Adam Block /NOAO /AURA/NSF; p. 315(top,a): Minnesota Astronomical Society; (top, b): NASA and The Hubble Heritage Team (STScI/AURA); (bottom, left to right): Dr. Raghvendra Sahai (JPL) and Dr. Arsen R. Hajian (USNO), NASA and The Hubble Heritage Team (STScI/AURA); NASA, A. Fruchter and the ERO team (STScI); J. P. Harrington and K. J. Borkowski (University of Maryland), HST, NASA; Bruce Balick (University of Washington), Vincent Icke (Leiden University, The Netherlands), Garrelt Mellema (Stockholm University), and NASA.

Chapter 13

Page 329 (both): Brad Farrington / Louisiana Delta Community College; p. 333 (top): N. Smith, J. A. Morse (U. Colorado) et al., NASA; (bottom, both): Anglo-Australian Observatory, photographs by David Malin; p. 336 (top): N. Levenson (Johns Hopkins), S. Snowden (USRA/GSFC); (bottom, left): Courtesy Nancy Levenson and colleagues, American Astronomical Society; (bottom, right): Jeff Hester, Arizona State University, and NASA; p. 340 (a): European Southern Observatory, ESO (inset): Hubble Space Telescope; (b-c): Jeff Hester, Arizona State University, and NASA; p. 352 ESA/Hubble.

Chapter 14

Page 359: The Hubble Heritage Team (AURA/STScI/NASA); p. 361 (all): Australian Astronomical Observatory, Photographs by David Malin; 363 (top): NOAO/AURA/NSF; (bottom, both): NASA; p. 364(top): Australian Astronomical Observatory, photograph by David Malin; (bottom): NASA and The Hubble Heritage Team (AURA/STScI); p. 365 (top, both): Johannes Schedler / Panther Observatory; (bottom, a): NASA/Swift/Stefan Immler (GSFC) and Erin Grand (UMCP); (bottom, b): Bill Schoening, Vanessa Harvey /REU program/ NOAO/AURA/NSF; p. 366 (a): Todd Boroson/NOAO/AURA/NSF; (b): Richard Rand, University of New Mexico; p. 369: NASA, ESA and the Hubble Heritage (STScI/AURA)-ESA/Hubble Collaboration Acknowledgment: M. West (ESO, Chile); p. 371 (all): Courtesy of Dr. Michael Brotherton; p. 374 (all): F. Owen, NRAO, with J. Biretta, STScI, and J. Eilek, NMIMT; p. 376 (both): Courtesy of Jeff Hester; p. 381 (both): WIYN/NOAO/NSF, WIYN Consortium, Inc.; p. 383: NASA/ESA/SAO, CXC, JPL-Caltech and STScI. J. DePasquale (Harvard-Smithsonian CfA) and B. Whitmore (StScI); p. 387: NASA, ESA, and the Hubble Heritage Team (STScI/AURA).

Chapter 15

Page 389: Frederick A. Ringwald; p. 390 (top): © Axel Mellinger; (bottom): C. Howk, University of Notre Dame, B. Savage, University of Wisconsin, N.S. Sharp (NOAO)/WIYN/NOAO/NSF; p. 391 (top): C. Jones, Harvard-Smithsonian Center for Astrophysics; (bottom, left): © Lynette Cook, all rights reserved; (bottom, right): NOAO/AURA/NSF; p. 392: NOAO/AURA/NSF; p. 393 (top): Courtesy of Jeff Hester; (bottom): The Electronic Universe Project; p. 395: ESA; p. 398 (top): ASPERA / G.Toma /A.Saftoiu; (bottom): Courtesy of Pierre Auger Observatory; p. 401 (top, from left to right): NASA/CXC/MIT/F.K. Baganoff et al./E. Slawik; Hubble: NASA, ESA, and Q.D. Wang (University of Massachusetts, Amherst); Spitzer: NASA, Jet Propulsion Laboratory, and S. Stolovy (Spitzer Science Center/Caltech); NRAO/AUI; (bottom): Keck/UCLA Galactic Center Group; p. 402: David A. Aguilar (CfA); p. 403 (top): This image was created by Prof.Andrea Ghez and her research team at UCLA and is from data sets obtained with the W.M. Keck Telescopes. Image creators include Andrea Ghez, Angelle Tanner, Seth Hornstein, and Jessica Lu; (bottom, both): Anglo-Australian Observatory, photographs by David Malin; p. 404: NASA, ESA, and Z. Levay (STScI).

Chapter 16

Page 411: © ESA and the Planck Collaboration; p. 418: Lucent Technologies' Bell Labs; p. 419 (both): Ann and Rob Simpson Nature Photography; p. 420 (a-c): NASA COBE Science Team (3); ©ESA and the Planck Collaboration; p. 428: From The Elegant Universe by Brian Greene ©1999 by Brian R. Greene. Used by permission of W.W. Norton & Company, Inc; p. 438 (both): Courtesy of the author.

Chapter 17

Page 441 (top): Thomas Jarrett (IPAC/Caltech); (left): Andre Karwath/Creative Common; p. 443 (top a): X-ray: NASA/CXC/MIT/E.-H Peng. Optical: NASA/STScI; (inset): NASA, N. Benitez (JHU), T. Broadhurst (The Hebrew University), H. Ford (JHU), M. Clampin(STScI), G. Hartig (STScI), G. Illingworth (UCO/Lick Observatory), the ACS Science Team and ESA; (b): NASA, ESA, D. Coe (NASA JPL/California Institute of Technology, and Space Telescope Science Institute), N. Benitez (Institute of Astrophysics of Andalusia, Spain), T. Broadhurst (University of the Basque Country, Spain), and H. Ford; (right, a): Max Tegmark and the Sloan Digital Sky Survey; (right, b): M. Blanton and the Sloan Digital Sky Survey; (right, c):Max Tegmark and the Sloan Digital Sky Survey; p. 449: Simulation: Matthew Turk, Tom Abel, Brian O'Shea Visualization: Matthew Turk, Samuel Skillman; p. 450 (a-b): NASA/JPL-Caltech/GSFC; NASA, ESA, G. Illingworth (University of California, Santa Cruz), R. Bouwens (University of California, Santa Cruz, and Leiden University), and the HUDF09 Team; p. 451 (top, a-b): Guedes, Javiera; Callegari, Simone; Madau, Piero; Mayer, Lucio The Astrophysical Journal, Volume 742, Issue 2, article id. 76 (2011); (bottom): Illustration Credit: NASA, ESA, and M. Kornmesser (ESO); p. 452 (top): X-ray (NASA/CXC/SAO/P. Green et al.), Optical (Carnegie Obs./Magellan/W.Baade Telescope/J.S.Mulchaey et al.); (center): NASA, ESA, J. Lotz (STScI), M. Davis (University of California, Berkeley), and A. Koekemoer (STScI); (bottom): NASA, ESA, the Hubble Heritage (STScI/AURA)-ESA/Hubble; Collaboration, and A. Evans (University of Virginia, Charlottesville/NRAO/Stony Brook University); p. 453 (top): NASA/CXC/CfA/M.Markevitch et al.; and NASA/STScI; Magellan/U.Arizona/D.Clowe et al.; Lensing Map: NASA/STScI; ESO WFI; Magellan/U. Arizona/D.Clowe et al.; (bottom): Courtesy Joel Primack and George Blumenthal.

Chapter 18

Page 461: Stromatollites. Photo by Paul Harrison. March 2005; http://creativecommons.org/licenses/by-sa/3.0/deed.en; p. 462 (top): Michael Perfit, University of Florida, Robert Embley/NOAA; (bottom): Woods Hole Oceanographic Institution, Deep Submergence Operations Group, Dan Fornari; p. 463: Gaertner/Alamy; p. 465 (top): Courtesy of National Science Foundation; (bottom): Kenneth E.Ward/Biografx/ Science Source; p. 466: With permission of the Royal Ontario Museum © ROM; p. 473 (both): NASA; p. 475: F. Drake (UCSC) et al., Arecibo Observatory (Cornell, NAIC); p. 476: Seti/NASA; p. 479: Seti/NASA.

INDEX

Page numbers followed by *f* and *t* refer to figures and tables, respectively.

A

A0 stars, 260
Abell 1689 (galaxy cluster), 443*f*
absolute magnitude, 250
absolute zero, 98
absorption, of light, 80
absorption lines
   and atomic energy levels, 255–57, 257*f*
   classifying stars based on, 259*f*
   defined, 83
   of galaxies, 376, 376*f*
   identifying atoms by, 258
   of Sun, 287
absorption spectra, 254
abundance
   defined, 187
   relative, 337, 337*f*, 394–95
accelerated frame of reference, 346–47
acceleration
   calculating, 56
   gravitational, 58
   and inertial frame of reference, 54*f*
   in Newton's second law, 55, 55*f*, 56
   of universe, 422–25, 423*f*
accretion, 159
accretion disks
   in binary systems, 321
   defined, 106
   formation of, 108–9, 108*f*
   large objects from, 109–10, 109*f*
   of Milky Way Galaxy, 402, 402*f*
   and supermassive black holes, 373–74
   temperature and composition in, 110–11, 110*f*
   of young stars, 105*f*, 109
achondrites, 236*f*, 237, 238
active comets, 226, 228–30, 228*f*
active galactic nuclei (AGNs)
   defined, 371
   interactions of galaxies with, 375
   in normal galaxies, 374–75
   production of, 373–74
   size of, 372–73, 372*f*
active regions, 293–95, 294*f*
adaptive optics, 89–90, 90*f*
Adams Ring, 208
Adrastea, 206
Advanced LIGO (Laser Interferometer Gravitational-Wave Observatory), 349–50
AGB stars, *see* asymptotic giant branch stars
age, of planetary surfaces, 131
AGNs, *see* active galactic nuclei
albedos, 162, 226
ALH84001 meteorite, 238
Allen, Paul, 476, 481
Allen Telescope Array (ATA), 476, 476*f*
Alpha Centauri, 116, 249, 264*f*, 382
alpha particles, 310
Alpher, Ralph, 417, 418
altitude, 24
Amalthea, 206
American Astronomical Society, 419
Ames Research Center, 120, 151
amino acids, 462
ammonia, 159, 221
Amor asteroids, 223–24
amplitude, 76
Andromeda Galaxy, 4
   distance to, 376
   in Local Group, 402, 442
   movement of Milky Way and, 378, 379
   spiral arms of, 365, 365*f*
Anglo-Australian Telescope, 443
angular momentum, 106–7, 107*f*, 351
angular resolution, 81
animals, on phylogenetic tree, 464
annular solar eclipse, 38
anomalies, 175
Antarctica, meteorites in, 236
Antarctic Circle, 33
Antennae galaxies, 382–83, 383*f*
anticyclonic motion, 171
antiparticles, 426–27, 429
aperture, 84
Apollo, 131, 138, 145, 146
Apollo 8, 128, 128*f*
Apollo 15, 6*f*, 146*f*
Apollo asteroids, 223
apparent daily motion, 22, 22*f*, 25–27, 27*f*
apparent magnitude, 250
apparent retrograde motion, 48–49, 48*f*
Aquarius, 30
Archaea, 464, 464*f*
arcminute (arcmin), 81, 248
arcsecond (arcsec), 89, 248
Arctic Circle, 33
Arcturus, 397
Arecibo radio telescope, 86, 86*f*, 473
argon-36, 177
Ariel, 223
A Ring (of Saturn), 206–9
Aristotle, 22
ash, 96, 305–8
A stars, 259
asteroid belt, 112, 113, 217
asteroids, 214, 223–25, 224*f*
   composition of, 224
   groups of, 223–24
   impacts of Earth by, 478–79
   meteorites from, 235–38
   NASA mission to sample, 239
   in Solar System, 4, 113
   visits to, 224–25
astrobiology, 469
astrometry, 118
astronauts
   acceleration by, 57–58, 57*f*
   orbiting by, 63, 63*f*
astronomers, tools of, 6–7
astronomical data, 80–83
   from charge-coupled devices, 82–83
   from human eye, 80–81
   on photographic plates, 81–82
   from spectrographs, 83
astronomical unit (AU), 28
astronomy, 2–15
   defined, 2
   demotion of Pluto in, 14–15
   mathematics in, 10–14
   observation and theory in, 285
   as science, 7–10
   study of planetary motion in, 48–51
   tools of, 6–7
   universal context from, 4–7
asymptotic giant branch (AGB) stars
   from horizontal branch stars, 312–13
   on H-R diagram, 313, 313*f*
   and post-AGB stars, 313–15
   stellar mass loss in, 313
   Sun as, 477–78
ATA (Allen Telescope Array), 476, 476*f*
Aten asteroids, 223
ATK, 91
Atlantic Ocean, 176
atmosphere(s), 156–77
   changes in, 158–61
   and climate, 174–76
   defined, 158
   evolution of, 158, 158*f*
   of giant planets, 186, 192–93, 192–93*f*, 197, 198
   greenhouse effect/gases in, 160–64, 174–76
   and planetary mass, 160
   in planet formation, 111
   primary, 111, 112, 158
   secondary, 111, 113, 159–65
   of terrestrial planets, 112–13
   of Titan, 221, 222
atmosphere, of Earth, 111–13, 164–71
   atmospheres of Venus and Mars *vs.*, 126, 156, 163–64, 172
   composition of, 164–65
   distortion of telescopic images by, 87
   evolution of, 158
   human impact on, 479
   and impact craters, 130
   layers of, 165–69, 166*f*
   and resolution of telescopes, 88–90, 90*f*
   as secondary atmosphere, 159–60
   and solar activity, 295–96
   in Urey-Miller experiment, 462, 462*f*
   and weather, 156
   winds of, 170–71
atmosphere, of Mars
   atmospheres of other planets *vs.*, 126, 156
   auroras in, 169
   composition and mass of, 158
   Curiosity rover's sampling of, 177
   evolution of, 158–60
   features of, 173–74
   and formation of planet, 113
   and possibility of extraterrestrial life, 470
atmosphere, of Sun, 285–97
   activity of, 289–97
   and atmosphere of Earth, 156, 167
   and atmospheres of planets, 158
   chromosphere and corona, 285*f*, 286–89, 286*f*, 288*f*, 289*f*
   photosphere, 285*f*, 286–87
   spectrum of, 287, 287*f*
atmosphere, of Venus
   atmospheres of other planets *vs.*, 126, 156
   auroras in, 169
   clouds in, 172*f*
   composition and mass of, 158
   evolution of, 158–60
   features of, 172–73
   and formation of planet, 112, 113
   and impact craters, 130
   impact of greenhouse effect on, 163–64
atmospheric greenhouses effect
   and climate, 174–76
   and surface temperature, 160–64, 161*f*
atomic energy levels, 252–57, 253*f*
atomic nuclei, 277, 277*f*, 310
atoms, 79
   in cooling universe, 429
   electrons in, 253*f*
   emission and absorption lines of, 257–58
   formation of, 6
   spectral lines of, 260–61
attraction, gravitational, 378, 379
AU (astronomical unit), 28
aurora australis, 169
aurora borealis, 169
auroras, 169, 169*f*, 203
autumnal equinox, 32
axes, on graphs, 11
axions, 447
axis of rotation
   of giant planets, 188
   tilt of Earth's, 30–33, 31*f*
   wobble in Earth's, 34–35, 34*f*

B

B0 stars, 260
B1 stars, 260
B9 stars, 260
Bacteria, 464, 464*f*
bar, defined, 166
barred spiral (SB) galaxies, 362, 363*f*, 391, 391*f*
basalt, 139, 146
Bell Laboratories, 417, 418*f*
belts, 189
Bennu, 239
beryllium, 6, 337
beryllium-8 nucleus, 310
Bessel, F. W., 249
Big Bang, 412*f*
   and acceleration of universe, 424
   defined, 412
   and expansion of infinite space, 414–15
   inflation after, 430–32
   moments after, 425–29, 429*f*
   and multiverses, 433
   as origin of universe, 6, 410
   Planck data about, 434
   radiation from, 417–21
Big Bang nucleosynthesis, 445–46, 445*f*
Big Island of Hawaii, 145
binary systems, 262
   center of mass in, 262–63, 262*f*
   flow of mass in, 320
   low-mass stars in, 319–22
   mass of stars in, 262–64
   orbits in, 262–63, 262*f*, 349
   star 2 in, 321–22
   stellar cataclysm in, 322
   stellar evolution in, 320–22, 320*f*
   total mass in, 264
   X-ray, 338, 338*f*

blackbodies, 101
blackbody radiation, 417–18, 418f
blackbody spectrum, 101, 101f
black hole(s), 350–52
    in Antenna galaxies, 383
    defined, 340
    from first stars, 449–50
    properties of, 350, 351
    Schwarzschild radii of, 350
    "seeing," 351–52
    in X-ray binary systems, 338, 338f
    see also supermassive black holes
Black Hole Era, 454
Bleriot, Louis, 209
Bleriot propeller, 209
"blueberries," 148
blue-green algae (cyanobacteria), 165, 463
Bode, Mike, 323
Boltzmann, Ludwig, 101
boron, 6, 337
Borrelly, Comet, 231
Borucki, Bill, 120
boson, Higgs, 426
Boss, Alan, 14
bound orbits, 65, 65f
boxcar experiment, 342, 343, 343f
brightness
    defined, 246
    and luminosity, 246, 250, 251f, 305
    of SS Cygni, 269
    of stars, 250, 251f
    of Sun, 251, 251f, 305
    and temperature of sunspots, 291
B Ring (of Saturn), 207
Brown, Mike, 14
Brown, Warren, 404
brown dwarfs, 104, 114
B stars, 259, 260
Bullet Cluster, 453f
burning
    carbon, 322, 331, 334
    defined, 277
    helium, 310–12, 331
    hydrogen, 277, 279, 321, 331
    hydrogen shell, 307–9, 331
    oxygen, 334
    silicon, 334
butterfly diagram, 292

C

Cabeus (lunar crater), 150, 151
Callisto, 113, 218
Cambrian explosion, 464
Cancer, 34
Cannon, Annie Jump, 259–60
Capricorn, 34
carbon
    abundance of, 337
    as basis for life, 468–69, 468f
    as catalyst, 330, 331
    in human body, 467–68
carbon-12, from triple-alpha process, 310, 310f
carbonaceous chondrites, 237
carbon burning, 322, 331, 334
carbon-nitrogen-oxygen (CNO) cycle, 330, 331
Carina spiral arm, 353
carrier particles, 426
Carter, Jimmy, 473
Cassini
    evidence of life on Titan from, 470
    Jupiter images, 189f, 190f
    lightning images, 194
    planetary ring images, 207f, 208, 209, 209f
    Saturn images, 190f, 191, 191f
    Titan images, 221, 221f, 222
    wind speed measurements, 195
Cassini Division, 205, 206
cataclysm, stellar, 322, 340
catalysts, 330, 331

Cat's Eye Nebula, 315f
CCDs (charge-coupled devices), 82–83, 82f
celestial equator, 23, 26
celestial sphere, 22–24, 22f, 23f, 27f
cell membranes, 466
cells, 466
Centaur rocket, 150
center of mass, 262f
    in binary systems, 262–63, 263f
    and radial velocity method, 117, 117f
centripetal force, 63–64, 64f
Cepheid variables, 331–32, 332f, 376, 380–81, 392
Ceres, 14, 15, 113, 216–17, 217f
CERN (European Organization for Nuclear Research), 7f
Chandrasekhar, Subrahmanyan, 322
Chandrasekhar limit, 322, 340
Chandra X-ray Observatory, 88, 382–83, 401, 452
chaos, defined, 111
charge-coupled devices (CCDs), 82–83, 82f
charged particles, in magnetic fields, 201–3
Charon, 14, 188, 216, 216f
Chelyabinsk, Russia, 233, 233f, 234, 479
chimneys, from disks, 397
chondrites, 236f, 237
chondrules, 236f, 237
chromatic aberrations, 85
chromosphere, 285–88, 285f, 286f, 288f, 293
circular velocity (v_{circ}), 64
circulation
    global, 171
    Hadley, 170–71, 171f
circumpolar stars, 26
civilizations, estimating number of, 474–75, 475f
climate, 174–76, 479
clocks, cosmic, 131–32
cloud cannibalism, 190
clouds
    on Earth, 166–67, 167f
    on giant planets, 186, 189–92
    structure of, 192–93
    on Venus, 172, 172f, 173
    viewing tops of, 189–92
    wind speed based on, 195
clusters
    galaxy, 442, 443, 443f, 452
    globular, 392–94, 392f, 393f, 396, 402
    and hierarchical clustering, 445
    open, 393–94, 393f
    star, 317–19, 318f, 319f
    super-, 442, 443
    see also specific clusters
CMB, see cosmic microwave background radiation
CNO (carbon-nitrogen-oxygen) cycle, 330, 331
COBE, see Cosmic Background Explorer
Colaprete, Anthony, 151
cold dark matter, 447, 448, 448f
color
    of spiral vs. elliptical galaxies, 364
    and surface temperature of stars, 252
    and wavelength of light, 250
coma, 228, 228f
Coma Cluster, 442, 442f
comet nuclei, 114, 228, 228f, 235
comets, 226–33, 226f
    active, 226, 228–30, 228f
    anatomy of, 228–30
    formation of, 111
    impacts by, 232–33, 479
    meteorites from, 235
    orbits of, 227, 227f
    origins of, 226
    in Solar System, 4
    visits to, 230–31
    see also specific comets
compass, 137
composite volcanoes, 144, 145f

compound lenses, 85
conduction, 281
cones (of eye), 80
conservation of angular momentum, 107, 107f
constellations, 30
    see also specific constellations
Contact (film), 481
convection, 140f
    energy transport by, 281–84
    and layers in atmosphere, 166, 167
    in tectonism, 140–42
convective zone, 283f, 284
conventional greenhouses effect, 160
cooling
    and interiors of terrestrial planets, 135
    neutrino, 334
    of universe, 429, 429f
Copernican Revolution, 49–50
Copernicus, Nicolaus, 49–50, 53
core(s)
    of Earth, 134
    of giant planets, 197, 198
    of high-mass stars, 330–31, 331f, 334
    of low-mass stars, see degenerate core
    molecular-cloud, 99–100, 100f
    of Moon, 134
    of Sun, 278, 280–85, 282f
core collapse, 334–36, 335f
Coriolis effects, 170–71, 170f, 171f, 194
Coriolis satellite, 323
corona, 285f, 286–89, 286f, 289f
coronae, 144
coronal holes, 289, 289f
coronal loops, 289, 289f
coronal mass ejections, 294, 294f, 295
COROT telescope, 472
Cosmic Background Explorer (COBE), 418–20, 419f, 420f, 434, 444
cosmic clocks, 131–32
cosmic microwave background radiation (CMB), 416–21, 419f
    emission of, 449
    and horizon problem, 430, 432
    in model of large-scale structure, 445
    and radiation from Big Bang, 417–18
    satellite data on, 418–19
    and structure of universe, 452–53
    variations in, 420–21, 444
cosmic rays, 398–99
cosmological constant, 423–24, 423f
cosmological principle, 8–9, 360, 432–33, 444
cosmology, 410
Crab Nebula, 340, 340f
Crater (constellation), 420
craters
    on Moon vs. Earth, 126
    secondary, 129
    see also impact craters
crescent Moon, 36, 37
C Ring (of Saturn), 207
critical density, 421–22, 422f
Crowther, Paul, 353
crust, of Earth, 134, 141
cryovolcanic plumes, 219, 219f
cryovolcanism, 219–20, 219f
C-type asteroids, 238
cube root, 52
cubing numbers, 52
Curiosity rover, 66–67, 149, 149f, 177, 469–70
Curtis, Heber D., 358
cyanobacteria (blue-green algae), 165, 463
cyclonic motion, 171
Cygnus Loop, 336f
Cygnus X-1, 351, 352, 352f

D

Dactyl, 224f, 225
Daphnis, 209
dark energy, 424

Dark Era, 454, 454f
dark matter, 445–48
    cold, 447, 448, 448f
    collapsing clumps of, 448
    composition of, 445–46
    defined, 368
    in galaxies, 367–70
    in galaxy clusters, 442
    hot, 447
    and mass of galaxies, 367
    in Milky Way Galaxy, 369–70, 399–400
    normal matter vs., 446, 447f
    observations of, 367–69
    types of, 369–70
    in ultrafaint dwarf galaxies, 451
dark matter halos, 369
dark matter mini-halos, 449, 450
daughter products, 132
Dawn spacecraft, 217, 225
days, 30
decay, 295
deep (term), 250
Deep Impact, 230–31, 231f
Deep Space 1, 230
degenerate core
    electron-degenerate matter in, 307
    helium burning in, 310–12
    of high-mass stars, 334, 337, 338
    of horizontal branch stars, 312
    mass and size of, 307–8
    of post-AGB stars, 314
    of red giants, 309, 309f
Degenerate Era, 454
Deimos, 225, 225f
Delta Airlines, 297
Delta Cephei, 331
density
    of asteroids, 224
    of chromosphere and corona, 286, 286f, 287
    critical, 421–22, 422f
    defined, 98
    of Earth, 133
    of giant planets, 186–87, 198–99
    and interior of planets, 133
    of normal matter, 446
    and temperature/pressure, 276f, 277
deoxyribonucleic acid (DNA), 463–66, 465f, 468
deuterium, 280, 445–46, 445f
Dialogue Concerning the Two Chief World Systems (Galileo), 53
diameters, of giant planets, 185, 186
Dicke, Robert, 418
differential rotation, 290
differentiation, 134, 198
diffraction, 88, 88f
diffraction limit, 88, 89
diffuse rings, 207
Dione, 223
direct imaging, of extrasolar planets, 118
disks
    galactic fountain model of, 395, 395f
    of Milky Way Galaxy, 392, 396–97, 396f
    star formation in, 365–67
    stars in halo vs., 397
    see also accretion disks
distance ladder, 380–81, 380f
DNA, see deoxyribonucleic acid
Donnelly, Mike, 239
Doppler blueshift, 115f, 116, 116f, 263
Doppler effect
    defined, 115
    finding extrasolar planets with, 115–17
    in helioseismology, 284
    testing rotation curve with, 368
Doppler redshift
    gravitational redshift vs., 349
    and Hubble's law, 376–77
    and motion in binary systems, 263
    in search for extrasolar planets, 115f, 116, 116f

INDEX I-3

Doppler shifts
　and expansion of universe, 416
　in helioseismology, 284
　and Hubble's law, 376–77
　and motion in binary systems, 263, 264
　and physical properties of stars, 261
　in search for extrasolar planets, 116
　and shape of Milky Way Galaxy, 399
　and supermassive black holes, 375
doubling time, 467
Drake, Frank, 474, 476
Draper, John W., 81
D Ring (of Saturn), 207
dust, 96, 364
dust storms, 174, 174f
dust tails, 228f, 229, 229f
dwarf galaxies
　in galaxy groups and clusters, 442
　giant galaxies vs., 365, 365f
　in Local Group, 402
　near Milky Way, 403, 403f, 450–51
　ultrafaint, 451
dwarf planet(s)
　formation of, 113
　number of, 216–17
　orbital data for, A-3
　physical data for, A-2
　Pluto's demotion to, 14–15
　in Solar System, 4
dynamic balance, of pressure and gravity, 102–3, 103f
Dysnomia, 216
Dyson, Freeman, 480
Dyson sphere, 480

E

Eagle Nebula, 102, 102f
Earth
　alignment of Mars, Sun, and, 66–67, 66f
　asteroids near, 223–24
　atmosphere of, see atmosphere, of Earth
　chemical composition of, 337
　density of, 133
　distance from Sun to, 75
　distance of galaxies from, 377
　erosion on, 147, 148
　forces of Moon and, 57
　formation of, 112, 113
　gravitational acceleration on, 58
　greenhouse effect on, 161, 163–64, 175
　impact craters on, 126, 130, 131
　interior of, 133–37
　life on, see life on Earth
　luminosity of, 101
　magnetic field of, 137–38, 168, 168f, 169
　motions of, 22–35, 38–41, 420
　orbital speed of, 129
　orbit of, 28, 29, 50, 51, 51f
　place in universe of, 4, 4f, 128
　revolution around Sun of, 28–35
　rotation of, see rotation, of Earth
　size and mass of Sun vs., 274
　size of giant planets and, 184
　solar activity's effect on, 294–97
　and solar eclipse for SDO, 40–41, 40f, 41f
　tectonism on, 139–42, 139f
　temperature of, 160, 163, 175, 175f, 176f, 296
　as terrestrial planet, 126
　tidal stresses on, 135–37, 135f, 136f
　time on surface of, 349
　volcanism on, 144–45
Earth-like planets, 472–73
earthquakes, 141–42, 141f
eccentricity (e), of ellipses, 50
eclipses, 38–41
　causes of, 38–39
　defined, 38
　lunar, 38–41, 38f, 39f
　for Solar Dynamics Observatory, 40–41
　see also solar eclipses
eclipsing binary systems, 264, 264f
Eddington, Arthur Stanley, 348
Efstathiou, George, 434
Einstein, Albert
　cosmological constant of, 423, 424
　equation for mass and energy by, 277
　in scientific revolution, 9, 9f
　theories of relativity of, 9, 342–44, 369
electric charge, of black holes, 350, 351
electromagnetic force, 425, 427, 429
electromagnetic spectrum, 77–79, 78f
electromagnetic waves, 77
electron-degenerate matter, 307–8
　see also degenerate core
electron neutrinos, 281
electrons
　atomic energy levels of, 252–55
　in atoms, 253f
　defined, 79
　energy states of, 254, 254f, 258, 258f
　light and, 80
　and positrons, 280, 426–27, 427f
electroweak force, 426
electroweak theory, 426
elements
　defined, 79
　emission and absorption lines of, 257–58
　formation of, 6, 337, 394
　in human body, 467–68
　parent, 132
　periodic table of, 187, A-1
　relative abundances of, 337, 337f, 394–95
　spectral lines of, 260–61
　see also heavy elements
ellipses, 50
elliptical galaxies
　defined, 362
　formation of, 452
　location of, 358
　mass of, 369
　orbits of stars in, 362–63, 362f
　spiral vs., 364–65
elliptical orbits
　of planets, 50, 50f
　precession of, 347
　velocity of objects with, 65, 65f
elliptic plane, 28f, 30
El Niño, 176
emission, of light, 80
emission lines, 83, 255, 257–58, 259f
emission spectra, 254
empirical laws, 50
Enceladus, 205–7, 219, 219f, 470
energy
　in core of Sun, 281–85, 282f
　of cosmic rays, 398–99
　dark, 424
　defined, 75
　and frequency, 77, 77f
　from high-mass stars, 334, 334f
　of impacts, 130
　kinetic, 75, 234
　and mass, 277, 344
　quantized, 77, 254
　from Sun, 75, 160–61, 302, 305
　thermal, 75, 100, 134, 194, 197, 198
　from transitions between states, 254, 254f
energy levels
　atomic, 252–57, 253f
　of photons, 254–56, 255–56f
energy states, of electrons, 254, 254f, 258, 258f
energy transport, 281–84, 282f
Englert, François, 426
EPOXI, 231, 231f
equal areas, law of, 51
equations
　acceleration, 56

boxcar experiment, 343
cosmic clocks, 132
density of Earth, 133
diameters of giant planets, 186
diffraction limit, 89
Doppler shift, 116
expansion and age of universe, 413, 413f
exponential growth, 467
Kepler's third law, 52
lifetime of Sun, 278
main-sequence lifetimes, 306
manipulation of, 29
mass of galaxies, 400
Newton's law of gravity, 60
parallax and distance, 249
radii of meteoroids, 234
redshift, 377
Schwarzschild radius, 350
Stefan-Boltzmann law and Wien's law, 101
temperatures of planets, 162–63
temperatures of sunspots, 291
thermal energy in giant planets, 198
units and scientific notation, 13–14
wavelength and frequency, 77
wind speeds on distant planets, 195
equator, 23, 26
equilibrium
　hydrostatic, 98, 276–77
　thermal, 158, 198
equinoxes, 32, 34, 35
equivalence principle, 345–46, 346f
E Ring (of Saturn), 204f, 205–7
Eris, 114, 216, 216f
Eros, 225, 225f
erosion, 147–48, 148f, 164
escape velocity (v_{esc}), 65
Eskimo Nebula, 315f
ESO (European Southern Observatory), 87
Eta Carinae, 6f, 333, 333f
Eukarya, 464, 464f
eukaryotes, 463, 463f
Eurasian Plate, 141
Europa, 113, 220, 220f, 221, 470
European Geosciences Union, 177
European Organization for Nuclear Research (CERN), 7f
European Southern Observatory (ESO), 87
European Space Agency, 118, 148, 295, 420, 434, 472
event horizons, 350, 351
events, in relativity theory, 342
EVLA (Expanded Very Large Array), 269
evolutionary track, of stars, 305
　see also stellar evolution
excited states, 254
exoplanets, see extrasolar planets
Expanded Very Large Array (EVLA), 269
explosions
　of high-mass stars, 328, 333–36
　of low-mass stars, 323
exponential growth, 467
extrasolar planets
　defined, 114
　discovery of, 115–19
　Earth-like, 472–73
　and planet candidates, 119, 119f
extraterrestrial life, 466–73
　and chemistry of life on Earth, 467–69
　and habitable zones of Milky Way Galaxy, 470–72
　intelligent, 473–77
　and search for Earth-like planets, 472–73
　in Solar System, 469–70
extremophiles, 463
eye, human, 80–81, 80f

F

facts, theories vs., 8
falsifiability, 7, 349, 350, 477
far side, of Moon, 35

faults, 141, 223
Fermi, Enrico, 475, 481
Fermi Gamma-ray Space Telescope, 401, 402f
Fermi paradox, 475, 481
51 Pegasi, 114
filters, optical, 252
first quarter Moon, 36
flash, helium, 310–12, 311f
flatness problem, 430–32
flat space, 415–16
fluid, 144
flux, 101, 291
focal planes, 84, 84f, 85f
focus, 50
Fomalhaut b, 118, 118f
force(s)
　action-reaction pairs of, 56–57, 57f
　centripetal, 63–64, 64f
　defined, 54, 55
　electromagnetic, 425, 427, 429
　electroweak, 426
　gravitational, 58–61, 99–100
　influence on motion of, 54–56
　net, 54, 98
　strong nuclear, 277, 425, 427, 429
　tidal, 205, 347, 351f
　weak nuclear, 425, 427, 429
formerly active moons, 222–23
47 Tucanae (star cluster), 319, 319f
fossils, 466, 466f
fragmentation of molecular clouds, 99–100
frame of reference
　accelerated, 346–47
　defined, 49
　inertial, 54, 54f, 344, 345, 345f
free fall, 63, 346–47
free float, 346–47
free neutrons, 337
frequency, 76–78, 77f
F stars, 260, 474
full Moon, 36
fusion, nuclear, 277–79

G

Gaia spacecraft, 118
galactic fountain model of disk, 395, 395f
galactic habitable zone, 472
Galatea, 208
galaxies, 358–83
　AGNs in normal, 374–75
　collisions of, 382–83
　dark matter in, 367–70
　defined, 360
　distance from Earth, 377
　distribution of, 360, 378, 445
　evolution of, 451–53
　formation of, 402–3, 448f, 450–51
　Hubble's law from study of, 375–82
　interactions between, 375
　mass of, 367, 400
　recession velocities of, 377
　shapes of, 361–65, 361f
　star formation in, 365–67
　supermassive black holes in center of, 371–75
　and Type II supernovae, 336–40
　see also specific galaxies
galaxy clusters, 442, 443, 443f, 452
galaxy groups, 402, 442
galaxy structures, 442–44
　components of, 442–43
　mapping, 443–44
　of most distant galaxies, 450–51, 450f
　types of, 442
Gale Crater, 67, 149, 469–70
Galilean moons, 51
Galilean relativity, 342
Galileo Galilei, 49, 51, 53, 59
Galileo spacecraft, 221
　comet impact images, 233
　Europa images, 220f

Io images, 218, 218f
planetary ring images, 206f
visits to asteroids, 224–25, 224f
Galle, Johann Gottfried, 182
gamma-ray bursts (GRBs), 352, 449
gamma-ray photons, 280, 283
gamma rays, 6, 78
gamma ray telescopes, 83, 84, 87
Gamow, George, 417, 418
Ganymede, 113, 222–23, 223f
gas giants, 186
  see also giant planets
gas halo, Milky Way, 397
Gaspra, 224
general relativistic time dilation, 348–49
general theory of relativity, 346–50
  cosmological constant in, 423
  and earliest moments in universe, 429
  and free fall/float, 346–47
  and geometry of spacetime, 347
  and gravitational lensing, 369
  observable consequences of, 347–50
  scientific revolution related to, 9
geocentrism, 48
geodesic, 346
geo-effective events, 297
geologically active moons, 218–20
geomagnetic storms, 296, 297
geothermal vents, 462, 462f
giant galaxies, 365, 365f
giant planets, 182–209
  atmospheres of, 186, 192–93, 192–93f, 197, 198
  clouds on, 189–92
  composition of, 186–88
  defined, 113
  diameters of, 185, 186
  interiors of, 197–99
  magnetic fields and magnetospheres of, 169, 199–203
  physical properties of, 184–86, 184t
  rings of, 203–9
  rotation of, 114, 182, 188, 195, 202
  temperatures of, 163
  terrestrial planets vs., 186
  weather on, 194–96
  winds on, 194–96
  see also specific planets
giant tube worms, 462f
gibbous Moon, 36, 37
Gingerich, Owen, 15
Giotto, 230, 230f
Gliese 581c, 473–74
Gliese 667C, 472
global circulation, 171
Global Oscillation Network Group (GONG), 284
global positioning system (GPS), 140, 349
globular clusters, 392–94, 392f, 393f, 396, 402
Gnedin, Oleg, 404
Goddard Space Flight Center, 239
GONG (Global Oscillation Network Group), 284
gossamer rings, 206
GPS (global positioning system), 140, 349
grand unified theories (GUTs), 427, 429
graphs, reading, 11–12
gravitational acceleration, 58
gravitational attraction, of galaxies, 378, 379
gravitational force, 58–61, 99–100
gravitational lensing, 118, 348, 369–70, 370f
gravitational redshift, 348–49, 349f
gravitational waves, 349–50
gravity, 58–61
  as distortion of spacetime, 345–50
  dynamic balance of pressure and, 102–3, 103f
  and evolution of high- vs. low-mass stars, 328

and expansion of universe, 415, 421–23, 423f
and flow of mass in binary systems, 320
and formation of stars and planets, 96
and grand unified theories, 427, 429
and Kepler's laws, 262
and large-scale structures, 444–48
and mass/weight, 58–59, 262
Newton's law of, 59–62, 61f, 185, 347
and orbits, 46, 62–63
of planetesimals, 109–10, 110f
and planet migration, 111
and pressure in low-mass stars, 305, 308, 311, 313
and pressure in protostars, 102–3
and pressure in Sun, 276, 276f
self-, 98, 98f, 99, 108
in spiral arms of galaxies, 366
and standard model, 425
and tides, 135–37, 135f
of white dwarfs, 321
GRBs (gamma-ray bursts), 352, 449
Great Dark Spot, 191, 194
Great Red Spot, 182, 189–90, 189f, 190f, 194, 195
Green Bank Telescope, 86f
greenhouse gases, 161, 174–76
greenhouses effect
  atmospheric, 160–64, 161f, 174–76
  conventional, 160
Gregorian calendar, 33
G Ring (of Saturn), 207
ground states, 253, 254
growth, exponential, 467
G stars, 260, 474
Guhathakurta, Madhulika, 296, 297
Gulf Stream, 176
Gusev (impact crater), 149
Guth, Alan, 431
GUTs (grand unified theories), 427, 429

H

habitable zones, 470–73
Hadley circulation, 170–71, 171f
Hale, George Ellery, 292, 293
Hale-Bopp, Comet, 228f
half-life, 132
Halley, Comet, 227f, 230, 230f
halo(s)
  dark matter, 369
  dark matter mini-halos, 449, 450
  of Milky Way Galaxy, 392, 393, 397
  stars in disk vs., 397
  studying formation of galaxies with, 402
halo ring, 206
Hα (hydrogen alpha) line, 288
Hartley 2, Comet, 231, 231f
Harvard-Smithsonian Center for Astrophysics, 443, 443f
Haumea, 216
Hawaiian Islands, 142, 144, 145
Hawking, Stephen, 352
Hawking radiation, 352, 352f, 454
Hayabusa, 225
HD 226868 (star), 351
HE 0437-5439 (star), 404, 404f
head (comet), 228
heating
  radioactive, 135
  tidal, 135–37
heavy elements
  creation of, 449, 450, 468
  defined, 187
  and evolution of Milky Way, 394–96
  on giant planets, 187, 188
heliocentrism, 49–50, 53
helioseismic waves, 284, 284f
helioseismology, 284–85
Helios spacecraft, 344
helium
  abundance of, 337
  atom of, 79, 79f

discovery of, 288
in early universe, 6, 445–46, 445f, 468
formation of stars from, 449
on giant planets, 198, 199
in main-sequence stars, 305, 306, 307f
in primary atmosphere, 111, 158
from proton-proton chain, 280
in protostars, 104
helium ash, 305–8
helium burning, 310–12, 331
helium core, of low mass-stars, see degenerate core
helium flash, 310–12, 311f
Hellekson, Bob, 91
heredity, 465, 465f, 466
Herschel, William, 182, 469
Herschel Space Observatory, 231
hertz, 76
Hertz, Paul, 434
Hertzsprung, Ejnar, 265
Hertzsprung-Russell (H-R) diagrams, 265–68, 265f
  AGB stars on, 313, 313f
  high-mass stars on, 330–33, 330f
  horizontal branch stars on, 312, 312f
  instability strip of, 331–33, 332f
  layout of, 266
  low-mass stars on, 308–9
  main sequence of, 266–67
  patterns in, 244
  post-AGB stars on, 314, 314f
  post-main sequence evolution on, 316–17, 317f
  protostars on, 304f, 305
  red giants on, 308–9, 311f
  for star clusters, 318–19, 318f, 319f
  stars not on main sequence of, 268
  Sun on, 302
  white dwarfs on, 315, 316, 316f
hierarchical clustering, 445
hierarchical merging, 452, 452f
Higgs, Peter, 426
Higgs boson, 426
Higgs field, 426
high-mass stars, 328–54
  black holes, 350–52
  core collapse in, 334–36, 335f
  evolution of, 304, 328
  explosions of, 333–36
  final days of, 333–34
  on H-R diagrams, 330–33, 330f
  on instability strip, 331–33
  neutron stars and pulsars from, 337–39
  nuclear reactions in cores of, 330–31, 334
  R136a1, 353–54
  relativistic physics for describing, 340–50
  supernovae from, 336–40
High Resolutions Fly's Eye Observatory, 398
Hi'iaka, 216
H II regions, 395
Himalayas, 141
Hipparchus, 250
Hipparcos satellite, 250, 266f
Hirschi, Raphael, 353
Holmes, Comet, 229, 229f
Homestake Mine experiment, 280–81
homogeneity, in distribution of galaxies, 360, 378, 445
horizon, defined, 23
horizon problem, 430, 430f, 432
horizontal branch stars, 312–13, 312f, 332
hot dark matter, 447
hot Jupiters, 119
hot spots, 142, 144, 145
Hounsell, Rebekah, 323
HR 8799 (star), 118, 118f
H-R diagrams, see Hertzsprung-Russell diagrams
HST, see Hubble Space Telescope
Hubble, Edwin, 358

classification of galaxies by, 362, 362f
discovery of expanding universe by, 423
discovery of Hubble's law by, 376, 377, 412
identification of Local Group by, 402
plots of galaxy velocity and distance by, 381f
Hubble classification of galaxies, 451–52, 451f
Hubble constant, 377, 380–81, 380f
Hubble's law, 375–82, 412–16
  and conditions of early universe, 412, 414
  discovery of, 376–77
  and distance ladder for Hubble constant, 380–81, 380f
  estimating age of universe with, 412–14, 413f
  and expansion of universe, 378–79, 414–16
  mapping galaxy structures with, 443–44, 443f
  mapping universe with, 382
  and redshift, 377, 416
  and size/shape of universe, 415–16
Hubble Space Telescope (HST), 13, 221
  atmospheric drag on, 295
  Ceres images, 217f
  discovery of Europa geysers, 470
  disk images, 105f, 109
  distance to SS Cygni measurement, 269
  Eta Carinae images, 333f
  galaxy collision data from, 383
  hyperfast star images, 404, 404f
  images of early galaxies, 450f
  magnitude of stars detected, 250
  Mars dust storm images, 174f
  merging galaxy images, 452, 452f
  nebulae images, 102f, 315f, 340f
  Neptune images, 191f
  observation of Fomalhaut b, 118
  planetary ring images, 203f
  Pluto images, 216f
  resolution in, 88–90
  Saturn images, 188f
  size of James Webb vs., 91
  supernovae images, 333f, 336f
  Uranus images, 196
  wavelengths detected, 88, 88f
  wind speed measurements, 195
Hubble time, 412, 412f, 413
humans
  evolution of, 464, 465
  fate of, 478–79
Huygens probe, 222, 222f
hydrogen
  absorption lines of, 260
  abundance of, 337
  Doppler velocities of, 399, 399f, 400
  in early universe, 6, 468
  energy states of, 258, 258f
  formation of stars from, 449
  fusion of, 279
  on giant planets, 199
  in human body, 467–68
  in main-sequence stars, 305, 306, 307f
  metallic, 198
  in Milky Way Galaxy, 391, 391f
  neutral, 391, 391f, 399, 399f, 400, 449
  in primary atmosphere, 158
  in protostars, 104
hydrogen alpha (Hα) line, 288
hydrogen burning
  in high-mass stars, 331
  in Sun, 277, 279
  in white dwarfs, 321
hydrogen fuel, exhaustion of
  in low-mass stars, 306–9
  in main-sequence stars, 302, 306
  in Sun, 278
hydrogen shell burning, 307–9, 331
hydrostatic equilibrium, 98, 276–77

hypervelocity stars, 404
hypotheses, 7–9, 424–25

I

ice, 111, 222, 222f
ice ages, 175
ice giants, 186
Ida, 224–25, 224f
ideas, 8
impact cratering, 129–31, 129f
impact craters, 129–31
    defined, 129
    on Earth, 126, 130, 131
    formation of, 129f
    on Ganymede, 223, 223f
    on Mars, 130–31, 130f
    on Mercury, 113f, 114, 130, 131
    on Moon, 126, 129f, 130, 131
    in Solar System, 113f, 114
    on terrestrial planets, 129–31
    on Venus, 130, 131
impacts
    by comets, 232–33
    cosmic clocks based on, 131–32
    and physical properties, 128–29, 128t
    and secondary atmosphere, 159
    on terrestrial planets, 128–32
index of refraction, 84, 84f
Indo-Australian Plate, 141
inert gases, 187
inertia, 54, 55
inertial frame of reference, 54, 54f, 344, 345, 345f
infinite universe, expansion of, 414–16
infinity, 350, 414
inflation, 430–32, 431f
infrared (IR) radiation, 7, 79
infrared telescopes, 83, 87, 88
inner halo, 397
inner Solar System
    comets near, 226f, 227, 229
    at death of Sun, 478
    formation of atmospheres in, 159
    planet formation in outer vs., 110–11
    see also terrestrial planets
instability strip, 331–33
integration time, 81, 82
intelligent life, 473–77
    attempting to detect signals from, 475–77
    estimating number of civilizations with, 474–75, 475f
    Kepler mission data in search for, 480–81
    sending messages into space for, 473–74
    very different forms of, 477
interferometers, 86
interferometric arrays, 86–87
interiors, of planets, see planetary interiors
International Astronomical Union, 15, 114
International Space Station, 74, 370
interstellar clouds, 98–99
interstellar medium, 289–90, 290f, 394–97
inverse proportions, 60, 249
inverse square law, 61
Io, 113, 202, 206, 218–19, 218f
ionization, in Cepheid variables, 331–32, 332f
ionosphere, 167
ions, 79
ion tails, 228, 228f, 229f
iron, 138, 334
iron meteorites, 236f, 237, 237f
IR (infrared) radiation, 7, 79
irregular galaxies, 362, 365, 452
irregular moons, 218
isotopes, 280, 446
isotropy, in distribution of galaxies, 360, 378, 445
Itokawa, 225

J

Jackson, Bernard, 323
James Webb Space Telescope, 88, 91
Japan Aerospace Exploration Agency, 225
Jenkins, Jon, 120
Jet Propulsion Laboratory (JPL), 66, 67, 177, 209, 434
jets, from accretion disks, 108f, 109
jet streams, 196, 196f
Jones, Heath, 454
JPL, see Jet Propulsion Laboratory
Jupiter, 189f
    asteroids near, 223, 224
    atmosphere of, 111, 192f
    characteristics of, 185
    clouds on, 189–90, 193
    comets near, 233
    composition of, 186–88
    formation of, 113
    as giant planet, 182
    habitability of, 471
    interior of, 197, 197f, 198
    magnetic field and magnetosphere of, 199–202, 201f
    moons of, 51, 74, 74f; see also specific moons
    orbit of, 51
    rings of, 203f, 206, 206f, 208
    rotation of, 188
    size of, 184
    thermal energy of, 198
    winds on, 194, 194f, 195

K

K5 stars, 306
KBOs (Kuiper Belt objects), 226
Keck Observatory, 89, 191f, 481
Kelvin temperature scale, 98
Kepler, Johannes, 49–52, 59–62
Kepler App, 119
Kepler's laws, 50–52
    first, 50
    and gravity, 262
    and Newton's laws, 62
    for planetary ring particles, 203
    second, 50–51, 51f
    third, 51, 52, 52f, 185, 264
Kepler space telescope
    finding Earth-like planets with, 472–73
    search for extrasolar planets with, 118–21, 480–81
    search for intelligent life with, 476, 480–81
kinetic energy, 75, 234
Kozyra, Janet, 297
K stars, 474
Kuiper, Gerard, 226
Kuiper Belt, 14, 15, 112, 113, 226, 227
Kuiper Belt objects (KBOs), 226

L

lambda peak, 101
La Niña, 176
Large Hadron Collider, 370, 426
Large Magellanic Cloud (LMC)
    globular clusters in, 393f
    LMC X-3 in, 352
    in Local Group, 442f
    location of Milky Way Galaxy and, 403f
    massive stars in, 353
    orbital motion of, 400, 400f
    studies of dark matter in, 370
    supernovae in, 333, 333f
large-scale structure, 443–48
    building testable models of, 445
    and dark matter, 445–48
    defined, 443
    and gravitational instabilities, 444
latitude, 25–27, 25f, 33
Lauretta, Dante, 239
law of equal areas, 51
Lawrence, Charles, 434
laws, scientific, 8
laws of motion, 46–67
    and alignment of Earth, Mars, and Sun, 66–67
    and Galileo's observations, 51, 53
    and gravity, 58–61
    Newton's, 53–58
    and orbits, 62–65
    and planetary motion in astronomy, 48–51
LCROSS (Lunar Crater Observation and Sensing Satellite), 150, 151
lead-206, 132
leap years, 33
Leavitt, Henrietta, 392
length contraction, 345
lensing, gravitational, 118, 348, 369–70, 370f
Leo, 30
Leonid meteor shower, 235f
life, defined, 462
life in universe, 460–80
    and life on Earth, 462–66, 477–79
    possibility of extraterrestrial life, 466–73
    search for intelligent life, 473–77, 480–81
life on Earth, 462–66, 477–79
    chemistry of, 467–69
    complexity of, 464–65
    and composition of atmosphere, 164–65, 165f
    and death of Sun, 477–78
    earliest organisms, 463
    evolution of, 465–66
    future of, 478–79
    origins of, 462–63
lifetime(s)
    of main-sequence stars, 305, 305t, 306, 318–19
    of Sun, 278
light, 72–91
    absorption of, 80
    from AGNs, 372–73
    from Cepheid variables, 331, 332
    collecting of, 84–91
    and Doppler effect, 115–16, 115f, 116f
    in electromagnetic spectrum, 77–79
    emission of, 80
    and mass in spiral galaxies, 367, 367f
    and matter, 79–80
    as particle, 76–77
    and recording of astronomical data, 80–83
    from Sun, 32, 159, 173, 295f
    unresolved points of, 372
    as wave, 75–76, 75f, 76f
    see also speed of light
lightning, 194
light-years (ly), 4–6, 5f, 75, 248
limb darkening, 287, 287f
limbs, of planets, 192
limestone, 164
lithium, 6, 337
lithosphere, 138
lithospheric plates, 140, 144
LMC, see Large Magellanic Cloud
LMC X-3, 352
Local Group, 4, 4f, 402, 402f, 442, 442f, 450–51
Lockheed Martin Space Systems, 239
Loihi, 145
long-period comets, 227, 229, 230, 233
Lowell, Percival, 469
Lowell Observatory, 376
low-mass stars, 302–23
    on asymptotic giant branch, 312–13
    in binary systems, 319–22
    defined, 304
    detection of explosions by, 323
    electron-degenerate matter in helium core of, 307
    evolution of high vs., 328
    exhaustion of hydrogen fuel in, 306–9
    helium burning in degenerate core of, 310–12
    on H-R diagram, 308–9
    hydrogen shell burning in, 307–8
    last stages of evolution for, 312–17
    main-sequence, 304–6
    post-AGB, 313–15
    and star clusters, 317–19
    stellar mass loss by, 313
    white dwarfs, 315–17
    see also Sun
LRO (Lunar Reconnaissance Orbiter), 150
luminosity
    of blackbody, 101
    and brightness, 246, 250, 251f, 305
    calculating, 251, 251f
    of Cepheid variables, 331
    defined, 100
    and distances to nearby stars, 246–50
    of galaxies, 364, 365
    of high-mass stars, 328
    on H-R diagram, 266
    of main-sequence stars, 305, 305f
    and mass, 305, 305f, 306
    of protostars, 100, 104f
    of quasars, 371
    of red giants, 308–9, 309f, 311–12
    and size of stars, 261, 261f
    of Sun, 277, 296, 305, 478
    and temperature, 101, 265
    of Type 1a supernovae, 322
    of white dwarfs, 316
    see also Hertzsprung-Russell (H-R) diagrams
luminosity class, 268
luminous (normal) matter, 367, 369, 446, 447f
Lunar Crater Observation and Sensing Satellite (LCROSS), 150, 151
lunar eclipses, 38–41, 38f, 39f
Lunar Reconnaissance Orbiter (LRO), 150
lunar tides, 136
ly, see light-years
Lyra, 332

M

M2-9 Nebula, 315f
M13 star cluster, 473, 473f
M51 galaxy, 81f, 328
M87 galaxy, 374
M92 globular cluster, 392, 392f
M104 galaxy, 364f
M109 galaxy, 391, 391f
MACHOs (massive compact halo objects), 369–70, 446
Magellan, 143, 143f, 172
magma, 144, 145f
magnetic bottles, 168, 168f
magnetic field(s)
    charged particles in, 201–3
    of Earth, 137–38, 168, 168f, 169
    of giant planets, 199–203, 200f
    and interiors of terrestrial planets, 137–38
    in Milky Way Galaxy, 397–98, 398f
    size and shape of, 199–201
    of Sun, 289–90, 292–93, 293f
magnetic highway, 290
magnetic south pole, 137
magnetosphere(s)
    of giant planets, 199–202, 201f
    as layer of atmosphere, 167–69
    and moons, 202
    of neutron stars, 338
    and solar wind, 200, 201, 295
magnitude, of stars, 250
main asteroid belt, 223
main sequence (on H-R diagram), 266–67, 266f
main-sequence stars, 304–6
    determining distance to, 266–67
    energy production in, 302
    exhaustion of hydrogen fuel in, 302, 306
    helium ash at center of, 305–8

high-mass stars' evolution beyond, 330–31
lifetimes of, 305, 305t, 306, 318–19
mass of, 267, 267f
nuclear fusion in, 277
structure of, 304–5
main-sequence turnoff, 319, 319f, 392
Makemake, 216
mammals, evolution of, 465
mantle, 134, 141, 143
Marcy, Geoff, 480, 481
Mare Imbrium, 146f
maria, 145–46
Mariana Trench, 141
Mariner 10, 146
Mars
    alignment of Earth, Sun, and, 66–67, 66f
    apparent retrograde motion of, 48, 48f
    asteroids near, 223–24
    atmosphere of, *see* atmosphere, of Mars
    climate on, 175
    erosion on, 148, 148f
    exploration of, 6
    formation of, 112
    greenhouse effect on, 161
    impact craters on, 130–31, 130f, 131
    interior of, 135
    magnetic field of, 138
    mass of, 160
    and meteorites, 237, 237f, 238
    moons of, 225, 225f
    search for life on, 469–70, 471
    seasons on, 174
    surface of, 173f
    tectonism on, 143, 143f
    temperature of, 163, 173, 174f
    as terrestrial planet, 126
    volcanism on, 146, 147f
    water on, 130–31, 148–50, 149f, 150f, 164, 470
Mars Atmosphere and Volatile Evolution Mission (MAVEN), 470
Mars Express, 67, 148, 225f
Marshall Space Flight Center, 239
Mars Odyssey, 148
Mars Reconnaissance Orbiter (MRO), 66, 67, 148f, 150, 225f
Mars Science Laboratory, 469
mass
    of AGB stars, 313
    and angular momentum, 106
    of asteroids, 224
    and atmosphere, 160
    of black holes, 350
    center of, 117, 117f, 262–63, 262f, 263f
    coronal ejections of, 294, 294f, 295
    of elliptical galaxies, 369
    and energy, 277, 344
    and expansion of universe, 421–22
    of first stars, 449
    flow of, 320
    of galaxies, 364–65, 367–69, 367f
    and geometry of spacetime, 347–49, 347f
    of giant planets, 185
    and gravity, 58–59, 262
    and inertia, 56
    loss of stellar, 313–15
    and luminosity, 305, 305f, 306
    of main-sequence stars, 267, 267f, 305, 305f, 318–19
    in Milky Way Galaxy, 400, 450–51
    of neutron stars, 340
    omega, 422–25, 424f, 430, 432, 445
    and planetary atmospheres, 160
    of protostars, 104
    in red giants, 313
    and size of degenerate core, 307–8
    of Solar System, 46, 185
    of stars in binary systems, 262–64
    and stellar evolution, 304
    of terrestrial planets, 160, 274

see also high-mass stars; low-mass stars
massive (term), 6
massive compact halo objects (MACHOs), 369–70, 446
mass-luminosity relationship, 306
mass transfer, 320–21, 321f
mathematics, 10–14
Mather, John, 419
Mathilde, 225
matter
    electron-degenerate, 307–8
    empty space in, 307
    and light, 79–80
    luminous (normal), 367, 369, 446, 447f
    see also dark matter
Mauna Kea Observatories (MKO), 87, 167, 167f
Maunder Minimum, 292, 292f
MAVEN (*Mars Atmosphere and Volatile Evolution Mission*), 470
McNaught, Comet, 229, 229f
medium, 75
    see also interstellar medium
Mercury, 287f
    atmosphere of, 156, 159
    erosion on, 147–48
    formation of, 112
    impact craters on, 113f, 114, 130, 131
    interior of, 135
    magnetic field of, 138
    orbit of, 347
    possibility of life on, 469
    tectonism on, 142–43
    temperatures of, 163
    as terrestrial planet, 126
    volcanism on, 146
    water on, 150
meridian, 23–24, 23f
mesosphere, 167
messages, for intelligent life, 473–74
Messenger, 131, 146, 150
metallic hydrogen, 198
Meteor Crater, 130, 130f
meteorites, 234–38
    asteroids as, 223
    defined, 130
    and history of Solar System, 105, 105f, 237–38
    and interior of planets, 133
    origins of, 235–36
    types of, 236–37, 236f
meteoroids, 130, 233, 233f, 234
meteors, 130
meteor showers, 235
methane, 193, 221, 222
Metis, 206
microlensing, 118
microwave radiation, 79
Mid-Atlantic Ridge, 144
migration, planet, 111
Milky Way Galaxy, 388–404
    age and chemical composition of stars in, 393–96
    cross section through disk of, 396–97, 396f
    dark matter in, 369–70, 399–400
    dust in plane of, 364, 364f
    and Earth's place in universe, 4, 4f
    effects of expanding universe on, 415
    estimating number of intelligent civilizations in, 474–75, 475f
    evolution of, 393–99, 394f–395f
    extrasolar planets in, 480
    Galileo's observation of, 51
    gravitational attraction of Andromeda and, 378, 379
    habitable zones of, 470–73
    halo of, 392, 393, 397
    hypervelocity stars from, 404
    in Local Group, 442, 442f
    magnetic fields and cosmic rays in, 397–99, 398f

mass in, 400, 450–51
measuring size of, 391–93
nearest spiral galaxy to, 360
planetary systems in, 98
shape of, 362, 390–91, 390f
spiral structure in, 390–91
studying galaxy formation from, 402–3
supermassive black hole in, 400–402
Miller, Stanley, 462
Miller-Jones, James, 269
Mimas, 114, 205
mini-Neptunes, 119
Miranda, 223
mirrors, in reflecting telescopes, 85
MKO, *see* Mauna Kea Observatories
model, defined, 48
molecular-cloud cores, 99–100, 100f
molecular clouds, 98–100
    angular momentum of, 106–7, 107f
    collapse of, 96, 98–100, 99f, 100f
    defined, 98–99
    fragmentation of, 99–100
    as interstellar clouds, 98–99
molecules, 79
momentum, angular, 106–7, 107f, 351
Moon
    atmosphere of, 156, 159
    eclipses of, 38–41, 38f, 39f
    erosion on, 147–48
    face of, 35
    forces of Earth and, 57
    formation of, 114, 134
    Galileo's observation of, 51
    gravitational acceleration on, 59
    impact craters on, 126, 129f, 130, 131
    interior of, 134, 134f, 135
    magnetic field of, 138
    meteorites from, 238
    motions of, 35–41
    orbit of, 35–37, 39, 39f, 41
    phases of, 35–37, 36f, 37f
    photographs of, 81, 81f
    physical properties of terrestrial planets *vs.*, 128t
    possibility of life on, 469
    and solar eclipse for SDO, 40–41, 40f, 41f
    surface of, 151
    tectonism on, 143
    and terrestrial planets, 112, 126
    and tides, 135–37
    visits to, 6
    volcanism on, 145–46, 146f
    water on, 150, 151
moonlets, 209
moons, 217–23
    of asteroids, 225
    of dwarf planets, 216
    estimating mass of planets from, 185
    formation of, 110, 113, 214
    formerly active, 222–23
    geologically active, 218–20
    of Jupiter, 51, 74, 74f
    of KBOs, 226
    physical properties of, A-3–A-4
    and planetary rings, 205, 206
    and planets' magnetospheres, 202
    possibly active, 220–22
    shepherd, 205
    see also specific moons
motion(s)
    apparent daily, 22, 22f, 25–27, 27f
    apparent retrograde, 48–49, 48f
    cyclonic and anticyclonic, 171
    of Earth, 22–35, 38–41
    Galileo's work on, 53
    influence of forces on, 54–56
    of Moon, 35–41
    planetary, 22–35, 38–41, 48–51
    of stars and galaxies, 362–63
    uniform circular, 63–64
    see also laws of motion

Mount Everest, 349
Mount Wilson, 376
MRO, *see* Mars Reconnaissance Orbiter
M stars, 260, 266, 474
M-type asteroids, 238
multicellular eukaryotes, 463
multiverses, 432–33
muon neutrinos, 281, 344, 344f
mutation, 465, 466

N

N (newtons), 56
Namaka, 216
nanometers, 79
National Aeronautics and Space Administration (NASA)
    on age of universe, 434
    effects of weather in space for, 296, 297
    images/video of galaxy collision from, 382, 383f
    NEO cataloging by, 233
    robotic exploration of outer Solar System by, 470
    structure simulations by, 453
    study of meteorite ALH84001 by, 238
    telescopes of, 88, 91
    see also specific missions and spacecraft
National Oceanic and Atmospheric Administration, 296
natural selection, 465, 466
NCP (north celestial pole), 23
neap tides, 137
near-Earth asteroids, 224
near-Earth objects (NEOs), 224, 232, 233
NEAR Shoemaker, 225, 225f
near side, Moon, 35
nebulae
    defined, 102
    planetary, 314f, 315, 315f, 340
    see also specific nebulae
negative curvature, 415
NEOs, *see* near-Earth objects
Neptune, 191f
    atmosphere of, 193f
    characteristics of, 185
    clouds of, 191–92
    composition of, 186–88
    discovery of, 182
    formation of, 113
    as giant planet, 182
    interior of, 198–99
    magnetic field and magnetosphere of, 199, 201
    orbit of, 15, 52
    rings of, 206, 208, 208f
    rotation of, 188
    thermal energy of, 198
    winds on, 194f, 196
net force, 54, 98
neutral hydrogen
    Doppler velocities of, 399, 399f, 400
    formation of first stars from, 449
    in Milky Way Galaxy, 391, 391f
neutrino cooling, 334
neutrinos
    defined, 280
    as hot dark matter, 447
    muon, 281, 344, 344f
    in photodisintegration, 335, 336
    from Sun, 279–81
neutrons
    defined, 79
    forces on, 277
    free, 337
    in photodisintegration, 335
neutron stars, 337–38, 339f, 340, 349, 383
New Frontiers Program, 239
New Horizons, 216, 226
new Moon, 36
Newton, Isaac, 9, 49, 53–62
Newtonian physics, 340–42, 344

newtons (N), 56
Newton's law of gravity, 59–62, 61*f*, 185, 347
Newton's laws of motion, 53–58
 combining, 57–58
 first, 54, 346
 and Kepler's laws, 62
 for orbits, 262
 second, 54–56, 55*f*, 58
 and speed of light, 341–42, 341*f*
 third, 56–57, 57*f*, 59
NGC 752 (star cluster), 319
NGC 891 (galaxy), 390*f*
NGC 1132 (galaxy), 369*f*
NGC 3198 (galaxy), 368, 368*f*
NGC 3603 (star cluster), 353
NGC 3877 (galaxy), 381*f*
NGC 6240 (galaxy), 452*f*
NGC 6530 (star cluster), 393*f*
nitrogen, 164, 221, 337, 467–68
normal galaxies, AGNs in, 374–75
normal (luminous) matter, 367, 369, 446, 447*f*
North American Plate, 141
north celestial pole (NCP), 23
Northern Hemisphere
 auroras in, 169
 comets in, 229, 229*f*
 seasons in, 10*f*, 11, 30
 tropical year in, 33
 view of Earth's rotation in, 25–26
north magnetic pole, 137
North Pole, 22, 24–25, 24*f*, 137
novae, 321–23
nuclear fusion, 277–79
nuclear reactions
 in cores of high-mass stars, 330–31, 331*f*, 334
 in early universe, 429, 445–46
nuclei
 atomic, 277, 277*f*, 310
 comet, 114, 228, 228*f*, 235
 see also active galactic nuclei (AGNs)
nucleic acids, 462

O
Obama, Barack, 151
oblate (term), 188
observable universe, 415, 428, 432
observation, theory and, 285
Occam's razor, 9
ocean(s)
 on Europa, 221
 on giant planets, 186
 origin of life in, 462, 462*f*
ocean currents, 176
ocean floor, formation of, 139–40, 139*f*
Odyssey rover, 66, 67
off-Earth colonization, 479
Olympus Mons, 146, 147*f*
omega lambda (Ω_λ), 423–25, 424*f*, 430, 432, 445
omega mass (Ω_{mass}), 422–25, 424*f*, 430, 432, 445
Oort, Jan, 226
Oort Cloud, 226
opacity, 282–85
open clusters, 393–94, 393*f*
open universe, 422
Opportunity rover, 66, 148, 149, 237, 237*f*
optics, adaptive, 89–90, 90*f*
orbit(s), 62–65
 in binary systems, 262–63, 262*f*, 349
 bound, 65, 65*f*
 centripetal force and circular velocity in, 63–64
 of comets, 227, 227*f*
 defined, 63
 of dwarf planets, 216, A-3
 of Earth, 28, 29, 50, 51, 51*f*
 elliptical, 50, 50*f*, 65, 65*f*, 347

and gravity, 46, 62–63
of Jupiter, 51
and laws of motion, 62–65
of Mercury, 347
of Moon, 35–37, 39, 39*f*, 41
of Neptune, 15, 52
Newton's and Kepler's laws, 62
of particles in rings, 203–5
of Pluto, 50, 51*f*, 216
of satellites, 64–65
shape of, 64–65
of stars in elliptical *vs.* spiral galaxies, 362–63, 362*f*
unbound, 65, 65*f*
orbital speed, 363
orbital velocity, 368, 368*f*
organic compounds, 111
organisms, earliest, 463
Origins-Spectral Interpretation Resource Identification Security Regolith Explorer (OSIRIS-REx), 239
Orion Spur, 391, 391*f*
O stars, 260, 266
outer halo, 397
outer Solar System
 planet formation in inner *vs.*, 110–11
 search for life in, 470
 see also giant planets
oxygen
 abundance of, 337
 and earliest organisms on Earth, 463
 in Earth's atmosphere, 164–65, 165*f*
 in human body, 467–68
oxygen burning, 334
ozone, 167, 465

P
Painted Desert, 12*f*
pair production, 426–27
Pan, 209
parallax
 defined, 248
 measuring distances with, 246–50, 248*f*, 249, 269
 spectroscopic, 267, 380
parallel universes, 431–32, 432*f*
Parasaurolophus, 466*f*
parent elements, 132
Parkes radio telescope, 7, 7*f*
parsec (pc), 248, 249
partial lunar eclipses, 39
partial solar eclipse, 38
particle, light as, 76–77
Pasachoff, Jay, 14, 15
patterns, 11, 20
pc (parsec), 248, 249
penumbra, 290*f*, 291
Penzias, Arno, 417–18, 418*f*
Peony nebula star, 353
period, 51
periodic table of elements, 187, A-1
period-luminosity relationship, 331
Perlmutter, Saul, 423
Perseus arm, 391*f*
perturbed (term), 159
phases
 of Moon, 35–37, 36*f*, 37*f*
 of Venus, 51, 52, 53*f*
Phobos, 225, 225*f*
Phoenix spacecraft, 150, 150*f*, 469
photinos, 447
photodisintegration, 334–35
photodissociation, 221–22
photographic plates, 81–82
photons, 77
 energy level structure and wavelengths of, 254–56, 255–56*f*
 energy transfer by, 282–83, 283*f*
 gamma-ray, 280, 283
photosphere, 285*f*, 286–87
photosynthesis, 165, 463
phylogenetic tree of life, 464, 464*f*

physical laws, 54
Piazzi, Giovanni, 15
Pierre Auger Observatory, 398, 398*f*
Pillan Patera, 218*f*
pinhole cameras, 274, 291
Pioneer 10, 473, 473*f*
Pioneer 11, 202, 473, 473*f*
pitch, wavelength and, 115
pixels, 82, 83
Planck era, 428
Planck mission
 CMB data from, 420, 421, 421*f*, 444
 data on age of universe from, 434
 data on "flatness" of space from, 425
 evidence of multiverses from, 433
planetary interiors, 133–38
 Earth, 133–34
 evolution of, 134–37
 giant planets, 197–99
 and interior of Moon, 134
 and magnetic fields, 137–38
planetary motion, 48–51
 Copernican revolution, 49–50
 in early astronomy, 48–49
 of Earth, 22–35, 48–51
 Kepler's laws on, 50–52
planetary nebulae, 314*f*, 315, 315*f*, 340
planetary rings, 203*f*
 composition of, 206
 formation and evolution of, 205
 orbits of particles in, 203–5
 propeller structures in, 208–9
 and structure of ring systems, 206–8
planetary systems, 114–21
 defined, 98
 and discovery of extrasolar planets, 115–19
 identifying other, 119
 planet formation in other, 114–21
 stability of, 470–71
 see also Solar System
planetesimals, 110*f*
 asteroids from, 237–38
 formation of planets from, 109–10
 impacts on Earth by, 232
 in Kuiper Belt and Oort Cloud, 226*f*
 in Solar System, 113–14
planet formation, 105–21
 accretion disk in, 108–9
 angular momentum in, 106–7
 atmosphere in, 111
 and chaotic encounters, 111
 and history of Solar System, 105–6
 in inner *vs.* outer Solar System, 110–11
 in other planetary systems, 114–21
 planetesimals in, 109–10, 110*f*
 protostars in, 105–10
 in Solar System, 111–14
 and temperature, 110–11
planet migration, 111
planets
 criteria for, 15
 defined, 98, 114, 216
 dwarf, *see* dwarf planet(s)
 extrasolar, 114–19, 119*f*, 472–73
 giant, *see* giant planets
 in habitable zones, 471–72, 471*f*
 orbits of, 50, 50*f*, A-3
 physical data for, A-2
 predicting temperatures of, 162–63
 search for Earth-like, 472–73
 terrestrial, *see* terrestrial planets
 see also specific planets
plasma, 167, 202, 202*f*
plate tectonics, 139–41, 139*f*, 141*f*
Pleiades supercomputer, 120
Pluto, 216*f*
 axis of rotation of, 188
 as dwarf planet, 14–15, 113, 216
 formation of, 114
 orbit of, 50, 51*f*, 216
Polaris, 30, 34

poles
 celestial, 23
 day length near, 33
 magnetic, 137
 view of Earth's rotation from, 24–25, 24*f*
Porco, Carolyn, 208
positive curvature, 415
positrons, 279, 280, 426–27, 427*f*
possibly active moons, 220–22
post-AGB stars, 313–15, 314*f*
Potts, Chris, 67
precession
 of elliptical orbits, 347
 of equinoxes, 34, 34*f*, 35
pressure
 defined, 98
 and density/temperature, 276*f*, 277
 dynamic balance of gravity and, 102–3, 103*f*
 and evolution of high- *vs.* low-mass stars, 328
 on giant planets, 197, 198
 and gravity of low-mass stars, 305, 308, 311, 313
 and gravity of protostars, 102–3
 and gravity of Sun, 276, 276*f*
 and interior of planets, 133–34, 137
 on Mars, 173, 174
 and temperature in atmospheric layers, 166, 166*f*
 and temperature in red giants, 311
primary atmospheres, 111, 112, 158
primary mirror, 85
primates, on phylogenetic tree, 464
Primordial Era, 454, 454*f*
primordial soup, 462
prokaryotes, 463, 463*f*
prominences, solar, 293–94, 293*f*, 294*f*
propeller belt, 208
propeller structures, 208–9
proportionality, 60, 249
proteins, 462
proton-proton chain, 279–80, 279*f*, 305
protons, 79, 80, 277
protostars, 100–110
 evolution of, 102–5
 gravity and pressure in, 102–3
 helium and hydrogen in, 104
 H-R diagram for, 304*f*, 305
 luminosity of, 100, 104*f*
 in planet formation, 105–10, 105*f*
 in star formation, 100–105
 stars *vs.*, 100, 102
Proxima Centauri, 249
Ptolemy (Claudius Ptolemaeus), 34, 49
pulsars, 338, 339*f*, 340
pulsating variable stars, 331–32

Q
QED (quantum electrodynamics), 426
quantized energy, 77, 254
quantum efficiency, 81–83
quantum electrodynamics (QED), 426
quantum mechanics, 10
quarks, 426
quasars, 371–73, 371*f*, 375

R
R136a1, 353–54
radar, on distance ladder, 380
radial velocity, 116
radial velocity method, 115–17, 119
radiant, 235, 235*f*
radiation, 252–61
 and atomic energy levels, 252–57
 from Big Bang, 417–18
 blackbody, 417–18, 418*f*
 classifying stars based on, 258–60
 and color/surface temperature of stars, 252
 cooling by, 135
 defined, 79

I-8 INDEX

energy transport by, 281–84, 283f
Hawking, 352, 352f, 454
infrared, 7, 79
microwave, 79
and normal matter vs. dark matter, 446, 447f
and sizes of stars, 261
spectral lines, 257–61
from Sun, 295
synchrotron, 202, 399
ultraviolet, 6, 78, 159, 295
see also cosmic microwave background radiation (CMB)
radiation belts, 202
radiative zone, 282–83
radioactive heating, 135
radioisotopes, 132
radiometric dating, 131, 131f
radio signals, from extraterrestrials, 476–77
radio stars, 371
radio telescopes, 83, 86–87
see also specific telescopes
radio waves, 77, 79
recession velocity, 377, 378, 381, 381f
recombination, 418, 419f, 448
red giant branch, 308–12, 309f, 311f
red giants, 308–12
evolution of, 312
helium burning in, 310–12
on H-R diagram, 308–9, 311f
hydrogen shell burning in, 308
luminosity of, 308–9, 309f, 311–12
size and structure of, 308f
stellar mass loss in, 313
redshift, 416f
and acceleration of universe expansion, 422–24
for blackbody radiation in early universe, 417–18, 417f
gravitational, 348–49, 349f
and Hubble's law, 376, 416
measuring distance with, 377, 382
of most distant galaxies, 450, 450f
for quasars, 371
redshift surveys, 443–44, 443f
red supergiants, 261, 331
reflecting telescopes, 84–85, 85f
refracting telescopes, 84–85, 85f
refraction, 84, 84f
refractory materials, 11, 110
reionization stage, 449
Reiss, Adam G., 423
relative abundance, 337, 337f, 394–95
relativistic effects, 344
relativity, 340–50
Galilean, 342
general theory of, 9, 346–50, 369, 423, 429
special theory of, 9, 341–46, 343f
resolution
angular, 81
of telescopes, 86, 88–90, 90f
rest wavelength, 116
retrograde rotation, 172–73
revolution, defined, 28
revolution of Earth, 28–35
manipulating equations related to, 29
and orbit, 28, 30
and seasons, 28, 30–33
and wobble in axis of rotation, 34–35, 34f
ribonucleic acid (RNA), 463
ringlets, 207
Ring Nebula, 315, 315f
rings, *see* planetary rings
ring systems, 204f, 206–8
RMC 136a, 353
RNA (ribonucleic acid), 463
Roche limit, 205
Roche lobes, 320
Rømer, Ole, 74, 75
Rose Center for Earth and Space, 15

rotation
differential, 290
of forming galaxies, 448
of giant planets, 114, 182, 188, 195, 202
of Moon, 35, 35f
of neutron stars, 338
retrograde, 172–73
in spiral galaxies, 366, 366f
of Sun, 290
synchronous, 35, 35f
of Venus, 172–73
rotation, of Earth, 22–27
and celestial sphere, 22–24, 22f, 23f, 27f
latitude and view of, 25–27, 25f
rotation of giant planets vs., 188
view of, from poles, 24–25, 24f
and wind, 170–71, 170f
rotational axis, *see* axis of rotation
rotation curves, 368–69, 368f, 400, 400f
RR Lyrae variables, 332, 392
RS Ophiuchi, 323
Rubin, Vera, 368
Russell, Henry Norris, 265

S

S0 galaxies, 362, 364, 365
Sagan, Carl, 481
Sagittarius, 34
Sagittarius A* region, 401, 401f
Sagittarius Dwarf, 403
San Andreas Fault, 141
sand dunes, 148
satellites
data on CMB from, 418–19
defined, 6, 63
orbits of, 64–65
see also specific satellites
Saturn, 188f, 190f, 191f
atmosphere of, 192f
characteristics of, 185
clouds of, 190–91
composition of, 186–88
formation of, 113
as giant planet, 182
interior of, 197, 197f, 198
magnetic field and magnetosphere of, 199, 201, 202
rings of, 203f, 204f, 205–9, 207f, 209f
rotation of, 188, 195, 202
thermal energy of, 198
winds on, 194f, 195–96
SB galaxies, *see* barred spiral galaxies
Schiaparelli, Giovanni, 469
Schmidt, Brian P., 423
Schmidt, Maarten, 371
Schultz, Peter, 151
Schwarzschild, Karl, 350
Schwarzschild radii, 350, 351f
science, 7–10
curiosity in, 244
falsifiability in, 349, 350
language of, 8–9
revolutions in, 9–10
testing ideas/explanations in, 7–8
scientific method, 7–8, 8f
scientific notation, 11–12
scientific revolutions, 9–10
scientific theory, 54
SCP (south celestial pole), 23
Scutum-Centaurus arm, 391f
SDO, *see* Solar Dynamics Observatory
Seager, Sara, 120–21
Search for Extraterrestrial Intelligence (SETI), 476, 481
seasons, 10f, 11
on giant planets, 188
and location on Earth, 33
on Mars, 174
passage of, 32–33
and revolution of Earth, 28, 30–33
secondary atmospheres, 159–64
around solid planets, 111

atmospheric greenhouse effect in, 160–64
formation of, 159
oxygen in, 165
and planetary mass, 160
of terrestrial planets, 113
secondary craters, 129
secondary mirrors, 85
second-generation stars, 450
seeing, astronomical, 89
seismic waves, 134
selection effect, 119
self-gravity, 98, 98f, 99, 108
self-replication, 465
semimajor axis, of planets' orbits, 50
SETI (Search for Extraterrestrial Intelligence), 476, 481
SETI@home, 476
Shapley, Harlow, 358, 392
shepherd moons, 205
shield volcanoes, 144, 145f, 146
Shoemaker-Levy 9, Comet, 233
short-period comets, 227
Siding Spring Observatory, 443, 454
signals, from intelligent life, 475–77
Sikhote-Alin region, 232, 479
silicates, 110
silicon, as basis for life, 468–69
silicon burning, 334
simulations
of formation of Milky Way–like galaxy, 451, 451f
of structure of universe, 452–53
singularity, 350
Sirius, 250
Six-degree-Field Galaxy Survey, 454
61 Cygni (star), 249
"slice of the universe" survey, 443, 443f
Slipher, Vesto, 376, 377
Sloan Digital Sky Survey, 443, 443f, 453, 453f
slope of line, 11–12
small bodies, 214–39
asteroids, 223–25, 239
comets, 226–33
dwarf planets, 216–17
meteorites, 234–38
moons, 217–23
Small Magellanic Cloud, 370, 393f, 403f, 442f
SMEI (Solar Mass Ejection Imager), 323
Smoot, George, 419
"Snowball Earth" hypothesis, 464, 465
SOFIA (Stratospheric Observatory for Infrared Astronomy), 87
Solar and Heliospheric Observatory (*SOHO*), 293f, 294f, 295, 296
Solar Dynamics Observatory (SDO), 40–41, 290f, 296
solar eclipses, 38f
annular, 38
causes of, 38–39
gravitational lensing during, 348, 348f
partial, 38
for SDO, 40–41, 40f, 41f
total, 38, 38f, 288, 288f, 289
viewing chromosphere during, 288, 288f
viewing corona during, 289
solar flares, 294–97, 294f
Solar Mass Ejection Imager (SMEI), 323
solar maximum, 66, 292
solar neutrino problem, 281, 284
Solar System
boundary of, 290
comets in, 226
defined, 98
determining distances within, 380
and Earth's place in universe, 4f
Earth's position in, 163–64
effects of expanding universe in, 415
extraterrestrial life in, 469–70

geocentric model of, 48
habitable zone of, 471, 471f
heliocentric model of, 49–50
history of, 105–6, 106f, 214, 237–38
layout of, 112f
mass of, 46, 185
meteorites as remnants of early, 234–38
moons of, 217–18, 217f
planet formation in, 110–14
primary atmospheres in, 158
Ptolemy's model of, 49, 49f
relative abundances of elements in, 337
ring systems in, 205
rotation and revolution of objects in, 22f
small bodies of, 214–39
terrestrial planets of, 126–51
volcanism in, 145–46
solar tides, 136–37, 136f
solar wind
boundary of, 290f
defined, 167
and layers of Earth's atmosphere, 167–69
and magnetic field of Sun, 289–90
and magnetosphere, 200, 201, 295
south celestial pole (SCP), 23
Southern Hemisphere
auroras in, 169
comets in, 229, 229f
seasons in, 31
South Pole, 23, 25, 137
spacecraft, orbiting by, 63, 63f
see also specific spacecraft
Space Radiation Group, 297
space shuttles, 46
space telescopes, 87–89
see also specific telescopes
spacetime, 344–50
Space Very Long Baseline Interferometer (SVLBI), 87
special theory of relativity, 9, 341–46
boxcar experiment in, 343, 343f
and free fall/float, 345–46
implications of, 344–45
and speed of light in vacuum, 341–42
time dilation in, 342–44
spectral lines, 257–61
defined, 116
emission, 83, 255, 257–58, 259f
and physical properties of stars, 260–61
see also absorption lines
spectral types, 260
spectrographs, 83
spectrometers, 83
spectroscopic binary systems, 264
spectroscopic parallax, 267, 380
spectroscopic radial velocity method, 116, 117, 119
spectroscopy, 83
spectrum(--a), 78f
absorption, 254
blackbody, 101, 101f
capturing, 83
classifying stars based on, 258–60, 259f, 260f
of cosmic microwave background radiation, 418–19, 419f
defined, 78
electromagnetic, 77–79, 78f
emission, 254
of galaxies, 376, 376f
of stars, 252–53
of Sun, 287, 287f
visible, 78–79, 78f
speed
and angular momentum, 106
equations involving, 29
graph of, 11f
orbital, 363
of waves, 76, 76f
of wind, 195

INDEX I-9

speed of light
  determining, 74–75, 74f
  and space travel, 477
  as ultimate speed limit, 345
  as unit in astronomy, 6
  in a vacuum, 74, 341–42, 341f
Spilker, Linda, 209
spiral arms, galaxy, 365–67, 365f, 366f, 391
spiral density waves, 367
spiral galaxies
  debate over location of, 358
  defined, 362
  elliptical vs., 364–65
  formation of, 448f
  mass of, 367–69, 367f
  merging of, 452
  Milky Way as, 390–91, 390f
  orbits of stars in, 363
  star formation in, 365–67
Spirit rover, 148, 173f
Spirograph Nebula, 315f
Spitzer Space Telescope, 88, 207, 383, 391, 450f
spring tides, 137
Sputnik 1 satellite, 6, 63
squaring numbers, 52
SS Cygni system, 269
standard candles, 331, 332, 380
standard model, 425–26, 427
star clusters, 317–19, 318f, 319f
Stardust, 230, 230f
star formation, 98–105
  accretion disk in, 108–9
  angular momentum in, 106–7
  in colliding galaxies, 382–83
  and evolution of Milky Way Galaxy, 393–96
  in evolving galaxies, 452
  of first stars, 449–50, 449f
  in galaxies, 365–67
  interstellar clouds in, 98–99
  in Milky Way, 403
  molecular clouds in, 98–100
  protostars in, 100–110
  star clusters and history of, 319
  and Stefan-Boltzmann law and Wien's law, 101
star maps, A-8–A-12
stars, 244–69
  age and chemical composition of, 393–96
  apparent daily motion of, 25–27, 27f
  brightest, A-7
  brightness of, 250, 251f
  classification of, 258–60
  color and surface temperature of, 252
  defined, 98, 114
  distances to nearby, 246–50
  Earth's orbit and view of, 28, 28f, 30
  evolution of, *see* stellar evolution
  galaxy shape and orbits of, 362–63
  habitable zones of, 471–72, 471f
  in halo vs. disk of galaxy, 397
  high-mass, *see* high-mass stars
  H-R diagrams of, 265–68
  hypervelocity, 404
  light from, 72
  low-mass, *see* low-mass stars
  luminosity of, 246–51
  mass of, 262–64
  nearest, A-5–A-6
  not on main sequence of H-R diagrams, 268
  protostars vs., 100, 102
  radiation from, 252–61
  second-generation, 450
  sizes of, 261
  in SS Cygni system, 269
  supergiant, 268
  see also specific stars
Stefan, Josef, 101
Stefan-Boltzmann constant, 101, 291

Stefan-Boltzmann law, 101, 162, 198, 252, 261, 291
stellar cataclysm, 322, 340
stellar evolution
  in binary systems, 320–22, 320f
  of high-mass stars, 328–54
  on H-R diagram, 265
  last stages of, for low-mass stars, 312–17
  of low-mass stars, 302–23
  and mass, 304
  star clusters as snapshots of, 317–18
stellar occultations, 185, 185f
Stelliferous Era, 454
stereoscopic vision, 246–48, 247f
stony-iron meteorites, 236f, 237, 238
stony meteorites, 236, 236f, 237
stratospheres, 167
Stratospheric Observatory for Infrared Astronomy (SOFIA), 87
strength, defined, 258
string theory, 428–29
stromatolites, 460, 463
strong nuclear force, 277, 425, 427, 429
structure, of universe, *see* universe, structure of
S-type asteroids, 238
subatomic particles
  and antiparticles, 426–27
  in cooling universe, 429
  in earliest moments of universe, 425–29
  in grand unified theories, 427
  in standard model, 425–26
  in theory of everything, 428–29
summer solstice, 32
Sun, 274–97
  age of "current" images from, 382
  alignment of Earth, Mars, and, 66–67, 66f
  angular momentum of, 107, 109
  apparent daily motion of, 22
  and celestial sphere, 24
  changes in, 290–94
  color of, 252
  composition of, 187, 306, 307f, 396
  core of, 278, 280–85, 282f
  death of, 477–78
  distance to, 75, 185
  empty space in, 307
  energy in core of, 281–85
  energy production in, 302, 305
  gravitational force of, 59–61
  gravity of, 46, 276, 276f
  H-R diagram for, 302
  hydrostatic equilibrium in, 276–77
  lifetime of, 278
  location of, 4
  luminosity and brightness of, 251, 251f, 305
  magnetic field of, 289–90, 292–93, 293f
  mass and size of Earth vs., 274
  in Milky Way Galaxy, 393, 393f
  motion of, 2, 420
  nuclear fusion in, 277–79
  observing neutrinos from, 280–81
  phases of Moon and position of, 35–37, 36f, 37
  post-AGB phase of, 315
  predicted evolution of, 316
  proton-proton chain in, 279–80, 279f
  revolution of Earth around, 28–35
  size and structure of red giant vs., 308f
  in Solar System, 112
  spectra of, 260
  structure of, 276–81, 276f, 304, 305
  thermal energy from, 75, 160–61
  and tides, 135–37
  and weather forecasting in space, 296–97
  see also atmosphere, of Sun
sunlight, 32, 159, 173, 295f
sunspot cycle, 291–94, 292f
sunspots, 287f, 290

and changes in Sun, 290–94
defined, 289
and magnetic field of Sun, 289–90
temperature and brightness of, 291
"super-Chandrasekhar" white dwarfs, 322
superclusters, 442, 443
super-Earths, 119
supergiant stars, 268
supermassive black hole(s)
  and accretion disks, 373–74
  and AGNs, 372–75
  at centers of galaxies, 371–75
  formation of, 449, 450, 452
  in Milky Way Galaxy, 400–402, 401f, 402f
  and quasars, 371–72, 371f
  and structure of universe, 455
Supernova 1987A, 333, 333f, 336
supernovae, 336–40
  and cosmic rays, 398
  and Crab Nebula, 340
  energetic and chemical legacies of, 336–37
  from high-mass stars, 335–40
  neutron stars and pulsars from, 337–39
  see also specific types
super red giant, 353
surface, defined, 252
SVLBI (Space Very Long Baseline Interferometer), 87
synchronous rotation, 35, 35f
synchrotron radiation, 202, 399

T

tails, comet, 228–29, 228f, 229f
Tarantula nebula, 353
tau neutrinos, 281
Taurus, 34, 340
Taylor, Lawrence, 151
tectonism, 138–44
  convection and, 140–42, 140f
  on other planets, 142–44
  theory of plate tectonics, 139–40
  volcanism and, 144–45
telescopes, 7, 84–91
  diffraction limit for, 89
  gamma ray, 83, 84, 87
  radio, 86–87
  reflecting, 84–85, 85f
  refracting, 84–85, 85f
  resolution of, 86, 88–90, 90f
  space, 87–89
  see also specific telescopes
Tempel 1, Comet, 230–31, 231f
temperature(s)
  and atmospheric greenhouse effect, 160–61, 160f
  and brightness of sunspots, 291
  of chromosphere and corona, 286, 286f, 287
  and color of stars, 252
  and composition in accretion disk, 110–11, 110f
  of cosmic microwave background radiation, 420
  and density/pressure, 276f, 277
  of Earth, 160, 163, 175, 175f, 176f, 296
  for formation of stars, 104
  on giant planets, 185
  in habitable zones, 471
  on H-R diagram, 266
  and interiors of planets, 134–37
  Kelvin scale, 98
  and luminosity, 101, 265
  of Mars, 163, 173, 174f
  of Mercury, 163
  and planet formation, 110–11
  of planets in forming Solar System, 113
  of post-AGB stars, 314
  predicting planets', 162–63
  and pressure in atmospheric layers, 166, 166f

of red giants, 308, 311
during seasons, 33
of stars, 260, 261, 261f
of Sun, 282, 283, 283f, 287–89
of Venus, 160, 163, 172, 172f
of white dwarfs, 316
Templeton Foundation, 480, 481
terrestrial planets, 126–51
  atmospheres of, 112–13, 156–77
  at death of Sun, 478
  erosion on, 147–48
  formation of, 112–13
  giant planets vs., 186
  greenhouse effect on, 161, 163–64
  impacts and characteristics of, 128–32, 128t
  interiors of, 133–38, 134f
  search for water on, 148–50
  and surface of Moon, 151
  and tectonism, 138–44
  and volcanism, 144–47
  wind/water on, 147–51
  see also specific planets
Tethys, 223
tetravalent atoms, 468
Thebe, 206
theories
  defined, 7, 8
  and observation, 285
theory of everything (TOE), 428–29
theory of plate tectonics, 139–40
thermal energy, 75, 100, 134, 194, 197, 198
thermal equilibrium, 158, 198
thermophiles, 466
thermosphere, 167
third quarter Moon, 36
Thomas, Reid, 67
tidal forces, 205, 347, 351f
tidal heating, 135–37
tide pools, 463, 463f
tides, 135–37, 135f, 136f
Tiktaalik roseae, 465f
time
  doubling, 467
  Hubble, 412, 412f, 413
  integration, 81, 82
  mass and geometry of, 348–49
  placing galaxies in, 382
  space-, 344–50
time dilation, 342–45, 348–49, 477
Tiscareno, Matthew, 209
Titan
  atmosphere of, 169
  Cassini images of, 221f
  evidence of life on, 470
  and magnetosphere of Saturn, 202
  as possibly active moon, 221–22
  spacecraft on, 6
  surface features of, 222f
TOE (theory of everything), 428–29
torus, 202, 374
total lunar eclipse, 39, 39f
total mass, in binary systems, 264
total solar eclipse, 38, 38f, 288, 288f, 289
trace elements, 468
transit method, 117–18, 117f
trans-Neptunian objects, 14
trends, on graphs, 11
Triangulum, 403
triple-alpha process, 310, 310f, 313, 337
Triton, 185, 188, 219, 220
Trojan asteroids, 223
tropical years, 33
Tropic of Cancer, 33
Tropic of Capricorn, 33
Tropics, 33
tropopause, 167
troposphere, 166, 167, 192, 193
Tunguska River, 232, 232f, 479
tuning fork diagram of galaxies, 362, 362f
Two-degree-Field Galaxy Redshift Survey, 443f

Tycho Brahe, 49, 50
type A stars, 332–33
Type Ia supernova
    from binary systems, 322
    determination of accelerating universe with, 423, 424, 424f
    on distance ladder, 381
    in NGC 3877, 381f
Type II supernova
    and Crab Nebula, 340
    creation of, 335f, 336
    energetic and chemical legacies of, 336–37
    from high-mass stars, 333
    neutron stars and pulsars from, 337–39
type O stars, 333
Tyson, Neil deGrasse, 15

U

UB 313 (Xena; dwarf planet), 14, 15
ultrafaint dwarf galaxies, 451
ultraviolet (UV) radiation, 6, 78, 159, 295
ultraviolet telescopes, 87, 88
umbra, 290f, 291
Umbriel, 218
unbounded universe, 416
unbound orbit, 65, 65f
uncertainty, observational, 250
uniform circular motion, 63–64
units, scientific, 11
universal gravitational constant, 61
universal law of gravitation, 61, 61f, 62
universe
    age of, 412, 413, 413f, 434
    conditions of early, 412, 414
    cooling, 429, 429f
    earliest moments of universe, 425–29
    Earth's place in, 4, 4f, 128
    observable, 415, 428, 432
    open, 422
    origin of, 6
    parallel, 431–32, 432f
    scale of, 4–6
    size/shape of, 415–16
    space and time mapping of, 382
    unbounded, 416
    see also life in universe
universe, evolution of, 6, 410–34
    and age of universe, 434
    and CMB confirmation of Big Bang theory, 416–21
    concept of inflation in, 430–32
    and evolution in multiverses, 432–33
    implications of Hubble's law for, 412–16
    and life in universe, 460
    predicting future, 453–54
    and rate of expansion, 421–25
    and structure of universe, 440
    and subatomic particles in earliest moments, 425–29, 429f
universe, expansion of, 375–82
    and age of universe, 413
    and gravity, 421–22
    and Hubble law, 378–79
    and Hubble's law, 376–77, 414–16
    rate of, 421–25
    testing hypothesis about, 424–25
    uniformity of, 378–79
universe, structure of, 440–55
    and evolution of galaxies, 451–53
    and formation of first galaxies, 450–51
    and formation of first stars, 449–50, 449f
    and galaxy structures, 442–44
    gravity and large-scale, 444–48
    influence of supermassive black holes on, 455
    and life in universe, 460
    predicting evolution of, 453–54
    supercomputer simulations of, 452–53
University of Arizona in Tucson, 239
University of California, 476
unresolved points of light, 372
uranium-238, 132
Uranus, 191f, 196f
    atmosphere of, 193f
    characteristics of, 185
    clouds of, 191
    composition of, 186–88
    discovery of, 182
    formation of, 113
    as giant planet, 182
    images of, 90f
    interior of, 198–99
    magnetic field and magnetosphere of, 199, 201
    rings of, 206, 208
    rotation of, 114, 188
    thermal energy of, 198
    winds on, 194f, 196
Urey, Harold, 462
Urey-Miller experiment, 462, 462f
U.S. Department of Defense, 323
UV radiation, see ultraviolet radiation

V

vacuum(s)
    defined, 74
    speed of light in, 74, 341–42, 341f
Valles Marineris, 143, 143f
vaporize, 130
v_{circ} (circular velocity), 64
Vega (star), 244, 252, 267
Vega 1, 230
Vega 2, 230
velocity(-ies)
    circular, 64
    defined, 55
    escape, 65
    and mass of binary stars, 262–63, 263f
    of objects with elliptical orbits, 65, 65f
    orbital, 368, 368f
    radial, 116
    recession, 377, 378, 381, 381f
Venus, 288f
    axis of rotation of, 188
    erosion on, 148
    exploration of, 6
    formation of, 112, 113
    greenhouse effect on, 161, 163–64, 175
    impact craters on, 130, 131
    interior of, 135
    magnetic field of, 138
    magnitude of, 250
    mass of, 160
    phases of, 51, 52, 53f
    possibility of life on, 469, 471
    tectonism on, 143–44, 143f
    temperature of, 160, 163, 172, 172f
    as terrestrial planet, 126
    transit of, 274
    volcanism on, 146
    water on, 150, 164
    see also atmosphere, of Venus
Venus Express, 164
vernal equinox, 32
Very Large Array (VLA), 86, 87f
Very Large Telescope (VLT), 87, 87f, 353
Very Long Baseline Array (VLBA), 86–87, 269
v_{esc} (escape velocity), 65
Vesta, 217, 225
Viking Orbiter, 143f
Viking spacecraft, 469
Virgo Cluster, 442, 442f
Virgo Supercluster, 4, 4f, 442
visible spectrum, 78–79, 78f
vision, stereoscopic, 246–48, 247f
visual binary systems, 264
VLA (Very Large Array), 86, 87f
VLBA (Very Long Baseline Array), 86–87, 269
VLT, see Very Large Telescope
voids, 443
volatile materials, 110–11, 192–93, 192f–193f
volcanism, 144–47
    cryo, 219–20, 219f
    on moons, 218–21
    and planetary rings, 206
    and secondary atmosphere, 111
    in Solar System, 145–46
    at tectonic plates, 141–42, 141f
    and tectonism, 144–45
volcanoes
    composite, 144, 145f
    shield, 144, 145f, 146
vortices, 190
Voyager, 218, 223f
Voyager 1
    and boundary of solar wind, 290f
    Earth images, 479, 479f
    in Jupiter's magnetosphere, 202
    lightning images, 194
    on magnetic highway, 290
    messages for intelligent life, 473
    planetary ring images, 207
    wind speed measurements, 195
Voyager 2
    Great Red Spot images, 190, 190f
    lightning images, 194
    magnetic field measurements, 199
    messages for intelligent life, 473
    observations of Uranus, 196
    Triton images, 220, 220f
    wind speed measurements, 195

W

waning Moon, 36
water
    on Earth, 164
    in habitable zones, 471
    on Mars, 130–31, 148–50, 149f, 150f, 164, 470
    on Mercury, 150
    on Moon, 150, 151
    search for, 148–50
    surface modifications by, 147–48
    on Venus, 150, 164
water erosion, 148, 164
wave(s)
    electromagnetic, 77
    gravitational, 349–50
    helioseismic, 284, 284f
    light as, 75–76, 75f, 76f
    radio, 77, 79
    seismic, 134
    spiral density, 367
wavelength, 76, 76f
    and color, 250
    and frequency, 77, 78
    of photons, 254–56, 255–56f
    and pitch, 115
    rest, 116
waxing Moon, 36
weakly interacting massive particles (WIMPs), 370
weak nuclear force, 425, 427, 429
weather
    and atmosphere of Earth, 156
    climate vs., 174
    on giant planets, 194–96
weather forecasting, in space, 296–97
weathering, 147–48
Wegener, Alfred, 139, 140
weight, 58–59
westerly winds, 195
white dwarfs, 315–17
    in binary systems, 321–22, 321f
    on H-R diagram, 315, 316, 316f
    luminosity class of, 268
    radii of, 261
    in X-ray binaries, 338, 338f
Wide-field Infrared Survey Explorer (WISE), 480
Wien, Wilhelm, 101
Wien's law, 101, 252, 260, 261
Wild 2, Comet, 230, 230f, 231
Wilkinson Microwave Anisotropy Probe (WMAP), 420, 420f, 423, 425, 434, 444
Williams, Iwan, 15
Wilson, Robert, 417–18, 418f
WIMPs (weakly interacting massive particles), 370
wind(s)
    and atmosphere of Earth, 170–71
    on Jupiter and Saturn, 194–96
    measuring speed of, 195
    on terrestrial planets, 148
    on Uranus and Neptune, 196
    on Venus, 173
    and weather on giant planets, 194
    westerly, 195
    zonal, 171
    see also solar wind
wind erosion, 148
wind streaks, 148
winter solstice, 32
WISE (Wide-field Infrared Survey Explorer), 480
WMAP, see Wilkinson Microwave Anisotropy Probe
Wright, Jason, 480

X

Xena (UB 313; dwarf planet), 14, 15
X-ray binary systems, 338, 338f
X-ray emissions, 289, 289f, 295
X-rays, 6, 78
X-ray telescopes, 84, 87, 88

Y

years, 28
    leap, 33
    light-, 4–6, 5f, 75, 248
    tropical, 33
Yellowstone Park, 145

Z

Zapp, Neal, 297
zenith, 23, 24
zero-age main sequence, 318
zodiac, 28f, 30
zonal winds, 171
zones, 189
Zorn, Torsten, 67